WASTEWATER MICROBIOLOGY

WILEY SERIES IN
ECOLOGICAL AND APPLIED MICROBIOLOGY

RECENT TITLES

MICROBIAL LECTINS AND AGGLUTININS:
Properties and Biological Activity
David Mirelman, Editor, 1986

THERMOPHILES: General,
Molecular, and Applied Microbiology
Thomas D. Brock, Editor, 1986

INNOVATIVE APPROACHES
TO PLANT DISEASE CONTROL
Ilan Chet, Editor, 1987

PHAGE ECOLOGY
Sagar M. Goyal, Charles P. Gerba,
Gabriel Bitton, Editors, 1987

BIOLOGY OF ANAEROBIC
MICROORGANISMS
Alexander J.B. Zehnder, Editor, 1988

THE RHIZOSPHERE
J.M. Lynch, Editor, 1990

BIOFILMS
William G. Characklis and
Kevin C. Marshall, Editors, 1990

ENVIRONMENTAL MICROBIOLOGY
Ralph Mitchell, Editor, 1992

BIOTECHNOLOGY IN PLANT
DISEASE CONTROL
Ilan Chet, Editor, 1993

ANTARCTIC MICROBIOLOGY
E. Imre Friedmann, Editor, 1993

WASTEWATER MICROBIOLOGY
Gabriel Bitton, 1994

WASTEWATER MICROBIOLOGY

GABRIEL BITTON

Department of Environmental Engineering Sciences
University of Florida, Gainesville

WILEY-LISS

A JOHN WILEY & SONS, INC., PUBLICATION
New York • Chichester • Brisbane • Toronto • Singapore

Address All Inquiries to the Publisher
Wiley-Liss, Inc., 605 Third Avenue, New York, NY 10158-0012

Printed in the United States of America.

Library of Congress Cataloging-in-Publication Data

Bitton, Gabriel
Wastewater microbiology / Gabriel Bitton.
p. cm. — (Wiley series in ecological and applied microbiology)
Includes bibliographical references and index.
ISBN 0-471-30985-0
ISBN 0-471-30986-9 (paperback)
1. Sanitary microbiology. 2. Water—Microbiology. 3. Sewage—Microbiology. I. Title. II. Series.
QR48.B53 1994
628.3'01'576—dc20 94-3875
CIP

The text of this book is printed on acid-free paper.

In memory of my father

Whenever one ceases to study, one forgets.
MAIMONIDES, *Book of Knowledge*

CONTENTS

PREFACE

Numerous colleagues and friends have encouraged me to prepare a second edition of *Introduction to Environmental Virology*, published by Wiley in 1980. Instead, I decided to broaden the topic by writing a text about the role of *all* microorganisms in water and wastewater treatment and the fate of pathogens and parasites in engineered systems.

In the 1960s, the major preoccupation of sanitary engineers was the development of wastewater treatment processes. Since then, new research topics have emerged and emphasis is increasingly placed on the biological treatment of hazardous wastes and the detection and control of new pathogens. The field of wastewater microbiology has blossomed during the last two decades as new modern tools have been developed to study the role of microorganisms in the treatment of water and wastewater. We have also witnessed dramatic advances in the methodology for detection of pathogenic microorganisms and parasites in environmental samples, including wastewater. New genetic probes and monoclonal antibodies are being developed for the detection of pathogens and parasites in water and wastewater. Environmental engineers and microbiologists are increasingly interested in toxicity and the biodegradation of xenobiotics by aerobic and anaerobic biological processes in wastewater treatment plants. Their efforts will fortunately result in effective means of controlling these chemicals. The essence of this book is an exploration of the interface between engineering and microbiology, which will hopefully lead to fruitful interactions between biologists and environmental engineers.

The book is divided into five main sections, which include fundamentals of microbiology, elements of public health microbioloby, process microbiology, biotransformations and toxic impact of chemicals in wastewater treatment plants, and the public health aspects of the disposal of wastewater effluents and sludges on land and in the marine environment. In the process microbiology section, each biological treatment process is covered from both the process microbiology and public health viewpoints.

This book provides a useful introduction to students in environmental sciences and environmental engineering programs and a source of information and references to research workers and engineers in the areas of water and wastewater treatment. It should serve as a reference book for practicing environmental engineers and scientists and for public health microbiologists. It is hoped that this information will be a catalyst for scientists and engineers concerned with the improvement of water and wastewater treatment and with the quality of our environment.

I am very grateful to all my colleagues and friends who kindly provided me with illustrations for this book and who encouraged me to write *Wastewater Microbiology*. I will always be indebted to them for their help, moral support, and good wishes. I am indebted to my graduate students who have contributed to my interest and knowledge in the microbiology of engineered systems. Special thanks are due to Dr. Ben Koopman for lending a listening ear to my book project and to Dr. Joseph Delfino for his moral support. I thank Hoa Dang-Vu Dinh for typing the tables for this book. Her attention to details is much appreciated.

Special thanks to my family, Nancy, Julie, and Natalie, for their love, moral support, and patience, and for putting up with me during the preparation of this book.

Gabriel Bitton
Gainesville, Florida
January 1994

FUNDAMENTALS OF MICROBIOLOGY

1

THE MICROBIAL WORLD

1.1. INTRODUCTION: THE PROTISTA

Protista are divided into two groups, the procaryotes (bacteria) and eucaryotes (fungi, protozoa, algae, and plant and animal cells). Viruses are obligate intracellular parasites that belong to neither of these two groups.

The main characteristics that distinguish procaryotes from eucaryotes are the following (Fig. 1.1):

1. Eucaryotic cells are generally more complex than procaryotic cells.
2. DNA is enclosed in a nuclear membrane and is associated with histones and other proteins only in eucaryotes.
3. Organelles are membrane-bound in eucaryotes.
4. Procaryotes divide by binary fission, whereas eucaryotes divide via mitosis.
5. Some structures are absent in procaryotes: for example, Golgi complex, endoplasmic reticulum, mitochondria, chloroplasts.

Other differences between procaryotes and eucaryotes are shown in Table 1.1.

We will now review the main characteristics of the procaryotic (bacteria, actinomycetes, cyanobacteria) and eucaryotic (fungi, protozoa, algae) microorganisms that are important from a process microbiology and public health viewpoints. We will also introduce the reader to environmental virology, the study of the fate of viruses of public health significance in wastewater, and other fecally contaminated environments.

1.2. CELL STRUCTURE

1.2.1. Cell Size

Except for filamentous bacteria, procaryotic cells are generally smaller than eucaryotic cells. Small cells have a higher growth rate than larger cells. This may be explained by the fact that small cells have a higher surface-to-volume ratio than larger cells. Thus, the higher metabolic activity of small cells is due to the additional membrane surface that is available for transport of nutrients into and waste products out of the cell.

1.2.2. Cytoplasmic Membrane (Plasma Membrane)

The cytoplasmic membrane is a 40- to 80-A thick semipermeable membrane that contains a phospholipid bilayer with proteins embedded within the bilayer (fluid mosaic model) (Fig. 1.2). The phospholipid bilayer is made of hydrophobic fatty acids oriented towards the inside of the bilayer and hydrophilic glycerol moieties oriented towards the outside of the bilayer. Certain cations such as Ca^{2+} and Mg^{2+} help stabilize the membrane structure. Sterols are other lipids that enter into the composition of plasma membranes of eucaryotic cells as well as some procaryotes such as mycoplasmas (these bacteria lack a cell wall).

Chemicals cross biological membranes by diffusion, active transport, and endocytosis.

Diffusion. Owing to the hydrophobic nature of the plasma membrane, lipophilic compounds diffuse better through the membrane than do io-

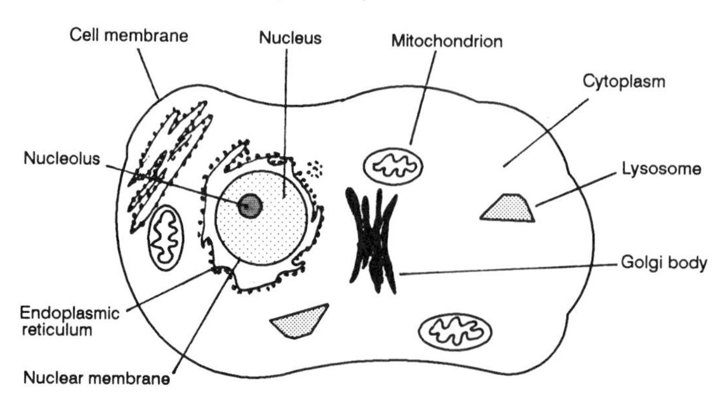

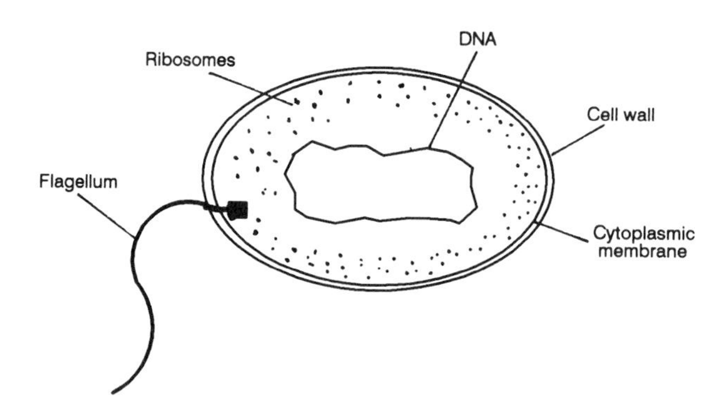

Fig. 1.1. Procaryotic and eucaryotic cells.

nized compounds. The rate of diffusion across cell membranes depends on their lipid solubility and concentration gradient across the membrane.

Active transport. Hydrophilic (i.e., lipid-insoluble) compounds can be transferred through the membrane by active transport. This transport involves highly specific carrier proteins, requires energy in the form of adenosine triphosphate (ATP) or phosphoenol pyruvate (PEP), and allows cells to accumulate chemicals against a concentration gradient. There are specific active transport systems for sugars, amino acids, and ions.

Toxic chemicals gain entry into cells mainly through diffusion, and some may use active transport systems similar to those used for nutrients.

Endocytosis. In eucaryotic cells, substances can cross the cytoplasmic membranes by endocytosis, in addition to diffusion and active transport. Endocytosis includes phagocytosis (uptake of particles) and pynocytosis (uptake of dissolved substances).

1.2.3. Cell Wall

All bacteria, except mycoplasma, have a cell wall. This structure confers rigidity to cells and maintains their characteristic shape, and it pro-

TABLE 1.1. Comparison of Procaryotes and Eucaryotes

Feature	Procaryotes (bacteria)	Eucaryotes (fungi, protozoa, algae, plants, animals)
Cell wall	Present in most procaryotes (absent in Mycoplasma); made of peptidoglycan	Absent in animal cells; present in plants, algae, and fungi
Cell membrane	Phospholipid bilayer	Phospholipid bilayer + sterols
Ribosomes	Size: 70S	Size: 80S
Chloroplasts	Absent	Present
Mitochondria	Absent; respiration associated with plasma membrane	Present
Golgi complex	Absent	Present
Endoplasmic reticulum	Absent	Present
Gas vacuoles	Present in some species	Absent
Endospores	Present in some species	Absent
Locomotion	Flagella composed of one fiber	Flagella or cilia composed of microtubules; amoeboid movement
Nuclear membrane	Absent	Present
DNA	One single molecule	Several chromosomes where DNA is associated with histones
Cell division	Binary fission	Mitosis

tects them from high osmotic pressures. It is composed of a mucopolysaccharide called peptidoglycan or murein (glycan strands cross-linked by peptide chains). Peptidoglycan is composed of N-acetylglucosamine and N--acetylmuramic acid and amino acids. A cell wall stain, called the Gram stain differentiates between gram-negative and gram-positive bacteria on the basis of cell wall composition. Peptidoglycan layers are thicker in gram-positive than in gram-negative bacteria. In addition to peptidoglycan, gram-positive bacteria contain techoic acids, which are made of an alcohol and a phosphate group.

Animal cells do not have cell walls but in other eucaryotic cells, the cell walls may be composed of cellulose (e.g., plant cells, algae, fungi), chitin (e.g., fungi), silica (e.g., diatoms), or polysachharides such as glucan and mannan (e.g., yeasts).

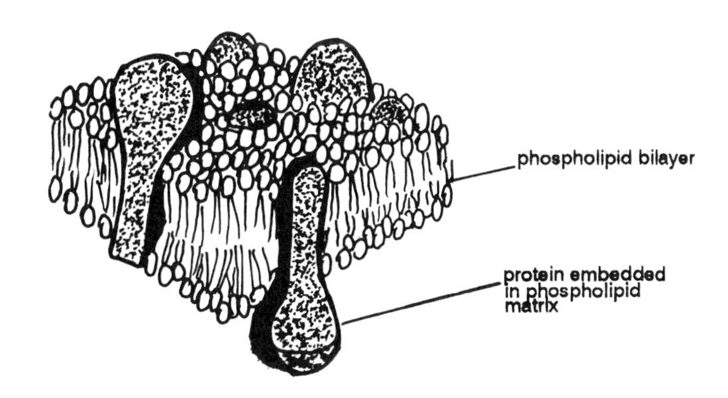

Fig. 1.2. Structure of cytoplasmic membrane. Adapted from Alberts et al. (1989).

1.2.4. Outer Membrane

The outermost layer of gram-negative bacteria contains phospholipids, lipopolysaccharides and proteins (Fig. 1.3). The lipopolysaccharides (LPS) constitute about 20% of the outer membrane by weight and consist of a hydrophobic region (lipid A) bound to an oligosaccharide core. The LPS molecules are held together with divalent cations. Proteins constitute about 60% of the outer membrane weight and are partially exposed to the outside. Some of the proteins form water-filled pores (porins) for the passage of hydrophilic compounds. Other proteins have a structural role as they help anchor the outer membrane to the cell wall. The outer membrane of gram-negative bacteria is an efficient barrier against hydrophobic chemicals (namely, some antibiotics and xenobiotics) but is permeable to hydrophilic compounds, some of which are essential nutrients.

Chemicals (e.g., EDTA, polycations) and physical treatments (e.g., heating, freeze-thawing, drying, and freeze-drying), as well as genetic alterations, can increase the permeability of outer membranes to hydrophobic compounds.

1.2.5. Glycocalyx

The glycocalyx is made of extracellular polymeric substance (EPS), which surrounds some microbial cells and is composed mainly of polysaccharides. In some cells, the glycocalyx is organized as a capsule (Fig. 1.4). Other cells produce loose polymeric materials, which are dispersed in the growth medium.

Extracellular polymeric substances are important from the medical and environmental viewpoints: (1) capsules contribute to pathogen virulence; (2) encapsulated cells are protected from phagocytosis in the body and in the environment; (3) EPS helps bacteria adsorb to surfaces such as teeth, mucosal surfaces, and environmentally important surfaces such as water distribution pipes (see chapter 15); (4) capsules protect cells against desiccation; (5) they play a role in metal complexation particularly in waste-

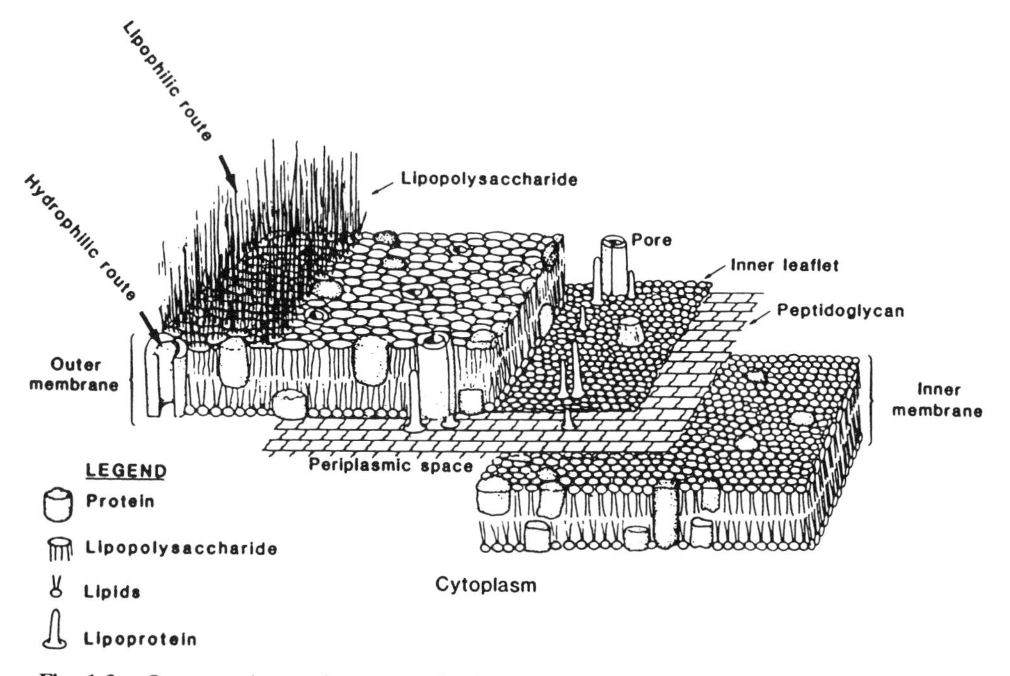

Fig. 1.3. Outer membrane of gram-negative bacteria. From Godfrey and Bryan (1984), with permission of the publisher.

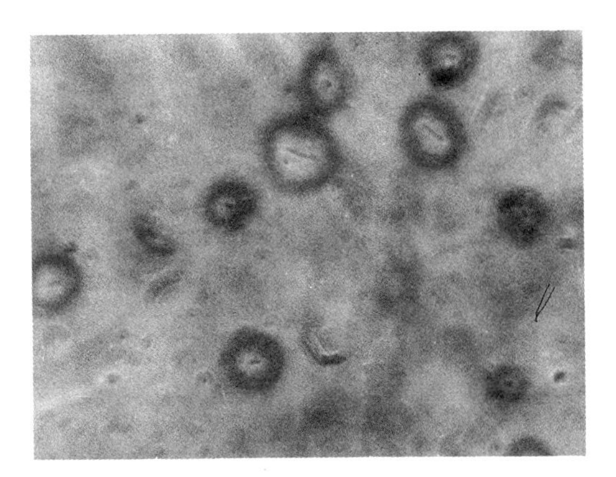

Fig. 1.4. Capsules of *Klebsiella aerogenes* (G. Bitton).

water treatment plants (see Chapter 18 and 19); (6) they play a role in microbial flocculation in the activated sludge process (see Chapter 8).

1.2.6. Cell Motility

Microbial cells can move by means of flagella, cilia, or pseudopods. Bacteria display various flagellar arrangements ranging from monotrichous (polar flagellum; e.g., *Vibrio coma*) and lophotrichous (bundle of flagella at one end of the cell; e.g., *Spirillum volutans*) to peritrichous (several flagella distributed around the cell; e.g., *E. coli*) (Fig. 1.5). The flagellum is composed of a protein called flagellin and is anchored by a hook to a basal body located in the cell envelope. Flagella enable bacteria to attain speeds of 50–100 μm/s. They enable cells to move towards food (chemotaxis), light (phototaxis), or oxygen (aerotaxis). Chemotaxis is the movement of a microorganism towards a chemical, generally a nutrient. It also enables the movement away from a harmful chemical (negative chemotaxis). Chemotaxis can be demonstrated by placing a capillary containing a chemical attractant into a bacterial suspension. Bacteria, attracted to the chemical, swarm around the tip and move inside the capillary. Two sets of proteins, chemoreceptors and transducers, are involved in triggering flagellar rotation and subsequent cell movement. From an ecological viewpoint, chemotaxis provides a selective advantage to bacteria, allowing them to detect carbon and energy sources. Toxicants (e.g., hydrocarbons, heavy metals) inhibit chemotaxis by blocking chemoreceptors, thus affecting food detection by motile bacteria as well as predator-prey relationships in aquatic environments.

Eucaryotic cells move by means of flagella, cilia, or cytoplasmic streaming (i.e., amoeboid movement). Flagella have a more complex structure than procaryotic flagella. Cilia are shorter

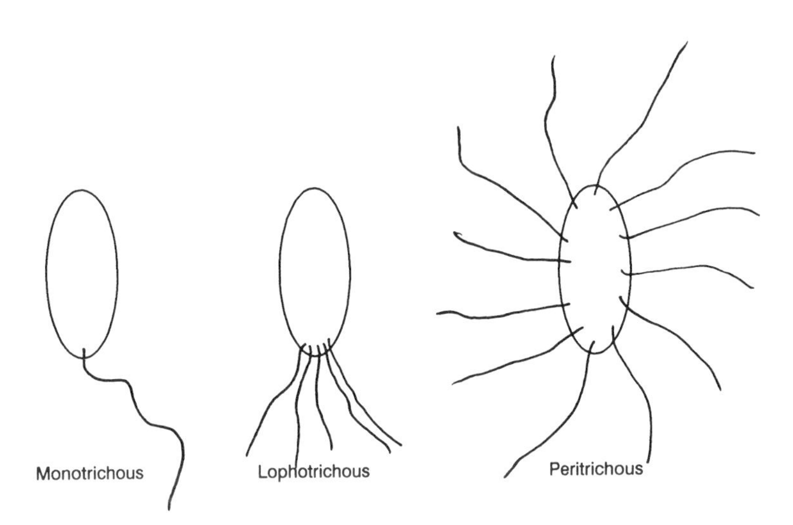

Fig. 1.5. Flagellar arrangements in bacteria.

and thinner than flagella. Ciliated protozoa use cilia for locomotion and for pushing food inside the cytostome, a mouthlike structure. Some eucaryotes (e.g., amoeba, slime molds) move by amoeboid movement by means of pseudopods (i.e., false feet), which are temporary projections of the cytoplasm.

1.2.7. Pili

Pili are structures that appear as short and thin flagella and are attached to cells in a manner similar to that of flagella. They play a role in cell attachment to surfaces, and in conjugation (involvement of a sex pilus), and they act as specific receptors for certain types of phages.

1.2.8. Storage Products

Cells may contain inclusions that contain storage products serving as a source of energy or building blocks. These inclusions, which can be observed under a microscope by means of special stains, include the following:

1. Carbon storage takes the form of glycogen, starch, and poly-β-hydroxybutyric acid (PHB), which stains with sudan black, a fat-soluble stain. PHB occurs exclusively in procaryotic microorganisms.
2. Volutin granules contain polyphosphate reserves. These granules, also called metachromatic granules, appear red when specifically stained with basic dyes such as toluidine blue or methylene blue.
3. Sulfur granules are found in sulfur filamentous bacteria (e.g., *Beggiatoa, Thiothrix*) and purple photosynthetic bacteria which use H_2S as an energy source and electron donor, respectively. H_2S is oxidized to S^0, which accumulates in sulfur granules that are readily visible under a light microscope. Upon depletion of the H_2S source, the elemental sulfur is further oxidized to sulfate.

1.2.9. Gas Vacuoles

Gas vacuoles are found in cyanobacteria, halobacteria (i.e., salt-loving bacteria), and photosynthetic bacteria. Electron microscope studies have shown that gas vacuoles are made of gas vesicles, which are filled with gas and surrounded by a protein membrane. Their role is to regulate cell buoyancy in the water column. Owing to this flotation device, cyanobacteria and photosynthetic bacteria sometimes form massive blooms at the surface of lakes or ponds.

1.2.10. Endospores

Endospores are formed inside bacterial cells and are released when cells are exposed to adverse environmental conditions. The location of the spore may vary. There are central, subterminal, and terminal spores. Physical and chemical agents trigger spore germination to form vegetative cells. Bacterial endospores are very resistant to heat and this is probably due to the presence of a dipicolinic acid–Ca complex in endospores. Endospores are also quite resistant to desiccation, radiation, and harmful chemicals. This is significant from a public health viewpoint, because they are much more resistant to chemical disinfectants than vegetative bacteria in water and wastewater treatment plants (Chapters 5 and 6).

1.2.11. Eucaryotic Organelles

Specialized structures, called organelles, are located in the cytoplasm of eucaryotic cells and carry out several important cell functions. We will now briefly review some of these organelles.

1.2.11.1. Mitochondria

Mitochondria (singular: mitochondrion) are oval or spherical structures surrounded by a double membrane. The outer membrane is very permeable to the passage of chemicals, and the inner membrane is folded to form shelves called cristae (singular: crista) (Fig. 1.6). They are the *site of cell respiration* and ATP production in eucary-

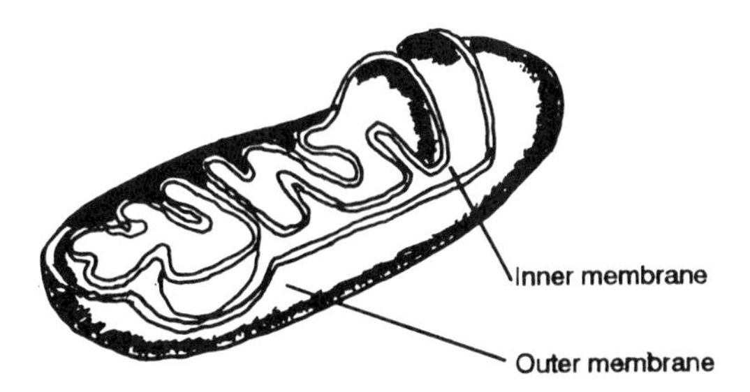

Fig. 1.6. Mitochondrion structure.

otic cells (Chapter 2). The number of mitochondria per cell varies with the type and the metabolic level of cells.

1.2.11.2. *Chloroplasts*

Chloroplasts are relatively large chlorophyll-containing structures found in plant and algal cells that also are surrounded by a double membrane. They are made of units called grana, which are interconnected by lamellae. Each granum is a stack of disks called thylakoids bathing into a matrix called stroma (Fig. 1.7). Chloroplasts are the sites of photosynthesis in plant and algal cells. The light and dark reactions of photosynthesis occur in the thylakoids and stroma, respectively (Chapter 2).

1.2.11.3. *Other Organelles*

Other important organelles that are found in eucaryotic cells but not in procaryotic cells are the following.

The *Golgi complex* consists of a stack of flattened membranous sacs, called saccules, which form vesicles that collect proteins, carbohydrates and enzymes.

The *endoplasmic reticulum* is a system of folded membranes attached to both the cell membrane and the nuclear membrane. The rough endoplasmic reticulum is associated with ribosomes and is involved in protein synthesis. The smooth endoplasmic reticulum is found in cells that make and store hormones, carbohydrates, and lipids.

Lysosomes are sacs that contain hydrolytic (digestive) enzymes and help in the digestion of phagocytized cells by eucaryotes.

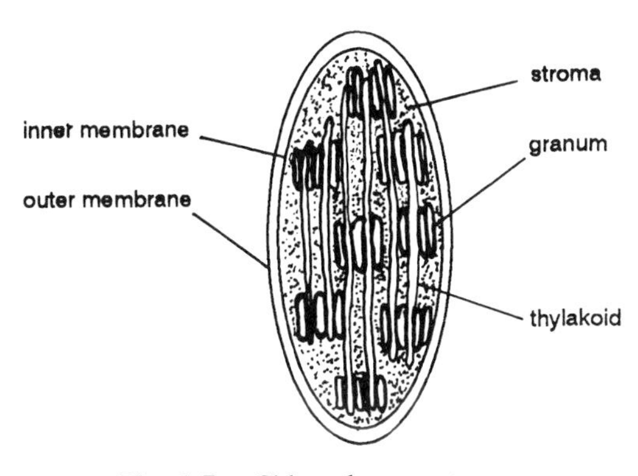

Fig. 1.7. Chloroplast structure.

1.3. CELL GENETIC MATERIAL

1.3.1. DNA Arrangements in Procaryotes and Eucaryotes

In procaryotes, DNA occurs as a single circular molecule that is tightly packed to fit inside the cell and is not enclosed in a nuclear membrane. Procaryotic cells may also contain small circular DNA molecules called plasmids.

Eucaryotes have a distinct nucleus surrounded by a nuclear membrane with very small pores that allow exchanges between the nucleus and the cytoplasm. DNA is present as chromosomes, which consist of DNA associated with histone proteins. Cells divide by mitosis, which leads to a doubling of the chromosome numbers. Each daughter cell has a full set of chromosomes.

1.3.2. Nucleic Acids

Deoxyribonucleic acid (DNA) is a double-stranded molecule that is made of several millions of units (e.g., approximately 4×10^6 base pairs in *E. coli* chromosome) called nucleotides. The double-stranded DNA is organized into a double helix (Fig. 1.8). Each nucleotide is made of a 5-carbon sugar (deoxyribose), a phosphate group, and a nitrogen-containing base linked to the carbon-5 and carbon-1 of the deoxyribose molecule, respectively. The nucleotides on a strand are linked together via a phos-

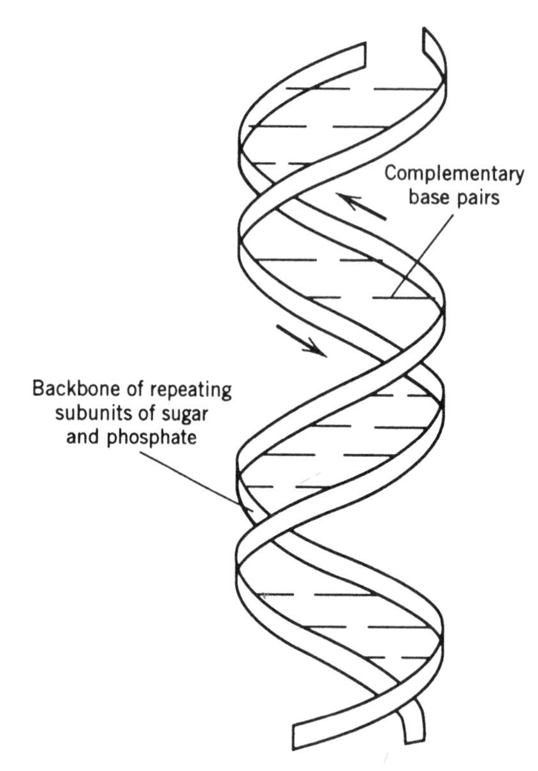

Fig. 1.8. DNA structure.

phodiester bridge. The hydroxyl group of a carbon-3 of a pentose (3′ carbon) is linked to the phosphate group on the carbon-5 (5′ carbon) of the next pentose. There are four different bases in DNA, two purines (adenine and guanine) and two pyrimidines (cytosine and thymine). A base on one strand pairs through hydrogen bonding with another base on the complementary strand. Guanine always pairs with cytosine, while adenine always pairs with thymine (Fig. 1.9). One strand runs in the $5' \rightarrow 3'$ direction, while the complementary strand runs in the $3' \rightarrow 5'$ direction. Physical and chemical agents cause DNA to unwind, leading to the separation of the two strands.

Ribonucleic acid (RNA) is generally single-stranded (some viruses have double-stranded RNA) and contains ribose in lieu of deoxyribose and uracil in lieu of thymine.

1.3.2.1. Replication and Transcription of DNA

Replication. The DNA molecule can make an exact copy of itself. The two strands separate and new complementary strands are formed. The double helix unwinds and each of the DNA strands acts as a template for a new complementary strand. Nucleotides move into the *replication fork* and align themselves against the complementary bases on the template. The addition of nucleotides is catalyzed by an enzyme called DNA polymerase.

Transcription. Transcription is the process of transfer of information from DNA to RNA. The complementary single-stranded RNA molecule is called messenger RNA (mRNA). mRNA carries the information from the DNA to the ribosomes, where it controls protein synthesis. Transcription is catalyzed by an enzyme called RNA polymerase. *Enzyme regulation* (repression or induction) occurs at the level of transcription. Sometimes, the product formed through the action of an enzyme represses the synthesis of that enzyme. The enzyme product acts as a corepressor which, along with a repressor, combines with the operator gene to block transcription and, therefore, enzyme synthesis. The synthesis of other enzymes, called inducible enzymes, occurs only when the substrate is present in the medium. Enzyme synthesis is induced because the substrate, called inducer, combines with the repressor to form a complex that has no affinity for the operator gene.

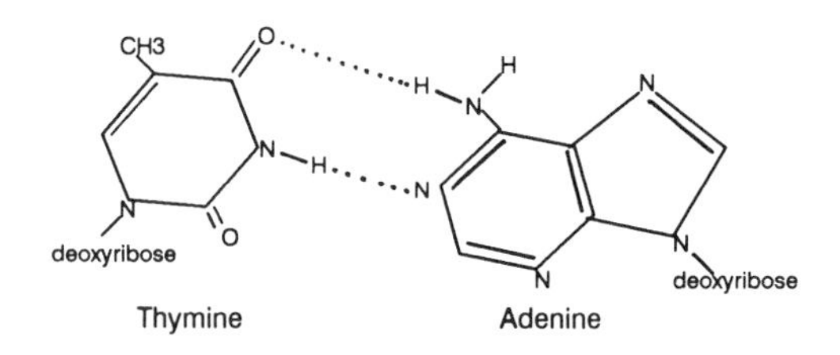

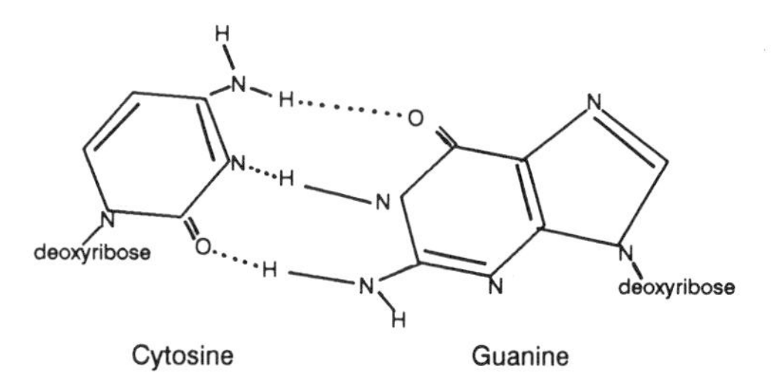

Fig. 1.9. Base-pairing in DNA.

1.3.2.2. *Translation*

Messenger RNA controls protein synthesis in the cytoplasm. This process is called translation. Another type of RNA is the transfer RNA (tRNA), which has attachment sites for both mRNA and amino acids and brings specific amino acids to the ribosome. Each combination of three nucleotides on the messenger RNA is called a codon or triplet, each of which codes for a specific amino acid. The sequence of codons determines that of amino acids in a protein. Some triplets code for the initiation and termination of amino acid sequences. There are 64 possible triplet codons.

1.3.3. Plasmids

A plasmid is a circular extrachromosomal DNA containing from 1,000 to 200,000 base pairs that reproduces independently from the chromosomal DNA. Plasmids are inherited by daughter cells following cell division. Plasmid replication can be inhibited by *curing* the cells with compounds such as ethidium bromide. Some of the plasmids may exist in a limited number (1–3) of copies (stringent plasmids) or in a relatively large number (10–220) of copies (relaxed plasmids). Relaxed plasmids are most useful as cloning vectors. Some plasmids cannot coexist and are thus incompatible with other plasmids in the same cell.

There are several types of plasmids:

1. *Conjugative plasmids* carry genes that code for their own transfer to other cells. F factors or sex factors are conjugative plasmids that can become integrated into the chromosomes. *E. coli* strains that possess the chromosome-integrated F factors are called *Hfr* (high frequency of recombination).

2. *Resistance transfer factors* (R factors) are plasmids that confer upon the host cell resistance to antibiotics (e.g., tetracycline, chloramphenicol, streptomycin) and heavy metals (e.g., mercury, nickel, cadmium). There is a great concern over these plasmids by the medical profession. The widespread use of antibiotics in medicine and agriculture results in the selection of multiple-drug-resistant bacteria with R factors.

3. *Col factors* are plasmids that code for production of colicins, which are proteinaceous bacterial inhibiting substances.

4. *Catabolic plasmids* code for enzymes that drive the degradation of unusual molecules such as camphor, naphtalene, and other xenobiotics found in environmental samples. They are important in the field of pollution control. Plasmids can be engineered to contain desired genes and can be replicated by introduction into an appropriate host (see section 1.3.6).

1.3.4. Mutations

Mutations, caused by physical and chemical agents, change the DNA code and impart new characteristics to the cell, allowing it, for example, to degrade a given xenobiotic or survive under high temperatures. Spontaneous mutations occur in one of 10^6 cells. However, the DNA molecule is capable of self-repair.

Conventional methods are used to obtain desired mutations in a cell. The general approach consists of exposing cells to a mutagen (e.g., ultraviolet light, chemicals) and then exposing them to desired environmental conditions. These conditions select for cells having the desired traits.

1.3.5. Genetic Recombinations

Recombination is the transfer of genetic material (plasmid or chromosomal DNA) from a donor cell to a recipient cell. There are three means by which DNA is transferred to recipient cells (Fig. 1.10) (Brock and Madigan, 1988).

1. *Transformation*. Exogenous DNA enters a *competent* recipient cell (that is, a cell capable of transformation by exogenous DNA) and becomes an integral part of a chromosome or plasmid. Cell competence is affected by the growth phase of the cells (i.e., physiological state of bacteria) as well as the composition of the growth medium. During transformation, the transforming DNA fragment attaches to the competent cell, is incorporated into the cell, and becomes single-stranded, and one strand is integrated into the recipient cell DNA while the other strand is broken down. Transformation efficiency is increased

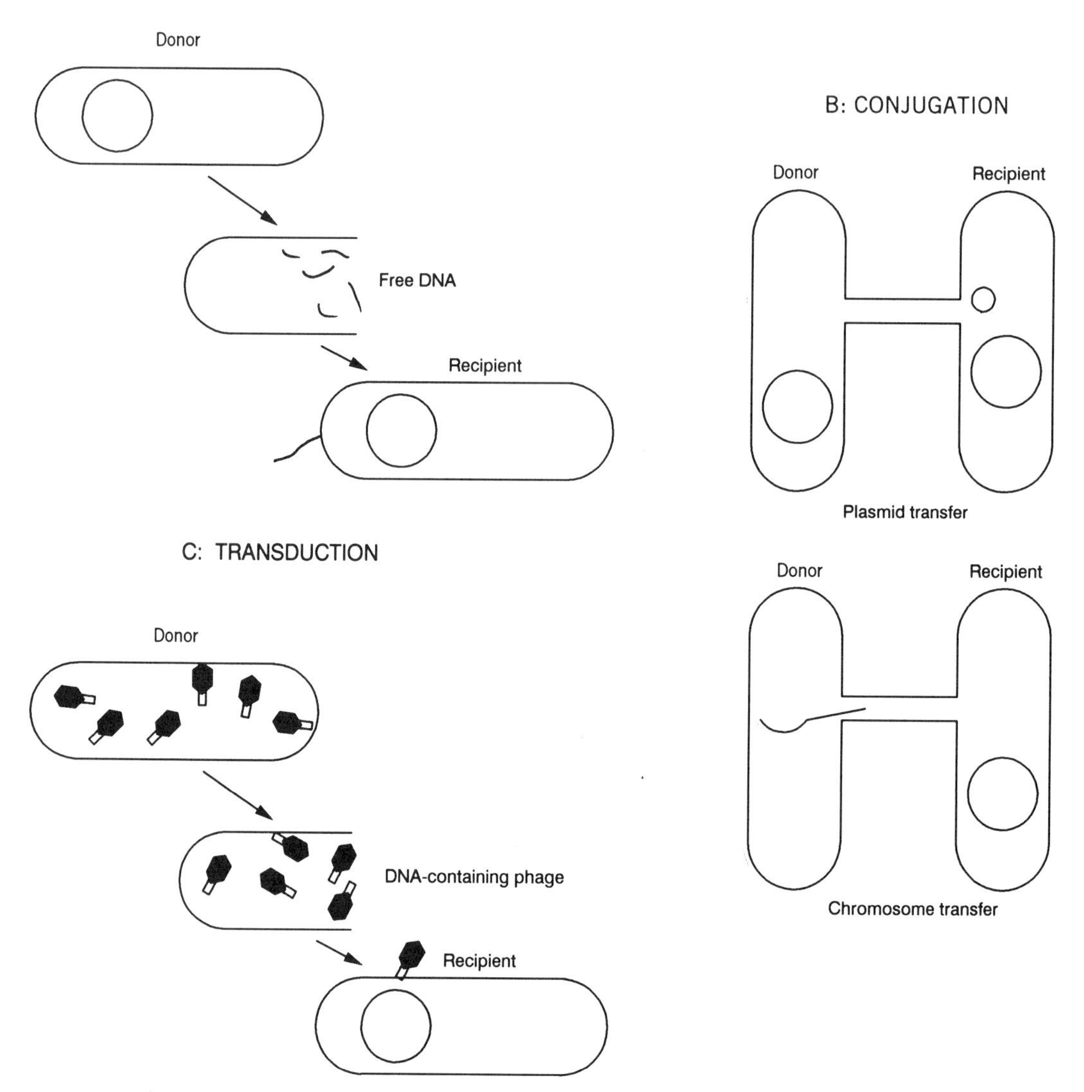

Fig. 1.10. Genetic recombinations among bacteria. **A.** Transformation. **B.** Conjugation. **C.** Transduction. Adapted from Brock and Madigan (1988).

by treating cells with high concentrations of calcium under cold conditions.

If the transforming DNA is extracted from a virus, the process is called *transfection*. DNA can be introduced into eucaryotic cells by electroporation (use of an electric field to produce pores in the cell membrane) or by the use of a particle gun to shoot DNA into the recipient cell.

2. *Conjugation*. This type of genetic transfer necessitates cell-to-cell contact. The genetic material (plasmid or a fragment of a chromosome mobilized by a plasmid) is transferred upon direct contact between a donor cell (F^+ or male cells) and a recipient cell (F^- or female cells). A special surface structure, called the *sex pilus*, of the donor cell forms a conjugation bridge and allows the transfer of the genetic material from the donor to the recipient cell.

Gene transfer by conjugation has been demonstrated in natural environments including wastewater, freshwater, seawater, sediments, and soils. Plasmids coding for antibiotic resistance can be transferred from environmental isolates to laboratory strains. Biotic and abiotic factors (e.g., cell type and density, temperature, oxygen, pH, surfaces) affect gene transfer by conjugation, but their impact under environmental conditions is not well known.

3. *Transduction*. Transduction is the transfer of genetic material from a donor to a recipient cell using a bacterial phage as a carrier. A small fragment of the DNA from a donor cell is incorporated into the phage particle. Upon infection of a donor cell, the DNA of the transducing phase particle may be integrated into the recipient cell DNA. The significance of transduction in gene transfer in the environment is not well known.

4. *Transposition*. Another recombination process is transposition, which consists of the movement (i.e., "jumping") of small pieces of plasmid or chromosomal DNA, called *transposons* ("jumping genes"), from one location to another on the genome. Transposons, which can move from one chromosome to another or from one plasmid to another, carry genes that code for the enzyme transposase, which catalyzes their transposition.

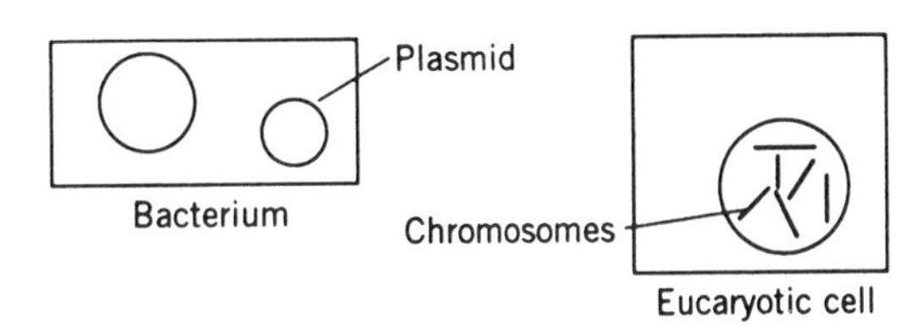

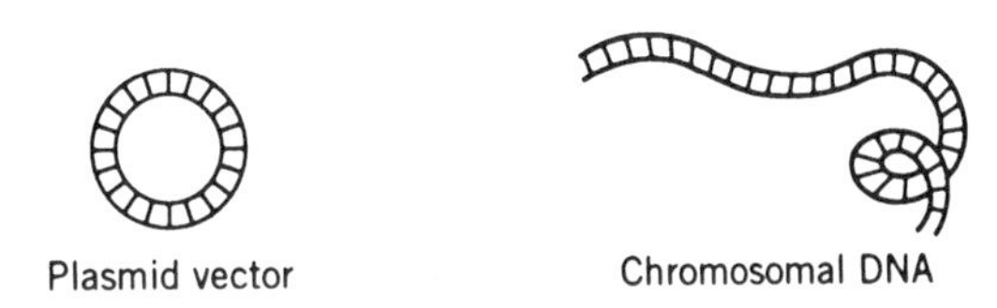

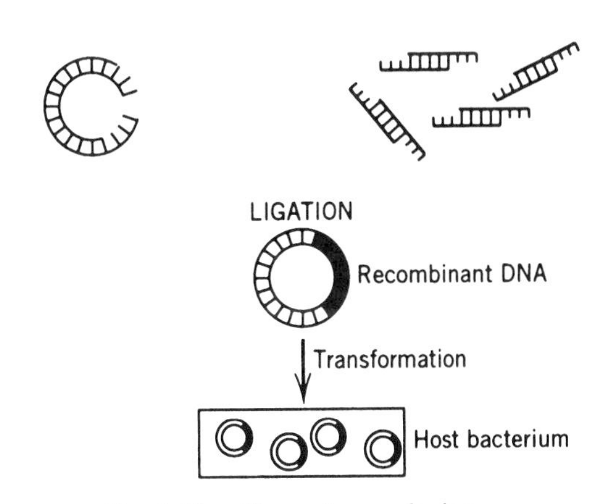

Fig. 1.11. Steps of gene cloning.

1.3.6. Recombinant DNA Technology

Recombinant DNA technology, commonly known as genetic engineering or gene cloning, is the deliberate manipulation of genes to produce useful gene products (e.g., proteins, toxins, hormones). There are two categories of recombination experiments: (1) *in vitro* recombination, which consists of using purified enzymes to break and then rejoin isolated DNA fragments in test tubes; and (2) *in vivo* recombination, which consists of encouraging DNA rearrangements that occur in living cells.

A typical gene cloning experiment consists of the following steps (Fig. 1.11):

1. *Isolation of the source DNA*. Several methods are used for the isolation of DNA from a wide range of cells.

2. *DNA fragmentation or splicing*. Restriction endonucleases are used to cleave the double-stranded DNA at specific sites (these enzymes normally help cells cope with foreign DNA and protect bacterial cells against phage infection). They are named after the microorganism from which they were initially isolated. For example, the restriction enzyme EcoRI was isolated from *E. coli*, whereas HindII enzyme was derived from *Haemophilus influenza*. EcoRI recognizes the following sequences on the double-stranded DNA:

-G-A-A-T-T-C-
-C-T-T-A-A-G-

and produces the following fragments:

```
-G-            A-A-T-T-C-
-C-T-T-A-A             -G-
```

The DNA fragments can be separated according to their size through electrophoresis.

3. *DNA ligation.* DNA fragments are joined to a cloning vector by using another enzyme called DNA ligase. Ligation is possible because both the source DNA and the cloning vector DNA have been cut with the same restriction enzyme. Commonly used cloning vectors are plasmids (e.g., pBR322) or phages (e.g., phage lambda).

4. *Incorporation of the recombinant DNA into a host.* The recombinant DNA is introduced into a cell for replication and expression. The recombinant DNA may be introduced into the host cell by transformation, for example. The most popular hosts are procaryotes such as *E. coli* or eucaryotes such as *Saccharomyces cerevisiae*. The host microorganism, now containing the recombinant DNA, will divide and make clones.

5. *Selection of the desirable clones.* Clones that have the desired recombinant DNA can be screened by using markers, like antibiotic resistance, that indicate the presence of the cloning vector in the cells. However, the selection of clones having the desired gene can be accomplished by utilizing nucleic acid probes (section 1.3.8) or by screening for the gene product. If the gene product is an enzyme (e.g., β-galactosidase), then clones are selected by looking for colonies that have the enzyme of interest (hosts cells are grown in the presence of the enzyme substrate).

1.3.7. Biotechnological Applications of Genetically Engineered Microorganisms

Biotechnological applications of genetically engineered microorganisms (GEMs) have been realized in various fields, including the pharmaceutical industry, agriculture, medicine, the food industry, energy, and pollution control. Notable applications are the production of human insulin and viral vaccines. In agriculture, research is focusing on the production of transgenic plants (i.e., genetically altered whole plants) that are resistant to insects, herbicides, or diseases.

The potential use of GEMs in pollution control is becoming increasingly attractive. It has been proposed to use GEMs to clean up hazardous waste sites and wastewaters by constructing microbial strains capable of degrading recalcitrant molecules. However, there are potential problems associated with the deliberate release of GEMs into the environment, given that, unlike chemicals, GEMs have the potential to grow and reproduce under *in situ* environmental conditions.

1.3.8. Nucleic Acid Probes

Nucleic acid probes are based on nucleic acid hybridization. The two strands of a DNA or RNA molecule are said to be complementary. Single complementary strands of DNA are produced by denaturation of DNA. Under appropriate conditions, the complementary strands hybridize (i.e., bind to each other). A probe is a single strand of DNA that contains specific sequences that, when combined with single-stranded target DNA, will hybridize with the complementary sequence in the target DNA and form a double-stranded structure. For easy detection, the probe is labeled with a radioisotope, an enzyme (e.g., β-galactosidase, peroxidase, or alkaline phosphatase) or a fluorescent compound.

The following are a few applications of nucleic acid probes.

1. *Detection of pathogens in clinical samples.* Probes have been developed for clinically important microorganisms such as *Legionella* spp., *Salmonella* species, enteropathogenic *E. coli, Neisseria gonorrhoea,* human immunodeficiency virus (HIV), and herpes viruses. Probe sensitivity can be increased by using polymerase chain reaction (PCR) technology (see Chapter 16 for details on this technology).

2. *Detection of metal resistance genes in envi-*

ronmental isolates. One example is a probe that was constructed for detecting the *mer* operon, which controls mercury detoxification.

3. *Tracking of specific bacteria in the environment*. Probes are useful in following the fate of specific environmental isolates and genetically engineered microbes in water, wastewater, sludge, and soils.

1.4. BRIEF SURVEY OF MICROBIAL GROUPS

1.4.1. Bacteria

1.4.1.1. Size and Shape of Bacteria

Except for filamentous bacteria (size may be greater than 100 μm) or cyanobacteria (size ranging approximately from 5 μm to 50 μm), the size of bacterial cells generally ranges between 0.3 μm (e.g., *Bdellovibrio bacteriovorus;* Mycoplasma) and 1–2 μm (e.g., *E. coli; Pseudomonas*). An unusually large bacterium was recently discovered by researchers at Indiana University, a gram-positive bacterium that sometimes grows longer than 500 μm and inhabits the intestines of marine surgeonfish.

Bacteria occur in three basic shapes: cocci (spherical; e.g., *Streptococcus,*), bacilli (rods-shaped; e.g., *Bacillus subtilis*) and spirilla (spiral; e.g., *Vibrio cholera; Spirillum volutans*) (Fig. 1.12). Because of their relatively small size, bacteria have a high surface-to-volume ratio, a critical factor in substrate uptake.

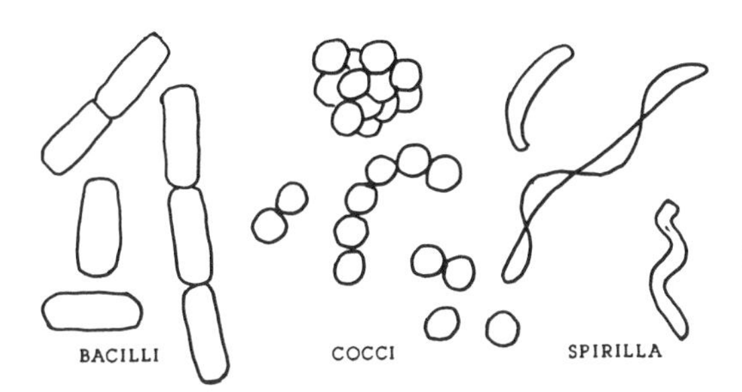

Fig. 1.12. Bacterial cell shapes. Adapted from Edmonds (1978).

1.4.1.2. Unusual Types of Bacteria (Fig. 1.13)

Sheathed bacteria. These bacteria are filamentous microorganisms surrounded by a tube-like structure called the sheath. The bacterial cells inside the sheath are gram-negative rods that become flagellated (swarmer cells) when they leave the sheath. The swarmer cells produce new sheaths at a relatively rapid rate. They are often found in polluted streams and in wastewater treatment plants. This group includes three genera: *Sphaerotilus, Leptothrix,* and *Crenothrix*. These bacteria have the ability to oxid-

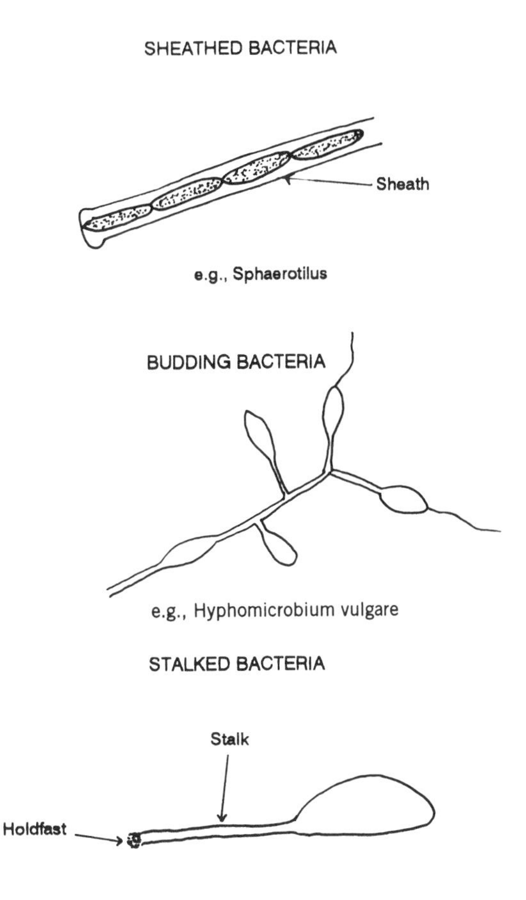

Fig. 1.13. Some unusual types of bacteria. Adapted from Lechevalier and Pramer (1971).

ize reduced iron to ferric hydroxide (e.g., *Sphaerotilus natans, Crenothrix*) or manganese to manganese oxide (e.g., *Leptothrix*). In Chapter 9, we will discuss the role of *Sphaerotilus natans* in activated sludge bulking.

Stalked bacteria. Stalked bacteria are aerobic, flagellated (polar flagellum), gram-negative rods that possess a stalk, a structure that contains cytoplasm and is surrounded by a membrane and a wall. At the end of the stalk is a holdfast, which allows the cells to adsorb to surfaces. The cells may adhere to one another and form rosettes. *Caulobacter* are typical stalked bacteria that are found in aquatic environments with low organic content. *Gallionella* (e.g., *G. ferruginea*) makes a twisted stalk, sometimes called a "ribbon," which consists of an organic matrix surrounded by ferric hydroxide. These bacteria are present in iron-rich waters and Fe^{2+} that have oxidized to Fe^{3+}. They are found in metal pipes in water distribution systems (Chapter 15).

Budding bacteria. Following attachment to a surface, some bacteria multiply by budding. They make filaments or hyphae, at the end of which a bud is formed. The bud acquires a flagellum (the cell is now called a swarmer), settles on a surface, and forms a new hypha with a bud at the top. *Hyphomicrobium* is widely distributed in soils and aquatic environments and requires one-carbon (e.g., methanol) compounds for growth. A phototrophic bacterium, *Rhodomicrobium,* is another example of budding bacteria.

Gliding bacteria. These filamentous gram-negative bacteria move by gliding, a slow motion on a solid surface. They resemble certain cyanobacteria except that they are colorless. *Beggiatoa* and *Thiothrix* are gliding bacteria that oxidize H_2S to S^0, which accumulates as sulfur granules inside the cells. *Thiothrix* filaments are characterized by their ability to form rosettes (more details are given in Chapters 3 and 9). Myxobacteria are another group of gliding microorganisms. They feed by lysing bacterial, fungal, or algal cells. Vegetative cells agregate to make "fruiting bodies," which lead to the formation of resting structures called myxospores. Under favorable conditions, myxospores germinate into vegetative cells.

Bdellovibrio (B. bacteriovorus). *Bdellovibrio* are small (0.2–0.3 μm), flagellated (polar flagellum) bacteria that are predatory on gram-negative bacteria. After attaching to the bacterial prey, *Bdellovibrio* penetrate inside the cells and multiply in the periplasmic space (i.e., space between the cell wall and the plasma membrane). Since they lyse their prey, they are able to form plaques on a lawn of the host bacterium. Some *Bdellovibrio* can grow independently on complex organic media.

Actinomycetes. Actinomycetes are gram-positive filamentous bacteria characterized by mycelial growth (i.e., branching filaments), which is analogous to growth by fungi. However, filaments are similar in diameter to bacteria (approximately 1 μm). Most actinomycetes are strict aerobes but a few of them require anaerobic conditions. Most of these microorganisms produce spores, and their taxonomy is based on these reproductive structures (e.g., single spores in *Micromonospora* or chains of spores in *Streptomyces*). They are commonly found in water, wastewater treatment plants and soils (with a preference for neutral and alkaline soils). Some of them (e.g., *Streptomyces*) produce a characteristic "earthy" odor that is due to the production of volatile compounds called geosmins (Chapter 15). They degrade polysaccharides (e.g., starch, cellulose), hydrocarbons, and lignin. Some of them produce antibiotics (e.g., streptomycin, tetracycline, chloramphenicol). Two well-known genera of actinomycetes are *Streptomyces* and *Nocardia* (Fig. 1.14). *Streptomyces* form a mycelium with conidial spores at the tip of the hyphae. These actinomycetes are important industrial microorganisms that produce hundreds of antibiotic substances. *Nocardia* are commonly found in water and wastewater and degrade hydrocarbons and other recalcitrant compounds. *Nocardia* are a significant

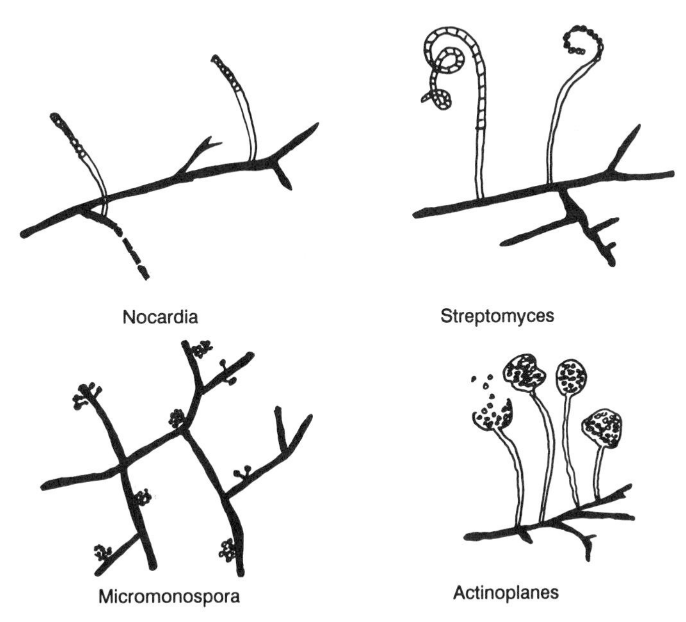

Fig. 1.14. Some common actinomycetes. Adapted from Cross and Goodfellow (1973).

constituent of foams in activated sludge units (Chapter 9).

Cyanobacteria. Cyanobacteria, often referred to as blue-green algae, are procaryotic organisms that differ from photosynthetic bacteria in the fact that they carry out oxygenic photosynthesis (see Chapter 2). (Fig. 1.15). They contain chlorophyll *a* and accessory pigments such as phycocyanin (blue pigment) and phycoerythrin (red pigment). The characteristic blue-green color exhibited by these organisms is due to the combination of chlorophyll *a* and phycocyanin. Cyanobacteria occur as unicellular, colonial, or filamentous organisms. They propagate by binary fission or fragmentation, and some may form resting structures, called akinetes, that germinate under favorable conditions into a vegetative form. Many of them contain gas vacuoles, which increases their buoyancy and helps them float to the top of the water column, where light is most available for photosynthesis. Some cyanobacteria (e.g., *Anabaena*) are able to fix nitrogen; the site of nitrogen fixation is a structure called a heterocyst.

Cyanobacteria are ubiquitous and, owing to their resistance to extreme environmental conditions (e.g., high temperatures, desiccation), they are found in desert soils and hot springs. They are responsible for algal blooms in lakes and other aquatic environments, and some are quite toxic (Chapter 15).

1.4.2. Fungi

Fungi are eucaryotic organisms that produce long filaments called hyphae, which form a mass called mycelium. Chitin is a characteristic component of the cell wall of hyphae. In most fungi, the hyphae are septate and contain cross-walls that divide the filament into separate cells containing one nucleus each. In some others, the hyphae are nonseptate and contain several nuclei. They are called coenocytic hyphae.

Fungi are heterotrophic organisms that include both macroscopic and microscopic forms. They use organic compounds as carbon source and energy and thus play an important role in nutrient recycling in aquatic and soil environments. Some fungi form traps that capture protozoa and nematodes. They grow well under acidic conditions (pH = 5) in foods, water, or wastewa-

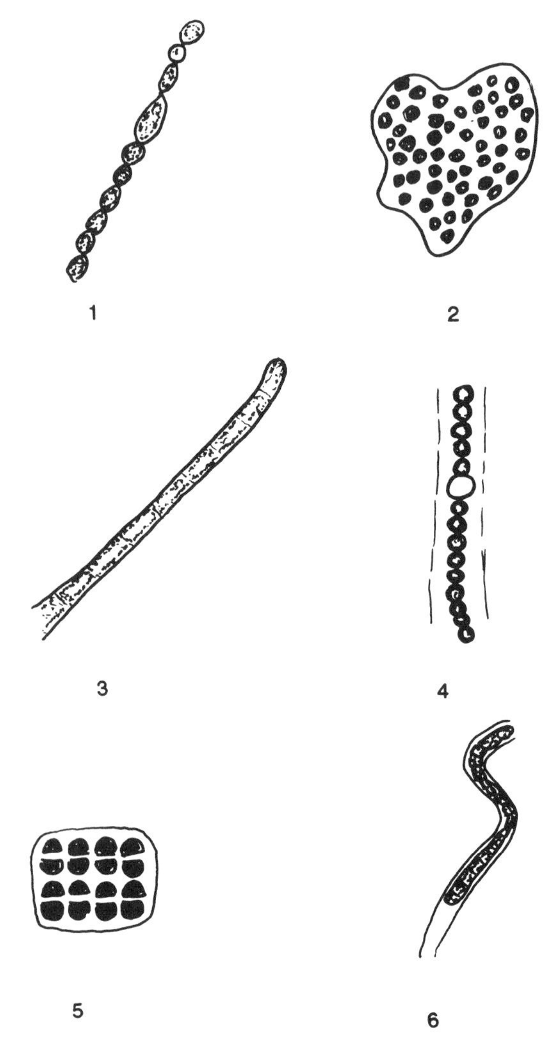

Fig. 1.15. Some common cyanobacteria (blue-green algae). (1) *Anabaena;* (2) *Anacystis;* (3) *Oscillatoria;* (4) *Nostoc;* (5) *Agmenellum;* (6) *Lyngbya.* Adapted from Benson (1973).

ter. Most fungi are aerobic, although some (e.g., yeasts) can grow under facultatively anaerobic conditions. Fungi are significant components of the soil microflora, and a great number of fungal species are pathogenic to plants, causing significant damages to agricultural crops. A limited number of species are pathogenic to humans and cause fungal diseases called mycoses. Airborne fungal spores are responsible for allergies in humans. Fungi are implicated in several industrial applications, such as fermentation processes and production of antibiotics (e.g., penicillin). We will discuss their role in composting in Chapter 12.

Identification of fungi is mainly based on the type of reproductive structure. Most fungi produce spores (sexual or asexual spores) for reproduction, dispersal, and resistance to extreme environmental conditions. Asexual spores are formed from the mycelium and germinate to give organisms that are identical to the parent. The nuclei of two mating strains fuse to give a diploid zygote, which gives haploid sexual spores following meiosis.

The Four Major Groups of Fungi (Fig. 1.16)

Phycomycetes. Phycomycetes are known as the *water molds* and occur on the surface of plants and animals in aquatic environments. They have nonseptate hyphae and reproduce by forming a sac, called a sporangium, that eventually ruptures to liberate zoospores, which settle and form a new organism. Some phycomycetes produce sexual spores. There are also terrestrial phycomycetes such as the common bread mold (*Rhizopus*) which reproduces asexually as well as sexually.

Ascomycetes. Ascomycetes have septate hyphae. Their reproduction is carried out by sexual spores (ascospores), which are contained in a sac called an ascus (eight or more ascospores in an ascus), or by asexual spores (conidia), which are often pigmented. *Neurospora crassa* is a typical ascomycete. Most of the yeasts (e.g., *Saccharomyces cerevisiae,* or baker's yeast) are classified as ascomycetes. They form relatively large cells that reproduce asexually by budding or fission, and sexually by conjugation and sporulation. Some of these organisms (e.g., *Candida albicans*) are pathogenic to humans. Yeasts, especially of the genus *Saccharomyces,* are important industrial microorganisms involved in making bread, wine, and beer.

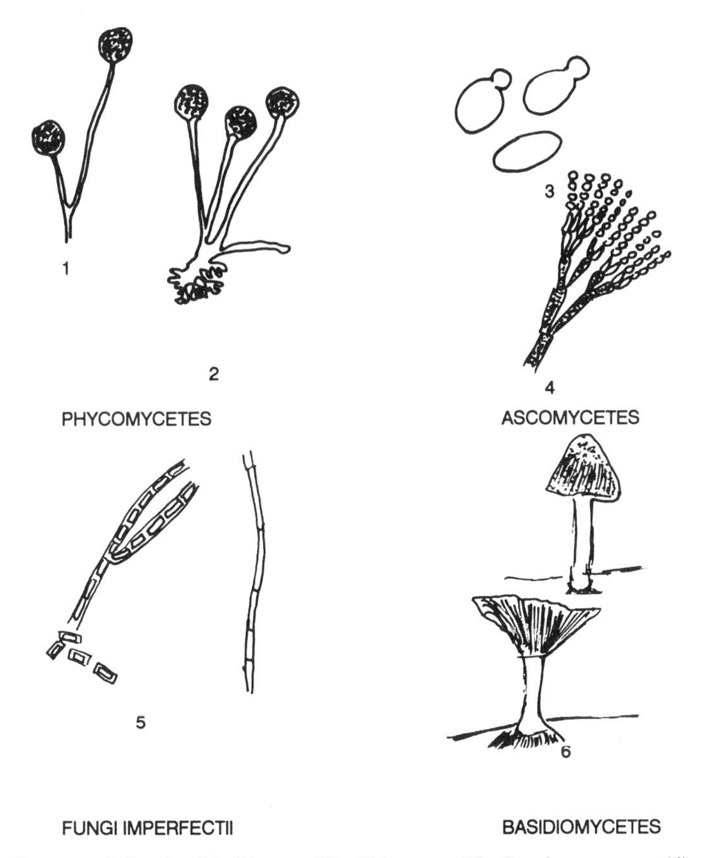

Fig. 1.16. Classes of fungi. (1) *Mucor;* (2) *Rhizopus;* (3) *Saccharomyces;* (4) *Penicillium;* (5) *Geotrichum;* (6) fruiting bodies of mushrooms (basidiomycetes).

Basidiomycetes. Basidiomycetes also have a septate mycelium. They produce sexual spores (basidiospores) on the surface of a structure called the basidium). Four basidiospores are formed on the surface of each basidium. Certain basidiomycetes, the *wood-rotting fungi,* play a significant role in the decomposition of cellulose and lignin. Common edible mushrooms (e.g., *Agaricus*) belong to the basidiomycete group. Unfortunately, some of them (e.g., *Amanita*) are quite poisonous.

Fungi imperfecti. Fungi Imperfecti have septate hyphae but no known sexual stage. Some of them (e.g., *Penicillium*) are used for the commercial production of important antibiotics. These fungi cause plant diseases and are responsible for mycoses in animals and humans (e.g., athlete's foot).

1.4.3. Algae

Most algae are floating unicellular microorganisms called phytoplankton. Many of them are unicellular, but some are filamentous (e.g., *Ulothix*) and others are colonial (e.g., *Volvox*). Although most of them are free-living organisms, some form symbiotic associations with fungi (lichens), animals (corals), protozoa, and plants.

Algae play the role of primary producers in aquatic environments, including oxidation ponds for wastewater treatment. Most are phototrophic microorganisms (Chapter 2). All algae contain chlorophyll *a;* some contain chlorophyll *b* and *c*, as well as other pigments such as xanthophyll and carotenoids. They carry out oxygenic photosynthesis (they use light as a source of energy and H_2O as electron donor) and grow in mineral media with vitamin supplements and with CO_2 as the carbon source. Under environmental conditions, vitamins are generally provided by bacteria. Some algae (e.g., euglenophyta) are heterotrophic and use organic compounds (simple sugars and organic acids) as a source of carbon and energy. Algae have either asexual or sexual reproduction.

Classification of Algae

Classification of algae is based mainly on the type of chlorophyll, the cell wall structure, and the nature of the carbon reserve material produced by the algae cells (Fig. 1.17).

Phylum chlorophyta (green algae). The green algae contain chlorophylls *a* and *b*, have a cellulose cell wall, and produce starch as a reserve material.

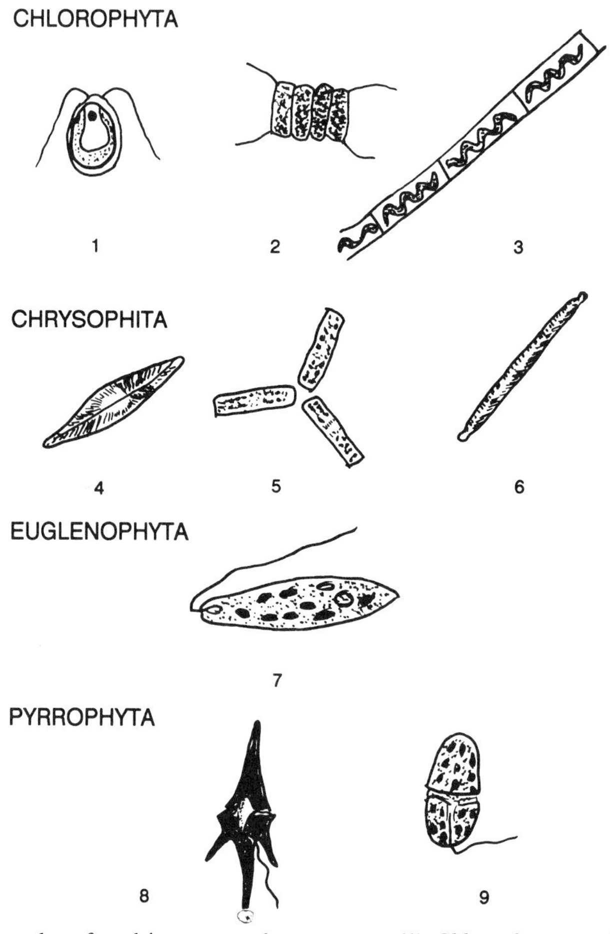

Fig. 1.17. Some algae found in water and wastewater. (1) *Chlamydomonas;* (2) *Scenedesmus;* (3) *Spyrogira;* (4) *Navicula;* (5) *Tabellaria;* (6) *Synedra;* (7) *Euglena;* (8) *Ceratium;* (9) *Gymnodynium.* Adapted from Benson (1973).

Phylum chrysophyta (golden-brown algae). This phylum contains an important group, the diatoms. They are ubiquitous, being found in marine and freshwater environments, sediments, and soils. They contain chlorophylls *a* and *c* and their cell walls typically contain silica (they are responsible for geological formations of diatomaceous earth); they produce lipids as reserve materials.

Phylum euglenophyta. The euglenophytes contain chlorophylls *a* and *b,* have no cell walls, and store reserves of paramylon, a glucose polymer. *Euglena* is a typical euglenophyte.

Phylum pyrrophyta (dinoflagellates). The pyrrophyta contain chlorophylls *a* and *c,* cellulose cell walls, and store starch.

Phylum rhodophyta (red algae). The red algae are found exclusively in a marine environment. They contain chlorophylls *a* and *d* and other pigments such as phycoerythrin, they store starch, and their cell walls are made of cellulose.

Phylum phaeophyta (brown algae). The cells of these exclusively marine algae contain chlorophylls *a* and *c* and xanthophylls; they store laminarin (β-1, 3-glucan) as reserve materials and also have cellulose walls.

1.4.4. Protozoa

Protozoa are unicellular organisms that are important to public health and process microbiology in water and wastewater treatment plants. The cells are surrounded by a cytoplasmic membrane that is covered by a protective structure called a pellicle. They form cysts under adverse environmental conditions, which are quite resistant to desiccation, starvation, high temperatures, lack of oxygen, and chemical insult, namely disinfection in water and wastewater treatment plants (Chapter 6). Protozoa are found in soils and aquatic environments, including wastewater. Some are parasitic to animals, including humans.

Protozoa are heterotrophic organisms that can absorb soluble food, which is transported across the cytoplasmic membrane. The holozoic protozoa are capable of engulfing particles such as bacteria. Ciliated protozoa use their cilia to move particles toward a mouthlike structure called the cytostome. They reproduce by binary fission, although sexual reproduction occurs in some species of protozoa (e.g., *Paramecium*).

Classification of Protozoa

Type of locomotion is the basis for classification of protozoa (Fig. 1.18). The medically important protozoa that can be transmitted by water and wastewater will be discussed in more detail in Chapter 4.

Sarcodina (amoebae). The sarcodina move by means of pseudopods (false feet). Movement by pseudopods is achieved by changes in the viscosity of the cytoplasm. Many of the amoebae are free-living but some are parasitic (e.g., *Entamoeba histolytica*). Amoebae feed by absorbing soluble food or by phagocytosis of the prey.

The foraminifera are sarcodina, found in the marine environment, that have shells called tests, which can be found as fossils in geologic formations.

Mastigophora (flagellates). The mastigophora move by means of flagella. Protozoologists include in this group the phytomastigophora, which are photosynthetic (e.g., *Euglena;* this alga is also heterotrophic). Another example of a flagellate is the human parasite *Giardia lamblia* (Chapter 4). Still another flagellate, *Trypanosoma gambiense,* transmitted to humans by the Tsetse fly, causes African sleeping sickness, which is characterized by neurological disorders.

Ciliphora (ciliates). These organisms use cilia for locomotion, but they also help in feeding. A well-known large ciliate is *Paramecium:* Some are parasitic to animals and humans. For example, *Balantidium coli* causes dysentery when cysts are ingested.

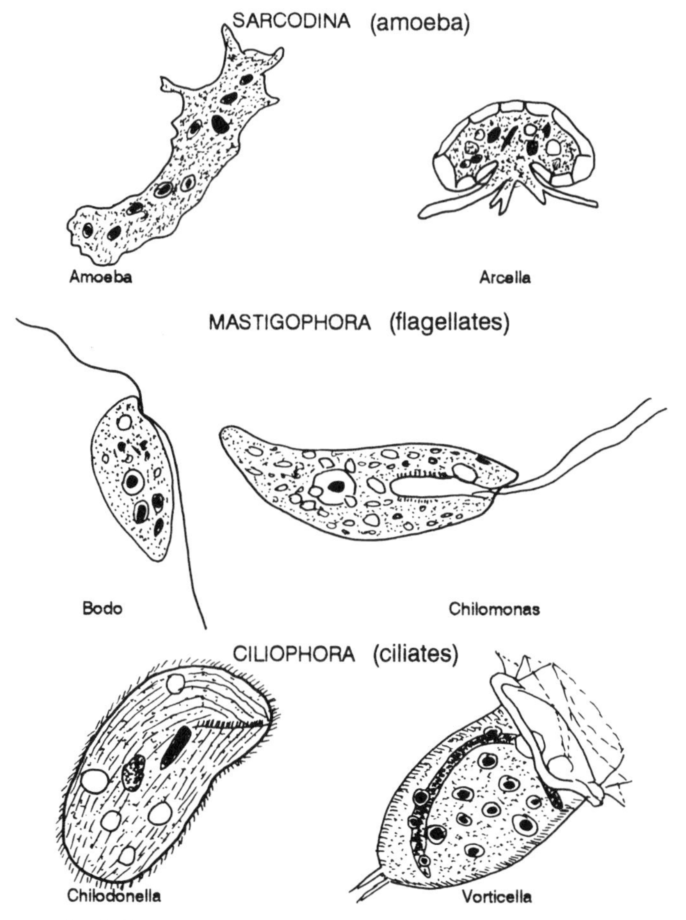

Fig. 1.18. Protozoa groups found in water and wastewater.

Sporozoa. The sporozoa have no means of locomotion and are exclusively parasitic. They feed by absorbing food and produce infective spores. A well-known sporozoan is *Plasmodium vivax,* which causes malaria. The infective agent, the sporozoite, is injected into humans by mosquito bites.

1.4.5. Viruses

Viruses belong neither to procaryotes nor to eucaryotes; they carry out no catabolic or anabolic function. Their replication occurs inside a host cell. The infected cells may be animal or plant cells, bacteria, fungi, or algae. Viruses are very small colloidal particles (25–350 nm) and most of them can be observed only with an electron microscope. Figure 1.19 shows the various sizes and shapes of viruses.

1.4.5.1. Virus Structure

A virus is made up of a core of nucleic acid (double-stranded or single-stranded DNA; double-stranded or single-stranded RNA) surrounded by a protein coat called a capsid. Capsids all composed of arrangements of various numbers of protein subunits known as capsomeres. The combination of capsid and nucleic acid core is called nucleocapsid. There are two main types of capsid symmetry: In helical symmetry, the capsid is a cylinder with a helical

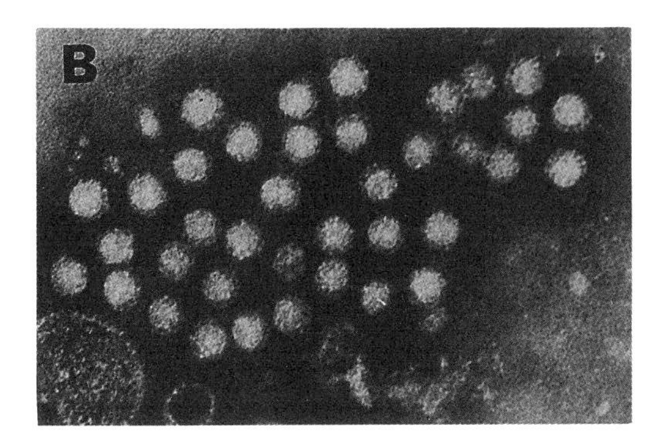

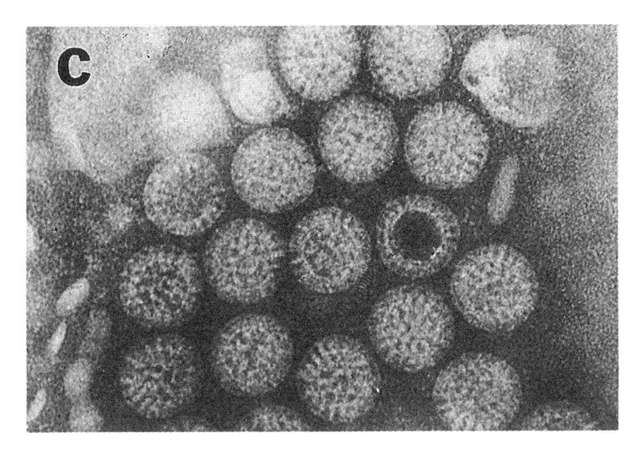

Fig. 1.19. Some enteric viruses of public health importance. **A.** Poliovirus 1. **B.** Hepatitis A virus. **C.** Rotavirus. Courtesy of R. Floyd and J.E. Banatvala.

structure (e.g., tobacco mosaic virus). In polyhedral symmetry, the capsid is an icosahedron that has 20 triangular faces, 12 corners, and 30 edges (e.g., poliovirus). Some viruses (e.g., bacterial phages) have more complex structures and some of them (e.g., influenza or herpes viruses) have an envelope that is composed of lipoproteins or lipids.

1.4.5.2. Virus Replication

Bacterial phages have been used as models to elucidate the phases involved in virus replication, which are the following (Fig. 1.20).

1. *Adsorption.* In order to infect the host cells, the virus particle must adsorb to receptors located on the cell surface. Animal viruses adsorb to surface components of the host cell. The receptors may be polysaccharides, proteins, or lipoproteins.
2. *Entry.* This step involves the entry of a virus particle or its nucleic acid into the host cell. Bacteriophages "inject" their nucleic acid into the host cell. For animal viruses the whole virion penetrates the host cell by endocytosis.
3. *Eclipse.* During this step, the virus particle is "uncoated" (i.e., the capsid is stripped away), and the nucleic acid is liberated.
4. *Multiplication.* This step involves the actual replication of the viral nucleic acid.
5. *Maturation.* The protein coat is synthesized and is combined with the nucleic acid to form a nucleocapsid.
6. *Release of mature virions.* Virus release generally results from the rupture of the host cell membrane.

1.4.5.3. Virus Detection and Enumeration

There are several approaches to virus detection and enumeration:

Animal inoculation. This was the traditional method for detecting viruses prior to the advent of tissue cultures. Newborn mice are infected with the virus and observed for symptoms of disease. Animal inoculation is essential for the detection of enteroviruses such as coxsackie A viruses.

Tissue cultures. Viruses are quantified by measuring their effect on established host cell lines which, under appropriate nutritional conditions, grow and form a monolayer on the inner

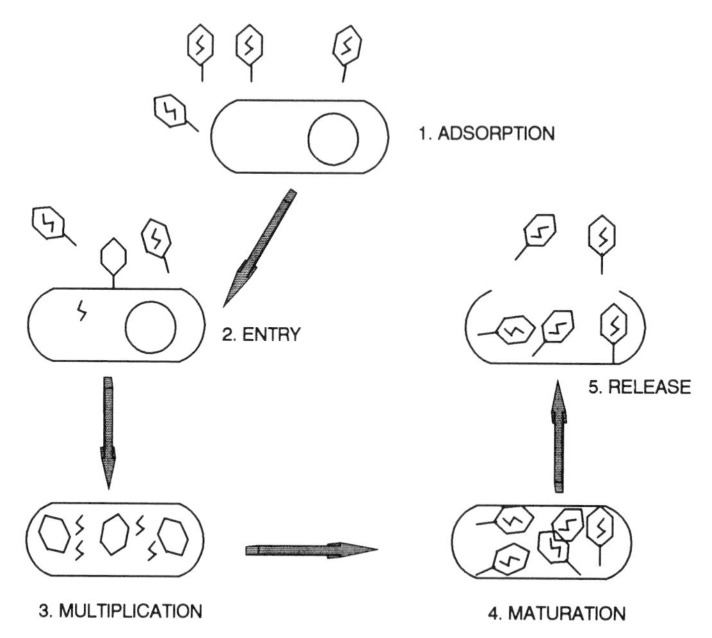

Fig. 1.20. Viral lytic cycle.

surface of glass or plastic bottles. There are two main types of host cell lines: (1) *primary cell lines* (these cells are removed directly from the host tissues and can be subcultured for only a limited number of times); and (2) *continuous cell lines* (animal cells, after serial subculturing, acquire characteristics that are different from the original cell line, allowing them to be subcultured indefinitely; they are derived from normal or cancerous tissues).

Many viruses (enteroviruses, reoviruses, adenoviruses) infect host cells and display a cytopathic effect. Others (e.g., rotaviruses, hepatitis A virus) multiply in the cells but do not cause a cytopathic effect. The presence of the latter needs to be confirmed by other tests, including immunological procedures, monoclonal antibodies, or nucleic acid probes. Other viruses (e.g., Norwalk-type virus) cannot yet be detected by tissue cultures.

Plaque assay. A viral suspension is placed on the surface of a cell monolayer and, following adsorption of viruses to the host cells, an overlay of soft agar or carboxymethylcellulose is poured onto the surface of the monolayer. Virus replication leads to localized areas of cell destruction called plaques. The results are expressed in numbers of plaque-forming units (PFU). Bacteriophages are also assayed by a plaque assay method based on similar principles. They form plaques that are zones of lysis of the host bacterial lawn (Fig. 1.21).

Serial dilution endpoint. Aliquots of serial dilutions of a viral suspension are inoculated into cultured host cells and, following incubation, the viral cytopathic effect (CPE) is recorded. The titer or endpoint is the highest viral dilution (i.e., smallest amount of viruses) capable of producing CPE in 50% of the cultures and is referred to as the tissue culture infectious dose ($TCID_{50}$).

Most probable number. Virus titration, using three dilutions of the viral suspension, is carried out in tubes or 96-well microplates. Virus-positive tubes or wells are recorded and the most probable number (MPN) is computed, from MPN tables.

1.4.5.4. Rapid Detection Methods

Immunoelectron microscopy. Viruses are incubated with specific antibodies and examined

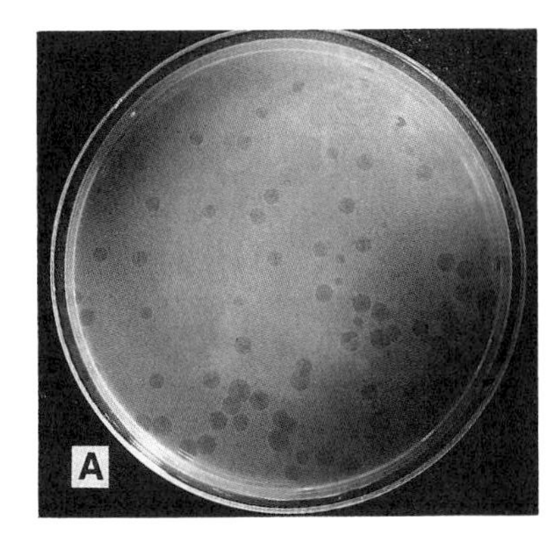

Fig. 1.21. Virus enumeration by plaque assay. **A.** Bacterial phage. **B.** Animal virus (poliovirus 1).

by electron microscopy for the presence of virus particles aggregated by the antibody. This is a useful technique for examining viruses such as the Norwalk-type agent.

Immunofluorescence. A fluorescent dye-labeled antibody is combined with the viral antigen, and the complex formed is observed with a fluorescence microscope. This method enables the detection of rotaviruses as fluorescent foci in tissue cultures.

Enzyme-linked immunosorbent assay (ELISA). A specific antibody is fixed on a solid support and the antigen (virus) is added to form an antigen–antibody complex. An enzyme-labeled specific antibody is then added to the fixed antigen. The presence of virus is detected by the formation of a colored product upon addition of the enzyme substrate. This enzymatic reaction can be conveniently quantified with a spectrophotometer.

Radioimmunoassay (RIA). This assay is also based on the binding of an antigen by a specific antibody. The antigen is quantified by labeling the antibody with a radioisotope (^{125}I) and measuring the radioactivity bound to the antigen–antibody complex.

When viruses growing in host cells are treated with a ^{125}I-labeled antibody, the radioactive foci can be enumerated following contact with a special film. This test, called radioimmunofocus assay (RIFA), is used for the detection of hepatitis A virus.

Nucleic acid probes. Gene probes are pieces of nucleic acid that help identify unknown microorganisms by hybridizing (i.e., binding) to the homologous organism's nucleic acid. For easy detection, the probes can be labeled with radioactive isotopes such as ^{32}P or with enzymes such as alkaline phosphatase, peroxidase, or β-galactosidase. Nucleic acid probes have been used for the detection of viruses (e.g., polioviruses, hepatitis A virus) in environmental samples (water, sediments, shellfish, etc.). Single-stranded RNA (ssRNA) probes have been utilized for the detection of hepatitis A virus in shellfish concentrates. Unfortunately, these probes are not sensitive and detect a minimum of 10^6 hepatitis A virus particles. Amplification of the target viral nucleic acid sequences by polymerase chain reaction (PCR) is being considered.

1.4.5.5. Virus Classification

Viruses can be classified on the basis of the host cell they infect. We will briefly discuss the classification of animal, algal, and bacterial phages.

Animal viruses. Animal viruses are classified mainly on the basis of their genetic material (DNA or RNA), presence of an envelope, capsid symmetry, and site of capsid assembly. All DNA viruses have double-stranded DNA except members of the parvovirus group, and all RNA viruses have single-stranded RNA except members of the reovirus group. Tables 1.2 and 1.3 show the major groups of animal viruses. Of great interest to us in this book is the enteric virus group, the members of which are encountered in water and wastewater. This group will be discussed in more detail in Chapter 4.

Retroviruses are a special group of RNA viruses. These viruses make an enzyme, called reverse transcriptase, that converts RNA into double-stranded DNA, which integrates into the host genome and controls viral replication. A notorious retrovirus is the human immunodeficiency virus (HIV), which causes aquired immunodeficiency syndrome (AIDS).

Algal viruses. A wide range of viruses or "virus-like particles" of eukaryotic algae have been isolated from environmental samples. The nucleic acid is generally double-stranded DNA but is unknown in several of the isolates. Viruses infecting *Chlorella* cells are large particles (125–200 nm diameter), with an icosahedral shape and a linear double-stranded DNA.

Cyanophages, discovered in 1960s, infect a number of cyanobacteria (blue-green algae). They are generally named after their hosts. For example, LPP1 cyanophage has a series of three cyanobacterial hosts: *Lyngbia, Phormidium,* and *Plectonema*. They range in size from 20 nm to 250 nm, and they all contain DNA. Cyanophages have been isolated around the world from oxidation ponds, lakes, rivers, and fishponds. Cyanophages have been proposed for use as biological control agents for the overgrowth (i.e., blooms) of cyanobacteria. Although some experiments have been successful on a relatively small scale, cyanobacteria control by cyanophages under field conditions remains to be demonstrated.

Bacterial phages. Bacteriophages infect a wide range of bacterial types. The various families of phages are illustrated in Figure 1.22. A typical T-even phage is made of a *head* (capsid), which contains the nucleic acid core; a *sheath* or "tail," which is attached to the head through a "neck"; and *tail fibers,* which help in the adsorption of the phage to its host cell. The genetic material is mostly double-stranded DNA but may be single-stranded DNA (e.g., ϕX174) or single-stranded RNA (e.g., f2, MS2). Phages adsorb to the host bacterial cell and initiate the

TABLE 1.2. Major Groups of Animal DNA Viruses[a,b]

Group	Parvoviruses	Papovaviruses	Adenoviruses	Herpesviruses	Poxviruses
Capsid symmetry	Cubic	Cubic	Cubic	Cubic	Complex
Virion: naked or enveloped	Naked	Naked	Naked	Enveloped	Complex coat
Site of capsid assembly	Nucleus	Nucleus	Nucleus	Nucleus	Cytoplasm
Reaction to ether (or other liquid solvent)	Resistant	Resistant	Resistant	Sensitive	Resistant
Diameter of virion (nm)	18–26	45–55	70–90	100	230–300

[a]Adapted from Melnick (1976).
[b]All DNA viruses of vertebrates have double-stranded DNA, except members of the parvoviruses, which have single-stranded DNA.

TABLE 1.3. Major Groups of Animal RNA Viruses[a,b]

	Picornavirus	Reovirus	Rotavirus	Rubella	Arbovirus	Myxovirus	Paramyxovirus	Rhabdovirus
Capsid symmetry	Cubic	Cubic	Cubic	Cubic	Cubic	Helical (or unknown)	Helical (or unknown)	Helical (or unknown)
Virion: naked or enveloped	Naked	Naked	Naked	Enveloped	Enveloped	Enveloped	Enveloped	Enveloped
Site of capsid assembly	Cytoplasm	Cytoplasm	Cytoplasm	Cytoplasm	Cytoplasm	Cytoplasm	Cytoplasm	Cytoplasm
Reaction to ether (or to other liquid solvent)	Resistant	Resistant	Resistant	Sensitive	Sensitive	Sensitive	Sensitive	Sensitive
Diameter of virion (nm)	20–30	75	64–66	60	40	80–120	150–300	60–180

[a]Adapted from Melnick (1976).

[b]All RNA viruses of vertebrates have single-stranded RNA, except members of the reovirus group, which are double-stranded.

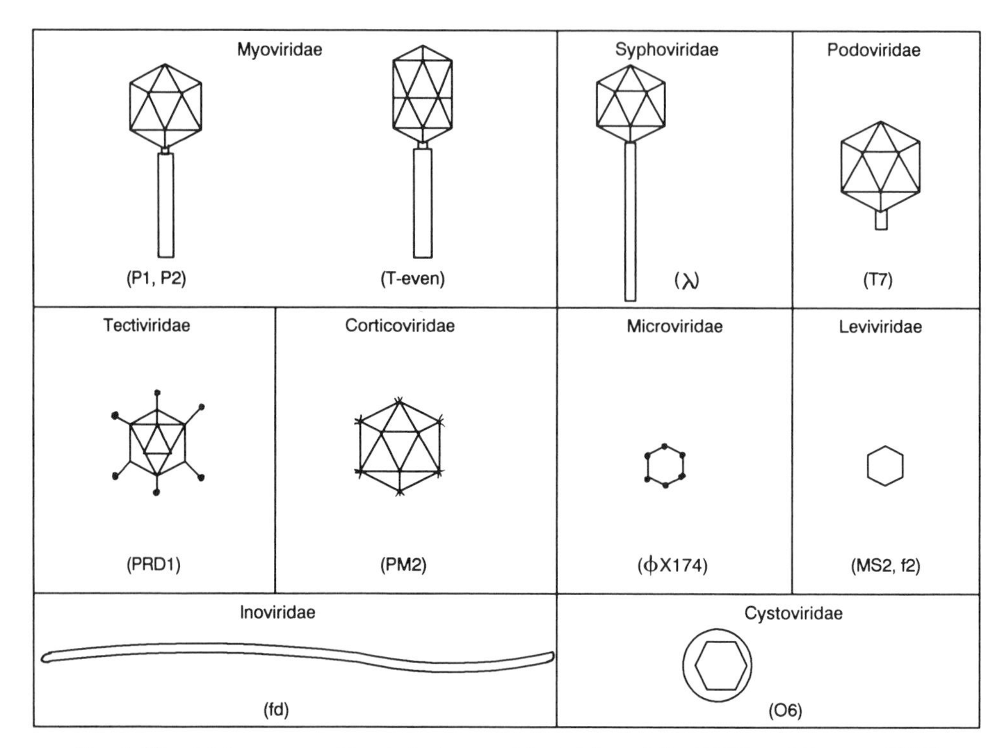

Fig. 1.22. Major groups of bacteriophages. Adapted from Jofre (1991) and Coetzee (1987).

lytic cycle, which results in the production of phage progeny and the destruction of the bacterial host cell. Sometimes the phage becomes incorporated into the host chromosome as a prophage. This process, called lysogeny, does not lead to the destruction of the host cell. Phages infecting *E. coli* are called coliphages. These phages are being considered as potential indicators of fecal contamination (Chapter 5).

1.5. FURTHER READING

Bitton, G. 1980. *Introduction to Environmental Virology.* John Wiley & Sons, New York, 326 pp.

Boyd, R.F. 1988. *General Microbiology,* 2nd ed. Times Mirror/Mosby Cool., St. Louis.

Brock, T.D., and M.T. Madigan. 1991. *Biology of Microorganisms,* 6th ed. Prentice-Hall, New York.

Chet, I., and R. Mitchell. 1976. Ecological aspects of microbial chemotactic behavior. Annu. Rev. Microbiol. 30: 221–239.

van Etten, J.L., L.C. Lane, and R.H. Meints. 1991. Viruses and viruslike particles of eucaryotic algae. Microbiol. Rev. 55: 586–620.

Gaudy, A.F., Jr., and E.T. Gaudy. 1988. *Elements of Bioenvironmental Engineering.* Engineering Press, San Jose, CA, 592 pp.

Goyal, S.M., C.P. Gerba, and G. Bitton, Eds. 1987. *Phage Ecology.* John Wiley & Sons, New York.

Halvorson, H.O., D. Pramer, and M. Rogul, Eds. 1985. *Engineered Organisms in the Environment: Scientific Issues.* American Society for Microbiology, Washington, DC.

Hancock, R.E.W. 1984. Alterations in outer membrane permeability. Annu. Rev. Microbiol. 38: 237–264.

Johnston, J.B., and S.G. Robinson. 1984. *Genetic Engineering and the Development of New Pollution Control Technologies.* U.S. EPA report #600/2-84-037. Washington DC.

Nakae, T. 1986. Outer membrane permeability of bacteria. Crit. Rev. Microbiol. 13: 1–62.

Rodriguez, R.L., and R.C. Tait. 1983. *Recombinant DNA Techniques: An Introduction.* Addison-Wesley, Reading, MA.

Schwartzbrod, L., Ed. 1991. *Virologie des Milieux Hydriques.* TEC & DOC–Lavoisier, Paris, 304 pp.

Sterritt, R.M., and J.N. Lester. 1988. *Microbiology for Environmental and Public Health Engineers.* E. & F.N. Spon, London.

Tortora, G.J., B.R. Funke, and C.L. Case. 1989. *Microbiology: An Introduction.* Benjamin/Cummings, Redwood City, CA, 810 pp.

2

MICROBIAL METABOLISM AND GROWTH

2.1. INTRODUCTION

In Chapter 1 we examined the microbial world with particular emphasis on the structure of microbial cells as well as the range of microorganisms found in aquatic environments. In the present chapter, we will examine the metabolism and growth kinetics of microbial populations.

2.2. ENZYMES AND ENZYME KINETICS

2.2.1. Introduction

Enzymes are protein molecules that serve as catalysts of biochemical reactions in animal, plant, and microbial cells. Commercial applications have been found for enzymes in the food (e.g., wine, cheese, beer), detergents, medical, pharmaceutical, and textile industries. However, no significant application has been proposed yet for wastewater treatment (see Chapter 16).

Enzymes do not undergo structural changes following their participation in chemical reactions and thus can be used repeatedly. They lower the activation energy and increase the rate of biochemical reactions. Enzymes may be intracellular or extracellular and are generally quite specific for their substrates. The substrate combines with the active site of the enzyme molecule to form an enzyme–substrate complex (ES). A new product (P) is formed and the unchanged enzyme E is ready to react again with the substrate:

$$E + S \underset{k_{-1}}{\overset{k_1}{\rightleftharpoons}} ES \xrightarrow{k_2} E + P$$

Some nonprotein groups or cofactors may become associated with enzyme molecules and participate in the catalytic activity of the enzyme. These include coenzymes (e.g., nicotinamide adenine dinucleotide, coenzyme A, flavin-adenine dinucleotide, and flavin mononucleotide) and metallic activators (e.g., K, Mg, Fe, Co, Cu, Zn, Mn, Mo). Dehydrogenase enzymes require coenzymes (e.g., FAD, FMN, NAD, coenzyme A, biotin) which accept the hydrogen removed from the substrate.

Enzymes are currently subdivided into six classes:

1. Oxidoreductases: responsible for oxidation and reduction processes in the cell.
2. Transferases: responsible for the transfer of chemical groups from one substrate to another.
3. Hydrolases: hydrolyse carbohydrates, proteins, and lipids into smaller molecules (for example, β-galactosidase hydrolyses lactose into glucose and galactose).
4. Lyases: catalyse the addition or removal of substituents groups.
5. Isomerases: catalyse isomer formation.
6. Ligases: catalyze the joining of two molecules, using an energy source such as ATP.

2.2.2. Enzyme Kinetics

The most important factors controlling enzymatic reactions are substrate concentration, pH, temperature, ionic strength, and the presence of toxicants.

At low substrate concentration, the enzymatic reaction rate V is proportional to the substrate concentration (first-order kinetics). At higher substrate concentrations, V reaches a plateau (zero-order kinetics). The enzymatic reaction rate V as a function of substrate concentration is given by the Michaelis–Menten equation (Eq. 2.1):

$$V = \frac{V_{\max}\,[S]}{K_m + [S]} \tag{2.1}$$

where V = reaction rate (units/time); $V_{\max}$ = maximum reaction rate (units/time); $[S]$ = substrate concentration (mol/L); and K_m = half-saturation constant (Michaelis constant); the substrate concentration at which $V = V_{\max}/2$.

The hyperbolic curve shown in Figure 2.1A can be linearized with the Lineweaver–Burke plot (Fig. 2.1B), which uses the reciprocals of both the substrate concentration and the reaction rate V:

$$\frac{1}{V} = \frac{K_m}{V_{\max}} \cdot \frac{1}{[S]} + \frac{1}{V_{\max}} \tag{2.2}$$

Plotting $1/V$ versus $1/[S]$ gives a straight line with a slope of $K_m/V_{\max}$, a y intercept of $1/V_{\max}$, and an x intercept of $-1/K_m$ (Fig. 2.1B). $V_{\max}$ and K_m can be obtained directly from the Lineweaver–Burke plot.

2.2.3. Effect of Inhibitors on Enzyme Activity

A wide range of toxicants are commonly found in industrial and municipal wastewater treatment plants. These inhibitors decrease the activity of enzyme-catalyzed reactions. There are three

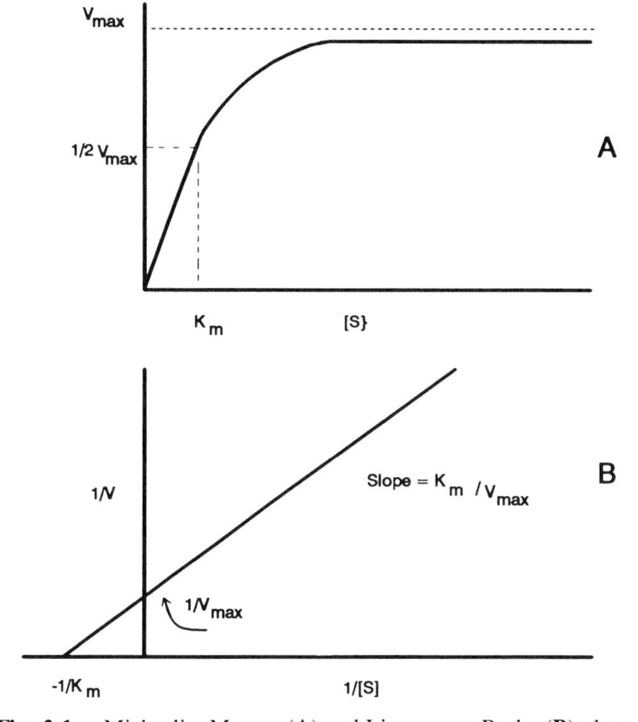

Fig. 2.1. Michaelis–Menten (**A**) and Lineweaver-Burke (**B**) plots.

types of enzyme inhibition: competitive, non-competitive, and uncompetitive (Fig. 2.2).

2.2.3.1. Competitive Inhibition

In competitive inhibition, the inhibitor (I) and the substrate (S) compete for the same reactive site on the enzyme (E). In the presence of a competitive inhibitor I, the reaction rate V is given by Equation 2.3:

$$V = \frac{V_{max}[S]}{[S] + K_m(1 + [I]/K_i)} \tag{2.3}$$

where $[S]$ = substrate concentration (mol/L); $[I]$ = inhibitor concentration (mol/L); K_i = inhibition coefficient; V_{max} = maximum enzyme reaction rate (t^{-1}); and K_m = Michaelis constant (mol/L).

The Lineweaver–Burke representation of Equation 2.3 is as follows:

$$\frac{1}{V} = \frac{K_m}{V_{max}}(1 + [I]/K_i)\frac{1}{[S]} + \frac{1}{V_{max}} \tag{2.4}$$

In competitive inhibition, V_{max} is unaffected,

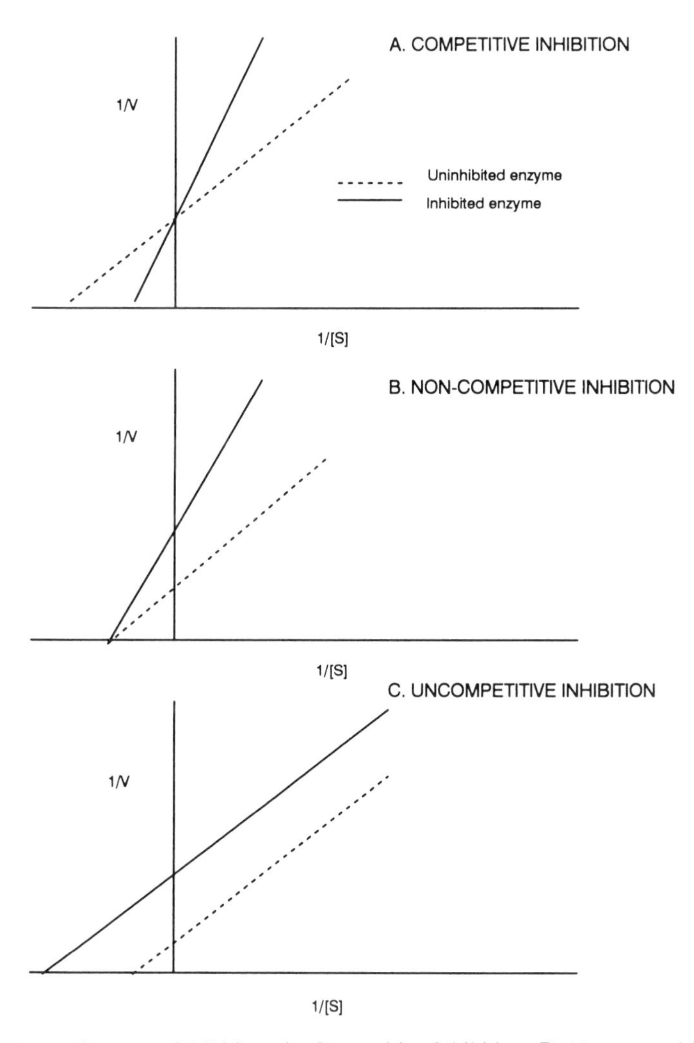

Fig. 2.2. Types of enzyme inhibition. **A.** Competitive inhibition. **B.** Noncompetitive inhibition. **C.** Uncompetitive inhibition. Adapted from Bitton and Koopman (1986), and Marison (1988b).

whereas K_m is increased by a factor of $(1 + [I]/K_i)$ (Fig. 2.2A).

2.2.3.2. *Noncompetitive Inhibition*

In noncompetitive inhibition, the inhibitor I can bind to both the enzyme E and the ES complex (Fig. 2.2B). According to Michaelis–Menten kinetics, the reaction rate is given by Equation 2.5:

$$V = \frac{V_{max}\,[S]}{(K_m + [S])\,(1 + [I]/K_i)} \quad (2.5)$$

The Lineweaver–Burke transformation of Equation 2.5 is

$$\frac{1}{V} = \frac{K_m}{V_{max}}(1 + [I]/K_i)\,\frac{1}{[S]} + \frac{1}{V_{max}}(1 + [I]/K_i) \quad (2.6)$$

In the presence of the inhibitor, the slope of the double-reciprocal plot is increased, whereas V_{max} is decreased. K_m remains unchanged.

2.2.3.3. *Uncompetitive Inhibition*

In uncompetitive inhibition the inhibitor binds to the enzyme–substrate complex but not to the free enzyme (Fig. 2.2C).

$$V = \frac{V_{max}\,[S]}{K_m + [S]\,(1 + [I]/K_i)} \quad (2.7)$$

The Lineweaver–Burke plot transformation gives Equation 2.8:

$$\frac{1}{V} = \frac{K_m}{V_{max}}\,\frac{1}{[S]} + \frac{1}{V_{max}}(1 + [I]/K_i) \quad (2.8)$$

In the presence of an inhibitor, the slope of the reciprocal plot remains the same but both V_{max} and K_m are affected by the inhibitor concentration (Fig. 2.2C).

2.3. MICROBIAL METABOLISM

2.3.1. Introduction

Metabolism is the sum of biochemical transformations that include interelated catabolic and anabolic reactions. Catabolic reactions are exergonic: They release energy derived from organic and inorganic compounds. Anabolic reactions (i.e., biosynthetic) are endergonic: they use the energy and chemical intermediates provided by catabolic reactions for biosynthesis of new molecules, cell maintainance, and growth. The relationship between catabolism and anabolism is shown in Figure 2.3.

The energy generated by catabolic reactions is transferred to energy-rich compounds such as adenosine triphosphate (ATP). (Fig. 2.4). This phosphorylated compound is composed of adenine, ribose (a five-carbon sugar), and three phosphates; it has two high-energy bonds that release chemical energy when hydrolyzed to adenosine diphosphate (ADP). Upon hydrolysis under standard conditions, each molecule of ATP releases approximately 7,500 calories.

$$\mathrm{A - P \sim P \sim P + H_2O \leftrightarrow A - P \sim P + Pi + energy} \quad (2.9)$$

(A = adenine + ribose).

The energy released is used for biosynthetic reactions, active transport, or movement, and some of it is dissipated as heat. Other energy-rich phosphorylated compounds are phosphoenolpyruvate (PEP) + ($\Delta G^0 = -14.8$ kcal/mol) and 1,3-diphosphoglycerate ($\Delta G^0 = -11.8$ kcal/mol).

ATP is generated by three mechanisms of phosphorylation.

1. *Substrate-level phosphorylation.* Substrate-level phosphorylation is the direct transfer of high-energy phosphate to ADP from an intermediate in the catabolic pathway. It produces all the energy in fermentative microorganisms but only a small portion of the energy in aerobic and anaerobic microorganisms. In fermentation, glucose is transformed to pyruvic acid through

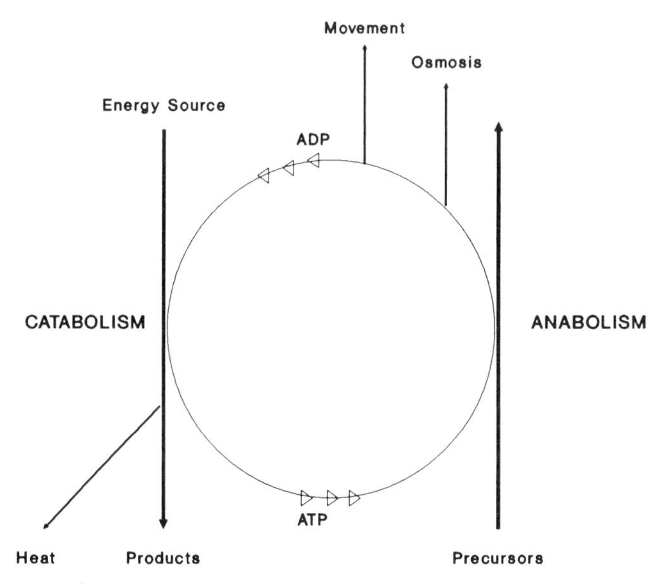

Fig. 2.3. Relationship between anabolic and catabolic reactions. From Scragg (1988), with permission of the publisher.

the Embden–Meyerhof pathway. Pyruvic acid is further transformed to several by-products by various microorganisms (e.g., alcohol by yeasts or lactic acid by *Streptococcus lactis*). Substrate-level phosphorylation results in ATP formation by the transfer of phosphate groups from a high-energy phosphorylated compound, such as 1,3-diphosphoglyceric acid, to ADP. In a typical fermentation, only 2 moles of ATP (equivalent to approximately 14,000 calories) are released per mole of glucose. The free energy ΔG^0 released by the combustion of a molecule of glucose being −686,000 cal, the efficiency of the process is only 2% (14,000/686,000 × 100 = 2%).

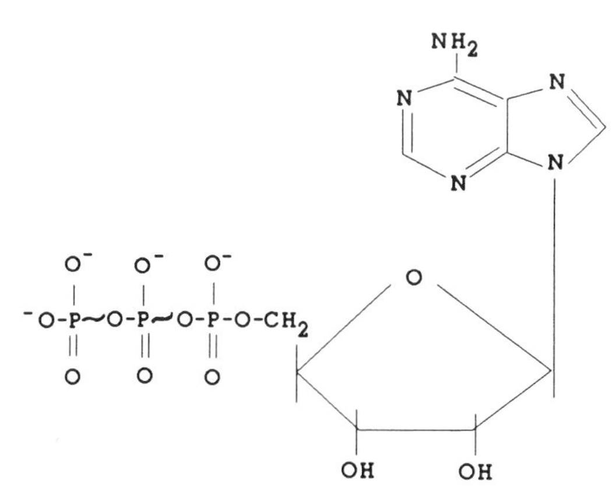

Fig. 2.4. Molecular structure of adenosine triphosphate (ATP).

2. *Oxidative phosphorylation* (electron transport system). ATP can also be generated by oxidative phosphorylation, in which process electrons are transported through the electron transport system (ETS) from an electron donor to a final electron acceptor, which may be oxygen, nitrate, sulfate, or CO_2. Further details are given in Section 2.3.2.

The electron transport system is located in the cytoplasmic membrane of procaryotes and in mitochondria of eucaryotes. Although the electron carriers used by bacteria are similar to those utilized in mitochondria, the former can use alternate final electron acceptors (e.g., nitrate or sulfate). It is assumed that the ATP yield for oxidative phosphorylation using alternate final electron acceptors is lower than in the presence of oxygen.

3. *Photophosphorylation*. Photophosphorylation is the process by which light energy is converted to chemical energy (ATP) and that occurs in both eucaryotes (e.g., algae) and procaryotes (e.g., cyanobacteria, photosynthetic bacteria). The ATP generated by photophosphorylation drives the reduction of CO_2 during the dark phase of photosynthesis. In photosynthetic organisms CO_2 serves as the source of carbon. The electron transport system of these organisms supplies both ATP and NADPH for the synthesis of cell food. Green plants and algae use H_2O as the electron donor and release O_2 as a by-product. Photosynthetic bacteria produce no oxygen from photosynthesis and use H_2S as the electron donor (more details are given in Section 2.3.2).

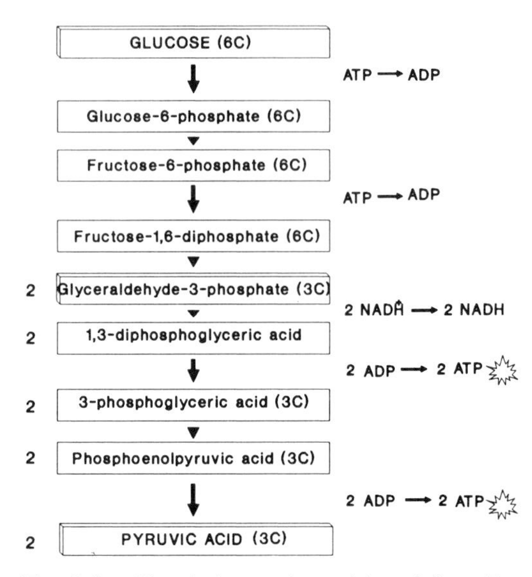

Fig. 2.5. Glycolysis reactions. Adapted from Tortora et al. (1989).

2.3.2. Catabolism

2.3.2.1. *Aerobic Respiration*

Respiration is an ATP-generating process that involves the transfer of electrons through the electron transport system. The substrate is oxidized, O_2 being used as the terminal electron acceptor. As indicated earlier, a small portion of ATP is also generated by substrate-level phosphorylation. The electron donor may be an organic compound (e.g., oxidation of glucose by heterotrophic microorganisms) or an inorganic compound (e.g., oxidation of H_2, Fe(II), NH_4 or S^0 by chemoautotrophic microorganisms).

Oxidation of Glucose. The breakdown of glucose involves the following steps.

GLYCOLYSIS. In glycolysis (Embden–Meyerhof–Parnas pathway) (Fig. 2.5) glucose is first phosphorylated and cleaved into two molecules of a key intermediate compound, glyceraldehyde-3-phosphate, which is then converted to pyruvic acid. Thus, glycolysis is the oxidation of one molecule of glucose to two molecules of pyruvic acid, a three-carbon compound. This pathway results in the production of two molecules of reduced nicotinamide-adenine dinucleotide (NADH) and a net gain of two molecules of ATP (4 ATP molecules produced − 2 molecules ATP used) per molecule of glucose.

Some microorganisms use the pentose phosphate pathway to oxidize pentoses. Glucose oxidation via this pathway produces 12 molecules of NADPH and one of ATP. Some procaryotic microorganisms use the Entner–Doudoroff pathway, which produces two molecules of NADPH and one of ATP per molecule of glucose oxidized.

TRANSFORMATION OF PYRUVIC ACID TO ACETYL COENZYME A. Pyruvic acid is decarboxylated (i.e., loss of one CO_2) to an acetyl group that combines with coenzyme A to give acetyl-CoA, a two-carbon compound. During this process, one molecule of NAD^+ is reduced to NADH.

$$\text{pyruvic acid} + NAD^+ + \text{coenzyme A} \rightarrow \text{acetyl-CoA} + NADH + CO_2 \quad (2.10)$$

KREBS CYCLE. The Krebs cycle (citric acid cycle or tricarboxylic acid cycle) (Fig. 2.6), in which pyruvate is completely oxidized to CO_2, releases more energy than glycolysis; it occurs in mitochondria in eucaryotes but is associated with the cell membrane in procaryotes. Acetyl-

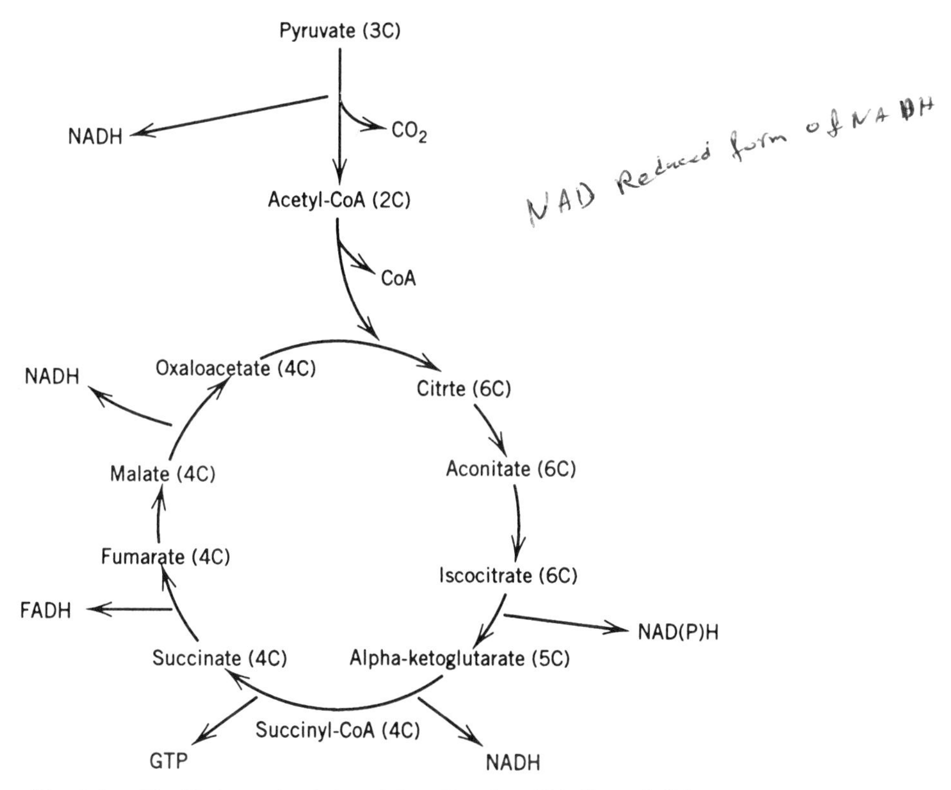

Fig. 2.6. The Krebs cycle. Adapted from Brock and Madigan (1991).

CoA enters the Krebs cycle and combines with oxaloacetic acid (a four-carbon compound) to form the six-carbon citric acid. The cycle consists of a series of biochemical reactions, each one being catalyzed by a specific enzyme. It results in the production of two molecules of CO_2 per molecule of acetyl-CoA that enters the cycle. The oxidation of compounds in the cycle releases electrons that reduce NAD^+ to NADH or flavine-adenine dinucleotide (FAD) to FADH. For each molecule of acetyl-CoA, 3 NADH and 1 FADH are formed. Moreover, one molecule of guanosine triphosphate (GTP) is formed by substrate level phosphorylation during the oxidation of α-glutaric acid to succinic acid.

(4) ELECTRON TRANSPORT SYSTEM. The electron transport system is a chain of electron carriers associated with the cytoplasmic membrane in procaryotes and with the mitochondria in eucaryotes. This system transfers the energy stored in NADH and FADH into ATP. The series consists of such electron carriers as flavin mononucleotide (FMN), coenzyme Q, iron-sulfur proteins, and cytochromes. For efficient trapping of energy, there is a stepwise release of ATP as the electrons are transported from one carrier to another to reach the final electron acceptor. Each molecule of NADH and of FADH generates three and two molecules of ATP, respectively. A series of steps are involved in the electron transport system (Fig. 2.7).

1. Electrons originating from the substrate are transferred to NAD (precursor: nicotinic acid), which is reduced to NADH.

2. Flavoproteins such as FMN and FAD accept hydrogen atoms and donate electrons.

3. Quinones (coenzyme Q) are nonprotein lipid-soluble carriers that also accept hydrogen atoms and donate electrons.

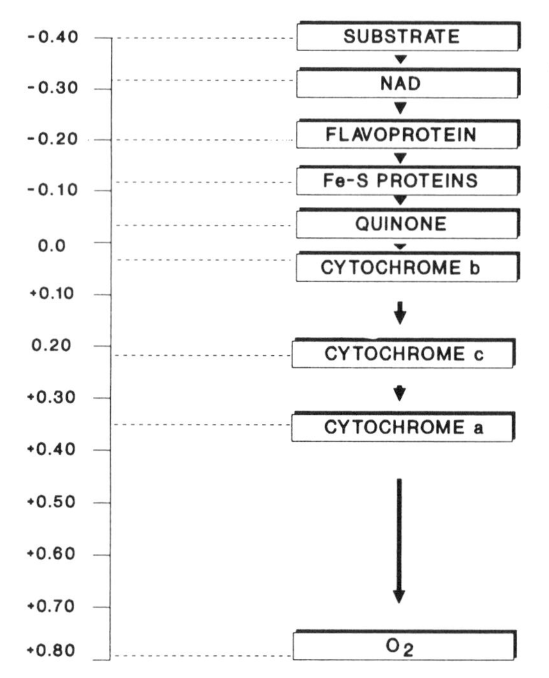

Fig. 2.7. Electron transport system. Adapted from Tortora et al. (1989).

4. Cytochromes are proteins with porphyrin rings that contain iron, which is reduced to Fe^{2+}. There are several classes of cytochromes designated by letters. The sequence of electron transport is the following: cytochrome b → cytochrome c → cytochrome a.

5. Oxygen is the final electron acceptor in aerobic respiration. Anaerobic respiration involves final electron acceptors other than oxygen. These electron acceptors may be NO_3, SO_4, CO_2 or some organic compounds. Anaerobic respiration releases less energy than aerobic respiration.

Inhibitors (e.g., cyanide, carbon monoxide), acting at various points of the ETS, can inhibit the electron carriers and thus interfere with both the electron flow and ATP synthesis. Other inhibitors (e.g., dinitrophenol), called uncouplers, inhibit only ATP synthesis.

Summary of ATP release by aerobic respiration. Aerobic respiration releases 38 ATP molecules per molecule of glucose oxidized to CO_2. A breakdown of ATP molecules released is as follows:

1. Glycolysis: Oxidation of glucose to pyruvic acid	
Yield of substrate-level phosphorylation	2 ATP
Yield of two molecules of NADH going through ETS if oxygen is available	6 ATP
2. Transformation of pyruvic acid to acetyl-coenzyme A	
Yields two molecules of NADH (there are two molecules of pyruvic acid/mole glucose) yield of 2 NADH	6 ATP
3. Krebs cycle	
For each acetyl-coenzyme A going through the cycle, 9 ATP are generated from NADH, 2 from FADH, and 1 from GTP; yield of 2 acetyl-coenzyme A molecules	24 ATP
Total ATP/molecule glucose	38 ATP

The efficiency of complete oxidation of glucose by aerobic respiration to CO_2 is approximately 38%.

2.3.2.2. Fermentation

Fermentation is the transformation of pyruvic acid to various products in the absence of a terminal electron acceptor. Both the electron donor and the acceptor are organic compounds and ATP is generated solely by substrate-level phosphorylation. Fermentation releases little energy (2 ATP/mole of glucose) and most of it remains in fermentation products. The latter depend on the type of microorganism involved in fermentation. For example, ethanol, lactic acid, and propionic acid are formed by *Saccharomyces, Lactobacillus,* and *Propionobacterium,* respectively.

The well-known process of alcoholic fermentation, carried out by yeasts, proceeds as follows:

$$\underset{CH_3COCOOH}{\text{pyruvic acid}} \xrightarrow[NADH_2 \rightarrow NAD]{CO_2} \underset{CH_3CH_2OH}{\text{ethyl alcohol}} \quad (2.11)$$

It involves two steps: (1) decarboxylation (i.e., removal of the carboxyl group)

$$\text{pyruvic acid} \xrightarrow{CO_2} \text{acetaldehyde} \quad (2.12)$$

and (2) reduction of acetaldehyde

$$\text{acetaldehyde} \xrightarrow[NADH_2 \rightarrow NAD]{} \text{ethyl alcohol} \quad (2.13)$$

The overall reaction for glucose fermentation to ethanol by yeasts is as follows:

$$\underset{\text{glucose}}{C_6H_{12}O_6} \longrightarrow \underset{\text{ethanol}}{2\ C_2H_5OH} + \underset{\text{carbon dioxide}}{2\ CO_2} \quad (2.14)$$

Yeasts can switch from aerobic respiration to fermentation, depending upon growth conditions.

In muscle tissues, pyruvate is converted to lactate, which accumulates during heavy exertion. Some of the lactate is washed away and carried to the liver, where it is metabolized.

2.3.3. Anabolism

Anabolism (biosynthesis) includes all the energy-consuming processes that result in the formation of new cells. It is estimated that 3,000 μmoles of ATP are required to make 100 mg of dry mass of cells. Moreover, most of this energy is used for protein synthesis (Brock and Madigan, 1991). Cells use energy (ATP) to make building blocks, synthesize macromolecules, repair damage to cells (maintainance energy), and maintain movement and active transport across the cell membrane. Most of the ATP generated by catabolic reactions is used for biosynthesis of biological macromolecules such as proteins, lipids, polysaccharides, purines, and pyrimidines. Most of the precursors of these macromolecules (amino acids, fatty acids, monosaccharides, nucleotides) are derived from intermediates formed during glycolysis, the Krebs cycle and other metabolic pathways (Entner–Doudoroff and pentose phosphate pathways). These precursors are linked together by specific bonds (e.g., peptide bond for proteins, glycoside bond for polysaccharides, phosphodiester bond for nucleic acids) to form cell biopolymers.

2.3.4. Photosynthesis

Photosynthesis is a process that converts light energy into chemical energy, using CO_2 as a carbon source and light as an energy source. Light is absorbed by chlorophyll molecules, which are found in algae, plants, and photosynthetic bacteria.

The general equation for photosynthesis is shown below:

$$6CO_2 + 12H_2O + \text{light} \rightarrow C_6H_{12}O_6 + 6H_2O + 6O_2 \quad (2.15)$$

Oxygenic photosynthesis consists of two types of reactions: (1) light reactions which result essentially in the conversion of light energy into chemical energy (ATP) and the production of NADPH; and (2) dark reactions in which NADPH is used to reduce CO_2.

2.3.4.1. Light Reactions

Most of the energy captured by photosynthetic pigments is contained in wavelengths between 400 nm (visible light) and 1,100 nm (near-infrared light). Algae, green photosynthetic bacteria, and purple photosynthetic bacteria absorb light at 670–685 nm, 735–755 nm, and 850–1,000 nm, respectively.

Light is absorbed by chlorophyll *a* (Fig. 2.8), a molecule made up of pyrrole rings surrounding a Mg atom and displaying two absorption peaks, the first peak being at 430 nm and the second one at 675 nm (Fig. 2.9). The pigments are organized in clusters, called photosystem I (p700) and pho-

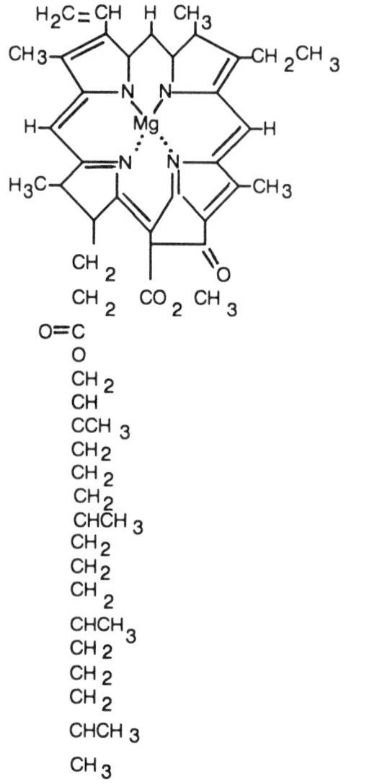

Fig. 2.8. Chemical structure of chlorophyll *a*.

tosystem II (p680), which play a role in the transfer of electrons from H_2O to NADP (photosynthetic bacteria have only photosystem I).

The flow of electrons in oxygenic photosynthesis has a path that resembles the letter Z and is called the noncyclic electron flow or the Z scheme (Fig. 2.10) (Boyd, 1988). Upon exposure of photosystem II to light, the energy released boosts an electron to a higher energy level as the oxidation–reduction potential becomes more negative. The electron hole is filled by an electron produced by photolysis of H_2O. ATP is produced as the boosted electron travels downhill, by electron carriers (quinones, cytochromes, plastocyanin), and is captured by photosystem I (p700). Upon exposure of p700 to light, the electron released is again boosted to a higher energy level. NADP is reduced to NADPH following downhill electron flow by other electron carriers (e.g., ferredoxin). Oxygen is a by-product of light reactions carried out in oxygenic photosynthesis by algae, cyanobacteria, and plants. Sometimes, the Z scheme displays a cyclic flow of electrons that results in the production of ATP but not NADPH.

2.3.4.2. Dark Reactions (Calvin–Benson Cycle)

The reducing compound (NADPH) and the energy (ATP) produced during the light reactions are

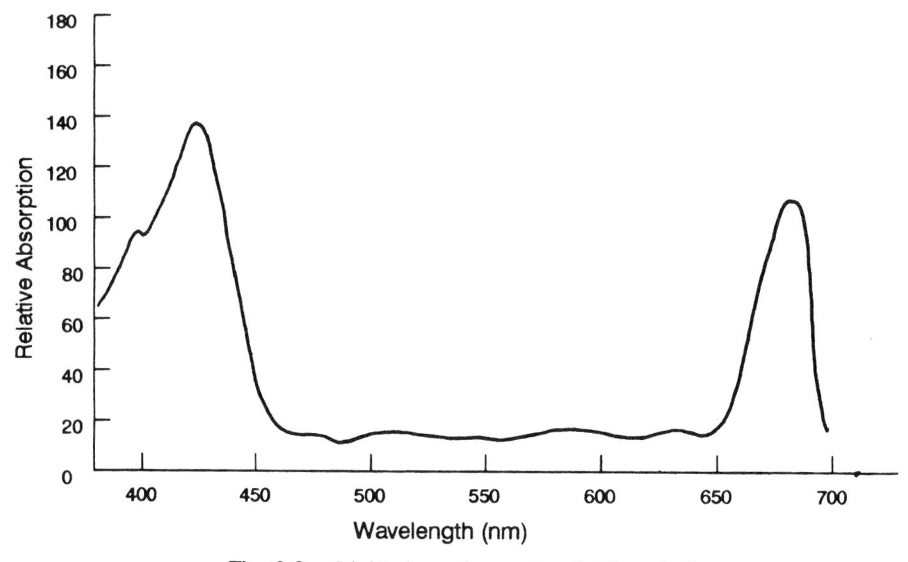

Fig. 2.9. Light absorption peaks of chlorophyll *a*.

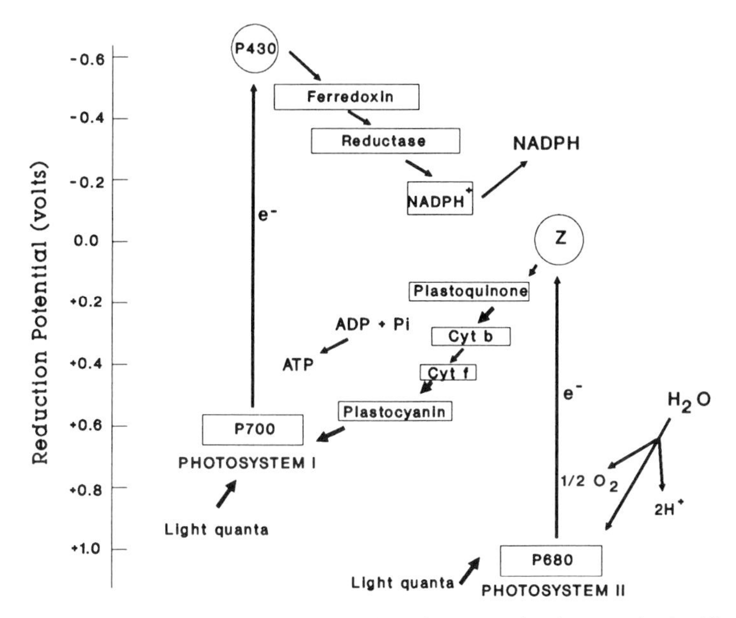

Fig. 2.10. Electron transfer during light reactions in oxygenic photosynthesis: The Z scheme. Adapted from Boyd (1988).

used to reduce CO_2 to organic compounds during the Calvin–Benson cycle, the first step of which (Fig. 2.11) is the combination of CO_2 with ribulose-1,5-diphosphate, a reaction catalyzed by the enzyme ribulose bisphosphate carboxylase. A subsequent series of reactions lead to the formation of one hexose molecule, fructose-6-phosphate.

The overall equation for the dark reactions is as follows:

$$6CO_2 + 18\ ATP + 12\ H_2O + 12\ NADPH \rightarrow C_6H_{12}O_6 + 12\ Pi + 18\ ADP + 12\ NADP^+ \quad (2.16)$$

2.3.5. Metabolic Classification of Microorganisms

The major elements that enter into the composition of a microbial cell are carbon, oxygen, nitrogen, hydrogen, phosphorus, and sulfur. Other nutrients that are necessary for biosynthesis of cell components include cations (e.g., Mg^{2+}, Ca^{2+}, Na^+, K^+, Fe^{++}); anions (e.g., Cl^-, SO_4^{2-}); trace elements (e.g., Co, Cu, Mn, Mo, Zn, Ni, Se), which serve as components or cofactors of several enzymes; and growth factors such as vitamins (e.g., riboflavin, thiamin, niacin, vitamin B_{12}, folic acid, biotin, vitamins B_6). A typical *E. coli* cell contains approximately 70% water, 3% carbohydrates, 3% amino acids, nucleotides, and lipids, 22% macromolecules (mostly proteins as well as RNA and DNA), and 1% inorganic ions.

Microorganisms need a carbon source (CO_2 or organic carbon) and an energy source (light or energy derived from the oxidation of inorganic or organic chemicals). The metabolic classification of microorganisms is based on two main criteria, energy source and carbon source.

2.3.5.1. *Phototrophs*

These microorganisms use light as the source of energy. They are subdivided into photoautotrophs and photoheterotrophs.

Photoautotrophs. This group includes algae, cyanobacteria, and photosynthetic bacteria, which are also called phototrophic bacteria. Photoautotrophs use CO_2 as a carbon source and

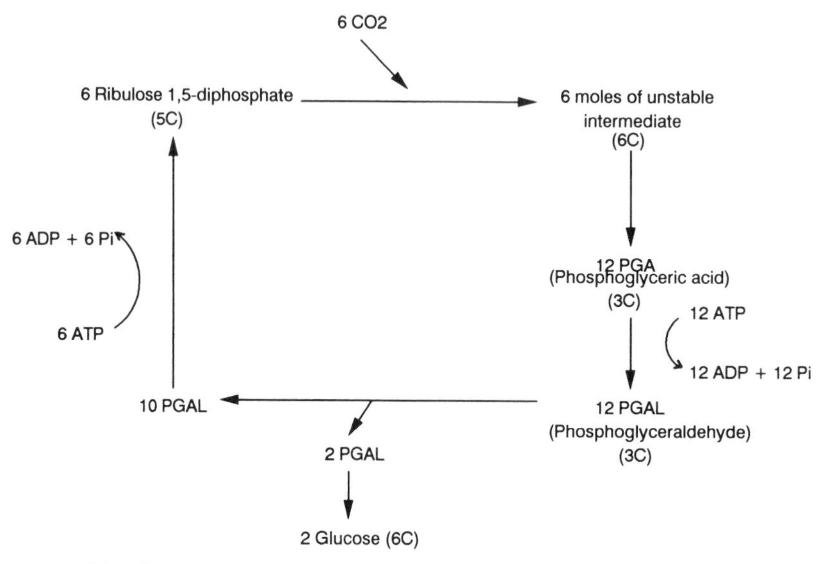

Fig. 2.11. The Calvin–Benson cycle. Adapted from Atlas (1986).

H_2O, H_2, or H_2S as electron donors. Photosynthetic bacteria carry out anoxygenic photosynthesis and most of them require anaerobic conditions. Oxygen is detrimental to the synthesis of their photosynthetic pigments, bacteriochlorophylls and carotenoids. For cyanobacteria and algae, the electron donor is H_2O, whereas for photosynthetic bacteria the donor is H_2S (some cyanobacteria are also able to use H_2S as electron donor, resulting in the deposition of S^0 outside their cells). The general taxonomy of phototrophic bacteria is shown in Table 2.1. There are approximately 60 species of phototrophic bacteria, broadly grouped into *purple* and *green* bacteria. Purple bacteria contain bacteriochlorophyll *a* (maximum absorption at 825–890 nm) and *b* (maximum absorption at about 1,000 nm), whereas green bacteria contain bacteriochlorophylls *c, d,* and *e,* which absorb light at wavelengths between 705 nm and 755 nm. These phototrophic bacteria (e.g., chromatiaceae, chlorobiaceae) use CO_2 as a carbon source, light as an energy source, and reduced sulfur compounds (e.g., H_2S, S^0) as electron donors.

Anoxygenic photosynthesis, using H_2S as the reductant source, is summarized as follows:

$$12H_2S + 6CO_2 \rightarrow C_6H_{12}O_6) + 6H_2O + 12S^0 \quad (2.17)$$

S^0 is deposited inside (e.g., in the case of green bacteria) or outside (e.g., in the case of purple bacteria) the cells of photosynthetic bacteria.

Photoheterotrophs (or photoorganotrophs). This group comprises all the facultative heterotrophs that derive energy from light or from organic compounds that serve as carbon sources and electron donors. The purple nonsulfur bacteria, the Rhodospirillaceae, use organic compounds as electron donors.

2.3.5.2. *Chemotrophs*

These microorganisms obtain their energy via oxidation of inorganic or organic compounds. They are subdivided into lithotrophs (chemoautotrophs) and heterotrophs (organotrophs).

Lithotrophs (chemoautotrophs). Lithotrophs use carbon dioxide as a carbon source (carbon fixation) and derive their energy (ATP)

TABLE 2.1. Recognized Genera of Anoxygenic Phototrophic Bacteria[a]

Taxonomic group	Morphology
	Purple bacteria
Purple sulfur bacteria (chromatiaceae and Ectothiorhodospiraceae)	
Amoebobacter	Cocci embedded in slime; contain gas vesicles
Chromatium	Large or small rods
Lamprocystis	Large cocci or ovoids with gas vesicles
Lamprobacter	Large ovals with gas vesicles
Thiocapsa	Small cocci
Thiocystis	Large cocci or ovoids
Thiodictyon	Large rods with gas vesicles
Thiospirillum	Large spirilla
Thiopedia	Small cocci with gas vesicles; cells arranged in flat sheets
Ectothiorhodospira	Small spirilla; do not store sulfur inside the cell
Purple nonsulfur bacteria (Rhodospirillaceae)	
Rhodocyclus	Half-circle or circle
Rhodomicrobium	Ovoid with stalked budding morphology
Rhodopseudomonas	Rods, dividing by budding
Rhodobacter	Rods and cocci
Rhodopila	Cocci
Rhodospirillum	Large or small spirilla
	Green bacteria
Green sulfur bacteria (Chlorobiaceae)	
Anacalochloris	Prosthecate spheres with gas vesicles
Chlorobium	Small rods or vibrios
Pelodictyon	Rods or vibrios, some form three-dimensional net; contain gas vesicles
Prosthecochloris	Spheres with prosthecae
Green gliding bacteria (Chloroflexaceae)	
Chloroflexus	Narrow filaments (multicellular), up to 100 μm long
Chloroherpeton	Short filaments (unicellular)
Chloronema	Large filaments (multicellular), up to 250 μm long; contain gas vesicles
Oscillochloris	Very large filaments, up to 2,500 μm long; contain gas vesicles

[a]From Madigan (1988), with permission of the publisher.

needs by oxidizing inorganic compounds such as NH_4, NO_2, H_2S, Fe^{2+}, or H_2. Most of them are aerobic.

Nitrifying bacteria are widely distributed in soils, water, and wastewater and oxidize ammonium to nitrate (see Chapter 3 for more details).

$$NH_4^+ \xrightarrow{\text{Nitrosomonas}} NO_2^- \xrightarrow{\text{Nitrobacter}} NO_3^- \quad (2.18)$$

Sulfur-oxidizing bacteria use hydrogen sulfide (H_2S), elemental sulfur (S^0), or thiosulfate ($S_2O_3^{2-}$) as energy sources. They are capable of growth in very acidic environments (pH = 2 or less). The oxidation of elemental sulfur to sulfate is as follows:

$$S + 3O_2 + 2H_2O \xrightarrow{(\textit{Thiobacillus thiooxidans})} 2H_2SO_4 + \text{energy} \quad (2.19)$$

Iron bacteria include bacteria (e.g., *Thiobacillus ferrooxidans*) that are acidophilic, derive energy from oxidation of Fe^{2+} to Fe^{3+}, and are also capable of oxidizing sulfur. Others oxidize ferrous iron at neutral pH (e.g., *Sphaerotilus natans*, *Leptothrix ochracea*, *Crenothrix*, *Clonothrix*, and *Gallionella ferruginea*).

Hydrogen bacteria (e.g., *Hydrogenomonas*) use H_2 as energy source and CO_2 as carbon source. Hydrogen oxidation is catalyzed by a hydrogenase enzyme. These bacteria are facultative lithotrophs, since they can grow also in the presence of organic compounds.

Heterotrophs (organotrophs). This is the most common nutritional group among microorganisms; it includes bacteria, fungi, and protozoa. Heterotrophic microorganisms obtain their energy from oxidation of organic matter. Organic compounds serve both as energy source and carbon source. This group includes the majority of bacteria, fungi, and protozoa in the environment.

2.4. MICROBIAL GROWTH KINETICS

Prokaryotic organisms such as bacteria reproduce mainly by binary fission (i.e., each cell gives two daughter cells). Growth of a microbial population is defined as an increase in numbers or an increase in microbial mass. The increase in the number of microbial cells or mass per unit time is termed the growth rate. The time required for a microbial population to double in numbers is the generation time or doubling time, which may vary from minutes to days.

Microbial populations can grow as batch cultures (closed systems) or as continuous cultures (open systems) (Marison, 1988a).

2.4.1. Batch Cultures

When a suitable growth medium is inoculated with cells, the growth of the microbial population follows the growth curve displayed in Figure 2.12, which shows four distinct phases.

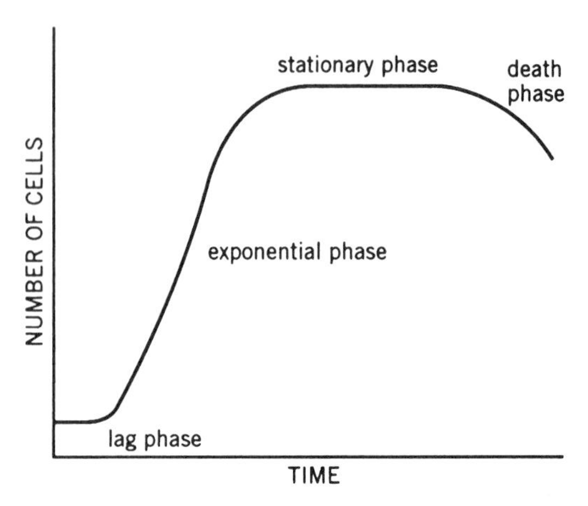

Fig. 2.12. Microbial growth curve.

2.4.1.1. *Lag Phase*

The lag phase is a period of cell adjustment to the new environment. Cells are involved in the synthesis of biochemicals and undergo enlargement. The duration of the lag phase depends on the cells' prior history (age, prior exposure to damaging physical or chemical agents, culture medium). For example, no lag phase is observed when an exponentially growing culture is transferred to a similar medium with similar growth conditions. Conversely, a lag period is observed when damaged cells are introduced into the culture medium.

2.4.1.2. *Exponential Growth Phase (Log Phase)*

The number of cells increases exponentially during the log phase. The exponential growth rate varies with the type of microorganism and growth conditions (e.g., temperature, medium composition). Under favorable conditions, the number of bacterial cells (e.g., *E. coli*) double every 15–20 min. The growth follows a geometric progression ($2^0 \rightarrow 2^1 \rightarrow 2^2 \rightarrow 2^n$).

$$X_t = X_0 \, e^{\mu t} \tag{2.20}$$

where μ = specific growth rate (hr^{-1}); X_t = cell biomass or numbers after time t; and X_0 = initial number or biomass of cells.

Using the natural logarithms on both sides of Equation 2.20, we obtain the following equation:

$$\ln X_t = \ln X_0 + \mu t \quad (2.21)$$

where μ is given by the following:

$$\mu = \frac{\ln X_t - \ln X_0}{t} \quad (2.22)$$

If n is the number of population doublings (i.e., number of generations) after time t, the doubling time t_d is given by

$$t_d = \frac{t}{n} \quad (2.23)$$

μ is related to the doubling time t_d by

$$\mu = \frac{\ln 2}{t_d} = \frac{0.693}{t_d} \quad (2.24)$$

Cells in the exponential growth phase are more sensitive to physical and chemical agents than those in the stationary phase.

2.4.1.3. *Stationary Phase*

The cell population reaches the stationary phase because microorganisms cannot grow indefinitely, mainly because of the lack of nutrients and electron acceptor, and the production and accumulation of toxic metabolites. Secondary metabolites (e.g., certain enzymes, antibiotics) are produced during the stationary phase.

There is no net growth (cell growth is balanced by cell death or lysis) of the population during the stationary phase.

2.4.1.4. *Death Phase*

During this phase, the death (decay) rate of the microbial population is higher than the growth rate. Cell death may be accompanied by cell lysis. The count of viable microorganisms decreases although the turbidity of the microbial suspension may remain constant. Culture turbidity may also decrease upon cell lysis.

2.4.2. Continuous Culture of Microorganisms

So far, we have described the growth kinetics of batch cultures. Maintenance of microbial cultures at the exponential growth phase over a long period of time can be achieved by growing the cells continuously in a completely mixed reactor where a constant volume is maintained. The most commonly used device is the chemostat (Fig. 2.13), which is essentially a complete-mix bioreactor without recycle. In addition to the flow rate of growth-limiting nutrient, environmental parameters such as oxygen level, temperature, and pH are also controlled. The substrate is added continuously at a flow rate Q to a reactor with a volume V containing concentration X of microorganisms. The dilution rate D, the reciprocal of the hydraulic retention time t, is given by the equation

$$D = \frac{Q}{V} = \frac{1}{t} \quad (2.25)$$

where D = dilution rate (time^{-1}); V = reactor volume (L); Q = flow rate of substrate (L/time); and t = time.

In continuous-flow reactors, microbial growth is described by the following equation:

$$dX/dt = \mu X - DX = X(\mu - D) \quad (2.26)$$

The above equation shows that the supply rate of the limiting nutrient controls the specific growth rate μ. At $D > \mu_{max}$, we observe a decrease in cell concentration and a washout of the population. Cell washout starts at the critical dilution rate D_c, which is approximately equal to μ_{max}.

The mass balance for X is given by the equation

$$V\frac{dX}{dt} = \mu XV - QX \quad (2.27)$$

$$= \frac{\mu_{max}[S]}{K_s + [S]}XV - QX \quad (2.28)$$

At steady state, $dX/dt = 0 \rightarrow \mu = D = Q/V = \dfrac{\mu_{max}[S]}{K_s + [S]}$ (2.29)

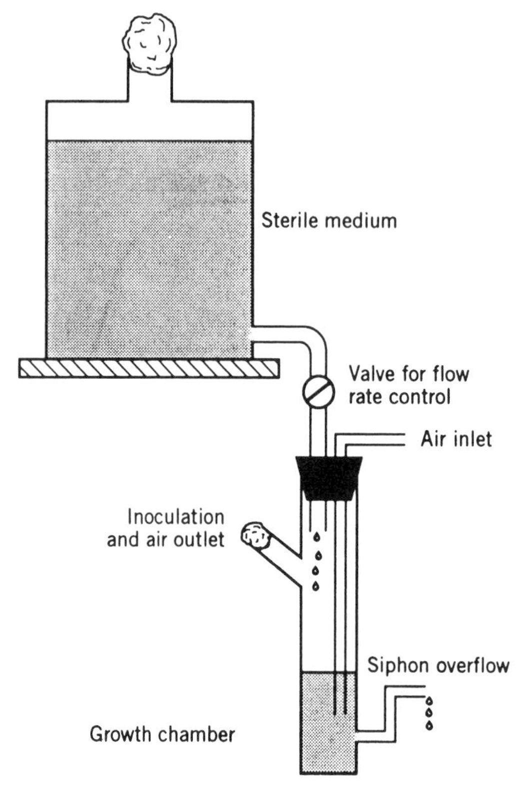

Fig. 2.13. Chemostat for continuous culture of microorganisms. From Marison (1988a), with permission of the publisher.

At steady state the substrate concentration $[S]$ and the cell concentration X in the reactor are given by Equations 2.30 and 2.31, respectively:

$$S = K_s \frac{D}{\mu_{max} - D} \quad (2.30)$$

$$X = Y(S_i - S_e) \quad (2.31)$$

where Y = growth yield; S_i = influent substrate concentration; and S_e = effluent substrate concentration. The steady state breaks down at very low or very high dilution rates.

2.4.3. Other Kinetic Parameters

There are three important parameters in microbial growth kinetics: growth yield Y, specific growth rate μ, and specific substrate uptake rate q.

The rate of increase of microorganisms in a culture (dX/dt) is proportional to the rate of substrate uptake/removal (dS/dt) by microbial cells:

$$dX/dt = Y\, dS/dt \quad (2.37)$$

where Y = the growth yield coefficient expressed as milligrams of cells formed per milligram of substrate utilized; dX/dt = the rate of increase in microorganism concentration (mg/L/day); and dS/dt = the rate of substrate removal (mg/L/day).

A more simplified equation showing the relationship between the three parameters is the following:

$$\mu = Yq \quad (2.38)$$

where μ = the specific growth rate (t^{-1}); and Y = the growth yield (milligrams of cells formed per milligram of substrate removed); q = substrate uptake rate (mg/L/day).

2.4.3.1. *Growth Yield*

As shown above, growth yield is the amount of biomass formed per unit amount of substrate removed. It reflects the efficiency of conversion of nutrients to cell material. The yield coefficient Y is obtained as follows:

$$Y = \frac{X - X_0}{S_0 - S} \quad (2.39)$$

S_0 and S = initial and final substrate concentrations, respectively (mg/L or mol/L); X_0 and X = initial and final microbial concentrations, respectively.

Several factors influence the growth yield: type of microorganisms, growth medium, substrate concentration, terminal electron acceptor, pH, and incubation temperature (Grady and Lim, 1980). Yield coefficients for several bacterial species are in the range of 0.4 to 0.6 (Heijnen and Roels, 1981).

For a pure microbial culture growing on a single substrate, the growth yield Y is assumed to be constant. However, in the natural environment, particularly in wastewater, there is a wide range of microorganisms, few of which are in the logarithmic phase. Many of them are in the stationary or in the declining phase of growth. Some of the energy will be used for cell maintenance. Thus, the growth yield Y must be corrected for the amount of cell decay occurring during the declining phase of growth. This correction will give the true growth yield coefficient, which is lower than the measured yield. Equation 2.38 becomes

$$\mu = Yq - k_d \quad (2.40)$$

where k_d is the endogenous decay coefficient (day^{-1}).

2.4.3.2. *Specific Substrate Uptake Rate q*

The specific substrate uptake (removal) is given by the following equation:

$$q = \frac{dS/dt}{X} \quad (2.41)$$

q (day^{-1}) is given by Monod's equation:

$$q = q_{max} \frac{[S]}{K_s + [S]} \quad (2.42)$$

2.4.3.3. *Specific Growth Rate* μ

μ is given by the following:

$$\mu = \frac{dX/dt}{X} \quad (2.43)$$

μ (day^{-1}) is given by Monod's equation:

$$\mu = \mu_{max} \cdot \frac{[S]}{K_s + [S]} \quad (2.44)$$

In waste treatment, the reciprocal of μ is the biological solid retention time

$$\mu = \frac{1}{\theta_c} \quad (2.45)$$

Thus,

$$\frac{1}{\theta_c} = Y q - k_d \quad (2.46)$$

2.4.4. Measurement of Microbial Growth

Various approaches are available for measuring the growth of microbial cultures in the laboratory or for determining microbial numbers or mass in an environmental sample.

2.4.4.1. *Total Number of Cells*

Total number of cells (live and dead cells) is measured by using special counting chambers such as the Petroff–Hauser chamber for bacterial counts or the Sedgewick–Rafter chamber for algal counts. The use of a phase-contrast microscope is required when nonphotosynthetic microorganisms are under consideration. Water and wastewater samples can also be filtered through membrane filters and, following proper staining of the cells, can be examined under a light microscope. The most popular method is the acridine orange direct count (AODC), by which cells are stained with acridine orange, retained on a membrane filter treated to suppress autofluorescence (use of Nuclepore filters treated with Irgalan black), and subsequently counted with an epifluorescent microscope. Electronic particle counters are also used for determining the total number of microorganisms in a sample. These methods do not differentiate, however, between live and dead microorganisms, and very small cells may be missed.

2.4.4.2. *Measurement of the Number of Viable Microbes on Solid Growth Media*

The purpose of this method is to measure the number of viable cells capable of forming colonies in a suitable growth medium (it is assumed that one viable cell will give one colony).

Plate count is determined by using the pour plate method (0.1–1 ml of microbial suspension is mixed with molten agar medium in a petri dish), or the spread plate method (0.1 ml of bacterial suspension is spread on the surface of an agar plate). The results of plates counts are expressed as colony forming units (CFU). CFU per plate should be between 30 and 300.

Membrane filters can also be used to determine microbe numbers in dilute samples. The sample is filtered and the filter is placed directly on a suitable agar medium. Viable cell counts are routinely used in aquatic and wastewater microbiology.

2.4.4.3. *Measurement of Dry Weight*

A sample is passed through a tared membrane filter (0.2 μm porosity) or centrifuged and dried at 105°C until a constant weight is reached. Cell dry weight is generally expressed in grams per liter. The average dry weight of a procaryotic cell varies between 10^{-15} and 10^{-11} g, whereas that of eucaryotic microorganisms varies between 10^{-11} and 10^{-7} g.

2.4.4.4. *Turbidity Measurement*

The turbidity of a cell suspension is determined with a spectrophotometer and is expressed as absorbance units. There is a close relationship between total microbe numbers and turbidity of bacterial suspensions. However, the cells must be dispersed by vigorous mixing or blender treatment prior to absorbance measurements.

2.4.4.5. *Determination of Cell Biochemicals*

Finally the growth of microbial cultures can be measured by determination of specific cell biochemicals such as ATP, DNA, RNA, or proteins.

2.4.5. Physical and Chemical Factors Affecting Microbial Growth

2.4.5.1. *Substrate Concentration*

The relationship between the specific growth rate μ and substrate concentration S is given by Monod's equation (Fig. 2.14A):

$$\mu = \mu_{max} \frac{[S]}{K_s + [S]} \tag{2.32}$$

where μ_{max} = maximum specific growth rate (hr^{-1}); S = substrate concentration (mg/L); and K_s = half-saturation constant (mg/L). K_s is the substrate concentration at which the specific growth rate is equal to $\mu_{max}/2$. K_s represents the affinity of the microorganism for the substrate.

μ_{max} and K_s are influenced by temperature, type of carbon source and other factors.

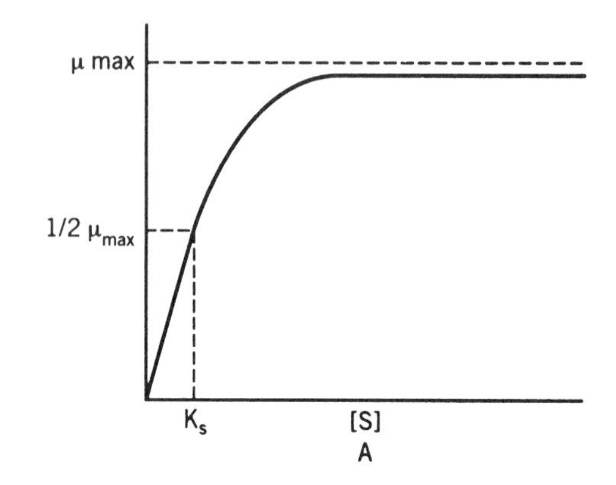

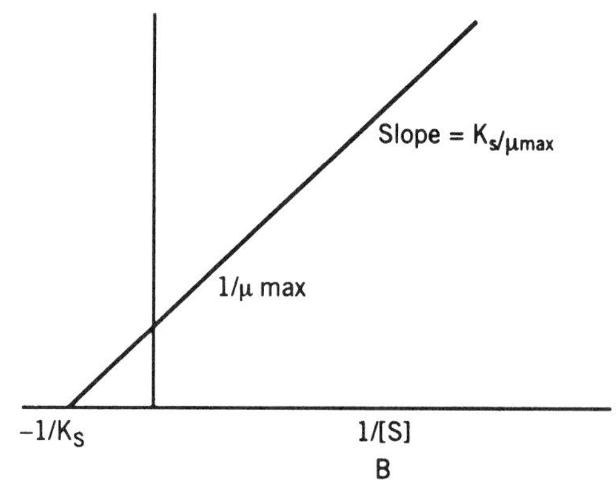

Fig. 2.14. Relationship between the specific growth rate μ and substrate concentration S: Monod's saturation curve.

Monod's equation can be linearized by using the Lineweaver–Burke equation:

$$\frac{1}{V} = \frac{K_s}{V_{max}[S]} + \frac{1}{V_{max}} \tag{2.33}$$

Figure 2.14B shows a plot of $1/\mu$ versus $1/S$. The slope, y intercept, and x intercept are (K_s/V_{max}), ($1/V_{max}$), and ($-1/K_s$), respectively. This plot allows the computation of K_s and μ_{max}. K_s values for individual chemicals found in wastewater are between 0.1 and 1.0 mg/L (Hanel, 1988).

2.4.5.2. *Temperature*

Temperature is one of the most important factors affecting microbial growth and survival. Microbial growth can occur at temperatures varying from below freezing to more than 100°C. Based on the optimum temperature for growth, microorganisms are classified as mesophiles, psychrophiles, thermophiles, or extreme thermophiles. Microbial growth rate is related to temperature by the Arrhenius equation:

$$\mu = Ae^{-E/RT} \tag{2.34}$$

where A = constant; E = activation energy (kcal/mole); R = gas constant; and T = absolute temperature (K).

Psychrophiles can grow at low temperatures because their cell membranes have a high content of unsaturated fatty acids, which helps maintain membrane fluidity, whereas a high content of saturated fatty acids helps thermophiles function at high temperatures. The decreased μ at high temperatures is due to the thermal denaturation of proteins, particularly enzymes, as well as changes in membrane structure, which leads to alterations in cell permeability.

2.4.5.3. *pH*

Biological treatment of wastewater occurs generally at neutral pH. In general, the optimum pH for growth of bacteria is around 7, although some may be obligately acidophilic (e.g., *Thiobacillus, Sulfolobus*) and thrive at pH below 2. Fungi prefer acidic environments, with a pH of 5 or lower. Cyanobacteria grow optimally at pH levels higher than 7. Bacterial growth generally results in a decrease of the pH of the medium by release of acidic metabolites (e.g., organic acids, H_2SO_4). Conversely, some microorganisms can increase the pH of their surrounding milieu (e.g., nitrifiers, algae).

pH affects the activity of microbial enzymes. It affects the ionization of chemicals and thus plays a role in the transport of nutrients and toxic chemicals into the cell.

2.4.5.4. *Oxygen Level*

Microorganisms can grow in the presence or in the absence of oxygen. They are divided into strict aerobes, facultative anaerobes (which can grow in the presence or in the absence of oxygen), and strict anaerobes. Aerobic microorganisms use oxygen as the terminal electron acceptor in respiration. Anaerobic counterparts use other electron acceptors such as sulfate, nitrate, or CO_2. Some microorganisms are microaerophilic and require low levels of oxygen for growth. Through their metabolism, aerobes may render the environment suitable for anaerobes by utilizing oxygen.

Upon reduction, oxygen forms toxic products such as superoxide (O_2^-), hydrogen peroxide (H_2O_2), or hydroxyl radicals. However, microorganisms have acquired enzymes to deactivate them. For example, H_2O_2 is destroyed by catalase and peroxidase enzymes, whereas O_2^- is deactivated by superoxide dismutase. The following are catalase- and superoxide dismutase-catalyzed reactions:

$$2O_2^- + 2H^+ \xrightarrow{\text{superoxide dismutase}} O_2 + H_2O_2 \tag{2.35}$$

$$2H_2O_2 \xrightarrow{\text{catalase}} 2H_2O + O_2 \tag{2.36}$$

2.5. FURTHER READING

Barnes, D., and P.J. Bliss. 1983. *Biological Control of Nitrogen in Wastewater Treatment.* E & F.N. Spon, London.

Bitton, G., and B. Koopman. 1986. Biochemical tests for toxicity screening, pp. 27–57, in: *Toxicity Testing Using Microorganisms,* G. Bitton and B. Dutka, Eds. CRC Press, Boca Raton, FL.

Boyd, R.F. 1988. *General Microbiology,* 2nd ed. Times Mirror/Mosby Cool., St. Louis.

Brock, T.D., and M.T. Madigan. 1991. *Biology of Microorganisms,* 6th ed. Prentice-Hall, Englewood Cliffs, NJ.

Gaudy, A.F., Jr., and E.T. Gaudy. 1988. *Elements of Bioenvironmental Engineering.* Engineering Press, San Jose, CA, 592 pp.

Grady, C.P.L., Jr., and H.C. Lim. 1980. *Biological Wastewater Treatment: Theory and Applications.* Marcel Dekker, New York, 963 pp.

Hammer, M.J. 1986. *Water and Wastewater Technology.* John Wiley and Sons, New York, 536 pp.

Hanel, K. 1988. *Biological Treatment of Sewage by*

the Activated Sludge Process. Ellis Horwood, Chichester, U.K., 299 pp.

Heijnen, J.J., and J.A. Roels. 1981. A macroscopic model describing yield and maintenance relationships in aerobic fermentation. Biotechnol. Bioeng. 23: 739–741.

Madigan, M.T. 1988. Microbiology, physiology, and ecology of phototrophic bacteria, pp. 39–111, in: *Biology of Anaerobic Microorganisms,* A.J.B. Zehnder, Ed. Wiley, New York.

Marison, L.W. 1988a. Growth kinetics, pp. 184–217, in: *Biotechnology for Engineers: Biological Systems in Technological Processes,* A. Scragg, Ed. Ellis Horwood, Chichester, U.K.

Marison, L.W. 1988b. Enzyme kinetics, pp. 96–119, in: *Biotechnology for Engineers: Biological Systems in Technological Processes,* A. Scragg, Ed. Ellis Horwood, Chichester, U.K.

Metcalf and Eddy, Inc. 1991. *Wastewater Engineering: Treatment, Disposal and Reuse.* McGraw-Hill, New York, 1334 pp.

Michal, G. 1978. Determination of Michaelis constant and inhibitor constants, pp. 29–42, in: *Principles of Enzymatic Analysis,* H.U. Bergmeyer, Ed. Verlag Chemie, Wienheim.

Rittmann, B.E., D.E. Jackson, and S.L. Storck. 1988. Potential for treatment of hazardous organic chemicals with biological processes, pp. 15–64, in: *Biotreatment Systems,* Vol. 3, D.L. Wise, Ed. CRC Press, Boca Raton, FL.

Scragg, A., Ed. *Biotechnology for Engineers: Biological Systems in Technological Processes.* Ellis Horwood, Chichester, U.K., 390 pp.

Segel, I.H. 1975. *Enzyme Kinetics.* Wiley, New York, 955 pp.

Shamat, N.A., and W.J. Maier. 1980. Kinetics of biodegradation of chlorinated organics. J. Water Pollut. Contr. Fed. 52: 2158–2166.

Stanier, R.Y., E.A. Adelberg, J.L. Ingraham, and M.L. Wheelis. 1979. *Introduction to the Microbial World.* Prentice-Hall, Englewood Cliffs, NJ, 468 pp.

Tortora, G.J., B.R. Funke, and C.L. Case. 1989. *Microbiology: An Introduction.* Benjamin/Cummings, Redwood City, CA, 810 pp.

3

ROLE OF MICROORGANISMS IN BIOGEOCHEMICAL CYCLES

This chapter is devoted to the examination of the biogeochemical cycles of nitrogen, phosphorus, and sulfur. We will discuss the microbiology of each cycle and emphasize the biotransformations and subsequent biological removal of these nutrients in wastewater treatment plants. Public health and technological aspects of each cycle will also be addressed.

A. NITROGEN CYCLE

3.1. INTRODUCTION

Nitrogen is essential to life and is a component of proteins and nucleic acids in microbial, animal, and plant cells. It is ironic that nitrogen gas is the most abundant gas (79% of the earth atmosphere) in the air we breath and yet it is a limiting nutrient in aquatic environments and in agricultural lands, leading to protein deficiency that is experienced by millions of people in developing countries. Unfortunately, nitrogen gas cannot be utilized by most organisms unless it is first converted to ammonia. This is because N_2 is a very stable molecule that will undergo changes only under extreme conditions (e.g., electrical discharge; high temperatures and pressures) (Barnes and Bliss, 1983).

3.2. MICROBIOLOGY OF THE NITROGEN CYCLE

Microorganisms play a major role in nitrogen cycling in the environment. The nitrogen cycle is displayed in Figure 3.1. We will now discuss the microbiology of five steps involved in nitrogen cycling: nitrogen fixation, assimilation, mineralization, nitrification, and denitrification (Alexander, 1977; Atlas and Bartha, 1987; Barnes and Bliss, 1983; Grady and Lim, 1980).

3.2.1. Nitrogen Fixation

Chemical reduction of nitrogen is very energy-intensive and expensive. As regards biological reduction of nitrogen, only a few species of bacteria and cyanobacteria (blue-green algae) are capable of carrying out nitrogen fixation, which ultimately results in the production of ammonia. The global biological nitrogen fixation is approximately 2×10^8 metric tons of N_2 per year. Agronomists have put considerable effort toward exploiting biological nitrogen fixation for maximizing crop yields.

3.2.1.1. Nitrogen-Fixing Microorganisms

There are two categories of nitrogen-fixing microorganisms (Table 3.1; Grant and Long, 1981). *Nonsymbiotic nitrogen-fixing microorganisms.* These include *Azotobacter* (e.g., *A.*

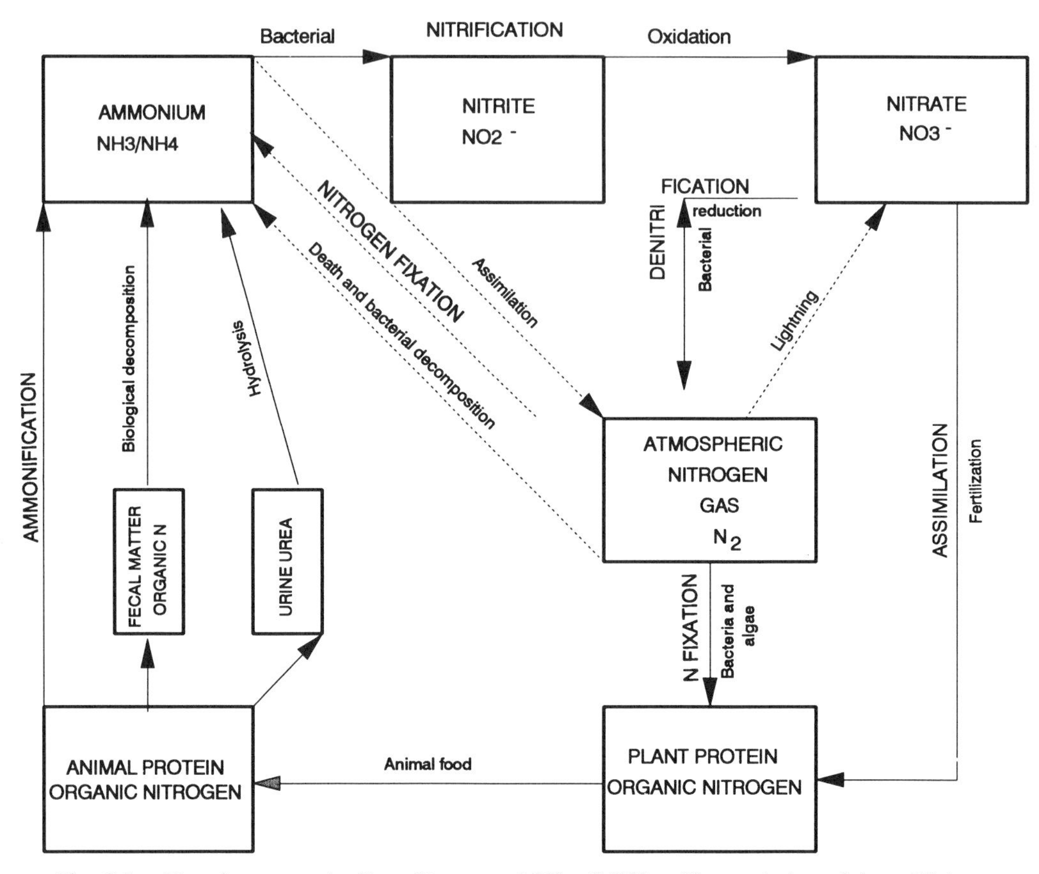

Fig. 3.1. The nitrogen cycle. From Barnes and Bliss (1983), with permission of the publisher.

TABLE 3.1. Nitrogen-Fixing Microorganisms[a]

Category	Microorganisms
A. Free-living nitrogen-fixing microorganisms	
Aerobes	*Azotobacter*
	Beijerinckia
Microaerophilic	*Azospirillum*
	Corynebacterium
Facultative anaerobes	*Klebsiella*
	Erwinia
Anaerobes	*Erwinia*
	Clostridium
	Desulfovibrio
B. Symbiotic associations	
Microbe–higher plants	Legume + *Rhizobium*
Cyanobacteria–aquatic weeds	*Anabaena-Azolla*
Others	Termites + enterobacteria

[a]Adapted from Grant and Long (1981).

agilis, A. chroococcum, A. vinelandii), gram-negative bacteria that forms cysts and fix nitrogen in soils and other environments. Other nitrogen-fixing microorganisms are *Klebsiella, Clostridium* (anaerobic, spore-forming bacteria active in sediments), and cyanobacteria (e.g., *Anabaena, Nostoc*). The latter fix nitrogen in natural waters and soils, and their fixation rate is ten times higher than free-nitrogen-fixing microorganisms in soils. The site of nitrogen fixation in cyanobacteria is a special cell called a heterocyst(Fig. 3.2). Cyanobacteria sometimes form associations with aquatic plants (e.g., *Anabaena–Azolla* association).

Symbiotic nitrogen-fixing microorganisms. Some prokaryotes may enter into a symbiotic relationship with higher plants to fix nitrogen. An example of significant agronomic importance is the legume–*Rhizobium* association. Upon infection of the root, *Rhizobium* form a nodule, which is the site of nitrogen fixation. Other examples are the association between *Frankia* and roots of woody perennial plants, and the association (with no nodule formation) between *Azospirillum* and the roots of maize and tropical grasses.

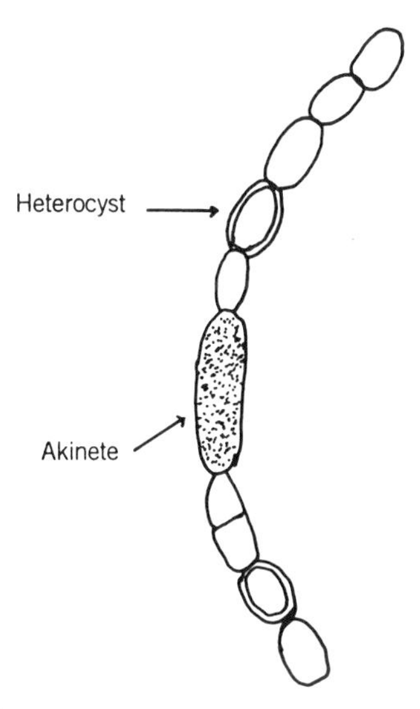

Fig. 3.2. Heterocysts, sites of nitrogen fixation in cyanobacteria (e.g., *Anabaena flos-aquae*).

3.2.1.2. Nitrogenase: The Enzyme Involved in Nitrogen Fixation

Nitrogen fixation is driven by an enzyme called nitrogenase which is made of iron sulfide and molybdo-iron proteins, both of which are sensitive to oxygen. This enzyme has the ability to reduce triple-bonded molecules such as N_2 and requires Mg^{2+} and energy in the form of ATP (15 to 20 ATP/N_2). The biosynthesis of this enzyme is controlled by the *nif* genes. Microorganisms have developed means to avoid inactivation of the oxygen-sensitive nitrogenase enzyme. For example, *Azotobacter* produces a copious amount of polysaccharides, which help reduce O_2 diffusion, protecting the enzyme from inactivation.

3.2.1.3. Nitrogen Fixation Methodology

Nitrogen fixation is determined by the acetylene reduction technique. This technique consists of measuring the reduction of C_2H_2 (acetylene) to C_2H_4 (ethylene).

The amount of nitrogen fixed is given by Equations 3.1 and 3.2.

$$\text{Moles of } N_2 \text{ fixed} = \frac{\text{moles } C_2H_2 \longrightarrow C_2H_4}{3} \tag{3.1}$$

$$\text{Grams of } N_2 \text{ fixed} = \frac{\text{moles } C_2H_2 \longrightarrow C_2H_4}{3} \times 28 \tag{3.2}$$

3.2.2. Nitrogen Assimilation

Heterotrophic and autotrophic microorganisms take up and assimilate NH_4^+ and NO_3^- after reduction to NH_4 (see more details in Section 3.2.5.1). Assimilation is responsible for some nitrogen removal in wastewater treatment plants. Plant and algal cells uptake nitrogen preferably in the form of NH_4^+. In soils, NH_4^+-based fertilizers are preferred to NO_3^--based ones (N-SERVE is used to inhibit nitrification in soils).

Cells convert NO_3^- or NH_4^+ to proteins and grow until nitrogen becomes limiting. For each 100 units of carbon assimilated, cells need approximately 10 units of nitrogen (C/N ratio = 10).

3.2.3. Nitrogen Mineralization (Ammonification)

Ammonification is the transformation of organic nitrogenous compounds to inorganic forms. This process is driven by a wide variety of microorganisms (bacteria, actinomycetes, fungi). In soils, some organic nitrogenous compounds become resistant to biodegradation because they form complexes with phenols and/or polyphenols.

Proteins are mineralized to NH_4^+ according to the following sequence: proteins → amino acids → deamination to NH_4.

An example is the transformation of urea to ammonium:

$$O{=}C\begin{matrix}\diagup NH_2\\ \diagdown NH_2\end{matrix} + H_2O \xrightarrow[\text{enzyme}]{\text{urease}} 2NH_3 + CO_2 \quad (3.3)$$

Proteins are converted to peptides or amino acids by extracellular proteolytic enzymes. Ammonium is produced following deamination (e.g., oxidative or reductive deaminations) of the amino acids according to reactions 3.4 and 3.5:

Oxidative deamination:

$$\underset{\text{amino acid}}{R{-}\underset{NH_2}{\underset{|}{C}H}{-}COOH} + \tfrac{1}{2}\,O_2 \rightarrow \underset{\text{keto acid}}{R{-}\underset{O}{\underset{\|}{C}}{-}COOH} + NH_4 \quad (3.4)$$

Reductive deamination:

$$R{-}\underset{NH_2}{\underset{|}{C}H}{-}COOH + 2H \rightarrow \underset{\text{acid}}{R{-}CH_2{-}COOH} + NH_4 \quad (3.5)$$

NH_4^+ predominates in acidic and neutral aquatic environments. As pH increases, NH_3 predominates and is released to the atmosphere.

$$NH_4^+ \rightarrow NH_3 + H^+ \quad (3.6)$$

3.2.4. Nitrification

3.2.4.1. *Microbiology of Nitrification*

Nitrification is the conversion of ammonium to nitrate by microbial action. This process is carried out by two categories of microorganisms.

1. Conversion of NH_4^+ to NO_2^-. *Nitrosomonas* (e.g., *N. europaea, N. oligocarbogenes*) oxidizes ammonium to nitrite via hydroxylamine (NH_2OH). Other ammonium oxidizers are *Nitrosospira, Nitrosococcus,* and *Nitrosolobus* (Focht and Verstraete, 1977).

$$2NH_4^+ + O_2 \rightarrow 2NH_2OH + 2H^+ \quad (3.7)$$

$$NH_4^+ + 1.5\,O_2 \rightarrow NO_2^- + 2H^+ + H_2O + 275\text{ kJ} \quad (3.8)$$

2. Conversion of NO_2^- to NO_3^-. *Nitrobacter* (e.g., *N. agilis, N. winogradski*) converts nitrite to nitrate:

$$NO_2^- + 0.5\,O_2 \rightarrow NO_3^- + 75\text{ kJ} \quad (3.9)$$

Other nitrite oxidizers are *Nitrospira* and *Nitrococcus* (Focht and Verstarete, 1977).

The oxidation of NH_4^+ to NO_2^- and then to NO_3^- is an energy-yielding process. Microorganisms utilize the generated energy to assimilate CO_2. Carbon requirements for nitrifiers are satisfied by carbon dioxide, bicarbonate, or carbonate. Nitrification is favored by the presence of oxygen and sufficient alkalinity to neutralize the hydrogen ions produced during the oxidation process. Theoretically, the oxygen requirement is 4.6 mg O_2/1 mg $NH_4^+{-}N$ oxidized to NO_3^- (U.S. EPA, 1975). Although they are obligate aerobes, nitrifiers have less affinity for oxygen than aerobic heterotrophic bacteria. The optimum pH for growth of *Nitrobacter* is between 7.2 and 7.8. Acid production resulting from nitrification can cause problems in poorly buffered wastewaters.

Although autotrophic nitrifiers are predominant in nature, nitrification may be carried out by heterotrophic bacteria (e.g., *Arthrobacter*) and fungi (e.g., *Aspergillus*) (Verstraete and Alexander, 1972). These microorganisms utilize organic carbon sources and oxididize ammonium to nitrate. However, heterotrophic nitrification is much slower than autotrophic nitrification and, probably, does not have any significant contribution.

3.2.4.2. *Nitrification Kinetics*

The growth rate of *Nitrobacter* is higher than that of *Nitrosomonas*. Thus, the rate-limiting step in nitrification is the conversion of ammonia to nitrite by *Nitrosomonas*. The following equation describes the growth according to Monod's model (Barnes and Bliss, 1983):

$$\mu = \frac{[NH_4{}^+]}{K_s + [NH_4{}^+]} \quad (3.10)$$

where μ = specific growth rate (days^{-1}); μ_{max} = maximum specific growth rate (days^{-1}); $[NH_4{}^+]$ = ammonium concentration (mg/L); and K_s = half saturation constant (ammonium substrate) (mg/L).

The ammonium oxidation rate q is related to the specific growth rate μ by the following equation (U.S. EPA, 1975):

$$q = \frac{\mu}{Y} \quad (3.11)$$

where Y = yield coefficient (see Chapter 2).

In wastewater treatment plants, oxygen is a limiting factor controlling the growth of nitrifiers. Therefore, Equation 3.10 is modified to take into account the effect of oxygen concentration:

$$\mu = \mu_{max} \frac{[NH_4^+]}{K_s + [NH_4^+]} \frac{[DO]}{K_o + [DO]} \quad (3.12)$$

where $[DO]$ = dissolved oxygen concentration (mg/L); and K_o = half-saturation constant (oxygen) (mg/L).

K_o has been estimated to be 0.15–2 mg/L, depending on temperature (U.S. EPA, 1975). Others have reported a range of K_o of 0.25–1.3 mg/L (Hawkes, 1983a; Stenstrom and Song, 1991; Verstraete and Vaerenbergh, 1986).

More complex equations expressing the growth kinetics of nitrifiers take into account the substrate ($NH_4{}^+$) concentration as well as environmental factors such as temperature, pH, and dissolved oxygen (Barnes and Bliss, 1983):

$$\mu_n = \mu_{max} \frac{[NH_4] - N}{0.4e^{0.118(T-15)} + [NH_4]} \times \frac{DO \cdot e^{0.095(T-15)}}{1 + DO} \times (1.83)(pH_{opt} - pH) \quad (3.13)$$

where μ_n = μ of nitrifiers; T = temperature (°C); pH_{opt} = optimum pH = 7.2; and μ_{max} = 0.3 day^{-1}.

The minimum residence time is a function of μ_n:

$$\text{min residence time} = 1/\mu_n \quad (3.14)$$

The μ_{max} of nitrifiers (0.006–0.035hr^{-1}) is much lower than the μ_{max} of mixed cultures of heterothrophs using glucose as a substrate (0.18–0.38 hr^{-1}) (Grady and Lim, 1980; Christensen and Harremoes, 1978; US EPA, 1977). The cell yield of nitrifiers is also lower than that of heterotrophic microorganisms (Rittmann, 1987). The maximum cell yield for *Nitrosomonas* is 0.29; the cell yield of *Nitrobacter* is much lower, around 0.08. However, experimental yield values are much lower and vary from 0.04 to 0.13 for *Nitrosomonas* and from 0.02 to 0.07 for *Nitrobacter* (Edeline, 1988; Painter, 1970). Thus, in nitrifying environments, *Nitrosomonas* is present in higher numbers than *Nitrobacter*. These relatively low yields have many implications for nitrification in wastewater treatment plants. The half-saturation constant (K_s for energy substrate is 0.05–5.6 mg/L for *Nitrosomonas* and 0.06–8.4 mg/L for *Nitrobacter* (Verstraete and Vaerenbergh, 1986).

Nitrification is favored in biologically treated effluents with low biochemical oxygen demand (BOD) and high ammonia content. A suspended-growth aeration process is the most favorable to nitrification of effluents. Nitrification occurs in the aeration tank (4–6 hr detention time), and sludge containing high numbers of nitrifiers is recycled to maintain high nitrifier activity.

3.2.4.3. Factors Controlling Nitrification

Several factors control nitrification in wastewater treatment plants. These factors are ammonia/nitrite concentration, oxygen concentration, pH, temperature, BOD_5/TKN ratio, and the presence of toxic chemicals (Grady and Lim, 1980; Hawkes, 1983a; Metcalf and Eddy, 1991):

Ammonia/nitrite concentration. Growth of *Nitrosomonas* and *Nitrobacter* follows Monod's kinetics and depends on ammonia and nitrite concentration, respectively (see Section 3.2.4.2).

Oxygen level. DO concentration remains one of the most important factors controlling nitrification. The half-saturation constant for oxygen (K_o) is 1.3 mg/L (Metcalf and Eddy, 1991). For nitrification to proceed, the oxygen should be well distributed in the aeration tank of an activated sludge system and its level should not be less than 2 mg/L.

$$NH_3 + 2O_2 \rightarrow NO_3^- + H^+ + H_2O \qquad (3.15)$$

To oxidize 1 mg of ammonia 4.6 mg of O_2 are needed (Christensen and Harremoes, 1978).

Pure culture studies have demonstrated the possible growth of *Nitrobacter* in the absence of dissolved oxygen, with NO_3 used as electron acceptor and organic substances as the source of carbon (Bock et al., 1988; Ida and Alexander, 1965; Smith and Hoare, 1968). The implications of these findings in wastewater treatment remain to be investigated.

Temperature. The growth rate of nitrifiers is affected by temperature in the range of 8–30°C. The optimum temperature has been reported to be around 30°C (Hitdlebaugh and Miller, 1981).

pH. The optimum pH for *Nitrosomonas* and *Nitrobacter* lies between 7.5 and 8.5 (U.S. EPA, 1975). Nitrification ceases at or below pH 6.0 (Painter, 1970; Painter and Loveless, 1983). Alkalinity is destroyed as a result of ammonia oxidation by nitrifiers. Theoretically, nitrification destroys alkalinity, as $CaCO_3$, in amounts of 7.14 mg/1 mg of NH_4^+−N oxidized (U.S. EPA, 1975). Therefore, there should be sufficient alkalinity in wastewater to balance the acidity produced by nitrification. The pH drop that results from nitrification can be minimized by aerating the wastewater to remove CO_2. Lime is sometimes added to increase wastewater alkalinity.

BOD_5/TKN ratio. The fraction of nitrifying organisms decreases as the BOD_5/TKN ratio increases. In combined carbon–oxidation nitrification processes, this ratio is greater than 5, whereas in separate-stage nitrification processes, the ratio is lower than 3 (Metcalf and Eddy, 1991).

Toxic inhibition. Nitrifiers are subject to product and substrate inhibition and are also quite sensitive to several toxic compounds found in wastewater (Bitton, 1983; Bitton et al., 1989). It appears that many of those compounds are more toxic to *Nitrosomonas* than to *Nitrobacter.* Organic matter in wastewater is not directly toxic to nitrifiers. Apparent inhibition by organic matter may be indirect and may be due to O_2 depletion by heterotrophes (Barnes and Bliss, 1983). The most toxic compounds to nitrifiers are cyanide, thiourea, phenol, anilines, and heavy metals (silver, mercury, nickel, chromium, copper, and zinc). The toxic effect of copper on *Nitrosomonas europea* increases as the substrate (ammonia) concentration is raised from 3 mg/L to 23 mg/L as N (Sato et al., 1988).

Table 3.2 summarizes the conditions necessary for optimal growth of nitrifiers (U.S. EPA, 1977).

3.2.5. Denitrification

Nitrification exerts an oxygen demand in a receiving body of water. Therefore, nitrate must be removed before discharge to receiving waters, particularly if the receiving stream serves as a source of drinking water.

TABLE 3.2. Optimal Conditions for Nitrification[a]

Characteristic	Design value
Permissible pH range (95% nitrification)	7.2–8.4
Permissible Temperatures (95% Nitrification), °C	15–35
Optimum temperature, °C (approximately)	30°
DO level at peak flow, mg/L	>1.0
MLVSS, mg/L	1,200–2,500
Heavy metals inhibiting nitrification (Cu, Zn, Cd, Ni, Pb, Cr)	<5 mg/L
Toxic organics inhibiting nitrification	
Halogen-substituted phenolic compounds	0 mg/L
Halogenated solvents	0 mg/L
Phenol and cresol	<20 mg/L
Cyanides and all compounds from which hydrocyanic acid is liberated on acidification	<20 mg/L
Oxygen requirement (stoichiometric, lb O_2/lb NH_3-N, plus carbonaceous oxidation demand)	4.6

[a]Adapted from U.S. EPA (1977).

3.2.5.1. *Microbiology of Denitrification*

The two most important mechanisms of biological reduction of nitrate are assimilatory and dissimilatory nitrate reduction (Tiedje, 1988):

Assimilatory nitrate reduction. By this mechanism nitrate is taken up and converted to nitrite and then to ammonium by plants and microorganisms. It involves several enzymes that convert NO_3^- to NH_3, which is then incorporated into proteins and nucleic acids. Nitrate reduction is driven by a wide range of assimilatory nitrate reductases, the activity of which is not affected by oxygen. Certain microorganisms (e.g., *Pseudomonas aeruginosa*) possess both an assimilatory nitrate reductase and a dissimilatory nitrate reductase that is oxygen-sensitive. The two enzymes are encoded by different genes (Sias et al., 1980).

Dissimilatory nitrate reduction (denitrification). This process is an anaerobic respiration by which NO_3 serves as an electron acceptor. NO_3^- is reduced to nitrous oxide, N_2O, and nitrogen gas, N_2. N_2 liberation is the predominant output of denitrification. However, N_2 has low water solubility and thus tends to escape as rising bubbles. The bubbles may interfere with the settling of sludge in a sedimentation tank (Dean and Lund, 1981). The microorganisms involved in denitrification are aerobic autotrophic or heterotrophic microorganisms that can switch to anaerobic growth when nitrate is used as the electron acceptor.

Denitrification is carried out according to the following sequence:

$$NO_3 \xrightarrow[\text{reductase}]{\text{Nitrate}} NO_2 \xrightarrow[\text{reductase}]{\text{Nitrite}} NO$$

$$\xrightarrow[\text{reductase}]{\text{Nitric oxide}} N_2O \xrightarrow[\text{reductase}]{\text{Nitrous oxide}} N_2$$

Denitrifiers belong to several physiological (organotrophs, lithotrophs, and phototrophs) and taxonomic groups (Tiedje, 1988) and can use various energy sources (organic or inorganic chemicals or light). Microorganisms that are capable of denitrification belong to the following genera: *Pseudomonas, Bacillus, Spirillum, Hyphomicrobium, Agrobacterium, Acinetobacter, Propionobacterium, Rhizobium, Corynebacteri-*

um, Cytophaga, Thiobacillus, and *Alcaligenes.* The most widespread genera are probably *Pseudomonas* (*P. fluorescens, P. aeruginosa, P. denitrificans*) and *Alcaligenes,* which are frequently found in soils, water, and wastewater (Painter, 1970; Tiedje, 1988).

Nitrous oxide (N_2O) may be produced during denitrification in wastewater, leading to incomplete removal of nitrate. This gas is a major air pollutant, the production of which must be prevented or at least reduced. Under certain conditions, up to 8% of nitrate is converted to N_2O, and favorable conditions for its production are low COD/NO_3-N, short solid retention time, and low pH (Hanaki et al., 1992).

3.2.5.2. Conditions for Denitrification

The main factors controlling denitrification in wastewater treatment plants and other environments are the following (Barnes and Bliss, 1983; Hawkes, 1983a,b).

Nitrate concentration. Since nitrate serves as an electron acceptor for denitrifying bacteria, the growth rate of denitrifiers depends on nitrate concentration and follows Monod-type kinetics (see Section 3.2.5.3).

Anoxic conditions. O_2 competes effectively with nitrate as a final electron acceptor in respiration. Glucose oxidation releases more free energy in the presence of oxygen (686 kcal/mole glucose) than in the presence of nitrate (570 kcal/mole glucose) (Delwiche, 1970). That is the reason why denitrification must be conducted in the absence of oxygen. Denitrification may occur inside activated sludge flocs and biofilms despite relatively high levels of oxygen in the bulk liquid. Thus, the presence of oxygen in wastewater may not preclude denitrification at the microenvironment level (Christensen and Harremoes, 1978).

Presence of organic matter. Denitrifying bacteria must have an electron donor to carry out the denitrification process. Several sources of electrons have been suggested and studied. The sources include pure compounds (e.g., acetic acid, citric acid, methanol), raw domestic wastewater, wastes from food industries (brewery wastes, molasses), and sludges. The preferred source of electrons, although more expensive, is methanol, which is used as a carbon source to drive denitrification (Christensen and Harremoes, 1978). Biogas, containing approximately 60% methane, can also serve as a sole carbon source in denitrification (Werner and Kaiser, 1991). It has long been known that methane can be used as a carbon source in denitrification (Harremoes and Christensen, 1971), as methanotrophic bacteria oxidize methane to methanol (Mechsner and Hamer, 1985; Werner and Kaiser, 1991).

$$6NO_3 + 5CH_3OH \rightarrow 3N_2 + 5CO_2 + 7H_2O + 6(OH)^- \qquad (3.16)$$

Thus, 5/6 mole of methanol is necessary for denitrifying one mole of NO_3. However, some of the methanol is used for cell respiration and cell synthesis. The maximum removal of nitrate is achieved when the ratio CH_3OH/NO_3 is approximately 2.5 (Fig. 3.3) (Christensen and Harremoes, 1977). In an anaerobic upflow filter, nearly complete nitrate removal (99.8%) was achieved at a ratio $\geq$ 2.65 (Hanaki and Polprasert, 1989). It has been suggested that a value of 3.0 should ensure complete denitrification (U.S. EPA, 1975).

pH. In wastewater, denitrification is most effective at pH between 7.0 and 8.5 and the optimum is around 7.0 (Christensen and Harremoes, 1977; Metcalf and Eddy, 1991). Alkalinity and pH increase following denitrification. Theoretically, denitrification produces 3.6 mg alkalinity as $CaCO_3$ per 1 mg nitrate reduced to N_2. However, in practice, this value is lower and a value of 3.0 has been suggested for design purposes (U.S. EPA, 1975). Thus, denitrification replaces about half of the alkalinity consumed during nitrification.

Temperature. Denitrification may occur between 35°C and 50°C. It also occurs at low temperatures (5–10°C) but at a slower rate.

Effect of trace metals. Denitrification is readily stimulated in the presence of molybdenum and selenium, which are active in the formation of formate dehydrogenase, one of the enzymes im-

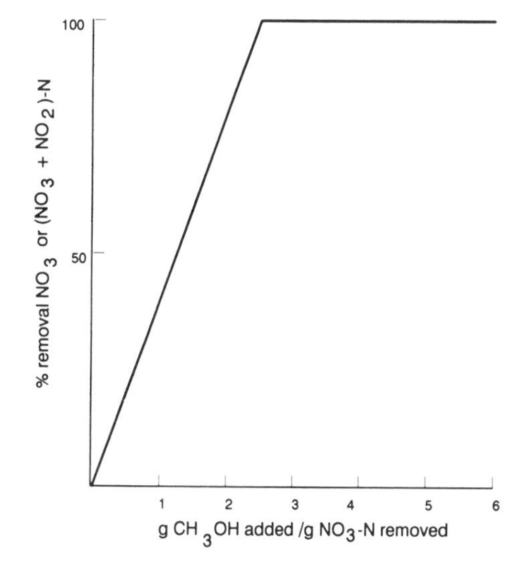

Fig. 3.3. Percentage removal of nitrate as a function of methanol/nitrate ratio. Adapted from Christensen and Harremoes (1977).

plicated in the metabolism of methanol. Molybdenum is essential to the synthesis of nitrate reductase (Chakrabarti and Jones, 1983).

Toxic chemicals. Denitrifying organisms are less sensitive to toxic chemicals than nitrifiers.

3.2.5.3. *Denitrification Kinetics*

The environmental factors discussed above have an effect on the growth kinetics of denitrifiers, as shown in Equation 3.17 (U.S. EPA, 1975):

$$\mu_D = \mu_{\max} \frac{D}{K_d + D} \frac{M}{K_m + M} \tag{3.17}$$

where μ_D = growth rate of denitrifiers; $\mu_{\max}$ = maximum growth rate of denitrifiers (as affected by nitrate and methanol concentrations, temperature, and pH); D = nitrate concentration (mg/L); K_d = half-saturation constant for nitrate (mg/L); M = methanol concentration (mg/L); and K_m = half-saturation constant for methanol (mg/L).

Denitrification rate is related to growth rate through Equation 3.18:

$$q_d = \mu_d / Y_d \tag{3.18}$$

where q_d = nitrate removal rate (mg NO_3-N/1 mg VSS/day); and Y_d = growth yield (mg VSS/1 mg NO_3-N removed).

In a completely mixed reactor, the solids retention time is given by Equation 3.19:

$$\frac{1}{\theta} = Y_d q_d - K_d \tag{3.19}$$

where θ = solids retention time (days); and K_d = decay coefficient (day^{-1}).

3.2.5.4. *Denitrification Methodology*

Denitrification can be measured by determining NO_3 disappearance, by measuring N_2 or N_2O formation, or by using ^{15}N. A popular method is the acetylene inhibition method, by which all nitrogen released is in the form of N_2O because of the specific inhibition of N_2O reductase by acetylene (Smith, et al., 1978). Chemical analysis is greatly simplified, since N_2O is a minor atmospheric constituent and can be assayed by gas chromatography.

3.3. NITROGEN REMOVAL IN WASTEWATER TREATMENT PLANTS

In domestic wastewater, nitrogen is found mostly in the form of organic nitrogen and ammonia. The average total nitrogen concentration in domestic sewage is approximately 35 mg/L. In this section we will mention only briefly the extent of, and means for, nitrogen removal in wastewater treatment plants. Specific processes for nitrogen removal will be covered in more detail in Chapters 8, 10, and 11.

3.3.1. Extent of Nitrogen Removal

Primary treatment of domestic wastewater removes approximately 15% of total nitrogen, mainly solids-associated organic nitrogen. Conventional biological treatment removes approximately another 10% of the nitrogen that is associated with cell biomass, which settles in the sedimentation tank. In biologically treated effluents, approximately 90% of the nitrogen is in the form of ammonia. Nitrogen removal can be lowered by the recycling of supernatants from sludge digesters into the wastewater treatment plant (Hammer, 1986). We will now discuss briefly additional means for nitrogen removal.

3.3.2. Means for Removal of Nitrogen

In sewage treatment plants there are several means for removing nitrogen from incoming wastewater.

Microbiological means: Nitrification–denitrification. Denitrification provides a means of removing nitrogen from a well-nitrified wastewater effluent. The overall efficiency of nitrification–denitrification can be as high as 95% (U.S. EPA, 1975).

Chemical and physical means. Liming results in a high pH (pH = 10 or 11), which converts NH_4^+ into NH_3, which can be removed from solution by air-stripping in packed cooling towers:

$$NH_4^+ + OH^- \rightarrow NH_4OH \xrightarrow{\text{air stripping}} NH_3 + H_2O \quad (3.20)$$

The first full-scale stripping tower was installed in the late 1970s in south Lake Tahoe in the United States. Some of the problems encountered are freezing in cold weather (ice formation on the packing material), scale formation (calcium carbonate scale), and air pollution with ammonia gas (Dean and Lund, 1981). However, ammonia can be recovered as an ammonium salt fertilizer.

Breakpoint chlorination, or superchlorination, oxidizes ammonium to nitrogen gas according to the reaction shown below (see Chapter 6 for more details).

$$3Cl_2 + 2NH_4^+ \rightarrow N_2 + 6\,HCl + 2H^+ \quad (3.21)$$

Breakpoint chlorination can remove 90%–100% ammonium (Metcalf and Eddy, 1991; U.S. EPA, 1975).

Nitrogen can also be removed by selective ion exchange, filtration, dialysis, and reverse osmosis.

3.4. ADVERSE EFFECTS OF NITROGEN DISCHARGES FROM WASTEWATER TREATMENT PLANTS

Wastewater treatment plants may discharge effluents with high ammonium or nitrate concentrations into receiving aquatic environments. This may lead to several environmental problems which are summarized as follows (Bitton, 1980a; Hammer, 1986; U.S. EPA, 1975).

3.4.1. Toxicity

Un-ionized ammonia is toxic to fish. At neutral pH, 99% of the ammonia occurs as NH_4^+, whereas NH_3 concentration increases at pH > 9.

$$NH_4^+\ OH^- \rightleftarrows NH_3 + H_2O \quad (3.22)$$

Therefore, ammonia toxicity is particularly important following discharge of alkaline wastewaters or rapid algal photosynthesis, which leads to high pH.

3.4.2. Oxygen Depletion in Receiving Waters

Ammonia may result in oxygen demand in receiving waters (recall that 1 mg ammonia exerts an oxygen demand of 4.6 mg O_2). The oxygen demand exerted by nitrifiers is called nitrogenous oxygen demand (NOD). Oxygen depletion adversely affects aquatic life.

3.4.3. Eutrophication of Surface Waters

Discharge of nitrogen into receiving waters may stimulate algal and aquatic plant growth. These, in turn, exert a high oxygen demand at nighttime, which adversely affects fish and other aquatic life and has a negative impact on the beneficial use of water resources for drinking or recreation. Nitrogen and phosphorus often are limiting nutrients in aquatic environments. Algal assay procedures help determine which of these two nutrients is the limiting one.

3.4.4. Effect of Ammonia on Chlorination-Efficiency

Chlorine combines with ammonium to form chloramines, which have a lower germicidal effect than free chlorine (see Chapter 6 for more details).

3.4.5. Corrosion

Ammonia, at concentrations exceeding 1 mg/L may cause corrosion of copper pipes (Dean and Lund, 1981).

3.4.6. Public Health Aspects of Nitrogen Discharges

Nitrate may be the cause of *methemoglobinemia* in infants and certain susceptible segments of the adult population (e.g., Navajos, Eskimos, and people with genetic deficiency of glucose-6-phosphate dehydrogenase or methemoglobin reductase have a higher incidence of methemoglobinemia), and it can lead to the formation of carcinogenic compounds (Bouchard et al., 1992; Craun, 1984b).

Methemoglobinemia ("blue babies" syndrome) is due to the conversion of nitrate to nitrite by nitrate-reducing bacteria in the gastrointestinal tract. Hemoglobin is converted to a brown pigment, methemoglobin, following oxidation, by nitrite, of Fe^{2+} in hemoglobin to Fe^{3+}.

$$\text{Hemoglobin } (Fe^{2+}) + O_2 \rightarrow \underset{\text{(red pigment)}}{\text{Oxyhemoglobin}} \quad (3.23)$$

$$\text{Hemoglobin } (Fe^{2+}) + NO_2^- \rightarrow \underset{\text{(brown pigment)}}{\text{Methemoglobin } (Fe^{3+})} \quad (3.24)$$

Since methemoglobin is incapable of binding molecular oxygen, the ultimate result is suffocation. Babies are more susceptible to methemoglobinemia because the higher pH in their stomachs allows a higher reduction of nitrate to nitrite by nitrate-reducing bacteria. Vitamin C offers a protective effect and helps maintain lower levels of methemoglobin. An enzyme, methemoglobin reductase, keeps methemoglobin at 1%–2% of the total hemoglobin in healthy adults.

Nitrite can also combine with secondary amines in the diet to form nitrosamines, which are known to be mutagenic and carcinogenic.

$$\underset{\text{Nitrite}}{NO_2^-} + \underset{\text{Secondary amine}}{\begin{matrix} R \\ R' \end{matrix}\!\!>\!NH} \rightarrow \underset{\text{Nitrosamine}}{\begin{matrix} R \\ R' \end{matrix}\!\!>\!N{-}N{=}O} \quad (3.25)$$

where R and R′ are alkyl and aryl groups.

The U.S. EPA interim drinking water standard for nitrate is 10 mg/L as NO_3-N. The World Health Organization (WHO) has also adopted a guideline of 10 mg/L NO_3-N. Public water supplies serving 1% of the U.S. population exceed the EPA limit (e.g., some Texas wells contain 110–690 mg/L NO_3-N) (Craun, 1984b).

B. PHOSPHORUS CYCLE

3.5. INTRODUCTION

Phosphorus is a macronutrient that is necessary to all living cells. It is an important component of adenosine triphosphate (ATP), nucleic acids (DNA and RNA), and phospholipids in cell membranes. P can be stored in intracellular volutin granules as polyphosphates in both procaryotes and eucaryotes. It is a limiting nutrient in regard to algal growth in lakes. The average concentration of total phosphorus (inorganic and organic forms) in wastewater is in the range of 10–20 mg/L.

Approximately 15% of the U.S. population contributes wastewater effluents to lakes, resulting in eutrophication of these water bodies. Eutrophication leads to significant changes in water quality and lowers the value of surface waters for fishing as well as for industrial and recreational uses. This can be controlled by reducing P inputs to receiving waters (Hammer, 1986).

3.6. MICROBIOLOGY OF THE PHOSPHORUS CYCLE

The major steps of the phosphorus cycle are displayed in Figure 3.4 and are described in the following discussion (Erlich, 1981).

3.6.1. Mineralization

Organic phosphorus compounds (e.g., phytin, inositol phosphates, nucleic acids, phospholipids) are mineralized to orthophosphate by a wide range of microorganisms that include bacteria (e.g., *B. subtilis, Arthrobacter*), actinomycetes (e.g., *Streptomyces*), and fungi (e.g., *Aspergillus, Penicillium*). Phosphatases are the enzymes responsible for degradation of phosphorus compounds.

3.6.2. Assimilation

Microorganisms assimilate phosphorus, which enters into the composition of several macromolecules in the cell. Some microorganisms have the ability to store phosphorus as polyphosphates in special granules. This topic will be covered in more detail in Section 3.8.

3.6.2. Precipitation of Phosphorus Compounds

The solubility of orthophosphate is controlled by the pH of the aquatic environment and by the presence of Ca^{2+}, Mg^{2+}, Fe^{3+} and Al^{3+}. When precipitation occurs, there is formation of insoluble compounds such as hydroxyapatite ($Ca_{10}(PO_4)_6(OH)_2$, vivianite $Fe_3(PO_4)_2{\cdot}8H_2O$ or variscite $AlPO_4{\cdot}2H_2O$ (Ehrlich, 1981).

3.6.3. Microbial Solubilization of Insoluble Forms of Phosphorus

Microorganisms, through their metabolic activity, help in the solubilization of P compounds. The mechanisms of solubilization are metabolic processes involving enzymes, production of organic and inorganic acids by microorganisms (e.g., succinic acid, oxalic acid, nitric and sulfuric acid), production of CO_2, which lowers pH, production of H_2S, which may react with

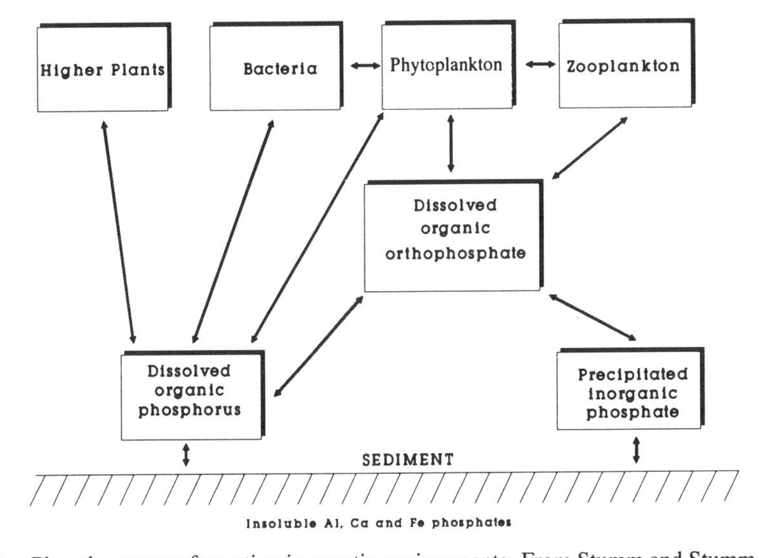

Fig. 3.4. Phosphorus transformation in aquatic environments. From Stumm and Stumm-Zollinger (1972), with permission of the publisher.

iron phosphate and liberate orthophosphate, and the production of chelators that can complex Ca, Fe, or Al.

3.7. PHOSPHORUS REMOVAL IN WASTEWATER TREATMENT PLANTS

The average concentration of total phosphorus (inorganic and organic forms) in wastewater is in the range of 10–20 mg/L, much of which comes from phosphate builders in detergents. Common forms of phosphorus in wastewater are orthophosphate (PO_4^{3-}) (50%–70% of phosphorus), polyphosphates, and phosphorus tied to organic compounds. Orthophosphate comprises approximately 90% of phosphorus in biologically treated effluents (Meganck and Faup, 1988). Since phosphorus is mainly responsible for eutrophication of surface waters, it must be removed by wastewater treatment processes before discharge of the effluents into surface waters. Most often, phosphorus is removed from wastewater by chemical means, but it can also be removed by biological processes (see Section 3.8). Several biological and chemical mechanisms are responsible for phosphorus removal in wastewater treatment plants (Arvin, 1985; Arvin and Kristensen, 1983):

1. Chemical precipitation, which is controlled by pH and cations such as Ca, Fe, and Al
2. Phosphorus assimilation by wastewater microorganisms
3. Polyphosphate accumulation by microorganisms
4. Microorganism-mediated enhanced chemical precipitation

Primary treatment of wastewaters removes only 5%–15% of phosphorus associated with particulate organic matter and conventional biological treatment does not remove a substantial amount of phosphorus (approximately 10%–25%) (Metcalf and Eddy, 1991). Most of the retained phosphorus is transferred to sludge (Hammer, 1986). Additional phosphorus removal can be achieved by adding iron and aluminum salts or lime to wastewater. Commercially available aluminum and iron salts are alum, ferric chloride, ferric sulfate, ferrous sulfate, and waste pickle liquor from the steel industry. These are generally added in excess to compete with natural alkalinity. Lime is less frequently used for phosphorus removal because of increased production of sludge as well as the operation and maintenance problems associated with its use (U.S. EPA, 1987b).

Aluminum reacts with phosphorus to form aluminum phosphate:

$$Al^{3+} + PO_4^{3-} \rightarrow AlPO_4 \qquad (3.26)$$

Ferric chloride reacts with phosphorus to form ferric phosphate:

$$FeCl_3 + PO_4^{3-} \rightarrow FePO_4 + 3\,Cl^- \qquad (3.27)$$

Other treatments for removal of phosphorus include adsorption to activated alumina, ion exchange, electrochemical methods, and deep-bed filtration (Meganck and Faup, 1988).

3.8. ENHANCED BIOLOGICAL REMOVAL OF PHOSPHORUS

In addition to chemical precipitation, phosphorus can also be removed by biological means. In the 1960s, Shapiro and his collaborators demonstrated the role of microorganisms in the uptake and release of phosphorus in the activated sludge system (Shapiro, 1967; Shapiro et al., 1967). Since then, two approaches have been taken to explain the mechanism of enhanced biological phosphorus removal in wastewater treatment plants: microorganism-mediated chemical precipitation and microorganism-mediated enhanced uptake of phosphorus.

3.8.1. Microorganism-Mediated Chemical Precipitation

According to this approach, precipitation of phosphate and its subsequent removal from wastewater is mediated by microbial activity in the aeration tank of the activated sludge process. At the head of a plug-flow aeration tank, microbial activity leads to low pH, which solubilizes phosphate compounds. At the end of the tank, a biologically mediated increase in pH leads to phosphate precipitation and incorporation into the sludge (Menard and Jenkins, 1970).

Biologically mediated phosphate precipitation also occurs inside denitrifying biofilms. Since denitrification produces alkalinity, denitrifier activity leads to an increase in pH and subsequent precipitation of calcium phosphate in biofilms (Arvin, 1985; Arvin and Kristensen, 1983). Precipitation of phosphorus can also be induced by the increase in phosphate concentration that results from release of phosphorus from the polyphosphate pool under anaerobic conditions.

3.8.2. Microorganism-Mediated Enhanced Uptake of Phosphorus

Enhanced removal of phosphorus is the result of microbial action in the activated sludge process (Toerien et al., 1990). Several mechanisms have been proposed to explain the enhanced uptake of phosphorus by microorganisms in wastewater.

Several microorganisms (e.g., *Acinetobacter, Pseudomonas, Aerobacter, Moraxella, E. coli, Mycobacterium, Beggiatoa*), called poly P bacteria, have the ability to accumulate phosphorus in excess of the normal cell requirement, which is around 1%–3% of the cell dry weight. Phosphorus is accumulated intracellularly in polyphosphate granules (e.g., volutin granules), which can be easily observed under bright-field or phase-contrast microscopy. Neisser's stain is used to observe these granules under a bright-field microscope (Meganck and Faup, 1988). Nuclear magnetic resonance (NMR) has been also recently used to detect polyphosphate granules in wastewater microorganisms (Florentz and Granger, 1983; Suresh et al., 1984). Polyphosphates serve as energy and phosphorus sources in microorganisms.

An enzyme, polyphosphate kinase, catalyzes polyphosphate biosynthesis in the presence of Mg^{2+} ions by transferring the terminal phosphoryl group from ATP to the polyphosphate chain. Polyphosphate degradation is driven by several enzymes (van Groenestijn et al., 1988b; Kornberg, 1957; Meganck and Faup, 1988; Ohtake et al., 1985; Szymona and Ostrowski, 1964) according to the following reactions:

$$\text{Polyphosphate}_n + \text{AMP} \xrightarrow[\quad]{\text{polyphosphate AMP phosphotransferase}} (\text{polyphosphate})_{n-1} + \text{ADP} \quad (3.28)$$

$$2\text{ADP} \underset{\quad}{\overset{\text{adenylate kinase}}{\rightleftarrows}} \text{ATP} + \text{AMP} \quad (3.29)$$

Polyphosphatases are other hydrolytic enzymes that are involved in polyphosphate degradation. In some bacteria, the hydrolysis of polyphosphates is driven by polyphosphate glucokinase and polyphosphate fructokinase, resulting in the phosphorylation of glucose and fructose, respectively.

In an anaerobic–aerobic activated sludge unit, inorganic phosphate is *released under anaerobic conditions and taken up by microorganisms under aerobic conditions* (Barnard, 1975). Figure 3.5 shows the release and uptake of phosphorus in a laboratory anaerobic–aerobic activated sludge unit (Hiraishi et al., 1989). Polyphosphate-accumulating aerobic bacteria such as *Acinetobacter* take up phosphorus under aerobic conditions, accumulate it as polyphosphate in granules, and release it under anaerobic conditions (Fuhs and Chen, 1975). For example, *Acinetobacter calcoaceticus* takes up phosphorus under aerobic conditions at a rate of 0.4–0.5 mmole/g dry cells per hour and releases it under anaerobic conditions at a rate of 0.015 mmole/g dry cells per hour (Ohtake et al., 1985). Magnesium acts as an important counterion of polyphosphates and is taken up and released simultaneously with phosphate (van Groenestijn et al., 1988a). Other studies have shown that, in addition to Mg^{2+}, K^+ and Ca^{2+} are also co-

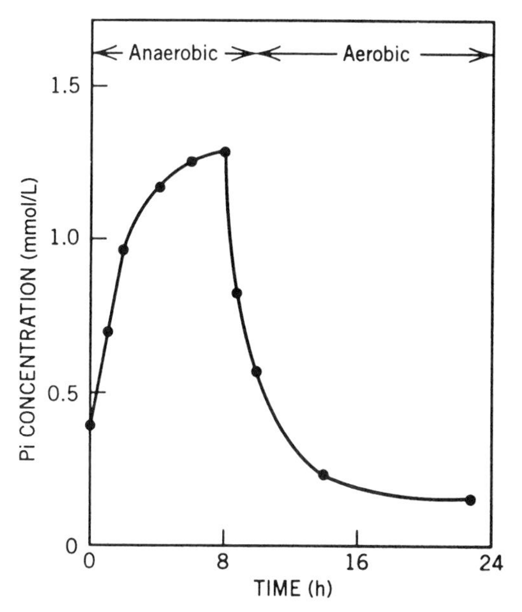

Fig. 3.5. P_i release and uptake by a laboratory anaerobic–aerobic activated sludge. From Hiraishi et al. (1989), with permission of the publisher.

transported with phosphate (Comeau et al., 1987).

There is a controversy on whether *Acinetobacter* is the predominant microorganism involved in enhanced phosphorus uptake. Some found no correlation between the number of *Acinetobacter* and the extent of phosphorus removal (Cloete and Steyn, 1988). The use of respiratory quinone profiles to characterize the bacterial population structure of the anaerobic–aerobic activated sludge system showed that *Acinetobacter* species were not important in this system (Hiraishi et al., 1989). However, it was shown that *Acinetobacter*, as detected by the biomarker diaminopropane, was the dominant organism only in wastewater treatment plants with low organic loading (Auling et al., 1991). Other poly P bacteria (*Pseudomonas, Aeromonas, Moraxella, Klebsiella, Enterobacter*) also contribute to removal of phosphate in activated sludge. Transmission electron microscopy indicated that the type of phosphate-accumulating bacteria depends on the wastewater composition and on the process used for phosphorus removal (Streichan et al., 1990).

Biochemical Model of Enhanced Phosphorus Uptake

Biochemical models have been proposed to explain the enhanced removal of phosphorus in activated sludge (Comeau et al., 1986; Wentzel et al., 1986). We will discuss the model of Comeau and collaborators, which is illustrated in Figure 3.6 (Comeau et al., 1986).

Phosphorus release under anaerobic conditions. Under anaerobic conditions (Fig. 3.6A),

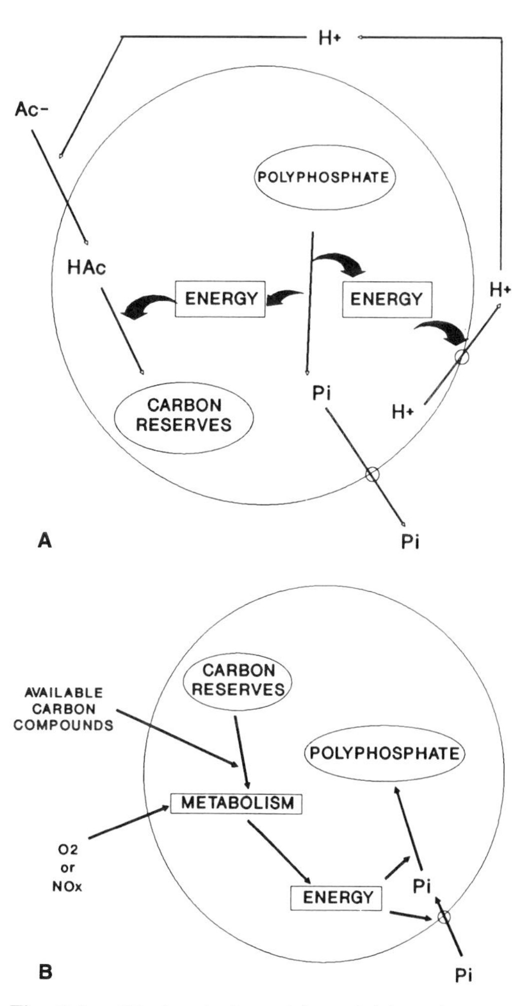

Fig. 3.6. Biochemical model explaining the enhanced P removal in activated sludge. **A.** Anaerobic conditions. **B.** Aerobic conditions. From Comeau et al. (1986), with permission of the publisher.

bacteria use energy derived from polyphosphate hydrolysis to take up carbon substrates, which are stored as poly-β-hydroxybutyrate (PHB) reserves, and to regulate the pH gradient across the cytoplasmic membrane. This phenomenon leads to the release of inorganic phosphorus. Simple organic substrates such as acetate are taken up by microorganisms and stored intracellularly as PHB, which subsequently are used as a carbon source during the aerobic phase. Acetate is converted to acetyl-CoA and the reaction is driven by energy supplied by the hydrolysis of accumulated intracellular polyphosphates. NADH, which provides the reducing power for PHB synthesis, is derived from the consumption of intracellular carbohydrate (Arun et al., 1988). Some investigators have suggested that the anaerobic zone serves as a fermentation milieu in which microorganisms such as *Aeromonas* (Brodisch and Joyner, 1983; Meganck et al., 1984) produce volatile fatty acids such as acetate, which are taken up by polyphosphate-accumulating organisms and stored as PHB (Fig. 3.7) (Meganck and Faup, 1988). To maximize carbon storage under anaerobic conditions, Comeau et al. (1987) suggested increasing the addition of simple carbon sources (e.g., septic wastewater, fermented primary sludge supernatant, acetate salts) and minimizing the addition of electron acceptors such as O_2 and NO_3. Readily biodegradable short-chain carbon substrates (butyric and isobutyric acids, valeric and isovaleric acids, ethanol, acetic acid, methanol, sodium acetate) can enhance removal of phosphorus (Abu-Ghararah and Randall, 1990; Jones et al., 1987).

Enhanced uptake under aerobic conditions. Under aerobic conditions (Fig. 3.6B), the energy derived from the metabolism of stored (e.g., PHB) or external carbon sources in the presence of O_2 or NO_3 is used for the accumulation of polyphosphates inside the cells. Under these conditions, inorganic phosphorus is taken up by the cells and stored as polyphosphates. Toxicants such as 2,4-dinitrophenol and H_2S have an adverse effect on phosphorus uptake under aerobic conditions (Comeau et al., 1987). Methods have been developed for distinguishing intracellular polyphosphate from extracellular precipitated orthophosphate. The use of these methods has indicated that polyphosphate is the predominant form of bioaccumulated phosphorus in activated sludge (de Haas, 1989).

Several proprietary processes based on release of phosphorus under anaerobic conditions and its uptake under aerobic conditions have been developed and marketed for phosphate removal in wastewater treatment plants. These processes will be discussed in Chapter 8.

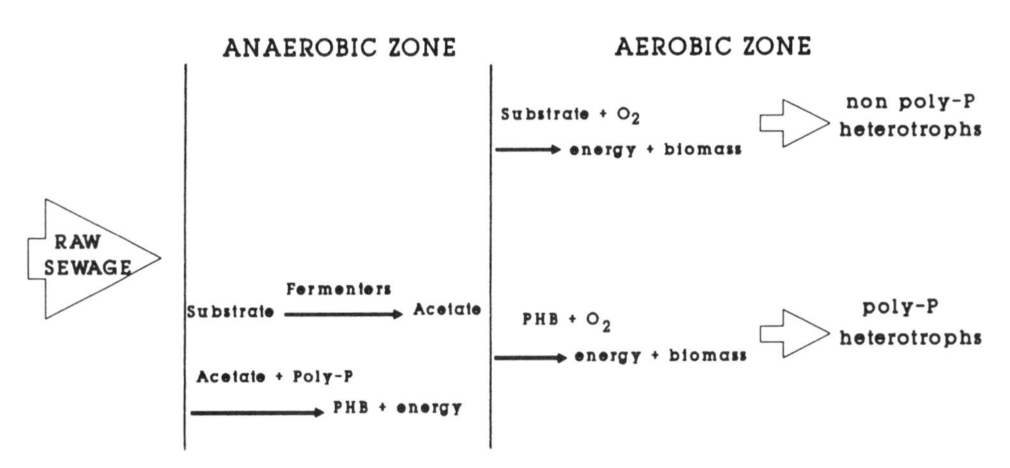

Fig. 3.7. Acetate formation by fermentation and subsequent storage as PHB by poly P bacteria. Adapted from Meganck and Faup (1988).

C. THE SULFUR CYCLE

3.9. INTRODUCTION

Sulfur is relatively abundant in the environment, and seawater is the largest reservoir of sulfate. Other sources include sulfur-containing minerals (e.g., pyrite, FeS_2 and chalcopyrite, $CuFeS_2$), fossil fuels, and organic matter. It is an essential element for microorganisms and enters into the composition of amino acids (cystine, cysteine, and methionine), cofactors (thiamine, biotin, and coenzyme A), ferredoxins, and enzymes (−SH groups.)

The sources of sulfur in wastewaters are organic sulfur, found in excreta, and sulfate, which is the most prevalent anion in natural waters.

3.10. MICROBIOLOGY OF THE SULFUR CYCLE

The steps involved in the sulfur cycle are summarized in Figure 3.8 (Sawyer and McCarty, 1967) and are as follows.

3.10.1. Mineralization of Organic Sulfur

Several types of microorganisms mineralize organic sulfur compounds, through aerobic and anaerobic pathways (Paul and Clark, 1989). Under aerobic conditions, sulfatase enzymes are involved in the degradation of sulfate esters to SO_4^{2-}:

$$R—O—SO_3^- + H_2O \xrightarrow{\text{sulfatase}} ROH + H^+ + SO_4^{2-} \qquad (3.30)$$

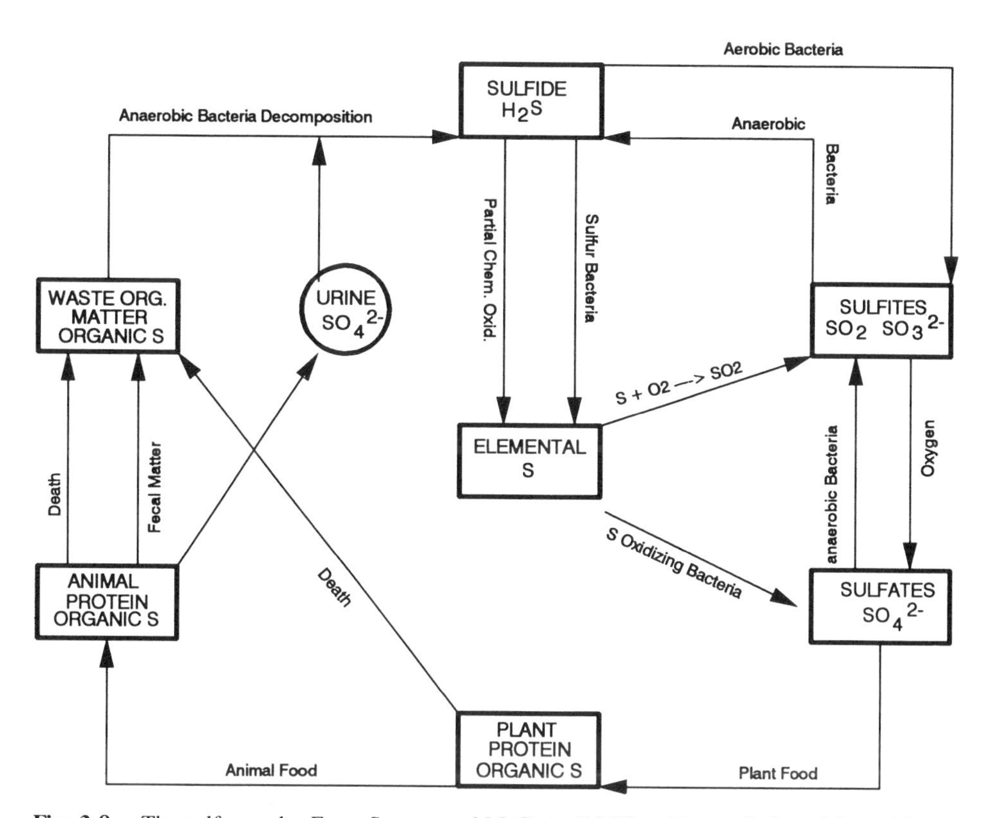

Fig. 3.8. The sulfur cycle. From Sawyer and McCarty (1967), with permission of the publisher.

Under anaerobic conditions, sulfur-containing amino acids are degraded to inorganic sulfur compounds or to mercaptans, which are odorous sulfur compounds.

3.10.2. Assimilation

Microorganisms assimilate oxidized as well as reduced forms of sulfur. Anaerobic microorganisms assimilate reduced forms such as H_2S, whereas aerobes utilize the more oxidized forms. The carbon-to-sulfur ratio is 100:1.

3.10.3. Oxidation Reactions

Several groups of microorganisms are involved in sulfur oxidation.

H_2S oxidation. H_2S is oxidized to elemental sulfur under aerobic and anaerobic conditions. Under aerobic conditions, *Thiobacillus thioparus* oxidizes S^{2-} to S^0:

$$S^{2-} + \frac{1}{2} O_2 + 2H^+ \rightarrow S^0 + H_2O \tag{3.31}$$

Under anaerobic conditions, oxidation is carried out by photoautotrophs (e.g., photosynthetic bacteria) and a chemoautotroph, *Thiobacillus denitrificans*. Photosynthetic bacteria use H_2S as an electron donor and oxidize H_2S to S^0, which is stored within the cells of chromatiaceae (purple sulfur bacteria) or outside the cells of chlorobiaceae (green sulfur bacteria). Filamentous sulfur bacteria (e.g., *Beggiatoa, Thiothrix*) also carry out H_2S oxidation to S, which is deposited in sulfur granules.

Oxidation of elemental sulfur. This reaction is carried out mainly by the aerobic, gram-negative, non-spore-forming thiobacilli (e.g., *Thiobacillus thiooxidans*) which grow at very low pH.

$$2S + 3O_2 + 2H_2O \longrightarrow 2H_2SO_4 \tag{3.32}$$

$$\underset{\text{thiosufate}}{Na_2S_2O_3} + 2O_2 + H_2O \rightarrow Na_2SO_4 + H_2SO_4 \tag{3.33}$$

Another sulfur oxidizer is *Sulfolobus*, which is a thermophilic acidophilic bacterium found in hot acidic spring waters (pH = 2–3; temperature = 55–85°C).

Sulfur oxidation by heterotrophs. Heterotrophs (e.g., *Arthrobacter, Micrococcus, Bacillus, Pseudomonas*) can also be responsible for sulfur oxidation in neutral and alkaline soils (Paul and Clark, 1989).

3.10.4. Sulfate Reduction

Sulfides are produced by assimilatory and dissimilatory sulfate reduction.

Assimilatory sulfate reduction. H_2S may result from the anaerobic decomposition by proteolytic bacteria (e.g., *Clostridia, Vellionella*) of organic matter containing sulfur amino acids such as methionine, cysteine, and cystine (Bowker et al., 1989).

Dissimilatory sulfate reduction. Sulfate reduction is the most important source of H_2S in wastewater. It is the reduction of sulfate by strict anaerobes, the sulfate-reducing bacteria.

$$SO_4^{2-} + \text{organic compounds} \rightarrow S^{2-} + H_2O + CO_2 \tag{3.34}$$

$$S^{2-} + 2H^+ \rightarrow H_2S \tag{3.35}$$

H_2S is toxic to animals and plants. It causes problems in paddy rice fields and is toxic to wastewater treatment plant operators (see Chapter 14).

In 1895, Beijerinck first isolated a bacterium capable of reducing sulfate to hydrogen sulfide. Sulfate-reducing bacteria belonging to the following genera have since been isolated from environmental samples (anaerobic sludge digesters, aquatic sediments, gastrointestinal tract):*Desulfovibrio, Desulfotomaculum, Desulfobulbus, Desulfomonas, Desulfobacter, Desulfococcus, Desulfonema, Desulfosarcina, Desulfobacterium,* and *Thermodesulfobacterium. Desulfotomaculum* is the only spore-forming genus among sulfate-reducing bacteria (Hamilton, 1985; Widdel, 1988).

In the absence of oxygen and nitrate, these strict anaerobic bacteria use sulfate as the terminal electron acceptor. They use low-molecular-weight carbon sources (e.g., electron donors) produced by the fermentation of carbohydrates, proteins, and other compounds. These carbon sources include lactate, pyruvate, acetate, propionate, formate, fatty acids, alcohols (ethanol, propanol), dicarboxylic acids (succinic, fumaric, and malic acids), and aromatic compounds (benzoate, phenol). H_2 is also used as electron donor. These bacteria have very low cell yields. (Hamilton, 1985; Rinzema and Lettinga, 1988; Widdel, 1988).

3.11. ENVIRONMENTAL AND TECHNOLOGICAL PROBLEMS ASSOCIATED WITH SULFUR OXIDATION AND REDUCTION

3.11.1. Problems Associated With Sulfate-Reducing Bacteria

3.11.1.1. Corrosion Problems

Pitting corrosion develops under strictly anaerobic conditions and involves sulfate-reducing bacteria such as *Desulfovibrio desulfuricans*. The pits are filled with iron sulfide. Questions remain regarding the exact role of sulfate-reducing bacteria in metal corrosion. Corrosion may be inherently linked to the metabolism of sulfate reducers or may be due to hydrogen sulfide or iron sulfides (Odom, 1990).

The development of biofilms on pipe surfaces leads to anaerobic conditions that are ideal for the growth of obligately anaerobic sulfate-reducing bacteria. These conditions are achieved when the biofilm reaches a certain thickness, which varies from 10 μm to 100 μm. A model for anaerobic microbial corrosion is displayed in Figure 3.9 (Hamilton, 1985).

The most accepted theory of anaerobic corrosion by sulfate-reducing bacteria is the cathodic depolarization theory of von Wolzogen-Kuhr and van der Vlugt (1934) (Fig. 3.10) (Widdel, 1988). A net oxidation of the metal is catalyzed by the activity of sulfate-reducing bacteria (SRB). At the anode, the metal is polarized by losing Fe^{2+}. The electrons released at the anode reduce protons (H^+) from water to molecular hydrogen (H_2) at the cathode. The molecular hydrogen is removed from the metal surface by sulfate-reducing bacteria (cathodic depolarization). The latter produce hydrogen sulfide, which combines with Fe^{2+} to form FeS, which accumulates on the metal surface.

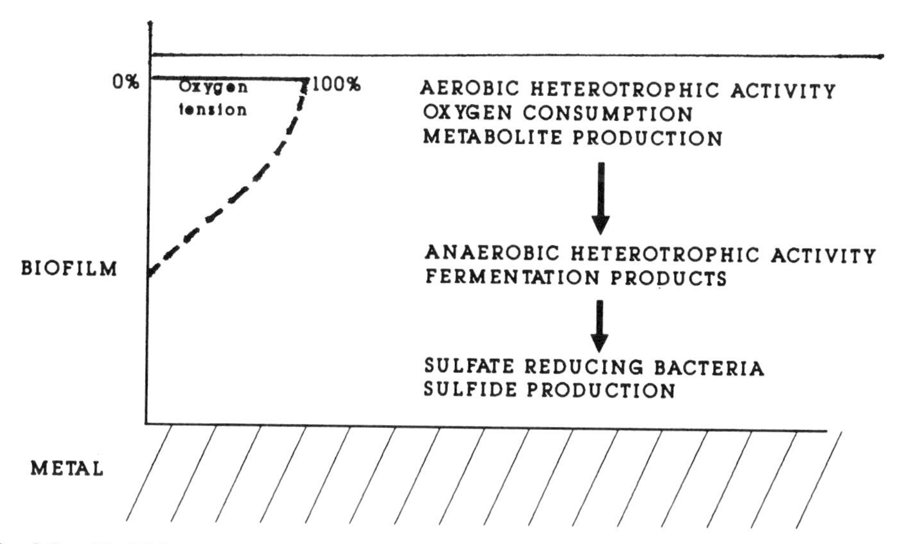

Fig. 3.9. Model for anaerobic microbial corrosion. From Hamilton (1985), with permission of the publisher.

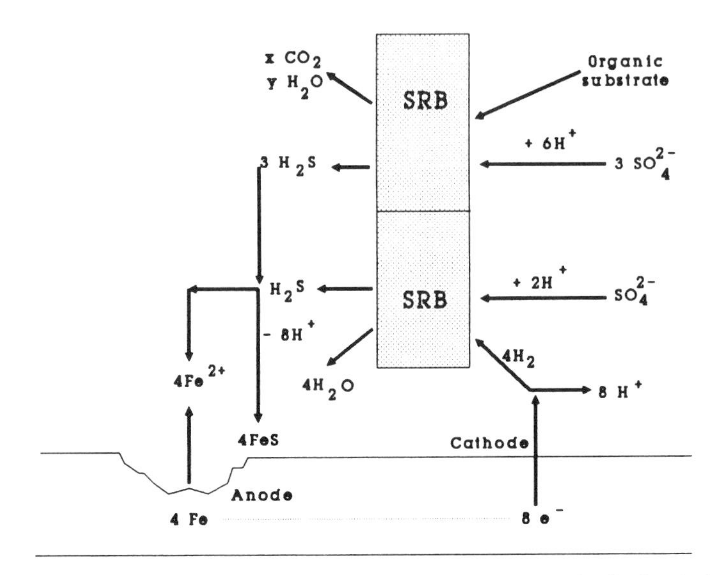

Fig. 3.10. Reactions during anaerobic corrosion as suggested by the cathodic depolarization theory of von Wolzogen-Kuhr and van der Vlugt. From Widdel (1988), with permission of the publisher.

Anaerobic corrosion of iron is summarized as follows (Ford and Mitchell, 1990; Hamilton, 1985):

Reaction at the anode:

$$4\ \mathrm{Fe(0)} \rightarrow 4\ \mathrm{Fe(II)} + 8e^- \qquad (3.36)$$

H_2O dissociation:

$$8H_2O \rightarrow 8H^+ + 8OH^- \qquad (3.37)$$

$$8H^+ + 8e^- \rightarrow 4H_2 \qquad (3.38)$$

Cathodic depolarization:

$$4H_2 + SO_4 \rightarrow S^{2-} + 4\ H_2O \qquad (3.39)$$

The presence of organic substrates (i.e., electron donors) helps in the production of additional sulfide. Hydrogenase enzyme of sulfate-reducing bacteria, which removes cathodic hydrogen, is involved in the initiation of the biocorrosion process, and its activity is a good indicator of biocorrosion (Bryant and Laishley, 1990; Bryant et al., 1991). A commercial kit is available for measuring hydrogenase activity (Costerton et al., 1989).

Sulfate-reducing bacteria can be controlled by preventing or reducing the input of organic matter or sulfate, changing environmental conditions by aeration or pH change, and by adding appropriate biocides (Widdel, 1988).

3.11.1.2. Impact of Sulfate Reduction on Anaerobic Digesters

In anaerobic digesters, H_2S is transferred to the biogas and causes corrosion problems, unpleasant odors, and inhibition of methanogenic bacteria (see Chapter 13) (Rinzema and Lettinga, 1988). Sulfate reducers and methanogens have similar characteristics. They both live under strict anaerobic conditions with similar pH and temperature ranges. Like methanogens, some sulfate reducers are able to oxidize H_2 and acetate and thus may compete with methanogens for these substrates (Fig. 3.11) (Rinzema and Lettinga, 1988). Competition for these substrates has been studied mostly in estuarine sediments, in which sulfate supply is more abundant than in freshwater sediments (Abram and Nedwell, 1978; Oremland and Polcin, 1982; Sorensen et al., 1981). Kinetic studies have shown that sulfate reducers generally have higher maximum growth rates and higher affinity for substrates (i.e., lower half-saturation constants, K_s) than

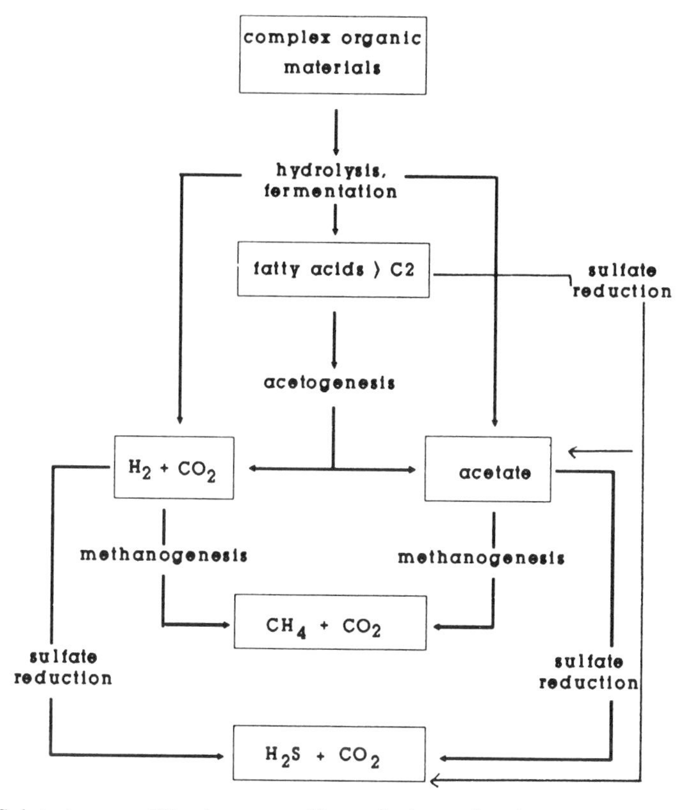

Fig. 3.11. Substrate competition between sulfate-reducing and methanogenic or acetogenic bacteria. From Rinzema and Lettinga (1988), with permission of the publisher.

methanogens. The half-saturation constant for hydrogen is 6.6 μM for methanogens, as compared with 1.3 μM for sulfate reducers. Similarly, the K_s values for acetate are 3 mM and 0.2 mM for methanogens and sulfate reducers, respectively (Speece, 1983). Thus, sulfate reducers may predominate over methanogens, providing that the sulfate supply is not limiting. However, despite their kinetic advantages, sulfate-reducing bacteria rarely predominate in anaerobic wastewater treatment. Future research should provide a satisfactory explanation for this phenomenon (Rinzema and Lettinga, 1988).

3.11.1.3. Aquatic Environments and Flooded Soils

The excessive production of hydrogen sulfide in anoxic waters and sediments may result in the poisoning of plants and animals, including fish. This detrimental effect is also observed in waterlogged soils (paddy fields) (Widdel, 1988).

3.11.1.4. Sulfate Reduction and the Oil and Paper-Making Industries

The environments of both industries are conducive to the formation of electron donors for sulfate-reducing bacteria. In the paper-making industry, cellulose fermentation produces electron donors for sulfate reduction. And aerobic degradation of hydrocarbons may release organic acids that can be used by sulfate reducers. The resulting hydrogen sulfide combines with iron and other heavy metals to form sulfides, which tend to plug injection wells. H_2S gas may also contaminate fuel gas and oil (Widdel, 1988).

3.11.2. Problems Associated with Sulfur-Oxidizing Bacteria

3.11.2.1. Acidic Mine Drainage

Thousands of miles of U.S. waterways, particularly in Appalachia, are affected by acidity. Acidic mine drainage results from water flowing through mines, or runoff from mounds of mine tailings. Iron pyrite (FeS_2) is commonly found associated with coal deposits. Under acidic conditions, iron- and sulfur-oxidizing bacteria drive the oxidation of pyrite to ferric sulfate and sulfuric acid, which cause the degradation of water quality and kill fish (Dugan, 1987a; 1987b).

$$FeS_2 + 3O_2 + 2H_2O \rightarrow 2H_2SO_4 + Fe^{3+} \quad (3.40)$$

Fe^{2+} is oxidized to Fe^{3+}, which forms ferric hydroxide and more acidity according to Equation 3.41:

$$Fe^{3+} + 3\ H_2O \rightarrow Fe(OH)_3 + 3H^+ \quad (3.41)$$

One of the means of biological control of acidic mine draining is inhibition of iron- and sulfur-oxidizing bacteria with anionic surfactants, benzoic acid, organic acids, alkyl benzene sulfonates, and sodium dodecyl sulfate (Dugan, 1987a). Some of these chemicals or their combinations (e.g., combination of sodium dodecyl sulfate and benzoic acid) were able to reduce acidic drainage from coal refuse under simulated field conditions (Dugan, 1987b). However, the application of these techniques under field conditions has not been attempted.

3.11.2.2. Removal of Pyritic Sulfur from Coal

Iron- and sulfur-oxidizing bacteria (*Thiobacillus ferrooxidans, Thiobacillus thiooxidans*) have been considered for removing sulfur from coal. The rate of biological sulfur removal increases as the particle size of the coal decreases (Tillet and Myerson, 1987).

3.11.2.3. "Microbial Leaching" (or "Microbial Mining")

Sulfur-oxidizing bacteria participate in recovery of metals (Cu, Ni, Zn, Pb) from low-grade ores. Mine tailings are piled up and the effluents from the piles are recirculated and become enriched in metals (they are called "pregnant" solutions). The effluents are rich in *Thiobacillus ferrooxidans*.

3.11.2.4. Role in Corrosion

Chemical corrosion is due mainly to oxidation and to the corrosive action of chlorine, acids, alkalies, and metal salts. Bacteria are also responsible for corrosion by producing inorganic (e.g., H_2SO_4) or organic acids (e.g., acetic acid, butyric acid) (Ford and Mitchell, 1990). Sulfuric acid, formed by oxidation of H_2S by sulfide-oxidizing bacteria (e.g., *Thiobacillus concretivorus*), is a major corrosive agent in distribution pipes, particularly concrete pipes. Another bacterium, *Thiobacillus ferrooxidans* is responsible for the corrosion of iron pipes and pumps. Other iron-oxidizing bacteria, *Gallionella* and *Sphaerotilus natans,* are associated with tubercle formation and corrosion of distribution pipes. H_2S can be removed by activated carbon or addition of iron salts or by using oxidants such as chlorine and hydrogen peroxide (Hamilton, 1985; Odom, 1990).

3.12. FURTHER READING

Barnes, D., and P.J. Bliss. 1983. *Biological Control of Nitrogen in Wastewater Treatment.* E. & F.N. Spon, London.

Christensen, M.H., and P. Harremoes. 1978. Nitrification and denitrification in wastewater treatment, pp. 391–414, in: *Water Pollution Microbiology,* Vol. 2, R. Mitchell, Ed. Wiley, New York.

Ehrlich, H.L. 1981. *Geomicrobiology.* Marcel Dekker, New York.

Ford, T., and R. Mitchell. 1990. The ecology of microbial corrosion. Adv. Microb. Ecol. 11: 231–262.

Meganck, M.T.J., and G.M. Faup. 1988. Enhanced biological phosphorus removal from waste waters, pp. 111–203, in: *Biotreatment Systems,* Vol. 3, D.L. Wise, Ed. CRC Press, Boca Raton, FL.

Widdel, F. 1988. Microbiology and ecology of sulfate- and sulfur-reducing bacteria, pp. 469–585, in: *Biology of Anaerobic Microorganisms,* A.J.B. Zehnder, Ed. Wiley, New York.

PUBLIC HEALTH MICROBIOLOGY

4

PATHOGENS AND PARASITES IN DOMESTIC WASTEWATER

4.1. ELEMENTS OF EPIDEMIOLOGY

4.1.1. Some Definitions

Epidemiology is the study of the spread of infectious diseases in populations. Infectious diseases are those that can be spread from one host to another. Epidemiologists play an important role in the control of these diseases.

Incidence of a disease is the number of individuals with the disease in a population, whereas prevalence is the percentage of individuals with the disease at a given time. A disease is *epidemic* when the incidence is high and *endemic* when the incidence is low. *Pandemic* refers to the spread of the disease across continents.

Infection is the invasion of a host by an infectious microorganism. It involves the entry (e.g., via the gastrointestinal and respiratory tracts, skin) of the pathogen into the host and its multiplication and establishment inside the host. Inapparent infection (or covert infection) is a subclinical infection with no apparent symptoms (i.e., the host reaction is not clinically detectable). Although it does not cause disease symptoms, it confers the same degree of immunity as an overt infection. For example, most enteric viruses cause inapparent infections. A person with inapparent infection is called a healthy carrier. However, carriers constitute a potential source of infection for others in the community (Finlay and Falkow, 1989; Jawetz et al., 1984).

Pathogenicity is the ability of an infectious agent to cause disease and injure the host. Pathogenic microorganisms may infect susceptible hosts, leading sometimes to overt disease, which results in the development of clinical symptoms that are easily detectable. The development of the disease depends on various factors, including infectious dose, pathogenicity, and host and environmental factors. Some organisms, however, are opportunistic pathogens and cause disease only in compromised individuals.

4.1.2. Chain of Infection

The potential for a biological agent to cause infection in a susceptible host depends on various factors, which are the following (Dean and Lund, 1981).

4.1.2.1. Types of Infectious Agent

Several infectious organisms may cause diseases in humans. These agents include bacteria, fungi, protozoa, metazoa (helminths), rickettsiae, and viruses (Fig. 4.1).

Evaluation of infectious agents is based on their *virulence* or their potential for causing diseases in humans. Virulence is related to the dose of infectious agent necessary for infecting the host and causing disease. The potential for causing illness also depends on the stability of the infectious agent in the environment. The *minimal infective dose* (MID) varies widely with the type of pathogen or parasite. For example, for *Salmonella typhi* or enteropathogenic *E. coli,* thousands to millions of organisms are necessary to establish infection, whereas the minimal infective dose for *Shigella* can be as low as 10

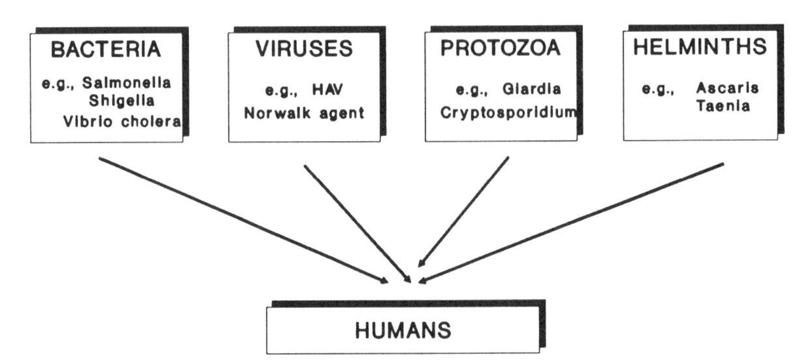

Fig. 4.1. Categories of organisms of public health significance.

cells. A few protozoan cysts or helminthic eggs may be sufficient to establish infection. For some viruses, only one or a few particles are sufficient for infecting individuals. For example, 17 infectious particles of echovirus 12 are sufficient for establishing infection (Table 4.1) (Bitton, 1980a; Bryan, 1977; Gunnerson et al., 1984; Schiff et al., 1984a, 1984b).

4.1.2.2. Reservoir of the Infectious Agent

A *reservoir* is a living or nonliving source of the infectious agent and allows the pathogen to survive and multiply. The human body is the reservoir for numerous pathogens, and person-to-person contact is necessary for maintaining the disease cycle. Domestic and wild animals also may serve as reservoirs for several diseases (e.g., rabies, brucellosis, tuberculosis, anthrax, leptospirosis, toxoplasmosis), called zoonoses, that can be transmitted from animals to humans. Nonliving reservoirs such as water, wastewater, food, or soils can also harbor infectious agents.

TABLE 4.1. Minimal Infective Doses for Some Pathogens and Parasites

Organism	Minimal infective dose
Salmonella spp.	10^4 to 10^7
Shigella spp.	10^1 to 10^2
Escherichia coli	10^6 to 10^8
Vibrio cholerae	10^3
Giardia lamblia	10^1 to 10^2 cysts
Cryptosporidium	10^1 cysts
Entamoeba coli	10^1 cysts
Ascaris	1–10 eggs
Hepatitis A virus	1–10 PFU

4.1.2.3. Mode of Transmission

Transmission involves the transport of an infectious agent from the reservoir to the host. It is the most important link in the chain of infection. Pathogens can be transmitted from the reservoir to a susceptible host by various routes (Sobsey and Olson, 1983).

Person-to-person transmission. The most common route of transmission of infectious agents is from person to person. The best examples of direct contact transmission are the sexually transmitted diseases, such as syphilis, gonorrhea, herpes, and AIDS. Coughing and sneezing discharge very small droplets containing pathogens within a few feet of the host (droplet infection). Transmission by these infectious droplets is sometimes considered an example of direct-contact transmission.

Waterborne transmission. The waterborne transmission of cholera was established in 1854 by John Snow, an English physician who noted a relationship between cholera epidemics and consumption of water from the Broad Street well in London. The waterborne route is not as important as the person-to-person contact route for the transmission of fecally transmitted diseases.

Waterborne disease outbreaks are reported to the U.S. Environmental Protection Agency (U.S. EPA) and to the Center for Disease Control (CDC) by local epidemiologists and health authorities; the system was started in the 1920s (Craun, 1986, 1988). During the period 1971–1985, 502 waterborne outbreaks of disease and 111,228 cases were reported. Figure 4.2 (Craun, 1988) shows that about three fourths of the outbreaks were due to untreated or inadequately treated groundwater and surface waters. Gastrointestinal illnesses of unidentified etiology and giardiasis are the most common waterborne diseases for groundwater and surface water systems (Table 4.2) (Craun, 1988). The outbreak rate (expressed as the number of outbreaks per 1,000 water systems) and the illness rate (expressed as numbers of cases/million person-years) decrease as the raw water is filtered and disinfected (Table 4.3) (Craun, 1988).

Foodborne transmission. Foods serve as a vehicle for the transmission of numerous infectious diseases caused by bacteria, viruses, protozoa, and helminth parasites. Food contamination results from unsanitary practices during production or preparation. Several pathogens and parasites have been detected in foodstuffs such as shellfish, vegetables, milk, eggs, and groundbeef. Their presence is of public health significance, particularly for foods that are eaten raw (e.g., shellfish). Vegetables contaminated with wastewater effluents are also responsible for disease outbreaks (e.g., typhoid fever, salmonellosis, amebiasis, ascariasis, viral hepatitis) (Bryan, 1977). Risks for transmission of these diseases are higher in countries that use poorly treated wastewater effluents for crop irrigation.

Shellfish (e.g., oysters, clams, mussels) are significant vectors of human diseases of bacterial, viral, protozoan, and helminthic origin. They are important in disease transmission for the following reasons (Bitton, 1980a): (1) They live in estuarine environments, which are often contaminated by domestic wastewater effluents; (2) being filter-feeders, they concentrate pathogens and parasites by pumping large quantities of estuarine water (4–20 L/hr); (3) they are often eaten raw or insufficiently cooked—it has been estimated that only one third of shellfish con-

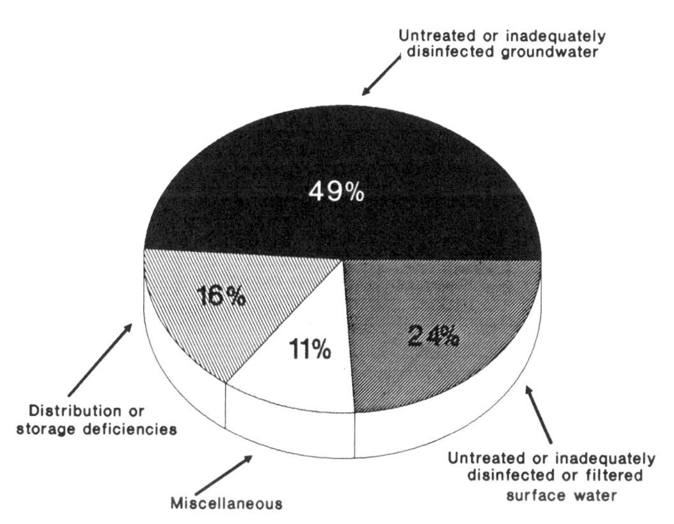

Fig. 4.2. Causes of 502 waterborne disease outbreaks: 1971–1985. From Craun (1988), with permission of the publisher.

TABLE 4.2. Etiology of Waterborne Outbreaks for Groundwater and Surface Water Systems: 1971–1985[a]

Illness	Number of outbreaks	Cases of illness
Gastroenteritis, undefined	251	61,478
Giardiasis	92	24,365
Chemical poisoning	50	3,774
Shigellosis	33	5,783
Hepatitis A	23	737
Gastroenteritis, viral	20	6,254
Campylobacterosis	11	4,983
Salmonellosis	10	2,300
Typhoid	5	282
Yersiniosis	2	103
Gastroenteritis, toxigenic *E. coli*	1	1,000
Cryptosporidiosis	1	117
Cholera	1	17
Dermatitis	1	31
Amebiasis	1	4
Total	502	111,228

[a]From Craun (1988), with permission of the publisher.

sumed every year in France are sufficiently cooked (Poggi, 1990). Other health hazards associated with shellfish consumption result from the ability of these mollusks to concentrate dinoflagellate toxins, heavy metals, hydrocarbons, pesticides, and radionuclides.

Airborne transmission. Some diseases (e.g., Q fever, some fungal diseases) can be spread by airborne transmission. This route is important in the transmission of biological aerosols generated by wastewater treatment plants or by spray irrigation with wastewater effluents. This topic is discussed thoroughly in Chapter 14.

Vector-borne transmission. The most common vectors for disease transmission are arthropods (fleas, insects) or vertebrates (e.g., rodents, dogs, and cats). The pathogen may or may not multiply inside the arthropod vector. Some vector-borne diseases are malaria (from *Plasmodium*), yellow fever and encephalitis (both from arboviruses), and rabies (from virus transmitted by the bite of rabid dogs or cats).

Fomites. Some pathogens can be transmitted by nonliving objects or fomites (e.g., clothes, utensils, toys).

4.1.2.4. Portal of Entry

Pathogenic microorganisms can gain access to the host mainly through the gastrointestinal tract (e.g., enteric viruses and bacteria), the respiratory tract (e.g., *Klebsiella pneumonae, Le-*

TABLE 4.3. Effect of Water Treatment on Outbreak and Illness Rates: 1971–85[a]

Type of community water system	Waterborne outbreaks per 1,000 water systems	Waterborne illness per million person-years
Untreated surface water	32.5	370.9
Disinfected-only surface water	40.5	66.3
Filtered and disinfected surface water	5.0	4.7

[a]From Craum (1988), with permission of the publisher.

gionella, myxoviruses), or the skin (e.g., *Aeromonas, Clostridium tetani, Clostridium perfringens*). Although the skin is a formidable barrier against pathogens, wounds or abrasions may facilitate their penetration into the host.

4.1.2.5. Host Susceptibility

Both the immune system and nonspecific factors play a role in the resistance of the host to infectious agents. Immunity to an infectious agent may be natural or acquired. *Natural immunity* is genetically specified and varies with species, race, age (the young and the elderly are more susceptible to infection), hormonal status, and physical and mental health of the host. People in poor health and the elderly are more susceptible to infectious agents than healthy adults. *Acquired immunity* develops as a result of exposure of the host to the infectious agent. Acquired immunity can be *passive* (e.g., a fetus acquiring the mother's antibodies) or *active* (e.g., active production of antibodies through contact with the infectious agent) (Jawetz et al., 1984).

The nonspecific factors include physiological barriers at the portal of entry (e.g., unfavorable pH, bile salts, production of digestive enzymes and other chemicals with antimicrobial properties, competition with the natural microflora in the colon) and destruction of the invaders by phagocytosis.

4.2. PATHOGENS AND PARASITES FOUND IN DOMESTIC WASTEWATER

Several pathogenic microorganisms and parasites are commonly found in domestic wastewater as well as in effluents from wastewater treatment plants. We will now review these infectious agents as well as some of the diseases they cause.

4.2.1. Bacterial Pathogens

Fecal matter contains up to 10^{12} bacteria per gram. The bacterial content of feces represents approximately 9% by wet weight (Dean and Lund, 1981). Wastewater bacteria have been characterized and belong to the following groups (Dott and Kampfer, 1988):

1. Gram-negative facultatively anaerobic bacteria (e.g., *Aeromonas, Plesiomonas, Vibrio, Enterobacter, Escherichia, Klebsiella,* and *Shigella*).
2. Gram-negative aerobic bacteria (e.g., *Pseudomonas, Alcaligenes, Flavobacterium,* and *Acinetobacter.*
3. Gram-positive spore-forming bacteria (e.g., *Bacillus* spp).
4. Non-spore-forming gram-positive bacteria (e.g., *Arthrobacter, Corynebacterium, Rhodococcus*).

A compilation of the most important bacteria that may be pathogenic to humans and that can be transmitted directly or indirectly by the waterborne route is shown in Table 4.4 (Dart and Stretton, 1980; Dean and Lund, 1981; Feachem et al., 1983; Harris, 1986; Sobsey and Olson, 1983). These pathogens cause enteric infections such as typhoid fever, cholera, and shigellosis. We will now review some of the most important bacterial pathogens found in wastewater.

4.2.1.1. *Salmonella*

Salmonellae are enterobacteriaceae that are widely distributed in the environment and include more than 2,000 serotypes. They are the most predominant pathogenic bacteria in wastewater and they cause typhoid and paratyphoid fever and gastroenteritis. Their numbers in wastewater range from a few to 8,000 organisms per 100 ml (Feachem et al., 1983). It has been estimated that 0.1% of the population excretes *Salmonella* at any given time. In the United States salmonellosis is primarily due to food contamination but its transmission by drinking water is still of great concern (Sobsey and Olson, 1983). *Salmonella typhi* is the etiological agent of typhoid fever, a deadly disease that has been brought under control as a result of the development of adequate water treatment processes (e.g., chlorination, filtration). This pathogen produces an endotoxin that causes fever, nausea and diarrhea

TABLE 4.4. Major Waterborne Bacterial Diseases[a]

Bacterial agent	Major disease	Major reservoir	Principal site affected
Salmonella typhi	Typhoid fever	Human feces	Gastrointestinal tract
Salmonella paratyphi	Paratyphoid fever	Human feces	Gastrointestinal tract
Shigella	Bacillary dysentery	Human feces	Lower intestine
Vibrio cholerae	Cholera	Human feces	Gastrointestinal tract
Enteropathogenic *E. coli*	Gastroenteritis	Human feces	Gastrointestinal tract
Yersinia enterocolitica	Gastroenteritis	Human/animal feces	Gastrointestinal tract
Campylobacter jejuni	Gastroenteritis	Human/animal feces	Gastrointestinal tract
Legionella pneumophila	Acute respiratory illness (legionnaire's disease)	Thermally enriched waters	Lungs
Mycobacterium tuberculosis	Tuberculosis	Human respiratory exudates	Lungs
Leptospira	Leptospirosis (Weil's disease)	Animal feces and urine	Generalized
Opportunistic bacteria	Variable	Natural waters	Mainly gastrointestinal tract

[a]Adapted from Sobsey and Olson (1983).

and may be fatal if not properly treated by antibiotics (Sterritt and Lester, 1988). Species implicated in food contamination are *S. paratyphi* and *S. typhimurium*. These species can grow readily in contaminated foods and cause food poisoning. Species such as *S. typhimurium* and *S. enteriditis* cause gastroenteritis, which is characterized by diarrhea and abdominal cramps.

4.2.1.2. *Shigella*

Shigella is the causal agent of bacillary dysentery, a diarrheal disease that produces bloody stools as a result of inflammation and ulceration of the intestinal mucosa. There are four pathogenic species of *Shigella: S. flexneri, S. dysenteriae, S. boydii,* and *S. sonnei*. These pathogens are transmitted by direct contact with an infected person who may excrete up to 10^9 shigellae per gram of feces. The infectious dose for *Shigella* is relatively small and can be as low as 10 organisms. Although person-to-person contact is the main mode of transmission of this pathogen, food-borne and waterborne transmission have also been documented. For example, groundwater was found to be responsible for a shigellosis outbreak in Florida that involved 1,200 people. However, *Shigella* persist less in the environment than do fecal coliforms. This pathogen is difficult to cultivate; thus, no quantitative data are available on its occurrence and removal in water and wastewater treatment plants.

4.2.1.3. *Vibrio cholerae*

Vibrio cholerae is a gram-negative curved rod bacterium that is almost exclusively transmitted by water. It releases an enterotoxin that causes mild to profuse diarrhea, vomiting, and a very rapid loss of fluids, which may result in death in a relatively short period of time (Sterritt and Lester, 1988). Although rare in the United States and Europe, this disease appears to be endemic in various areas throughout Asia. This pathogen is found in wastewater, and levels of 10 to 10^4 organisms per 100 ml of wastewater during a cholera epidemic have been reported (Kott and Betzer, 1972). Explosive epidemics of cholera and typhoid fever have been documented in Peru and Chile and have been associated with the consumption of sewage-contaminated vegetables (Shuval, 1992). This bacterium is also naturally present in the environment and attaches to solids, including zooplankton (e.g., copepods) and phytoplankton (e.g., *Volvox*). These plankton-associated bacteria may occur in the nonculturable state and can be observed under the microscope, by means of a fluorescent-monoclonal-antibody technique (Brayton et al., 1987; Huk et al., 1990).

4.2.1.4. *E. coli*

Several strains of *E. coli,* many of which are harmless, are found in the gastrointestinal tract of humans and warm-blooded animals. There are several categories of *E. coli* strains, however, that bear virulence factors and cause diarrhea. There are enterotoxigenic (ETEC), enteropathogenic (EPEC), enterohemorrhagic (EHEC), and enteroinvasive strains of *E. coli* (Levine, 1987). Enterotoxigenic *E. coli* causes gastroenteritis with profuse watery diarrhea accompanied by nausea, abdominal cramps, and vomiting (Harris, 1986). Approximately 2%–8% of the *E. coli* present in water have been found to be enteropathogenic *E. coli,* which causes traveler's diarrhea. Food and water are important in the transmission of this pathogen. However, the infective dose for this pathogen is relatively high, being in the range of 10^6 to 10^9 organisms.

Some of these diarrheagenic strains of *E. coli* have been detected in treated water with genetic probes, and they can represent a health risk to consumers (Martins et al., 1992). During a 1989–1990 survey of waterborne disease outbreaks in the United States, the etiologic agent in one of 26 outbreaks was enterohemorrhagic *E. coli* 0157:H7, an agent that causes bloody diarrhea, particularly among the very young and very old members of the community (Herwaldt et al., 1992). This pathogen was isolated from a water reservoir in Philadelphia (McGowan et al., 1989). Two outbreaks of *E. coli* O157:H7 have been shown to be associated with waterborne transmission. One occurred in Scotland in the summer of 1990 (Dev et al., 1991): the second outbreak occurred in Cabool, Missouri, in the winter of 1990, following disturbances in the water distribution network; it resulted in 243 documented cases of diarrhea and four deaths among elderly citizens (Geldreich et al., 1992; Swerdlow et al., 1992).

4.2.1.5. *Yersinia*

Y. enterocolitica is responsible for acute gastroenteritis with invasion of the terminal ileum. Swine are a major animal reservoir but many domestic and wild animals can also serve as reservoirs for this pathogen. Food-borne (e.g., milk, tofu) outbreaks of yersiniosis have also been documented in the United States. The role of water is uncertain but there are instances where this pathogen was suspected to be the cause of waterborne transmission of gastroenteritis (Schiemann, 1990). This psychrotrophic organism thrives at temperatures as low as 4°C, is mostly isolated during the cold months but is poorly correlated with traditional bacterial indicators (Sobsey and Olson, 1983; Wetzler et al., 1979). This organism has been isolated from wastewater effluents, river water and from drinking water (Bartley et al., 1982; Meadows and Snudden, 1982; Stathopoulos and Vayonas-Arvanitidou; 1990; Wetzler et al., 1979).

4.2.1.6. *Campylobacter* (e.g., *C. fetus* and *C. jejuni*)

This pathogen is known to infect humans as well as wild and domestic animals. It causes acute gastroenteritis (fever, nausea, abdominal pains, diarrhea, vomiting) and is transmitted to humans by contaminated food, mainly poultry, and contaminated water. This pathogen is the cause of several outbreaks of gastroenteritis in the United States and elsewhere (Blaser and Peller, 1981; Blaser et al., 1983). Municipal water supplies as well as water from mountain streams were implicated as sources of infection (Mentzing, 1981; Palmer et al., 1983; Taylor et al., 1983). The first outbreak in the United States was reported in 1978 in Vermont, where 2,000 out of a population of 10,000 were affected. The seasonal occurrence of *Campylobacter* in surface waters has been documented (Carter et al., 1987). *Campylobacter* has been detected in surface waters, potable water, and wastewater, but no organisms have been recovered from digested sludge (Andrin and Schwartzbrod, 1992; Stathopoulos and Vayonas-Arvanitidou, 1990; Stelzer, 1990). Recovery from surface waters is highest in the fall

(55% of samples positive) and winter (39% of samples positive). Numbers of *Campylobacter* organisms did not display any correlation with heterotrophic plate counts, total and fecal coliforms, or fecal streptococci.

4.2.1.7. *Leptospira*

Leptospira is a small spirochete that can gain access to the host through abrasions of the skin or through mucous membranes. It causes *leptospirosis,* which is characterized by the dissemination of the pathogen in the patient's blood and the subsequent infection of the kidneys and the central nervous system (Sterrit and Lester, 1988). The disease can be transmitted from animals (rodents, domestic pets, and wildlife) to humans coming into contact (e.g., by bathing) with waters polluted with animal wastes. This zoonotic disease may strike sewage workers. However, this pathogen is not of major concern because it does not appear to survive well in wastewater (Rose, 1986).

4.2.1.8. *Legionella pneumophila*

This bacterium is the etiological agent of *legionnaire's disease*, first described in 1976 in Philadelphia, Pennsylvania. This disease is a type of acute pneumonia with a relatively high fatality rate; it may also involve the gastrointestinal and urinary tracts as well as the nervous system. *Pontiac fever* is another syndrome associated with *Legionella* infection. People manifesting this syndrome have fever, headaches, and muscle aches but may recover without treatment. This organism is transmitted mainly by aerosolization (Muraca et al., 1988).

Aerosolization. Outbreaks of legionnaire's disease are associated with exposure to microbial aerosols from cooling towers, evaporative condensers (Cordes et al., 1980; Dondero et al., 1980), humidifiers, and shower heads. Natural draft cooling towers are used to cool the hot water generated by power-generating plants. These towers generate microbial aerosols, including *Legionella,* at the top of the structure. It is postulated that the source of *Legionella* is the water drawn from nearby surface waters or the potable water supply to replace the moisture lost during the cooling cycle (Muraca et al., 1988).

Ingestion. Legionella pneumophila serogroup 1 has been detected in drinking water samples (Hsu et al., 1984; Tobin et al., 1986), but so far no outbreak has been attributed to consumption of contaminated drinking water.

Nosocomial legionnaire's disease can be contracted via exposure to *Legionella* from the water distribution system in hospitals (Best et al., 1984a). Potable water has been shown to be the source of epidemic legionnaire's disease in a hospital: The number of cases increased following a pressure drop in the distribution system, which probably caused the release of *Legionella* cells associated with biofilms growing in the distribution pipes. However, the number of cases dropped following hyperchlorination (>2 mg/L free residual chlorine) (Fig. 4.3) (Shands et al., 1985).

This bacterium appears to be ubiquitous and has been isolated from wastewater, soil, and natural aquatic environments including tropical waters (Fliermans et al., 1979, 1981; Ortiz-Roque and Hazen, 1987). Its presence in wastewater has been linked, on one occasion at least, with increased levels of antibodies among wastewater irrigation workers (Bercovier et al., 1984). However, the epidemiological significance of this finding is unclear at the present time. In the natural environment this pathogen can thrive in association with other bacteria, green and blue-green algae (Berendt, 1980; Popes et al., 1982; Stout et al., 1985; Tison et al., 1980), amoebae (Barbaree et al., 1986; Barker et al., 1992; Thyndall and Domingue, 1982), or ciliates (Fields et al., 1984; Smith-Somerville et al., 1991), resulting in increased resistance to biocides, chlorination, low pH, and high temperatures (Barker et al., 1992; States et al., 1990). At a site of legionellosis outbreak, protozoa able to sustain the intracellular growth of *L. pneumophila* were isolated from a cooling tower water. Multiplication of this pathogen in coculture with protozoa is shown in Figure 4.4 (Barbaree et al., 1986). Some bacteria are able to support the growth of *Legionella pneumophila* on media lacking L-cystein. Conversely, 16% to 32% of hetero-

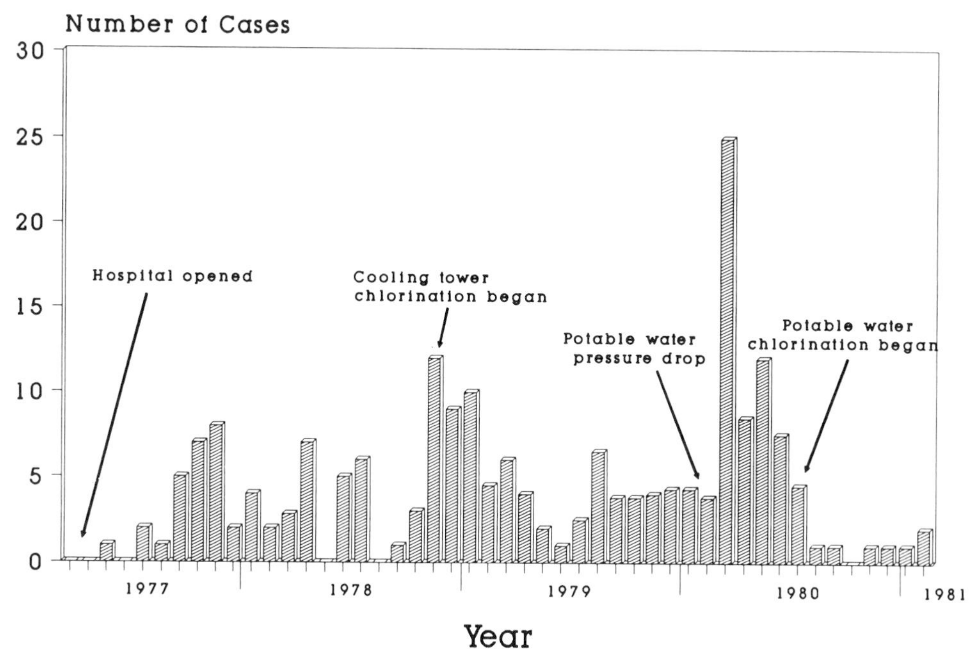

Fig. 4.3. Reduction of legionnaire's disease cases by water superchlorination. From Shands et al. (1985), with permission of the publisher.

trophic plate count bacteria isolated from chlorinated drinking water were found to inhibit *Legionella* species (Toze et al., 1990).

4.2.1.9. *Bacteroides fragilis*

Enterotoxin-producing strains of this anaerobic bacterium may be involved in causing diarrhea in humans. This pathogen has been found in wastewater at levels ranging from 6.2×10^4 to 1.1×10^5 CFU/ml, 9.3% of which were enterotoxigenic (Shoop et al., 1990).

4.2.1.10. Opportunistic Bacterial Pathogens

This group includes heterotrophic gram-negative bacteria belonging to the following genera (Sobsey and Olson, 1983): *Pseudomonas, Aeromonas, Klebsiella, Flavobacterium, Enterobacter, Citrobacter, Serratia, Acinetobacter, Proteus*, and *Providencia*. Segments of the population particularly at risk of infection with opportunistic pathogens are newborn babies and elderly and sick people. These organisms may occur in high numbers in institutional (e.g., hospital) drinking water and attach to water distribution pipes, and some of them may grow in finished drinking water (see Chapter 15). However, their public health significance with regard to the population at large is not well known. Other opportunistic pathogens are the nontubercular mycobacteria that cause pulmonary and other diseases. The most frequently isolated nontubercular mycobacteria belong to the species *Mycobacterium avium-intracellulare* (Wolinsky, 1979). Potable water, particularly hospital water supplies, can support the growth of these bacteria, which may be linked to nosocomial infections (duMoulin et al., 1981).

4.2.2. Antibiotic Producing Bacteria

Patients receiving antibiotic therapy harbor a large number of antibiotics-resistant bacteria in their intestinal tracts. These bacteria are excreted in large numbers in feces and eventually reach

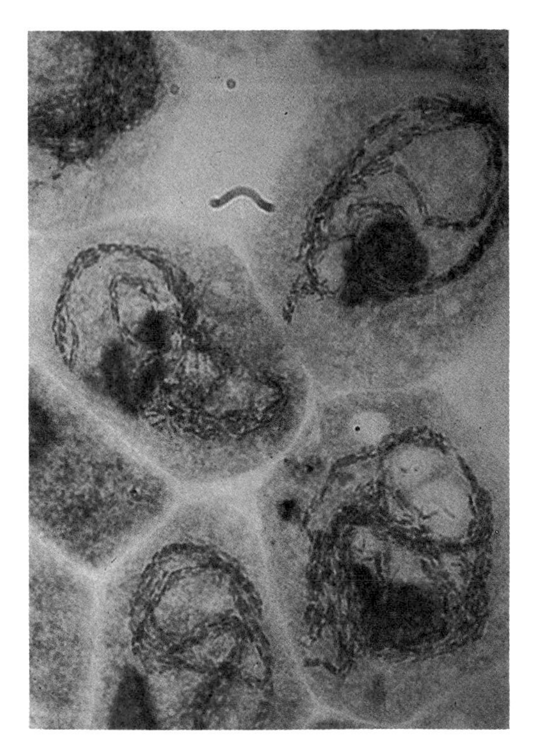

Fig. 4.4. *Tetrahymena pyriformis* infected with *Legionella pneumophila* serotype 1. Courtesy of Barry S. Fields.

the community wastewater treatment plant. The genes coding for resistance to antibiotics are often located on plasmids (R factors) and, under appropriate conditions, can be transferred to other bacteria through conjugation, which requires cell-to-cell contact. If the recipient bacteria are potential pathogens, they may be of public health concern as a result of their acquisition of resistance to antibiotics. Drug-resistant microorganisms produce nosocomial and community-acquired human infections, which can lead to increased morbidity, mortality, and disease incidence. Drug resistance can in turn complicate and increase the cost of therapy based on administration of antibiotics to patients who have been exposed to pathogens of environmental origin. Strategies for tackling this serious problem include the reduced use of antibiotics in humans and animals and preventive measures against the transmission of infectious diseases (Cohen, 1992; Neu, 1992). Bacterial resistance to antibiotics has been demonstrated in terrestrial and aquatic environments, particularly those contaminated with wastes from hospitals (Grabow and Prozesky, 1973).

Gene transfer between microorganisms is known to occur in natural environments as well as in engineered systems such as wastewater treatment plants (Colwell and Grimes, 1986; McClure et al., 1990). Investigators have used survival chambers to demonstrate the transfer of R plasmids among bacteria in domestic wastewater. The mean transfer frequency in wastewater varied between 4.9×10^{-5} and 7.5×10^{-5}. The highest transfer frequency (2.7×10^{-4}) was observed between *Salmonella enteritidis* and *E. coli* (Mach and Grimes, 1982). Nonconjugative plasmids (e.g., pBR plasmids) can also be transferred, but this necessitates the presence of a mobilizing bacterial strain to mediate the transfer (Gealt et al., 1985). Several indigenous mobilizing strains have been isolated from raw wastewater. Each of these strains is capable of aiding in the transfer of the plasmid pBR325 to a recipient *E. coli* strain (McPherson and Gealt, 1986). Under laboratory conditions, plasmid mobilization from genetically engineered bacteria to environmental strains has also been demonstrated under low-temperature and low-nutrient conditions in drinking water (Sandt and Herson, 1989).

The occurrence of multiple-antibiotic-resistant (MAR) indicator and pathogenic (e.g., *Salmonella*) bacteria in water and wastewater treatment plants has been documented (Alcaid and Garay, 1984; Armstrong et al., 1981, 1982; Walter and Vennes, 1985). In untreated wastewater, the percentage of multiple-antibiotic-resistant coliforms varies between less than 1% to about 5% of the total coliforms (Fig. 4.5) (Walter and Vennes, 1985). Chlorination appears to select for resistance to antibiotics in wastewater treatment plants (Staley et al., 1988). However, other observers (Murray et al., 1984) reported that chlorination increased the bacterial resistance to some antibiotics (e.g., ampicilin, tetracycline) but not to others (e.g., chloramphenicol, gentamicin). The proportion of bacteria carrying R factors seems to increase follow-

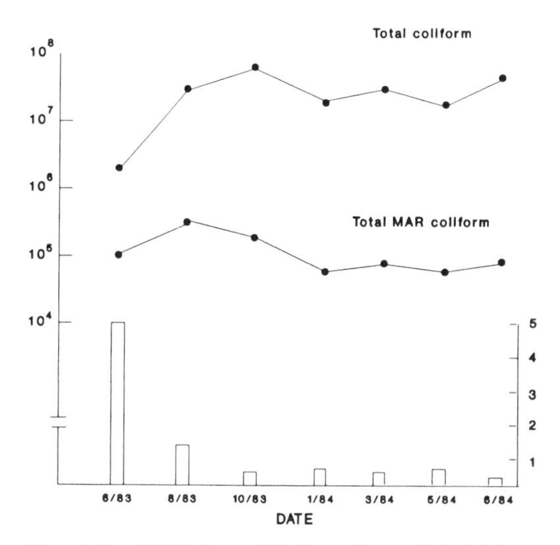

Fig. 4.5. Multiple-antibiotic-resistant (MAR) bacteria in domestic wastewater. From Walter and Vennes (1985), with permission of the publisher.

ing water and wastewater treatment (Armstrong et al., 1981; Bell, 1978; Bell et al., 1981; Calomiris et al., 1984). For example, in one study MAR was expressed by 18.6% of heterotrophic plate count bacteria in untreated water as compared to 67.8% for bacteria in the distribution system (Armstrong et al., 1981). Similarly, in a water treatment plant in Oregon, the percentage of MAR bacteria rose from 15.8% in untreated (river) water to 57.1% in treated water (Armstrong et al., 1982). MAR is furthermore associated with resistance to heavy metals (e.g., Cu^{2+}, Pb^{2+}, Zn^{2+}). This phenomenon has been observed both in drinking water (Calomiris et al., 1984) and wastewater (Varma et al., 1976). The public health significance of this phenomenon is not well known.

4.2.3. Viral Pathogens

Water and wastewater may become contaminated by approximately 140 types of enteric viruses. These viruses enter into the human body orally, multiply in the gastrointestinal tract, and are excreted in large numbers in the feces of infected individuals. Table 4.5 (Bitton, 1980a; Jehl-Pietri, 1992; Schwartzbrod, 1991; Schwartzbrod et al., 1990) shows a list of enteric viruses found in aquatic environments that are pathogenic to humans. Many of the enteric viruses cause inapparent infections that are difficult to detect. They are responsible for a broad spectrum of diseases that range from skin rash, fever, respiratory infections, and conjunctivitis to gastroenteritis and paralysis. Their presence in a community's wastewater reflects virus infections among the population.

Enteric viruses are present in relatively small numbers in water and wastewater. Therefore, environmental samples of 10–1,000 L must be concentrated in order to detect these pathogens. A number of approaches have been considered for accomplishing this task (Gerba, 1987b; Goyal and Gerba, 1982a; Farrah and Bitton, 1982). The most widely used approach is based on the adsorption of viruses to microporous filters of various compositions (nitrocellulose, fiberglass, charge modified cellulose, epoxy-fiberglass, cellulose + glass fibers). This step is followed by elution of the adsorbed viruses from the filter surface. Further concentration of the sample can be obtained by membrane filtration or organic flocculation. The concentrate is then assayed with animal tissue cultures or immunological or genetic probes (see Chapter 1). Table 4.6 (Gerba, 1987b) shows a compilation of most of the methods available for concentrating viruses from water and wastewater.

From an epidemiological standpoint, enteric viruses are mainly transmitted by person-to-person contacts. However, they may also be communicated by water transmission either directly (drinking water, swimming, aerosols) or indirectly through contaminated food (e.g., shellfish, vegetables). Waterborne transmission of enteric viruses is illustrated in Figure 4.6 (Gerba et al., 1975a). The infection process depends on the minimal infective dose (MID) and on host susceptibility, which involves host factors (e.g., specific immunity, sex, age) and environmental factors (e.g., socioeconomic level, diet, hygienic conditions, temperature, humidity). Although the MID for viruses is controversial, it is generally relatively low compared with bacterial pathogens. Experiments with human volunteers have shown an MID of 17 PFU

TABLE 4.5. Some Human Enteric Viruses[a]

Virus group	Serotypes	Some diseases caused
A. Enteroviruses		
Poliovirus	3	Paralysis, aseptic meningitis
Coxsackievirus		
A	23	Herpangia, aseptic meningitis, respiratory illness, paralysis, fever
B	6	Pleurodynia, aseptic meningitis, pericarditis, myocarditis, congenital heart disease, anomalies, nephritis, fever
Echovirus	34	Respiratory infection, aseptic meningitis, diarrhea, pericarditis, myocarditis, fever, and rash
Enteroviruses (68–71)	4	Meningitis, respiratory illness
Hepatitis A virus (HAV)		Infectious hepatitis
B. Reoviruses	3	Respiratory disease
C. Rotaviruses	4	Gastroenteritis
D. Adenoviruses	41	Respiratory disease, acute conjunctivitis, gastroenteritis
E. Norwalk agent (calicivirus)	1	Gastroenteritis
F. Astroviruses	5	Gastroenteritis

[a]Adapted from Bitton (1980), Jehl-Pietri (1992), and Schwartzbrod et al. (1990).

TABLE 4.6. Methods Used for Concentrating Viruses From Water[a]

Method	Initial volume of water	Applications	Remarks
Filter Adsorption-Elution Negatively charged filters	Large	All but the most turbid waters	Only system shown useful for concentrating viruses from large volumes of tapwater, sewage, seawater, and other, natural waters; cationic salt concentration and pH must be adjusted before processing.
Positively charged filters	Large	Tapwater, sewage, seawater	No preconditioning of water necessary at neutral or acidic pH level.
Adsorption to metal salt precipitate, aluminum hydroxide, ferric hydroxide	Small	Tapwater, sewage	Have been useful as reconcentration methods.
Charged filter aid	Small	Tapwater, sewage	40-L Volumes tested, low cost; used as a sandwich between prefilters.
Polyelectrolyte PE60	Large	Tapwater, lake water, sewage	Because of its unstable nature and lot-to-lot variation in efficiency for concentrating viruses, method has not been used in recent years.

(*continued*)

TABLE 4.6. (*Continued*)

Method	Initial volume of water	Applications	Remarks
Bentonite	Small	Tapwater, sewage	
Iron oxide	Small	Tapwater, sewage	
Talcum powder	Large	Tapwater, sewage	Can be used to process up to 100-L volumes as a sandwich between filter paper support.
Gauze pad	Large		First method developed for detection of viruses in water, but not quantitative or very reproducible
Glass powder	Large	Tapwater, seawater	Columns containing glass powder have been made that are capable of processing 400-L volumes
Organic flocculation	Small	Reconcentration	Widely used method for reconcentrating viruses from primary filter eluates
Protamine sulfate	Small	Sewage	Very efficient method for concentrating reoviruses and adenoviruses from small volumes of sewage
Polymer two-phase	Small	Sewage	Processing is slow; method has been used to reconcentrate viruses from primary eluates
Hydroextraction	Small	Sewage	Often used as a method for reconcentrating viruses from primary eluates
Ultrafiltration			
Soluble filters	Small	Clean waters	Clogs rapidly even with low turbidity
Flat membranes	Small	Clean waters	Clogs rapidly even with low turbidity
Hollow fiber or capillary	Large	Tapwater, lake water	Up to 100 L may be processed, but water must often be prefiltered
Reverse osmosis	Small	Clean waters	Also concentrates cytotoxic compounds that adversely affect assay methods

[a]Adapted from Gerba (1987b).

(plaque-forming units) for echovirus 12 (Schiff et al., 1984a, 1984b). Several epidemiological surveys have shown that enteric viruses are responsible for 4.7%–11.8% of waterborne epidemics (Cliver, 1984; Craun, 1988; Lippy and Waltrip, 1984). Epidemiological investigations have definitely proven the waterborne and foodborne transmission of viral diseases such as hepatitis and gastroenteritis.

4.2.3.1. Hepatitis

Hepatitis is caused mainly by the following viruses (Jehl-Pietri, 1992; Pilly, 1990):

Infectious hepatitis is caused by hepatitis A virus (HAV), a 27-nm RNA enterovirus (enterovirus type 72, belonging to the family picornaviridae) with a relatively short incubation period (2–6 weeks) and a fecal–oral transmission

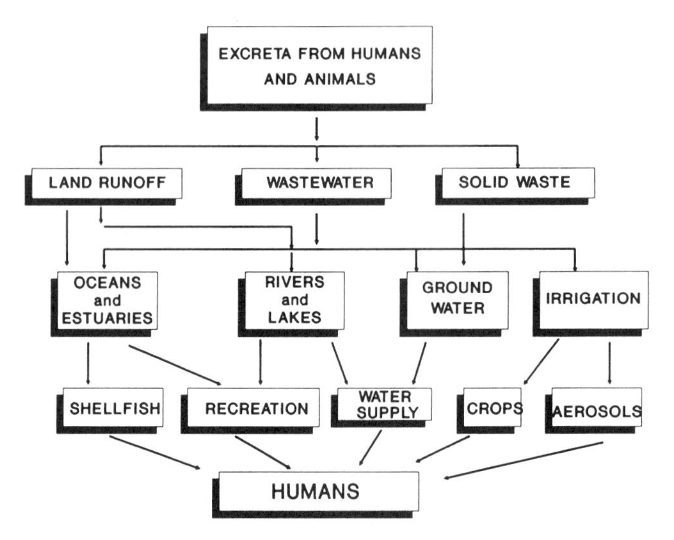

Fig. 4.6. Waterborne transmission of enteric viruses. From Gerba et al. (1975a), with permission of the publisher.

route. Although it can be replicated on primary and continuous human or animal tissue cultures, it is hard to detect because it does not always have a cytopathic effect. Other means for detection of HAV include genetic probes and immunological methods (immunoelectron microscopy, radioimmunoassay, enzyme immunoassay, radioimmunofocus assay; see Chapter 1 for more details).

Serum hepatitis is caused by hepatitis B virus (HBV), a 42-nm DNA virus displaying a relatively long incubation time (4–12 weeks). This virus is transmitted by contact with infected blood or by sexual contact. Mortality (1%–4%) is higher than for infectious hepatitis (<0.5%).

Non-A, non-B viral hepatitis is caused by two viruses: A 50- to 60-nm flavivirus with clinical and epidemiological characteristics similar to hepatitis B virus and a 32- to 34-nm calicivirus with characteristics similar to hepatitis A virus.

Chronic delta hepatitis is caused by a 28- to 35-nm RNA virus that has not been fully characterized. Its epidemiological and clinical patterns are similar to those of hepatitis B virus.

HAV causes liver damage with necrosis and inflammation. After the onset of infection, the incubation period may last up to 6 weeks. One of the most characteristic symptoms is jaundice.

Hepatitis A is transmitted by the fecal–oral route either by person-to-person direct contact or by waterborne or food-borne transmission (Jehl-Pietri, 1992). This disease is distributed worldwide and the prevalence of HAV antibodies is higher among lower socioeconomic groups and increases with the age of the infected individual. Direct-contact transmission has been documented mainly in nurseries (especially among infants wearing diapers), mental institutions, prisons, and in military camps. Waterborne transmission of infectious hepatitis has been conclusively demonstrated and documented worldwide on several occasions. It has been estimated that 4% of hepatitis cases observed between 1975 and 1979 in the United States were the result of waterborne transmission (Cliver, 1985). These hepatitis cases are due to consumption of improperly treated water. However, swimming in recreational waters has not been associated with hepatitis A infection. Food-borne transmission of HAV appears to be more important than waterborne transmission. Consumption of shellfish grown in wastewater-contaminated waters accounts for numerous hepatitis and gastroenteritis outbreaks documented worldwide (Table 4.7) (Jehl-Pietri, 1992; Schwartzbrod, 1991). The most recent hepatitis epidemic to result from the consumption of contaminated shellfish occurred in 1988 in Shanghai and was responsible for 292,000 cases (Hu et al., 1989).

4.2.3.2. *Viral Gastroenteritis*

Gastroenteritis is probably the most frequent waterborne illness; it is caused by protozoan parasites and by bacterial and viral pathogens (e.g., rotaviruses and Norwalk-type agent) (Williams and Akin, 1986). In this section we will examine rotaviruses and Norwalk-type virus as causal agents of gastroenteritis.

Rotaviruses. Rotaviruses, belonging to the family Reoviridae, are 70-nm particles containing double-stranded RNA surrounded with a double-shelled capsid. Rotaviruses are the major cause of infantile acute gastroenteritis in children under 2 years of age (Gerba et al., 1985). This disease largely contributes to childhood mortality in developing countries and is responsible for millions of childhood deaths per year in Africa, Asia, and Latin America. It is also responsible for outbreaks among adult populations (e.g., the elderly) and is a major cause of travelers' diarrhea. Up to 10^{10} rotavirus particles can be detected in patients' stools. The virus is spread mainly by the fecal–oral route but a respiratory route has also been suggested (Flewett, 1982; Foster et al., 1980). There have been several outbreaks of gastroenteritis, in which rotaviruses originating from wastewater have been implicated (Gerba et al., 1985). Some waterborne outbreaks associated with rotaviruses are summarized in Table 4.8 (Gerba et al., 1985; Williams and Akin, 1986).

TABLE 4.7. Some Viral Hepatitis Outbreaks Due to Shellfish Consumption[a]

Year	Shellfish	Country	Number of cases
1953	Oysters	USA	30
1955	Oysters	Sweden	600
1961	Oysters	USA	84
1962	Clams	USA	464
1963/1966	Clams/oysters	USA	180
1964	Clams	USA	306
1964	Oysters	USA	3
1966	Clams	USA	4
1968/1971	Clams/oysters	Germany	34
1971	Clams	USA	17
1972	Mussels	France	13
1973	Oysters	USA	265
1976	Mussels	Australia	7
1978	Mussels	England	41
1979	Oysters	USA	8
1980	Oysters	Philippines	7
1980/1981	Cockle	England	424
1982	Various shellfish	England	172
1982	Oysters	USA	204
1982	Clams	USA	150
1984	Cockle	Malaysia	322
1984	Mussels	Yugoslavia	51
1985	Clams	USA	5
1988	Clams	China	292,000

[a]Adapted from Jehl-Pietri (1992) and Schwartzbrod (1991).

TABLE 4.8. **Rotavirus Waterborne Outbreaks**[a]

Year	Location	Number ill	Remarks
1977	Sweden	3,172	Small town water supply contaminated with sewage effluent
1980	Brazil	~900	Contamination of private school's water
1980	Norfolk Island	—	Contamination of ground water supply
1981	Russia	173	Contamination of community water system
1981	Colorado	1,500	Source contamination (chlorinator and filtration failure)
1982	Israel	~2,000	Reservoir contaminated by children
1981–1982	East Germany	11,600	Floodwater contamination of wells
1982–1983	China	13,311	Contaminated water supply
1991	Arizona	900	Well water contaminated by sewage at a resort

[a]Adapted from Gerba et al. (1985) and Williams and Akin (1986).

Detection of rotaviruses in wastewater and other environmental samples is accomplished by electron microscopy, enzyme-linked immunosorbent assays (ELISA kits are commercially available), or tissue cultures (a popular cell line is MA-104, which is derived from fetal rhesus monkey kidney). Detection in cell cultures includes methods such as plaque assay, cytopathic effect (CPE), and immunofluorescence. Information on the fate of rotaviruses in the environment is available mostly for simian rotaviruses (e.g., strain SA-11) and little is known concerning the four known human rotavirus serotypes.

Norwalk-type agent. This small, 27-nm virus, first discovered in 1968 in Norwalk, Ohio, is a major cause of waterborne disease and is also implicated in food-borne outbreaks. It causes diarrhea and vomiting and appears to attack the proximal small intestine, but the mechanism of pathogenicity is not known (Harris, 1986). Since this virus cannot be propagated in tissue cultures, little is known about its structure and nucleic acid content. The tools used for its detection in clinical samples, immune electron microscopy and radioimmunoassay techniques, are not sensitive enough for environmental samples. The development of gene probes for detecting this virus awaits determination of its nucleic acid. The Norwalk virus plays a major role in waterborne gastroenteritis (Table 4.9) (Gerba et al., 1985; Williams and Akin, 1986), but it also appears to play a role in travelers' diarrhea (Keswick et al., 1982). Forty two percent of outbreaks of nonbacterial gastroenteritis have been attributed to Norwalk virus (Kaplan et al., 1982).

4.2.4. Protozoan Parasites

Most protozoan parasites produce cysts that are able to survive outside their host under adverse environmental conditions. Encystment is triggered by factors such as lack of nutrients, accumulation of toxic metabolites, and host immune response. Under appropriate conditions, a new trophozoite is released from the cyst. This process is called excystment (Rubin et al., 1983).

The major waterborne pathogenic protozoa affecting humans are the following (Table 4.10).

4.2.4.1. *Giardia lamblia*

This flagellated protozoan parasite has a pear-shaped trophozoite (9–21 μm long) and an ovoid cyst stage (8–12 μm long and 7–10 μm wide) (Fig. 4.7). An infected individual may shed up to (1–5) × 10^6 cysts per gram feces (Jakubowski and Hoff, 1979; Lin, 1985). Domestic wastewater is a significant source of *Giardia* and wild and domestic animals act as important reservoirs of *Giardia* cysts. This parasite is endemic in mountainous areas in the United States and infects both humans and domestic and wild animals (e.g., beavers, muskrats, dogs, cats). Infection is caused by ingestion of cysts found in water. In humans, infections may last from months to

TABLE 4.9. Some Waterborne Outbreaks of Norwalk Virus[a]

Year	Location	Number ill	Remarks
1978	Pennsylvania	350	Drinking water with insufficient chlorination (attack rate 17%–73%)
1978	Tacoma, WA	600	Well 51.4 m deep (attack rate 72%)
1979	Arcata, CA	30	Sprinkler irrigation system not for human consumption
1980	Maryland	126	Well 95 ft. (attack rate 64%), disinfected water
1980	Rome, GA	1,500	Spring (attack rate 72%)
1982	Tate, GA	500	Springs and well; springs possibly contaminated from rainfall events

[a]Adapted from Gerba et al. (1985, and Williams and Akin (1986).

years. The infectious dose for Mongolian gerbils is more than 100 cysts (Schaefer et al., 1991), but it is fewer than 10 cysts in humans (Rendtorff, 1954). Passage through the stomach appears to promote the release of trophozoites, which attach to the epithelial cells of the upper small intestine and reproduce by binary fission. They may coat the intestinal epithelium and interfere with absorption of fats and other nutrients. They encyst as they travel through the intestines and reach the large intestine (AWWA, 1985a).

Giardia has an incubation period of 1–8 weeks. It causes diarrhea, abdominal pains, nausea, fatigue, and weight loss, but giardiasis is rarely fatal. Although its usual mode of transmission is the person-to-person route, *Giardia* is recognized as one of the most important etiological agents in waterborne disease outbreaks (Craun, 1979, 1984b). During the 1971–1985 period more than 50% of the outbreaks resulting from the use of surface waters were caused by *Giardia* (Fig. 4.8) (Craun, 1988). The first major documented outbreak of giardiasis occurred in 1974 in Rome, New York, and was associated with the presence of *Giardia* in the water supply. It affected approximately 5,000 people (10% of the town's population). The outbreak occurred as

TABLE 4.10. Major Waterborne Diseases Caused by Protozoa[a]

Organism	Disease (site affected)	Major reservoir
Giardia lamblia	Giardiasis (GI tract)	Human and animal feces
Entamoeba histolytica	Amoebic disentery (GI tract)	Human feces
Acanthamoeba castellani	Amoebic meningoencephalitis (central nervous system)	Soil and water
Naeleria gruberi	Amoebic meningoencephalitis (central nervous system)	Soil and water
Balantidium coli	Dysentery/intestinal ulcers (GI tract)	Human feces
Cryptosporidium	Profuse and watery diarrhea; weight loss; nausea; low-grade fever (GI tract)	Human and animal feces

[a]Adapted from Sobsey and Olson (1983).

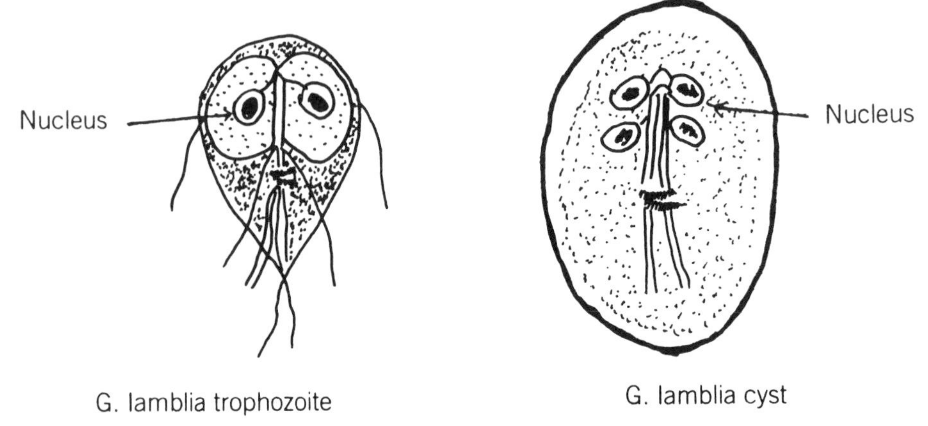

Fig. 4.7. *Giardia lamblia* trophozoite and cyst. From Lin (1985), with permission of the publisher.

a result of consumption of water that had been chlorinated but not filtered. Recent amendments of the Safe Drinking Water Act (Surface Water Treatment Rule) in the United States mandates the U.S. EPA to require filtration and disinfection for all surface waters and groundwater under the direct influence of surface water to control the transmission of *Giardia* spp. and enteric viruses. However, exceptions (e.g., effective disinfection) to this requirement are being considered (Clark et al., 1989; U.S. EPA, 1989d). Other outbreaks have been reported in Colorado, Vermont, New Hampshire, Utah, and Washington. Approximately 80 outbreaks of giardiasis were recorded in the United States from 1965 to 1983 (AWWA, 1985a). *Giardia* was the etiologic agent for 7 of 26 (27%) outbreaks of waterborne diseases reported in the United States in 1989–1990 (Herwaldt et al., 1992).

Most giardiasis outbreaks are associated with the consumption of untreated or unsuitably treated water (water chlorinated but not filtered;

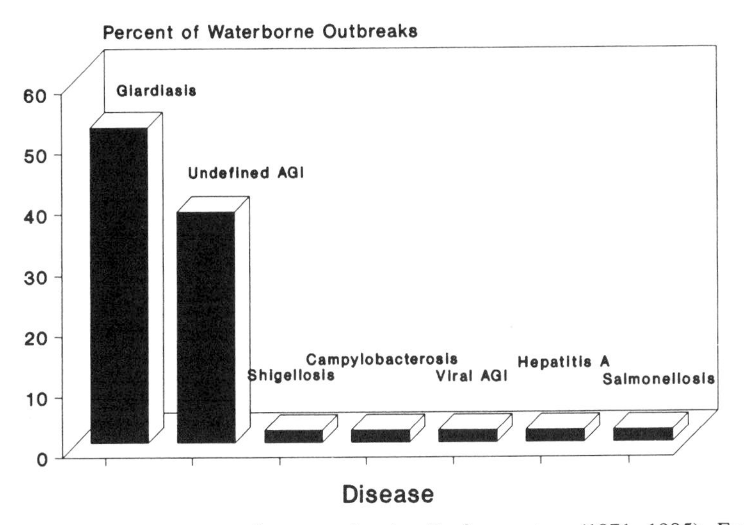

Fig. 4.8. Etiology of waterborne disease outbreaks: Surface waters (1971–1985). From Craun (1988), with permission of the publisher.

interruption of disinfection). Faulty design or construction of filters may lead to breakthrough of *Giardia lamblia* and subsequent contamination of drinking water. Traditional bacterial indicators are not suitable for indicating the presence of *Giardia* cysts in water and other environmental samples (Rose et al., 1991). Good correlation has been reported between the occurrence of *Giardia* cysts and *Cryptosporidium* oocysts as well as some traditional parameters of water quality (turbidity) (LeChevallier et al., 1991b).

Giardia lamblia cysts can be concentrated from water and wastewater, by means of ultrafiltration cassettes or adsorption to polypropylene cartridges (APHA, 1985; Hibler and Hancock, 1990; Isaac-Renton et al., 1986). Since *Giardia lamblia* cannot be cultured in the laboratory, the cysts must be detected and identified by immunofluorescence with polyclonal or monoclonal antibodies or by phase-contrast microscopy (Sauch, 1989). Moreover, cysts exposed to a chlorine concentration of 1–11 mg/L, although fluorescing, could not be confirmed by phase-contrast microscopy because they lost their internal structures (Sauch and Berman, 1991).

Cysts can also be selectively concentrated from water samples by an antibody–magnetite procedure. Following exposure to a mouse anti-*Giardia* antibody, cysts are allowed to react with anti-mouse antibody-coated magnetite particles and then concentrated by high-gradient magnetic separation (Bifulco and Schaefer, 1993). Fluorogenic dyes such as fluorescein diacetate (FDA) can be used to indicate cyst viability. FDA uptake and degradation releases fluorescein, which renders the cysts fluorescent. Although cysts responding to FDA correlate well with infectivity to animals, this stain may overestimate their viability (Labatiuk et al., 1991). Conversely, there is a negative correlation between cyst staining with propidium iodide and infectivity (Schupp and Erlandsen, 1987; Sauch et al., 1991). Recently, a cDNA probe was constructed for the detection of *Giardia* cysts in water and wastewater concentrates (Abbaszadegan et al., 1991; Nakhforoosh and Rose, 1989). However, this technology does not provide information on cyst viability (Rose et al., 1991). Amplification of the *giardin* gene DNA by polymerase chain reaction (PCR) has been used to detect *Giardia*. Distinction of live from dead cysts was made possible by measuring the amount of RNA before and after excystation (Mahbubani et al., 1991). However, several substances (e.g., humic acids) that are present in environmental samples interfere with pathogen and parasite detection by PCR technology (Rodgers et al., 1992). These interferences are under study.

A survey of raw wastewater from several states in the United States showed that *Giardia* cyst numbers varied from hundreds to thousands of cysts per liter (Casson et al., 1990; Sykora et al., 1990) but cyst concentration may be as high as 10^5 cysts per liter (Jakubowski and Eriksen, 1979). In Arizona, *Giardia* was detected at concentrations of 48 cysts per 40L of activated sludge effluent (Rose et al., 1989b). This concentration decreased to 0.3 cysts/40L following sand filtration. It was suggested that sewage examination for *Giardia* cysts may serve as a means for determining the prevalence of giardiasis in a given community (Jakubowski et al., 1990). A survey of a sewage treatment plant in Puerto Rico showed that 94%–98% of this parasite is removed following passage through the plant (Correa et al., 1989). This parasite is more resistant than bacteria to chlorine (Jarrol et al., 1984). *Giardia* cysts have been detected in 16% of potable water supplies (lakes, reservoirs, rivers, springs, groundwater) in the United States at an average concentration of 3 cysts per 100L (Rose et al., 1991). Another survey of surface water supplies in the United States and Canada showed that cysts occurred in 81% of the samples (LeChevallier et al., 1991b).

4.2.4.2. *Cryptosporidium*

The coccidian protozoan parasite *Cryptosporidium* was first described at the end of the last century. It was known to infect animal species (calves, lambs, chickens, turkeys, mice, pigs, dogs, cats) but infection of humans was reported only in the 1970s. *Cryptosporidium parvum* is the major species responsible for infections in humans and animals (Current, 1987; Rose et al., 1985; Rose, 1990).

The infective stage of this protozoan is the oocyst (5–7 μm), which readily persists under environmental conditions. Following ingestion by a suitable host, the oocyst undergoes excystation and releases infective sporozoites, which parasitize epithelial cells mainly in the host's gastrointestinal tract. The life cycle of *Cryptosporidium* is illustrated in Figure 4.9 (Fayer and Ungar, 1986). Some observers have suggested that the minimum infective dose for *Cryptosporidium* in humans may be less than or equal to 1,000 oocysts (Blewett et al., 1993; Hart et al., 1984; Peeters et al., 1989) but other animal models show that as few as one to ten oocysts may initiate infection (Kwa et al., 1993; Miller et al., 1986; Rose, 1988). The parasite causes a profuse and watery diarrhea that is often associated with weight loss and sometimes nausea, vomiting, and low-grade fever (Current, 1988). The duration of the symptoms and the outcome depend on the immunological status of the patient. The diarrhea generally lasts 1–10 days in immunocompetent patients but may persist for longer periods (more than a month) in immunodeficient patients (e.g., AIDS patients, cancer patients undergoing chemotherapy). Examination of thousands of human fecal samples in the United States, Canada, and Europe has revealed that the prevalence of human cryptosporidiosis ranges from 1% to 5% (Ongerth and Stibbs, 1987).

Person-to-person, waterborne, food-borne, and zoonotic routes are all involved in the transmission of *Cryptosporidium*. Person-to-person transmission is the major route, especially in daycare centers. The zoonotic route, the transmission of the pathogen from infected animals to humans, is suspected to be greater for *Cryptosporidium* than for *Giardia* (AWWA, 1988). Studies are being conducted on the prevalence of this protozoan parasite in the environment, since some of the outbreaks resulted from waterborne

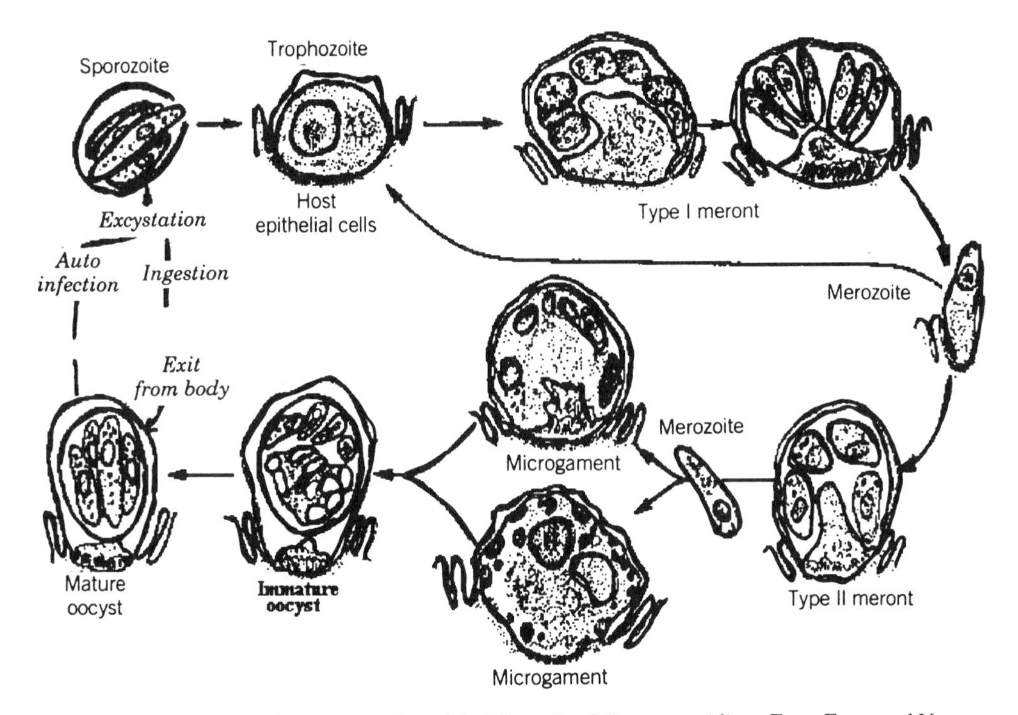

Fig. 4.9. Diagrammatic representation of the life cycle of *Cryptosporidium*. From Fayer and Ungar (1986), with permission of the publisher.

transmission. Recently, cryptosporidiosis outbreaks have occurred in Texas and Georgia. An outbreak in Carrollton, Georgia, affected approximately 13,000 people and was epidemiologically associated with consumption of drinking water from a water treatment plant where rapid sand filtration was part of the treatment process. The problems discovered in the plant were ineffective flocculation and restarting of sand filter without backwashing. *Cryptosporidium* was identified in 39% of the stools of patients during the outbreak and in samples of treated water, while no other traditional indicator was identified in the samples (Hayes et al., 1989).

Other outbreaks of cryptosporidiosis have been reported in Europe (Rush et al., 1990). Although the Carrollton outbreak has shown that *Cryptosporidium* may result from waterborne transmission, the public health significance of this route is still unknown. The pathogen is not efficiently removed or inactivated by traditional water treatment processes such as sand filtration or chlorination, although lime treatment for water softening can partially inactivate *Cryptosporidium* oocysts (Robertson et al., 1992; Rose, 1990).

Concentration techniques have been developed for the recovery of this parasite but they are still at the developmental stage. The methods used involve the retention of oocysts on polycarbonate filters (Ongerth and Stibbs, 1987) or polypropylene cartridge filters (Musial et al., 1987). Following elution with a detergent solution from the filters, the oocysts are detected in the concentrates by using polyclonal or, more recently, monoclonal antibodies combined with epifluorescence microscopy (Rose et al., 1989c), DNA probe hybridization, and detection by polymerase chain reaction (Johnson et al., 1993). However, the recovery efficiency of these concentration techniques is relatively low (less than 10% in river water to 59% for tapwater (Rose, 1988). Oocyst viability is determined by *in vitro* excystation, mouse infectivity, or staining with fluorogenic vital dyes such as DAPI (4′, 6-diamidino-2-phenylindole) or propidium iodide (Campbell et al., 1992). This methodology allowed the detection of this hardy parasite in wastewater and in surface and drinking water. Oocysts occur in raw wastewater at levels varying between 850 and 13,700 oocysts per liter. The range of oocyst concentrations in wastewater effluents varies between 4 and 3,960 cysts per liter (Madore et al., 1987; Musial et al., 1987). A survey of potable water supplies in the United States showed that oocysts were present in 55% of the samples at an average concentration of 43 oocysts per 100L (Rose et al., 1991). Another survey indicated the presence of oocysts in 87% of surface water samples (LeChevallier et al., 1991b). Analysis of river water in western United States showed *Cryptosporidium* oocysts in each of the 11 rivers examined, at concentrations ranging from 2 to 112 oocysts per liter (Ongerth and Stibbs, 1987). This parasite has also been detected on one occasion in drinking water (Rose et al., 1986).

4.2.4.3. *Entamoeba histolytica*

E. histolytica forms infective cysts (10–15 μm in diameter) that are shed for relatively long periods by asymptomatic carriers; it persists well in water and wastewater and may be subsequently ingested by new hosts. Level of cysts in raw wastewater can be as high as 5,000 cysts per liter.

This protozoan parasite is transmitted to humans mainly by contaminated water and food. It causes amebiasis, or amoebic dysentery, which is a disease of the large intestine. Symptoms vary from diarrhea alternating with constipation to acute dysentery. It may cause ulceration of the intestinal mucosa, resulting in diarrhea and cramps. It is a cause of morbidity and mortality mostly in developing countries and is acquired mostly by consumption of contaminated drinking water in tropical and subtropical areas. Waterborne transmission of this protozoan parasite, however, is rare in the United States.

4.2.4.4. *Naegleria*

Naegleria are free-living protozoa that have been isolated from wastewater, surface waters, swimming pools, soils, domestic water supplies, and thermally polluted effluents (Marciano-Cabral,

1988). *Naegleria fowleri* is the causal agent for primary amoebic meningoencephalitis (PAME), which was first reported in Australia in 1965. It is most often fatal, 4–5 days after entry of the parasite into the body. The protozoan enters the body through the mucous membranes of the nasal cavity and migrates to the central nervous system. The disease has been associated with swimming and diving, mostly in warm lakes in southern states of the United States (Florida, South Carolina, Georgia, Texas, and Virginia). Another concern is the fact that *Naegleria* may harbor *Legionella pneumophila* and other pathogenic microorganisms (Newsome et al., 1985). The implications of this association with regard to human health are under investigation. There are rapid identification techniques (e.g., cytometry; the API ZYM system, which is based on detection of enzyme activity) that can distinguish between *Naegleria fowleri* and other *Naegleria* species (Kilvington and White, 1985).

4.2.5. Helminth Parasites

Although helminth parasites are not generally studied by microbiologists, their presence in wastewater, along with bacterial and viral pathogens and protozoan parasites, is nonetheless of great concern in regard to human health. The ova constitute the infective stage of parasitic helminths; they are excreted in feces and spread by wastewater, soil, or food. The ova are very resistant to environmental stresses and to chlorination in wastewater treatment plants (Little, 1986). The most important parasites are the following (Table 4.11).

4.2.5.1. *Taenia spp.*

Taenia saginata (beef tapeworm) and *Taenia solium* (pig tapeworm) are now relatively rare in the United States. These parasites develop in an intermediate host to reach a larval stage called *cysticercus* and they may finally reach humans, who serve as final hosts. Cattle ingest the infective ova while grazing and serve as intermediate hosts for *Taenia saginata,* pigs being the intermediate hosts for *Taenia solium*. The cysticerci invade muscles, eyes, and brain. These parasites cause enteric disturbances, abdominal pains, and weight loss.

4.2.5.2. *Ascaris lumbricoides (Roundworms)*

The life cycle of this helminth (Fig. 4.10) includes a phase in which the larvae migrate through the lungs and cause pneumonitis (Loeffler's syndrome). This disease can be acquired through ingestion of only a few infective eggs. Infected individuals excrete a large quantity of

TABLE 4.11. Major Parasitic Helminths

Organism	Disease (main site affected)
Nematodes (roundworms)	
Ascaris lumbricoides	Ascariasis—intestinal obstruction in children (small intestine)
Trichuris trichiura	Whipworm—(trichuriasis) (intestine)
Hookworms	
Necator americanus	Hookworm disease (GI tract)
Ancylostoma duodenale	Hookworm disease (GI tract)
Cestodes (tapeworms)	
Taenia saginata	Beef tapeworm—abdominal discomfort, hunger pains, chronic indigestion (GI tract)
Taenia solium	Pork tapeworm (GI tract)
Trematodes (flukes)	
Schistosoma mansoni	Schistosomiasis (complications in liver [cirrhosis], bladder, and large intestine)

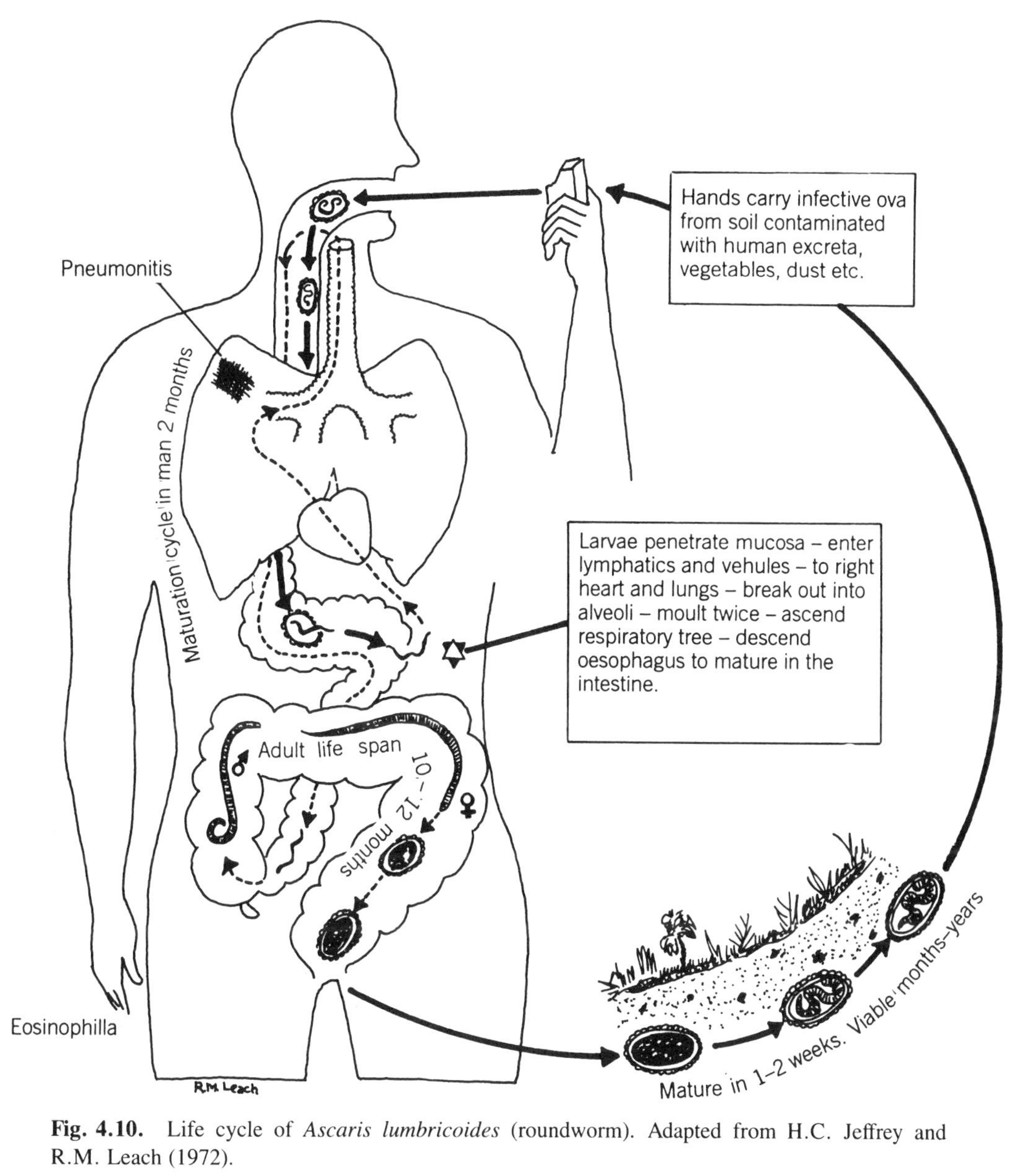

Fig. 4.10. Life cycle of *Ascaris lumbricoides* (roundworm). Adapted from H.C. Jeffrey and R.M. Leach (1972).

eggs, and each female *Ascaris* can produce approximately 200,000 eggs per day (Little, 1986). The eggs are dense and thus are readily removed by sedimentation in wastewater treatment plants. Although they are effectively removed by the activated sludge treatment, they are quite resistant to chlorine action (Rose, 1986).

4.2.5.3. *Toxocara canis*

This parasite mainly infects children with habits of eating dirt. In addition to causing intestinal disturbances, the larvae of this parasite can migrate into the eyes, causing severe ocular damage, sometimes resulting in loss of the eye.

4.2.5.4. *Trichuris trichiura*

Trichuris trichiura causes whipworm infections in humans. The eggs are dense and settle quite readily in sedimentation tanks.

4.2.6. Other Problem-Causing Microorganisms

Surface waters that feed water treatment plants may harbor large concentrations of blue-green algae such as *Anabaena flos-aquae, Microcystis aeruginosa,* and *Schizothrix calcicola*. These algae produce exotoxins (peptides and alkaloids) as well as endotoxins (lipopolysaccharides) that may be responsible for syndromes such as gastroenteritis (Carmichael, 1981a, 1989). However, because of the lack of knowledge about the occurrence and potential for removal of these toxins in water and wastewater treatment plants, their health risks have not yet been evaluated (see also Chapter 15).

4.3. FURTHER READING

Bitton, G. 1980a. *Introduction to Environmental Virology.* Wiley, New York, 326 pp.

Cliver, D.O. 1984. Significance of water and environment in the transmission of virus disease. Monogr. Virol. 15: 30–42.

Craun, G.F., Ed. 1986. *Waterborne Diseases in the United States.* CRC Press, Boca Raton, FL, pp. 295.

Feachem, R.G., D.J. Bradley, H. Garelick, and D.D. Mara. 1983. *Sanitation and Disease: Health Aspect of Excreta and Wastewater Management.* John Wiley and Sons, Chichester, U.K.

Gerba, C.P., S.N. Singh, and J.B. Rose. 1985. Waterborne gastroenteritis and viral hepatitis. CRC Crit. Rev. Environ. Control 15: 213–236.

Rose, J.B. 1990. Occurrence and control of *Cryptosporidium* in drinking water, pp. 294–321, in: *Drinking Water Microbiology,* G.A. McFeters, Ed. Springer Verlag, New York.

Schwartzbrod, L., Ed. 1991. *Virologie des Milieux Hydriques.* TEC & DOC Lavoisier, Paris, 304 pp.

Sobsey, M.D., and B. Olson. 1983. Microbial agents of waterborne disease, in: *Assessment of Microbiology and Turbidity Standards for Drinking Water,* P.S. Berger and Y. Argaman, Eds. EPA Report # EPA 570-9-83-001.

5

INDICATOR MICROORGANISMS

5.1. INTRODUCTION

The direct detection of pathogenic bacteria and viruses and cysts of protozoan parasites requires costly and time-consuming procedures, and well-trained labor. These requirements led to the concept of indicator organisms of fecal pollution. As early as 1914, the U.S. Public Health Service adopted the coliform group as an indicator of fecal contamination of drinking water (Gerba, 1987a). Later on, various microorganisms have been used for indicating the occurrence of fecal contamination, treatment efficiency in water and wastewater treatment plants, and the deterioration and postcontamination of water in distribution systems (Olivieri, 1983).

The criteria for an ideal indicator organism are the following:

1. It should be a member of the intestinal microflora of warm-blooded animals.
2. It should be present when pathogens are present and absent in uncontaminated samples.
3. It should be present in greater numbers than the pathogen.
4. It should be at least equally resistant as the pathogen to environmental insults and to disinfection in water and wastewater treatment plants.
5. It should not multiply in the environment.
6. It should be detectable by means of easy, rapid, and inexpensive methods.
7. The indicator organism should be nonpathogenic.

5.2. REVIEW OF INDICATOR MICROORGANISMS

Proposed or commonly used microbial indicators are the following (APHA, 1989; Berg, 1978; Ericksen and Dufour, 1986; Olivieri, 1983):

5.2.1. Total Coliforms

The total group of coliforms includes the aerobic and facultative anaerobic, gram-negative, nonspore-forming, rod-shaped bacteria that ferment lactose with gas production within 48 hr at 35°C (APHA, 1989).

This group includes *E. coli, Enterobacter, Klebsiella,* and *Citrobacter.* These coliforms are discharged in high numbers (2×10^9 coliforms per day per capita) in human and animal feces, but not all of them are of fecal origin. These indicators are useful for determining the quality of potable water, shellfish harvesting waters, and recreational waters. They are less sensitive, however, than viruses or protozoan cysts to environmental factors and to disinfection. Some members of this group (e.g., *Klebsiella*) may sometimes grow under environmental conditions in industrial and agricultural wastes. In water treatment plants, total coliforms are one of the best indicators of the treatment efficiency of the plant. This group has also been found useful for assessing the safety of reclaimed wastewater in the Windhoek reclamation plant in Namibia (Grabow, 1990).

5.2.2. Fecal Coliforms

Fecal coliforms include all coliforms that can ferment lactose at 44.5°C. The fecal coliform group comprises bacteria such as *Escherichia coli* and *Klesiella pneumonae*. The presence of fecal coliforms indicates the presence of fecal material from warm-blooded animals. However, human and animal contaminations cannot be differentiated. Some have suggested the sole use of *E. coli* as an indicator of fecal pollution, since it can be easily distinguished from the other members of the fecal coliform group (e.g., absence of urease and presence of β-glucuronidase). Fecal coliforms display a survival pattern similar to that of bacterial pathogens, but their usefulness as indicators of protozoan or viral contamination is limited. They are much less resistant to disinfection than viruses or protozoan cysts. Coliform standards are unreliable with respect to viral pollution of shellfish and overlying waters. They may also regrow in water and wastewater under appropriate conditions. Several methodological modifications have been proposed to improve the recovery of these indicators, particularly injured fecal coliforms (Eriksen and Dufour, 1986). Moreover, the detection of *E. coli* in pristine sites in a tropical rain forest suggests that it may not be a reliable indicator of fecal pollution in tropical environments (Bermudez and Hazen, 1988; Hazen, 1988).

5.2.3. Fecal Streptococci

This group comprises *Streptococcus faecalis, S. bovis, S. equinus,* and *S. avium*. Since they commonly inhabit the intestinal tract of humans and warm-blooded animals, they are used to detect fecal contamination in water. Members of this group persist well but do not reproduce in the environment. A subgroup of the fecal streptococci group, the enterococci (*S. faecalis* and *S. faecium*), has been suggested as useful for indicating the presence of viruses, particularly in sludge and seawater. The fecal coliform to fecal streptococci ratio (FC/FS ratio) serves as an indicator of the origin of pollution of surface waters. A ratio of 4 or more indicates a contamination of human origin, whereas a ratio below 0.7 is indicative of animal pollution (Geldreich and Kenner, 1969). This ratio is only valid for recent (24 hours) fecal pollution. However, some investigators have questioned the usefulness of this ratio (Pourcher et al., 1991).

5.2.4. Anaerobic Bacteria

The main anaerobic bacteria that have been considered as indicators are the following.

Clostridium perfringens. This microorganism is an anaerobic gram-positive, spore-forming, rod-shaped bacterium that produces spores that are quite resistant to environmental stresses and to disinfection. The hardy spores make this bacterium too resistant to be useful as an indicator organism. It has been suggested nonetheless to use this microorganism as an indicator of past pollution and as a tracer to follow the fate of pathogens. This organism also appears to be a reliable indicator for tracing fecal pollution in the marine environment (e.g., marine sediments impacted by sludge dumping off the coast of New Jersey) (Burkhardt and Watkins, 1992; Hill et al., 1993).

Bifidobacteria. The *Bifidobacteria* are anaerobic, non-spore-forming, gram-positive bacteria that have been suggested as fecal indicators. Since some of them (e.g., *B. bifidum, B. adolescentis, B. infantis*) are primarily associated with humans, they may help distinguish between human and animal contamination. However, they require the development of suitable methods for their detection.

Bacteroides spp. These anaerobic bacteria occur in the intestinal tract at concentrations in the order of 10^{10} cells per gram of feces, and the survival of *B. fragilis* in water is lower than that of *E. coli* or *S. faecalis*. A fluorescent antiserum test for this microorganism was suggested as a useful method for indicating the fecal contamination of water (Fiksdal et al., 1985; Holdeman et al., 1976).

5.2.5. Bacteriophages

Bacteriophages are similar to the enteric viruses but are more easily and rapidly detected in environmental samples and are found in higher numbers than enteric viruses in wastewater and other environments (Bitton, 1980a; Goyal et al., 1987; Grabow, 1986). Several investigators have suggested the potential use of coliphages as water quality indicators in estuaries (O'Keefe and Green, 1989), seawater (good correlation between coliphage and *Salmonella*) (Borrego et al., 1987), recreational freshwater (Dutka et al., 1987), and potable water (Ratto et al., 1989). Of all the indicators examined, coliphages exhibited the best correlation with enteric viruses in polluted streams in South Africa. The incidence of both enteric viruses and coliphages were inversely correlated with temperature (Fig. 5.1) (Geldenhuys and Pretorius, 1989). They may also serve as indicators for assessing the removal efficiency of water and wastewater treatment plants (Bitton, 1987). In an activated sludge system, coliphages that formed plaques greater than 3 mm were significantly correlated with numbers of enteroviruses (Funderburg and Sorber, 1985).

In water treatment plants, coliphages help provide information concerning the performance of water treatment processes such as coagulation, flocculation, sand filtration, adsorption to activated carbon, and disinfection (Table 5.1) (Payment, 1991). Some coliphages, particularly RNA phage f2, are more resistant to chlorination than enteroviruses such as poliovirus type 1. However, MS2, another RNA phage, was not suitable as a surrogate for enteric viruses in determination of ozone disinfection efficiency (Finch and Fairbairn, 1991). It is thus doubtful that bacterial phages can be used as indicators for enteric viruses in all situations (Gerba, 1987a).

F-specific bacteriophages (male-specific phages) enter a host bacterial cell by adsorbing to the F or sex pilus of the cell. Since F-specific phages are infrequently detected in human fecal matter and show no direct relationship with fecal pollution level, they cannot be considered as indicators of fecal pollution (Havelaar et al., 1990; Morinigo et al., 1992b). However, their presence in high numbers in wastewaters and their relatively high resistance to chlorination contribute to their consideration as an index of wastewater contamination (Debartolomeis and Cabelli, 1991; Havelaar et al., 1990; Nasser et al., 1993; Yahya and Yanko, 1992). Monitoring of postchlorinated effluents after rainfalls showed the fecal coliforms and enterococci were much more sensitive to chlorine than male-specific bacteriophages (Rippey and Watkins, 1992). As regards shellfish contamination, male-specific bacteriophages

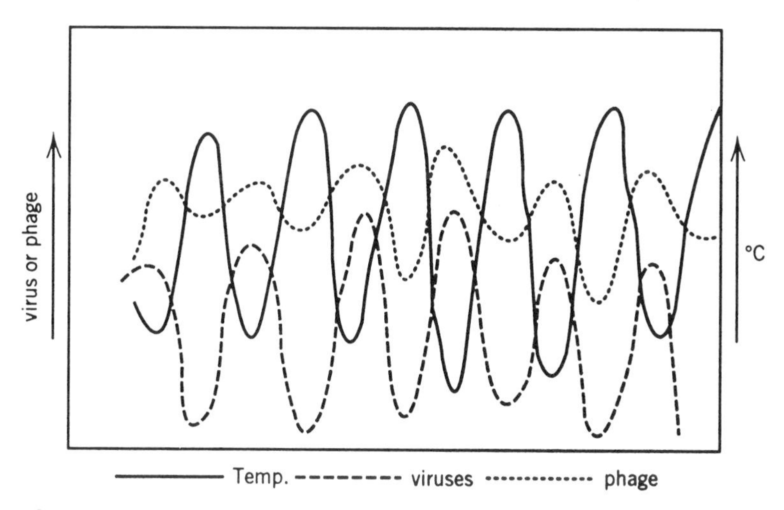

Fig. 5.1. Cyclical seasonal pattern of temperature and growth of viruses and coliphages. From Geldenhuys and Pretorius (1989), with permission of the publisher.

TABLE 5.1. Monitoring of Coliphages, Clostridia, and Viruses at Various Stages of the Pont-Viau, Canada, Water Filtration Plant[a]

Organisms	Type of water	Geometric mean	Percentage reduction	Percentage positive
Human enteric viruses (mpniu/100 L)	Raw[b]	79	na	91.0
	Settled[b]	0	>99	0.0
	Filtered[c]	0	>99.9	6.5
	Finished[c]	0	<99.9	0.0
Coliphages (PFU/100 L)	Raw[b]	565	na	100.0
	Settled[b]	3.1	99.953	50.0
	Filtered[c]	0.5	99.992	30.3
	Finished[c]	0.0	99.99997	0.6
Clostridia (CFU/100 L)	Raw[b]	11349	na	100.0
	Settled[b]	83.8	99.262	93.8
	Filtered[c]	1.2	99.989	51.5
	Finished[c]	0.0	99.9982	1.9

[a]Adapted from Payment (1991).
[b]100-L sample.
[c]1,000-L sample.

survive at least for 7 days in hard-shelled clams at ambient seawater temperatures and do not undergo replication with or without added host cells (Burkhardt et al., 1992).

The potential of bacteriophage of *Bacteroides* ssp. to serve as indicators of viral pollution was also explored (Tartera and Jofre, 1987). Phages active against *Bacteroides fragilis* HSP 40 were detected in feces (found in 10% of human fecal samples but not in animal feces), sewage, and other polluted aquatic environments (river water, seawater, groundwater, sediments) but were absent in nonpolluted sites (Cornax et al., 1990; Tartera and Jofre, 1987) (Table 5.2). These indicators do not appear to multiply in environmental samples (Tartera et al., 1989) and are more resistant to chlorine than are bacterial indicators (*S. faecalis, E. coli*) or viruses (poliovirus type 1, rotavirus SA11 and coliphage f2) (Fig. 5.2) (Bosch et al., 1989). However, they are less re-

TABLE 5.2. Levels of Bacteriophages Active Against *B. fragilis* HSP 40 in Water and Sediments[a]

Samples	No. of samples	% Samples positive for phages	Maximum value/100 ml	Minimum value/100 ml	Mean value/100 ml
Sewage	33	100	1.1×10^5	7	6.2×10^3
River water[b]	22	100	1.1×10^5	93	1.6×10^4
River sediment[b]	5	100	4.6×10^5	90	1.08×10^5
Seawater[b]	22	77.2	1.1×10^3	<3	1.2×10^2
Marine sediment[b]	12	91.0	43	<3	13.4
Groundwater[b]	19	21.0	NK[d]	0	
Nonpolluted[c]	50	0			

[a]Adapted from Tartera and Jofre (1987).
[b]Samples from areas with sewage pollution.
[c]Water and sediments from areas without known sewage pollution.
[d]NK, not known.

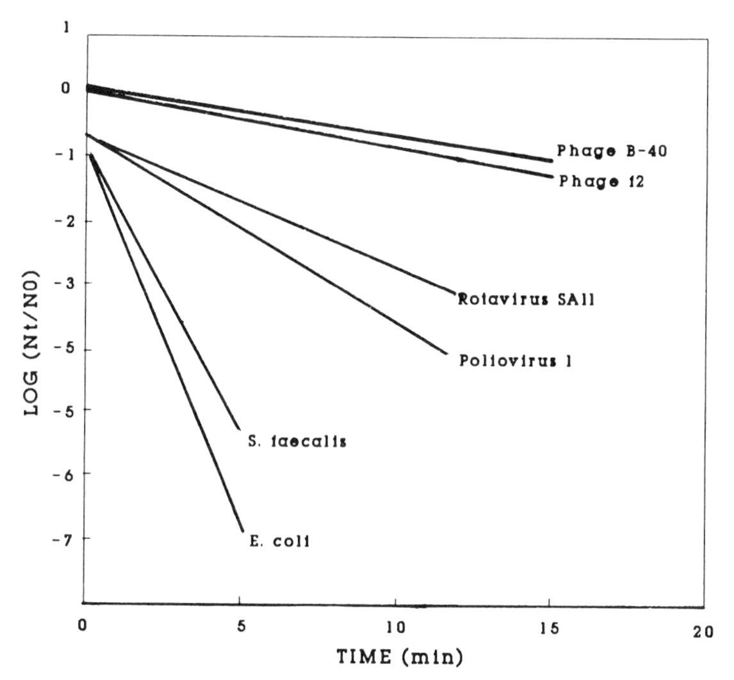

Fig. 5.2. Regression line of the inactivation of bacteriophages active against *Bacteroides fragilis* (B-40) and other microorganisms by chlorine. Adapted from Bosch et al. (1989).

sistant than coliphage f2 to UV irradiation (Bosch et al., 1989). Thus, these organisms may be suitable indicators of human fecal pollution, and their use makes possible the distinction between human and animal fecal pollution. They display a positive correlation with enteroviruses and rotaviruses (Jofre et al., 1989) and their persistence is similar to that of hepatitis A virus in seawater (Chung and Sobsey, 1993). Their suitability as surrogates for viral pollution has not been fully demonstrated.

5.2.6. Yeasts and Acid-Fast Organisms

Some investigators have proposed yeasts and acid-fast mycobacteria (*Mycobacterium fortuitum* and *M. phlei*) as indicators of disinfection efficiency (Grabow et al., 1980; Haas et al., 1985). The acid-fast bacterium *Mycobacterium fortuitum* is more resistant to free chlorine and ozone than is E. coli or poliovirus type 1 (Engelbrecht and Greening, 1978; Engelbrecht et al., 1974; Farooq and Akhlaque, 1983).

5.2.7. Heterotrophic Plate Count (HPC)

HPC represents the aerobic and facultative anaerobic bacteria that derive their carbon and energy from organic compounds. The number of recovered bacteria depends on medium composition, period of incubation (1–7 days), and temperature of incubation (20–35°C) (Reasoner, 1990). This group includes gram-negative bacteria belonging to the following genera: *Pseudomonas, Aeromonas, Klebsiella, Flavobacterium, Enterobacter, Citrobacter, Serratia, Acinetobacter, Proteus, Alcaligenes, Enterobacter, Moraxella.* The heterotrophic plate count microorganisms found in chlorinated distribution water are shown in Table 5.3 (LeChevallier et al., 1980). Some members of this group are opportunistic pathogens (e.g., *Aeromonas, Flavobacterium* (see Chapter 4), but little is known about the effects of high numbers of HPC bacteria on human health. In drinking water, the number of HPC bacteria may vary from less than 1 CFU/ml to more than 10^4 CFU/ml, and they are influenced mainly by tem-

TABLE 5.3. Identification of HPC Bacteria in Distribution Water and Raw Water[a]

	Distribution water		Raw water	
Organism	Total[a]	% of total	Total	% of total
Actinomycete	37	10.7	0	0
Arthrobacter spp.	8	2.3	2	1.3
Bacillus spp.	17	4.9	1	0.6
Corynebacterium spp.	31	8.9	3	1.9
Micrococcus luteus	12	3.5	5	3.2
Staphylococcus aureus	2	0.6	0	0
S. epidermidis	18	5.2	8	5.1
Acinetobacter spp.	19	5.5	17	10.8
Alcaligenes spp.	13	3.7	1	0.6
F. meningosepticum	7	2.0	0	0
Group IVe	4	1.2	0	0
Group M5	9	2.6	2	1.3
Group M4	8	2.3	2	1.3
Moraxella spp.	1	0.3	1	0.6
Pseudomonas alcaligenes	24	6.9	4	2.5
P. cepacia	4	1.2	0	0
P. fluorescens	2	0.6	0	0
P. mallei	5	1.4	0	0
P. maltophilia	4	1.2	9	5.7
Pseudomonas spp.	10	2.9	0	0
Aeromonas spp.	33	9.5	25	15.9
Citrobacter freundii	6	1.7	8	5.1
Enterobacter agglomerans	4	1.2	18	11.5
Escherichia coli	1	0.3	0	0
Yersinia enterocolitica	3	0.9	10	6.4
Group IIK biotype I	0	0	1	0.6
Hafnia alvei	0	0	9	5.7
Enterobacter aerogenes	0	0	1	0.6
Enterobacter cloacae	0	0	1	0.6
Klebsiella pneumoniae	0	0	0	0
Serratia liquefaciens	0	0	1	0.6
Unidentified	65	18.7	28	17.8
Total	347	100.0	157	99.7

[a]Adapted from LeChevallier et al. (1980).

perature, presence of a chlorine residual, and level of assimilable organic matter. HPC level should not exceed 500 organisms per ml (LeChevallier et al., 1980). HPC was found to be the most sensitive indicator for the removal and inactivation of microbial pathogens in reclaimed wastewater.

HPC is useful to water treatment plant operators with regard to the following (AWWA, 1987; Grabow, 1990; Reasoner, 1990):

1. Assessing the efficiency of various treatment processes, including disinfection, in a water treatment plant
2. Monitoring the bacteriological quality of the finished water during storage and distribution
3. Determining bacterial growth on surfaces of materials used in treatment and distribution systems

4. Determining the potential for regrowth or aftergrowth in treated water in distribution systems.

5.2.8. Chemical Indicators of Water Quality

Fecal sterols (coprostanol, coprosterol, cholesterol, and coprostanone). Some investigators have reported a correlation between fecal sterols and fecal contamination. However, fecal sterols may be degraded following water and wastewater treatment operations and may not be affected by chlorination.

Free chlorine residual. Free chlorine residual is a good indicator for drinking-water quality.

Levels of endotoxins. Endotoxins are lipopolysaccharides present in the outer membrane of gram-negative bacteria. Their concentration in environmental samples is conveniently measured by the *Limulus* amoebocyte lysate (LAL) assay. This test is based on the reaction of white blood cells of the horseshoe crab with endotoxins. This reaction leads to an increase in sample turbidity that is conveniently measured with a spectrophotometer. A number of studies have been undertaken to establish a relationship of endotoxin levels in wastewater and drinking water with the levels of total and fecal coliform bacteria (Evans et al., 1978; Haas et al., 1983; Jorgensen et al., 1979). Although a statistically significant association was noted, endotoxin level was not recommended as a surrogate indicator (Haas et al., 1983).

A panel on measurement of the microbial quality of drinking water made the specific recommendations shown in Table 5.4 (Olivieri, 1983).

5.3. DETECTION OF SOME INDICATOR MICROORGANISMS

Several methods are available for the detection of indicator microorganisms in environmental samples, including wastewater (Ericksen and Dufour, 1986; Seidler and Evans, 1983). Some of the procedures have been standardized and are routinely used by government and private laboratories (APHA, 1989). In this section we will focus on the detection of total and fecal coliforms and bacteriophages, and determination of heterotrophic plate count, stressing only some of the methodological advances made in the past few years.

5.3.1. Detection of Total and of Fecal Coliforms

As noted above, the total coliform group includes all the aerobic and facultative anaerobic, gram-negative, non-spore-forming, rod-shaped bacteria that ferment lactose with gas production within 48 hr at 35°C. Total coliforms are detected by the most probable numbers (MPN) technique or by the membrane filtration method. These procedures are described in detail in *Standard Methods for the Examination of Water and*

TABLE 5.4. Recommended Surrogate Measurements for Water Treatment Systems[a]

Class of test	Source water	Treatment train	Distribution system
Routine	1. Chlorine demand	1. Free chlorine residual	1. Free chlorine residual
	2. Turbidity	2. Turbidity	2. Turbidity
	3. Fecal indicator	3. Total coliform	3. Total coliform
		4. Plate count	4. Plate count
Periodic	—	1. Enterococci	1. Enterococci
		2. *C. perfringens*	2. *C. perfringens*
Diagnostic	Sanitary survey	1. Sanitary survey	1. Sanitary survey
		2. Microbial identification	2. Microbial identification

[a]From Olivieri (1983), with permission of the publisher.

Wastewater (APHA, 1989). MPN generally overestimates coliform numbers in tested samples; the overestimation depends on the number of total coliforms present in the water sample and on the number of tubes per dilution.

Fecal coliforms are defined as those bacteria that produce gas when grown in EC broth at 44.5°C or produce blue colonies when grown in m-FC agar at 44.5°C. A 7-hr test is also available for detecting this indicator group (Reasoner et al., 1979). A recent evaluation of this test showed more than 90% agreement between this method and the traditional MPN five-tube test (Barnes et al., 1989). The 7-hr test would be useful in emergency situations.

Several factors influence the recovery of coliforms, among them the type of growth medium, the diluent and membrane filter used, the presence of noncoliforms, and the sample turbidity (one should use the MPN approach when turbidity exceeds 5 nephelometric turbidity units [NTU]). Heterotrophic plate count bacteria are also able to reduce the number of coliform bacteria, presumably by competing successfully for limiting organic carbon (McFeters et al., 1982; LeChevallier and McFeters, 1985a). Another important factor affecting the detection of coliforms in water and wastewater is the occurrence of injured bacteria in environmental samples. Injury is due to physical (e.g., temperature; light), chemical (e.g., toxic metals and organic toxicants; chlorination), and biological factors. These debilitated bacteria do not grow well in the selective detection media used (presence of selective ingredients such as bile salts and deoxycholate) under temperatures much higher than those encountered in the environment (Bissonnette et al., 1975; 1977; Domek et al., 1984; McFeters et al., 1982; Zaske et al., 1980). In gram-negative bacteria, injury causes damage to the outer membrane, which becomes more permeable to selective ingredients such as deoxycholate. The injured cells can, however, undergo repair when grown on a nonselective nutrient medium. The low recovery of injured coliforms in environmental samples may underestimate the presence of fecal pathogens in the samples. Copper- and chlorine-induced injuries have been studied under *in vitro* and *in vivo* conditions to learn more about the pathogenicity of injured pathogens (LeChevallier et al., 1985; Singh et al., 1985; Singh et al., 1986a). Sublethally injured pathogens display a temporary reduction in virulence, but under suitable *in vivo* conditions they may regain their pathogenicity. Copper- and chlorine-stressed cells retain their full pathogenic potential even after exposure to the low pH of the stomach of orally infected mice (Singh and McFeters, 1987).

A growth medium, m-T7 agar, was proposed for the recovery of injured microorganisms (LeChevallier et al., 1983). The recovery of fecal coliforms on m-T7 agar is greatly improved when the samples are preincubated at 37°C for 8 hr. Recovery of fecal coliforms from wastewater samples was 3 times higher than with the standard m-FC method (LeChevallier et al., 1984b). Chlorine-stressed *Escherichia coli* displays reduced catalase, leading to its inhibition from the accumulated hydrogen peroxide. Thus, improved detection of chlorine-stressed coliform bacteria can also be accomplished by incorporating catalase and/or pyruvate into the growth medium to block and degrade hydrogen peroxide (Calabrese and Bissonnette, 1989, 1990; McDonald et al., 1983).

5.3.2. Rapid Methods for Coliform Detection

5.3.2.1. *Enzymatic Assays*

Enzymatic assays constitute an alternative approach for detecting indicator bacteria, namely total coliforms and *E. coli,* in water and wastewater. These assays are specific, sensitive, and rapid. In most tests, the detection of total coliforms consists of observing β-galactosidase activity, which is based on the hydrolysis of the substrate *o*-nitrophenyl-β-D-galactopyranoside (ONPG) to yellow nitrophenol, which absorbs light at 420 nm. A fluorogenic compound, 4-methylumbelliferone-β-D-galactoside, can also be used as a substrate for β-galactosidase (Berg and Fiksdal, 1988). The detection of total coliforms by β-galactosidase assay can be improved by incorporating isopropyl-β-

D-thiogalactopyranoside (IPTG), a gratuitous inducer of β-galactosidase production, in the growth medium (Diehl, 1991).

Rapid assays for detection of *E. coli* are based on the hydrolysis of a fluorogenic substrate, 4-methylumbelliferone glucuronide (MUG), by β-glucuronidase, an enzyme found in *E. coli*. The end product is fluorescent and can be easily detected with a long-wave UV lamp. These tests have been used for the detection of *E. coli* in clinical and environmental samples (Berg and Fiksdal, 1988; Trepeta and Edberg, 1984). β-Glucuronidase is an intracellular enzyme that is found in *E. coli* as well as some *Shigella* species (Feng and Hartman, 1982). A most-probable-number fluorogenic assay based on β-glucuronidase activity has been used for the detection of *E. coli* in water and food samples (Feng and Hartman, 1982; Robison, 1984). The assay consists of incubating the sample in lauryl-tryptose broth amended with 100 mg/L MUG, and observing the development of fluorescence within 24 hr incubation at 35°C. This assay can be adapted to membrane filters, since β-glucuronidase-positive colonies are fluorescent or have a fluorescent halo when examined under a long-wave UV light. This test can detect the presence of one viable *E. coli* cell within 24 hr. A similar miniaturized fluorogenic assay, using MUG as the substrate, was considered for the determination of *E. coli* numbers in marine samples. This assay displayed a 87.3% confirmation rate (Hernandez et al., 1990; 1991).

A commercial test, the Autoanalysis Colilert (AC) test, also called the minimal media ONPG-MUG (MMO-MUG), was recently developed to enumerate simultaneously in 24 hr both total coliforms and *E. coli* in environmental samples (Covert et al., 1989; Edberg et al., 1988, 1989, 1990). The test is performed by adding the sample to tubes that contain powdered ingredients consisting mainly of salts and specific enzyme substrates, which also serve as the only carbon source for the target microorganisms. The enzyme substrates are ONPG for detecting total coliforms and MUG for specifically detecting *E. coli*. Thus, according to the manufacturer, ONPG and MUG serve as enzyme substrates as well as food sources for the microorganisms. After 24-hr incubation, samples positive for total coliforms turn yellow, whereas *E. coli*-positive samples fluoresce under long-wave UV illumination in the dark. It appears that *Escherichia* species other than *E. coli* are not detected by the Colilert test (Rice et al., 1991). Examination of human and animal (cow, horses) fecal samples revealed that 95% of *E. coli* isolates were β-glucuronidase-positive after 24-hr incubation (Rice et al., 1990). Several surveys concerning coliform detection in drinking water have shown that the AC test had a sensitivity similar to that of the standard multiple-tube fermentation method, and the membrane filtration method for drinking water (Edberg et al., 1988; Katamay, 1990). This test yielded numbers of chlorine-stressed *E. coli* in wastewater similar to or higher than those of the U.S. EPA-approved EC-MUG test (Covert et al., 1992; McCarty et al., 1992). A new version of Colilert, Colilert-MW, was developed to detect total coliforms and *E. coli* in marine waters (Palmer et al., 1993).

It has been found, however, that some *E. coli* strains are nonfluorogenic; one third of *E. coli* isolates from fecal samples from human volunteers were found to be nonfluorogenic (Chang et al., 1989). Furthermore, a certain percentage of *E. coli* isolates producing virulence factors (e.g., enterotoxigenic or enterohemorrhagic *E. coli*) are not recovered on MMO-MUG medium (Martins et al., 1992). Thus, some investigators do not recommend the implementation of the AC test as a routine procedure for *E. coli* in environmental samples (Lewis and Mak, 1989). Furthermore, the AC test disagreed with the standard membrane filtration fecal coliform test for treated but not for untreated water samples. This was due to the presence of false-negative results obtained by the AC test (Clark et al., 1991).

A MUG-based solid medium was proposed for the detection of *E. coli* after only a 7.5-hr incubation. Testing this method in water gave a specifity of 96.3% (Sarhan and Foster, 1991). Another chromogenic substrate, 5-bromo-4-chloro-3-indoxy-β-glucuronide (BCIG), is also potentially useful for the rapid and specific identification of *E. coli* on solid media. *E. coli* colonies turn blue following incubation of samples for 22–24 hr at 44.5°C. Ninety-nine percent of the BCIG-positive blue colonies were also MUG positive (Watkins et al., 1988). As to the detection of total and fecal coliforms by enzyme-based techniques, only two methods have been approved, as of 1991, by the U.S. EPA for detection of fecal coliforms: EC-MUG and nutrient agar plus MUG. The MMO-MUG was approved only for detecting total coliforms but it can be used in combination with EC-MUG for *E. coli* detection (Pontius, 1992).

Similar enzymatic methods have been developed for the detection of fecal streptococci. These indicators can be detected by incorporating a fluorogenic substrate, 4-methylumbelliferone-β-D-glucoside (MUD), into a selective medium. Miniaturized tests, using microtitration plates and MUD, were successful in the selective detection of this group in fecal, freshwater, wastewater, and marine samples (Hernandez et al., 1990, 1993; Pourcher et al., 1991). The enterococci group can be rapidly detected by a fluorogenic-chromogenic enzymatic assay. This test is based on the detection of the activity of two specific enzymes, pyroglutamyl aminopeptidase and β-D-glucosidase (Manafi and Sommer, 1993).

5.3.2.2. *Other Methods for Coliform Detection*

Monoclonal antibodies. E. coli can be detected by using monoclonal antibodies directed against outer membrane proteins (e.g., OmpF protein) or against alkaline phosphatase, an enzyme localized in the cell periplasmic space (Joret et al., 1989). Although some monoclonal antibodies are specific for *E. coli* and *Shigella,* some have questioned their specifity and affinity, and further research is needed to demonstrate the application of this tool to routine *E. coli* detection in field samples (Kfir et al., 1993).

Polymerase-chain-reaction/gene-probe detection method. In this method, specific genes (e.g., *LacZ, lamB* genes) in *Escherichia coli* are amplified by the polymerase chain reaction (PCR) (see Chapter 16 for further details) and subsequently detected with a gene probe. With this method one can detect 1–5 cells of *E. coli* per 100 ml of water (Atlas et al., 1989; Bej et al., 1990). Another genetic probe involves the *uid* genes, which code for β-glucuronidase in *E. coli* and *Shigella* species. This probe, when combined with PCR, can detect as low as one or two cells but cannot distinguish between *E. coli* and *Shigella.* (Bej et al., 1991a; Cleuziat and Robert-Baudouy, 1990). The PCR/gene probe method appears to be more sensitive than the AC method in the detection of *E. coli* in environmental samples (Bej et al., 1991c). This is probably due to the presence of approximately 15% of β-glucuronidase-negative strains in environmental samples.

5.3.3. Heterotrophic Plate Count (HPC)

The heterotrophic plate count in water and wastewater is defined as the total number of bacteria that can grow following incubation of the sample on plate count agar at 35°C for 48 hr. These bacteria may interfere with the detection of coliforms in water samples. HPC is greatly influenced by the temperature and length of incubation, the growth medium, and the plating method (pour plate versus spread plate). A growth medium, designated R2A, was recently developed for use in heterotrophic plate counts. This medium is recommended for use with an incubation period of 5–7 days at 28°C (Reasoner and Geldreich, 1985).

HPC should not exceed 500 colonies per 1 ml. Numbers above this limit generally signal a deterioration of water quality in distribution systems.

An original approach is the use of recombinant *lux*+ phages for the detection of indicator bacteria within 1–5 hr. Enteric bacteria become bioluminescent following infection with recombinant phage and can be measured with a bioluminometer. This procedure awaits further development (Kodikara et al., 1991).

5.3.4. Bacteriophages

Domestic wastewater harbors a wide range of phage strains that can be detected with a variety of host bacteria. Their levels in raw wastewater are in the range of 10^5 to 10^7 phage particles per liter, but decrease significantly following waste treatment operations (Bitton, 1987). The detection of phage in water and wastewater effluents comprises the following steps (Goyal, 1987):

Phage concentration. Phage can be concentrated from large volumes of water by adsorption to negatively or positively charged membrane filters. This step is followed by elution of adsorbed phage from the membrane surface with glycine at high pH (pH = 11.5), beef extract at pH = 9.0, or casein at pH = 9. If necessary, a reconcentration step is included to obtain a low-volume concentrate (Goyal and Gerba, 1983; Goyal et al., 1980; Logan et al., 1980). Phage can also be concentrated from 2- to 4-L samples by the magnetite–organic flocculation (Bitton et al., 1981a). The sample, after addition of casein and magnetite, is flocculated at pH = 4.5–4.6. The flocs, with the trapped viruses, are pulled down with a magnet, resolubilized, and assayed for phage.

Decontamination of concentrate. Indigenous bacteria that interfere with the bacterial phage assay must be inactivated or removed from the concentrate by chloroform extraction, membrane filtration, addition of antibiotics, or use of selective media (e.g., nutrient broth modified with sodium dodecyl sulfate) (Goyal, 1987; Kennedy et al., 1985). Treatment of the concentrate with hydrogen peroxide followed by plating on a medium supplemented with crystal violet is also useful for the inactivation of interfering bacteria (Ashgari et al., 1992).

Phage assay. Concentrates are assayed for phage by the double-layer-agar method (Adams, 1959) or the single-layer-agar procedure (Grabow and Coubrough, 1986; Havelaar and Hogeboom, 1983). The phage numbers may also be obtained by a most-probable-number procedure (Kott, 1966). The assay of male-specific phages requires the use of specific host cells such as *Salmonella typhimurum* strain WG49 or *Escherichia coli* strain HS[*pFamp*]R, but it may be complicated by the growth of somatic phages on the host cells. Somatic phages can be suppressed by treating the sample with lipopolysaccharides from the host cells (Handzel et al., 1993).

We have reviewed the characteristics and detection methodology for the traditional and less traditional microbial indicators used for assessing contamination of aquatic and other environments by pathogenic microorganisms. Most, if not all, of these indicators are not ideal, because some are more sensitive (e.g., vegetative bacterial indicators) or more resistant (e.g., bacterial spores) than viruses or protozoan parasites to environmental stresses and disinfection. There is probably no universal ideal indicator microorganism that is suitable for various environments under variable conditions. A search for the elusive ideal indicator is still going on in many laboratories the world over. The advent of gene probes and polymerase-chain-reaction technology has given hope for the development of rapid and simple methods for detecting small numbers of bacterial or viral pathogens and protozoan parasites in wastewater, wastewater effluents, sludges, food, drinking water, and other environmental samples. Furthermore, multiplex PCR can be used to detect a wide range of pathogenic microorganisms and parasites in the same sample (see Chapter 16). The road is open to direct, rapid, and possibly inexpensive methods for detecting pathogens and parasites in wastewater and other environmental samples.

5.4. FURTHER READING

Berg, G., Ed. 1978. *Indicators of Viruses in Water and Food.* Ann Arbor Science Publishers, Ann Arbor, MI.

Ericksen, T.H., and A.P. Dufour. 1986. Methods to identify water pathogens and indicator organisms, pp. 195–214, in: *Waterborne Diseases in the United States,* G.F. Craun, Ed. CRC Press, Boca Raton, FL.

Gerba, C.P. 1987a. Phage as indicators of fecal pollution, pp. 197–209, in: *Phage Ecology,* S.M. Goyal, C.P. Gerba, and G. Bitton, Eds. Wiley-Interscience, New York.

Goyal, S.M., C.P. Gerba, and G. Bitton, Eds. 1987. *Phage Ecology.* Wiley-Interscience, New York, 321 pp.

6

WATER AND WASTEWATER DISINFECTION

6.1. INTRODUCTION

Disinfection is the destruction of microorganisms capable of causing diseases. Disinfection is an essential and final barrier against human exposure to disease-causing pathogenic microorganisms, including viruses, bacteria, and protozoan parasites. Chlorination was initiated at the beginning of this century to provide an additional safeguard against pathogenic microorganisms. The destruction of pathogens and parasites by disinfection helps considerably in the reduction of waterborne and food-borne diseases. However, in recent years, the finding that chlorination can lead to the formation of by-products that can be toxic or genotoxic to humans and animals has led to a quest for safer disinfectants. It was also realized that some pathogens or parasites are indeed quite resistant to disinfectants and that the traditional indicator microorganisms are sometimes not suitable for ensuring safe water.

In addition to their use for pathogen and parasite destruction, some of the disinfectants (e.g., ozone, chlorine dioxide) are also employed for oxidation of organic matter, iron, and manganese and for controlling taste and odor problems and algal growth.

This chapter deals with the disinfectants most used by the water and wastewater treatment industries.

6.2. FACTORS INFLUENCING DISINFECTION

Several factors control the disinfection of water and wastewater (Lippy, 1986; Sobsey, 1989).

6.2.1. Type of Disinfectant

Disinfection efficacy depends on the type of chemical used. Some disinfectants (e.g., ozone, chlorine dioxide) are stronger oxidants than others (e.g., chlorine).

6.2.2. Type of Microorganism

There is a wide variation in the resistance of various microbial pathogens to disinfectants. Spore-forming bacteria are generally more resistant to disinfectants than vegetative bacteria. Resistance to disinfectants varies also among vegetative bacteria and among strains belonging to the same species (Ward et al., 1984). For example, *Legionella pneumophila* is much more resistant to chlorine than is *E. coli* (Kuchta et al., 1983; Muraca et al., 1987). In general, resistance to disinfection goes along the following order: vegetative bacteria < enteric viruses < spore-forming bacteria < protozoan cysts.

6.2.3. Disinfectant Concentration and Contact Time

Inactivation of pathogens with disinfectants increased with time and ideally should follow first-order kinetics. Inactivation versus time follows a straight line when data are plotted on log-log paper.

$$N_t/N_0 = e^{-kt} \tag{6.1}$$

where N_0 = number of microorganisms at time 0; N_t = number of microorganisms at time t; k = decay constant (time $^{-1}$); and t = time.

However, field inactivation data actually show a deviation from first-order kinetics (Fig. 6.1) (Hoff and Akin, 1986). Curve C in Figure 6.1 shows deviation from first-order kinetics. The tailing off of the curve results from the survival of a resistant subpopulation within a heterogenous population or from protection of the pathogens by interfering factors (see below). Microbial clumping explains the "shoulder" of survival curves obtained when microorganisms are exposed to chlorine action (Rubin et al., 1983).

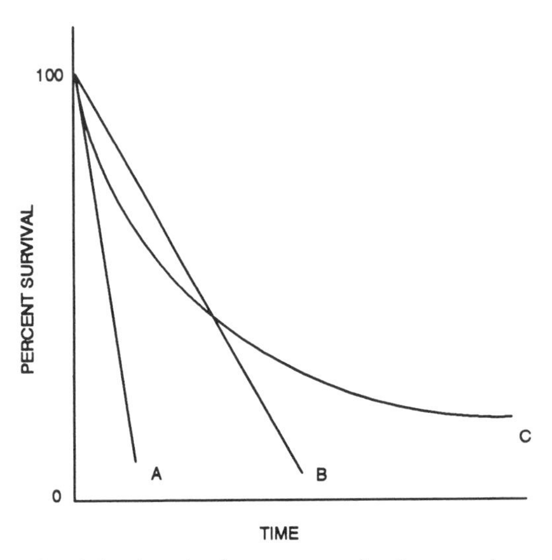

Fig. 6.1. Inactivation curves of microorganisms following disinfection. **A.** Sensitive homogeneous population. **B.** More resistant homogeneous population. **C.** Heterogeneous population or one partially protected by aggregation. From Hoff and Akin (1986), with permission of the publisher.

Disinfectant effectiveness can be expressed as $C \cdot t$, C being the disinfectant concentration and t the time required to inactivate a certain percentage of the population under specific conditions (pH and temperature). The relationship between disinfectant concentration and contact time is given by the Watson's law (Clark et al., 1989):

$$K = C^n t \tag{6.2}$$

where K = constant for a given microorganism exposed to a disinfectant under specific conditions; C = disinfectant concentration (mg/L); t = time required to kill a certain percentage of the population (min); and n = a constant called the "coefficient of dilution."

When t is plotted against C on a double-logarithmic paper, n is the slope of the straight line (see Fig. 6.2) (Clark et al., 1989). The value of n determines the importance of the disinfectant concentration or contact time in microorganism inactivation. If $n < 1$, disinfection is more affected by contact time than by disinfectant concentration. If $n > 1$, the disinfectant level is the predominant factor controlling disinfection (Rubin et al., 1983). However, n is often close to unity.

Determination of Ct values can also take into account the temperature and pH of the suspending medium. For example, an equation was developed to predict *Giardia lamblia* inactivation of cysts following chlorine treatment (Clark et al., 1989; Hibler et al., 1987).

$$C \cdot t = 0.9847\ C^{0.1758}\ pH^{2.7519}\ T^{-0.1467} \tag{6.3}$$

where C = chlorine concentration ($C < 4.23$ mg/L); t = time to inactivate 99.99% of the cysts; pH = pH (range is between 6 and 8); and T = temperature (range is between 0.5 and 5.0°C).

Ct values for a range of pathogenic microorganisms are shown in Table 6.1 (Hoff, 1986; Hoff and Akin, 1986). The order of resistance to chlorine is the following: protozoan cysts > viruses > vegetative bacteria.

Another way to express the efficiency of a given disinfectant is the lethality coefficient,

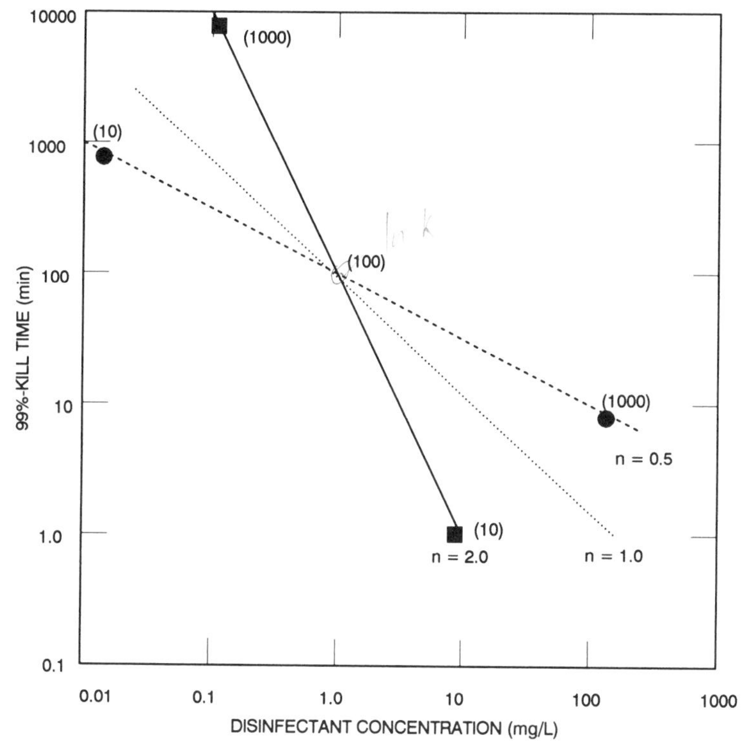

Fig. 6.2. Effect of *n* value on Ct at various disinfectant concentrations (Ct values given in parentheses). From Clark et al. (1989), with permission of the publisher.

given by the following equation (Morris, 1975):

$$\lambda = 4.6/Ct_{99} \tag{6.4}$$

where 4.6 = natural log of 100; C = residual concentration of disinfectant (mg/L); and t_{99} = contact time (min) for 99% inactivation of microorganisms.

The values of λ for destroying 99% of a range of microorganisms by ozone in 10 min at 10–15°C vary from 5 for *Entamoeba histolytica* to 500 for *E. coli* (Chang, 1982).

6.2.4. Effect of pH

As regards disinfection with chlorine, pH controls the amount of HOCl (hypochlorous acid)

TABLE 6.1. Microbial Inactivation by Chlorine: Ct Values (Temperature = 5°C; pH = 6.0)[a]

Microorganism	Chlorine concn., mg/L	Inactivation time, min	Ct
E. coli	0.1	0.4	0.04
Poliovirus 1	1.0	1.7	1.7
E. histolytica cysts	5.0	18	90
G. lamblia cysts	1.0	50	50
	2.0	40	80
	2.5	100	250
G. muris cysts	2.5	100	250

[a]Adapted from Hoff and Akin (1986).

and OCl^- (hypochorite) in solution (see Section 6.3.1.). HOCl is 80 times more effective than OCl^- for *E. coli*. For disinfection with chlorine, Ct increases with pH (Lippy, 1986). Conversely, bacterial, viral, and protozoan cyst inactivation by chlorine dioxide is generally more efficient at higher pH values (Berman and Hoff, 1984; Chen et al., 1985; Sobsey, 1989). The effect of pH on microbial inactivation by chloramine is not well established owing to conflicting results. The effect of pH on pathogen inactivation by ozone is also not well established.

6.2.5. Temperature

Pathogen and parasite inactivation increases (i.e., Ct decreases) as temperature increases.

6.2.6. Chemical and Physical Interference with Disinfection

Chemical compounds that interfere with disinfection are inorganic and organic nitrogenous compounds, iron, manganese, and hydrogen sulfide. Dissolved organic compounds also exert a chlorine demand and their presence results in reduced disinfection efficiency (see section on chloramination).

Turbidity in water is composed of inorganic (e.g., silt, clay, iron oxides) and organic matter as well as microbial cells. It is measured by determining light scattering by particulates present in the water. It interferes with the detection of coliforms in water (Geldreich et al., 1978; LeChevallier et al., 1981) and it may also reduce the disinfection efficiency of chlorine and other disinfectants. A turbidity of one nephelometric unit (NTU) is allowed in drinking water. Turbidity needs to be removed because particle-associated microorganisms are more resistant to disinfection than freely suspended microorganisms. Total organic carbon (TOC) associated with turbidity exerts a chlorine demand and thus interferes with the maintenance of a chlorine residual in water. Microorganisms associated with fecal material, cell debris, and wastewater solids are also protected from disinfection (Hoff, 1978; Berman et al., 1988; Foster et al., 1980; Harakeh, 1985; Hejkal et al., 1979). These findings are particularly important for communities that treat their water solely by chlorination. Figure 6.3 shows the protective effect of turbidity on coliform bacteria (LeChevallier et al., 1981). It was also shown that the protective effect of particulates in water and wastewater depends on the nature and the size of the particles. Hence, cell-associated poliovirus is protected from chlorine inactivation, whereas bentonite and aluminum phosphate do not offer such protection to the virus (Hoff, 1978). Viruses and bacterial indicators are not protected from ozone inactivation by bentonite (Boyce et al., 1981). An examination of the protective effect of particulates in primary effluents showed that the >7.0-μm fraction was responsible for the protective effect (Berman et al., 1988). A study of solids-associated viruses under field conditions showed that they are more

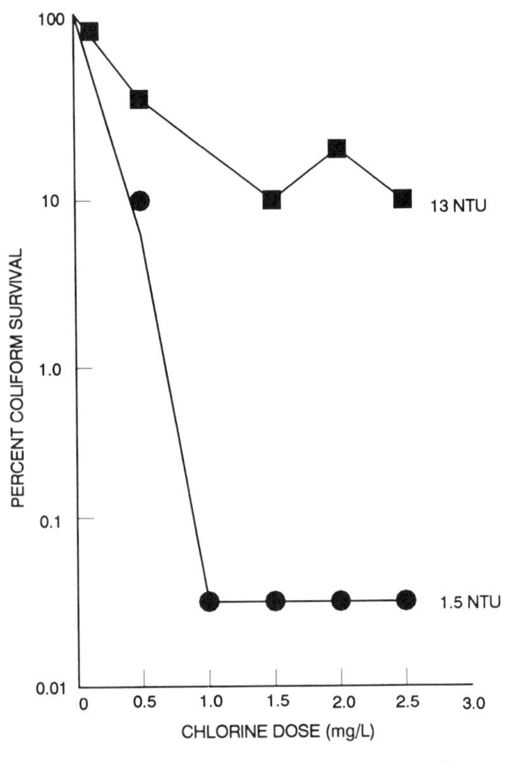

Fig. 6.3. Effect of turbidity on coliform persistence in chlorinated water. Adapted from LeChevallier et al. (1981).

resistant to chlorination than "free" virions. Reducing turbidity to less than 0.1 NTU could be a preventive measure for counteracting the protective effect of particulate matter during disinfection.

6.2.7. Protective Effect of Macroinvertebrates

Macroinvertebrates may enter and colonize water distribution systems (Levy et al., 1984; Small and Greaves, 1968). The public health significance of the presence of the animals in water distribution systems has been addressed. Nematodes may ingest viral and bacterial pathogens and thus protect these microorganisms from chlorine action (Chang et al., 1960). *Hyalella azreca,* an amphipod, protects *E. coli* and *Enterobacter cloacae* from chlorination. In the presence of 1 mg/L chlorine the decay rate of macroinvertebrate-associated *E. cloacae* was $k = 0.022\ hr^{-1}$, whereas the decay rate of unassociated *E. cloacae* was $k = 0.93\ hr^{-1}$ (Fig. 6.4.) (Levy et al., 1984). Enteropathogenic bacteria are also protected from chlorine action when ingested by protozoa (King et al., 1988).

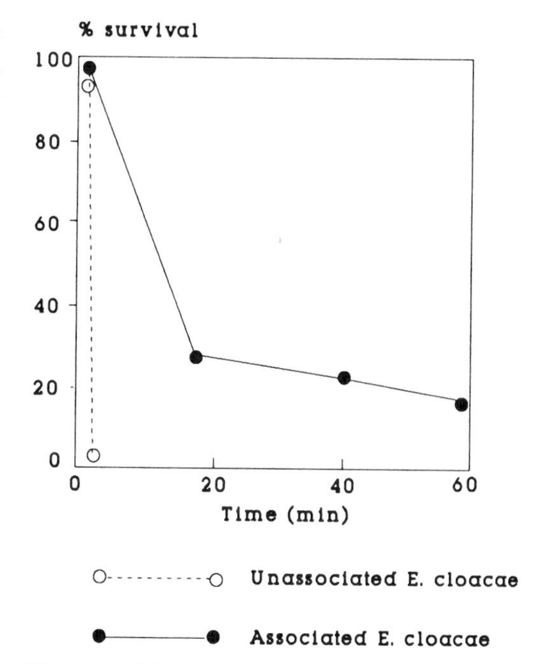

Fig. 6.4. Effect of chlorination on *Enterobacter cloacae* associated with a macroinvertebrate. Adapted from Levy et al. (1984).

6.2.8. Other Factors

Several studies have shown that laboratory-grown pathogenic and indicator bacteria are more sensitive to disinfectants than those occurring in the natural aquatic environment. Hence, a naturally occurring *Flavobacterium* sp. was 200 times more resistant to chlorine than when subcultured under laboratory conditions (Wolfe and Olson, 1985). *Klebsiella pneumoniae* was more resistant to chloramine when grown under low-nutrient conditions (Stewart and Olson, 1992a). The increased resistance to chloramine is due to several physiological factors, namely increased cell aggregation and extracellular polymer production, alteration of membrane lipids, and decreased oxidation of sulfhydryl groups (Stewart and Olson, 1992b). The greater resistance of "environmental" bacterial strains to nutrient limitation and to deleterious agents such as disinfectants may also be due to the synthesis of stress proteins, the role of which is not well understood (Matin and Harakeh, 1990). This phenomenon raises questions regarding the usefulness of laboratory disinfection data to predict pathogen inactivation under field conditions (Hoff and Akin, 1986).

Prior exposure may also increase microbial resistance to disinfectants. Repeated exposure of microorganisms to chlorine results in the selection of bacteria and viruses that are resistant to disinfection (Bates et al., 1977; Leyval et al., 1984; Ridgeway and Olson, 1982). Clumping or aggregation of pathogenic microorganisms generally reduces the disinfectant efficiency. Bacterial cells, viral particles, and protozoan cysts inside the aggregates are well protected from disinfectant action (Chen et al., 1985; Sharp et al., 1976).

6.3. CHLORINE

6.3.1. Chlorine Chemistry

Chlorine gas (Cl_2) introduced into water hydrolizes according to the following equation:

$$Cl_2 \text{ (chlorine gas)} + H_2O \leftrightarrow HOCl \text{ (hypochlorous acid)} + H^+ + Cl^- \qquad (6.5)$$

Hypochlorous acid dissociates in water according to the following:

$$HOCl \text{ (hypochlorous acid)} \leftrightarrow H^+ + OCl^- \text{ (hypochlorite ion)} \qquad (6.6)$$

Figure 6.5 shows that the proportion of HOCl and OCl^- depends on the pH of the water. Chlorine, as HOCl or OCl^-, is defined as *free available chlorine*. HOCl combines with ammonia and organic nitrogen compounds to form chloramines, which are combined available chlorine (see Section 6.4 for more details).

6.3.2. Inactivation of Microorganisms by Chlorine

Of the three chlorine compounds (HOCl, OCl^-, and NH_2Cl), hypochlorous acid is the most effective for the inactivation of microorganisms in water and wastewater. The presence of interfering substances in wastewater reduces the disinfection efficacy of chlorine, and relatively high concentrations of chlorine (20–40 ppm) are required for adequate reduction of viruses. In wastewater effluents, no free-chlorine species are present after some seconds of contact.

Chlorine, specifically HOCl, is generally quite efficient in inactivating pathogenic and indicator bacteria. Water treatment with 1 mg/L or less for about 30 min is generally efficient in significantly reducing numbers of bacteria. *Campylobacter jejuni* displays more than 99% inactivation in the presence of 0.1 mg/L free chlorine (contact time = 5 min) (Blaser et al., 1986). Although there is wide variation in the resistance of enteric viruses to chlorine, these pathogens are generally more resistant to this disinfectant than vegetative bacteria. That explains why viruses are frequently detected in secondarily treated effluents. Chloramines are much less efficient than free residual chlorine (about 50 times less efficient) at inactivation of viruses. Protozoan cysts (e.g., *Giardia lamblia, Entamoeba histolytica, Naegleria gruberi*) are more resistant to chlorine than both bacteria and viruses. In the presence of HOCl at pH = 6, the Ct for *E. coli* is 0.04 as compared to a Ct of 1.05 for poliovirus type 1 and a Ct of 80 for *G. lamblia* (Logsdon and Hoff, 1986).

Cryptosporidium is extremely resistant to disinfection. A chlorine or monochloramine concentration of 80 mg/L is necessary to cause 90% inactivation following 90-min contact time (Korich et al., 1990). This parasite is not completely inactivated in a 3% solution of sodium hypochlorite (Campbell et al., 1982) and the oocysts can remain viable for 3–4 months in 2.5% potassium dichromate solution (Current, 1988). This parasite would thus be extremely resistant to disinfection as carried out in water and wastewater treatment plants (Korich et al., 1989).

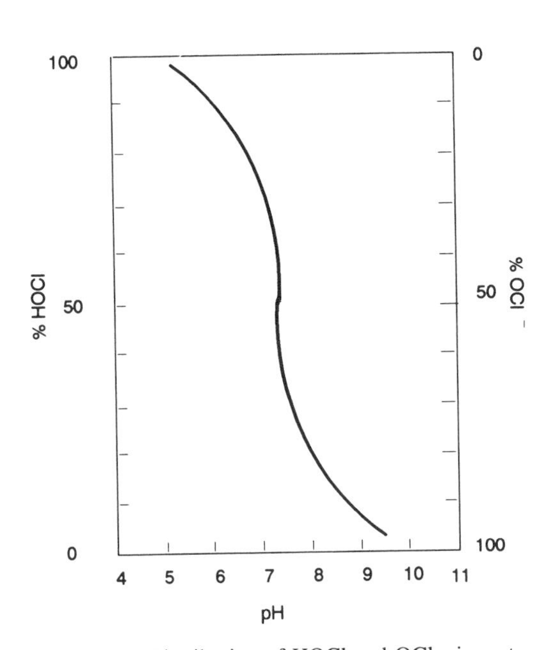

Fig. 6.5. Distribution of HOCl and OCl^- in water as a function of pH. From Bitton (1980a), with permission of the publisher.

6.3.3. Cell Injury by Chlorine

Physical (heat, freezing, sunlight) and chemical agents (chlorine, sublethal levels of heavy metals such as copper) can cause injury to bacterial cells (LeChevallier and McFeters, 1985b). Injury caused by environmental agents can lead to cell size reduction, damage to cell barriers, and altered cell physiology and virulence (Singh and McFeters, 1990).

Chlorine and copper appear to cause significant injury to coliform bacteria in drinking water (Camper and McFeters, 1979; Domek et al., 1984). The injured bacteria fail to grow in the presence of selective agents (e.g., sodium lauryl sulfate, sodium deoxycholate) traditionally incorporated in growth media designed for the isolation of indicator and pathogenic bacteria (Bissonnette et al., 1975; Busta, 1976). However, chlorine- and copper-injured pathogens (e.g., enterotoxigenic *E. coli*) retain their potential for enterotoxin production (Singh and McFeters, 1986) and are able to recover in the small intestine of animals, retaining their pathogenicity. This suggests that cells injured by chlorine treatment are still of potential health significance (Singh et al., 1986a). Injury by chlorine can affect a wide variety of pathogens, including enterotoxigenic *E. coli, Salmonella typhimurium, Yersinia enterocolitica,* and *Shigella spp.* The extent of injury by chlorine depends on the type of microorganism involved.

6.3.4. Potentiation of the Killing Action of Free Chlorine

The killing action of free chlorine can be potentiated by adding salts such as KCl, NaCl, or CsCl (Berg et al., 1990; Haas et al., 1986; Sharp et al., 1980). Following chlorination, viruses are more effectively inactivated in drinking water (e.g., Cincinnati drinking water) than in purified water (Fig. 6.6) (Berg et al., 1989). The mechanism of the potentiating effect of salts is not fully understood.

The disinfecting ability of chlorine can also be enhanced in the presence of heavy metals. The inactivation rate of pathogenic (e.g., *Legionella pneumophila*) bacteria and viruses (e.g., poliovirus) is increased when free chlorine

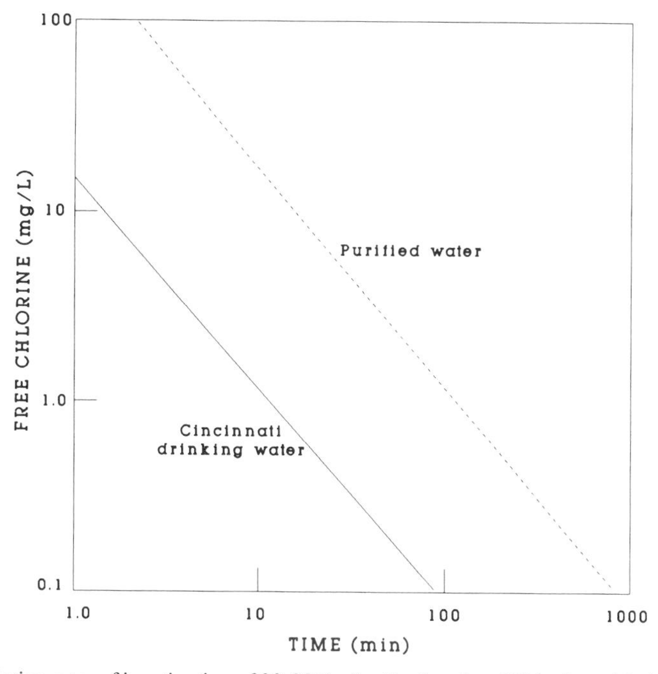

Fig. 6.6. Relative rates of inactivation of 99.99% of poliovirus 1 at 5°C by free chlorine at pH 9.0 in purified water and in Cincinnati drinking water. From Berg et al. (1989), with permission of the publisher.

is modified with electrolytically generated copper and silver (400 and 40 μg/L, respectively) (Yahya and Gerba, 1990; Landeen et al., 1989) (Fig. 6.7). This phenomenon was also demonstrated for indicator bacteria in swimming pool water (Yahya et al., 1989). This process does not, however, completely eliminate certain enteric viruses (e.g., hepatitis A virus) from water (Bosch et al., 1993).

6.3.5. Mechanism of Action of Chlorine

Chlorine causes two types of damage to bacterial cells.

Disruption of cell permeability. Free chlorine disrupts the integrity of the bacterial cell membrane, thus leading to loss of cell permeability and to the disruption of other cell functions (Berg et al., 1986; Haas and Engelbrecht, 1980; Venkobachar et al., 1977). Exposure to chlorine leads to a leakage of proteins, RNA, and DNA (Venkobachar et al., 1977). Cell death is the result of a release of TOC and UV-absorbing materials, a decrease in potassium uptake, and a reduction in protein and DNA synthesis (Haas and Engelbrecht, 1980). Permeability disruption has also been implicated as the cause of damage of chlorine to bacterial spores (Kulikovsky et al., 1975).

Damage to nucleic acids and enzymes. Chlorine also damages the bacterial nucleic acids (Hoyana et al., 1973; Shih and Lederberg, 1974) as well as enzymes (e.g., catalase). One of the consequences of reduced catalase activity is inhibition by the accumulated hydrogen peroxide (Calabrese and Bissonnette, 1990).

As to viruses, the mode of action of chlorine may depend on the type of virus. Nucleic acid damage is the primary mode of inactivation for bacterial phage f2 (Dennis et al., 1979; Olivieri et al., 1980) or poliovirus type 1 (O'Brien and Newman, 1979). The protein coat appears to be the target site for other types of viruses (e.g., rotaviruses) (Vaughn and Novotny, 1991).

6.3.6. Toxicology of Chlorine and Chlorine By-products

In general, the risks from chemicals in water are not as well defined as those from pathogenic microorganisms. This is due to the lack of data on disinfection by-products. The toxicology of chlorine and by-products is of obvious importance, since it is estimated that 79% of the U.S.

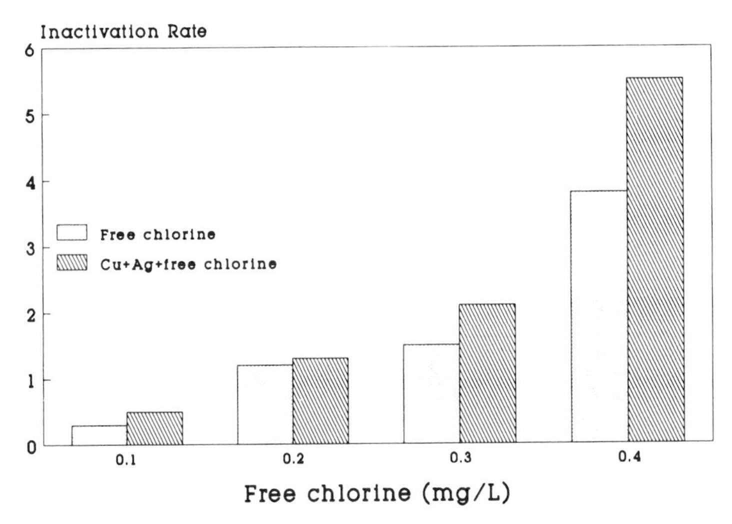

Fig. 6.7. Inactivation of *Legionella pneumophila* by exposure to electrolytically generated copper and silver and/or various concentrations of free chlorine. From Landeen et al. (1989), with permission of the publisher.

population is exposed to chlorine (U.S. EPA, 1989). There is evidence of an association between chlorination of drinking water and increased risk of colon and bladder cancers. This association is stronger for consumers who have been exposed to chlorinated water for more than 15 years (Craun, 1988; Jolley et al., 1985; Larson, 1989). Trihalomethanes (THM) such as chloroform, dichloromethane, bromodichloromethane, dibromochloromethane, bromoform, 1,2-dichloroethane, and carbon tetrachloride are chlorinated compounds produced by chlorination of water and wastewater and are suspected carcinogens. There is also the possibility of an association of the chlorination of water with increased risk of cardiovascular diseases but this awaits further investigation (Craun, 1988). These findings led the U.S. EPA to establish a maximum contaminant level (MCL) of 100 μg/L for THM. This level will be reduced further in the near future. Because water treatment with chloramines does not produce any trihalomethanes, consumers drinking chloraminated water appear to experience less bladder cancer than those consuming chlorinated water (Zierler et al., 1987).

The main approaches to reducing or controlling THM in drinking water are the following (Wolfe et al., 1984): (1) Removal of THM precursors prior to chlorination—there is a strong relation between total THM formation potential (TTHMFP) and total organic carbon in water (Fig. 6.8) (LeChevallier et al., 1992); (2) removal of THM; and (3) use of alternative disinfectants that do not generate THM (e.g., chloramination, ozone, or UV irradiation).

6.3.7. Chloramination

Chloramination is the disinfection of water with chloramine in lieu of free chlorine. The Denver Water Department has been using successfully chloramination for more than 70 years for water treatment (Dice, 1985). Chloramines do not react with organics to form THM. Although they are less effective disinfectants than free chlorine, they appear to be more effective in controlling biofilm microorganisms because they interact poorly with capsular polysaccharides. Thus, use of free chlorine has been suggested as a primary disinfectant in water distribution systems with conversion of the residual to monochloramine if biofilm control is the goal (LeChevallier et al., 1990).

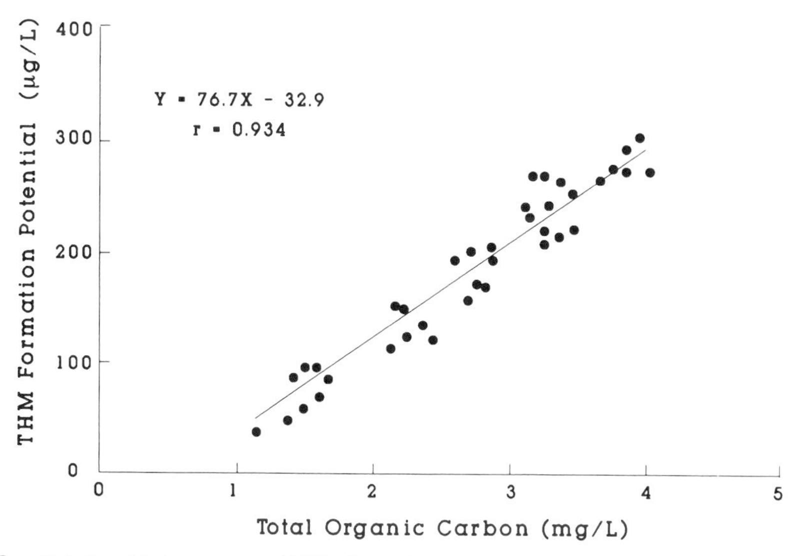

Fig. 6.8. Relationship between total THM formation potential (TTHMFP) and total organic carbon (TOC). From LeChevallier et al. (1992), with permission of the publisher.

6.3.7.1. *Chloramine Chemistry*

In aqueous solutions, HOCl reacts with ammonia and forms inorganic chloramines according to the following equations (Snoeyink and Jenkins, 1980):

$$NH_3 + HOCl \rightarrow \underset{\text{monochloramine}}{NH_2Cl} + H_2O \quad (6.7)$$

$$NH_2Cl + HOCl \rightarrow \underset{\text{dichloramine}}{NHCl_2} + H_2O \quad (6.8)$$

$$NHCl_2 + HOCl \rightarrow \underset{\text{trichloramine}}{NCl_3} + H_2O \quad (6.9)$$

The proportion of the three forms of chloramines greatly depends on the pH of the water. Monochloramine is predominant at pH > 8.5. Monochloramine and dichloramine coexist at pH between 4.5 and 8.5, and trichloramine is formed at pH < 4.5. Figure 6.9 (Wolfe et al., 1984) shows the effect of pH on the distribution of chloramines. Monochloramine is the predominant kind of chloramine formed at the pH range usually encountered in water and wastewater treatment plants (pH = 6–9). In water treatment plants the formation of monochloramine is desirable because dichloramine and trichloramines impart unpleasant taste to the water.

The mixing of chlorine and ammonia produces a chlorine dose–residual curve displayed in Figure 6.10 (Kreft et al., 1985). In the absence of chlorine demand, a chlorine dose of 1 mg/L produces a chlorine residual of 1 mg/L. However, in the presence of ammonia, the chlorine residual reaches a peak (formation of mostly monochloramine, at a ratio of chlorine to ammonia-N between 4:1 and 6:1) and then decreases to a minimum called the breakpoint. The breakpoint, where the chloramine is oxidized to nitrogen gas, occurs when the ratio of chlorine to ammonia-N is between 7.5:1 and 11:1.

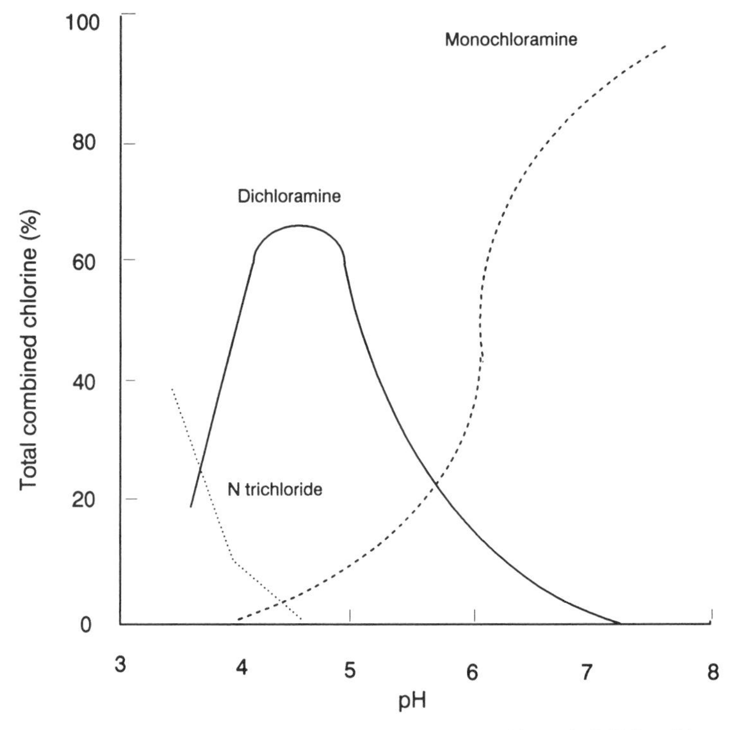

Fig. 6.9. Distribution of types of chloramine with pH. From Wolfe et al. (1984), with permission of the publisher.

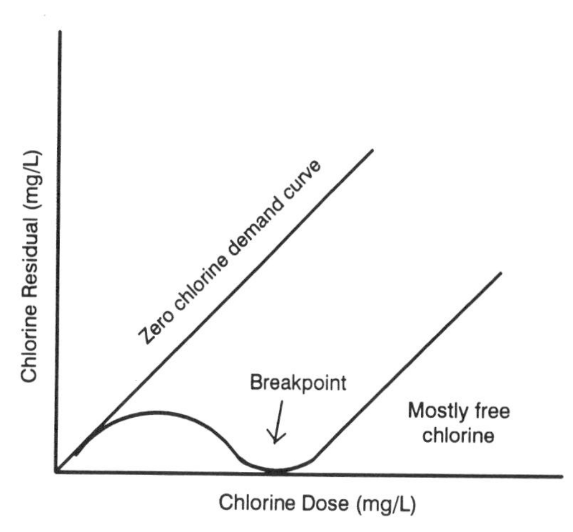

Fig. 6.10. Dose–demand curve for chlorine-ammonia reaction after 1 hr at pH 7–8. From Kreft et al. (1985), with permission of the publisher.

$$2NH_3 + 3HOCl \rightarrow N_2 + 3H_2O + 3HCl \qquad (6.10)$$

Addition of chlorine beyond breakpoint chlorination insures the existence of a free available residual.

6.3.7.2. *Biocidal Effect of Inorganic Chloramines*

In the 1940s, Butterfield and collaborators showed that free chlorine inactivates enteric bacteria much faster than inorganic chloramines. Furthermore, the bactericidal activity of chloramines was found to increase with temperature and hydrogen ion concentration (Watie and Butterfield, 1944). Similar observations were made with regard to viruses (Kelly and Sanderson, 1958) and protozoan cysts (Stringer and Kruse, 1970). Mycobacteria, some enteric viruses, (e.g., hepatitis A virus, rotaviruses), and protozoan cysts are quite resistant to chloramines (Pelletier and DuMoulin, 1988; Rubin, 1988; Sobsey et al., 1988). Thus, it was recommended that drinking water should not be disinfected with only chloramines unless the source water is of good quality (Sobsey, 1989).

The inactivation of pathogens and parasites with chloramines is summarized in Table 6.2 (Sobsey, 1989).

TABLE 6.2. Inactivation of Health-Related Microorganisms in Water by Chloramines: Ct Values[a]

Microbe	Water	°C	pH	Est Ct
Bacteria				
E. coli	BDF[b]	5	9.0	113
Coliforms, *S. typhimurium*, and *S. sonnei*	Tap + 1% sewage	20	6.0	8.5
M. fortuitum	BDF	20	7.0	2,667
M. avium	BDF	17	7.0	ND[c]
M. Intracellulare	BDF	17	7.0	ND
Viruses				
Polio 1	BDF	5	9.0	1,420
Polio 1	primary effluent	25	7.5	345
Hepatitis A	BDF	5	8.0	592
Coliphage MS2	BDF	5	8.0	2,100
Rotavirus SA11				
Dispersed	BDF	5	8.0	4,034
Cell-associated	BDF	5	8.0	6,124
Protozoan cysts				
G. muris	BDF	3	6.5–7.5	430–580
G. muris	BDF	5	7.0	1,400

[a]Adapted from Sobsey (1989).
[b]BDF = buffered-demand free water.
[c]No data available.

6.3.7.3. *Toxicological Aspects of Chloramines*

Dichloramine and trichloramine have offensive odors and have a threshold odor concentration of 0.8 and 0.02 mg/L, respectively (Kreft et al., 1985). Chloramines cause hemolytic anemia in kidney hemodialysis patients (Eaton et al., 1973), but no effect was observed in animals or humans ingesting chloramines by mouth. Although chloramines are mutagenic to bacteria and initiate skin papillomas in mice, study of the potential carcinogenic effects in humans awaits further research.

In the aquatic environment, chloramines are toxic to fish and invertebrates. At 20°C, the 96-hr LC_{50} (50% lethal concentration) of monochloramine ranges from 0.5 to 1.8 mg/L. One of the mechanisms of toxicity to fish is the irreversible oxidation of hemoglobin to methemoglobin, which has a lower oxygen-carrying capacity (Groethe and Eaton, 1975; Wolfe et al., 1984).

6.4. CHLORINE DIOXIDE

6.4.1. Chemistry of Chlorine Dioxide

Chlorine dioxide (ClO_2) does not appear to form trihalomethanes nor does it react with ammonia to form chloramines. Therefore, its use as a disinfectant in water treatment is becoming widespread. Because it cannot be stored in compressed form in tanks, chlorine dioxide must be generated at the site. ClO_2 is generated from the reaction of chlorine gas with sodium chlorite:

$$2\ NaClO_2 + Cl_2 \rightarrow 2\ ClO_2 + 2\ NaCl \qquad (6.11)$$

ClO_2 does not hydrolyze in water but exists as a dissolved gas. In alkaline solutions, it forms chlorite and chlorate:

$$2\ ClO_2 + OH^- \rightarrow ClO_2^- + ClO_3^- + H_2O \qquad (6.12)$$

In water treatment plants, chlorite is the predominant form generated. To reduce THM formation, ClO_2 is used as a preoxidant and a primary disinfectant and is followed by the addition of chlorine to maintain a residual (Aieta and Berg, 1986).

6.4.2. Effect of Chlorine Dioxide on Microorganisms

Chlorine dioxide is a fast-acting and effective microbial disinfectant and is equal or superior to chlorine in inactivating bacteria and viruses in water and wastewater (Aieta and Berg, 1986; Bitton, 1980a; Longley et al., 1980; Narkis and Kott, 1992). It is also effective in the destruction of cysts of pathogenic protozoa such as *Naegleria gruberi* (Chen et al., 1985). As shown for human and simian rotaviruses, the virucidal efficiency of chlorine dioxide increases as the pH is increased from 4.5 to 9.0 (Chen and Vaughn, 1990). Bacteriophage f2 inactivation is also much higher at pH 9.0 than at pH 5.0 (Fig. 6.11) (Noss and Olivieri, 1985).

Inactivation of health-related microorgan-

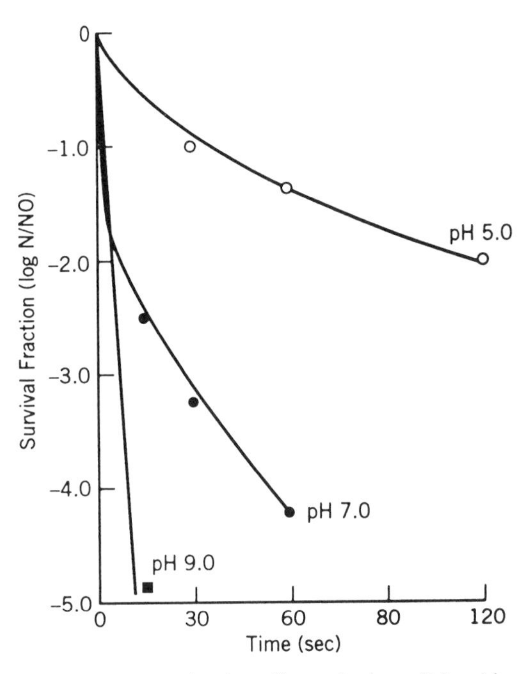

Fig. 6.11. Inactivation of bacteriophage f2 by chlorine dioxide as a function of pH. Adapted from Noss and Olivieri (1985).

isms (Ct values) by chlorine dioxide is summarized in Table 6.3 (Sobsey, 1989).

6.4.3. Mode of Action of Chlorine Dioxide

The primary mode of action of chlorine dioxide involves the disruption of protein synthesis in bacterial cells (Benarde et al., 1967). The disruption of the outer membrane of gram-negative bacteria has also been observed (Berg et al., 1986).

An examination of the mechanisms of virus inactivation by chlorine dioxide has revealed contradicting results (Vaughn and Novotny, 1991). Work with f2 bacterial phage has shown that the protein coat is the primary target of the lethal action of chlorine dioxide (Noss et al., 1986; Olivieri et al., 1985). The loss of attachment of this phage to its host cell paralleled virus inactivation (Noss et al., 1986). Specifically, degradation of tyrosine residues within the protein coat appears to be the primary site of action of chlorine dioxide in f2 phage (Noss et al., 1986). The disruption of the viral protein coat was suggested in regard to other viruses such as

TABLE 6.3. Inactivation of Health-Realted Microorganisms in Water by Chlorine Dioxide: Ct Values[a]

Microbe	Water	ClO_2 residual (mg/L)	Temp. (°C)	pH	Time (min.)	% Reduction	Ct
Bacteria							
E. coli	BDF[b]	0.3–0.8	5	7.0	0.6–1.8	99	0.48
Fecal coliforms	effl.	1.9	?	?	10	99.94	ND
Fecal strep.	effl.	1.9	?	?	10	99.5	ND
C. perfringens	effl.	1.9	?	?	10	0	ND
L. pneumophila	BDF	0.5–0.35	23	?	15	99.9–99.99	ND
K. pneumonia	BDF	0.12	23	?	15	99.3–99.7	ND
Viruses							
Coliphage f2	BDF	1.5	5	7.2	2	99.994	ND
Polio 1	BDF	0.4–14.3	5	7.0	0.2–11.2	99	0.2–6.7
Polio 1	effl.	1.9	?	?	10	99.4	ND
Rota SA11:							
Dispersed	BDF	0.5–1.0	5	6.0	0.2–0.6	99	0.2–0.3
Cell-associated	BDF	0.45–1.0	5	6.0	1.2–4.8	99	1.0–2.1
Cell-associated	BDF	0.46–0.52	5	10.0	0.3–0.4	99	0.16–0.2
Hepatitis A	BDF	0.14–0.23	5	6.0	8.4	99	1.7
Hepatitis A	BDF	0.2	5	9.0	<0.33	>99.9	<0.04
Coliphage MS2	BDF	0.15	5	6.0	34	99	5.1
Coliphage MS2	BDF	0.15	5		<0.33	>99.95	<0.03
Protozoan cysts							
N. gruberi	BDF	0.8–1.95	5	7.0	7.8–19.9	99	15.5
N. gruberi	BDF	0.46–1.0	25	5.0	5.4–13.2	99	6.35
N. gruberi	BDF	0.42–1.1	25	9.0	2.5–6.7	99	2.91
G. muris	BDF	0.1–5.55	5	7.0	1.3–168	99	10.7
G. muris	BDF	0.26–1.2	25	5.0	4.0–24	99	5.8
G. muris	BDF	0.21–1.12	25	7.0	3.3–28.8	99	5.1
G. muris	BDF	0.15–0.81	25	9.0	2.1–19.2	99	2.7

[a]Adapted from Sobsey (1989).

[b]BDF = buffered demand free water.

poliovirus (Brigano et al., 1979). Some observers have concluded that the primary site of action of chlorine dioxide is the viral genome (Alvarez and O'Brien, 1982; Taylor and Butler, 1982).

6.4.4. Toxicology of Chlorine Dioxide

Chlorine dioxide interferes with the thyroid function and produces high serum cholesterol in animals fed a diet low in calcium and rich in lipids (Condie, 1986). Chlorine dioxide has two inorganic by-products, chlorite (ClO_2^-) and chlorate (ClO_3^-). Chlorite is of greater health concern than chlorate and both may combine with hemoglobin to cause methemoglobinemia.

6.5. OZONE

6.5.1. Introduction

Ozone is produced by passing dried air between electrodes separated by an air gap and a dielectric and by applying an alternating current with the voltage ranging from 8,000 to 20,000 V (Fig. 6.12). Ozone was first introduced as a strong oxidizing agent for the removal of taste, color, and odors. The first water treatment plant to use ozone started operations in 1906 in Nice, France. This oxidant is now used as a primary disinfectant to inactivate pathogenic microorganisms and for the oxidation of iron and manganese, taste- and odor-causing compounds, color, refractory organics, and THM precursors. Preozonation also lowers THM formation potential and promotes particle coagulation during water treatment (Chang and Singer, 1991); it is also used in combination with activated carbon treatment. In the United States more than 40 wastewater treatment plants are presently using ozone mostly as an oxidant, and a few of them use it as a disinfectant. Ozone can be applied at various points of a conventional water treatment plant, depending on the type of use (Rice, 1989; AWWA, 1985b). Its effectiveness as a disinfectant is not controlled by pH, and it does not interact with ammonia.

Ozone is more expensive than chlorination or even UV disinfection in respect to construction as well as operation and maintenance costs. Power use is the most expensive operating cost item. Since ozone does not leave any residual in water, ozone treatment is sometimes combined with postchlorination. Ozone breaks down complex compounds into simpler ones, some of which may serve as substrates for microbial growth in water distribution systems (Bancroft et al., 1984).

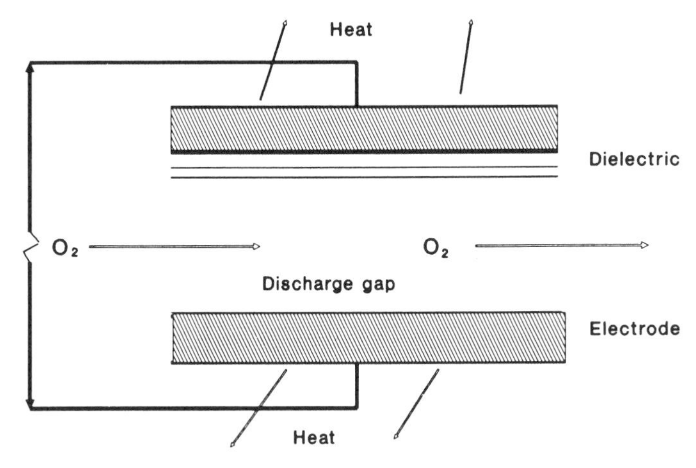

Fig. 6.12. Basic ozonator configuration. From McCarthy and Smith (1974), with permission of the publisher.

6.5.2. Effect of Ozone on Indicator and Pathogenic Microorganisms

Ozone is a much more powerful oxidant than chlorine. The threshold ozone concentration above which bacterial inactivation is very rapid is only 0.1 mg/L (Bitton, 1980a). The Ct values for 99% inactivation are very low and range from 0.001 to 0.2 for *E. coli* and from 0.04 to 0.42 for enteric viruses (Engelbrecht, 1983; Hall and Sobsey, 1993).

Ozone appears to be more effective against human and simian rotaviruses than chlorine, monochloramine, or chlorine dioxide (Chen and Vaughn, 1990; Korich et al., 1990). The ozone concentration required to inactivate 99.9% of enteroviruses in water (25°C, pH = 7.0) in 10 min varies between 0.05 and 0.6 mg/L (Engelbrecht, 1983). However, some bacterial pathogens (e.g., *Mycobacterium fortuitum*) appear to be more resistant than viruses to ozone. The resistance of a number of microorganisms to ozone has been found to be as follows: *Mycobacterium fortuitum* > poliovirus type 1 > *Candida parapsilosis* > *E. coli* > *Salmonella typhimurium* (Fig. 6.13) (Farooq and Akhlaque, 1983). Suspended solids (e.g., clays, sludge solids) significantly reduce viral inactivation by ozone. The protective effect of solids is illustrated in Figure 6.14 (Kaneko, 1989).

We have seen that *Cryptosporidium* oocysts are very resistant to chlorination. Ozone, at a concentration of 1.1 mg/L totally inactivates *Cryptosporidium parvum* oocysts in 6 min at levels of 10^4 oocysts per 1 ml (Peeters et al., 1989). Cysts of *Giardia lamblia* and *G. muris* are also effectively inactivated by ozone (Fig. 6.15) (Wickramanayake et al., 1985). At pH 7, at 5°C, more than two-log reduction of the viability of *G. lamblia* cysts is achieved in a few minutes with less than 0.5 mg ozone per 1 L. The effectiveness of ozonation varies greatly with temperature. The resistance of *G. lamblia* cysts to ozone increased when the temperature was lowered from 25°C to 5°C (Wickramanayake et al., 1985). A similar phenomenon was observed for *Cryptosporidium* oocysts (Joret et al., 1992).

The Peroxone process, using a mixture of

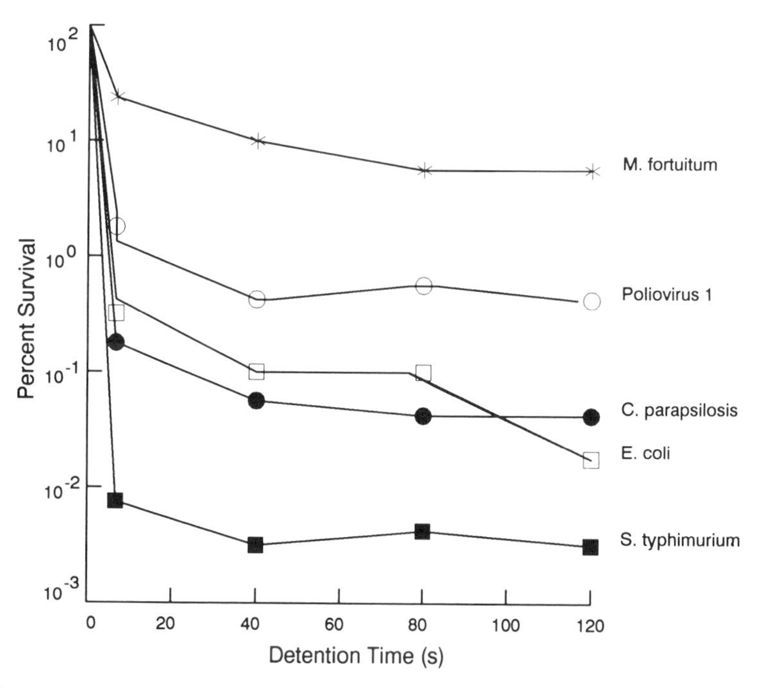

Fig. 6.13. Inactivation of various microorganisms by ozone in an activated sludge effluent. Adapted from Farooq and Akhlaque (1983).

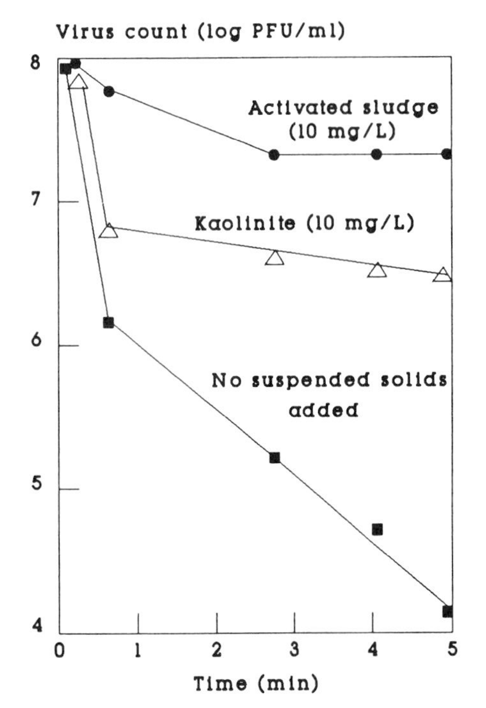

Fig. 6.14. Effect of suspended solids on poliovirus inactivation by ozone. Adapted from Kaneko (1989).

ozone and hydrogen peroxide, has been investigated for controlling taste and odors, disinfection by-products, and microbial pathogens. It appears that the inactivation efficiency of Peroxone (H_2O_2:O_3 = 0.3 or less) is similar to that of ozone alone. However, Peroxone is superior to ozone for the oxidation of taste and odor-causing compounds (Ferguson et al., 1990).

6.5.3. Mechanisms of Ozone Action

In aqueous media ozone produces free radicals that inactivate microorganisms. Ozone affects the permeability, enzymatic activity, and DNA of bacterial cells (Hamelin et al., 1978; Ishizaki et al., 1987), and guanine and/or thymine residues appear to be the most susceptible targets of ozone (Ishizaki et al., 1984). Ozone treatment also leads to the conversion of closed circular plasmid DNA (ccDNA) of *E. coli* to open circular DNA (ocDNA) (Ishizaki et al., 1987).

As shown in poliovirus, ozone inactivates viruses by damaging the nucleic acid core (Roy et al., 1981). The protein coat is also affected (De Mik and De Groot, 1977; Riesser et al., 1977; Sproul et al., 1982), but the damage to the protein coat may be small and may not significantly affect the adsorption of poliovirus to its host cell (VP4, a capsid polypeptide responsible for attachment to host cell, was not affected by ozone). For rotaviruses, ozone alters both the capsid and the RNA core (Chen et al., 1987).

6.5.4. Public Health Import of Ozonation By-products

We have discussed the formation of mutagenic/carcinogenic compounds following chlorination of water and wastewater. Less is known regarding ozonation by-products. Aldehydes are by-products of potential concern, but their health significance is not presently known (U.S. EPA, 1989). Recent studies show that water treated with an ozone dose of 1 mg/L may show some increase in mutagenicity. Mutagenicity was reduced, however, at higher ozone levels (>3

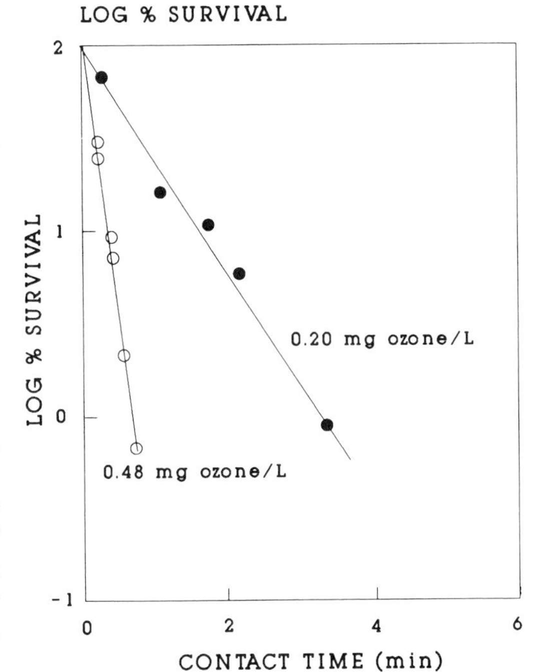

Fig. 6.15. Inactivation of *Giardia lamblia* cysts by ozone. Adapted from Wickramanayake et al. (1985).

mg/L). The mutagenic compounds can be removed by granular activated carbon (GAC) treatment (Bourbigot et al., 1986; van Hoof et al., 1985; Matsuda et al., 1992; Rice, 1989).

6.6. ULTRAVIOLET LIGHT

6.6.1. Introduction

UV disinfection was first used at the beginning of this century to treat water in Henderson, Kentucky, but it was abandoned in favor of chlorination. Owing to technological improvements, this disinfection alternative is now regaining popularity, particularly in Europe (Wolfe, 1990).

UV disinfection systems use low-pressure mercury lamps enclosed in quartz tubes. The tubes are immersed in flowing water in a tank and allow passage of UV radiation at the germicidal wavelength of 2,537 A. However, transmission of UV by quartz decreases upon continuous use. Therefore, the quartz lamps must be regularly cleaned, by mechanical, chemical, and ultrasonic cleaning methods. Teflon has been proposed as an alternative to quartz but its transmission of UV radiation is lower than in quartz systems.

6.6.2. Mechanism of UV Damage

Studies with viruses have demonstrated that the initial site of UV damage is the viral genome, followed by structural damage to the virus coat (Rodgers et al., 1985).

Ultraviolet radiation damages microbial DNA at a wavelength of approximately 260 nm. It causes thymine dimerization, which blocks DNA replication and effectively inactivates microorganisms.

6.6.3. Inactivation of Pathogens by UV Radiation

Microbial inactivation is proportional to the UV dose, which is expressed in microwatt-seconds per square centimeter. The inactivation of microorganisms by UV radiation can be represented by the following equation (Luckiesh and Holladay, 1944; Severin, 1980):

$$N/N_0 = e^{-KP_d t} \qquad (6.13)$$

where N_0 = initial number of microorganisms (#/mL); N = number of surviving microorganisms (#/mL); K = inactivation rate constant (μW·s/cm^2); P_d = UV light intensity reaching the organisms (μW/cm^2); and t = exposure time in seconds.

The above equation is subject to several assumptions, one of which is that the logarithm of the survival fraction should be linear with regard to time (Severin, 1980). In environmental samples, however, the inactivation kinetics is not linear with time, which may be due to resistant organisms among the natural population and to differences in flow patterns.

The efficacy of UV disinfection depends on the type of microorganism under consideration. In general, the resistance of microorganisms to UV follows the same pattern as with chemical disinfectants, which is as follows (Fig. 6.16) (Chang et al., 1985): protozoan cysts > bacterial spores > viruses > vegetative bacteria. This trend is supported by Table 6.4 (Wolfe, 1990), which shows the approximate UV dose (in μW-s/cm^2) for 90% inactivation of microorganisms. A virus such as hepatitis A virus requires a UV dose of 2,700 μW-s/cm^2 for one-log inactivation (Wolfe, 1990) but necessitates 20,000 μW-sec/cm^2 to achieve a three-log reduction (Battigelli et al., 1993).

6.6.4. Variables Affecting UV Action

Many variables (e.g., suspended particles, chemical oxygen demand, color) in wastewater effluents affect UV transmission in water and, thus, the dose necessary for disinfection (Harris et al., 1987; Severin, 1980). Several organic compounds (e.g., humic substances, phenolic compounds, lignin sulfonates from the pulp and paper mill industry, ferric iron) interfere with UV transmission in water.

Indicator bacteria are partially protected from the harmful UV radiation when embedded with-

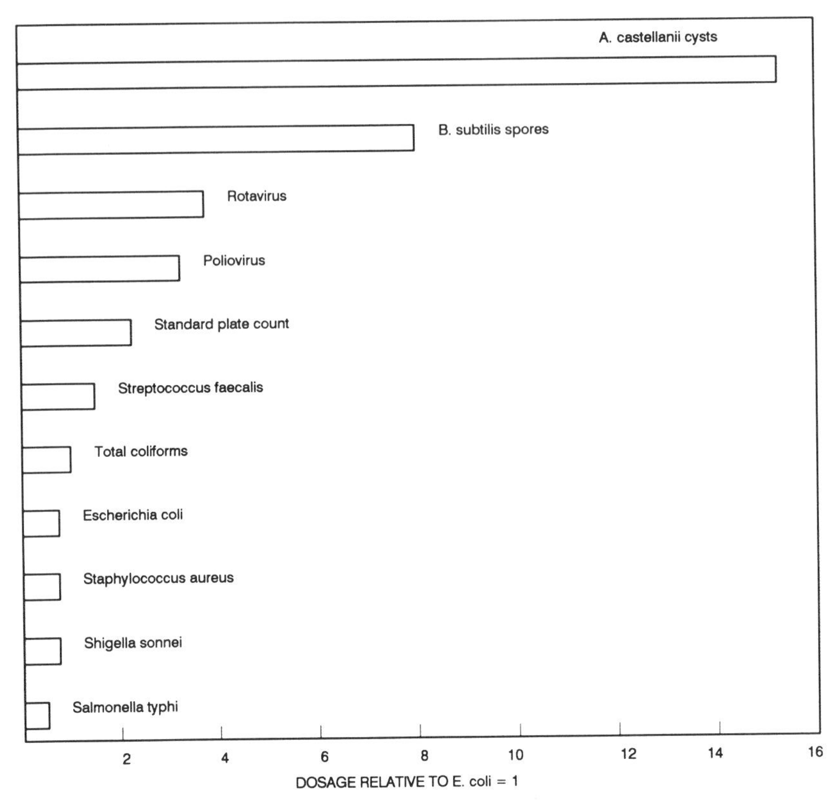

Fig. 6.16. Relative UV dose required to inactive 99.9% of various microorganisms compared to that for *E. coli*. From Chang et al. (1985), with permission of the publisher.

in particulate matter (Oliver and Cosgrove, 1977; Harris et al., 1987; Qualls et al., 1983, 1985). Suspended solids protect microorganisms only partially from the lethal effect of UV radiation. This is because suspended particles in water and wastewater absorb only a portion of the UV light (Bitton et al., 1972). Wastewater solids absorb 75% of the light, and scattering accounts for the remaining 25% (Qualls et al., 1983, 1985). Most clay minerals do not afford much protection to microorganisms because they scatter most of the UV light. It is known that their protective effect depends on their specific absorption and scattering of UV radiation, and it decreases with an increase of light scattering (Bitton et al., 1972). Thus, flocculation followed by filtration of effluents through sand or anthracite beds to remove interfering substances should improve UV disinfection efficiency (Dizer et al., 1993). This is illustrated in Figure 6.17 (Severin, 1980), which shows the effect of sand filtration on inactivation of coliforms with UV radiation.

Photoreactivation may occur following exposure of UV-damaged microbial cells to visible light at wavelengths between 300 nm and 500 nm (Jagger, 1958). The potential for UV-damaged bacteria to undergo repair after UV irradiation has been demonstrated (Carson and Petersen, 1975; Meschner et al., 1990). DNA damage can also be repaired in the dark by the cell excision repair system. The UV-damaged DNA segment is excised and replaced by a newly synthesized segment. Photoreactivation was demonstrated in a full-scale wastewater treatment plant using UV disinfection. However, although total and fecal coliforms were photoreactivated, fecal streptococci failed to be (Harris et al., 1987; Whitby et al., 1984). Figure 6.18 shows the higher survival of *E. coli* subjected to photoreactivation after UV irradiation (Harris et al., 1987). *Legionella pneumophila* treated by UV irradiation was also photoreactivated following exposure to visible light (indirect sunlight). Thus, UV-treated water should not be exposed to visible light during storage (Knudson, 1985).

6.6.5. UV Disinfection of Water and Wastewater

6.6.5.1. *Potable Water*

UV disinfection is particularly useful for potable water (Wolfe, 1990). Since groundwater accounts for about 50% of waterborne diseases in the United States, UV is a potentially useful disinfectant for this resource. This disinfectant is particularly efficient against viruses, which are major agents of waterborne diseases in groundwater (Craun, 1986b). A drinking water plant using UV disinfection in London treats up to 14.5 MGD. In hospitals, chlorine is added to maintain a disinfectant residual following UV irradiation. Continuous UV irradiation rapidly inactivates *Legionella* in plumbing systems and in circulating hot tubs and whirlpools (Gilpin, 1984; Muraca et al., 1987). At a dose of 30,000 μW-s/cm^2, UV reduces *Legionella pneumophila* by 4–5 logs within 20 min in a hospital water distribution system (Muraca et al., 1987). Although this dose is ineffective in inactivating *Giardia lamblia* cysts, which require a dose of 63,000 μW-s/cm^2 for a 1-log reduction in cyst viability (Rice and Hoff, 1981), it is much higher than the minimum dose of 16,000 μW-s/cm^2 recommended by the U.S. Public Health Service for disinfection of water with ultraviolet irradiation (U.S. DHEW, 1967).

TABLE 6.4. Approximate Dosage for 90% Inactivation of Selected Microorganisms by UV[a]

Microorganism	Dosage μW-s/cm^2
Bacteria	
E. coli	3,000
Salmonella typhi	2,500
Pseudomonas aeruginosa	5,500
Salmonella enteritis	4,000
Shigella dysenteriae	2,200
Shigella paradysenteriae	1,700
Shigella flexneri	1,700
Shigella sonnei	3,000
Staphylococcus aureus	4,500
Legionella pneumophila	380
Vibrio cholerae	3,400
Viruses	
Poliovirus 1	5,000
Coliphage	3,600
Hepatitis A virus	3,700
Rotavirus SA 11	8,000
Protozoan cysts	
Giardia muris	82,000
Giardia lamblia	63,000
Acanthamoeba castellanii	35,000

[a]Adapted from Wolfe (1990) and Rice and Hoff (1981).

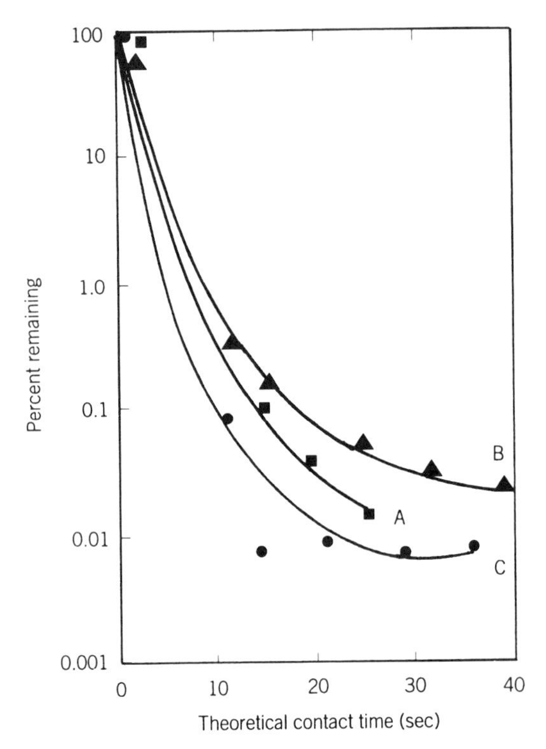

Fig. 6.17. Effect of suspended solids on the inactivation of fecal coliforms with UV light. **A.** Activated sludge effluent. **B.** Activated sludge effluent with waste solids added. **C.** Sand-filtered effluent. Adapted from Severin (1980).

6.6.5.2. *Wastewater Effluents*

UV disinfection of wastewater effluents is now an economically competitive alternative to chlorination and does not generate toxic by-products as shown for chlorination. Several UV disinfection systems have been built or are in the planning stage (Qualls et al., 1985; Severin, 1980; Whitby et al., 1984). Problems encountered include the difficulty of measuring the UV dose necessary for disinfection of wastewater effluents. Models as well as bioassays utilizing *Bacillus subtilis* spores or RNA phage MS2 have been proposed for the determination of the effective UV dose for disinfection (Havelaar et al., 1990b; Qualls and Johnson, 1983; Qualls et al., 1989; Wilson et al., 1992). MS2 displays higher resistance to UV irradiation than enteric viruses (e.g., hepatitis A virus, rotavirus) (Battigelli et al., 1993; Wiedenmann et al., 1993).

The U.S. PHS minimum dose of 16,000 $\mu W\text{-}s/cm^2$ leads to more than 3-log reduction of coliforms in wastewater effluents. The wide variation observed in coliform inactivation by UV in wastewater is probably due to the varying proportions of solids-associated coliforms in the effluents (Qualls et al., 1985). UV disinfection, investigated in a full-scale plant in Ontario, was shown to be as efficient as chlorination in respect to the inactivation of total coliforms, fecal col-

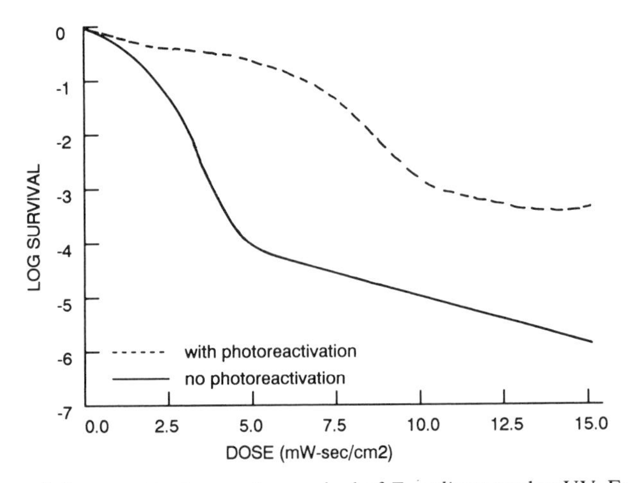

Fig. 6.18. Effect of photoreactivation on the survival of *E. coli* exposed to UV. From Harris et al. (1987), with permission of the publisher.

iforms, and fecal streptococci (Whitby et al., 1984). Moreover, it was superior to chlorination for the inactivation of *Clostridium perfringens* and coliphages.

Simultaneous addition of ozone and UV does not contribute more inactivation of fecal coliforms in wastewater effluents. However, inactivation of the bacterial indicators is increased if UV irradiation precedes or follows ozonation (Venosa et al., 1984).

6.6.6. Some Advantages and Disadvantages of UV Disinfection

The following are some advantages of disinfecting water and wastewater with UV irradiation (Sobsey, 1989; Wolfe, 1990):

1. Efficient inactivation of bacteria and viruses in potable water (higher doses are required for protozoan cysts)
2. No production of any known undesirable carcinogenic or toxic by-products (no adverse effects were observed by Oliver and Carey (1976) in rainbow trout exposed to UV-treated effluents)
3. No taste and odor problems
4. No need to handle and store toxic chemicals
5. Small space requirement by UV units

The following are some disadvantages of UV disinfection.

1. No disinfectant residual in treated water; (therefore, a postdisinfectant like chlorine or ozone should be added)
2. Difficulty in determining UV dose
3. Biofilm formation on the lamp surface (however, modern UV units are designed to prevent fouling by microorganisms)
4. Problems in maintenance and cleaning of UV lamps
5. Potential problem due to photoreactivation of UV-treated microbial pathogens
6. Higher cost of UV disinfection than chlorination (however, the cost is lower than for ozonation)

6.6.7. Other Technologies Based on Photoinactivation

6.6.7.1. Use of Solar Radiation for Disinfection of Drinking Water

In remote areas with no access to treated drinking water, solar radiation has been considered for disinfecting water. A solar radiation intensity of at least 600 W/m^2 for 5 hr is necessary for an adequate reduction of bacterial indicators and pathogens (e.g., *Vibrio cholerae, Salmonella typhi*) (Odeymi, 1990). No data are available on the inactivation of viruses or parasites.

6.6.7.2. Photodynamic Inactivation

Photodynamic inactivation (photochemical disinfection) consists of using visible light or sunlight as the energy source, O_2, and a sensitizer dye such as methylene blue. This approach was studied under laboratory conditions (Acher and Juven, 1977; Gerba et al., 1977) and in pilot plants (Acher et al., 1990; Eisenberg et al., 1987). Photochemical disinfection of wastewater effluents under field conditions achieved microbial reductions of 1.8 log for poliovirus 1, 3.0 log for coliforms, 3.1 log for fecal coliforms, and 3.7 log for enterococci in approximately 1 hr under alkaline conditions. These reductions necessitate a light intensity of 700–2,100 $\mu E\ m^{-2}\ s^{-1}$, a methylene blue level of 0.8–0.9 ppm, pH = 8.7–8.9, and a dissolved oxygen level of 4.5–5.5 mg/L (Acher et al., 1990).

6.7. WASTEWATER IRRADIATION

In Chapter 12, we discuss the treatment of wastewater sludges by gamma radiation (cesium-137 or cobalt-60) and high-energy electron beams. Research is now focusing on the effectiveness of this technology in the disinfection of wastewater as well as in the removal of organic matter (BOD

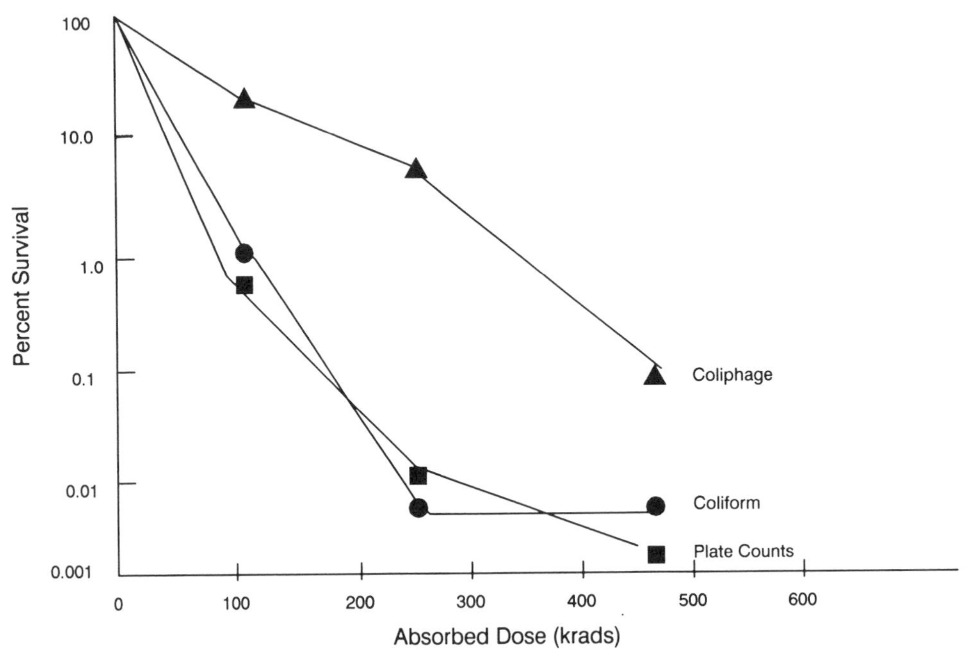

Fig. 6.19. Disinfection of raw wastewater with gamma irradiation. From Farooq et al. (1992), with permission of the publisher.

and COD). Disinfection of raw wastewater with gamma radiation (^{60}cobalt source), at a dose of 463 krad, resulted in 3 log inactivation for coliphage and 4–5 log inactivation for coliforms and heterotrophic plate count (Fig. 6.19) (Farooq et al., 1992). Wastewater treatment with high energy electrons was less effective than gamma radiation and resulted in 2–3 log inactivation for the three categories of microbial indicators (Fig. 6.20) (Farooq et al., 1992).

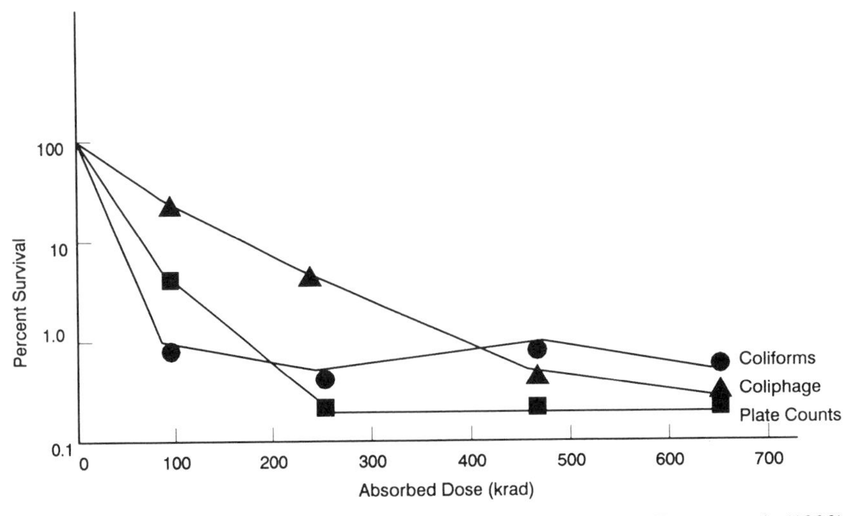

Fig. 6.20. Disinfection of raw wastewater with electron beams. From Farooq et al. (1992), with permission of the publisher.

6.8. FURTHER READING

Aieta, E.M., and J.D. Berg. 1986. A review of chlorine dioxide in drinking water treatment. J. Am. Water Works Assoc. 78: 62–72.

Hoff, J.C., and E.W. Akin. 1986. Microbial resistance to disinfectants: Mechanisms and significance. Environ. Health Perspect. 69: 7–13.

Sobsey, M.D. 1989. Inactivation of health-related microorganisms in water by disinfection processes. Water Sci. Technol. 21: 179–195.

Wolfe, R.L., N.R. Ward, and B.H. Olson. 1984. Inorganic chloramines as drinking water disinfectants: A review. J. Am. Water Works Assoc. 76: 74–88.

WASTEWATER AND WATER TREATMENT MICROBIOLOGY

7

INTRODUCTION TO WASTEWATER TREATMENT

7.1. INTRODUCTION

In the middle of the nineteenth century in England, waterborne diseases such as cholera were rampant and several epidemics in London resulted in thousands of victims. Increasing awareness of the role of microorganisms in diseases led to an enhanced demand for wastewater treatment. This has resulted in the passage of legislation that encouraged the construction of wastewater treatment plants (Guest, 1987). Wastewater treatment practice started in the beginning of the twentieth century. By the end of the last century, the British Royal Commission on Sewage Disposal proposed that the goal of wastewater treatment should be to produce a final effluent of 30 mg/L of suspended solids and 20 mg/L of biochemical oxygen demand (BOD) (Sterritt and Lester, 1988).

Today there are approximately 15,000 wastewater treatment facilities in the United States, 80% of which are small plants (<1 MGD). These plants treat approximately 37 billion gallons of wastewater per day. Approximately 75% of the facilities have secondary treatment or greater (Ouellette, 1991; U.S. EPA, 1989f). These plants have been constructed to treat both domestic and industrial wastes. Nontoxic wastes are contributed mainly by the food industry and by domestic sewage, whereas toxic wastes are contributed by the coal processing (phenolics, ammonia, cyanide), petrochemical, (oil, petrochemicals, surfactants), pesticide, pharmaceutical, and electroplating (toxic metals such as cadmium, copper, nickel, zinc) industries (Kumaran and Shivaraman, 1988). The biotransformation and toxicity of these chemicals will be discussed in Chapters 17 and 18. Since many of these plants may fail to meet the desired criteria, pretreatment steps are required at industrial sites prior to the entry of the waste into municipal wastewater treatment plants.

The major contaminants found in wastewater are biodegradable organics, volatile organic compounds, recalcitrant organics, toxic metals, suspended solids, nutrients (nitrogen and phosphorus), microbial pathogens, and parasites (Fig. 7.1). Research efforts are now being focused on the fate of toxic substances following passage through wastewater treatment plants.

The objectives of waste treatment processes are the following:

1. Reduce the organic content of wastewater (i.e., reduction of BOD); this category also includes the removal/reduction of trace organics which are recalcitrant to biodegradation and may be toxic or carcinogenic (see Chapters 17 and 18).

2. Removal/reduction of nutrients (N, P) to reduce pollution of receiving surface waters or groundwater if the effluents are applied onto land (see Chapters 8, 10 and 11).

3. Removal or inactivation of pathogenic microorganisms and parasites.

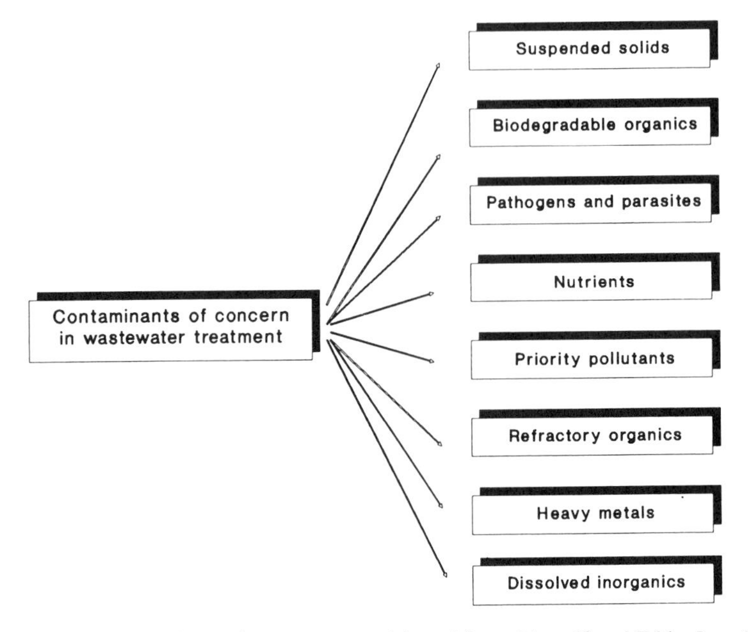

Fig. 7.1. Major contaminants in wastewater. Adapted from Metcalf and Eddy, Inc. (1991).

7.2. COMPOSITION OF DOMESTIC WASTEWATER

Domestic wastewater is a combination of human and animal excreta (feces and urine) and grey water resulting from washing, bathing, and cooking. People excrete 100–500 g wet weight of feces and 1–1.3 L of urine per capita per day. Each person contributes 15–20 g BOD_5/day (Feachem et al., 1983; Gotaas, 1956; Sterritt and Lester, 1988). Other characteristics of human feces and urine are displayed in Table 7.1 (Feachem et al., 1983; Gotaas, 1956; Polprasert, 1989). The chemical characteristics of untreated domestic wastewater are displayed in Table 7.2 (Metcalf and Eddy, 1991).

Domestic wastewater is composed mainly of proteins (40%–60%), carbohydrates (25%–

TABLE 7.1. Composition of Human Feces and Urine[a]

Component	Feces	Urine
Quantity (wet) per person per day	100–400 g	1.0–1.31 kg
Quantity (dry solids) per person per day	30–60 g	50–70 g
Moisture content	70%–85%	93%–96%
Approximate composition (percent dry weight) organic matter	88–97	65–85
Nitrogen (N)	5.0–7.0	15–19
Phosphorus (as P_2O_5)	3.0–5.4	2.5–5.0
Potassium (as K_2O)	1.0–2.5	3.0–4.5
Carbon (C)	44–55	11–17
Calcium (as CaO)	4.5	4.5–6.0
C/N ratio	_6–10	1
BOD_5 content per person per day	15–20 g	10 g

[a]From Polprasert (1989), with permission of the publisher.

TABLE 7.2. Typical Characteristics of Domestic Wastewater (All Values in mg/L)[a]

Parameter	Concentration		
	Strong	Medium	Weak
BOD_5	400	220	110
COD	1,000	500	250
Organic N	35	15	8
NH_3-N	50	25	12
Total N	85	40	20
Total P	15	8	4
Total solids	1,200	720	350
Suspended solids	350	220	100

[a]From Metcalf and Eddy, Inc. (1991), with permission of the publisher.

50%), fats and oils (10%), urea derived from urine, and a large number of trace organics, which include pesticides, surfactants, phenols, and priority pollutants. The latter category comprises nonmetals (As, Se), metals (e.g., Cd, Hg, Pb), benzene compounds (e.g., benzene, ethylbenzene), and chlorinated compounds (e.g., chlorobenzene, lindane, tetrachloroethene, trichloroethene) (Metcalf and Eddy, 1991). The bulk of organic matter in domestic wastewater is easily biodegradable and consists mainly of carbohydrates, amino acids, peptides and proteins, volatile acids, and fatty acids and their esters (Giger and Roberts, 1978; Painter and Viney, 1959).

In domestic wastewaters, organic matter is found as dissolved organic carbon (DOC) and particulate organic carbon (POC). POC represents approximately 60% of organic carbon and some of it is of sufficient size to be removed by sedimentation (Fig. 7.2) (Rickert and Hunter, 1971). In fixed-film processes, DOC is directly absorbed by the biofilm, whereas POC is adsorbed to the biofilm surface to be subsequently hydrolyzed by microbial action (Sarner, 1986).

Three main tests are used for the determination of organic matter in wastewater. These include biochemical oxygen demand (BOD), total organic carbon (TOC), and chemical oxygen demand (COD). Trace organics are detected and measured by means of sophisticated instruments such as gas chromatography and mass spectroscopy.

7.2.1. Biochemical Oxygen Demand (BOD_5)

BOD is the amount of dissolved oxygen (DO) consumed by microorganisms for the biochemi-

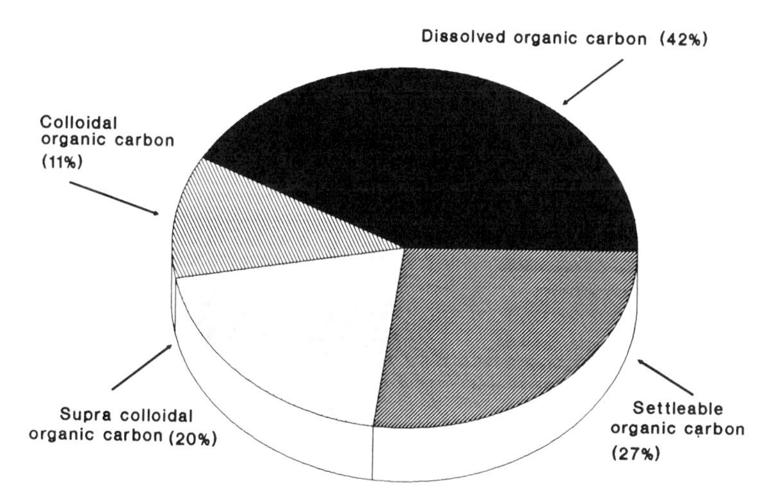

Fig. 7.2. Percentage distribution of total organic carbon (TOC) by particle size in domestic wastewater. Estimated particles sizes: settleable, > 100 μm; supracolloidal, 1–100 μm; colloidal, 1 nm to 1 μm; dissolved, < 1 nm. Adapted from Rickert and Hunter (1971).

cal oxidation of organic (carbonaceous BOD) and inorganic matter (autotrophic or nitrogenous BOD). The sample must be diluted if the BOD exceeds 8 mg/L.

7.2.1.1. *Carbonaceous BOD (CBOD)*

CBOD is the amount of oxygen utilized by a mixed population of heterotrophic microorganisms to oxidize organic compounds in the dark at 20°C over a period of 5 days. This is described by the following general equations:

$$\text{organic compounds} \xrightarrow[\text{heterotrophs}]{O_2} CO_2 + H_2O + NH_4 + \text{bacterial mass}$$

$$\text{bacterial biomass} \xrightarrow[\text{protozoa}]{O_2} \text{protozoan biomass} + CO_2$$

The conversion of bacterial biomass to protozoan biomass proceeds with a yield of 0.78 mg protozoa per 1 mg bacteria (Edeline, 1988).

The BOD test was developed originally to predict the effect of wastewater on receiving streams and to determine their capacity to assimilate organic matter (Gaudy, 1972; Gaudy and Gaudy, 1988). Much has been written on the kinetics of BOD exertion in wastewater. Figure 7.3 (Gaudy, 1972) shows the existence of a plateau and its length depends on the relationship between prey (bacteria) and predator (protozoa) microorganisms. Many reasons have been offered for the existence of a plateau during oxygen uptake: predation by protozoa, lag period between exogenous respiration on the substrate and the onset of endogenous respiration, substrate made of various carbon sources that are metabolized sequentially, and metabolism of a metabolic product only after an acclimation period.

Aliquots of wastewater are placed in a 300-ml BOD bottle and diluted in phosphate buffer (pH = 7.2) containing other inorganic elements (N, Ca, Mg, Fe) and saturated with oxygen. Sometimes acclimated microorganisms or dehydrated cultures of microorganisms, sold in capsule form, are added to municipal and industrial wastewaters, which may not have a sufficient microflora to carry out the BOD test. The micro-

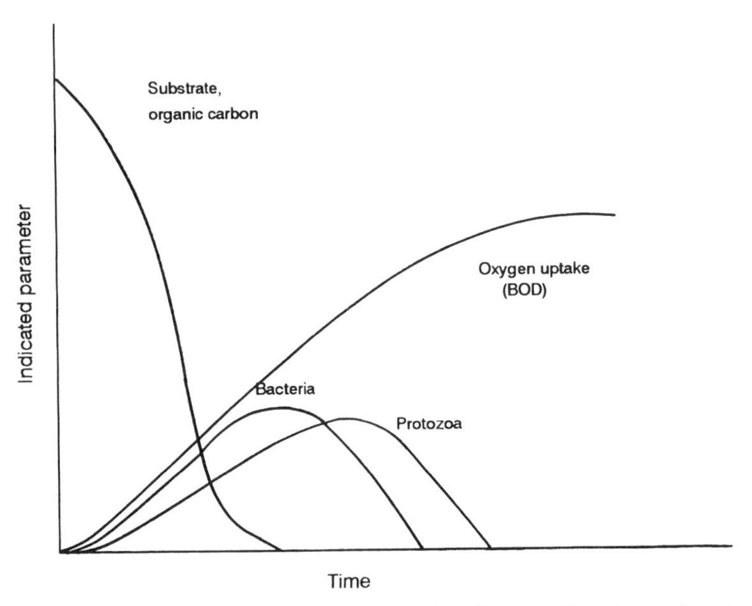

Fig. 7.3. Relationship between oxygen consumption, microbial growth, and organic carbon removal. Adapted from Gaudy (1972).

bial seed is usually composed of microorganisms commonly found in the wastewater environment (e.g., *Pseudomonas, Bacillus, Nocardia, Streptomyces*). A nitrification inhibitor is sometimes added to the sample to determine the carbonaceous BOD, which is exerted solely by heterotrophic microorganisms (Fitzmaurice and Gray, 1989). Dissolved oxygen concentration is determined at time 0 and after 5-day incubation, by means of an oxygen electrode, chemical procedures (e.g., Winkler test), or a manometric BOD apparatus (Hammer, 1986). The BOD test is carried out on a series of dilutions of the sample, the dilution depending on the source of the sample. When dilution water is not seeded, the BOD value is expressed in milligrams per liter, according to the following equation (APHA, 1989; Hammer, 1986):

$$BOD\ (\text{mg/L}) = \frac{D_1 - D_5}{P} \quad (7.1)$$

where D_1 = initial DO; D_5 = DO at day 5; and P = decimal volumetric fraction of wastewater utilized.

If the dilution water is seeded,

$$BOD\ (\text{mg/L}) = \frac{(D_1 - D_2) - (B_1 - B_2)f}{P} \quad (7.2)$$

where D_1 = initial DO of the sample dilution (mg/L); D_2 = final DO of the sample dilution (mg/L); P = decimal volumetric fraction of sample used; B_1 = initial DO of seed control (mg/L); B_2 = final DO of seed control (mg/L); and f = ratio of seed in sample to seed in control = (% seed in D_1)/(% seed in B_1).

The microbial biomass (bacteria and protozoa) consumes oxygen during the 5-day incubation period (carbonaceous BOD). Because of depletion of the carbon source, the carbonaceous BOD reaches a plateau (Fig. 7.4) called the ultimate carbonaceous BOD. The relationship between BOD at any time and ultimate carbonaceous BOD is given by the following equation (Hammer, 1986):

$$BOD\ (\text{at any time } t) = \text{ultimate } BOD\ (1 - 10^{-kt}) \quad (7.3)$$

k = BOD rate constant, with a value of approximately 0.1/day for domestic wastewaters. k depends on the following (Davis and Cornwell, 1985):

Type of waste. k varies from 0.15–0.30 in raw wastewater to 0.05–0.10 in surface water.

Acclimation of microorganisms to the type of waste analyzed. The BOD test should be conducted with acclimated microbial seeds.

Temperature. The BOD test is conducted at the standard temperature of 20°C. At other temperatures, k is given by the following formula:

$$k_T = k_{20}\, \theta^{T-20} \quad (7.4)$$

where T = temperature in °C; k_T = BOD rate constant at a given temperature T (day^{-1}); k_{20} = rate constant at 20°C; and θ = temperature coefficient = 1.135 at $4 < T < 20$°C and = 1.056 at $20 < T < 30$°C.

Toxic substances present in wastewater interfere with the BOD test. Another disadvantage of the BOD test is that it overlooks recalcitrant compounds.

7.2.1.2. Nitrogenous Oxygen Demand (NOD)

Autotrophic bacteria such as nitrifying bacteria also require oxygen to oxidize NH_4^+ to nitrate (see Chapter 3). The oxygen demand exerted by nitrifiers is called autotrophic BOD or nitrogenous oxygen demand (NOD) (Fig. 7.4). During the determination of the BOD of wastewater samples, nitrifiers exert an oxygen demand and their activity leads to higher BOD values, sometimes resulting in noncompliance of the wastewater treatment plant with federal or state regulations. It has been estimated that about 60% of the compliance violations in the United States are due to nitrification occurring during the BOD test (Dagues, 1981; Hall and Foxen, 1983). This phenomenon is significant in nitrified wastewater effluents, in which nitrifying bacteria may account for 24%–86% of the total BOD (Hall and Foxen, 1983; Washington et al., 1983). The theoretical nitrogenous oxygen demand is 4.57 g oxygen used per 1 g of ammonium oxidized to nitrate. However, the actual value is somewhat

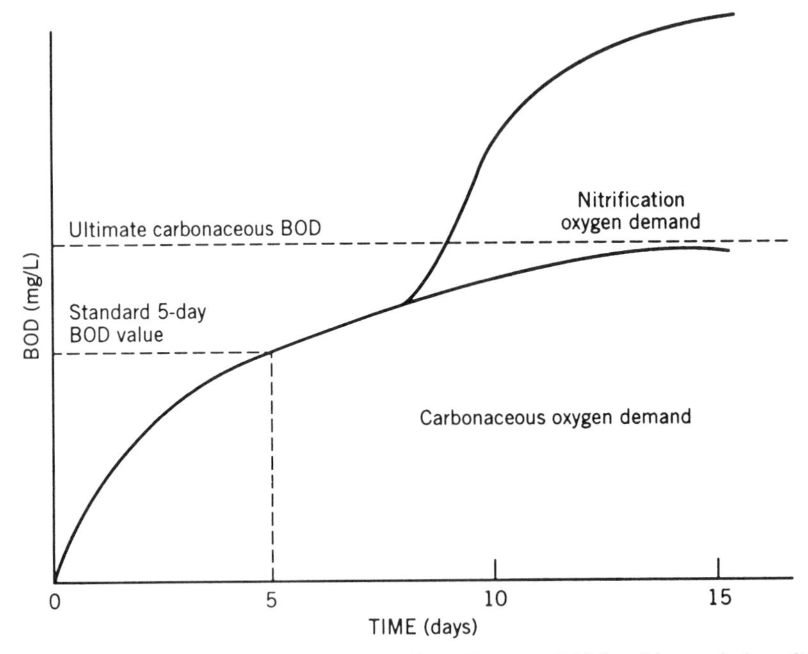

Fig. 7.4. Carbonaceous and autotrophic BOD. From Hammer (1986), with permission of the publisher.

lower because some of the nitrogen is incorporated into the bacterial cells (Davis and Cornwell, 1985). The nitrogenous oxygen demand (NOD) is as follows (Verstraete and van Vaerenbergh, 1986):

$$\text{NOD (mg/L)} = (\text{available N} - \text{assimilated N}) \times 4.33 \tag{7.5}$$

It is thus necessary to carry out an inhibited BOD test to distinguish between carbonaceous and nitrogenous BOD (APHA, 1989). Addition of 2-chloro-6(trichloro methyl) pyridine at a final concentration of 10 mg/L is recommended for inhibition of nitrification. It has been shown that this chemical does not suppress the oxidation of organic matter (Young, 1983).

BOD removal may vary from 70% in high-rate activated sludge to up to 95% in extended aeration activated sludge.

7.2.2. Chemical Oxygen Demand (COD)

COD is the amount of oxygen necessary to oxidize the organic carbon completely to CO_2 and H_20 and ammonia (Sawyer and McCarty, 1978).

COD is measured by oxidation with potassium dichromate ($K_2Cr_2O_7$) in the presence of sulfuric acid and silver and is expressed in milligrams per liter. In general, 1 g of carbohydrate or 1 g of protein is approximately equivalent to 1 g of COD. If the COD value is much higher than the BOD value, it means that the sample contains large amounts of organic compounds that are not easily biodegraded.

In untreated domestic wastewater, COD ranges between 250 and 1,000 mg/L. For typical untreated domestic wastewater, the BOD_5/COD ratio varies from 0.4 to 0.8 (Metcalf and Eddy, 1991).

The BOD_5, COD, and BOD_5/COD ratios of some typical wastewaters are shown in Table 7.3 (Verstraete and van Vaerenbergh, 1986).

7.2.3. Total Organic Carbon (TOC)

TOC represents the total organic carbon in a given sample and is independent of the oxidation state of the organic matter. TOC is determined by oxidation of the organic matter with heat and

TABLE 7.3. COD, BOD_5, and BOD_5/COD Ratios of Selected Wastewaters[a]

	COD (in mg/L)	BOD_5 (in mg/L)	BOD_5/COD
Domestic sewage			
Raw	500	300	0.60
After biological treatment	50	10	0.20
Slaughterhouse wastewater	3,500	2,000	0.57
Distillery vinasse	60,000	30,000	0.50
Dairy wastewater	1,800	900	0.50
Rubber factory	5,000	3,300	0.66
Tannery wastewater	13,000	1,270	0.10
Textile-dying			
Raw	1,360	660	0.48
After biological treatment	116	5	0.04
Draft mill effluent			
Raw	620	226	0.36
Biologically stabilized	250	30	0.12

[a]Adapted from Verstraete and Vaerenbergh (1986).

oxygen (aeration step is eliminated if volatile organic compounds are present in the sample) or chemical oxidants, followed by the measurement of the CO_2 liberated with an infrared analyzer (Hammer, 1986; Metcalf and Eddy, 1991).

7.3. OVERVIEW OF WASTEWATER TREATMENT

Physical forces as well as chemical and biological processes drive the treatment of wastewater. Treatment methods that rely on physical forces are called *unit operations*. These include screening, sedimentation, filtration, and flotation. Treatment methods based on chemical and biological processes are called *unit processes*. Chemical unit processes include disinfection, adsorption, and precipitation. Biological unit processes involve microbial activity which is responsible for degradation of organic matter and removal of nutrients (Metcalf and Eddy, 1991).

Wastewater treatment comprises the following four steps:

1. Preliminary treatment. The objective of this operation is to remove debris and coarse materials that may clog equipment in the plant.
2. Primary treatment. Treatment is brought about by physical processes (unit operations) such as screening and sedimentation.
3. Secondary treatment. Biological (e.g., activated sludge, trickling filter, oxidation ponds) and chemical (e.g., disinfection) unit processes are used to treat wastewater. Removal of nutrients also generally occurs during secondary treatment of wastewater.
4. Tertiary or advanced treatment. Unit operations and chemical unit processes are used to further remove BOD, nutrients, pathogens and parasites, and sometimes toxic substances.

The third part of this book will address the following aspects for each unit process:

- **Process microbiology**
  We will examine the types and activity of microorganisms involved in each process.
- **Public health microbiology**
  Some of the unit processes were specifically designed to remove pathogenic microorganisms and parasites. We will examine the effectiveness of each process in removing or inactivating microbial pathogens and parasites.

8

ACTIVATED SLUDGE PROCESS

8.1. INTRODUCTION

Activated sludge is a suspended-growth process that was started in England at the turn of the century. Since then, this process has been adopted worldwide as a secondary biological treatment for domestic wastewaters.

This process consists essentially of an aerobic treatment that oxidizes organic matter to CO_2 and H_2O, NH_4, and new cell biomass. Air is provided by diffused or mechanical aeration. The microbial cells form flocs that are allowed to settle in a clarification tank.

8.2. DESCRIPTION OF THE ACTIVATED SLUDGE PROCESS

8.2.1. Conventional Activated Sludge System

A conventional activated sludge process (Fig. 8.1) includes the following:

Aeration tank. Aerobic oxidation of organic matter is carried out in this tank. Primary effluent is introduced and mixed with return activated sludge (RAS) to form the mixed liquor, which contains 1,500–2,500 mg/L of suspended solids. Aeration is provided by mechanical means. An important characteristic of the activated sludge process is the recycling of a large proportion of the biomass. This makes the mean cell residence time (i.e., sludge age) much greater than the hydraulic retention time (Sterritt and Lester, 1988). This practice helps maintain a large number of microorganisms that effectively oxidize organic compounds in a relatively short time. The detention time in the aeration basin varies between 4 hr and 8 hr.

Sedimentation tank. This tank is used for the sedimentation of the microbial flocs (sludge) produced during the oxidation phase in the aeration tank. As mentioned earlier, a portion of the sludge in the clarifier is recycled back to the aeration basin and the remainder is "wasted" to maintain a proper food-to-microorganism (F/M) ratio.

We will now define some operational parameters commonly used in activated sludge (Davis and Cornwell, 1985; Verstraete and van Vaerenbergh, 1986).

Mixed-liquor suspended solids (MLSS). The content of the aeration tank in an activated sludge system is called mixed liquor. MLSS is the total amount of organic and mineral suspended solids, including microorganisms, in the mixed liquor. It is determined by filtering an aliquot of mixed liquor, drying the filter at 105°C, and determining the weight of solids in the sample.

Mixed-liquor volatile suspended solids (MLVSS). The organic portion of MLSS is represented by MLVSS, which comprises nonmicrobial organic matter as well as dead and live microorganisms and cellular debris (Nelson and Lawrence, 1980). MLVSS is determined following heating of dried filtered samples at 600–650°C, and represents approximately 65 to 75% of MLSS.

Food-to-microorganism ratio. This parameter indicates the organic load into the activated sludge system and is expressed in kilograms BOD per kilogram MLSS per day (Curds and

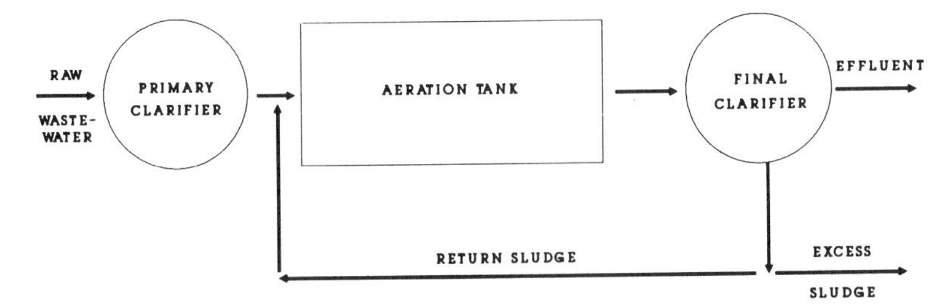

Fig. 8.1. Conventional activated sludge system.

Hawkes, 1983; Nathanson, 1986). It is expressed as follows:

$$F/M = \frac{Q \times BOD}{MLSS \times V} \qquad (8.1)$$

where Q = flow rate of sewage in million gallons per day (MGD); BOD = 5-day biochemical oxygen demand (mg/L); $MLSS$ = mixed liquor suspended solids (mg/L); and V = volume of aeration tank (gallons).

F/M is controlled by the rate of activated sludge wasting. The higher the wasting rate the higher the F/M ratio. For conventional aeration tanks the F/M ratio is 0.2–0.5 lb BOD_5/day/lb MLSS, but it can be higher (up to 1.5) for activated sludge when high-purity oxygen is used (Hammer, 1986). A low F/M ratio means that the microorganisms in the aeration tank are starved, leading to a more efficient wastewater treatment.

Hydraulic retention time (*HRT*). The hydraulic retention time is the average time spent by the influent liquid in the aeration tank of the activated sludge process; it is the reciprocal of the dilution rate D (Sterritt and Lester, 1988).

$$HRT = \frac{1}{D} = \frac{V}{Q} \qquad (8.2)$$

where V = volume of the aeration tank; Q = flow rate of the influent wastewater into the aeration tank; and D = dilution rate.

Sludge age. Sludge age is the mean residence time of microorganisms in the system. While the hydraulic retention time may be in the order of hours, the mean cell residence time may in the order of days. This parameter is the reciprocal of the microbial growth rate μ (see Chapter 2). Sludge age is given by the following formula (Hammer, 1986; Curds and Hawkes, 1983):

$$\text{Sludge age (days)} = \frac{MLSS \times V}{SS_e \times Q_e + SS_w \times Q_w} \qquad (8.3)$$

where $MLSS$ = mixed-liquor suspended solids (mg/L); V = volume of aeration tank (L); SS_e = suspended solids in wastewater effluent (mg/L); Q_e = quantity of wastewater effluent (m^3/day); SS_w = suspended solids in wasted sludge (mg/L); Q_w = quantity of wasted sludge (m^3/day).

Sludge age may vary from 5 to 15 days in conventional activated sludge. It varies with the season of the year and is higher in the winter than in the summer season (U.S. EPA, 1987a).

The important parameters controlling the operation of an activated sludge are organic loading rates, oxygen supply, and control and operation of the final settling tank. This tank has two functions: clarification and thickening. For routine operation, one must measure sludge settleability by determining the sludge volume index (SVI) (Forster and Johnston, 1987) (see Chapter 9 for more details).

8.2.2. Some Modifications of the Conventional Activated Sludge Process

There are several modifications of the conventional activated sludge process (Nathanson, 1986; U.S. EPA, 1977) (Fig. 8.2.).

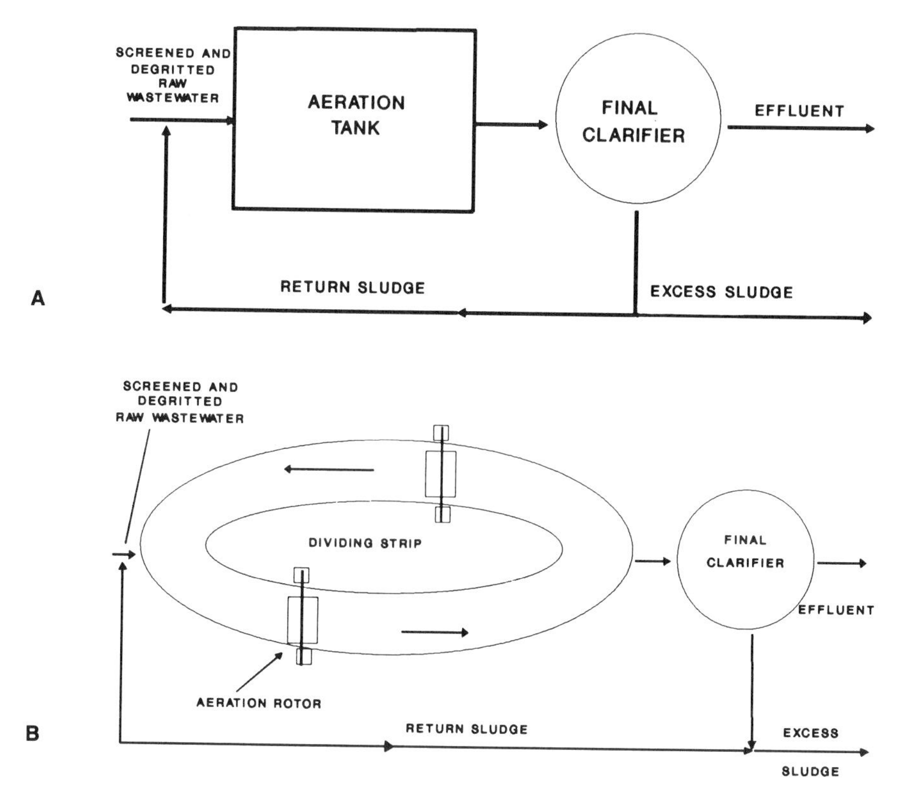

Fig. 8.2. Modifications of the activated sludge process. **A.** Extended aeration system. **B.** Oxidation ditch. From U.S. EPA (1977), with permission of the publisher.

8.2.2.1. Extended Aeration System (Fig. 8.2A)

This process, used in package treatment plants, has the following features:

1. The aeration time is much longer (about 30 hr) than in conventional systems. The sludge age is also longer and can be extended to more than 15 days.
2. The wastewater influent entering the aeration tank has not been treated by primary settling.
3. The system operates at much lower F/M ratio (generally < 0.1 lb BOD/day/lb MLSS) than conventional systems (0.2–0.5 lb BOD/day/lb).
4. This system requires less aeration than conventional treatment and is mainly suitable for small communities that use package treatment.

8.2.2.2. Oxidation Ditch (Fig. 8.2B)

The oxidation ditch consists of an aeration oval channel with one or more rotating rotors for wastewater aeration. This channel receives screened wastewater and has a hydraulic retention time of approximately 24 hr.

8.2.2.3. Step Aeration

The primary effluent enters the aeration tank through several points, thus improving its distribution into the tank and making more efficient use of oxygen. This increases the treatment capacity of the system.

8.2.2.4. *Contact Stabilization*

Following contact of the wastewater with the sludge in a small contact tank for a short period of time (20–40 min), the mixture flows to a clarifier and the sludge is returned to a stabilization tank with a retention time of 4–8 hr. This system produces less sludge.

8.2.2.5. *Completely Mixed Aerated System*

A completely mixed system allows a more uniform aeration of the wastewater in the aeration tank. This system can sustain shock and toxic loads.

8.2.2.6. *High-Rate Activated Sludge*

This system is used for the treatment of high-strength wastes and is operated at much higher BOD loadings than those encountered in the conventional activated sludge process. This results in shorter hydraulic retention periods (i.e., shorter aeration periods). The system is operated at higher MLSS concentrations.

8.2.2.7. *Pure Oxygen Aeration*

The pure oxygen aeration system is based on the principle that the rate of transfer of oxygen is higher for pure oxygen than for atmospheric oxygen. This results in higher availability of dissolved oxygen, leading to improved treatment and reduced production of sludge.

8.3. THE BIOLOGY OF ACTIVATED SLUDGE

The two main goals of the activated sludge system are (1) *oxidation* of the biodegradable organic matter in the aeration tank (soluble organic matter is thus converted to new cell mass); and (2) *flocculation,* that is, separation of the newly formed biomass from the treated effluent.

8.3.1. Survey of Organisms Present in Activated Sludge Flocs

The activated sludge flocs contain bacterial cells as well as inorganic and organic particles. Floc size varies between < 1 μm (the size of some bacterial cells) to 1,000 μm or more (Parker et al., 1971; U.S. EPA, 1987a) (Fig. 8.3). Viable cells in the floc, as measured by ATP analysis and dehydrogenase activity, would account for 5%–20% of the total cells (Weddle and Jenkins, 1971). Some investigators maintain that the active fraction of bacteria in activated sludge flocs represent only 1%–3% of total bacteria (Hanel, 1988).

The following are the microorganisms to be observed in activated sludge flocs.

8.3.1.1. *Bacteria*

Bacteria constitute the major component of activated sludge flocs. More than 300 strains of bacteria thrive in activated sludge. They are responsible for the oxidation of organic matter and for nutrient transformations, and they produce polysaccharides and other polymeric materials that aid in the flocculation of microbial biomass. The major genera found in the flocs are *Zooglea, Pseudomonas, Flavobacterium, Alcaligenes, Bacillus, Achromobacter, Corynebacterium, Comomonas, Brevibacterium,* and *Acinetobacter,* as well as filamentous microorganisms. Some examples of filamentous microorganisms are the sheathed bacteria (e.g., *Sphaerotilus*) and gliding bacteria (e.g., *Beggiatoa, Vitreoscilla*), which are responsible for sludge bulking (see Chapter 9 for more details on these bacteria). As the oxygen level in the flocs is diffusion-limited, the number of active aerobic bacteria decreases as the floc size increases (Hanel, 1988). The inner region of relatively large flocs favors the development of strictly anaerobic bacteria such as methanogens. It has been proposed that the presence of methanogens can be explained by the formation of several anaerobic pockets inside the flocs or by the tolerance of certain methanogens to oxygen (Fig. 8.4) (Wu et al., 1987). Thus, activated sludge could be a convenient and suitable seed material for starting anaerobic reactors.

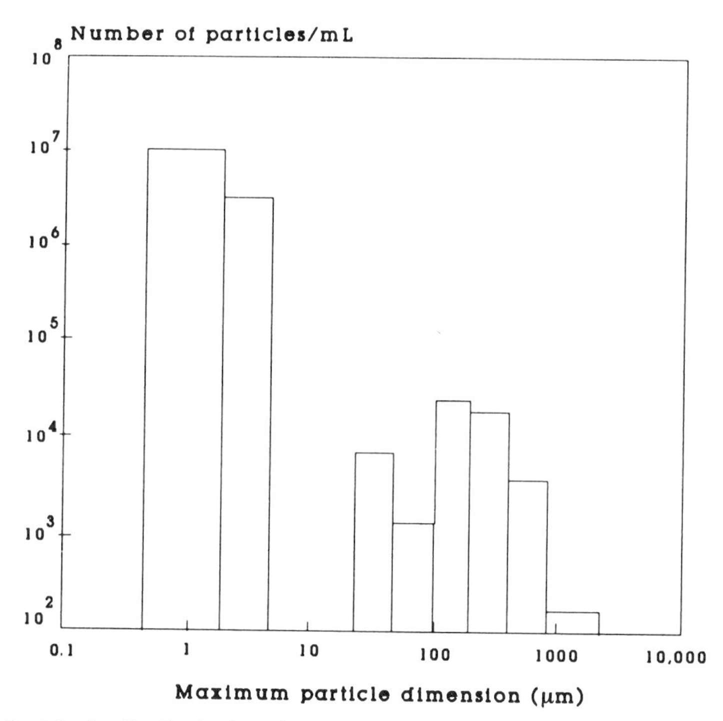

Fig. 8.3. Particle size distribution in activated sludge. From Parker et al. (1971), with permission of the publisher.

Total aerobic bacterial counts in standard activated sludge are in the order of 10^8 CFU/mg of sludge. Table 8.1 displays some bacterial genera found in standard activated sludge. The majority of the bacterial isolates were identified as *Comamonas–Pseudomonas* species. An analy-

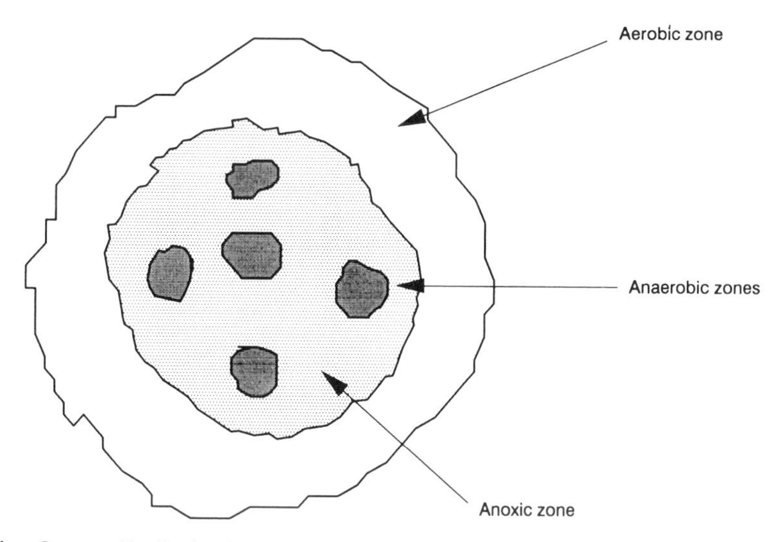

Fig. 8.4. Oxygen distribution in an activated sludge floc. From Wu et al. (1987), with permission of the publisher.

TABLE 8.1. Distribution of Aerobic Heterotrophic Bacteria in Standard Activated Sludge[a]

Genus or group	Percentage of total isolates
Comamonas–Pseudomonas	50.0
Alcaligenes	5.8
Pseudomonas (fluorescent group)	1.9
Paracoccus	11.5
Unidentified (gram-negative rods)	1.9
Aeromonas	1.9
Flavobacterium–Cytophaga	13.5
Bacillus	1.9
Micrococcus	1.9
Coryneform	5.8
Arthrobacter	1.9
Aureobacterium–Microbacterium	1.9

[a]Adapted from Hiraishi et al. (1989).

sis of the quinone composition of activated sludge revealed that ubiquinone Q-8 was the predominant quinone (Hiraishi et al., 1989). *Caulobacter,* a stalked bacterium generally found in organically poor waters, has also been isolated from wastewater treatment plants in general and activated sludge in particular (MacRae and Smit, 1991).

Zoogloea are exopolysaccharide-producing bacteria that produce typical finger-like projections and are found in wastewater and other organically enriched environments (Fig. 8.5) (Norberg and Enfors, 1982; Unz and Farrah, 1976; Williams and Unz, 1983). They are isolated by using enrichment media containing *m*-butanol, starch, or *m*-toluate as the carbon source. They are found in various stages of wastewater treatment but their numbers comprise only 0.1%–1% of the total bacterial numbers in the mixed liquor (Williams and Unz, 1983). The relative importance of these bacteria in wastewater treatment needs further investigation.

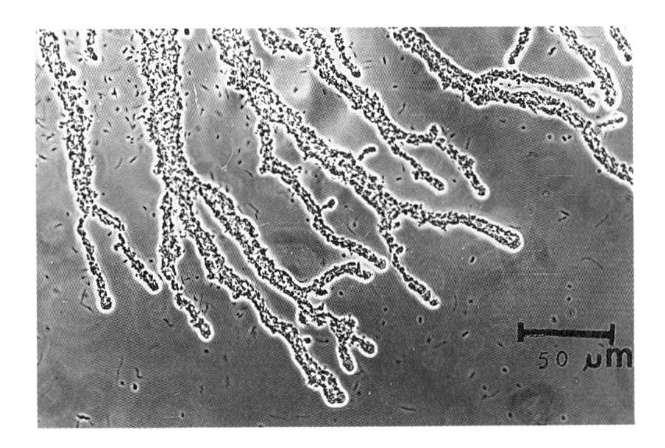

Fig. 8.5. *Zooglea ramigera.* Courtesy of Samuel R. Farrah.

Activated sludge flocs also harbor autotrophic bacteria such as nitrifiers (*Nitrosomonas, Nitrobacter*), which convert ammonium to nitrate (see Chapter 3) and phototrophic bacteria such as the purple nonsulfur bacteria (Rhodospirillaceae), which are detected at concentrations of approximately 10^5 cells per ml. The purple and green sulfur bacteria are found at much lower levels. However, phototrophic bacteria probably play a minor role in BOD removal in activated sludge (Madigan, 1988; Siefert et al., 1978). Figure 8.6 is a scanning electron micrograph of an activated sludge floc.

8.3.1.2. *Fungi*

Activated sludge does not usually favor growth of fungi although some fungal filaments are oc-

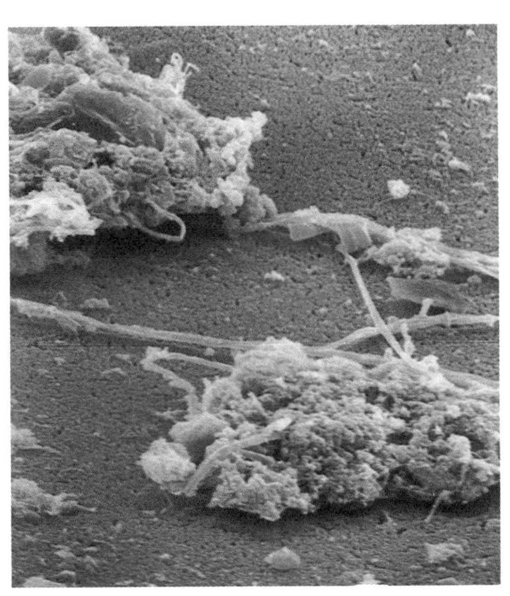

Fig. 8.6. Activated sludge flocs observed via scanning electron microscopy. Courtesy of R.J. Dutton and G. Bitton.

casionally observed in activated sludge flocs. Fungi may grow abundantly under specific conditions of low pH, toxicity, and nitrogen-deficient wastes. The predominant genera found in activated sludge are *Geotrichum, Penicillium, Cephalosporium, Cladosporium,* and *Alternaria* (Pipes and Cooke, 1969; Tomlinson and Williams, 1975).

Sludge bulking (see Chapter 9) may result from the abundant growth of *Geotrichum candidum,* which is favored by low pH from acid wastes.

8.3.1.3. Protozoa

Protozoa are significant predators of bacteria in activated sludge as well as in natural aquatic environments (Curds, 1982; Drakides, 1980; Fenchel and Jorgensen, 1977; LaRiviere, 1977). Protozoan grazing on bacteria can be experimentally determined by measuring the uptake of ^{14}C- or ^{35}S-labeled bacteria or fluorescently labeled bacteria (Hoffmann and Atlas, 1987; Sherr et al., 1987). Such grazing can be significantly reduced in the presence of toxicants. For example, *Aspidisca costata* grazing on bacteria in activated sludge is reduced in the presence of cadmium (Hoffmann and Atlas, 1987). The protozoa most often found in activated sludge have been thoroughly described (Fig. 8.7) (Dart and Stretton, 1980; Edeline, 1988; Eikelboom and van Buijsen, 1981).

Ciliates. The cilia, which give these organisms their names, are used for locomotion and for pushing food particles into the mouth. They are subdivided into *free, creeping,* and *stalked* ciliates. Free ciliates feed on free-swimming

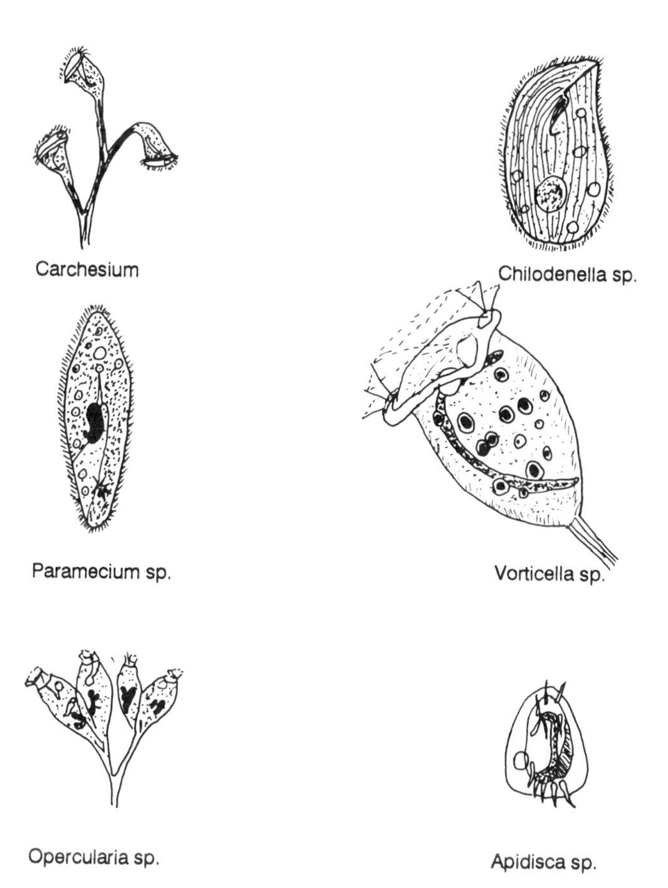

Fig. 8.7. Some protozoa found in wastewater, particularly in activated sludge. Adapted from Dart and Stretton (1980).

bacteria. The most important genera found in activated sludge are *Chilodonella, Colpidium, Blepharisma, Euplotes, Paramecium, Lionotus, Trachelophyllum,* and *Spirostomum.* Creeping ciliates graze on bacteria on the surface of activated sludge flocs. Two important genera are *Aspidisca* and *Euplotes.* Stalked ciliates are attached by their stalk to the flocs. The stalk has a muscle (myoneme) that allows it to contract. The predominant stalked ciliates are *Vorticella, Carchesium, Opercularia,* and *Epistylis.*

Flagellates. These protozoa move by one or several flagella. They take up food by mouth or by absorption through their cell wall. Some important flagellates found in wastewater are *Bodo* spp., *Pleuromonas* spp., *Monosiga* spp., *Hexamitus* spp., and the colonial protozoa, *Poteriodendron* spp.

Rhizopoda (amoebae). Amoebae move slowly by pseudopods ("false feet"), which are temporary projections of the cells. This group is subdivided into amoeba (e.g., *Amoeba proteus*) and thecamoeba, which are surrounded by a shell (e.g., *Arcella*).

Flagellated protozoa and free-swimming ciliates are usually associated with high bacterial concentrations (> 10^8 cells per 1 ml), whereas stalked ciliates occur at low bacterial concentrations (< 10^6 per 1 ml). Protozoa contribute significantly to the reduction of BOD, suspended solids, and numbers of bacteria, including pathogens (Curds, 1975). The protozoan species composition of activated sludge may indicate the BOD removal efficiency of the process. For example, the presence of large numbers of stalked ciliates and rotifers indicates a low BOD. The ecological succession of microorganisms during activated sludge treatment is illustrated in Figure 8.8 (Bitton, 1980a).

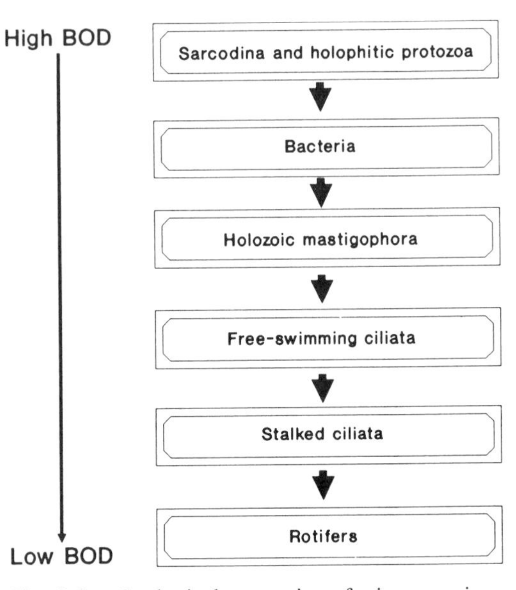

Fig. 8.8. Ecological succession of microorganisms in activated sludge. From Bitton (1980a), with permission of the publisher.

8.3.1.4. Rotifers

Rotifers are metazoa (i.e., multicellular organisms) with sizes varying from 100 μm to 500 μm. Their body, anchored to a floc particle, frequently "stretches out" from the floc surface (Doohan, 1975; Eikelboom and van Buijsen, 1981). The rotifers found in wastewater treatment plants belong to two main orders, Bdelloidea (e.g., *Philodina* spp., *Habrotrocha* spp.) and Monogononta (e.g., *Lecane* spp., *Notommata* spp.). The four most common rotifers found in activated sludge and trickling filters are illustrated in Figure 8.9 (Curds and Hawkes, 1975). The role of rotifers in activated sludge is twofold: (1) They help in the removal of freely suspended bacteria (i.e., nonflocculated bacteria); (2) they contribute to floc formation by producing fecal pellets surrounded by mucus.

The presence of rotifers at later stages of activated sludge treatment is due to the fact that these animals display a strong ciliary action that helps in feeding on reduced numbers of suspended bacteria (their ciliary action is stronger than that of protozoa).

8.3.2. Organic Matter Oxidation in the Aeration Tank

Domestic wastewater has a C:N:P ratio of 100:5:1, which satisfies the C, N, and P requirements of a wide variety of microorganisms. Organic matter in wastewater occurs as soluble, colloidal, and particulate fractions (see Chapter 7). The soluble organic matter serves as a food

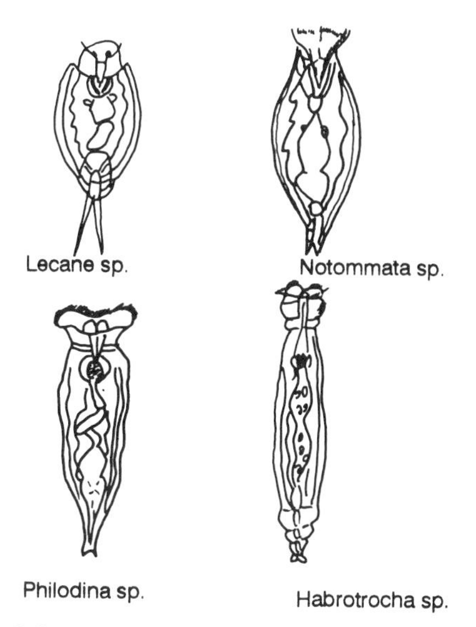

Fig. 8.9. Rotifers most commonly found in activated sludge. From Curds and Hawkes (1975), with permission of the publisher.

source for the heterotrophic microorganisms in the mixed liquor. It is quickly removed by adsorption and coflocculation, as well as absorption and oxidation by microorganisms. Aeration for only a few hours leads to the transformation of soluble BOD into microbial biomass. Aeration serves two purposes: (1) supplying oxygen to the aerobic microorganisms, and (2) keeping the activated sludge flocs in constant agitation to provide an adequate contact between the flocs and the incoming wastewater. An adequate dissolved oxygen concentration is also necessary for the activity of heterotrophic and autotrophic microorganisms, especially nitrifying bacteria. The dissolved oxygen level must be in the 0.5–0.7 mg/L range. Nitrification ceases when DO is below 0.2 mg/L (Dart and Stretton, 1980).

Figure 8.10 (Curds and Hawkes, 1983) summarizes the degradation and biosynthetic reactions that occur in the aeration tank of an activated sludge process.

8.3.3. Sludge Settling

The mixed liquor is transferred from the aeration tank to the settling tank, where the sludge separates from the treated effluent. A portion of the sludge is recycled to the aeration tank and the remaining sludge is wasted and transferred to an aerobic or anaerobic digester for further treatment.

Microbial cells occur in aggregates or flocs, the density of which is sufficient for settling in the clarifying tank. Sludge settling depends on the F/M ratio and sludge age. Good settling occurs when the sludge microorganisms are in the

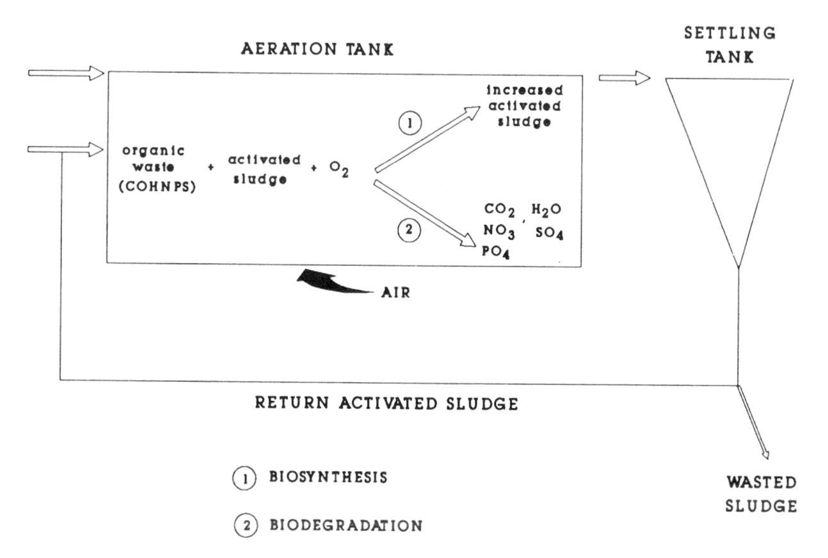

Fig. 8.10. Removal of organic matter by the activated sludge process. Adapted from Curds and Hawkes (1983).

endogenous phase, which occurs when carbon and energy sources are limited and when the microbial specific growth rate is low. Good sludge settling with subsequent efficient BOD removal occurs at low F/M ratios (i.e., high MLSS concentrations). Conversely, a high F/M ratio is conducive to poor sludge settleability. In municipal wastewaters, the optimum F/M ratio is between 0.2 and 0.5. (Gaudy and Gaudy, 1988; Hammer, 1986). A mean cell residence time of 3–4 days is necessary for effective settling (Metcalf and Eddy, 1991). Poor settling can also be caused by sudden changes in physical parameters (e.g., temperature, pH), absence of nutrients (e.g., N, P, micronutrients), and presence of toxicants (e.g., heavy metals), which can cause a partial deflocculation of activated sludge (Chudoba, 1989).

A model explaining the structure of an activated sludge floc has been proposed. According to this model, filamentous microorganisms form backbones, to which zoogleal (floc-forming) microorganisms attach to form strong flocs (Sezgin et al., 1978). This model does not explain, however, the absence of filamentous backbones in well-flocculating activated sludges (Chudoba, 1989). It is now accepted that polymers produced by some activated sludge microorganisms are mainly responsible for floc formation. It was first thought that the production of an intracellulaı storage product, poly-β-hydroxybutyric acid, was responsible for bacterial aggregation. It is now generally accepted that extracellular slimes produced by *Zooglea ramigera* and other activated sludge microorganisms play a leading role in bacterial flocculation and floc formation (Friedman et al., 1969; Harris and Mitchell, 1973; Norberg and Enfors, 1982; Pavoni et al., 1972; Tenney and Stumm, 1965; Vallom and McLoughlin, 1984). These polymers are produced during the endogenous phase of growth and help bridge the microbial cells to form a three-dimensional matrix (Parker et al., 1971; Pavoni et al., 1972; Tago and Aida, 1977). Figure 8.11 (Pavoni et al., 1972) shows the correlation between microbial flocculation and exocellular polymer production by activated sludge microorganisms. The polymer concentration increases as the solid retention time increases (Chao and Keinath, 1979). Polymer production by activated sludge microorganisms has been studied in the laboratory under batch and continuous culture conditions. The polymers have a high molecular weight (MW > 10,000) and constitute 36%–62% of the total COD of the cultures. They are composed of sugars (e.g., glucose, galactose), amino sugars, uronic acids (glucuronic and galacturonic acids), and amino acids, indicating their heteropolysaccharidic character. These heteropolysaccharides are refractory to biodegradation (Hejzlar and Chudoba, 1986; Horan and Eccles, 1986). Cations (e.g., Ca, Mg) may bridge the carboxyl groups of extracellular polysaccharides of floc-forming bacteria. A possible structure of the activated sludge floc is displayed in Figure 8.12 (Forster and Dallas-Newton, 1980). Excessive production of extracellular polysaccharides can be responsible for nonfilamentous bulking, a condition consisting of loose flocs that do not settle well. This condition contrasts with filamentous bulking, which is caused by the excessive growth of filamentous bacteria. Microbial flocculation can be enhanced by adding commercial polyelectrolytes or by adding iron and aluminum salts as coagulants (see Chapter 9). Some investigators have suggested that hydrophobic interactions are also involved in the flocculation of activated sludge (Urbain et al., 1993).

The conventional way of monitoring for sludge settleability is by determining the sludge volume index (SVI). Mixed liquor drawn from the aeration tank is introduced into a 1-L graduated cylinder and allowed to settle for 30 min. Sludge volume is recorded. SVI, which is the volume occupied by one gram of sludge, is given by the following formula:

$$SVI\ (\text{ml/g}) = \frac{SV \times 1{,}000}{MLSS} \tag{8.4}$$

where SV = volume of the settled sludge in the graduated cylinder (ml), and $MLSS$ = mixed-liquor suspended solids (mg/L).

In a conventional activated sludge plant (with MLSS < 3,500 mg/L) the normal range of SVI is 50–150 ml/g. This topic will be discussed further in Chapter 9.

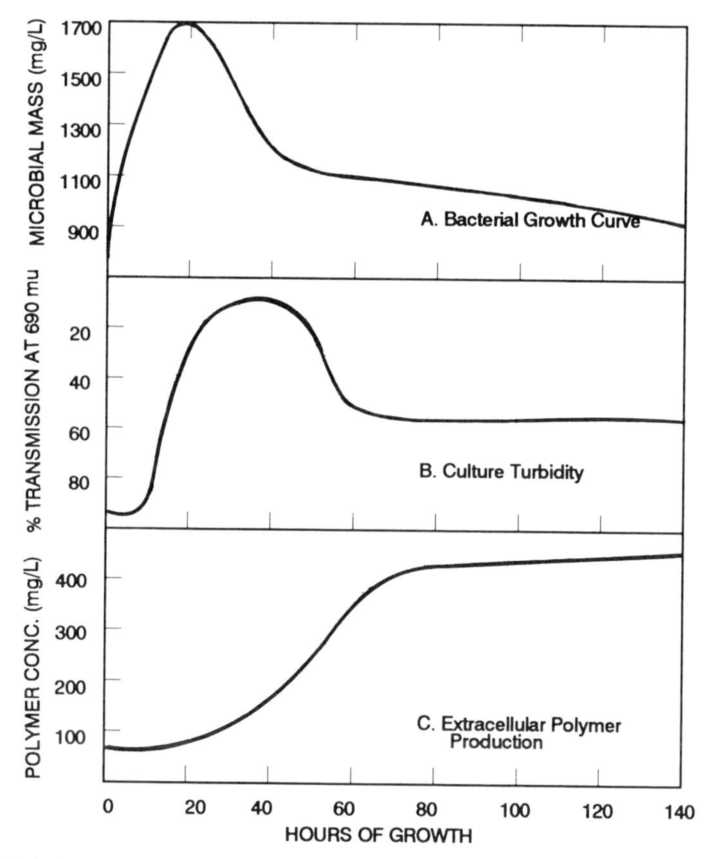

Fig. 8.11. Relationship between bacterial growth (**A**), bacterial flocculation (**B**), and polysaccharide production (**C**). Adapted from Pavoni et al. (1972).

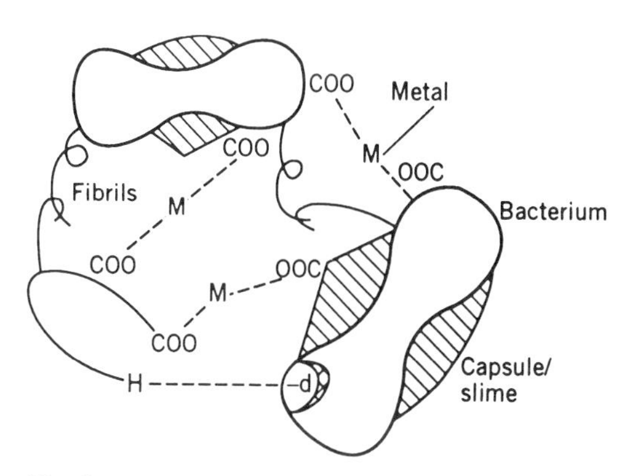

Fig. 8.12. Possible structure of an activated sludge floc. From Forster and Dallas-Newton (1980), with permission of the publisher.

8.4. NUTRIENT REMOVAL BY THE ACTIVATED SLUDGE PROCESS

8.4.1. Nitrogen Removal

We have seen in Chapter 3 that nitrogen can be removed from incoming wastewater by chemical –physical means (e.g., breakpoint chlorination or air stripping to remove ammonia) or by biological means, which consist of nitrification followed by denitrification. We will now discuss these two biological processes in suspended-growth bioreactors.

8.4.1.1. Nitrification in Suspended-Growth Reactors

Several factors control nitrification kinetics in activated sludge plants. These factors, discussed in detail in Chapter 3, are ammonia/nitrite concentration, oxygen concentration, pH, temperature, BOD_5/TKN ratio, and the presence of toxic chemicals (Grady and Lim, 1980; Hawkes, 1983; Metcalf and Eddy, 1991).

The establishment of a nitrifying population in activated sludge depends on the wastage rate of the sludge and therefore on the BOD load, MLSS, and retention time. The growth rate of nitrifying bacteria (μ_n) must be higher than the growth rate (μ_h) of heterotrophs in the system (Barnes and Bliss, 1983; Christensen and Harremoes, 1978; Hawkes, 1983).

Since $\mu_h = 1/\theta \rightarrow \mu_n >= 1/\theta$, θ = detention time.

In reality, the growth rate of nitrifiers is lower than that of heterotrophs in sewage; therefore a long sludge age is necessary for the conversion of ammonia to nitrate. Nitrification is expected at a sludge age above 4 days (Hawkes, 1983).

$$\mu_h = 1/\theta = Y_h\, q_h - K_d \qquad (8.5)$$

where Y_h = heterotrophic yield coefficient (lb BOD obtained/lb BOD removed/day; q_h = rate of substrate removal (lb BOD removed/lb VSS/day; μ_h = specific growth rate of heterotrophs (day^{-1}); K_d = decay coefficient (day^{-1}).

Y_h and K_d are assumed to be constant. Therefore, μ_h is reduced by decreasing q_h (U.S. EPA, 1975).

There are two nitrification systems in suspended-growth reactors (Fig. 8.13) (U.S. EPA, 1977):

1. *Combined carbon–oxidation nitrification (single-stage nitrification system).* This process is characterized by a high BOD_5/TKN ratio and a low population of nitrifiers—most of the oxygen requirement is exerted by heterotrophs.

2. *Two-stage nitrification.* Nitrification proceeds well in two-stage activated sludge systems. BOD is removed in the first stage, while nitrifiers are active in the second stage.

8.4.1.2. Denitrification in Suspended-Growth Reactors

Nitrification must be followed by denitrification to remove nitrogen from wastewater. Since the nitrate concentration is often much greater (>1 mg/L) than the half-saturation constant for denitrification (K_s = 0.08 mg/L), the rate of denitrification is independent of nitrate concentration but depends on the concentration of biomass and electron donor (e.g., methanol) in wastewater (Barnes and Bliss, 1983). The conventional activated sludge system can be modified to encourage denitrification. The systems used are the following (Fig. 8.14) (Curds and Hawkes, 1983).

Single sludge system (Fig. 8.14A). This system comprises a series of aerobic and anaerobic tanks in lieu of a single aeration tank.

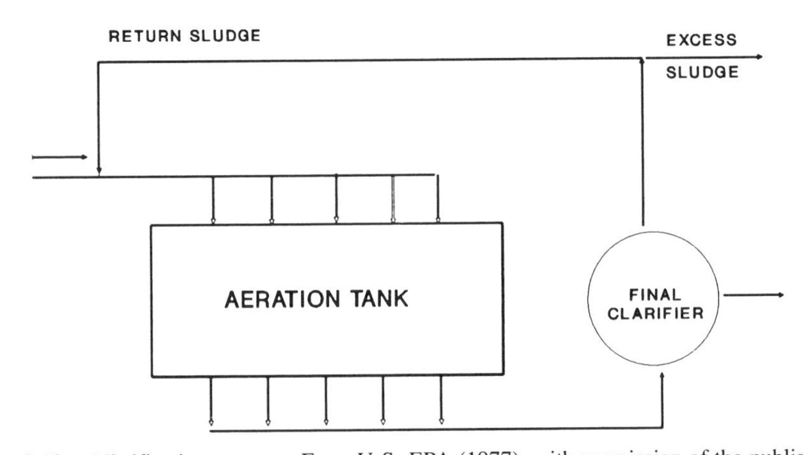

Fig. 8.13. Nitrification systems. From U.S. EPA (1977), with permission of the publisher.

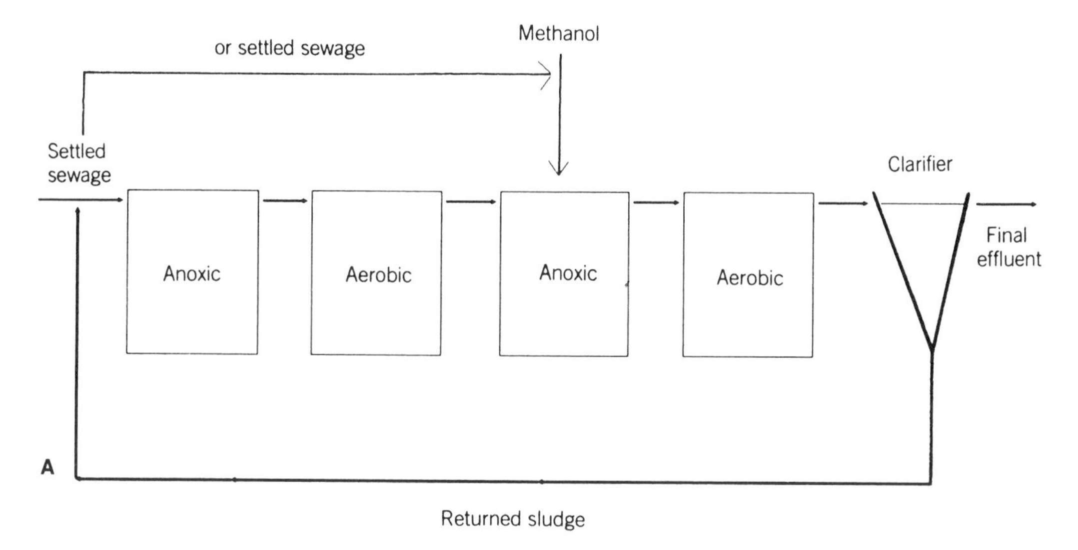

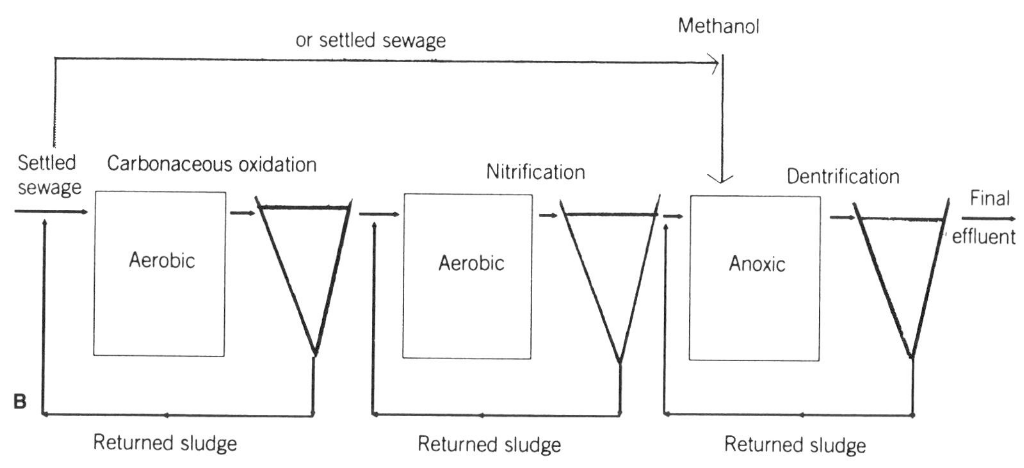

Fig. 8.14. Denitrification systems: **A.** Single-sludge system **B.** Multisludge system. From Curds and Hawkes (1983), with permission of the publisher.

Multisludge system (Fig. 8.14B). Carbonaceous oxidation, nitrification, and denitrification are carried out in three separate systems. Methanol or settled sewage serves as the source of carbon for denitrifiers.

Bardenpho Process. This process, illustrated in Figure 8.15, was developed by Barnard in South Africa (U.S. EPA, 1975). The process consists of two aerobic and two anoxic tanks followed by a sludge settling tank. Tank 1 is anoxic and is used for denitrification, with wastewater used as a carbon source. Tank 2 is an aerobic tank utilized for both carbonaceous oxidation and nitrification. The mixed liquor from this tank, which contains nitrate, is returned to Tank 1. The anoxic Tank 3 removes the nitrate remaining in the effluent by denitrification. Finally, Tank 4 is an aerobic tank used to strip the nitrogen gas that results from denitrification, thus improving mixed liquor settling.

8.4.2. Phosphorus Removal

In wastewater treatment plants, phosphorus is removed by chemical means (e.g., phosphorus precipitation with iron or aluminum) and by mi-

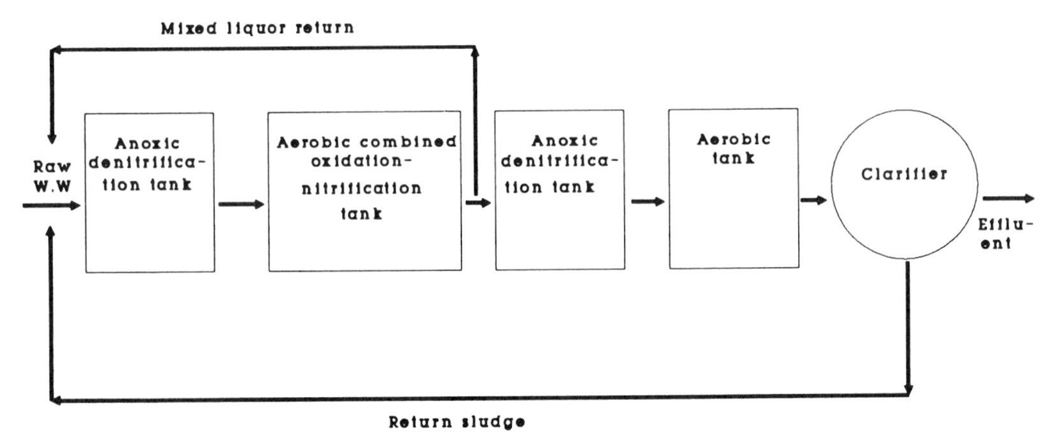

Fig. 8.15. Denitrification systems: Bardenpho process. From U.S. EPA (1975), with permission of the publisher.

crobiological means. In Chapter 3, we discussed the mechanisms of microorganism-mediated chemical precipitation and uptake of phosphorus. We will now briefly describe the proprietary phosphate removal processes. All of the processes incorporate aerobic and anaerobic stages and are based on phosphorus uptake during the aerobic stage and its subsequent release during the anaerobic stage (Manning and Irvine, 1985; Meganck and Faup, 1988; U.S. EPA, 1987b). The commercial systems can be divided into mainstream and sidestream processes. The most popular are described below.

8.4.2.1. Mainstream Processes

Mainstream phosphorus removal processes are illustrated in Figure 8.16.

A/O (aerobic/oxic) process. The A/O process consists of a modified activated sludge system that includes an anaerobic zone (detention time = 0.5–1 hr) upstream of the conventional aeration tank (detention time = 1–3 hr) (U.S. EPA, 1987b). Figure 8.17 (Deakyne et al., 1984) illustrates the microbiology of the A/O process: During the anaerobic phase, inorganic phosphorus is released from the cells as a result of polyphosphate hydrolysis. The energy liberated is used for the uptake of BOD from wastewater. Removal efficiency is high when the BOD/phosphorus ratio exceeds 10 (Metcalf and Eddy, 1991). During the aerobic phase, soluble phosphorus is taken up by bacteria, which synthesize polyphosphates, using the energy released from BOD oxidation.

The A/O process results in the removal of phosphorus and BOD from effluents and produces a phosphorus-rich sludge. The key features of this process are the relatively low SRT (solid retention time) and high organic loading rates (U.S. EPA, 1987b).

Bardenpho process. This system, described in section 8.4.1.2, removes nitrogen as well as phosphorus by a nitrification–denitrification process (Meganck and Faup, 1988; Barnard, 1974).

UCT process. The treatment train in the UCT process includes three tanks (anaerobic → anoxic → aerobic), followed by a final clarifier. In order to have strictly anaerobic conditions in the anaerobic tank, sludge is not recycled from the final clarifier to the first tank (Meganck and Faup, 1988; U.S. EPA, 1987b).

8.4.2.2. Sidestream Processes

PhoStrip is a sidestream process that is designed for phosphorus removal by biological as well by chemical means (Fig. 8.18) (Tetreault et al., 1986). A sidestream return activated sludge is diverted to an anaerobic tank called the anaerobic phosphorus stripper, where phosphorus is

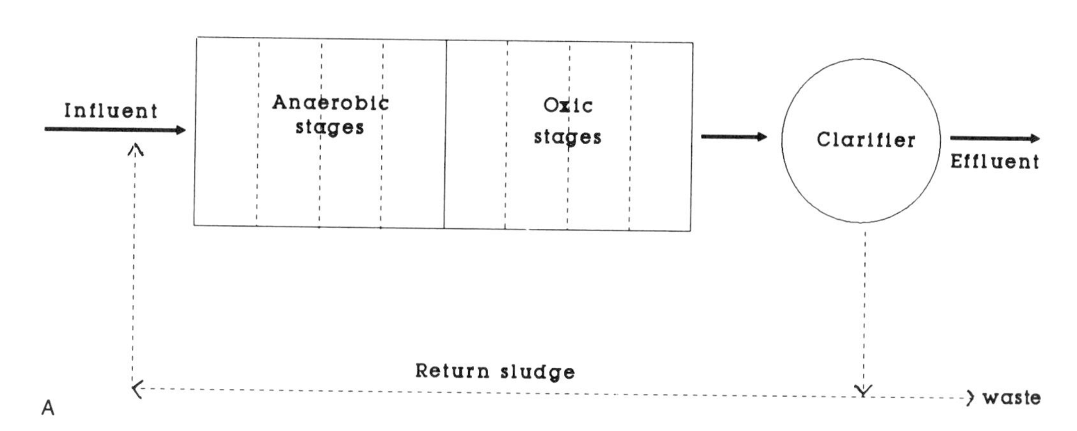

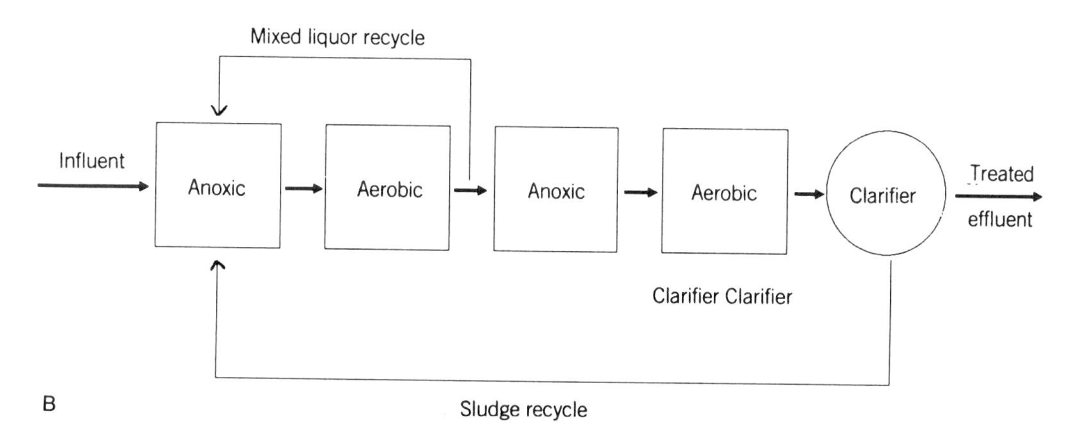

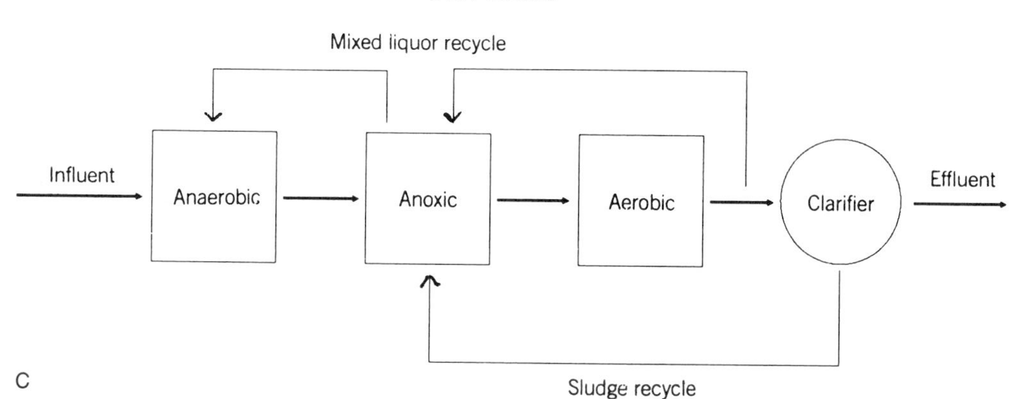

Fig. 8.16. Propietary biological phosphorus removal processes: Mainstream processes. **A.** A/O process. **B.** Bardenpho process. **C.** UCT process. Adapted from Meganck and Faup (1988) and U.S. EPA (1987b).

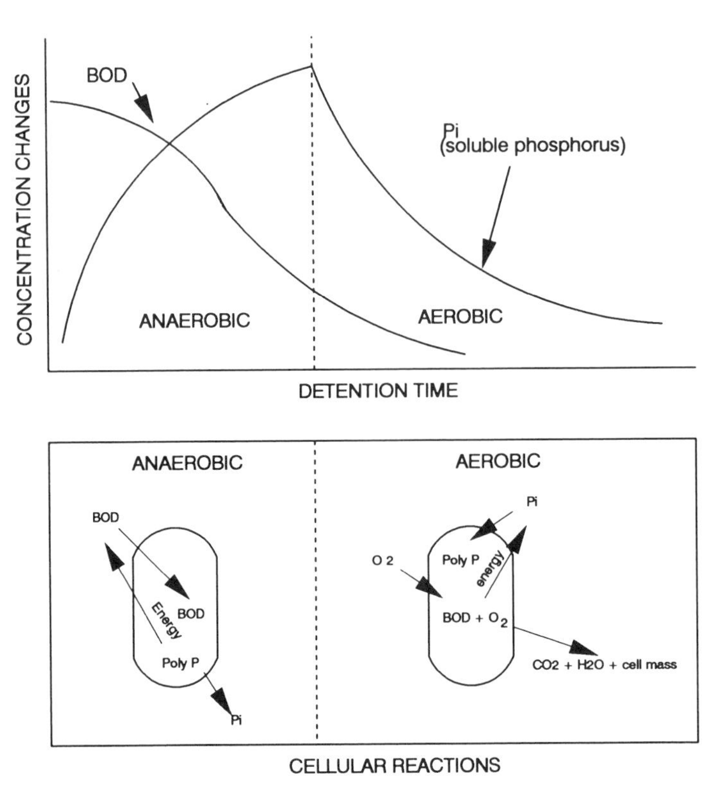

Fig. 8.17. Microbiology of the A/O process. From Deakyne et al. (1984), with permission of the publisher.

released from the sludge. The phosphorus-rich supernatant is treated with lime to remove phosphorus by chemical precipitation. The solid retention time in the anaerobic tank is 5–20 hr. Phosphorus uptake in the aeration basin is ensured when the DO level is higher than 2 mg/L. The PhoStrip process can help achieve an effluent total P concentration of less than 1 mg/L if the soluble BOD_5:soluble P is low (between 12 and 15) (Tetreault et al., 1986).

8.5. PATHOGEN REMOVAL BY ACTIVATED SLUDGE

Both components (aeration and sedimentation tanks) of the activated sludge process affect to some extent the removal/inactivation of pathogens and parasites. During the aeration phase, environmental factors (e.g., temperature, sunlight) and biological factors (e.g., inactivation by antagonistic microorganisms), and possibly aeration, have an impact on pathogen/parasite survival. Floc formation during the aeration phase is also instrumental in removing undesirable microorganisms. During the sedimentation phase, certain organisms (e.g., parasites) undergo sedimentation while floc-entrapped microbial pathogens settle readily in the tank. As compared with other biological treatment processes, activated sludge is relatively efficient in removing pathogenic microorganisms and parasites from incoming primary effluents.

8.5.1. Bacteria

Activated sludge is generally more efficient than trickling filters for removing indicator (e.g., coliforms) and pathogenic (*Salmonella*) bacteria. The removal efficiency may vary from 80% to

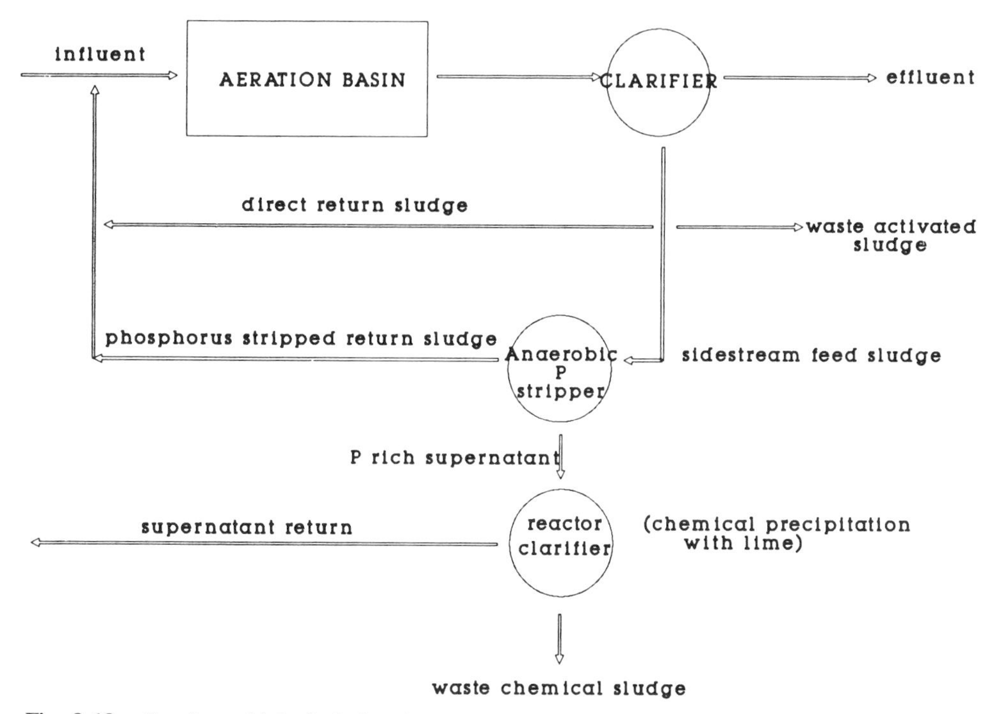

Fig. 8.18. Propietary biological phosphorus removal processes: Sidestream processes (PhoStrip). From Tetreault et al. (1986), with permission of the publisher.

more than 99%. Bacteria are removed by inactivation, grazing by ciliated protozoa (grazing is particularly effective for free-swimming bacteria), and adsorption to sludge solids and/or encapsulation within sludge flocs followed by sedimentation (Feachem et al., 1983; Omura et al., 1989; Yaziz and Lloyd, 1979).

8.5.2. Viruses

The activated sludge process is the most efficient biological process for removal of virus from sewage. It appears that most of the virus particles (>90%) are solids-associated and are ultimately transferred to sludge (Bitton, 1980a; Omura et al., 1989; Rao et al., 1986). The ability of activated sludge to remove viruses is related to the ability to remove solids. Thus, many of the viruses that escape into the effluents are solids-associated (Gerba et al., 1978). Viruses are also inactivated by environmental and biological factors. Attempts have been made to estimate the contribution of association with solids and of inactivation to removal of viruses in activated sludge. After 10 hr of aeration, 25% are removed by adsorption to sludge flocs and 75% are removed by inactivation (Glass and O'Brien, 1980). Therefore, inactivation alone is not sufficient for removing most of the viruses with a retention time varying from 6 hr to 12 hr.

Field studies in India showed that 90%–99% of enteric viruses were removed by a conventional activated sludge process (Table 8.2) (Rao et al., 1977). Rotaviruses are removed to the same extent as enteroviruses, and a 93%–99% removal was observed in a 1.5-MGD plant in Houston, Texas (Rao et al., 1986). Removal of enterovirus at lower levels was observed in some instances. For example, removal of enterovirus at Haut de Seine and Nancy wastewater treatment plants in France was only 48% and 69% respectively (Schwartzbrod et al., 1985).

Coliphage has been used as model viruses for studying virus removal by an activated sludge process (Bitton, 1987). These studies generally show that the activated sludge process removes 90%–99% of phage (Safferman and Morris, 1976). Studies on indigenous coliphage in wastewater treatment plants showed that while 12%–30% of coliphage is associated with suspended solids in raw wastewater, more than 97% is solids-associated in the aeration tank. Most of the adsorbed coliphage was RNA F-specific phages (Ketratanakul and Ohgaki, 1989).

In summary, virus removal/inactivation by activated sludge may be due to the following:

1. Virus adsorption to or encapsulation within sludge solids (this results in the transfer of viruses to sludge)
2. Virus inactivation by sewage bacteria (some activated sludge bacteria may have some antiviral activity)
3. Virus ingestion by protozoa (ciliates) and small metazoa (nematodes).

8.5.3. Parasitic Protozoa

Protozoan cysts such as those of *Entamoeba histolytica* are not inactivated under the conditions prevailing in the aeration tank of an activated sludge process. They are, however, entrapped in sludge flocs and are thus transferred to sludge following sedimentation. Similar removals have been observed for both *Entamoeba histolytica* and *Giardia* cysts (Panicker and Krishnamoorthi, 1978). More than 98% of *Giardia* cysts are removed and become concentrated in sludge. Numbers of cysts at various treatment stages of a California wastewater treatment plant are shown in Table 8.3 (Casson et al., 1990). More than

TABLE 8.2. Removal of Enteric Viruses by Primary and Activated Sludge Treatments at the Dadar Sewage Treatment Plant, Bombay, India[a]

Season	Months	Raw sewage: (virus concentration, PFU/L)	% Reduction after primary treatment	% Reduction after activated sludge treatment
Rainy				
	June 1972	1,000	33.5	97.9
	July 1972	1,250	24.1	97.0
	June 1973	1,200	29.7	95.5
	July 1973	837	29.8	98.9
Autumn				
	September 1972	300	64.7	98.0
	October 1972	312	66.0	91.7
	October 1973	572	73.0	96.4
	November 1973	1,087	56.0	90.0
Winter				
	January 1973	587	41.4	96.8
	February 1973	468	83.4	95.5
	January 1974	605	47.0	99.6
Summer				
	March 1973	812	57.0	97.6
	April 1973	875	59.7	98.6
	May 1973	731	66.0	93.5
	March 1974	250	68.8	98.0
	June 1974	694	74.7	99.0

[a]Adapted from Rao et al. (1977).

TABLE 8.3. ***Giardia* Cyst Removal in a California Wastewater Treatment Plant**[a]

		Sampling point						
		1	2	3	4	5	6	7
Date	Sampling period	Raw wastewater	Primary effluent	Secondary effluent	Postchlorination and filtration	Cholrine contact chamber effluent	Final effluent	Return activated sludge
3/13/89	7:00–15:00	260	380	1	1	1	2	270
	15:00–23:00	360	380	4	1	1	4	530
	23:00–7:00	140	620	2	1	1	1	670
Calculated flow composite		276	427	2	1	1	3	450
3/14/89	7:00–15:00	310	100	2	1	1	1	270
	15:00–23:00	470	660	1	4	1	2	270
	23:00–7:00	110	220	1	1	1	1	530
Calculated flow composite		326	351	1	2	1	1	331

[a]From Casson et al. (1990), with permission of the publisher.

99% of *Giardia* cysts were removed by the activated sludge treatment and most of them were transferred to sludge. Under laboratory conditions, the activated sludge process removes 80%–84% of *Cryptosporium parvum* oocysts (Villacorta-Martinez de Maturana et al., 1992).

8.5.4. Helminth Eggs

Because of their size and density, eggs of helminth parasites (e.g., *Taenia, Ancylostoma, Necator*) are removed by sedimentation during primary treatment of wastewater and during the activated sludge treatment; thus they are largely concentrated in sludges (Dean and Lund, 1981). Following a survey of wastewater treatment plants in Chicago, parasite eggs were not detected in unchlorinated sewage effluents (Fitzgerald, 1982). Some observers, however, have reported that activated sludge does not completely eliminate parasite eggs of *Ascaris* and *Toxocara* from wastewater (Schwartzbrod et al., 1989).

8.6. FURTHER READING

Barnes, D., and P.J. Bliss. 1983. *Biological Control of Nitrogen in Wastewater Treatment*. E. & F.N. Spon, London.

Curds, C.R., and H.A. Hawkes. 1983. *Ecological Aspects of Used-Water Treatment*, Vol. 2. Academic Press, London.

Edeline, F. 1988. *L'Epuration Biologique des Eaux Residuaires: Theorie et Technologie*. Editions CEBEDOC, Liege, Belgium, 304 pp.

Forster, C.F., and D.W.M. Johnston. 1987. Aerobic processes, pp. 15–56, in: *Environmental Biotechnology*, C.F. Forster and D.A.J. Wase, Eds. Ellis Horwood, Chichester, U.K.

Hanel, L. 1988. *Biological Treatment of Sewage by the Activated Sludge Process*. Ellis Horwood, Chichester, U.K.

Metcalf and Eddy, Inc. 1991. *Wastewater Engineering: Treatment, Disposal and Reuse*, McGraw-Hill, New York, 1334 pp.

U.S. EPA. 1977. *Wastewater Treatment Facilities for Sewered Small Communities*. Report # EPA--625/1-77-009.

9

BULKING AND FOAMING IN ACTIVATED SLUDGE PLANTS

9.1. INTRODUCTION

Since the introduction of continuous-flow reactors, sludge bulking has been one of the major problems affecting biological waste treatment (Sykes, 1989). There are several types of problems regarding solid separation in activated sludge. These problems are summarized in Table 9.1 (Eikelboom and van Buijsen, 1981; Hawkes, 1985; Jenkins, 1992; Jenkins et al., 1984).

Dispersed growth. In a well-operated activated sludge, bacteria that are not associated with the flocs are generally consumed by protozoa. Their presence in high numbers as dispersed cells results in a turbid effluent. Dispersed growth is associated with the failure of floc-forming bacteria to bioflocculate. This phenomenon occurs as a result of high BOD loading and oxygen limitation. Toxicity (e.g., metal toxicity) may also cause a dispersed growth of activated sludge bacteria.

Nonfilamentous bulking. This phenomenon is sometimes called "zoogleal bulking" and is caused by excess production of exopolysaccharides by bacteria (e.g., *Zooglea*) found in activated sludge. This results in reduced settling and compaction. This type of bulking is rare and is corrected by chlorination (Chudoba, 1989).

Pinpoint flocs. Pinpoint flocs are caused by the disruption of sludge flocs into very small fragments that may pass into the activated sludge effluent. Some observers believe that filamentous bacteria constitute the backbone of activated sludge flocs and therefore that their occurrence in low numbers may cause the flocs to loose their structure, resulting in poor settling and release of turbid effluents.

Rising sludge. Rising sludge is the result of excess denitrification, which results from anoxic conditions in the settling tank. Sludge particles attach to rising nitrogen bubbles and form a sludge blanket at the surface of the clarifier. The final outcome is a turbid effluent with an increased BOD_5. One solution to rising sludge is to lower the sludge retention time (i.e., increase the recirculation rate of activated sludge) in the settling tank.

Filamentous bulking. Bulking is a problem that consists of slow settling and poor compaction of solids in the clarifier of the activated sludge system (Jenkins and Richard, 1985). Filamentous bulking is usually caused by the excessive growth of filamentous microorganisms. This common problem will be discussed in more detail.

Foaming/scum formation. The problem of scum formation, which is due to the proliferation of *Nocardia* and *Microthrix* in aeration tanks of activated sludge units, will be discussed in Section 9.6.

9.2. FILAMENTOUS BULKING

Bulking is due to the overgrowth of filamentous bacteria in activated sludge. These bacteria are normal components of activated sludge flocs but may outcompete the floc-forming bacteria under specific conditions.

TABLE 9.1. **Causes and Effects of Activated Sludge Separation Problems**[a]

Name of problem	Cause of problem	Effect of problem
Dispersed growth	Microorganisms do not form flocs but are dispersed, forming only small clumps or single cells.	Turbid effluent. No zone settling of sludge.
Slime (jelly); viscous bulking (also possibly has been referred to as nonfilamentous bulking).	Microorganisms are present in large amounts of exocellular slime.	Reduced settling and compaction rates. Virtually no solids separation, in severe cases, resulting in overflow of sludge blanket from secondary clarifier.
Pin floc (or pinpoint floc)	Small, compact, weak, roughly spherical flocs are formed, the larger of which settle rapidly. Smaller aggregates settle slowly.	Low sludge volume index (SVI) and a cloudy, turbid effluent.
Bulking	Filamentous organisms extend from flocs into the bulk solution and interfere with compaction and settling of activated sludge.	High SVI—very clear supernatant.
Risking sludge (blanket rising)	Dentrification in secondary clarifier releases poorly soluble N_2 gas, which attaches to activated sludge flocs and floats them to the secondary clarifier surface.	A scum of activated sludge forms on surface of secondary clarifier.
Foaming/scum formation	Caused by (1) nondegradable surfactants and (2) the presence of *Nocardia* sp. and sometimes by (3) the presence of *Microthrix parvicella*.	Foams float large amounts of activated sludge solids to surface of treatment units. Foams accumulate and can putrify. Solids can overflow into secondary effluent or overflow tank freeboard onto walkways.

[a]Adapted from Jenkins et al. (1984).

9.2.1. Measurement of Sludge Settleability

Sludge settleability is determined by measuring the sludge volume index, which is given by the following formula:

$$SVI = \frac{V \times 1{,}000}{MLSS} \qquad (9.1)$$

where V = volume of settled sludge after 30 min (mL/L); and $MLSS$ = mixed-liquor suspended solids (mg/L).

SVI is expressed in mL per gram and is thus the volume occupied by 1 g of sludge (Nathanson, 1986). A high SVI ($>$ 150 ml/g) indicates bulking conditions, whereas an SVI below 70 ml/g indicates the predominance of pin (small) flocs (U.S. EPA, 1987a).

9.2.2. Relation Between Filamentous and Floc-Forming Bacteria

Based on the relationship between floc-forming and filamentous bacteria, three types of flocs are observed in activated sludge:

Normal flocs. A balance between floc-forming and filamentous bacteria results in strong flocs that keep their integrity in the aeration basin and settle well in the sedimentation tank.

Pinpoint flocs. In these flocs, filamentous bacteria are absent or occur in low numbers. This results in small flocs that do not settle well. The secondary effluent is turbid despite the low SVI.

Filamentous bulking. Filamentous bulking is caused by the predominance of filamentous organisms. The filaments interfere with sludge settling and compaction. Poor sludge settling, as expressed by SVI, is observed when the length of extended filaments exceeds 10^7 μm/mg suspended solids (Fig. 9.1) (Palm et al., 1980; Sezgin, 1982; Sezgin et al. 1978, 1980). More than 25 types of filamentous microorganisms have been identified in bulking activated sludge.

There are major physiological differences between floc-forming and filamentous bacteria. The differences are summarized in Table 9.2 (Sykes, 1989). Filamentous bacteria have a higher surface-to-volume ratio than their floc-forming counterparts, and this helps them survive under low oxygen concentrations and low-nutrient conditions. They also have a low half-saturation constant (K_s in Monod's equation; see Chapter 2) and have a high affinity for substrates, thus behaving as oligotrophs and surviving well under starvation conditions. Filamentous bacteria are able to predominate under low dissolved oxygen, low-F/M, low-nutrient conditions, and high sulfide levels. However, it appears that low F/M is the predominant cause of bulking in wastewater treatment plants. These differences between filamentous and floc-forming bacteria can be exploited to control filamentous bulking in activated sludge (see Section 9.3.2).

9.2.3. Types of Filamentous Microorganisms

Some 25–30 types of filamentous microorganisms are responsible for bulking in activated sludge. A survey of bulking activated sludge plants in the United States has revealed that approximately 15 major types of filamentous microorganisms are responsible for bulking, the most predominant ones being *Nocardia* (an actinomycete also responsible for foaming; see Section 9.6) and Type 1701 (Jenkins and Richard, 1985) (Table 9.3). Type 021N was found in 19%

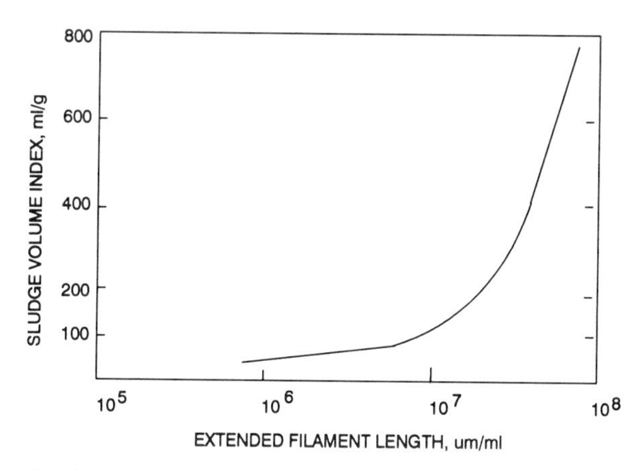

Fig. 9.1. Relationship between numbers of filamentous bacteria and sludge volume index. From Sezgin (1982), with permission of the publisher.

TABLE 9.2. Comparison of Physiological Characteristics of Floc-Formers and Filamentous Organisms[a]

Characteristic	Bacteria	
	Floc-former	Filamentous
Maximum substrate uptake rate	High	Low
Maximum specific growth rate	High	Low
Endogenous decay rate	High	Low
Decrease in specific growth rate from low substrate concentration	Significant	Moderate
Resistance to starvation	Low	High
Decrease in specific growth rate from low DO	Significant	Moderate
Potential to sorb organics when excess is available	High	Low
Ability to use nitrate as an electron acceptor	Yes	No
Exhibits abundant uptake of phosphorus	Yes	No

[a]From Sykes (1989), with permission of the publisher.

TABLE 9.3. Filamentous Organisms Predominant in U.S. Bulking Activated Sludges[a]

Rank	Filamentous organism	Percentage of treatment plants with bulking sludge where filament was observed to be dominant[b]
1	*Nocardia* spp.	31
2	Type 1701	29
3	Type 021N	19
4	Type 0041	16
5	*Thiothrix* spp.	12
6	*Sphaerotilus natans*	12
7	*Microthrix parvicella*	10
8	Type 0092	9
9	*Haliscomenobacter hydrossis*	9
10	Type 0675	7
11	Type 0803	6
12	*Nostocoida limicola*	6
13	Type 1851	6
14	Type 0961	4
15	Type 0581	3
16	*Beggiatoa* spp.	<1
17	Fungi	<1
18	Type 0914	<1
	All others	<1

[a]From Jenkins and Richard (1985), with permission of the publisher.
[b]Percentage of 525 samples from 270 treatment plants with bulking problems.

of more than 400 bulking sludge samples in the United States. Of all the bulking episodes due to Type 021N 80% were associated with the treatment of industrial wastes or mixtures of industrial and domestic wastes (Richard et al., 1985b). Another survey of 17 wastewater treatment plants in Pennsylvania showed that the four most frequently encountered filamentous microorganisms were Type 0041, Type 1701, *Haliscomenobacter hydrossis,* and Type 021N (Williams and Unz, 1985b). Type 021N was also detected with a relative frequency of 13% and 21% in two wastewater treatment plants in Berlin (Ziegler et al., 1990).

Less frequently found bulking microorganisms include an actinomycete isolate (# NRRL B 16216) that caused bulking in a bench-scale reactor dedicated to treating coke plant wastewater (White et al., 1986), and a cyanobacterium, *Schizothrix calcicola,* which was responsible for numerous bulking episodes in an Ohio wastewater treatment plant (Rozich et al., 1982).

9.2.4. Techniques for the Isolation and Identification of Filamentous Microorganisms

Filamentous bacteria have long been considered "unusual" microorganisms in classical microbiology textbooks. The pioneering work of Eikelboom and van Buijsen has led to the development of methods for the isolation and identification of these organisms (Eikelboom, 1975; Eikelboom and van Buijsen, 1981). The most recent techniques concern the isolation of filamentous sulfur bacteria in activated sludge (*Thiothrix, Beggiatoa,* and Type 021N) (Williams and Unz, 1985b, 1989; Richard et al., 1985a). These bacteria grow on media low in organic carbon with reduced sulfur sources (sulfide or thiosulfate). Ammonium and 2- to 4-carbon organic acids support the growth of all filamentous sulfur bacteria (Williams and Unz, 1989). *Microthrix parvicella* has also been cultivated successfully in a chemically defined medium containing mainly Tween 80 and reduced nitrogen and sulfur compounds (Slijkhuis, 1983; Slijkhuis and Deinema, 1988).

The application of conventional techniques for the identification of filamentous microorganisms is difficult and time-consuming. Other problems are their slow growth and difficulties in obtaining pure cultures from activated sludge samples (Wanner and Grau, 1989). Thus, filamentous microorganisms are characterized mainly by microscopic examination, mostly with a phase-contrast microscope. For such identification, information about the following characteristics should be obtained (Fig. 9.2) (Eikelboom and van Buijsen, 1981).

Filament shape. The filaments can be straight, curved, mycelial, or twisted.

Size and shape of cells within the filament (e.g., rods, cocci).

Branching. Fungi and actinomycetes such as *Nocardia* have branched filaments; Some filamentous bacteria such as *Sphaerotilus natans* display false branching.

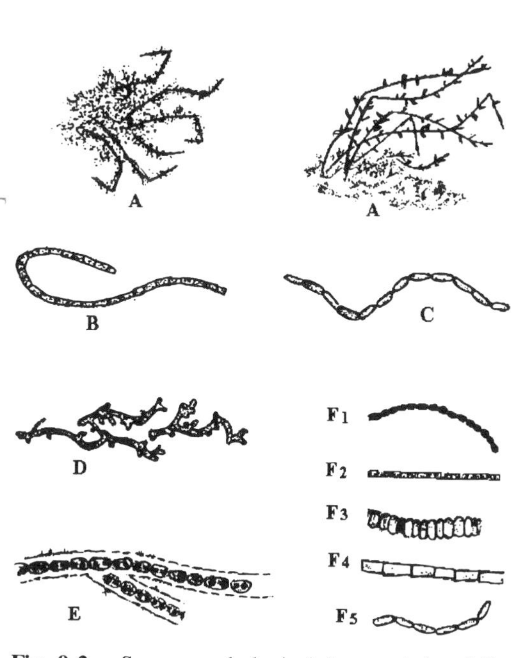

Fig. 9.2. Some morphological characteristics of filamentous bacteria. Adapted from Eikelboom and van Buijsen (1981).

Filament motility. *Beggiatoa,* for example, move by *gliding* on a surface.

Presence of a sheath. Some filamentous bacteria produce a sheath, a tubular structure that encloses the cells. Although difficult to see, this structure can be observed in preparations in which cells are missing. Staining with 0.1% crystal violet is a common method for detecting sheaths. However, for some filamentous bacteria such as *Thiothrix,* sheaths may be present in some strains but absent in others (Williams et al., 1987).

Presence of epiphytic bacteria on filament surfaces. Several bacterial cells become attached to the surface of some filamentous bacteria.

Filament size and diameter can be useful for distinguishing between fungi and branched filamentous actinomycetes such as *Nocardia.*

Presence of granules. Granules are inclusions in filamentous bacteria that are used to store food reserves. The presence of sulfur granules is indicated by the observation of bright yellow granules following addition of sodium sulfide or thiosulfate to the sample. *Beggiatoa, Thiothrix,* and type 021N have sulfur granules (Richard et al., 1985b). The electron micrographs displayed in Figure 9.3 show sulfur granules in filaments of both *Thiothrix* and 021N (Williams and Unz, 1985a; Williams et al., 1987). The sulfur granules are surrounded by a single-layered envelope within invaginations of the cytoplasmic membrane. *Thiothrix* was first observed by Winogradsky in 1888 but its isolation in pure culture was not undertaken until more than a hundred years later. This sheathed bacterium makes filaments that produce motile gonidia (Larkin, 1980). Other tests help detect the presence of polyphosphate and polyhydroxybutyric (PHB) granules.

Staining of activated sludge dry smears. The following tests are performed, with ordinary transmitted light.

Gram stain: This test distinguishes between gram-positive and gram-negative bacteria and is based on the chemical composition of bacterial cell walls.

Neisser stain: This staining technique helps in

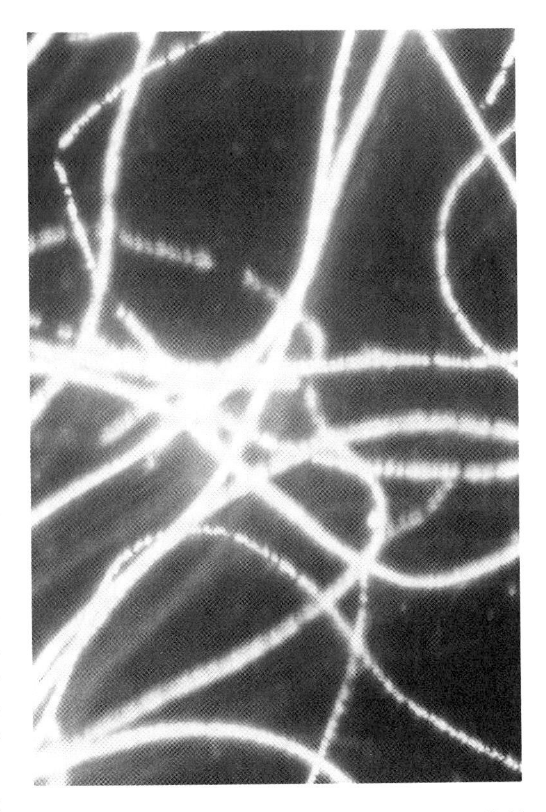

Fig. 9.3. Sulfur granules of *Thiothrix* (dark-field microscopy; × 1,000). Courtesy of R. Brigmon and G. Bitton.

Fig. 9.4. Rosette formation in *Thiothrix.* Dark-field microscopy; × 1,000. Courtesy of R. Brigmon and G. Bitton.

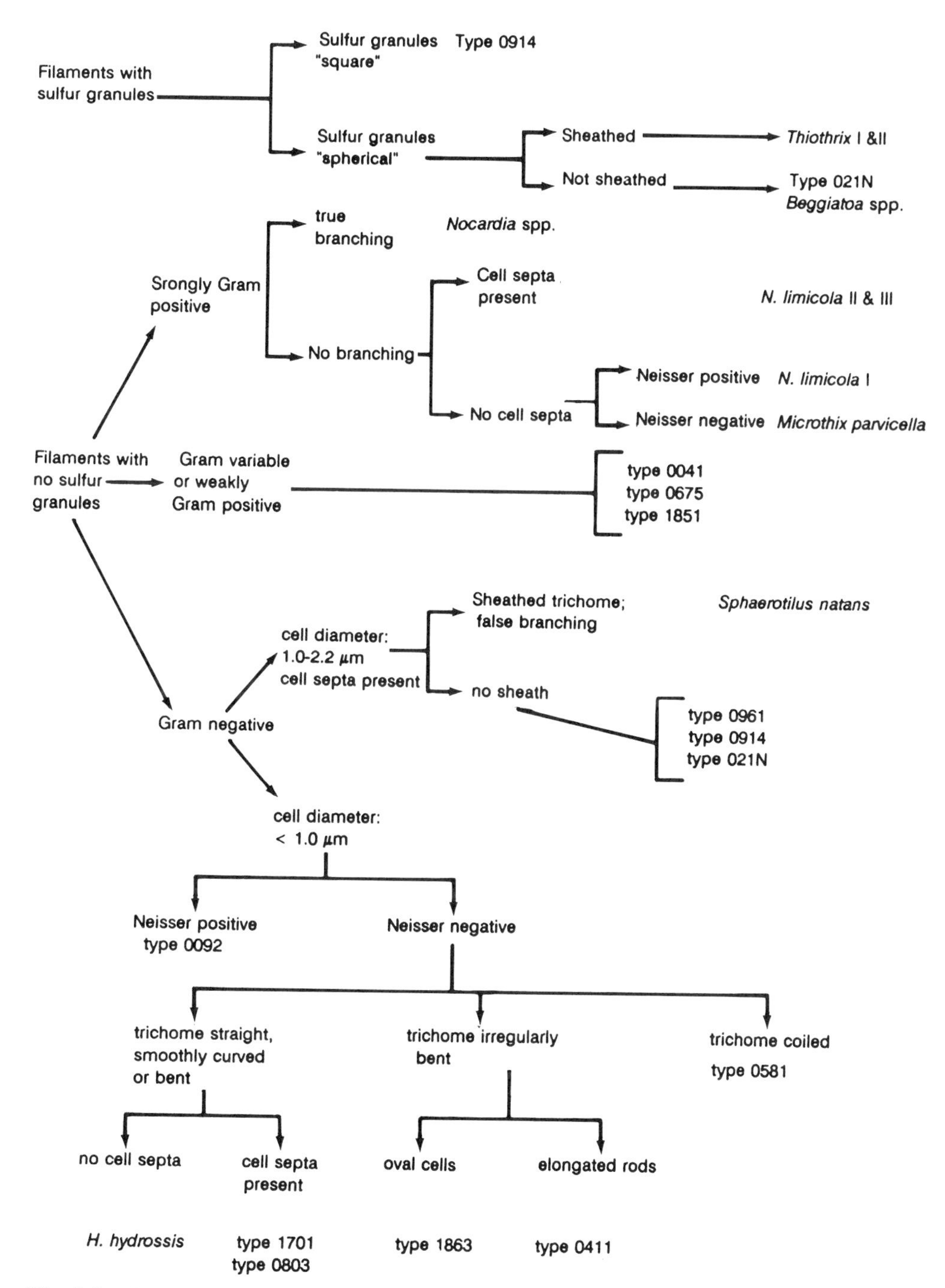

Fig. 9.5. Simplified dichotomous key for identification of filamentous microorganisms. Adapted from Jenkins et al. (1984) and Eikelboom (1975).

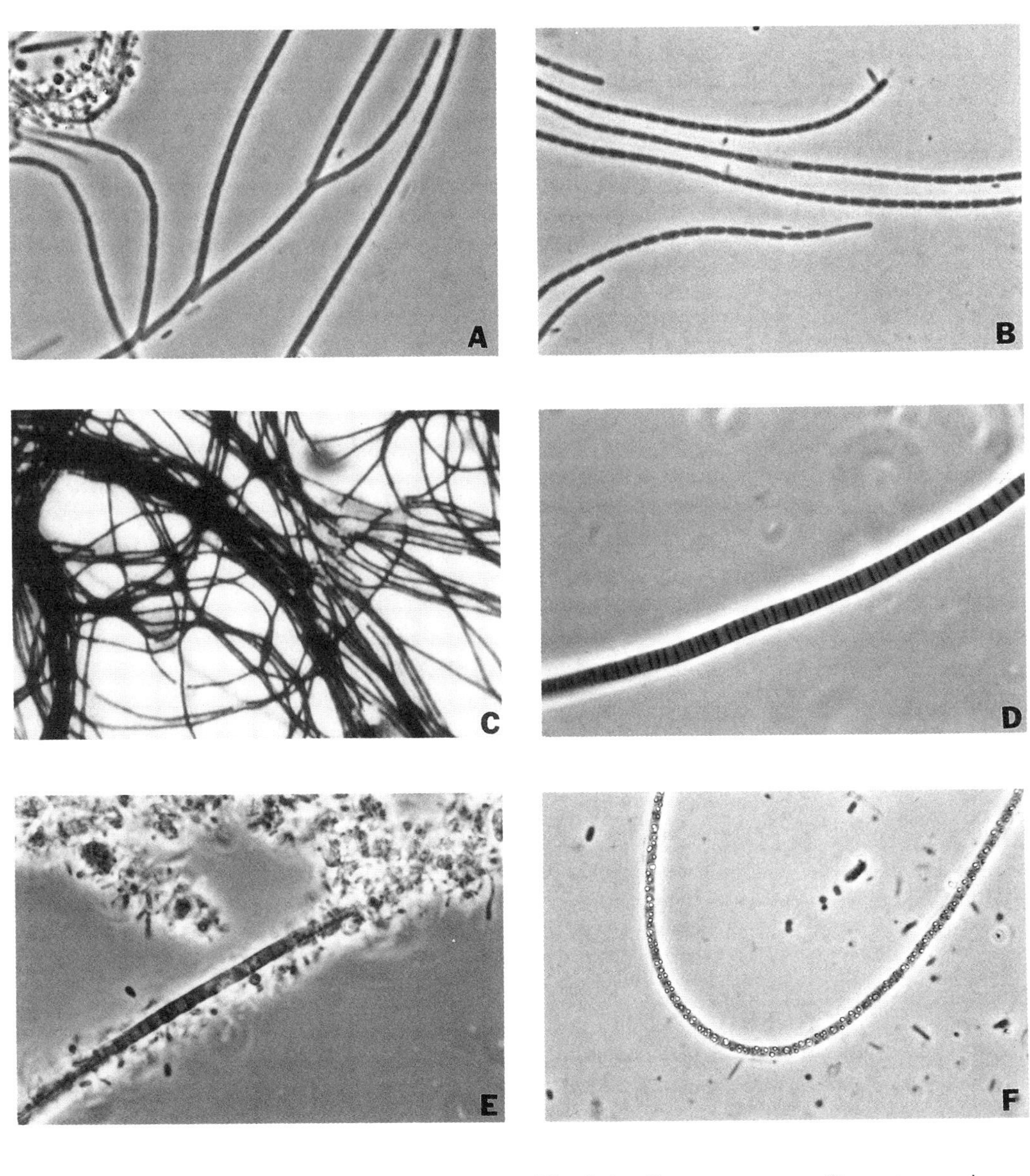

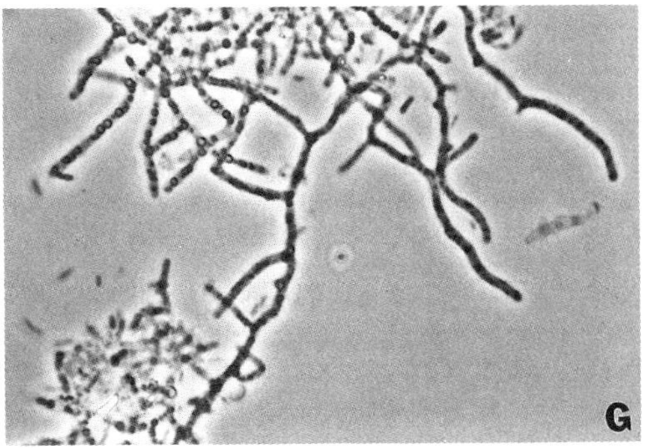

Fig. 9.6. Some common filamentous microorganisms observed in bulking sludges. **A.** *Sphaerotilus natans*. **B.** Type 1701. **C.** *Microthrix parvicella*. **D.** Type 021N. **E.** Type 0041. **F.** *Thiothrix* II. **G.** *Nocardia* sp. Except for *M. parvicella,* which was gram-stained, all others were observed by phase-contrast microscopy at × 1,000. Courtesy of M. Richards.

the observation of polyphosphate granules in filamentous bacteria. Cells are stained with a mixture of methylene blue and crystal violet, and then counterstained with chrysoidin Y (Eikelboom and van Buijsen, 1981). Neisser-negative bacteria appear as light brown to yellowish filaments, whereas Neisser-positive bacteria display dark polyphosphate granules.

Other characteristics. Rosettes are sometimes observed in *Thiothrix* and Type 021N (Fig. 9.4) (Williams et al., 1987). In both organisms the basal cells at the end of the filaments are held together by a holdfast material that is stained with ruthenium red, suggesting the presence of acidic polysaccharides (Costerton, 1980).

Using these characteristics, one then attempts to identify the filamentous bacteria according to the dichotomous key shown in Figure 9.5 (Eikelboom, 1975; Eikelboom and van Buijsen, 1981; Jenkins et al., 1984). Figure 9.6 displays micrographs of some common filamentous bacteria.

A fluorescent antibody technique has also been used for the detection of *Sphaerotilus natans* in activated sludge. The antiserum did not display any reaction with other tested filamentous microorganisms in activated sludge (*Haliscomenobacter hydrossis, Microthrix parvicella,* Type 1701, Type 021N, Type 0041, *Thiothrix*) (Howgrave-Graham and Steyn, 1988). Monoclonal antibodies can also be useful tools for the rapid identification of filamentous microorganisms in activated sludge. A monoclonal antibody has been made against *Thiothrix* and tested in wastewater treatment plants as well as in sulfur spring waters in Florida (Brigmon, Bitton, and Zam, unpublished data).

9.3. SOME FACTORS CAUSING FILAMENTOUS BULKING

Filamentous microorganisms are normal components of the activated sludge microflora. Their overgrowth may be due to one or a combination of the following factors.

9.3.1. Waste Composition

High-carbohydrate wastes (e.g., brewery and corn wet-milling industries) appear to be conducive to sludge bulking. Carbohydrates composed of glucose, maltose, and lactose, but not galactose, support the growth of filamentous bacteria (Chudoba, 1985; Chudoba et al., 1985). Some filaments (e.g., *S. natans, Thiothrix* spp., Type 021N) appear to be favored by readily biodegradable organic substrates, whereas others (e.g., *M. parvicella,* Type 0041) are able to use slowly biodegradable substrates (Jenkins, 1992).

9.3.2. Substrate Concentration

This is one of the most prevalent cause of filamentous bulking. Filamentous microorganisms are slow-growing organisms and have lower half-saturation constant K_s and μ_{max} than floc-formers. A study of the interaction between Type 021N (a filamentous bacterium) and *Zooglea ramigera* (a typical floc-forming bacterium) showed that, under low substrate concentration (low F:M ratio), Type 021N outpaces *Z. ramigera* owing to its higher affinity for the substrate (i.e., low K_s) and lower decay rate. Conversely, under high substrate concentration, *Z. ramigera* outpaces the filamentous bacterium because of its higher maximum growth rate (van Niekerk et al., 1987). Thus at low substrate concentrations, filamentous microorganisms have a higher substrate removal rate than the floc-formers, which prevail at high substrate concentrations (Fig. 9.7) (Chudoba, 1985, 1989; Chudoba et al., 1973). A high substrate concentration can be established by means of a biological selector (see section 9.5.4).

9.3.3. Sludge Loading and Sludge Age

These two parameters are related by the following formula (Chudoba, 1985):

$$\frac{1}{\theta} = Y B - k_d \qquad (9.2)$$

where θ = sludge age; Y = yield coefficient; B = sludge loading; and K_d = decay rate of total biomass.

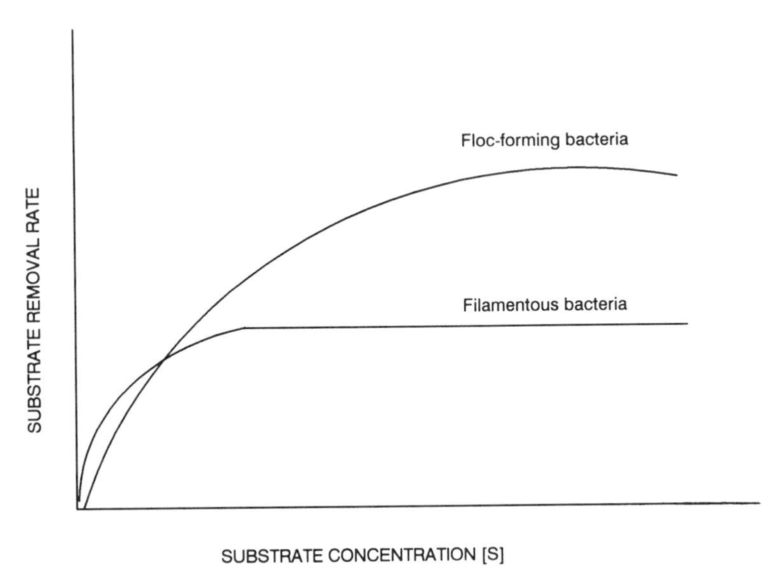

Fig. 9.7. Substrate removal rate (r_x) of floc-forming and filamentous bacteria. From Chudoba (1985), with permission of the publisher.

The relationships depend on whether the reactor is a completely mixed or plug-flow system. In completely mixed systems, increasing sludge loading leads to a decrease of SVI and thus to a decrease of filamentous microorganisms. At high B values (low sludge age values), filamentous microorganisms are washed out and this leads to poor quality effluents. In the plug-flow pattern, floc-forming bacteria predominate at an optimum B value of approximately 0.3 g g^{-1} day $^{-1}$ as BOD_5. Increasing B leads to an increase in SVI. However, at very high B values, filamentous microorganisms are washed out as in the completely mixed systems (Chudoba, 1985).

Some filamentous microorganisms (e.g., *Thiothrix*, Type 1701, *S. natans*) are found over a wide range of sludge age (i.e., of MCRT, or mean cell retention time) values while others occur only at low (e.g., Type 1863) or high (e.g., *M. parvicella*, Type 0092) values (Fig. 9.8) (Jenkins, 1992).

9.3.4. pH

The optimum pH in the aeration tank is 7–7.5. pH values below 6 may favor the growth of fungi (e.g., *Geotrichum*, *Candida*, *Trichoderma*) and cause filamentous bulking (Pipes, 1974). In laboratory activated sludge units, bulking caused by the excessive growth of fungi occurred after 30 days at pH = 4.0 and pH = 5.0 (Hu and Strom, 1991). A survey of wastewater treatment plants in Pennsylvania showed that fungi were encountered in 10% of the samples (Williams and Unz, 1985b).

9.3.5. Sulfide Concentration

High sulfide concentration in the aeration tank causes the overgrowth of sulfur filamentous bacteria such as *Thiothrix*, *Beggiatoa*, and Type 021N. These microorganisms use sulfide as a source of energy and oxidize it to elemental sulfur, which is stored as intracellular sulfur granules. *Beggiatoa* growth occurs mostly in fixed-film bioreactors (Jenkins, 1992).

9.3.6. Dissolved Oxygen Level

The growth of certain filamentous bacteria (e.g., *Sphaerotilus natans*, Type 1701, *Haliscomenobacter hydrossis*) is favored by relatively low dissolved oxygen levels in the aeration tank (Lau et al., 1984a; Palm et al., 1980; Sezgin et al., 1978; Travers and Lovett, 1984). Aeration tanks should be operated with a minimum of 2 mg O_2/L to avoid predominance of specific filamentous microorganisms, namely *Sphaerotilus natans* (Chudoba, 1985). The growth kinetics of *Sphaerotilus natans* and its interaction with a floc-forming bacterium (*Citrobacter* sp.) have been studied under laboratory conditions, by using continuous culture techniques. It was essentially shown that a low level of dissolved oxygen is a major factor contributing to the proliferation of this filamentous bacterium in activated sludge. *Sphaerotilus* has a lower K_{DO} (K_{DO} = 0.01 mg/L) than the floc forming bacterium (K_{DO} = 0.15 mg/L) and, thus, would thrive at low DO in the mixed liquor (Lau et al., 1984a; 1984b). However, no relationship between oxygen level and filament numbers was found when the dominant filamentous bacteria were *Microthrix parvicella* or Type 0041 (Forster and Dallas-Newton, 1980).

9.3.7. Nutrient Deficiency

Deficiencies in nitrogen, phosphorus, iron, or trace elements may cause bulking. This factor has not received much attention. The growth of *S. natans, Thiothrix,* and Type 021N can be associated with nitrogen and phosphorus deficiencies. It has been suggested that the C/N/P ratio should be 100/5/1 (U.S. EPA, 1987a). It has also been suggested that iron and trace element deficiencies may cause bulking.

An integrated hypothesis for sludge bulking has been proposed. According to this hypothesis, activated sludge consists of three categories of "model" microorganisms (Chiesa and Irvine, 1985): (1) fast-growing "zoogleal" type microorganisms; (2) slow-growing filamentous organisms with high substrate affinity (i.e., low K_s), and (3) fast-growing filamentous organisms with a high affinity for dissolved oxygen (i.e., low K_{DO}). At high substrate concentrations, category 1 is favored as long as there is enough dissolved oxygen. Low substrate concentrations below a critical concentration S^* favor the proliferation of category 2. Category 3 prevails under low DO conditions. An intermittent feeding pattern creates favorable conditions for the development of nonfilamentous microorganisms, which have high substrate uptake rates during periods

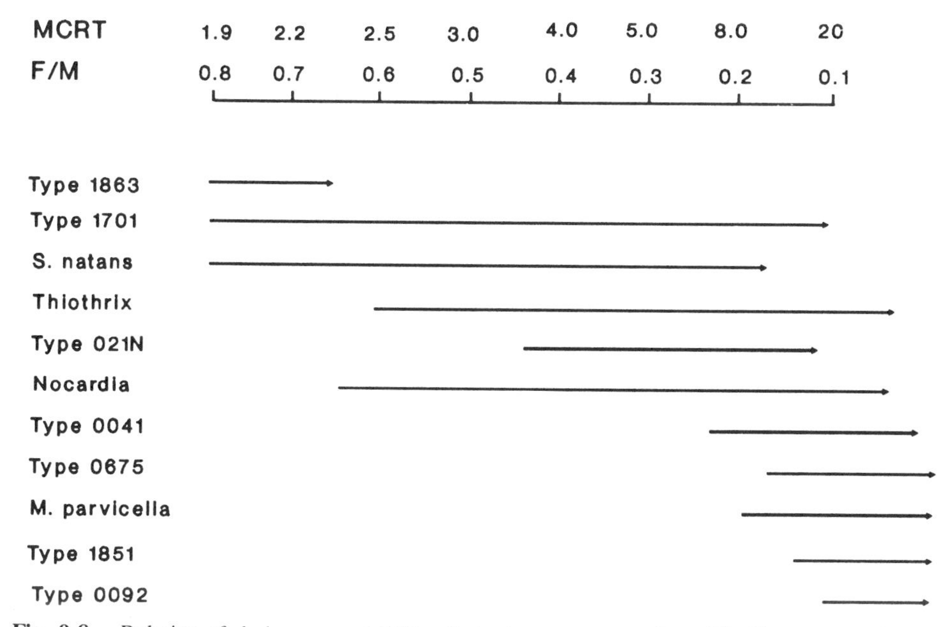

Fig. 9.8. Relation of sludge age and F/M ratio to the occurrence of specific filamentous microorganisms in activated sludge. From Jenkins (1992), with permission of the publisher.

of high substrate concentration and a capacity to store reserve materials during periods of starvation (endogenous metabolism).

9.4. USE OF FILAMENTOUS MICROORGANISM IDENTIFICATION AS A TOOL FOR DIAGNOSING THE CAUSE(S) OF BULKING

Excessive growth of specific filamentous microorganisms is indicative of specific operational problems in the plant, such as low DO, low F/M ratio (i.e., low organic loading rate), high concentration of sulfides in wastewater, nitrogen and phosphorus deficiencies, and low pH (Table 9.4) (Richard et al., 1982, 1985a; Strom and Jenkins, 1984). Therefore identification of the causative organism is recommended, although some investigators question their use as indicator organisms in the assessment of sludge bulking (Wanner and Grau, 1989; Williams and Unz, 1985b).

In a survey of 89 U.S. wastewater treatment plants with bulking problems, Type 1701 was found to be the predominant filamentous microorganism in 33% of the plants (Richard et al., 1985a). The overgrowth of these filamentous bacteria in activated sludge systems is linked to low dissolved oxygen in the aeration tank. This bacterium has a $k_{DO} = 0.014$ mg/L as compared to $k_{DO} = 0.073$ mg/L for a floc-forming microorganism (Richard et al., 1985a). It thrives in activated sludge plants that treat complex carbohydrates such as brewing wastes and starches. Excessive growth of fungi is indicative of low pH in the aeration basin. The predominance of *Sphaerotilus natans*, Type 021N, Type 1701, and *Thiothrix* (also an indicator of high sulfide levels) is indicative of low DO in the aeration basin.

Bulking due to sulfur-oxidizing *Thiothrix* is associated with the presence of sulfides in septic wastes (Farquhar and Boyle, 1972). Bulking caused by Type 021N is associated with the treatment of septic domestic wastes containing readily degradable carbonaceous substrates such as simple sugars and organic acids in systems operated at low F/M ratio (< 0.3). This type of bulking is also associated with nutrient deficiencies (N or P deficiencies) or high sulfide levels (Richard et al., 1985a). The predominance of *Microthrix parvicella*, Types 0041, 0675, 0961, 0803, and 0092 is associated with low F:M ratios (Nowak et al., 1986; U.S. EPA, 1987a; Daigger et al., 1985). These filamentous microorganisms prevail over floc-forming bacteria under alternating anoxic–aerobic conditions (Casey et al., 1992). The occurrence of *Nostocoida lumicola* is also associated with low organic loading (Nowak and Brown, 1990).

Low temperature also encourages the excessive growth of *M. parvicella*. These microorganisms maintain a higher growth rate than

TABLE 9.4. Dominant Filament Types That Are Indicative of Activated Sludge Operational Problems[a]

Suggested causative conditions	Indicative filament types
Low F/M	*M. parvicella, Nocardia* sp., *H. hydrossis,* 0041, 0675, 0092, 0581, 0961, 0803
Low DO	1701, *S. natans*; possibly 021N and *Thiothirx* sp.
Presence of sulfides	*Thiothrix* sp., *Beggiatoa* sp., possibly 021N
Low pH	Fungi
Nutrient deficiency (N and/or P)	*Thiothrix* sp., possibly 021N

[a]From Richard et al. (1985a), with permission of the publisher.

floc-forming microorganisms at low organic concentrations. One cure for this type of bulking is the incorporation of a selector that promotes the growth of floc-forming organisms at the expense of the filamentous types (see Section 9.5.4). *Microthrix parvicella* also displays excessive growth in systems (activated sludge or oxidation ditch) fed wastewater with excess fatty acids and low DO concentration (Slijkhuis and Deinema, 1988). The growth of Type 0961 is affected by the feed pattern in the aeration basin. Higher organic loading leads to the loss of their selective advantage over floc-forming bacteria (Nowak et al., 1986).

9.5. CONTROL OF SLUDGE BULKING

Various approaches are available for controlling sludge bulking in wastewater treatment plants. Some of them are described below.

9.5.1. Treatment with Oxidants

Filamentous bacteria can be controlled by treating the return sludge with chlorine or hydrogen peroxide to "selectively" kill filamentous microorganisms.

Bulking control by chlorination was proposed over 50 years ago and this practice is probably the most widely used for this purpose. Chlorine may be added to the aeration tank or to the return activated sludge (Fig. 9.9) (Jenkins et al., 1984). The method of choice is chlorine addition to the RAS line as chlorine gas or sodium hypochlorite about three times per day. Chlorine concentration should be 10–20 mg/L (concentrations above 20 mg/L may cause deflocculation and formation of pinpoint flocs). Chlorine dosage for bulking control can be rapidly estimated, by means of a short-term enzymatic bioassay based on reduction of a tetrazolium salt by dehydrogenases (Logue et al., 1983).

Hydrogen peroxide is generally added to the return activated sludge at concentrations of 100–200 mg/L. However, as shown for chlorine, excessive levels of hydrogen peroxide can be deleterious to floc-forming bacteria (Cole et al., 1973). In addition to its role as an oxidizing agent, hydrogen peroxide may also act as a source of oxygen in the aeration tank. Ozone has also been proposed for curing filamentous bulking (Colignon et al., 1986).

9.5.2. Treatment with Flocculants

Synthetic organic polymers, lime, and iron salts can be added to the mixed liquor to promote

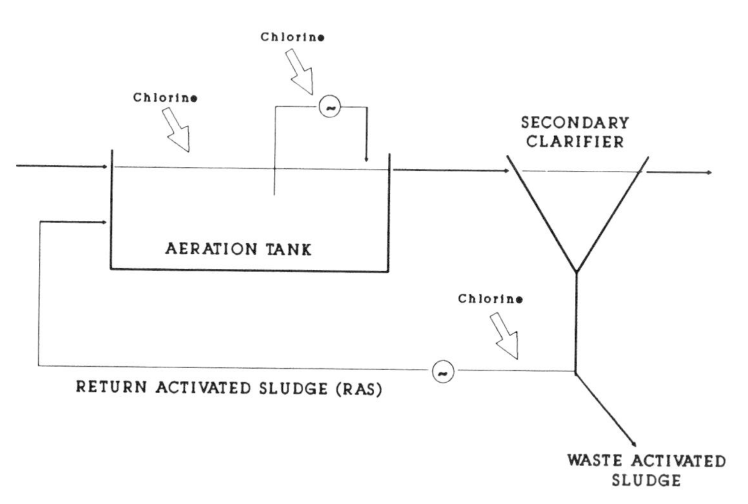

Fig. 9.9. Bulking control by chlorination: Chlorine dosing points in an activated sludge system. From Jenkins et al. (1984), with permission of the publisher.

sludge settling. However, the addition of lime and iron salts increases the solids load, and the use of polymers is costly. The addition of cationic polymers at concentrations of 15–20 mg/L has resulted in a successful control of bulking in wastewater from the brewing industry.

9.5.3. Manipulation of RAS Flow Rates

The clarifier in the activated sludge process has two functions: clarification (i.e., floc removal to obtain a clear effluent) and thickening of the sludge.

The degree of thickening achieved in the clarifier is given by Equation:

$$\frac{X_u}{X} = \frac{(Q + Q_r)}{Q_r} \tag{9.3}$$

where X_u = RAS suspended solids concentration (w/v); X = MLSS in aeration tank (w/v); Q = influent flow rate (v/t); and Q_r = RAS flow rate (v/t).

Bulking interferes essentially with sludge thickening in the clarifier and results in a decrease of the RAS suspended solids concentration (X_u). This decrease must be compensated by an increase in RAS flow rate (Q_r). Thus, increasing RAS flow rate helps prevent failure of the clarifier. Reduction of MLSS concentration in the clarifier feed can also help control bulking. This reduction can be achieved by decreasing the mixed-liquor solids inventory, which is obtained by increasing the sludge wasting rate (Jenkins et al., 1984).

9.5.4. Biological Selectors

Biological selectors are alternative process configurations that favor the growth of floc-forming bacteria over filamentous bacteria and, thus, help control bulking. A selector is a tank or compartment in which certain parameters (e.g., F/M ratio, electron acceptor) can be manipulated to discourage the overgrowth of undesirable filamentous microorganisms. The incoming wastewater and the return activated sludge are mixed in the selector tank under the desired conditions prior to entering the aeration basin (Sykes, 1989).

There are three categories of biological selectors: aerobic, anoxic, and anaerobic.

Aerobic selectors. The concept of kinetic selection was introduced in the 1970s (Chudoba et al., 1973). The kinetics are based on Monod's equation (see Chapter 2). At relatively high substrate concentrations ($S > K_s$), the specific growth rate is controlled by μ_{max}. However, at low substrate concentrations ($S < K_s$), the specific growth rate is controlled mainly by K_s. Because of their low K_s, filamentous bacteria predominate at low substrate concentrations (Fig. 9.10) (Chudoba et al., 1973). In contrast, floc-forming bacteria predominate under high substrate concentrations.

Thus, devising an aerobic selector consists of creating a substrate concentration gradient (F/M gradient) across the reactor. This gradient can be created by using, for example, several reactors in series or establishing an F/M gradient within the same tank (Fig. 9.11A) (Chudoba et al., 1973; Sykes, 1989). This configuration gives a selective advantage to floc-forming bacteria, which take up most of the soluble substrate at the head of the reactor.

Anoxic selectors. An anoxic condition is defined as the absence of oxygen and the presence of nitrate as the electron acceptor. Pilot plant studies have demonstrated the positive effect of anoxic conditions on sludge settleability (Chambers, 1982; Price, 1982). This approach consists of setting up an anoxic reactor followed by an aerobic one. In the anoxic reactor, the floc-forming bacteria predominate over the filamentous bacteria because they can take up organic substrates, using nitrate as an electron acceptor. Some filamentous microorganisms (e.g., *S. Natans,* 021N) cannot use nitrate or nitrite as electron acceptors (Wanner et al., 1987a). Nitrate is provided by the recycling of return activated sludge as well as mixed liquor (Fig. 9.11B); (Barnard, 1973; Sykes, 1989). The subsequent low organic substrate concentration in the aerobic reactor is not sufficient to sustain the growth of filamentous bacteria. The use of an anoxic selector has been found to be successful

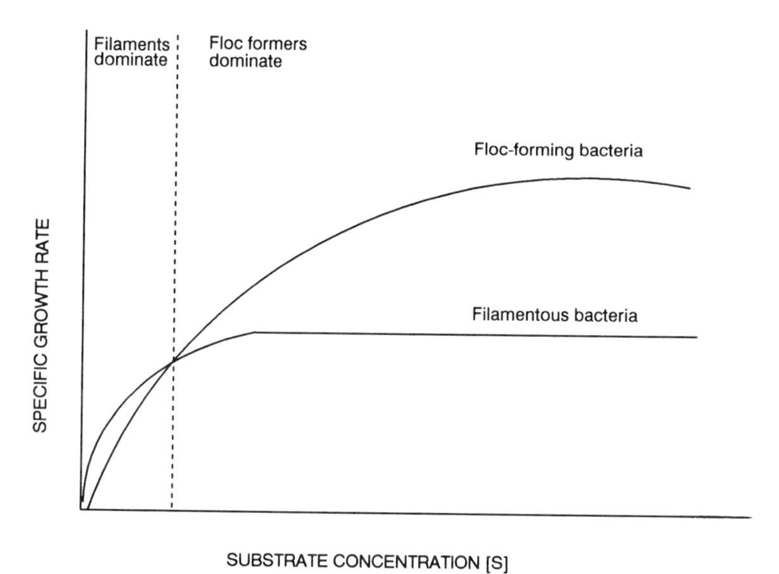

Fig. 9.10. Kinetic selection of filamentous microorganisms based on Monod's equation. From Chudoba et al. (1973), with permission of the publisher.

in controlling the growth of *Nostocoida lumicola* (Nowak and Brown, 1990).

Anaerobic selectors. Anaerobiosis conditions are defined as the lack of oxygen as well as nitrate as electron acceptors. Anaerobic conditions suppress the growth of filamentous bacteria such as *Sphaerotilus natans* and Type 021N. An anaerobic selector is based on the ability of floc-forming bacteria to accumulate polyphosphates (i.e., abundant uptake of phosphorus by poly-P bacteria; see Chapter 3) under aerobic conditions and to use them as a source of energy for uptake of soluble organic substrates under anaerobic conditions. Under anaerobic conditions, filamentous bacteria are not able to take up organic substrates at a rate comparable to the action of poly-P bacteria (Wanner et al., 1987b). Thus, processes designed for removal of biological phosphorus help select for floc-forming bacteria in the anaerobic reactor (Figure 9.11C; Sykes, 1989). The accumulated carbohydrates will be metabolized in the aerobic reactor. However, an excessive growth of *Thiothrix* has been observed under anaerobic conditions, under which this filamentous microorganism is able to utilize organic substrates by dissimilatory sulfate reduction and subsequently to cause bulking (Wanner et al., 1987b).

9.5.5. Biological Control

Microorganisms, mainly bacteria and actinomycetes, isolated from various sources (e.g., activated sludge, compost, soil) are capable of lysing filamentous bacteria. An active lytic microorganism against Type 021N was isolated from soil (Yagushi et al., 1991). Predatory ciliated protozoa (e.g., *Trithigmostoma cucullulus*) are also able to ingest filamentous microorganisms, and their growth in aeration tanks is followed by a decrease of the sludge volume index, an indication of their controlling effect on sludge bulking. Inoculation of bulking activated sludge with these protozoa also results in a decrease of SVI (Fig. 9.12) (Inamori et al., 1991). This approach needs further exploration under field conditions.

9.5.6. Other Specific Methods

Preaeration of wastewater to remove sulfides helps control the growth of *Thiothrix* but higher

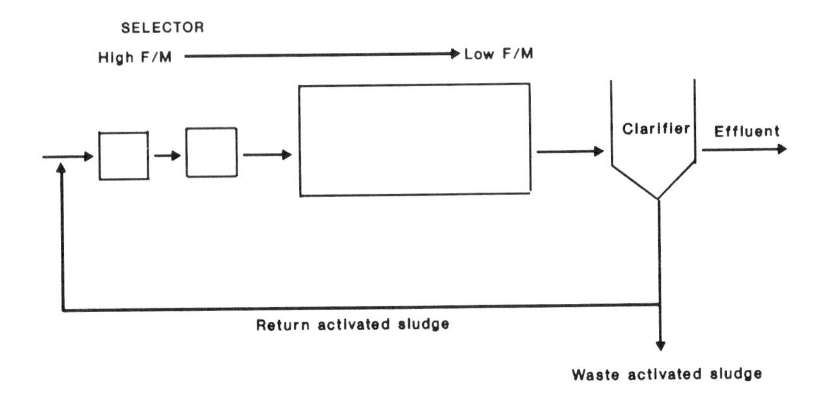

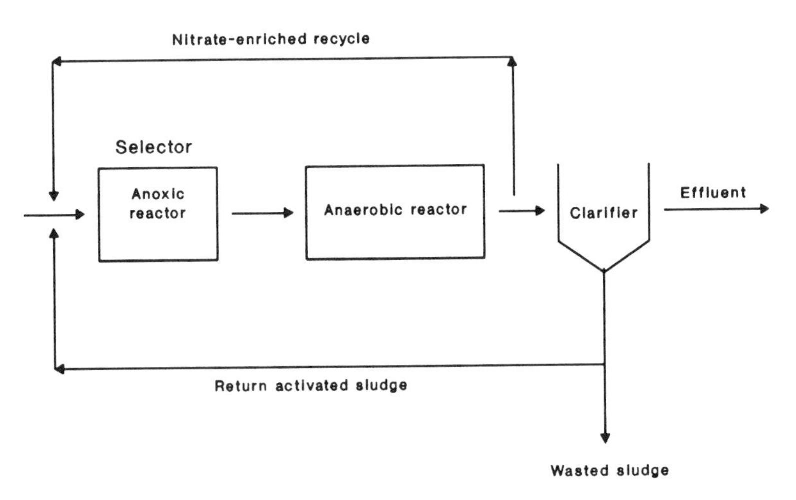

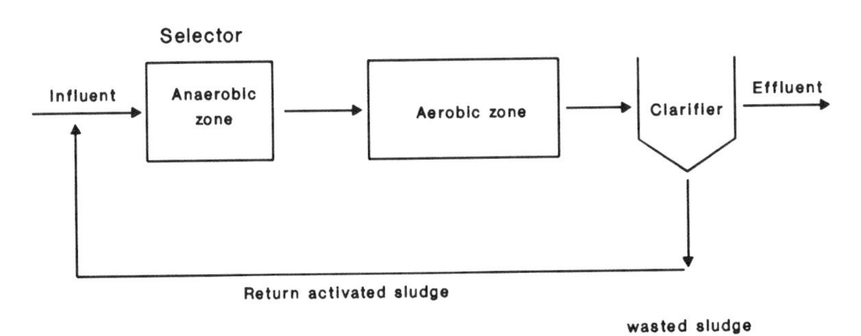

Fig. 9.11. Biological selectors for bulking control. **A.** Aerobic selector. **B.** Anoxic selector. **C.** Anaerobic selector. From Chudoba et al. (1973), with permission of the publisher.

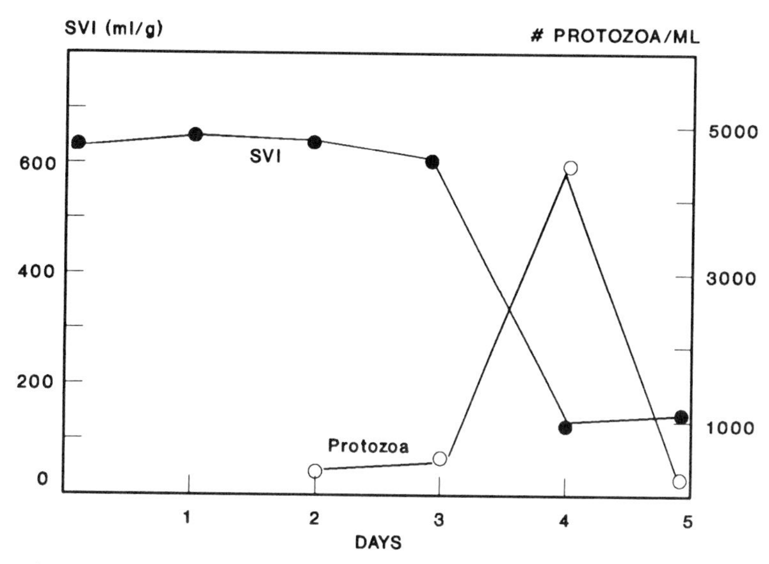

Fig. 9.12. Control of filamentous bacteria by predatory protozoa. From Inamori et al. (1991), with permission of the publisher.

dissolved oxygen concentrations did not reduce Type 0961 growth (Farquhar and Boyle, 1972; Nowak et al. 1986). Filamentous sulfur bacteria (*Thiothrix, Beggiatoa,* Type 021N) are unable to grow well under low pH conditions (Williams and Unz, 1989). Thus, adjustment and maintenance of wastewater at low pH (e.g., pH 5.5) may help control the growth of filamentous sulfur bacteria, although the possibility exists that the low pH might promote the growth of fungal filaments (Unz and Williams, 1988).

Iron compounds (e.g., ferrous sulfate, potasium ferrate, Fe-cystein) strongly inhibit the respiration of filamentous bacteria such as *Sphaerotilus, Thiothrix,* and *Type 021N* (Chang et al., 1979; Kato and Kazama, 1991; Lee, Koopman, and Bitton, unpublished results). These chemicals deserve further exploration.

Some investigators recommend a two-phase approach to solving a bulking episode: (1) chlorination to kill the extended filamentous microorganisms; and (2) identification of the causative microorganism(s) in order to make specific design or operational changes (Jenkins et al., 1984).

9.6. FOAMING IN ACTIVATED SLUDGE

9.6.1. Introduction

Foaming is a common problem encountered in many wastewater treatment plants around the world. A recent survey indicated that two thirds of the 114 U.S. plants surveyed experienced foaming at one time or another (Pitt and Jenkins, 1990). A 40% incidence has been observed in South Africa (Blackbeard et al., 1986). Half to more than 92% of the plants surveyed in Australia experienced a foaming problem during the sampling period (Blackall et al., 1988; 1991; Seviour et al., 1990) and 20% of 6,000 activated sludge plants surveyed in France were affected by foaming (Pujol et al., 1991).

The following are types of foams and scums that are found in activated sludge (Jenkins et al., 1984):

1. Undegraded surface-active organic compounds
2. Poorly biodegradable detergents that produce white foams

3. Scums due to rising sludge that results from denitrification in the clarifier
4. Brown scums due to excessive growth of actinomycetes.

Figure 9.13 shows the excessive scum growth in the aeration basins at a wastewater treatment plant in Atlanta. This type of scum appears to be the most troublesome in activated sludge and will be covered in some detail.

9.6.2. Problems Caused by Foams

Scum formation in activated sludge plants causes the following problems:

1. Excess scum can overflow into walkways (Fig. 9.13) and cause slippery conditions, leading to hazardous situations for plant workers.
2. Excess scum may pass into activated sludge effluent, resulting in increased BOD and suspended solids in the effluent.
3. Foaming causes problems in anaerobic digesters.
4. Foaming can produce the nuisance of odors, especially in warm climates.
5. Foam carries the potential of infection of wastewater workers with opportunistic pathogenic actinomycetes such as *Nocardia asteroides*.

Fig. 9.13. Scum covering the aeration basins at a wastewater treatment plant in Atlanta. Courtesy of Mesut Sezgin.

9.6.3. Foam Microbiology

Actinomycetes are the most important agents responsible for the brown viscous scum in activated sludge. These organisms generally proliferate in aeration basins operated at high mean cell residence time, but they may also be found in the clarifier.

Microorganisms identified in foams are *Nocardia amarae, N. rhodochrous, N. asteroides, N. caviae, N. pinensis, Streptomyces* spp., *Microthrix parvicella, Micromonospora*, Type 0675, and *Rhodococcus* spp. (Goddard and Forster, 1987a; Lechevalier and Lechevalier, 1974, 1975; Lemmer and Kroppenstedt, 1984; Pujol et al., 1991; Seviour et al., 1990; Sezgin and Karr, 1986; Sezgin et al., 1988). Figure 9.14 shows micrographs of *Nocardia* foams in mixed liquor. Other microorganisms that cause foaming problems are *Nostocoida limicola* and Type 0041; like *Nocardia amarae*, they have the ability to produce biosurfactants (Goddard and Forster, 1987b; Sutton, 1992; Wanner and Grau, 1989). Several other filamentous microorganisms have been reported in foams but their association with foam formation is not clear.

Nocardia amarae and, to a lesser extent, *N. pinensis* are the major organisms found in foams examined by U.S. and Australian investigators. Nuisance foams are produced when *Nocardia* occurs in mixed liquor at levels exceeding 26 mg *Nocardia* per 1 g VSS (Jenkins, 1992). *Rhodococcus* sp. is the dominant organism in foams examined in Europe. *Microthrix parvicella*, a nonbranching filamentous organism that is implicated in bulking, is sometimes a dominant organism in foams, as observed in surveys conducted in France and South Africa (Blackall et al., 1988; 1991; Blackbeard et al., 1986; Pujol et al., 1991).

Foam microorganisms use several growth substrates varying from sugars to high-molecular-weight polysaccharides, proteins, and aromatic compounds (Lemmer, 1986; Lemmer and Kroppenstedt, 1984). It has been suggested that scum actinomycetes survive in the aeration tank because they are able to switch from K-strategy (i.e., ability to grow at low substrate concentrations because of low K_s) to μ_{max}-strategy (i.e.,

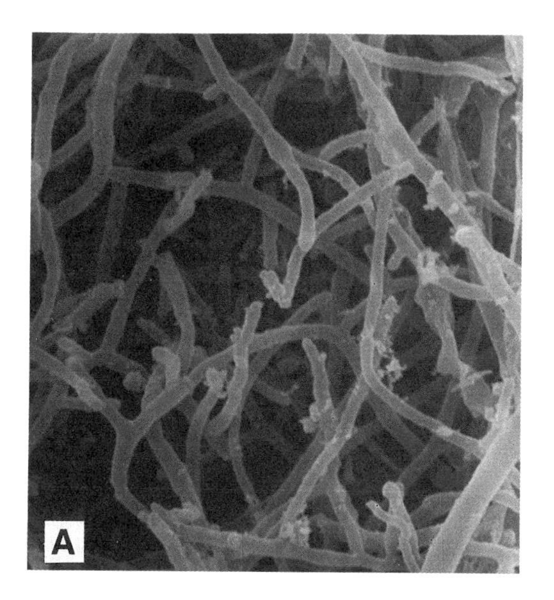

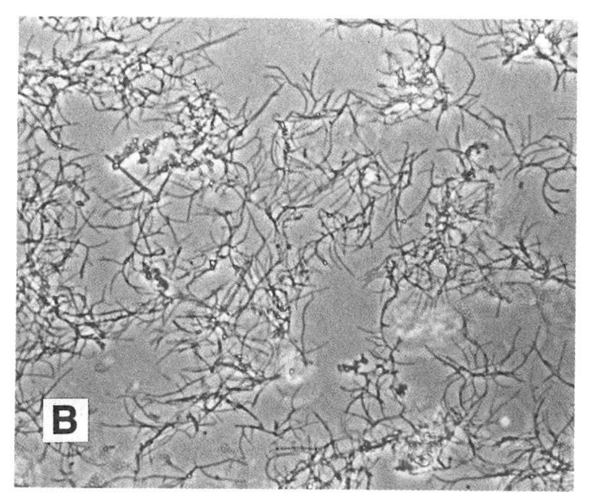

Fig. 9.14. *Nocardia* sp. in mixed liquor. **A.** Scanning electron micrograph of *Nocardia* foam at the University of Florida activated sludge process. Courtesy of J. Awong and G. Bitton. **B.** Phase-contrast micrograph of *Nocardia* sp. Courtesy of Mesut Sezgin.

ability to produce high biomass when sufficient nutrients are present) (Lemmer, 1986). However, study of the growth kinetics of *N. amarae* showed that this actinomycete has a relatively low μ_{max} (0.087 h^{-1}) and a relatively high K_s (675 mg/L). These values lead to the conclusion that excessive growth of this actinomycete in activated sludge cannot be associated with favorable growth kinetics (Baumann et al., 1988) but can be explained by the production of biosurfactants and selective utilization of hydrophobic compounds such as hydrocarbons by *N. amarae*, which has a hydrophobic surface (Lemmer and Baumann, 1988a). These characteristics are essential for foam production and transport of the cells to the bubble phase (Blackall and Marshall, 1989; Blackall et al., 1988). Anionic surfactants and their biodegradation products can significantly enhance foaming in *Nocardia*-containing activated sludge (Ho and Jenkins, 1991). Other advantages of actinomycetes over other wastewater bacteria are their higher resistance to desiccation and UV irradiation and their ability to store polyphosphates and poly-β-hydroxybutyric acid (Lemmer and Baumann, 1988b).

9.6.4. Mechanisms of Foam Production

The causes and mechanism(s) of foam production are not well understood. Some possible mechanisms are the following (Soddell and Seviour, 1990):

1. Gas bubbles produced by aeration or metabolism (e.g., N_2) may assist in flotation of foam microorganisms.
2. The hydrophobic nature of the cell walls of foam microorganisms help their transport to the air–water interface.
3. Biosurfactants produced by foam microorganisms assist in foam formation.
4. Foams are associated with relatively long retention times (> 9 days), with warm temperatures (> 18°C) (Pipes, 1978), and with wastewaters rich in fats (Eikelboom, 1975).

9.6.5. Foam Control

Numerous measures for controlling foams in activated sludge have been proposed. Several of these cures are not always successful and have not been rigorously tested under field conditions. The control measures that have been proposed are the following (Jenkins et al., 1984; Soddell and Seviour, 1990):

1. Chlorination of foams (chlorine sprays) or return activated sludge. Some operators have reported success following chlorination of return activated sludge. However, excessive chlorine levels may cause dispersion of floc and deterioration of effluent. This practice has not been successful in controlling scums in an Atlanta plant (Sezgin and Karr, 1986).

In Phoenix, *Nocardia* foaming was controlled by spraying the foam with a high concentration (2,000–3,000 mg/L) of chlorine (Albertson and Hendricks, 1992).

2. Increase in sludge wasting. Since one of the causes of foaming is relatively long mean cell retention time, foam can be controlled by increasing sludge wasting (i.e., reducing sludge age), thus causing *Nocardia* to be washed out. *Nocardia* numbers declined to undetectable levels at an MCRT of 2.2 days at 16°C and 1.5 days at 24°C (Fig. 9.15) (Cha et al., 1992). A survey indicated that MCRT reduction was the most common strategy used by U.S. wastewater treatment plants, with a success rate of 73% (Pitt and Jenkins, 1990). However, this control measure is not always successful in eliminating foaming in full-scale wastewater treatment plants (Mori et al., 1992) and is not desirable when long retention times are required (e.g., nitrification process). At the R.M. Clayton plant in Atlanta, scum was eliminated by reducing sludge age from 10 days to less than 3 days for a period of approximately 25 days (Sezgin and Karr, 1986).

3. Use of biological selectors. The use of an anoxic selector was sometimes successful in controlling the establishment of scum in the wastewater treatment plant in Georgia (Sezgin and Karr, 1986). Bench-scale studies showed than an aerobic selector can control *Nocardia* populations at an MCRT of 5 days (Cha et al., 1992).

4. Reducing air flow in the aeration basin. By this means scum accumulation can be controlled because these filamentous microorganisms are strict aerobes (Sezgin and Karr, 1986).

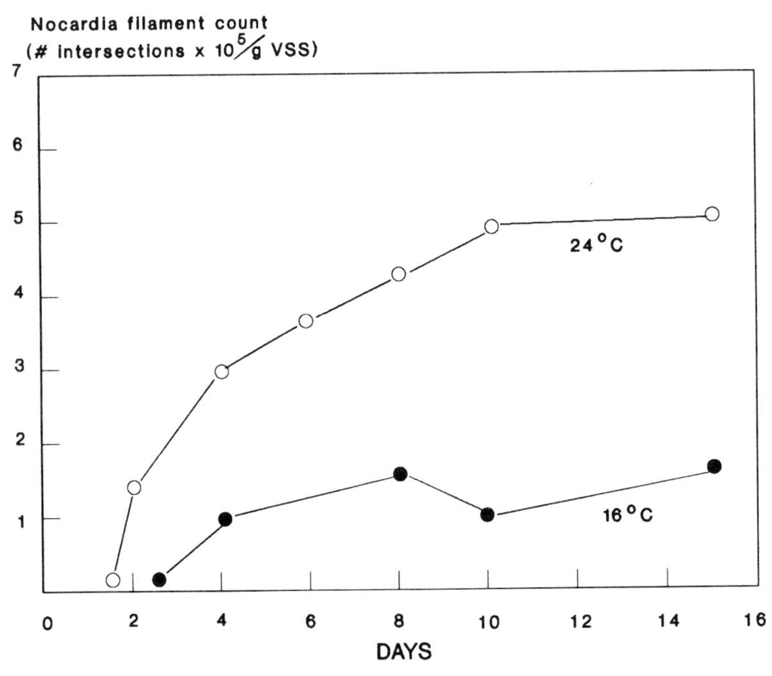

Fig. 9.15. Effect of MCRT and temperature on *Nocardia* populations in bench-scale activated sludge units. From Cha et al. (1992), with permission of the publisher.

5. Reduction in pH and in oil and grease levels. This approach also resulted in a decrease in scum accumulation (Sezgin and Karr, 1986).

6. Addition of anaerobic digester supernatant to wastewater. This supernatant is toxic to pure cultures of *Nocardia* (Lechevalier et al., 1977; Lemmer and Kroppenstedt, 1984). However, under field conditions, the addition of this supernatant to wastewater is not always successful in controlling actinomycetes foams (Blackall et al., 1991). The toxic agent found in anaerobic digester supernatant needs to be fully characterized.

7. Water sprays to control foam buildup. This approach does not result in a complete mechanical collapse of the foam.

8. Antifoam agents and iron salts. Their use has produced mixed results. Under laboratory conditions, it was shown that a suspension of montmorillonite (a three-layer clay mineral), at a concentration of 100 μg/ml, prevents the formation of stable foam by *N. amarae* (Blackall and Marshall, 1989).

9. Some foam accumulation was observed in low-turbulence zones in the aeration tank. This problem can be avoided by proper location of the aerators (Blackall et al., 1991).

10. Physical removal of the foam. The skimmed-off scum should not be recycled into the primary clarifier or the aeration tank, but should be wasted (Lemmer and Baumann, 1988a, 1988b).

11. Use of antagonistic microflora. The use of bacteria and predatory protozoa to control foam actinomycetes has not been successful (Soddell and Seviour, 1990).

9.7. FURTHER READING

Eikelboom, D.H. 1975. Filamentous organisms observed in bulking activated sludge. Water Res. 9: 365–388.

Eikelboom, D.H., and H.J.J. van Buijsen. 1981. *Microscopic Sludge Investigation Manual.* Report # A94a. TNO Research Institute, The Netherlands.

Jenkins, D., M.G. Richard, and G.T. Daigger. 1984. *Manual on the Causes and Control of Activated Sludge Bulking and Foaming.* Water Research Commission, Pretoria, South Africa.

Soddell, J.A., and R.J. Seviour. 1990. Microbiology of foaming in activated sludge foams. J. Appl. Bacteriol. 69: 145–176.

10

PROCESSES BASED ON ATTACHED MICROBIAL GROWTH

10.1. INTRODUCTION

Biofilm reactors comprise trickling filters, rotating biological contactors (RBC), and submerged filters (down-flow and up-flow filters). These reactors are used for oxidation of organic matter, nitrification, denitrification, or anaerobic digestion of wastewater (Fig. 10.1) (Harremoes, 1978).

In a fixed-film biological process, microorganisms are attached to a solid substratum in which they reach relatively high concentrations. The support materials include gravels, stones, plastic, sand, and activated carbon particles. Two important factors influencing microbial growth on the support material are the flow rate of wastewater and the size and geometric configuration of particles. Several rate-limiting phenomena are involved in biofilms. Electron donors and acceptors must diffuse inside the biofilms and reaction products must be transported out of the biofilm (Fig. 10.2) (Harremoes, 1978).

Fixed-biofilm processes offer the following advantages (Rittmann et al., 1988):

1. They allow the development of microorganisms with relatively low specific growth rates (e.g., methanogens).
2. They are less subject to variable or intermittent loadings.
3. They are suitable for small reactor sizes.

Two types of fixed-film reactors (trickling filters and rotating biological contactors) are described in this chapter. Anaerobic fixed-film bioreactors are described in Chapter 13. Fixed-film kinetics are discussed in Chapter 17.

10.2. TRICKLING FILTERS: PROCESS DESCRIPTION

The trickling or percolating filter was introduced in 1890 and is one of the earliest systems for biological waste treatment. It has four major components (Fig. 10.3).

1. A circular or rectangular tank containing the filter medium with a bed depth of approximately 1.0–2.5 m. The filter medium provides a large surface area for microbial growth. The ideal filter medium should provide a large surface area to maximize microbial attachment and growth. It should also provide sufficient void space for air diffusion as well as allowing sloughed microbial biofilm to pass through. It should not be toxic to microorganisms and should be chemically and mechanically stable (Grady and Lim, 1980).

The filter media used in trickling filters are stones (crushed limestone and granite), ceramic material, treated wood, hard coal, or plastic media. Selection of filter media is based on factors such as specific surface area, void space, unit weight, media configuration and size, and cost (U.S. EPA, 1977). The smaller the size, the

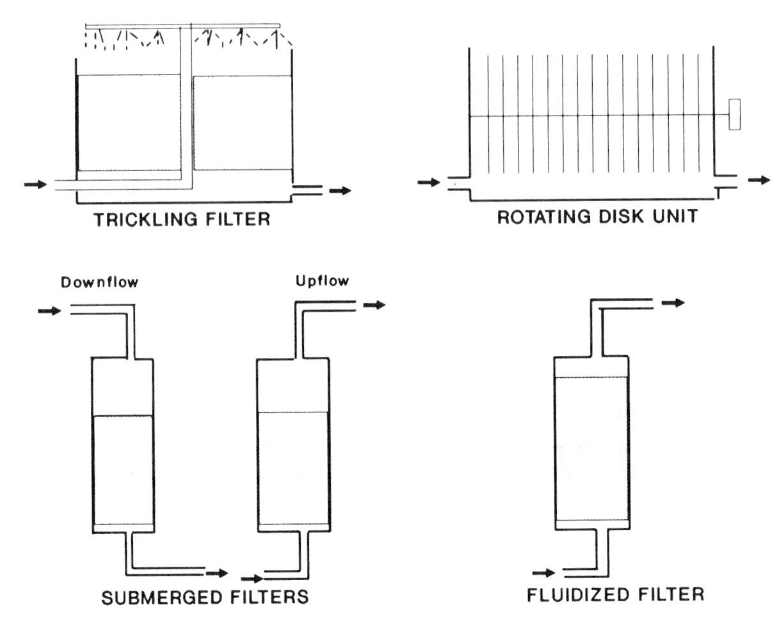

Fig. 10.1. Biofilm reactors used for wastewater treatment. Adapted from Harremoes (1978).

higher the surface area for microbial attachment and growth, but the lower the percentage void space. Plastic media, introduced in the 1970s, are made of PVC or polypropylene and are mainly used in high-rate trickling filters. They have a low bulk density and offer optimum surface area (85–140 m^2/m^3) and much higher void space (up to 95%) than other filter media. Thus, filter clogging is considerably minimized when these media are used. Plastic is also a light material that requires less heavily reinforced concrete tanks than do stone media. Therefore, biological tower reactors containing these materials can be as high as 6–10 m.

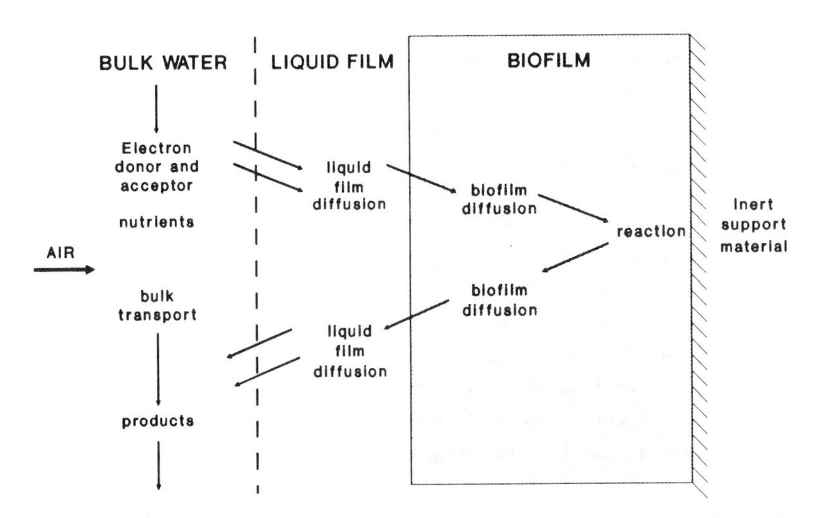

Fig. 10.2. Potentially rate-limiting phenomena involved in biofilm reactions. From Harremoes (1978), with permission of the publisher.

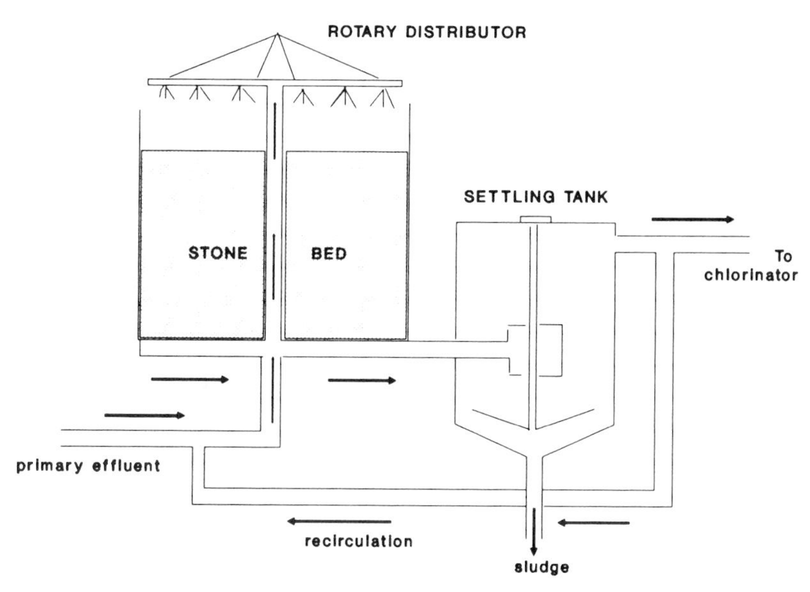

Fig. 10.3. The trickling filter process.

2. A wastewater distributor that allows a uniform hydraulic load over the filter material. It has one to four arms and its configuration and speed depend on the filter media used. Hydraulic load varies from less than 5 m^3/m^2/day for low-rate filters to more than 25 m^3/m^2/day for high-rate filters (Nathanson, 1986). Wastewater is percolated or trickled over the filter and provides nutrients for the growth of microorganisms on the filter surface.

3. An underdrain system for collection of liquids and introduction of air. The underdrain collects treated wastewater as well as biological solids (i.e., microbial biomass) that have been sloughed off the biofilm material.

4. A final clarifier, also called the humus tank, for separation of solids from the treated wastewater.

Trickling filters can operate in different modes (Forster and Johnston, 1987):

The single-pass mode. The single-pass mode handles an organic loading rate of 0.06–0.12 kg BOD m^{-3} day $^{-1}$.

The alternating double-filtration (ADF) mode. The ADF mode involves the alternate use (1–2 weeks interval) of two sets of filters and humus tanks and allows higher organic loading rates with no problems of filter clogging.

The recirculation mode. Trickling filter effluents are partially recirculated through the filter to increase the treatment efficiency of the filter media (Metcalf and Eddy, 1991). A portion of the treated effluent is returned to the filter. The recirculation ratio R is the ratio of the flow rate of recirculated effluent to the flow rate of the wastewater influent (Nathanson, 1986):

$$R = Q_R/Q \tag{10.1}$$

where Q_R = the flow rate of recirculated trickling filter effluent; and Q = the flow rate of wastewater influent.

Changes in the quality and quantity of wastewater can be handled by adjusting the rate of recirculation. Recirculation improves contact between wastewater and the filter material, helps dilute high-BOD or toxic wastewater, increases dissolved oxygen for biodegradation of organics and for tackling odor problems, improves distribution of the influent on the filter surface, prevents the filter from drying out during the night when the wastewater flow is low, and avoids ponding (i.e., puddles at the surface of the filter) (Davis and Cornwell, 1985; U.S. EPA, 1977).

10.3. BIOLOGY OF TRICKLING FILTERS

A trickling filter essentially converts soluble organic matter to biomass, which is further removed by settling in the final clarifier. Organic loading is the rate at which BOD is applied to the trickling filter, and is expressed in kilograms BOD applied per cubic meter of filter per day. A typical organic loading is 0.5 kg/m^3·d and may vary from 0.1–0.4 kg/m^3/d for low-rate filters to 0.5–1 kg/m^3/d for high-rate filters. This parameter is important to the performance of the trickling filter and may dictate the hydraulic loading onto the filter. BOD removal by trickling filters is approximately 85% for low-rate filters and 65%–75% for high-rate filters (U.S. EPA, 1977).

10.3.1. Biofilm Formation

Light microscopy and transmission and scanning electron microscopy serve as tools for studying the formation of biofilms in wastewater (Eighmy et al., 1983). The biofilm that forms on the surface of the filter media in trickling filters is called the zoogleal film; it is composed of bacteria, fungi, algae, protozoa, and other life forms (Fig. 10.4). The processes involved in biofilm formation in wastewater are similar to those that occur in natural aquatic environments. After conditioning of the substratum with organic materials, the surface is colonized with gram-negative bacteria followed by filamentous bacteria. Permanent adsorption to the substratum requires the formation of a polysaccharide-containing matrix, named the glycocalyx. These polysaccharides help anchor the biofilm microorganisms to the surface of the filter material (Bitton and Marshall, 1980). The glycocalyx also provides a surface that is rich in polyanionic compounds that complex metal ions (Eighmy et al., 1983).

Biofilm microorganisms degrade the organic matter present in wastewater. The increase in biofilm thickness, however, leads to limited oxygen diffusion to the deeper layers of the biofilm, thus creating an anaerobic environment near the filter media surface. Microorganisms in the deeper layers face a reduced supply of organic substrates and enter into the endogenous phase of growth. They are subsequently sloughed off the surface. Sloughing is followed by the formation of a new biofilm (Metcalf and Eddy, 1991).

10.3.2. Organisms Present in Biofilms

Trickling filters are notable for the diversity of life forms participating in wastewater treatment, making this process more complex than the activated sludge process. In addition to procaryotic and eucaryotic microorganisms, trickling filters contain higher life forms such as rotifers, nematodes, annelid worms, snails, and many insect larvae. The groups of organisms encountered in trickling filters are the following:

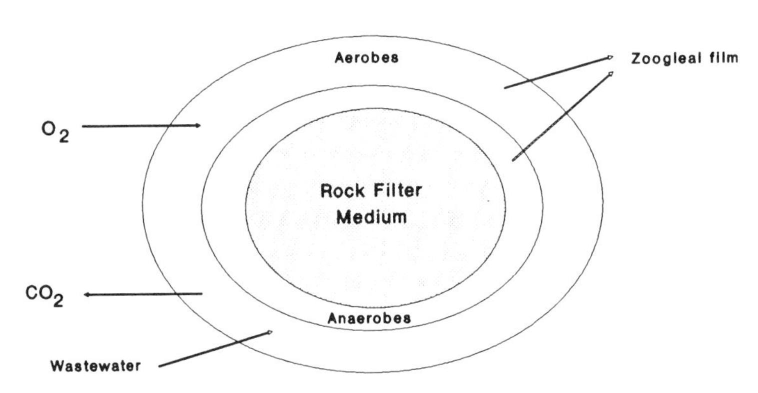

Fig. 10.4. Zoogleal film formation on packing media in trickling filters. Adapted from Zajik (1971).

Bacteria. Bacteria are active in the uptake and degradation of soluble organic matter. Colloidal organic matter is also trapped in the filter and degraded by extracellular enzymes (Grady and Lim, 1980). Some of the bacterial genera that are active in trickling filters are *Zooglea, Pseudomonas, Flavobacterium, Achromobacter, Alcaligenes*, filamentous bacteria (e.g., *Sphaerotilus*), and nitrifying bacteria (*Nitrosomonas* and *Nitrobacter*).

Nitrifying bacteria convert ammonia into nitrate (see Chapter 3). Nitrification efficiency is operationally defined as the percentage of ammonia removed during treatment. This definition does not take into account other nitrogen transformations such as nitrogen assimilation and denitrification (Parker and Richards, 1986). As discussed in regard to activated sludge, the extent of nitrification in trickling filters depends on a variety of factors, including temperature, dissolved oxygen, pH, presence of inhibitors, filter depth and media type, loading rate, and wastewater BOD (Balakrishnan and Eckenfelder, 1969; Parker and Richards, 1986; U.S. EPA, 1977). Low-rate trickling filters allow the development of a high-nitrifying population. Conversely, high-rate filters, owing to higher loading rates and continuous sloughing of the biofilm, do not allow nitrification to proceed (U.S. EPA, 1977). For rock media filters, organic loading should not exceed 10 lb BOD_5/1,000 ft^3/day (U.S. EPA, 1975). The decrease of nitrification at higher loading rates is due to the predominance of heterotrophs in the biofilm (Parker and Richards, 1986). Higher loading rates (22 lbs BOD_5/1,000 ft^3/day) are allowable in plastic-media trickling filters because of the higher surface area of the plastic media (Stenquist et al., 1974). Figure 10.5 (Parker and Richards, 1986) shows the relationship of BOD_5 level and nitrification in a 16-ft plastic media tower (combined carbon-oxidation nitrification system) in Garland, Texas and Atlanta, Georgia. Nitrification was initiated when the BOD_5 level was less than 20 mg/L and occurred only at the bottom of the tower, where the BOD_5 level was relatively low and where the competition between heterotrophs and nitrifiers was low (Wanner and Gujer, 1984). If two filters are used, heterotrophic growth occurs in the first filter and nitrification in the second filter.

Fungi. Fungi are also active in the biofilm in connection with waste stabilization. They predominate, however, only under low pH conditions, a situation that can be created by the introduction of some acidifying industrial wastes. Some examples of fungi found in trickling filters are *Fusarium, Penicillium, Aspergilus, Mucor, Geotrichum*, and yeasts. The hyphae growth is helpful for the transfer of oxygen to the lower depths of the biofilm.

Algae. Many types of algae also grow on the biofilm surface (e.g., *Ulothrix, Phormidium, Anacystis, Euglena, Chlorella*). They produce oxygen during the daytime following photosynthesis, and some species of blue-green algae are also able to fix nitrogen. In contrast to activated sludge, algae and fungi are important components of biofilms in trickling filters.

Protozoa. These unicellular procaryotic organisms feed on biofilm bacteria. Continuous removal of bacteria by protozoa helps maintain a high decomposition rate (Uhlmann, 1979). The protozoa found in biofilms are flagellates (e.g., *Bodo, Monas*), ciliates (e.g., *Colpidium, Vorticella*), and amaeba (e.g., *Amaeba, Arcella*).

Rotifers. Rotifers (e.g., *Rotaria*) are also encountered in biofilms.

Macroinvertebrates. Several groups of macroinvertebrates (nematodes, lumbricidae, collembola, and diptera) are found in trickling filters. Insect larvae (e.g., chiromonids, *Psychoda alternata, P. severini, Sylvicola fenestralis*) feed on the biofilm and help control its thickness, thus avoiding clogging of the filter by microbial exopolymers. These larvae develop into adult insects ("filter flies") in 2–3 weeks and may be a nuisance, particularly to wastewater treatment plant operators. Flies in numbers as high as 30,000/m^2·day have been reported (Edeline, 1988). Insects are controlled by increasing the wetting of the filter surface, since *Psychoda* larvae emerge only in dry filters, and by chemical control with insecticides (Forster and Johnston, 1987). Commercial preparations of *Bacillus thuringiensis* var. *israelensis* can also be applied

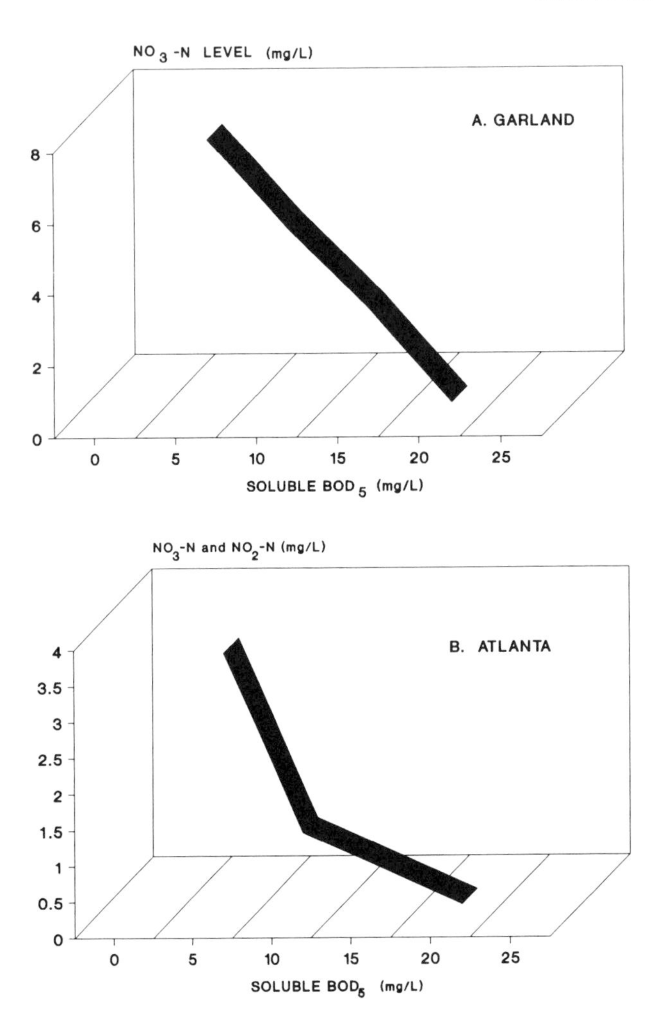

Fig. 10.5. Relationship between nitrification and soluble BOD_5. **A.** Garland, Texas, plant. **B.** Atlanta, Georgia, plant. Adapted from Parker and Richards (1986).

to filter beds to control insect larvae. The spores of this entomogenous pathogen contain a protoxin that is activated in the insect gut, causing its death (Houston et al., 1989a, 1989b).

Cold temperatures and toxicants slow down predator (protozoa and macroinvertebrates) activity and thus increase the chances of filter clogging with subsequent ponding. This may adversely affect the performance of the humus tank in the spring when predator activity resumes. Excess biofilm ("spring sloughing") overloads the settling tank, with subsequent increase of solids in the final effluent (Forster and Johnston, 1987). Nitrification is also affected by low winter temperatures.

More work is needed for a more thorough understanding of the microbial ecology of trickling filters.

10.3.3. Biofilm Kinetics

The proper functioning of a trickling filter depends on the growth kinetics within the biofilm that develops on the filter material. Biofilm growth is described by the following equation (La Motta, 1976; Uhlmann, 1979):

$$dX/dt = \mu X - kX \tag{10.2}$$

where μX = growth; kX = loss; X = number of microorganisms (microbial biomass); μ = growth rate constant (hr^{-1}); and k = decay rate constant (hr^{-1}).

The growth rate μ depends on the wastewater loading rate, while k depends on factors such as removal by grazing, decay of bacterial and fungal biomass, and hydraulic load.

At steady state

$$dX/dt = 0 \tag{10.3}$$

which implies that

$$\mu X = kX \tag{10.4}$$

This situation is desirable for the proper functioning of the trickling filter. Biofilm thickness depends on the strength (i.e., BOD_5) of the incoming wastewater and controls the removal of substrate by the filter. Above a critical value, biofilm thickness no longer controls the removal of substrate (Fig. 10.6) (Uhlmann, 1979). Since oxygen diffusion within the biofilm is limited to approximately 0.3 mm or less, the thickness of the active portion of the biofilm is controlled by the extent of oxygen diffusion within the film.

BOD removal by the active portion of the biofilm proceeds according to the Monod equation and thus depends on substrate concentration (see Chapter 2).

Fig. 10.6. Effect of biofilm thickness on substrate removal rate. Adapted from Uhlmann (1979).

10.3.4. Some Advantages and Disadvantages of Trickling Filters

Some advantages and problems related to trickling filters are the following.

Advantages. Trickling filters are attractive to small communities because of easy operation, low maintenance costs, and reliability. They are used to treat toxic industrial effluents and are able to withstand shock loads of toxic imputs. Furthermore, the sloughed biofilms can be easily removed by sedimentation.

Disadvantages. High organic loading may lead to filter clogging as a result of excessive growth of slime bacteria in biofilms. Excessive biofilm growth can also cause odor problems in trickling filters. Clogging restricts air circulation, thus resulting in low availability of oxygen to biofilm microorganisms. However, modifications have helped improve the BOD removal of trickling filters (Best et al., 1985; Grady and Lim, 1980). The following are some of these improvements:

1. Alternating double filtration (ADF), which consists of alternating two filters for receiving the waste
2. Slowing down wastewater distribution
3. Use of plastic materials in the filter for increased surface area and improvement of air circulation
4. Management of odor problems by increasing air flow by means of forced ventilation

10.4. REMOVAL OF PATHOGENS AND PARASITES BY TRICKLING FILTERS

The removal of pathogens and parasites by trickling filters is generally low and erratic.

Removal of bacteria is inconsistent and varies from 20% to above 90%, depending on the operation of the trickling filter. Removal of *Salmonella* by trickling filters is lower than by the activated sludge process and may vary from 75% to 95% (Feachem et al., 1983).

The removal of viruses by trickling filters is also generally low and erratic. Filtration rates affect the removal of viruses, and probably other pathogenic microorganisms. Laboratory experiments have shown that at a medium filtration rate (10 MGD per acre), the removal of viruses, coliforms, and fecal streptococci, as well as BOD and COD reduction, was greater than at a higher rate (23 MGD per acre) (Clarke and Chang, 1975). At the Kerrville, Texas wastewater treatment plant, the mean removal of total and fecal coliforms by a trickling filter was 92% and 95%, respectively. However, removal of enterovirus was erratic and varied from 59% to 95% (Moore et al., 1981). Similarly, removal of bacterial phage is inconsistent and varies between 40% and 90%, depending on the season of the year (Kott et al., 1974). In New Zealand, two trickling filter systems removed between 0% and 20% of viruses despite a significant reduction of fecal coliforms (> 90% removal) (Lewis et al., 1986). Similar observations were made in Japan; the removal efficiency of trickling filters is lower for viruses than for indicator bacteria. The mechanism of removal of viruses by trickling filters is not well known. Some investigators have suggested that viruses are removed by adsorption to the biofilm material (Omura et al., 1989).

Trickling filter plants in India removed 74%–91% of *Entamoeba histolytica*. The removal of *Giardia* cysts is similar to that of *Entamoeba* (Panicker and Krishnamoorthi, 1978). A trickling filter effluent in Maryland harbored from 4–44 cysts per liter. The removal efficiency of trickling filters is generally lower than that of activated sludge (Casson et al., 1990).

10.5. ROTATING BIOLOGICAL CONTACTORS

10.5.1. Process Description

A rotating biological contactor is another example of a fixed film bioreactor. This process was conceived in Germany at the beginning of this century and introduced in the United States in the 1920s but was marketed only in the 1960s (Huang and Bates, 1980; U.S. EPA, 1977).

An RBC consists of a series of disks that are mounted on a horizontal shaft and rotate slowly in the wastewater. The disks are approximately 40% submerged in wastewater (Fig. 10.7). At any time, the submerged portion of the disk removes BOD as well as dissolved oxygen. The rotation of the disks provides aeration as well as the shear force that causes sloughing of the biofilm from the disk surface. Increased rotation improves oxygen transfer and enhances the contact between attached biomass and wastewater (Antonie, 1976; Hitdlebaugh and Miller, 1981; March et al., 1981). The advantages offered by RBC are short residence time, low operation and maintenance costs, and production of a readily dewatered sludge that settles rapidly (Weng and Molof, 1974).

10.5.2. RBC Biofilms

As discussed for trickling filters, there is an initial adsorption of microorganisms to the disk surface to form a 1- to 4-mm-thick biofilm that is

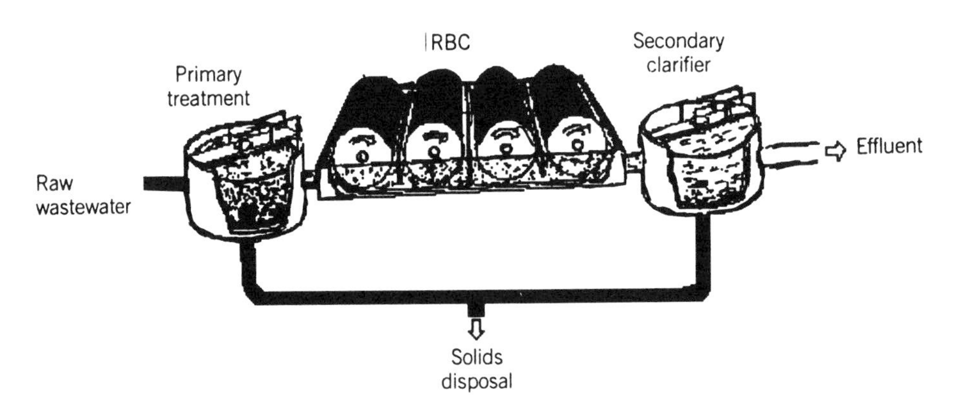

Fig. 10.7. Rotating biological contactor. From U.S. EPA (1975), with permission of the publisher.

responsible for BOD removal in rotating biological contactors. The rotating disks provide a large surface area for the attached biomass.

Biofilms developing on RBC comprise a complex and diverse microbial community made of eubacteria, filamentous bacteria, protozoa, and metazoa. Commonly observed filamentous organisms include *Sphaerotilus, Beggiatoa, Nocardia,* and filamentous algae such as *Oscillatoria* (Hitdlebaugh and Miller, 1981; Kinner et al., 1983; Pescod and Nair, 1972; Pretorius, 1971; Torpey et al., 1971). Biofilm examination by transmission electron microscopy shows that *Sphaerotilus* contains many poly-β-hydroxybutyrate inclusions, an indication of storage of excess carbon by the bacteria. These inclusions may account for 11% to more than 20% of the dry weight of this bacterium (Rouf and Stokes, 1962). Scanning electron microscopy shows that the RBC biofilm is composed of two layers: an outer whitish layer containing *Beggiatoa* filaments and an inner black layer (due to ferrous sulfide precipitation) containing *Desulfovibrio,* a sulfate-reducing bacterium (Alleman et al., 1982). Figure 10.8 (Alleman et al., 1982) explains the relationship between *Desulfovibrio* and *Beggiatoa* within the RBC biofilm:

Anaerobic zone. In this layer, fermentative bacteria provide the end products (organic acids, alcohols) used by sulfate-reducing bacteria.

Aerobic zone. Hydrogen sulfide, produced by sulfate-reducing bacteria in the anaerobic zone, diffuses into the aerobic zone and is readily used by *Beggiatoa* as an electron donor. H_2S is oxidized to elemental sulfur.

The organism succession on RBC surfaces is similar to that observed in activated sludge (Kinner and Curds, 1989): Bacterial colonization is followed by protozoan flagellates and small amoebae → free-swimming bacteriovorous ciliates (e.g., *Colpidium*) → nematodes → stalked ciliates (e.g., *Vorticella*) → rotifers.

After reaching a certain thickness, the biofilm sloughs off and the sloughed material ultimately reaches the final clarifier.

The first stages of an RBC mostly remove organic materials (i.e., BOD_5 removal), whereas subsequent stages remove NH_4 as a result of nitrification, when the BOD_5 is low enough.

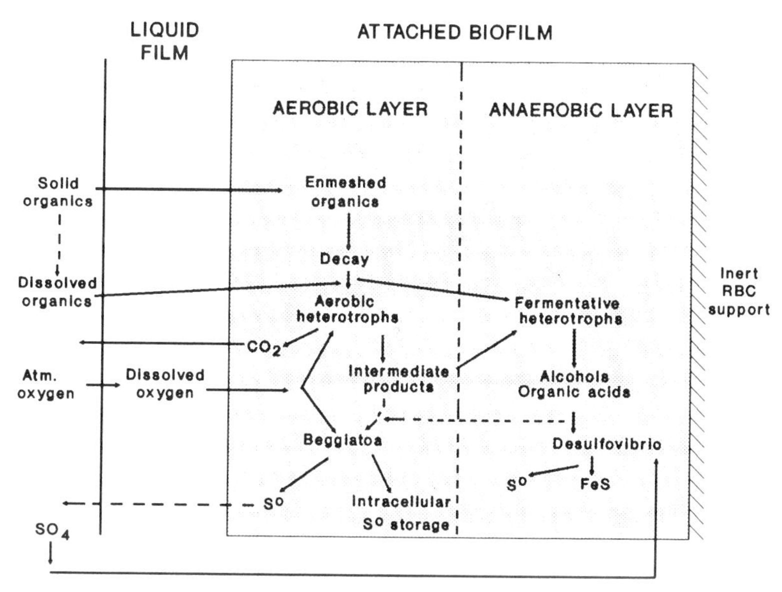

Fig. 10.8. Relationship between *Beggiatoa* and *Desulfovibrio* in an RBC biofilm. Adapted from Alleman et al. (1982).

Ammonia oxidizers cannot effectively compete with the faster-growing heterotrophs that oxidize organic matter. Nitrification occurs only when the BOD is reduced to approximately 14 mg/L, and increases with rotational speed (Weng and Molof, 1974). RBC performance is negatively affected by low dissolved oxygen in the first stages and by low pH in the later stages where nitrification occurs (Hitdlebaugh and Miller, 1981).

RBCs have some of the advantages of trickling filters (e.g., low cost, low maintenance, resistance to shock loads) but not some of the disadvantages (e.g., filter clogging, filter flies).

10.5.3. Pathogen Removal in RBC

Not much is known about the removal of indicator and pathogenic microorganisms in RBCs. RBCs appear to be fairly efficient in removing indicator bacteria. One log or more of fecal coliforms are removed by this process (Sagy and Kott, 1990).

10.6. FURTHER READING

Edeline, F. 1988. L'Epuration Biologique des Eaux Residuaires: Theorie et Technologie. Editions CEBEDOC, Liege, Belgium, 304 pp.

Forster, C.F., and D.W.M. Johnston. 1987. Aerobic processes, pp. 15–56, in: *Environmental Biotechnology,* C.F. Forster and D.A.J. Wase, Eds. Ellis Horwood, Chichester, U.K.

Grady, C.P.L., Jr., and H.C. Lim. 1980. *Biological Waste Treatment.* Marcel Dekker, New York, 963 pp.

Harremoes, P. 1978. Biofilm kinetics, pp. 71–109, in: *Water Pollution Microbiology,* Vol. 2, R. Mitchell, Ed. Wiley, New York.

U.S. EPA. 1977. *Wastewater Treatment Facilities for Sewered Small Communities.* EPA Report # EPA-625/1-77-009.

11

WASTE STABILIZATION PONDS

11.1. INTRODUCTION

Treatment of wastewater in ponds is probably the most ancient means of waste treatment known to humans. Oxidation ponds are also called stabilization ponds or lagoons and they serve mostly small rural areas where land is readily available at relatively low cost. They are used for secondary treatment of wastewater or as polishing ponds. The various types of stabilization ponds will be discussed in the following sections.

Waste stabilization ponds are classified as facultative, aerobic, anaerobic, aerated, high-rate aerated, and maturation ponds (Hammer, 1986; Hawkes, 1983b; Nathanson, 1986; Reed et al., 1988).

11.2. FACULTATIVE PONDS

Facultative ponds are the most common type of lagoons for domestic wastewater treatment. Waste treatment is provided by both aerobic and anaerobic processes. These ponds range from 1 m to 2.5 m in depth and are subdivided in three layers: an upper aerated zone, a middle facultative zone, and a lower anaerobic zone. The detention time varies between 5 days and 30 days (Hammer, 1986; Negulescu, 1985; Hawkes, 1983b). Among the advantages of these ponds are low initial cost and ease of operation. Some disadvantages include odor problems associated mainly with growth of algae, and with mosquitoes, which are a public health concern.

11.2.1. Biology of Facultative Ponds

Waste treatment in oxidation ponds is the result of natural biological processes carried out mainly by bacteria and algae. Waste treatment is carried out by a mixture of aerobic, anaerobic, and facultative microorganisms. These ponds allow the accumulation of solids, which are degraded anaerobically at the bottom of the pond. Many categories of organisms play a role in the treatment process. These include mainly algae, heterotrophic bacteria, and zooplankton. The microbiological processes in facultative ponds are summarized in Figure 11.1 (U.S. EPA, 1977).

11.2.1.1. Activity in the Photic Zone

In the photic zone, photosynthesis is carried out by a wide range of algal species (mostly green and blue-green algae, diatoms), producing 10–66 g algae/m^2/day (Edeline, 1988). The most common species encountered in oxidation ponds are *Chlamydomonas, Euglena, Chlorella, Scenedesmus, Microactinium, Oscillatoria,* and *Microcystis* (Hawkes, 1983b). The type of algae that predominates is determined by a variety of factors. For example, diatoms prevail at lower temperatures than blue-green algae (Edeline, 1988). Algal photosynthesis depends on available temperature and light. In the presence of high numbers of algae, light penetration is limited to the first 2 ft of the water column. Mixing is important for the maintenance of aerobic conditions within the pond and for providing the exchange of nutrients and gases between phototrophs and heterotrophs. This exchange is impeded when the pond becomes stratified, a phe-

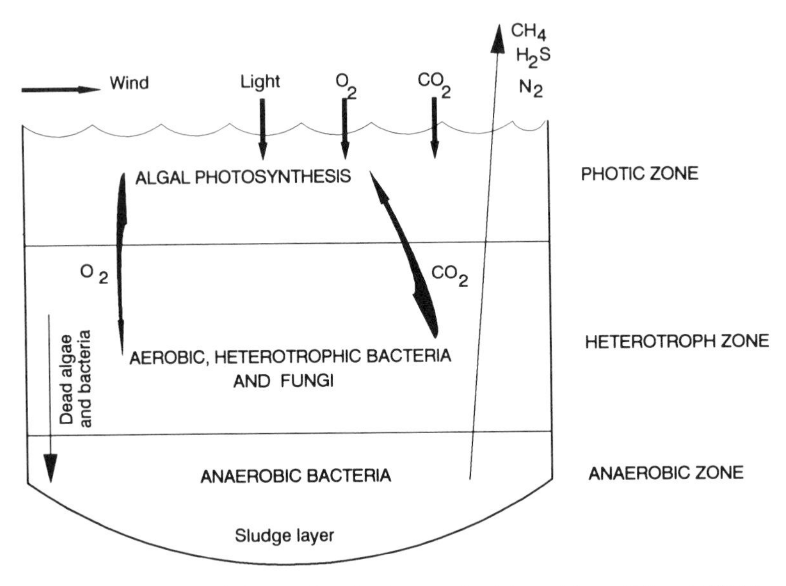

Fig. 11.1. Microbiology of facultative ponds. Adapted from Bitton (1980a).

nomenon that occurs under warm conditions in the absence of natural circulation. Stratification is due to the establishment of a temperature difference between the warm upper layer, or epilimnion, and the lower and colder layer, called the hypolimnion. The zone between the epilimnion and hypolimnion is called the thermocline; it is characterized by a sharp decrease in temperature (Fig. 11.2) (Curds and Hawkes, 1983).

Algae are also involved in uptake of nutrients, mainly nitrogen and phosphorus. Some algae are able to fix nitrogen (e.g. blue-green algae), while most others utilize ammonium or nitrate. Photosynthesis leads to an increase in pH, particularly if the treated wastewater has a low alkalinity, and this may create conditions for removal of nutrients. At high pH, phosphorus precipitates as calcium phosphate, and ammonium ion may be lost as ammonia. Furthermore, algal photosynthesis produces oxygen, which is used by heterotrophic microorganisms. Other photosynthetic microorganisms in oxidation ponds are the photosynthetic bacteria, which use H_2S as electron donor instead of H_2O. Therefore, their main role is the removal of H_2S, which causes odor problems (see Chapter 2). The numbers of both algae and photosynthetic bacteria decrease with increased organic loading (Houghton and Mara, 1992).

11.2.1.2. Heterotrophic Activity

Bacterial heterotrophs are the principal agents responsible for degradation of organic matter in facultative ponds. The role of fungi, as well as that of protozoa, appears to be less significant. Heterotrophic activity results in the production of CO_2 and micronutrients necessary for the growth of algae. Algae, in return, provide the oxygen that is necessary for aerobic heterotrophs. Surface reaeration is another source of oxygen for heterotrophs.

Dead bacterial and algal cells and other solids settle to the bottom of the pond, where they undergo anaerobic decomposition (Fig. 11.1). Anaerobic microbial activity results in the production of gases such as methane, hydrogen sulfide, carbon dioxide, and nitrogen. Hydrogen sulfide production may encourage the growth of photosynthetic bacteria, namely the purple sulfur bacteria (*Chromatium, Thiocapsa, Thiopedia*) (Edeline, 1988; Holms and Vennes, 1970; Houghton and Mara, 1992). Blooms of Rhodospirillaceae (nonsulfur purple bacteria) in sew-

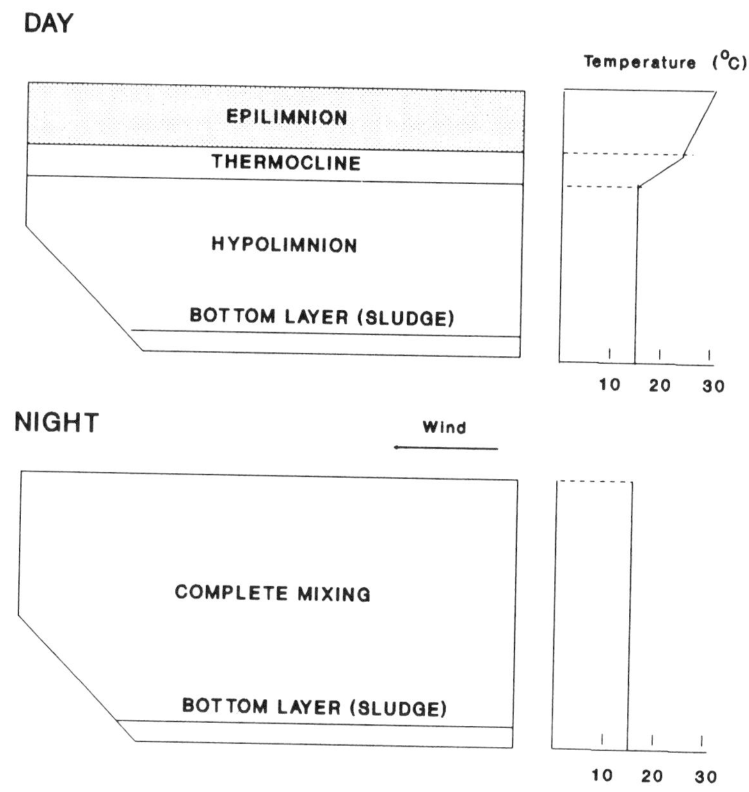

Fig. 11.2. Stratification in oxidation ponds. From Curds and Hawkes (1983), with permission of the publisher.

age lagoons have also been documented (Jones, 1956).

Although some of the carbon is lost to the atmosphere as CO_2 or CH_4, most of it is converted to microbial biomass. Unless the microbial cells are removed by sedimentation or some other solids removal process (e.g., intermittent sand filtration), little carbon reduction is obtained in oxidation ponds. Furthermore, the microbial cells in pond effluents may exert an oxygen demand in receiving waters (U.S. EPA, 1977).

11.2.1.3. Zooplankton Activity

Zooplankton (rotifera, cladocera, and copepoda) prey on algal and bacterial cells and can play a significant role in controlling these populations. Their activity is thus important to the operation of the pond. Cladocera (e.g., *Daphnia*) are filter feeders that feed mostly on bacterial cells and detrital particles but less on filamentous algae. They are therefore helpful in reducing the turbidity of pond effluents.

11.2.2. Effect of Temperature on Pond Operation

Temperature plays an important role in the activity of phototrophs and heterotrophs in wastewater ponds. It also significantly affects anaerobic waste degradation in the pond sediments. No methanogenic activity and subsequent reduction of sludge volume occurs at temperatures below 15°C. BOD loading varies from 2.2 g/m^2·day in cold climates to 5.6 g/m^2·day in warmer climates (Hammer, 1986). During the colder months, the ponds go anaerobic because of the lack of solar radiation and, hence, photosynthesis. One of the empirical equations developed for expressing the loading rate of a pond is the Gloyna equation, which gives the pond volume as a function of the prevailing temperature (Gloyna, 1971):

$$V = (3.5 \times 10^{-5})\, NgL_a \times 1.085^{\,(35-T_m)} \quad (11.1)$$

where V = pond volume (m^3); N = number of people contributing the waste; g = per capita waste contribution (L d^{-1}); L_a = BOD (mg L^{-1}); and T_m = average water temperature of coldest month (°C).

In warm climates, the effluent has a BOD of less than 30 mg/L. However, the suspended solid concentration may be high because of the presence of algal cells in the pond effluents.

11.2.3. Removal of Suspended Solids, Nitrogen, and Phosphorus by Ponds

Oxidation pond effluents often have a high level of suspended solids that are composed mostly of algal cells and wastewater solids (Reed et al., 1988) and can be treated by intermittent sand filters or microstrainers.

Nitrogen is removed by ponds by a number of mechanisms, including nitrification/denitrification, volatilization as ammonia, and algal uptake (see Chapter 3). Ponds remove approximately 40%–80% of nitrogen. Phosphorus removal by ponds is low and can be increased by in-pond treatment with iron and aluminum salts or with lime.

11.3. OTHER TYPES OF PONDS

Aerobic ponds. Aerobic ponds are shallow ponds (0.3–0.5 m deep) and are generally mixed to allow the penetration of light necessary for growth of algae and subsequent generation of oxygen. The detention time of wastewater is generally 3–5 days.

Aerated lagoons. Aerated lagoons are 2–6 m deep with a detention time of less than 10 days. They are used to treat high-strength domestic wastewater. They are mechanically aerated with air diffusers or mechanical aerators. Treatment (i.e., BOD removal) depends on aeration time, temperature, and type of wastewater. At 20°C, there is an 85% BOD removal with an aeration period of 5 days. Faulty operation of the aerated lagoon may result in foul odors.

Anaerobic ponds. Anaerobic ponds (Hammer, 1986) have a depth of 2.5–9 m and a relatively long detention time of 20–50 days (Metcalf and Eddy, 1991; Reed et al., 1988). These ponds serve as a pretreatment step for high-BOD organic wastes with high protein and fat content (e.g., meat wastes) and with high concentration of suspended solids. Organic matter is biodegraded under anaerobic conditions to CH_4, CO_2, and other gases such as H_2S (see Chapter 13 for more details on the microbiology of anaerobic digestion of wastes). These ponds do not require expensive mechanical aeration and generate small amounts of sludge. Some problems associated with these ponds are the production of odorous compounds (e.g., H_2S), sensitivity to toxicants, and the requirement of relatively high temperatures. Anaerobic digestion of wastewater is virtually halted below 10°C. These lagoons are not suitable for domestic wastewaters, which characteristicly have a low BOD.

Maturation or tertiary ponds. Maturation ponds are 1–2 m deep and serve as tertiary treatment for wastewater effluents from activated sludge or trickling filters. The detention time is approximately 20 days. Oxygen provided by surface reaeration and algal photosynthesis is used for nitrification. Their role is to further reduce BOD, suspended solids, and nutrients (nitrogen and phosphorus) and to further inactivate pathogens.

11.4. PATHOGEN REMOVAL BY OXIDATION PONDS

Removal or inactivation of pathogens in oxidation ponds is controlled by a variety of factors among which are temperature, sunlight, pH, lytic action of bacteriophages, predation by macroorganisms, and attachment to settleable solids.

11.4.1. Factors Controlling Removal of Bacteria

Oxidation ponds remove a significant percentage (90%–99%) of indicator and pathogenic bacteria. Coliform die-off in ponds increases with an increase in temperature, retention time, and pH, but decreases with an increase in BOD_5 and pond depth (Saqqar and Pescod, 1992). Other factors include aeration, antibacterial extracellular algal compounds, depletion of nutrients, sunlight intensity, and growth of algae (Fernandez et al., 1992; Qin et al., 1991). The possible reasons for the efficient removal of bacteria are the following:

1. Long detention time used in oxidation ponds.

2. High pH generated as a result of photosynthesis. Fecal coliform decline is higher in waste stabilization ponds where pH exceeds 9 (Parhad and Rao, 1974; Pearson et al., 1987). A study of the interaction between *Chlorella* and *E. coli* showed that the decline of the latter was due to the high pH (10–10.5) generated as a result of algal photosynthesis and growth. However, *E. coli* and *Chlorella* are able to grow together when the wastewater is buffered to pH = 7.5. This phenomenon is illustrated in Figure 11.3 (Parhad and Rao, 1974).

3. Predation by zooplankton. We have seen that these organisms play a significant role in the control of bacterial populations in ponds.

4. Inactivating effect of sunlight. The UV-B spectrum (wavelength between 280 nm and 320 nm) of sunlight may play a significant role in the destruction of coliforms in lagoons (Moeller and Calkins, 1980). Since ultraviolet light does not penetrate well into the water column, visible light could affect bacterial die-off. Its impact is increased at high dissolved oxygen concentrations as well as high pH levels (Curtis et al., 1992a, 1992b).

11.4.2. Bacterial Die-off Kinetics

Several kinetic models have been proposed for the prediction of bacterial die-off in waste stabilization ponds. As shown previously, there are several factors that affect bacterial decay rates in ponds. These factors include mainly temperature, solar radiation, predation, and antibiosis that results from growth of algae in the pond. Several investigators have proposed die-off models based mainly on the effects of temperature and solar radiation.

A simple model was proposed by Marais (1974):

$$N_e/N_i = 1 + K\theta \quad (11.2)$$

where N_e = coliform numbers in the pond effluent (numbers per 100 ml); N_i = coliform numbers in pond influent (numbers per 1 ml); K = first-order die-off coefficient (day^{-1}); and θ = theoretical retention time (days).

Temperature is a primary factor in bacterial die-off. There is a relationship between the decay constant K and temperature:

$$K_T = K_{20}\, C^{T-20} \quad (11.3)$$

where K_{20} = decay constant at 20°C; C = constant; and T = mean water temperature in pond (°C).

In a completely mixed pond within a temperature range of 5°C to 21°C, K_T is given by the following equation:

$$K_T = 2.6\,(1.19)^{T-20} \quad (11.4)$$

However, decay rates are sometimes lower than those predicted by Equation 11.4. For example, the relationship between decay rates and temperature for a number of waste stabilization ponds in Kenya gave the following equation (Mill et al., 1992):

$$K_T = 0.712\,(1.166)^{T-20} \quad (11.5)$$

In a series of n ponds N_e, the bacterial number in the pond effluent, is given by Equation 11.6:

$$N_e = \frac{N_i}{(KR_1 + 1)(KR_2 + 1)\,.\,.\,.\,(KR_n + 1)} \quad (11.6)$$

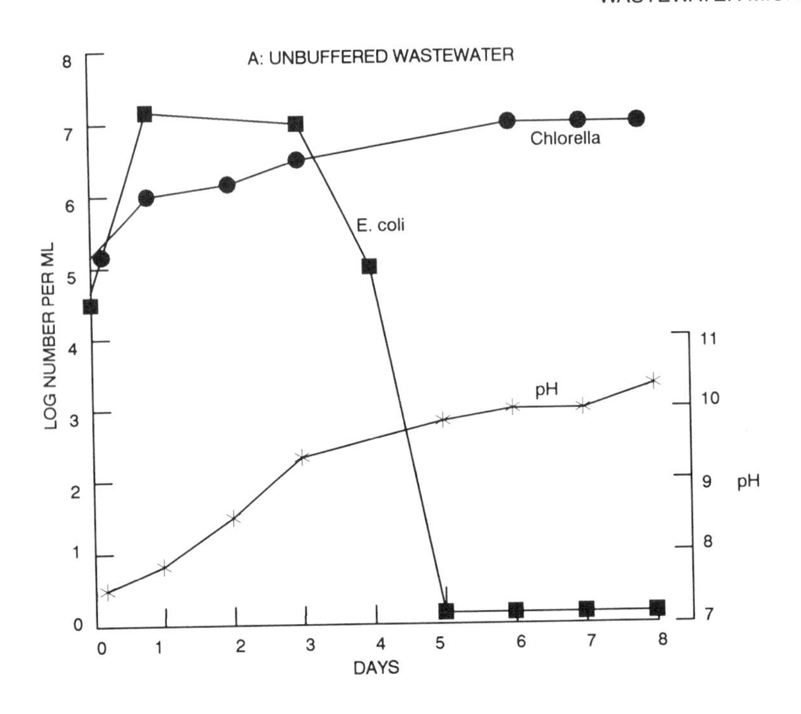

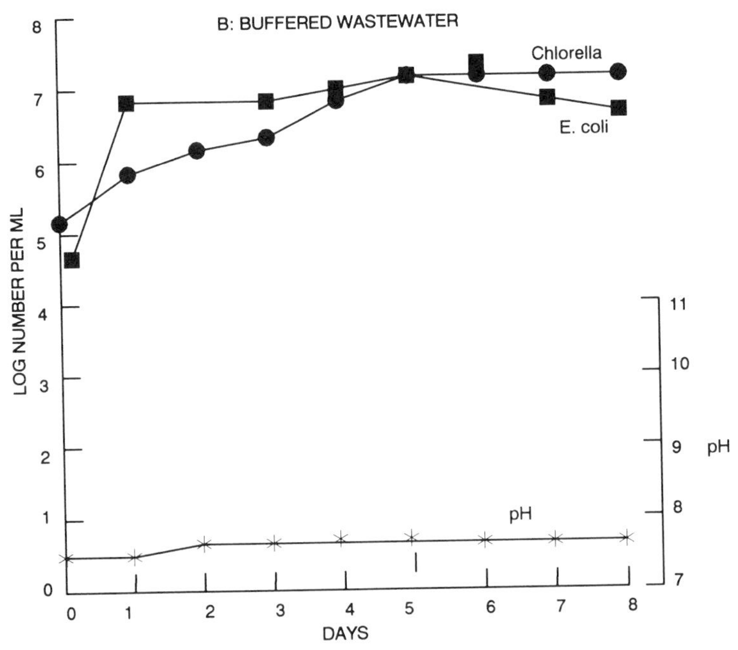

Fig. 11.3. Growth of *Escherichia coli* and *Chlorella* in unbuffered and buffered wastewater. From Parhad and Rao (1974), with permission of the publisher.

where n = number of ponds in the series. It is assumed that all ponds have the same size. R_1, R_2, and R_n are the retention times in pond 1, 2, . . . , and n, respectively. For a fixed total retention time, the removal efficiency increases as the number of ponds in the series increases.

Solar radiation greatly influences the rate of coliform die-off in oxidation ponds (Sarikaya and Saarci, 1987). The relationship between the decay constant K and light intensity I is given by Equation 11.7:

$$K = K_d + K_s(I) \tag{11.7}$$

where K_d = decay rate constant in the dark for I = 0 (hr^{-1}); K_d is temperature-dependent. K_s = decay rate constant due to the effect of light (hr^{-1}) and I = light intensity (cal/cm^2/hr).

The relationship between K and I at temperatures ranging from 25°C to 30°C is given by Equation 11.8 (Sarikaya and Saarci, 1987).

$$K = 0.018 + 0.012\,(I) \tag{11.8}$$

Therefore, the change in bacterial concentration N with time is given by Equation 11.9:

$$dN/dt = -KN = -(K_d + K_s\,I)N \tag{11.9}$$

Pond depth also plays a role in bacterial decay in waste stabilization ponds. The effect of pond depth on the coliform decay rate constant is given by Equation 11.10 (Sarikaya et al., 1987).

$$K = 1.156 + 5.244 \times 10^{-3}\,\frac{S_0}{kH}\,(1 - e^{-KH}) \tag{11.10}$$

where K = decay rate constant (day^{-1}); S_0 = daily solar radiation (cal/cm^2·day); k = light attenuation coefficient (m^{-1}); and H = pond depth (m).

The above equation shows that bacterial decay rates in shallow ponds are higher than in deeper ponds.

Concentration of algae can also be incorporated into die-off kinetic models. A multiple linear regression equation gives the bacterial die-off rate K as a function of temperature as well as algae concentration and influent COD loading rate (Polprasert et al., 1983):

$$K = f(T, C_s, OL) \tag{11.11}$$

where T = Temperature (°C); C_s = algae concentration (mg/L); and OL = COD loading rate (kg COD/ha·d).

K is given by the following equation (Polprasert et al., 1983):

$$e^K = 0.6351\,(1.0281)^T\,(1.0016)^{C_s}\,(0.9994)^{OL} \tag{11.12}$$

More recently, a model was developed for a waste stabilization pond system in Jordan. The model shows that the coliform reduction rate increases with increasing temperature and pH and with decreasing soluble BOD_5 (Saqqar and Pescod, 1992).

$$K_b = 0.50\,(1.02)^{T-20}\,(1.15)^{(pH-6)2}\,(0.99784)^{SBOD-100} \tag{11.13}$$

where K_b = fecal coliform reduction rate (day^{-1}); T = water temperature (°C); and $SBOD$ = soluble BOD_5 (mg/L).

11.4.3. Virus Removal and/or Inactivation in Oxidation Ponds

Temperature and solar radiation are important factors that control virus persistence in oxidation ponds (Bitton, 1980a). Removal of virus is expected to be high under a hot and sunny climate in the upper layer of the pond. The survival of a poliovirus Type 1 in a model pond is displayed in Figure 11.4 (Funderburg et al., 1978). During the summer season, it takes 5 days to obtain a 2-log reduction of viruses, while in winter 25 days are required to achieve a similar reduction. Biological factors are also involved in virus inactivation in ponds. As shown for bacteria, high pH resulting from heavy growth of algae may increase inactivation of virus in oxidation ponds.

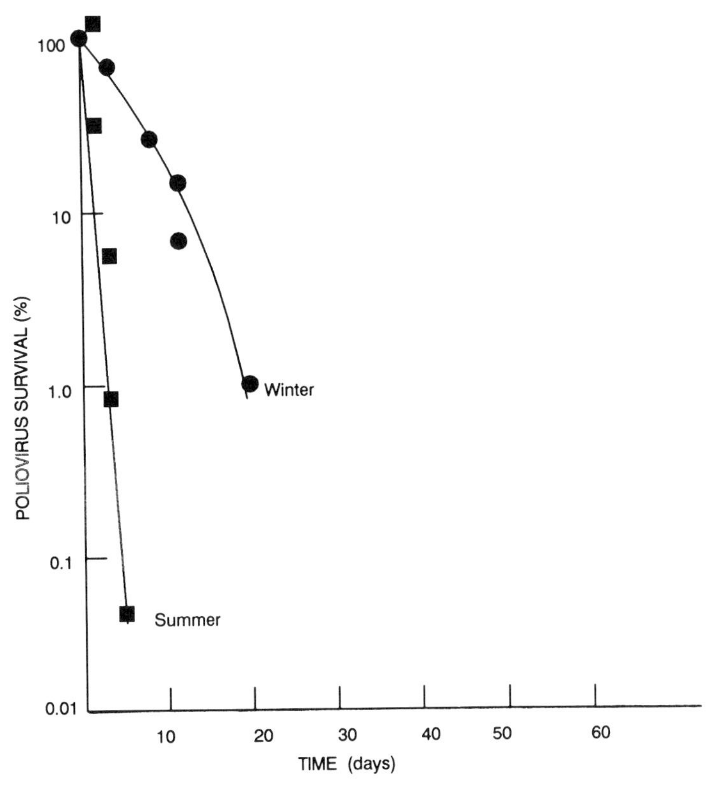

Fig. 11.4. Effect of the season of the year on the survival of poliovirus in a model oxidation pond. Adapted from Funderburg et al. (1978).

Virus adsorbed to suspended solids settle in pond sediments, where they survive for longer periods than in the water column. Disturbance of the contaminated sediments, however, may help in the resuspension of viral and bacterial pathogens.

11.4.4. Protozoan Cysts and Helminth Eggs

Protozoan cysts and helminth eggs are effectively removed because of the relatively long detention times in oxidation ponds. The percentage removal of *Giardia* and *Entamoeba histolytica* in an aerated lagoon varies from 67% to 100% (Panicker and Krishnamoorthi, 1981). Sedimentation is responsible for the removal of helminth parasite eggs. The percentage removal of parasites (*Ascaris lumbricoides*, hookworm) in an aerated lagoon was found to vary between 50% and 100% (Panicker and Krishnamoorthi, 1981). In Morocco, 100% removal of helminth was obtained following treatment by two oxidation lagoons in series, with a total detention time of 10 days. Most of the eggs were transferred to the lagoon sediments (Schwartzbrod et al., 1989). The percentage removal of nematode eggs follows Equation 11.14 (Ayres et al., 1992):

$$R = 100\,[1 - 0.14\exp(-0.38\theta)] \quad (11.14)$$

where R = percentage nematode removal; and θ = retention time (days).

For design purposes, use of the lower 95% confidence limit in Equation 11.14 was recommended to meet the World Health Organization (1989) guideline of less than one nematode egg per liter for irrigation.

Pond sludges therefore contain pathogenic microorganisms, protozoan cysts, and parasite eggs. They should be properly treated prior to disposal.

11.5. FURTHER READING

Hawkes, H.A. 1983b. Stabilization ponds, in: *Ecological Aspects of Used-Water Treatment,* Vol. 2, C.R. Curds and H.A. Hawkes, Eds. Academic Press, London.

U.S. EPA. 1977. *Wastewater Treatment Facilities for Sewered Small Communities.* Report # EPA-625/1-77-009.

12

SLUDGE MICROBIOLOGY

12.1. INTRODUCTION

Sludge is mostly composed of solids generated during wastewater treatment processes. Sludge treatment and disposal are probably the most costly operations in wastewater treatment plants. The U.S. Environmental Protection Agency has reported that approximately 7 million dry tons of sludge are produced annually in the United States. Sludge production is expected to increase in the future and to double by the year 2000. Member states of the European Community produce approximately 5.5 million dry tons of solids per year (Bowden, 1987).

The types of sludges that must be treated or disposed of are the following (Davis and Cornwell, 1985):

Grit. Grit is a mixture of coarse and dense materials such as sand, bone chips, and glass. These materials are collected in the grit chamber and are directly disposed of in a landfill.

Primary sludge. This is the sludge generated by the primary treatment and accumulated in the primary clarifier. It contains 3%–8% solids.

Secondary sludge. Secondary sludge is generated by biological treatment processes (e.g., activated sludge, trickling filter). The solids are mostly organic and range from 0.5% to 2% for activated sludge and up to 5% for trickling filter sludge.

Tertiary sludge. This sludge results from tertiary (chemical) treatment of wastewater.

12.2. SLUDGE PROCESSING

We will now review the various physical, chemical, and biological treatments available for processing sludges generated by biological treatment processes (Fig. 12.1) (Forster and Senior, 1987; Davis and Cornwell, 1985; Metcalf and Eddy, 1991).

12.2.1. Screening

Raw sludge is sometimes screened to remove coarse materials that may cause problems such as blockage of pipes and pumps.

12.2.2. Thickening

The goal of thickening is to increase the solid concentration of sludge. Thickening can be accomplished in tanks, where the solids are allowed to settle to the bottom or centrifuged in the presence or absence of chemical conditioners. This process can increase the solids concentration of primary sludge to approximately 12%. This operation reduces the transportation cost of sludge to the disposal site.

12.2.3. Dewatering

Sludges are dewatered by filtration or application to drying beds. Filtration helps achieve a higher solids concentration than thickening. The most common filtration device is the filter press, which is made of a series of plates fitted with a filter cloth or, more recently, membrane filters. Filtration is carried out by applying positive pressure or vacuum. Vacuum filtration can be applied to raw sludge prior to incineration or to stabilized sludge prior to disposal on land. In the presence of conditioners, filtration helps achieve a solids content of 20%–40%, depending on

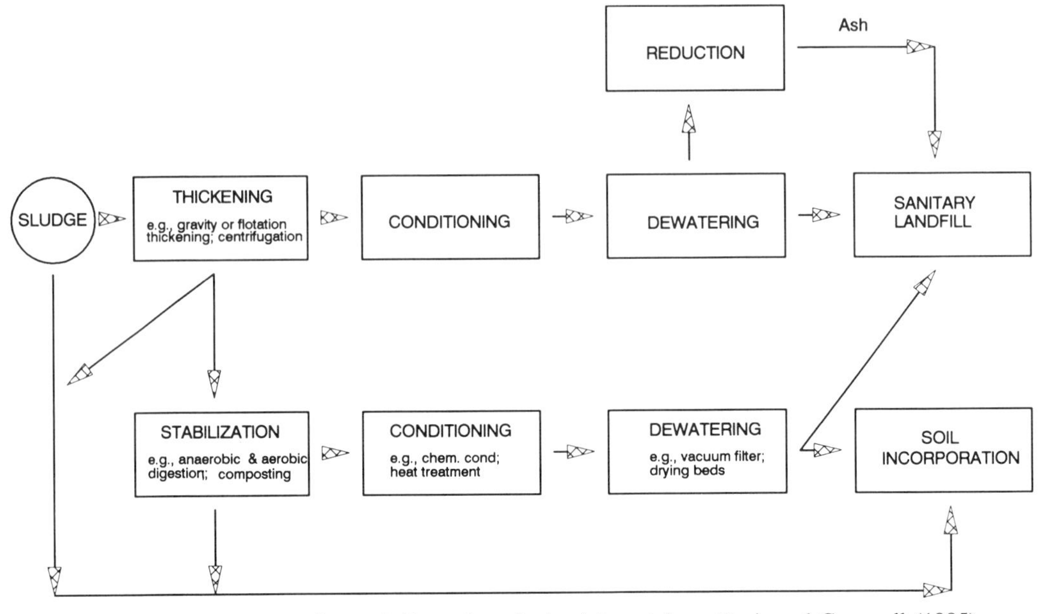

Fig. 12.1. Sludge processing and disposal methods. Adapted from Davis and Cornwell (1985).

sludge type and filtration mode. The filtrate has a high BOD, however, and must be recycled through the treatment plant.

Sludge may also be applied on drying beds to reach a solids content of approximately 40% following a drying period ranging from 10 to 60 days, depending on weather conditions. Sand drying beds are commonly used, especially in small plants. The sludge is removed mechanically or manually after the drying period.

12.2.4. Conditioning

Conditioning facilitates the separation of solids from the liquid phase. This is accomplished by chemical or heat treatment. Chemical treatment consists of adding conditioners to help improve sludge filterability. These chemicals may be inorganic salts (alum, ferrous and ferric salts, lime) or the increasingly popular synthetic organic polymers called polyelectrolytes. Sludge can also be heated under pressure (1,000–2,000 kPa) at temperature ranging from 175°C to 230°C (Davis and Cornwell, 1985). This treatment also reduces the affinity of sludge for water.

12.2.5. Stabilization

The purpose of stabilization is to break down the organic fraction of the sludge in order to reduce its mass and to obtain a product that is less odorous as well as safer from a public health standpoint. Sludge stabilization includes the following technologies: anaerobic digestion, aerobic digestion, composting, lime stabilization, and heat treatment (Metcalf and Eddy, 1991). Lime and heat stabilization are covered in Section 12.3.9.

12.2.5.1. Anaerobic Digestion

A detailed discussion of anaerobic digestion is presented in Chapter 13.

12.2.5.2. Aerobic Digestion

Aerobic digestion consists of adding air or oxygen to sludge contained in a 10- to 20-ft-deep open tank. A typical circular aerobic digester is shown in Figure 12.2 (U.S. EPA, 1977). Oxygen concentration must be maintained above 1 mg/L to avoid production of foul odors. The detention time in the digester is 12–30 days, depending on the prevailing temperature. Micro-

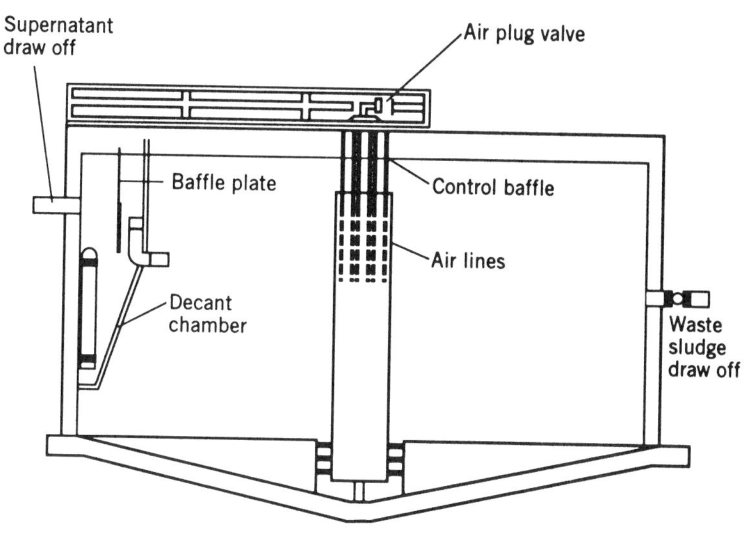

Fig. 12.2. Aerobic sludge digester. From U.S. EPA (1977), with permission of the publisher.

organisms degrade aerobically available organic substrates. They enter into the endogenous phase of growth when the biodegradable organic matter is depleted. The prevailing conditions in the tank also promote nitrification, which converts NH_4 to NO_3. The end result is a reduction in sludge solids. The stabilized sludge is allowed to settle and the supernatant is generally recycled to the head of the plant because it has a high BOD and high N and P levels (BOD = 100–500 mg/L; nitrate concentration = 200–500 mg/L; total P concentration = 50–200 mg/L) (U.S. EPA, 1977).

An important innovation in aerobic digestion of sludges has been the development and implementation of autoheated thermophilic aerobic digestion. This process generates heat as a result of free energy released by the aerobic degradation of organic matter by sludge microorganisms. Much of the energy that results from the oxidation of organic compounds is released as heat. The heat energy produced by the aerobic digestion of primary and secondary sludges is around 25 kcal/L (Metcalf and Eddy, 1991). The rise in temperature is a function of the level of organic matter and is given by Equation 12.1 (Forster and Senior, 1987; Jewell and Kabrick, 1980):

$$\Delta T = 2.4\ \Delta(\text{COD}) \qquad (12.1)$$

where ΔT = increase in temperature in °C; and ΔCOD = COD (mg/L) oxidized.

Most of the heat generated during biological oxidation is lost during sludge aeration. Therefore to achieve autoheating, the heat resulting from organic matter oxidation can be conserved by providing enough biodegradable organic matter, by insulating the bioreactor, and by increasing oxygen transfer efficiency to levels exceeding 10% (Jewell and Kabrick, 1980).

The advantages of aerobic digestion are low capital cost, easy operation, and production of odorless stabilized sludge. Some disadvantages are the high consumption of energy necessary for supplying oxygen, dependence upon weather conditions, and an end product with relatively low dewatering capacity.

12.2.6. Composting

The main objectives of composting are to produce stabilized organic matter accompanied by reduction of odors and to destroy pathogens and parasites.

12.2.6.1. Process Description

Composting consists essentially of mixing sludge with a bulking agent, stabilizing the mixture in the presence of air, curing, screening to recover the bulking agent, and storing the resulting compost material. There are three main types of composting systems (Benedict et al., 1988; Pedersen, 1983; Reed et al., 1988):

1. *Aerated static pile process* (Beltsville system) (Fig. 12.3) (Benedict et al., 1988). This process is the most widely used in the United States (Goldstein, 1988); it consists of mixing dewatered sludge (raw, anaerobically, or aerobically digested sludge) with a bulking agent (new or recycled) such as wood chips, leaves, corncobs, bark, peanut and rice hulls, or dried sludge. Wood chips are the most commonly used bulking agent in composting. Bulking materials offer structural support and favor aeration during composting. The pile is covered with screened compost to reduce or remove odors and to maintain high temperatures inside the pile. Aeration is provided by blowers and air diffusers during the 21-day active composting period in the pile. Afterwards, the compost is cured for at least 30 days, dried, and screened to recycle the bulking agent.

2. *Windrow process* (Fig. 12.4) (Benedict et al., 1988). Dewatered sludge is mixed with the bulking agent and stacked in 1- to 2-m-high rows called windrows. The composting period lasts approximately 30–60 days. Aeration is provided by turning the windrows two or three times per week. Additional induced aeration can be provided in the aerated windrow process.

3. *Enclosed systems.* These systems are enclosed to ensure a better control of temperature, oxygen concentration, and odors during composting. They require little space and minimize odor problems. Their cost, however, is higher than that of open systems.

12.2.6.2. Process Microbiology

The composting process is dominated by the degradation of organic matter by microorganisms under aerobic, moist, and warm conditions. Composting results in the production of a stable product that is used as a soil conditioner or a fertilizer, or as feed for fish in aquaculture. Furthermore, this process inactivates pathogens and transforms organic forms of nitrogen and phosphorus into inorganic forms, which are more bioavailable for uptake by agricultural crops (Polprasert, 1989; Walker and Wilson, 1973).

Microbial succession during composting. Composting consists of several temperature-controlled phases, each of which is driven by specific groups of microorganisms (Fig. 12.5) (Polprasert, 1989; Fogarty and Tuovinen, 1991):

1. Latent phase, for the acclimatization of microbial populations to the compost environment.
2. Mesophilic phase, dominated by bacteria that raise the temperature following decomposition of organic matter.
3. Thermophilic phase, which is characterized by the growth of thermophilic bacteria, fungi, and actinomycetes (organic matter is degraded at high rates during this phase).
4. Cooling phase, during which temperature decreases again to the mesophilic range (thermophilic microorganisms are replaced by mesophilic ones).
5. Maturation phase, in which the temperature drops to ambient levels, allowing the establishment of biota from other trophic levels (protozoa, rotifers, beetles, mites, nematodes) as well as other important processes such as nitrification, which is sensitive to high temperatures.

The composting mass may be regarded as a microbial ecosystem. Microbial succession during composting is relatively rapid, since it may take only a few days to reach 55°C. Community diversity during the thermophilic stage is relatively low (Finstein et al., 1980). An examination of the thermophilic microflora of two composting plants treating refuse or refuse–sludge mixtures showed that 87% of the isolates belong to the genus *Bacillus* (e.g., *Bacillus stearothermophilus*), a spore-forming bacterium) (Duvoort-

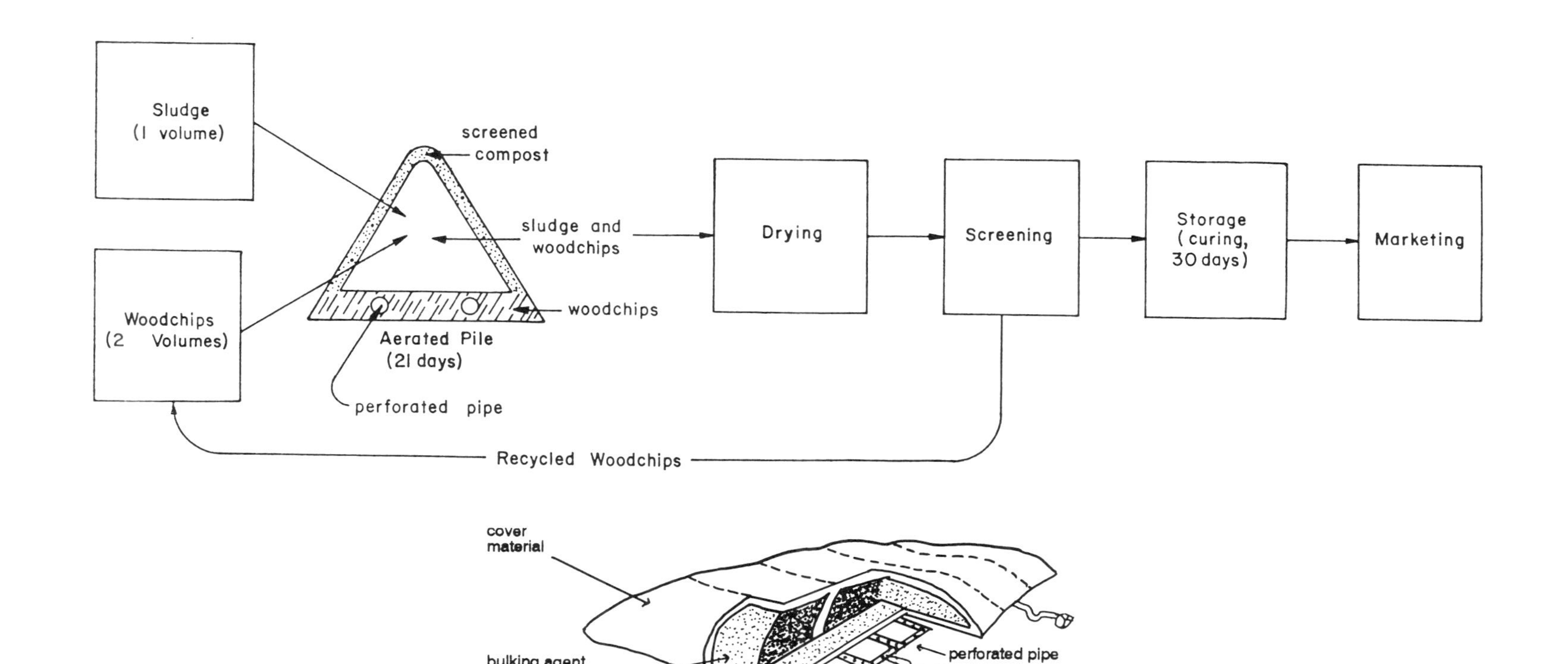

Fig. 12.3. Composting: Aerated static pile process. From Benedict et al. (1988), with permission of the publisher.

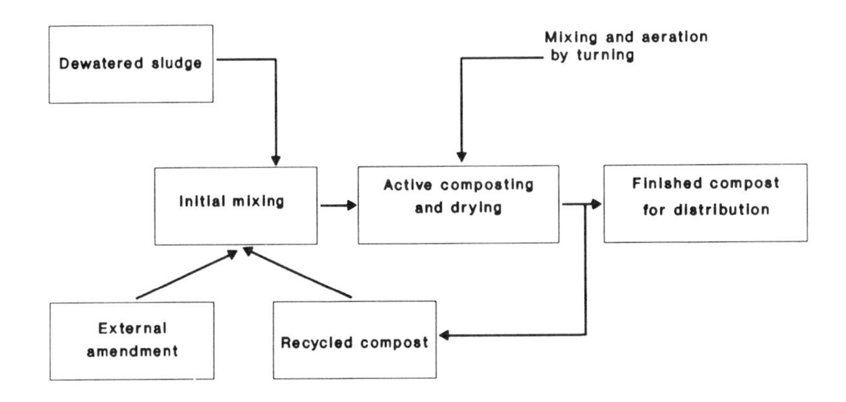

Fig. 12.4. Composting: Conventional windrow process. From Benedict et al. (1988), with permission of the publisher and W. D. Burge.

van Engers and Coppola, 1986; Strom, 1985a, 1985b). Thermophilic bacteria play a role in the degradation of carbohydrates and proteins, whereas actinomycetes (e.g., *Streptomyces* spp. and *Thermoactinomyces* spp.) and fungi (e.g., *Aspergillus fumigatus*) play a role in the degradation of more complex organics such as cellulose and lignin (Polprasert, 1989).

12.2.6.3. *Factors Affecting Composting*

Several physical and chemical parameters control the activity of microorganisms during composting: temperature, aeration, moisture, pH, and carbon and nitrogen content, as well as the type of material being composted and the composting system used.

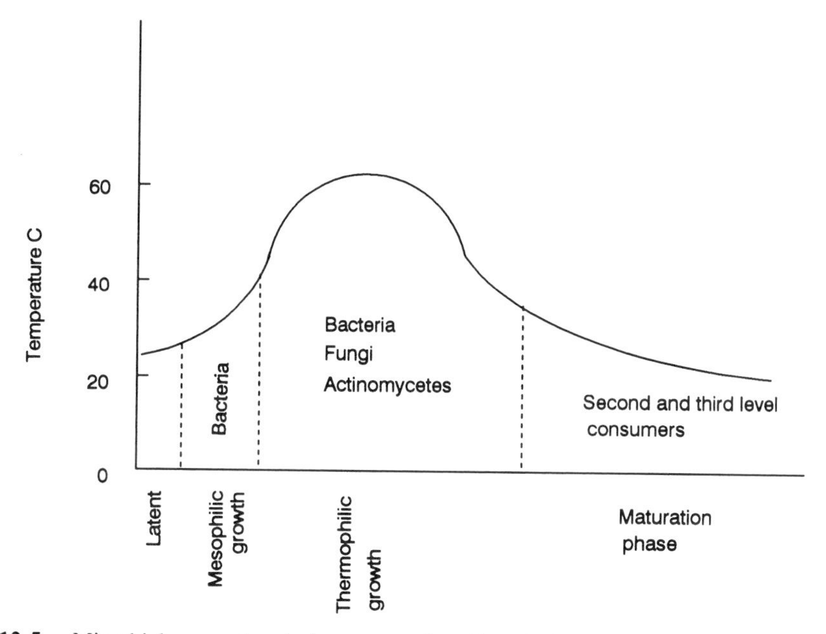

Fig. 12.5. Microbial succession during composting. From Polprasert (1989), with permission of the publisher.

Temperature is probably the most important factor affecting microbial activity in composts and can be controlled by adjusting the moisture and aeration levels (McKinley and Vestal, 1985; Polprasert, 1989). The optimum temperature for composting sewage sludge, as measured by microbial activity, is 55–60°C and should not exceed 60°C (Bach et al., 1984; Fogarty and Tuovinen, 1991; McKinley et al., 1985; Nakasaki et al., 1985a; Strom, 1985a). Although the high temperatures achieved during composting inhibit microbial activity (Kuter et al., 1985), they are instrumental in pathogen inactivation (see Section 12.3.7). In the static pile process, a temperature of 55°C or above must be maintained for at least 3 days to ensure effective pathogen inactivation (Benedict et al., 1988).

Aeration provides oxygen to the aerobic compost microorganisms, helps control the compost temperature, and removes excess moisture as well as gases (Fogarty and Tuovinen, 1991).

Moisture control is also an important factor during composting. The optimum moisture content should be 50%–60% and should not exceed 65% in enclosed composting systems.

Finally, as to *chemical content,* the optimum C/N ratio should be 25:1; and levels of toxic compounds (heavy metals, organic toxicants, salts) should be monitored to avoid inhibition of compost microorganisms and to comply with disposal regulations (Metcalf and Eddy, 1991).

12.2.6.4. Advantages and Disadvantages of Composting

Composting produces a stable end product that is a good source of nutrients and serves as a useful soil additive in agriculture and horticulture. A further advantage is the ability of composting to inactivate pathogenic microorganisms. However, compost is labor-intensive and is sometimes the source of foul odors.

12.3. PATHOGEN REMOVAL DURING SLUDGE TREATMENT

We have previously shown that biological wastewater treatment processes do not completely remove or inactivate pathogenic and parasitic organisms. Many of these organisms become bound to solids following wastewater treatment and are merely transferred to wastewater sludge. Further treatment is necessary to eliminate them or at least reduce their numbers significantly. We will now review the removal/inactivation capacity of some of the sludge treatment processes.

12.3.1. Pathogens and Parasites Found in Sludge

Bacterial pathogens of primary concern in sludge include *Salmonella, Shigella, Campylobacter, Yersinia, Leptospira,* and enteropathogenic *E. coli*. Primary sludge collected in Nancy, France, was found to harbor 97–1,230 PFU of enteroviruses per liter of sludge. Five of six samples contained hepatitis A antigen as determined by radioimmunoassay (Albert et al., 1990). The types of virus detected in anaerobically digested sludges are polioviruses, coxsackie A and B viruses, echoviruses, and reoviruses. The virus load reported is as high as 15 $TCID_{50}$/ml (TCID = tissue culture infective dose) (Lydholm and Nielsen, 1981). Similar types of virus are generally detected in aerobically digested sludge but in smaller numbers than in anaerobically digested sludge (Bitton et al., 1985; Goddard et al., 1981; Goyal et al., 1984; Scheuerman, 1984).

The parasites most often found in sludge are *Ascaris* species such as *A. lumbricoides* (human intestinal roundworm) and *A. suum* (pigs' roundworm), *Taenia saginata, Toxocara* (e.g., *T. cati,* the roundworm of cats, and *T. canis,* the roundworm of dogs), and *Trichuris* (e.g., *T. trichiura,* the human whipworms) (Little, 1986). The numbers of parasitic eggs detected per kilogram of anaerobically digested sludge (35°C; 42-day detention time; 25% solids) from the Caen–Mondeville agglomeration in France were 2,200–2,400 for *Taniidae,* 410–1,200 for *Ascaris* spp., 350–410 for *Toxocara* spp., and 4,910–7,250 for the family *Trichuridae*. Eight percent of the parasitic eggs were viable (Barbier et al., 1990). However, because of their low specific gravity, few protozoan cysts are found in municipal sludges (Little, 1986).

U.S. federal regulations have set up two categories of treatments to reduce pathogen levels in sludges (Appleton et al., 1986; Pedersen, 1981, 1983; U.S. EPA, 1979; Venosa, 1986; Yeager and O'Brien, 1983):

1. *Processes to significantly reduce pathogens* (PSRP). This category includes aerobic digestion (60 days at 15°C to 40 days at 20°C), anaerobic digestion (60 days at 20°C to 15 days at 35–55°C), lime stabilization (pH = 12 after a 2-hr contact time), mesophilic composting, air drying, and low-temperature composting (minimum: 40°C for at least 5 days). Most facilities are equipped with processes that qualify as PSRP. The use of sludge in the PSRP category for land application is allowed. However, there are restrictions with regard to crop production (no food crops should be grown within 18 months), animal grazing (prevention of grazing for at least 1 month by animals that provide products that are consumed by humans), and public access to the treated site (public access must be controlled for at least 12 months).

2. *Processes to further reduce pathogens* (PFRP). This category includes heat treatment (at 180°C for 30 min), irradiation, high-temperature composting, thermophilic aerobic digestion (10 days at 55–60°C) and heat drying. Some PFRP (e.g., high-temperature composting, thermophilic aerobic digestion, heat drying) do not require prior PSRP. PFRP processes are required if edible crops are exposed to sludge. There are no restrictions on the use of PFRP sludge on land.

We will now examine the effect of some of the sludge treatment processes on inactivation of pathogens and parasites.

12.3.2. Anaerobic Digestion

Detention time and temperature are the major factors that affect pathogen survival during anaerobic digestion of sludge. The sludge-feeding protocol also influences the reduction of microbial pathogens. The draw/fill mode (the digested sludge is withdrawn before feeding of the digester) achieves greater pathogen reduction than the fill/draw mode (feed is added before withdrawal of product). The difference in pathogen reduction between the two modes is smaller for viruses than for bacteria (Farrel et al., 1988; Pedersen, 1983; Venosa, 1986).

Anaerobic digestion achieves 1- to 3-log reduction in pathogenic and indicator bacteria. *Salmonella* is inactivated under both mesophilic and thermophilic digestion. However, at an SRT as low as 10 days, mesophilic two-phase anaerobic digestion achieves a greater reduction of bacterial indicators (TC, FC, FS) than conventional anaerobic digestion. Reduction of enterovirus is similar in both processes (Lee et al., 1989).

Anaerobic digestion appears to be efficient in protozoan cyst inactivation, which essentially depends on digestion temperature. It achieves a 3-log reduction in viability of cysts of *Giardia muris* (as measured by *in vivo* cyst excystation) in 7.9 days, in 19 hr, and in 13.6 min at 21.5°C, 37°C, and 50°C, respectively. It was shown that protozoan cysts (*Giardia muris, Giardia lamblia, Cryptosporidium parvum*) are eliminated within 24 hr at 37°C (Gavaghan et al., 1992; van Praagh et al., 1993). As regards the parasitic ova, their levels in digested and lagooned sludges from Chicago are shown in Table 12.1 (Arther et al., 1981). Embryonation tests showed that 55% of the ova in anaerobically digested sludge are viable (Arther et al., 1981). Thus, parasites (e.g., *Ascaris, Trichuris, Toxocara*) readily survive the mesophilic anaerobic digestion process. *Ascaris* ova are destroyed only by thermophilic digestion or by heating at 55°C for 15 min (Pike et al., 1988). Thermophilic anaerobic digestion at 53°C reduces this parasite to undetectable levels (Lee et al., 1989).

Varying levels of viruses are detected in anaerobically digested sludge. In France, they were found at levels varying from 50 to 130 PFU/L, and 44% of the digested sludge samples were positive for viruses (Schwartzbrod and Mattieu, 1986). Thus, anaerobic digestion does not completely eliminate viruses. Enteric viruses survive in sludge that has been subjected to mesophilic (30–32°C) or even thermophilic anaerobic digestion (50°C). Poliovirus 1, embedded in sludge flocs, survives more than 30 days of anaerobic digestion (Moore et al., 1976). Temperature plays an important role in virus inactivation in sludge (Berg and Berman, 1980; Eisenhardt et al., 1977; Ward et al., 1976). Another known virucidal agent in digesting sludge is the uncharged form of ammonia, which predominates at higher pH values. Nucleic acids are the main target of inactivation by ammonia (Ward and Ashley, 1977a).

12.3.3. Aerobic Digestion

Although much of the sludge produced in the United States is treated by anaerobic digestion, many small communities utilize the less complex aerobic digestion, a process considered by the U.S. EPA to significantly reduce pathogen numbers in sludge prior to land application (U.S.

TABLE 12.1. Levels of Parasitic Nematode Ova in Digested and Lagooned Sludge[a]

Date collected	Numbers of ova per 100 g dry sludge			
	Ascaris spp.	*Toxocara* spp.	*Toxascaris leonina*	*Trichuris* spp.
May 1976	192	64	64	32
June 1976	171	150	43	43
June 1977	231	331	66	33
Oct. 1977	218	146	18	36
Mean	203	172.7	47.7	36
Std. Dev.	26.8	112.7	22.4	4.9

[a]From Arther et al. (1981), with permission of the publisher.

EPA, 1979). *Salmonella* spp. (mostly *S. enteriditis*) was detected at levels of 0.8–33 MPN/g in aerobically digested sludges from three wastewater treatment plants in Florida (Table 12.2) (Farrah and Bitton, 1984). Reduction in enteric bacteria and viruses during aerobic digestion of sludge depends on both detention time and temperature (Martin et al., 1990; Scheuerman et al., 1991). Bacterial indicators (fecal coliforms, fecal streptococci) die off increasingly as temperature is raised from 20°C to 40°C (Fig. 12.6) (Kuchenrither and Benefield, 1983). Temperature was also found to be the single most important factor influencing the survival of enteric viruses (poliovirus 1, coxsackie B3, echovirus 1, rotavirus SA-11) during aerobic digestion of sludge (Scheuerman et al., 1991). Mesophilic aerobic digestion appears to be ineffective, however, in reducing the number of parasite eggs such as *Ascaris* (Pedersen, 1983). Figure 12.7 (Kabrick and Jewell, 1982) shows that thermophilic aerobic digestion when temperature ≥ 45°C achieves a much greater destruction of pathogens (*Salmonella, P. aeruginosa,* viruses) than traditional mesophilic anaerobic digestion. A similar trend is observed for parasites (Kabrick et al., 1979; Kuchenrither and Benefield, 1983). Thus, autoheated aerobic digestion can produce virtually pathogen- and parasite-free sludge.

12.3.4. Dual Digestion System (DDS)

The dual digestion system involves an aerobic and an anaerobic stage (Appleton and Venosa, 1986; Appleton et al., 1986): Step 1 includes a covered aerobic digester that uses pure oxygen, with a 1-day detention time; heat (temperature around 55°C) is generated during oxidation of organic matter in the aerobic digester. Step 2 involves treatment in an anaerobic digester with an 8-day detention time.

The destruction of pathogenic and indicator organisms is higher in Step 1 than in Step 2. Reduction of microbial indicators by DDS is greater than that achieved with other stabilization processes such as high-rate anaerobic digestion, lime stabilization, or mesophilic composting. Pathogen reductions are greater than or equal to those achieved by thermophilic aerobic digestion (Appleton et al., 1986).

12.3.5. Sludge Lagooning

Inactivation of pathogens in sludge lagoons depends on weather conditions and type of pathogen. At temperatures exceeding 20°C, it appears that the storage time for 1-log reduction of pathogens should be 1, 2, and 6 months for bacteria, viruses, and parasite eggs, respectively (Pedersen, 1983). Virus monitoring in a sludge lagoon in Jay, Florida, showed that the levels dropped to undetectable levels in approximately 1–2 months following addition of fresh sludge to the lagoon (Fig. 12.8) (Farrah et al., 1981). Longer storage periods are necessary for pathogen and parasite inactivation in cold climates.

TABLE 12.2. Detection of *Salmonella* spp. in Aerobically Digested Sludge[a]

Treatment plant	Sludge source	Percentage of samples positive for *Salmonella* spp.	Mean MPN/g
Kanapaha	Undigested	27	4.6
	Digested	36–40	0.9–2.3
Main Street	Undigested	29	7.5
	Digested	14–25	0.4–0.8
Tallahassee	Undigested	40	30
	Digested	100	33

[a]Adapted from Farrah and Bitton (1984).

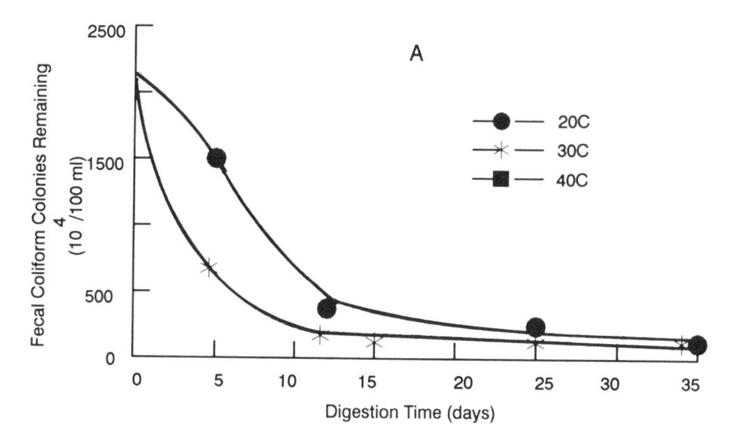

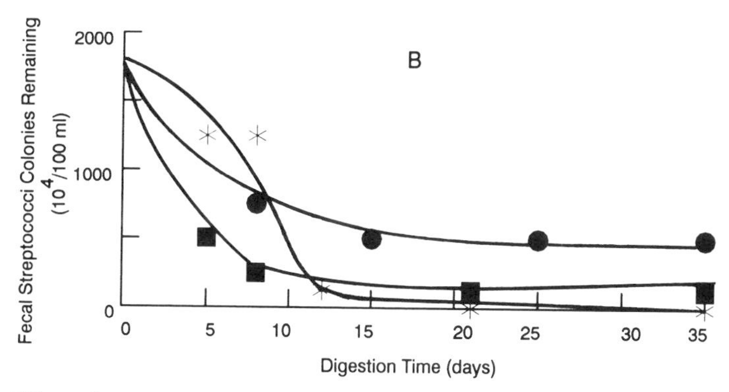

Fig. 12.6. Effect of temperature on inactivation of indicator bacteria during aerobic digestion of sludge. **A.** Fecal coliforms. **B.** Fecal streptococci. From Kuchenrither and Benefield (1983), with permission of the publisher.

12.3.6. Pasteurization

The process of pasteurization consists in subjecting the sludge to heat treatment at 70°C for 30 min. Although energy-intensive, pasteurization effectively destroys helminth eggs and most bacterial and viral pathogens. Some have proposed heating sludge more economically at a lower temperature for a longer time period. *Ascaris* ova are destroyed when pasteurization is carried out at temperatures above 55°C for approximately 2 hr (Carrington, 1985). Pasteurization at 70°C for 30 min also ensures the destruction of more that 99% of *Taenia saginata* ova. This process achieves a complete inactivation of *Salmonella* and enteroviruses (Pike et al., 1988; Saier et al., 1985).

The mechanism of heat inactivation of viruses in sludge has been investigated. A heat-stable ionic detergent, associated with sludge solids, appears to protect enteroviruses but not reoviruses from heat inactivation. This protective effect can be overcome by ammonia, a virucidal agent associated with the liquid portion of digested sludge at pH > 8.5. It has been concluded that heat treatment of sludge under alkaline conditions is effective against viruses (Ward and Ashley, 1977b, 1978; Ward et al., 1976).

12.3.7. Composting

From a public health standpoint, there are two concerns about the production and use of compost material: (1) pathogenic microorganisms in

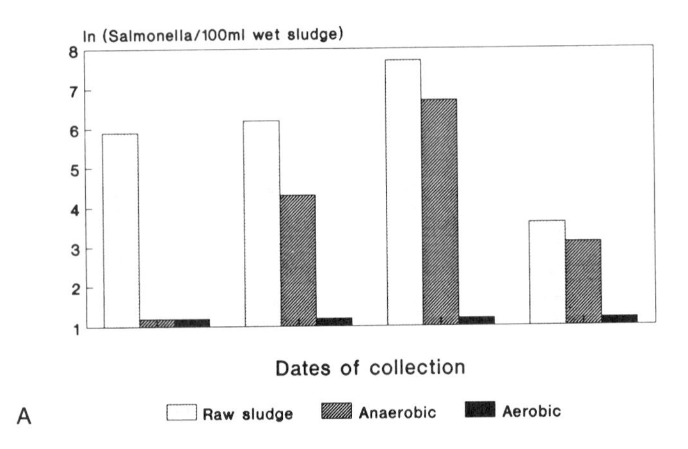

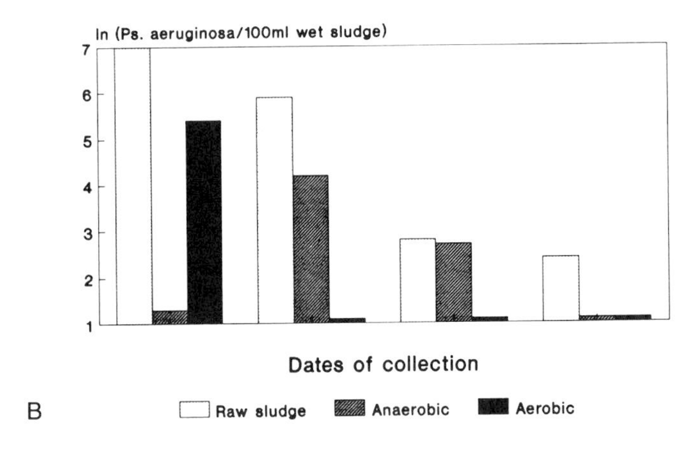

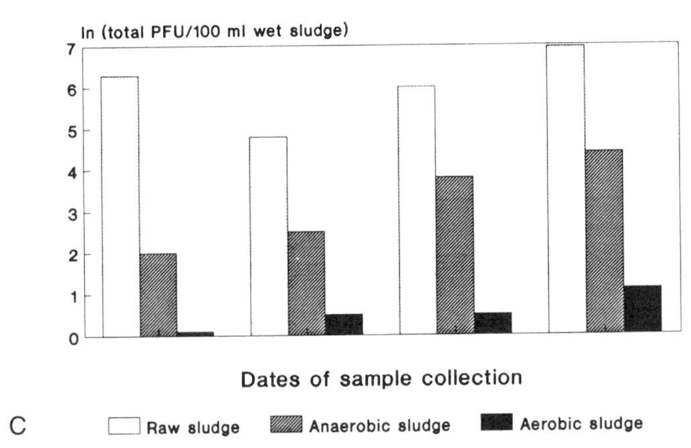

Fig. 12.7. Inactivation of pathogenic microorganisms by thermophilic aerobic sludge digestion. **A.** *Salmonella*. **B.** *Pseudomonas aeruginosa*. **C.** viruses. From Kabrick and Jewell (1982), with permission of the publisher.

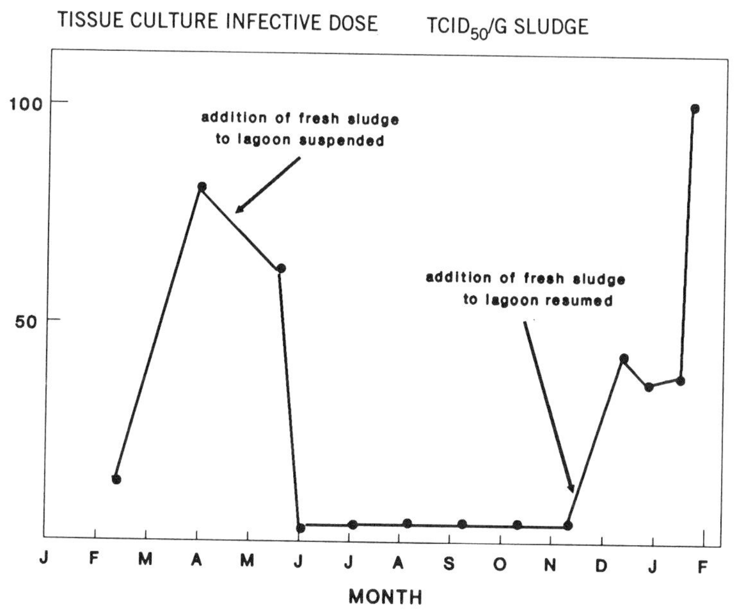

Fig. 12.8. Survival of enteroviruses in a sludge lagoon in Florida. From Farrah et al. (1981), with permission of the publisher.

the waste being composted; and (2) health hazards of secondary pathogens (fungi, actinomycetes) to compost workers.

Temperature is the key factor controlling survival of pathogens in sludge undergoing composting. However, since the compost material is quite heterogenous, it is difficult to maintain a uniform temperature within the compost biomass. This may explain the difficulties in achieving a complete inactivation of pathogens. It has been suggested that compost must be maintained at 55°C for at least 3 days at static pile facilities and for at least 15 days at windrow facilities to ensure the safety of this material (Benedict et al., 1988; Epstein, 1979). This process is effective in reducing bacterial indicators (except possibly for fecal streptococci) and bacterial pathogens such as *Salmonella* and *Shigella*. However, regrowth of pathogens (e.g., *Salmonella*) has been observed in composted sludge (Brandon et al., 1977; Hussong et al., 1985; Russ and Yanko, 1981). Twelve percent of sewage sludge compost from 30 municipalities in the United States was found to harbor *Salmonella* at densities up to 10^4 cells per gram. *Salmonella* regrowth in composted sludge occurs when the moisture content is above 20%, temperatures are in the mesophilic range (20–40°C) and the C/N ratio excedes 15:1 (Russ and Yanko, 1981). Some investigators argue that temperature should be maintained below 60° to achieve high microbial activity during composting. Inactivation of pathogens could be easily accomplished during curing, when temperature can reach 70°C (Kuter et al., 1985). Laboratory studies show evidence of effective inactivation of viruses during composting and curing periods (Bitton, 1980a; Pedersen, 1983). Temperature of 55°C or above readily inactivate protozoan cysts and helminth eggs in composting sludge. Figure 12.9 (Brandon, 1978) shows that *Ascaris* eggs are readily inactivated at 55°C or 60°C but not at 45°C.

Survival of pathogens in compost is also affected by moisture and indigenous compost microflora. Although the level of nutrients is sufficient for supporting growth, the indigenous microflora limits pathogen regrowth (Hussong et al., 1985).

Aspergillus fumigatus is an opportunistic

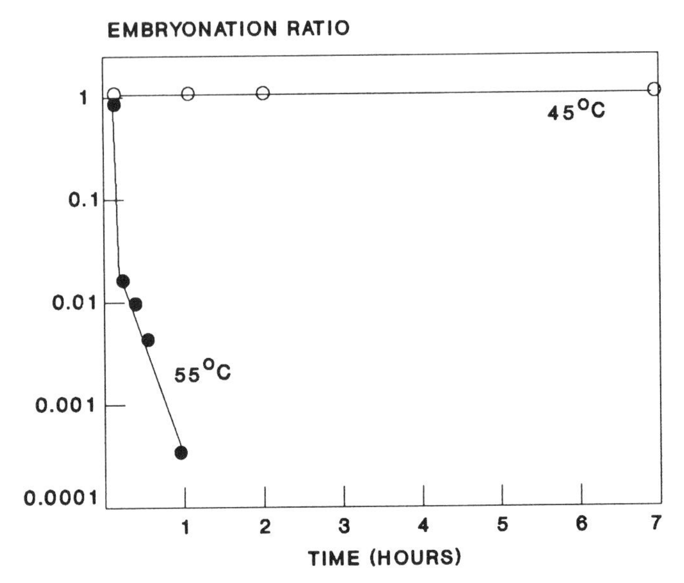

Fig. 12.9. Heat inactivation of *Ascaris* eggs in composted sludge. Adapted from Brandon (1978).

pathogen that causes allergies when inhaled during composting operations (Clark et al., 1984). This pathogen has been isolated from compost workers and can cause serious lung damage. *Thermoactinomyces* has been implicated in farmer's lung disease (Blyth, 1973). Thus, proper preventive measures (e.g., wearing protective masks) should be taken to avoid spore inhalation by compost workers (see Chapter 13).

12.3.8. Air Drying

Digested sludge is dewatered on sand beds in small wastewater treatment plants. Liquid sludge is placed on a 1-ft layer of sand and allowed to dry. Water is removed by drainage and evaporation and most of it is collected within the first few days in an underdrain system and usually returned to the plant. Afterwards, water is removed mostly by evaporation. Both processes remove approximately 45%–75% of the water (U.S. EPA, 1977). Sludge drying to 95% solids brings about a reduction in bacterial pathogens ranging from 0.5 to almost 4 logs (Yeager and O'Brien, 1983; Ward et al., 1981).

Viruses are generally detected in dried sludge; thus, sludge drying is not a reliable process for complete inactivation of viruses.

12.3.9. Chemical Inactivation

12.3.9.1. Lime Stabilization

Lime, added as $Ca(OH)_2$ or CaO, is an inexpensive chemical that is used as a flocculating agent and for odor control in wastewater treatment plants. The lime stabilization process consists of adding a lime slurry to liquid sludge to achieve a pH higher than 12. Modifications of this process (Westphal and Christensen, 1983) include (1) the addition of lime slurry to iron-conditioned sludge prior to vacuum filtration and (2) the addition of dry lime to dewatered conditioned sludge cake.

In order to be effective, lime stabilization must achieve a pH of 12 or above for at least 2 hours. This treatment leads to 3- 6-log reduction of bacterial indicators (Venosa, 1986). Figure 12.10 (Farrel et al., 1974; Pedersen, 1983) shows that pH should exceed 10.5 to achieve effective bacterial inactivation. In Germany it was shown that lime treatment of raw sludge at pH = 12.8 completely inactivated *Salmonella*

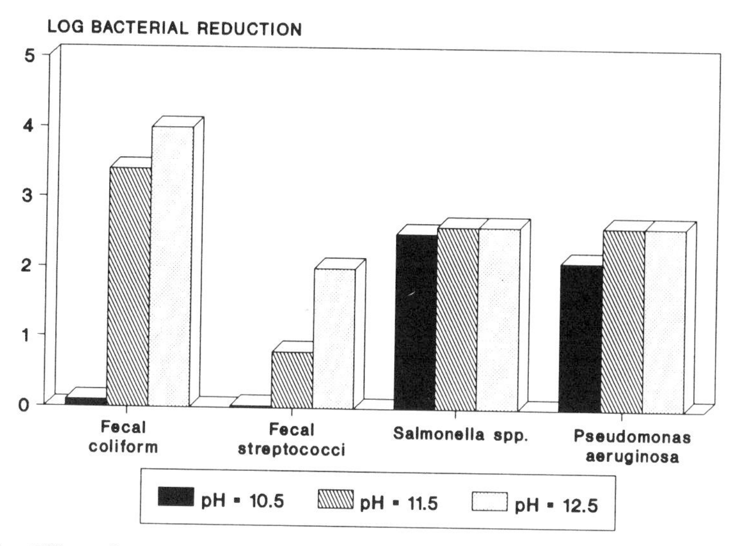

Fig. 12.10. Effect of pH on inactivation of pathogenic and indicator bacteria in sludge (0.5 hr detention time). From Farrel et al. (1974), with permission of the publisher.

senftenberg within 3 hr (Pfuderer, 1985). Viruses are not detected in limed sludge after 12 hr of contact time (Sattar et al., 1976). Only 1 of 10 sludge samples conditioned with lime and ferric chloride and subsequently dehydrated was positive for viruses (Schwartzbrod and Mattieu, 1986). However, liming does not seem to be completely effective for reduction of parasites.

A new process, the N-Viro soil, has been accepted as a PFRP by the U.S. EPA. It consists of treating dewatered sludge with lime and then mixing it with cement kiln dust. The sludge is then allowed to cure for 3–7 days. The liming and the resulting high alkalinity of the product may be responsible for efficient inactivation of pathogenic microorganisms.

12.3.9.2. *Ammonification*

Addition of ammonia or ammonium sulfate to digested sludge effectively reduces parasites in sludge but the treatment is expensive (Reimers et al., 1986).

12.3.9.3. *Ozonics Process*

This process consists of treating sludge for 30–90 min with a high ozone concentration (200 ppm) at low pH (pH = 2.5–3.5) and under pressure (60 psi). This process is effective against bacterial pathogens and viruses but not against parasites (Reimers et al., 1986).

12.3.10. Irradiation

This treatment consists of subjecting sludge to isotopes emitting gamma radiation (cesium-137 or cobalt-60) or to high-energy electron beams. This process essentially destroys the microbial genetic material. The energy level of the irradiation is relatively low and thus does not result in the production of radioactive sludge (Ahlstrom and Lessel, 1986).

The first sludge irradiator was built in Germany at the beginning of the 1970s. Other irradiators were built subsequently in the United States. In 1976, an experimental electron beam irradiator was installed in Deer Island, Massachusetts. The electron beams are generated by a 750-kV electron accelerator. This unit, which treats a thin layer (approx. 2 mm) of liquid sludge spread on a rotating drum, is illustrated in Figure 12.11 (Ahlstrom and Lessel, 1986). This system was designed to treat approximately 100,000 gallons of sludge per day. In 1978, a sludge irradiator was built at Sandia National

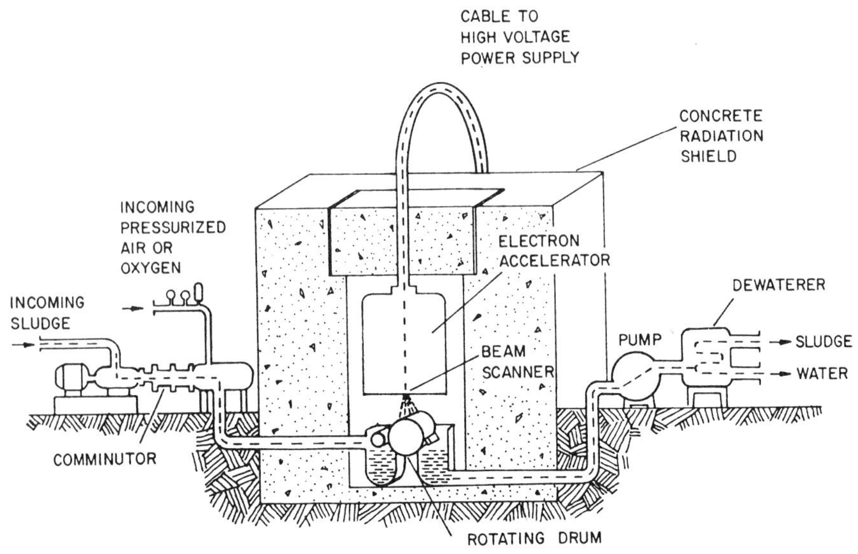

Fig. 12.11. Sludge irradiation process used at Deer Island wastewater treatment plant. From Ahlstrom and Lessel (1986), with permission of the publisher.

Laboratory in Albuquerque, New Mexico. This unit is driven by cesium-137 and treats dried or dewatered sludge (Fig. 12.12). More recently, in 1984, a larger electron beam irradiator was constructed near Miami, Florida. This unit handles thin layers of liquid sludge (2% solids) at an applied dose of 350–400 krads and at a rate of 120 GPM (Ahlstrom and Lessel, 1986). Afterwards, the sludge is conditioned with polymers, dewatered, and later stored in a sludge lagoon to be subsequently marketed as "Daorganite" for agricultural use.

Ionizing radiation inactivates microorganisms either directly or indirectly by production of free radicals. Nucleic acids are the main target of ionizing radiation. The sensitivity of microorganisms to sludge irradiation varies with the type of pathogen or parasite. D_{10}, the radiation dose necessary for inactivation of 90% of parasites/pathogens, increases as sludge solids increase. Bacteria (e.g., *Salmonella typhimurium, Klebsiella* sp.) are protected from ionizing radiation at reduced sludge moisture levels (Ward et al., 1981). Bacterial pathogens are efficiently reduced (6–8 log) at radiation doses ranging from 200 to 300 krad. Pathogens such as *Salmonella* can regrow in irradiated sludge, but this problem is not encountered in sludge irradiated at doses exceeding 400 krad. The D_{10} values (expressed in krad; 1 rad = 1 joule/kg = 100 erg/g) for reduction of selected bacterial pathogens and parasites have been compiled (Ahlstrom and Lessel, 1986; Forster and Senior, 1987) and are shown in Table 12.3. It appears that viruses are the pathogens most resistant to irradiation (Yeager and O'Brien, 1983). It was recommended to use 0.5 Mrad for liquid sludges and 1.0 Mrad for dry sludges. A 1-Mrad dose is sufficient for elimination of bacterial, fungal, and viral pathogens as well as parasites (Yeager and O'Brien, 1983).

With regard to pathogen/parasite inactivation, the following processes are more efficient than irradiation alone:

Thermoradiation. Combination of irradiation with heat treatment at 47°C results in higher destruction of pathogens. For example, thermoradiation, with a radiation dose of 150–200 krad and a temperature of 51°C leads to a 5-log reduction of poliovirus in 5 min (Bitton, 1980a). This process also improves sludge settling and filterability.

Oxiradiation. Irradiation in the presence of air or oxygen results in more inactivation of pathogens. This process is being developed in Germany.

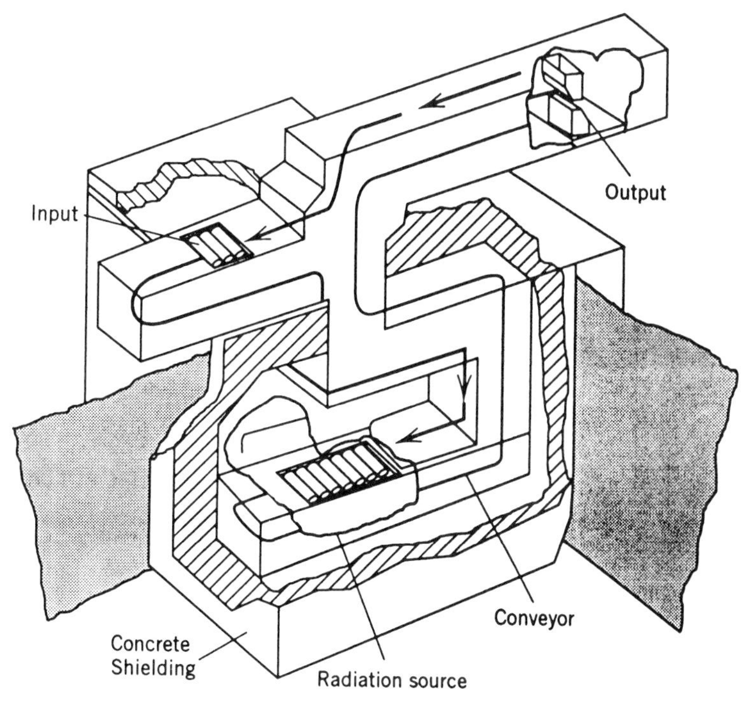

Fig. 12.12. Sandia irradiator for dried sludge (Albuquerque, NM). From Ahlstrom and Lessel (1986), with permission of the publisher.

TABLE 12.3. Sludge Irradiation: D Values of Selected Pathogens and Parasites[a]

Organism	D Value (krad)
Bacteria	
E. coli	<22–36
Micrococcus spp.	14
Klebsiella spp.	36–92
Enterobacter spp.	34–62
Salmonella typhimurium	<50–140
Proteus mirabilis	<22–50
Streptococcus faecalis	110–250
Viruses	
Poliovirus	350
Coxsackievirus	200
Echovirus	170
Reovirus	165
Adenovirus	150
Parasites	
Ascaris spp.	<66
Fungi	
Aspergillus fumigatus	50–60

[a] From Ahlstrom and Lessel (1986), with permission of the publisher.

12.3.11. Heat Treatment

This process involves heating sludge, under pressure, at temperatures up to 260°C for approximately 30 min. This treatment stabilizes sludge and improves its dewatering. Microbial pathogens and parasites should be effectively destroyed following heat treatment.

12.3.12. Microwave

Microwave treatment has been proposed as an alternative pasteurization method for the destruction of pathogens and parasites. Pathogens such as *Salmonella senftenberg* are effectively inactivated (in drinking water and in liquid manure) by microwave treatment at temperatures between 67°C and 69°C with an average holding time of about 7 s (Niederwohrmeier et al., 1985).

12.4. EPIDEMIOLOGICAL SIGNIFICANCE OF PATHOGENS IN SLUDGE

Prospective epidemiological studies have been conducted to assess the potential health risks associated with sludge treatment and disposal.

An epidemiological study of workers at composting facilities showed that workers did not experience any ill effects in comparison with a control group. Minor effects were observed, which consisted of skin irritation and inflammation of nose and eyes, that resulted from exposure to dust and to the thermophilic fungus *Aspergillus fumigatus* (Clark et al., 1984). Another prospective epidemiological study was conducted in Ohio to determine the health risks of sludge handling to farm workers. Sludge, soil, forage, and fecal samples were analyzed for bacterial, viral, and parasitic pathogens. Serum samples were tested for antibodies to coxsackieviruses, echoviruses, and hepatitis A virus. This study has essentially shown that sludge handling did not result in any adverse health effects on the study population in comparison with the control group (Ottolenghi and Hamparian, 1987).

The presence of *Salmonella* spp., *Shigella* spp., and *Campylobacter* spp. in sludge applied to land was monitored in Ohio. No *Shigella* or *Campylobacter* were detected in sludge, but 21 serotypes of *Salmonella* were isolated, the dominant serotype being *Salmonella infantis*. From an examination of antibodies to salmonellae it was concluded that the risk of infection to farm populations was minimal.

The prospective epidemiological approach may not be the most suitable one for assessing the health risks associated with sludge production and disposal. There are several confounding factors associated with this approach. Sludge harbors a myriad of pathogenic microorganisms and parasites that can cause a myriad of symptoms and diseases. Moreover, the transmission of the pathogenic agents may not be limited to sludge, since other transmission routes (e.g., person-to-person, water, food) may be implicated (Jakubowski, 1986). There are several other problems associated with the assessment of the epidemiological significance of pathogenic microorganisms in wastewater and sludge, among them the following (Block, 1983):

1. Methodological problems in determining the number of pathogens in sludge
2. High variability of pathogen densities in sludges
3. Lack or fragmentary nature of information about the minimum infective doses for pathogens, which may vary from less than 10 to 10^{10} organisms (Block, 1983; Bryan, 1977)

12.5. FURTHER READING

Bitton, G., B.L. Damron, G.T. Edds, and J.M. Davidson, Eds. 1980. *Sludge-Health Risks of Land Application*. Ann Arbor Science, Ann Arbor, MI, 366 pp.

Block, J.C., A.H. Havelaar, and P. l'Hermite, Eds. 1986. *Epidemiological Studies of Risks Associated with the Agricultural Use of Sewage Sludge:*

Knowledge and Needs. Elsevier Applied Science, London, 168 pp.

Davis, M.L., and D.A. Cornwell. 1985. *Introduction to Environmental Engineering*. PWS Engineering, Boston.

Strauch, D., A.H. Havelaar, and P. l'Hermite, Eds. *Inactivation of Microorganisms in Sewage Sludge by Stabilization Processes*. Elsevier Applied Science, London.

Venosa, A.D. 1986. Detection and significance of pathogens in sludge, in: *Control of Sludge Pathogens,* C.A. Sorber, Ed. Water Pollution Control Federation, Washington, DC.

13

ANAEROBIC DIGESTION OF WASTEWATER AND SLUDGE

13.1. INTRODUCTION

Anaerobic digestion consists of a series of microbiological processes that convert organic compounds to methane. Methane production is a common phenomenon in several diverse natural environments ranging from glacier ice to sediments, marshes, termites, rumen, and oil fields. The microbiological nature of methanogenesis was discovered more than a century ago (Koster, 1988). While several types of microorganisms are implicated in aerobic processes, anaerobic processes are driven mostly by bacteria.

Anaerobic digestion has long been used for the stabilization of wastewater sludges. In recent years, however, it has also been considered for the treatment of industrial wastewater. This was made possible through a better understanding of the microbiology of this process and through improved reactor designs. Its advantages over aerobic processes are the following (Lettinga et al., 1980; Sahm, 1984; Sterritt and Lester, 1988; Switzenbaum, 1983):

1. Anaerobic digestion uses readily available CO_2 as an electron acceptor. It requires no oxygen, the supply of which adds substantially to the cost of wastewater treatment.

2. Anaerobic digestion produces lower amounts of sludge (3–20 times less than aerobic processes), since the energy yields of anaerobic bacteria are relatively low. Most of the energy derived from substrate breakdown is found in the final product, CH_4. As regards cell yields, 50% of organic carbon is converted to biomass under aerobic conditions, whereas only 5% is converted into biomass under anaerobic conditions. The net amount of cells produced per metric ton of COD destroyed is 20–150 kg, as compared to 400–600 kg for aerobic digestion (Speece, 1983; Switzenbaum, 1983).

3. Anaerobic digestion produces a useful gas, methane. This gas contains about 90% of the energy, has a calorific value of approximately 9,000 kcal/m^3, and can be burned on site to provide heat for digesters or to generate electricity. Little energy (3%–5%) is wasted as heat. Methane production contributes to the BOD reduction in digested sludge.

4. Energy required for wastewater treatment is reduced.

5. Anaerobic digestion is suitable for high-strength industrial wastes.

6. It is possible to apply high loading rates to the digester.

7. Anaerobic systems can biodegrade xenobiotic compounds such as chlorinated aliphatic hydrocarbons (e.g., trichloroethylene, trihalomethanes) and recalcitrant natural compounds such as lignin (see Chapter 16).

Some disadvantages of anaerobic digestion are the following:

1. It is a slower process than aerobic digestion.

2. It is more sensitive to upsets by toxicants.

3. Start-up of the process requires long periods.

4. As regards biodegradation of xenobiotic compounds by cometabolism, anaerobic pro-

cesses require relatively high concentrations of primary substrates (Rittmann et al., 1988).

13.2. PROCESS DESCRIPTION

13.2.1. Single-Stage Digestion

Anaerobic digesters are large fermentation tanks provided with mechanical mixing, heating, gas collection, sludge addition and withdrawal ports, and supernatant outlets (Fig. 13.1) (Metcalf and Eddy, 1991). Sludge digestion and settling occur simultaneously in the tank. Sludge stratifies and forms the following layers from the bottom to the top of the tank: digested sludge, actively digesting sludge, supernatant, a scum layer, and gas. Higher sludge loading rates are achieved in the high-rate version, in which sludge is continuously mixed and heated.

13.2.2. Two-Stage Digestion

This process consists of two digesters (Fig. 13.2) (Metcalf and Eddy, 1991), one tank continuously mixed and heated for sludge stabilization and the other one for thickening and storage prior to withdrawal and ultimate disposal. Although conventional high-rate anaerobic digestion and two-stage anaerobic digestion achieve comparable methane yield and COD stabilization efficiency, the latter process allows operation at much higher loading rates and shorter hydraulic retention times (Ghosh et al., 1985).

13.3. PROCESS MICROBIOLOGY

Consortia of microorganisms, mostly bacteria, are involved in the transformation of complex high-molecular-weight organic compounds to methane. Furthermore, there are synergistic interactions between the various groups of bacteria implicated in anaerobic digestion of wastes. The overall reaction is the following (Polprasert, 1989):

$$\text{Organic matter} \rightarrow CH_4 + CO_2 + H_2 + NH_3 + H_2S \qquad (13.1)$$

Although some fungi and protozoa can be found in anaerobic digesters, bacteria are undoubtedly the dominant microorganisms. Large numbers of strict and facultative anaerobic bacteria (e.g., *Bacteroides, Bifidobacterium, Clostridium, Lactobacillus, Streptococcus*) are involved in the hydrolysis and fermentation of organic compounds.

There are four categories of bacteria that are involved in the transformation of complex materials into simple molecules such as methane and carbon dioxide. These bacterial groups operate in

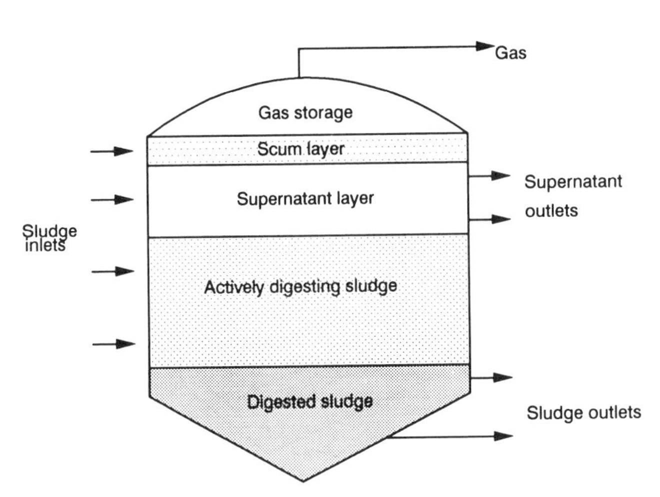

Fig. 13.1. Conventional single-stage anaerobic digester. Adapted from Metcalf and Eddy (1991).

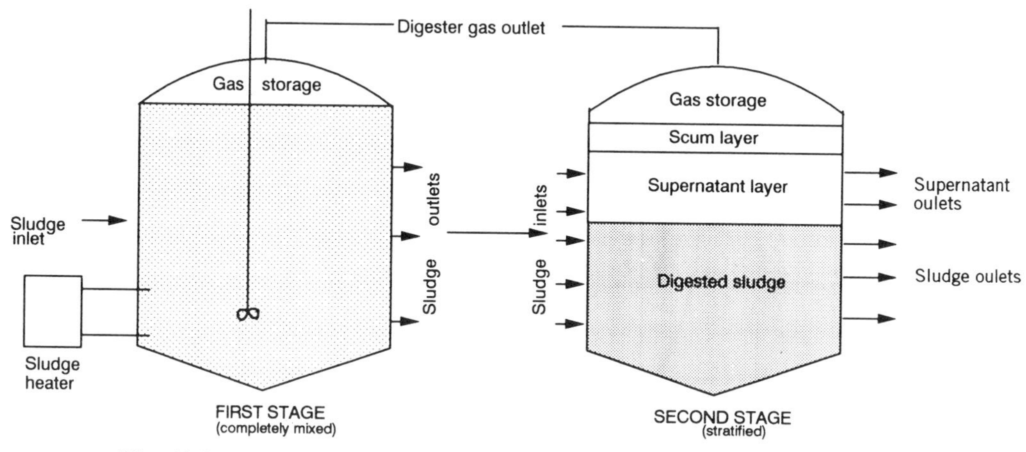

Fig. 13.2. Two-stage anaerobic digester. Adapted from Metcalf and Eddy (1991).

a synergistic relationship (Archer and Kirsop, 1991; Barnes and Fitzgerald, 1987; Sahm, 1984; Sterritt and Lester, 1988; Zeikus, 1980): (Fig. 13.3) (Koster, 1988).

13.3.1. Group 1: Hydrolytic Bacteria

Consortia of anaerobic bacteria break down complex organic molecules (proteins, cellulose, lignin, lipids) into soluble monomer molecules such as amino acids, glucose, fatty acids, and glycerol. The monomers are directly available to the next group of bacteria. Hydrolysis of the complex molecules is catalyzed by extracellular enzymes such as cellulases, proteases, and lipases. However, the hydrolytic phase is relatively slow and can be limiting in anaerobic digestion of wastes such as raw cellulolytic wastes, which contain lignin (Polprasert, 1989; Speece, 1983).

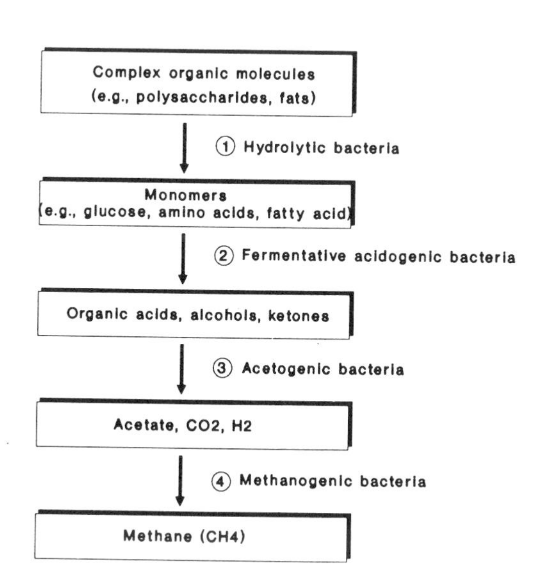

Fig. 13.3. Metabolic bacterial groups involved in anaerobic digestion of wastes.

13.3.2 Group 2: Fermentative Acidogenic Bacteria

Acidogenic (i.e., acid-forming) bacteria (e.g., *Clostridium*) convert sugars, amino acids, and fatty acids to organic acids (e.g., acetic, propionic, formic, lactic, butyric, or succinic acids), alcohols and ketones (e.g., ethanol, methanol, glycerol, acetone), acetate, CO_2, and H_2. Acetate is the main product of carbohydrate fermentation. The products formed vary with the type of bacteria as well as with culture conditions (temperature, pH, redox potential).

13.3.3. Group 3: Acetogenic Bacteria

Acetogenic bacteria (acetate and H_2-producing bacteria) such as *Syntrobacter wolinii* and *Syntrophomonas wolfei* (McInernay et al., 1981) convert fatty acids (e.g., propionic acid, butyric acid) and alcohols into acetate, hydrogen, and carbon dioxide, which are used by the methanogens. This group requires low hydrogen tensions for fatty acid conversion; and therefore a close monitoring of hydrogen concentration is necessary. Under relatively high H_2 partial pressure, acetate formation is reduced and the substrate is converted to propionic acid, butyric acid, and ethanol rather than methane. There is a symbiotic relationship between acetogenic bacteria and methanogens. Methanogens help achieve the low hydrogen tension required by acetogenic bacteria.

Ethanol, propionic acid, and butyric acid are converted to acetic acid by acetogenic bacteria according to the following reactions:

$$\underset{\text{ethanol}}{CH_3CH_2OH} + CO_2 \rightarrow \underset{\text{acetic acid}}{CH_3COOH} + 2H_2 \quad (13.2)$$

$$\underset{\text{propionic acid}}{CH_3CH_2COOH} + 2H_2O \rightarrow \underset{\text{acetic acid}}{CH_3COOH} + CO_2 + 3H_2 \quad (13.3)$$

$$\underset{\text{butyric acid}}{CH_3CH_2CH_2COOH} + 2H_2O \rightarrow \underset{\text{acetic acid}}{2CH_3COOH} + 2H_2 \quad (13.4)$$

Acetogenic bacteria grow much faster than methanogenic bacteria. The former group has a μ_{max} of approximately 1 hr^{-1}, whereas the μ_{max} of the latter is around 0.04 hr^{-1} (Hammer, 1986).

13.3.4. Group 4: Methanogens

Anaerobic digestion of organic matter in the environment releases 500–800 million tons of methane per year into the atmosphere and this represents 0.5% of the organic matter derived from photosynthesis (Kirsop, 1984; Sahm, 1984). The fastidious methanogenic bacteria occur naturally in deep sediments or in the rumen of herbivores. This group is composed of both gram-positive and gram-negative bacteria with a wide variety of shapes. Methanogenic microorganisms grow slowly in wastewater and their generation times range from 3 days at 35°C to as high as 50 days at 10°C.

Methanogens are subdivided into two subcategories.

Hydrogenotrophic methanogens (i.e., hydrogen-utilizing chemolithotrophs) convert hydrogen and carbon dioxide into methane.

$$CO_2 + 4H_2 \rightarrow \underset{\text{methane}}{CH_4} + 2H_2O \quad (13.5)$$

The hydrogen-utilizing methanogens help maintain the very low-level partial pressures necessary for the conversion of volatile acids and alcohols to acetate (Speece, 1983).

Acetotrophic methanogens, also called acetoclastic bacteria or acetate-splitting bacteria, convert acetate into methane and CO_2.

$$CH_3COOH \rightarrow CH_4 + CO_2 \quad (13.6)$$

The acetoclastic bacteria grow much more slowly (generation time = a few days) than the acid-forming bacteria (generation time = a few hours). This group comprises two main genera: *Methanosarcina* (Smith and Mah, 1978) and *Methanothrix* (Huser et al., 1982). During thermophilic (58°C) digestion of lignocellulosic waste, *Methanosarcina* was the dominant acetotrophic bacteria encountered in the bioreactor. After 4 months, *Methanosarcina* (μ_{max} = 0.3 day $^{-1}$; K_s = 200 mg/L) was displaced by *Methanothrix* (μ_{max} = 0.1 day $^{-1}$; K_s = 30 mg/L). It was postulated that the competition in favor of *Methanothrix* was due to the lower acetate K_s value of this organism (Gujer and Zehnder, 1983; Koster, 1988; Zinder et al., 1984).

About two thirds of methane is derived from acetate conversion by acetotrophic methanogens. The other third is the result of carbon dioxide reduction by hydrogen (Mackie and Bryant, 1981).

From a taxonomic viewpoint, methanogens belong to a separate kingdom, the archaebacteria, and differ from procaryotes in the following ways (Sahm, 1984):

1. They differ in cell wall composition; for example, the cell wall of methanogens lacks muramic acid.

2. Methanogens have a specific coenzyme, F_{420}, a 5-deazaflavin analog that acts as an electron carrier in metabolism. Its oxidized form absorbs light at 420 nm (Cheeseman et al., 1972). This blue-green fluorescent coenzyme has been proposed for use in quantifying methanogens in mixed cultures (van Beelen et al., 1983). F_{420} determination in cell extracts is carried out by extraction followed by fluorescence measurement or by high-performance liquid chromatography with fluorimetric detection (Peck, 1989). Methanogenic colonies can be distinguished from nonmethanogenic ones by fluorescence microscopy (Edwards and McBride, 1975; Kataoka et al., 1991). However, it was found that the use of F_{420} for methanogenic consortia can be misleading as regards the determination of acetoclastic methanogenic activity (Dolfing and Mulder, 1985). Another nickel-containing coenzyme, F_{430}, is also unique to methanogens.

3. Methanogens have ribosomal RNA sequences that differ from those of other procaryotes.

A tentative general classification of methanogens is presented in Table 13.1 (Balch et al., 1979). Methanogens are grouped into three orders: Methanobacteriales (e.g., *Methanobacterium, Methanobrevibacter, Methanothermus*), Methanomicrobiales (e.g., *Methanomicrobium, Methanogenium, Methanospirillum, Methanosarcina,* and *Methanococcoides*), and Methanococcales (e.g., *Methanococcus*). At least 49 species of methanogens have been described (Vogels et al., 1988). Table 13.2 (Koster, 1988) is a compilation of some of the methanogens that have been isolated and their respective substrates.

13.4. METHODS FOR DETECTION OF METHANOGENS

During the past decade, progress has been made in the development of procedures for determining the numbers and activity of methanogens in anaerobic digesters. Isolation of methanogens from environmental samples is a difficult task but the development of relatively sophisticated techniques has facilitated the search for and subsequent isolation of new species of methanogenic bacteria.

Standard microbiological enumeration tech-

TABLE 13.1. Classification of Methanogens[a]

Order	Family	Genus	Species
Methanobacteriales	Methanobacteriaceae	Methanobacterium	*M. formicicum* *M. bryanti* *M. thermoautotrophicum* *M. ruminantium*
		Methanobrevibacter	*M. arboriphilus* *M. smithii* *M. vannielli*
Methanococcales	Methanococcaceae	Methanococcus	*M. voltae*
		Methanomicrobium	*M. mobile*
Methanomicrobiales	Methanomicrobiaceae	Methanogenium	*M. cariaci* *M. marisnigri*
		Methanospirillum	*M. hungatei* *M. barkeri*
	Methanosarcinaceae	Methanosarcina	*M. mazei*

[a]From Balch et al. (1979), with permission of the publisher.

TABLE 13.2. Isolated Methanogens and Their Substrates[a]

Bacteria	Substrate	Bacteria	Substrate
Methanobacterium bryantii	H_2	*Methanogenium cariaci*	H_2 and HCOOH
M. formicicum	H_2 and HCOOH	*M. marisnigri*	H_2 and HCOOH
M. thermoautotrophicum	H_2	*M. tatii*	H_2 and HCOOH
M. alcaliphilum	H_2	*M. olentangyi*	H_2
		M. thermophilicum	H_2 and HCOOH
Methanobrevibacter arboriphilus	H_2	*M. bourgense*	H_2 and HCOOH
M. ruminantium	H_2 and HCOOH	*M. aggregans*	H_2 and HCOOH
M. smithii	H_2 and HCOOH		
		Methanococcoides methylutens	CH_3NH_2 and CH_3OH
Methanococcus vannielii	H_2 and HCOOH		
M. voltae	H_2 and HCOOH	*Methanothrix soehngenii*	CH_3COOH
M. deltae	H_2 and HCOOH	*M. concilii*	CH_3COOH
M. maripaludis	H_2 and HCOOH		
M. jannaschii	H_2	*Methanothermus fervidus*	H_2
M. thermolithoautotrophicus	H_2 and HCOOH		
M. frisius	H_2, CH_3OH, CH_3NH_2, and $(CH_3)_3N$	*Methanolobus tindarius*	CH_3OH, CH_3NH_2, $(CH_3)_2NH$, and $(CH_3)_3N$
Methanomicrobium mobile	H_2 and HCOOH	*Methanosarcina barkeri*	CH_3OH, CH_3COOH, H_2, CH_3NH_2, $(CH_3)_2NH$ and $(CH_3)_3N$
M. paynteri	H_2		
Methanospirillum hungatei	H_2 and HCOOH	*Methanosarcina thermophila*	CH_3OH, CH_3COOH, H_2 CH_3NH_2, $(CH_3)_2NH$, and $(CH_3)_3N$
Methanoplanus limicola	H_2 and HCOOH		
M. endosymbiosus	H_2		

[a]From Koster (1988), with permission of the publisher.

niques are not suitable for methanogenic bacteria. Methanogens are fastidious and occur as microbial consortia. They are difficult to culture in the laboratory. Immunological analysis, with polyclonal (Archer, 1984; Macario and Macario, 1988) or monoclonal (Kemp et al., 1988) antibodies, is now being used as a tool for determining the numbers and identity of methanogens in anaerobic digesters. Indirect immunofluorescence (IIF) and slide immunoenzymatic assays (SIA) have shown that the methanogenic microflora of anaerobic digesters is more diverse than previously thought. The predominant species detected were *Methanobacterium formicum* and *Methanobrevibacter arboriphilus* (Macario and Macario, 1988).

Microbial activity in anaerobic digesters is usually determined by measuring volatile fatty acids (VFA) or methane. Lipid analysis has been used to determine the biomass, community structure, and metabolic status in experimental digesters. Microbial biomass, community structure, and metabolic stress are indicated by determining the total lipid phosphate, phospholipid fatty acids, and poly-β-hydroxybutyric acid, respectively (Henson et al., 1989; Martz et al., 1983; White et al., 1979). Microbial activity in anaerobic sludge can also be determined by measuring ATP and INT-dehydrogenase activity. These parameters correlate well with traditional ones such as gas production rates (Chung and Neethling, 1989). ATP determination responds to pulse feeding of the digester and to addition of toxicants (Chung and Neethling, 1988). Tests are available for the estimation of the amount of acetotrophic bacteria in sludge (Valcke and Verstraete, 1983; van der Berg et al., 1974). One of these tests measures the capacity of the sludge to convert acetate into methane. The test gives information on the percentage of acetotrophic methanogens in anaerobically digested sludge.

Phosphatase activity has also been proposed as a biochemical tool to predict digester upset or failure. An increase in acid and alkaline phosphatases can predict instability of the digestion process well in advance of conventional tests (pH, VFA, gas production) (Ahley and Hurst, 1981).

13.5. FACTORS CONTROLLING ANAEROBIC DIGESTION

Anaerobic digestion is affected by temperature, retention time, pH, chemical composition of wastewater, competition of methanogens with sulfate-reducing bacteria, and the presence of toxicants.

13.5.1. Temperature

Methane production has been documented under a wide range of temperatures ranging between 0°C and 97°C. Although psychrophilic methanogenic bacteria have not been isolated, thermophilic strains operating at an optimum range of 50–75°C are found in hot springs. *Methanothermus fervidus* has been found in a hot spring in Iceland and grows at 63–97°C (Sahm, 1984).

In municipal wastewater treatment plants, anaerobic digestion is carried out in the mesophilic range at temperatures from 25°C to up to 40°C with an optimum at approximately 35°C. Thermophilic digestion operates at temperature ranges of 50–65°C. It allows higher loading rates and is also conducive to greater destruction of pathogens. One drawback is its higher sensitivity to toxicants (Koster, 1988).

Because of their slower growth as compared with acidogenic bacteria, methanogenic bacteria are very sensitive to small changes in temperature. As to utilization of volatile acids by methanogenic bacteria, a decrease in temperature leads to a decrease of the maximum specific growth rate while the half-saturation constant increases (Lawrence and McCarty, 1969). Thus, mesophilic digesters must be designed to operate at temperature between 30°C and 35°C for their optimal functioning.

13.5.2. Retention Time

The hydraulic retention time (HRT), which depends on wastewater characteristics and environmental conditions, must be long enough to allow metabolism by anaerobic bacteria in digesters. Digesters based on attached growth have a lower HRT (1–10 days) than those based on dispersed growth (10–60 days) (Polprasert, 1989). The retention times of mesophilic and thermophilic digesters range between 25 and 35 days but can be lower (Sterritt and Lester, 1988).

13.5.3. pH

Most methanogenic bacteria function in a pH range between 6.7 and 7.4, but optimally at pH = 7.0–7.2, and the process may fail if the pH is close to 6.0. Acidogenic bacteria produce organic acids, which tend to lower the pH of the bioreactor. Under normal conditions, this pH reduction is buffered by the bicarbonate that is produced by methanogens. Under adverse environmental conditions, the buffering capacity of the system can be upset, eventually stopping the production of methane. Acidity is more inhibitory to methanogens than to acidogenic bacteria. An increase in volatile acids level thus serves as an early indicator of system upset. Monitoring the ratio of total volatile acids (as acetic acid) to total alkalinity (as calcium carbonate) has been suggested to ensure that it remains below 0.1 (Sahm, 1984). One method for restoring the pH balance is to increase alkalinity by adding chemicals such as lime, anhydrous ammonia, sodium hydroxide, or sodium bicarbonate.

13.5.4. Chemical Composition of Wastewater

Methanogenic bacteria can produce methane from carbohydrates, proteins, and lipids, as well as from complex aromatic compounds (e.g., ferulic, vanilic, and syringic acids). However, a few compounds such as lignin and *n*-paraffins are hardly degraded by anaerobic bacteria.

Wastewater must be nutritionally balanced (nitrogen, phosphorus, sulfur) to maintain an adequate anaerobic digestion. The C:N:P ratio for anaerobic bacteria is 700:5:1 (Sahm, 1984). However, some observers argue that the C/N ratio for optimal gas production should be 25–30:1 (Polprasert, 1989). Methanogens use ammonia and sulfide as nitrogen and sulfur sources, respectively. Although un-ionized sulfide is toxic to methanogens at levels exceeding 150–200 mg/L, it is required by methanogens as a major source of sulfur (Speece, 1983). Moreover, trace elements such as iron, cobalt, molybdenum, and nickel are also necessary. Nickel, at concentrations as low as 10 μM, significantly increases methane production in laboratory digesters (Williams et al., 1986). Addition of nickel can increase the acetate utilization rate of methanogens from 2 g to as high as 10 g acetate per gram VSS per day (Speece et al., 1983). Nickel enters into the composition of the cofactor F_{430}, which is involved in biogas formation (Diekert et al., 1981; Whitman and Wolfe, 1980).

13.5.5. Competition of Methanogens with Sulfate-Reducing Bacteria

Methanogens and sulfate-reducing bacteria may compete for the same electron donors, acetate and H_2. The study of the growth kinetics of these two bacterial groups shows that sulfate-reducing bacteria have a higher affinity for acetate (K_s = 9.5 mg/L) than methanogens (K_s = 32.8 mg/L). This means that sulfate-reducing bacteria will outcompete methanogens under low acetate concentrations (Shonheit et al., 1982; Oremland, 1988; Yoda et al., 1987). This competitive inhibition results in the shunting of electrons from methane generation to sulfate reduction (Lawrence et al., 1966; McFarland and Jewell, 1990). Sulfate reducers and methanogens are very competitive at COD/SO_4 ratios of 1.7–2.7. An increase of this ratio is favorable to methanogens, whereas a decrease is favorable to sulfate reducers (Choi and Rim, 1991).

13.5.6. Toxicants

A wide range of toxicants are responsible for the occasional failure of anaerobic digesters. Inhibition of methanogenesis is generally indicated by reduced methane production and increased concentration of volatile acids. The following are some of the toxicants.

Oxygen. Methanogens are obligate anaerobes and are adversely affected by trace levels of oxygen (Oremland, 1988; Roberton and Wolfe, 1970).

Ammonia. Un-ionized ammonia is quite toxic to methanogenic bacteria. However, since the production of un-ionized ammonia is pH-dependent (i.e., more un-ionized ammonia forms at high pH), little toxicity is observed at neutral pH. Ammonia is inhibitory to methanogens at levels of 1,500–3,000 mg/L. Toxicity caused by continuous addition of ammonia decreases as the solids retention time is increased (Bhattacharya and Parkin, 1989).

Chlorinated hydrocarbons. Chlorinated aliphatics are much more toxic to methanogens than to aerobic heterotrophic microorganisms (Blum and Speece, 1992). Chloroform is very toxic to methanogenic bacteria and leads to their complete inhibition, as measured by methane production and hydrogen accumulation, at concentrations above 1 mg/L (Fig. 13.4) (Hickey et al., 1987). Acclimation to this compound increases the tolerance of methanogens to up to 15 mg/L of chloroform. Methanogen recovery depends on biomass concentration, solids retention time, and temperature (Yang and Speece, 1986).

Benzene ring compounds. Pure cultures of methanogens (e.g., *Methanothix concilii, Methanobacterium espanolae, Methanobacterium bryantii*) are inhibited by benzene ring compounds (e.g., benzene, toluene, phenol, pentachlorophenol). Pentachlorophenol is the most toxic of all the benzene ring compounds tested (Patel et al., 1991). Among the phenolic compounds, the order of inhibition of methanogenesis is nitrophenols $>$ chlorophenols $>$ hydroxyhyphenols (Wang et al., 1991).

Formaldehyde. Methanogenesis is severely inhibited at a formaldehyde concentration of 100 mg/L but appears to recover at lower formaldehyde concentrations (Fig. 13.5) (Hickey et al., 1988; Parkin and Speece, 1982).

Volatile acids. If the pH is maintained near neutrality, volatile acids such as acetic or butyric acid appear to be only a little toxic to methanogenic bacteria. Propionic acid, however, displays toxicity to both acid-forming and methanogenic bacteria.

Long-chain fatty acids. The long-chain fatty acids (e.g., caprylic, capric, lauric, myristic, and oleic acids) inhibit the activity of acetoclas-

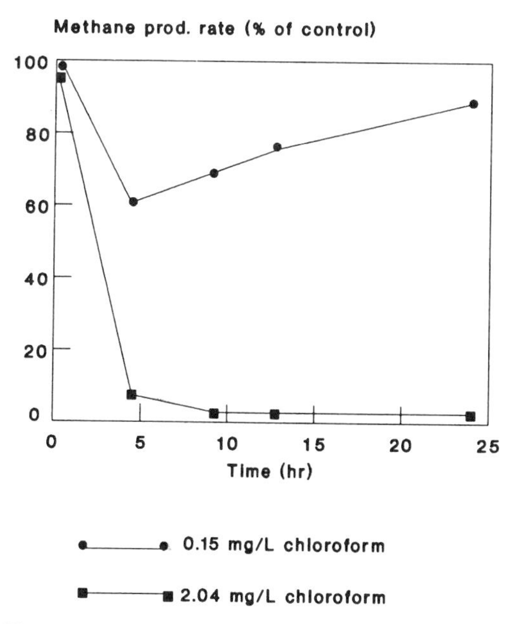

Fig. 13.4. Effect of chloroform on methane production. From Hickey et al. (1987), with permission of the publisher.

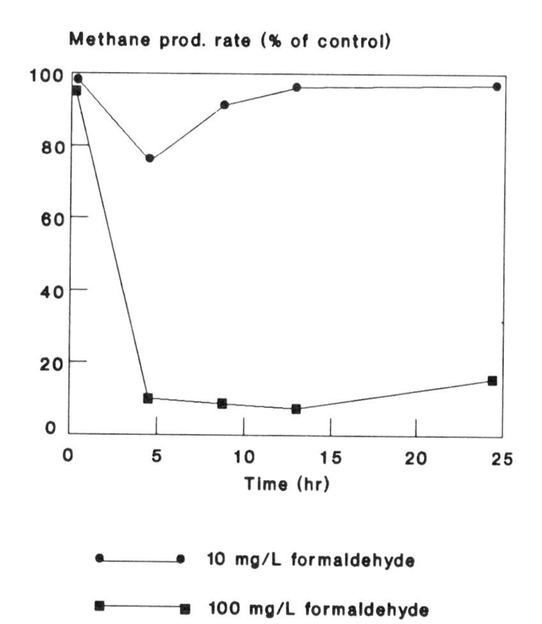

Fig. 13.5. Effect of formaldehyde on methanogenesis. From Hickey et al. (1987), with permission of the publisher.

tic methanogens (e.g., *Methanothrix* spp.) in acetate-fed sludge (Koster and Cramer, 1987).

Heavy metals. Heavy metals (e.g., Cu^{++}, Pb^{++}, Cd^{++}, Ni^{++}, Zn^{++}, Cr^{+6}) found in wastewaters and sludges from industrial sources are inhibitory to anaerobic digestion (Lin, 1992; Mueller and Steiner, 1992). The sequence of the extent of inhibition of anaerobic digestion of municipal sludge was found to be Ni $>$ Cu $>$ Cd $>$ Cr $>$ Pb. Toxicity increased as the heavy metal affinity for sludge decreased. Thus, lead is the least toxic heavy metal, owing to its high affinity for sludge (Mueller and Steiner, 1992). Metal toxicity is reduced following reaction with hydrogen sulfide, which leads to the formation of insoluble heavy metal precipitates. However, as discussed previously, some metals (e.g., nickel, cobalt, and molybdenum), at trace concentrations, may stimulate methanogenic bacteria (Murray and van den Berg, 1981; Shonheit et al., 1979; Whitman and Wolfe, 1980).

Cyanide. Cyanide is used in industrial processes such as metal cleaning and electroplating. While cyanide toxicity is concentration- and time-dependent, methanogen recovery depends on biomass concentration, solids retention time, and temperature (Fedorak et al., 1986; Yang and Speece, 1985).

Sulfide. Sulfide is one of the most potent inhibitors of anaerobic digestion (Anderson et al., 1982). Since diffusion through the cell membrane is more rapid for un-ionized than ionized hydrogen sulfide, sulfide toxicity is very dependent on pH (Koster et al., 1986). Sulfides are toxic to methanogenic bacteria when their levels exceed 150–200 mg/L. Acid-forming bacteria are less sensitive to hydrogen sulfide than methanogens. Within the latter group, hydrogen-oxidizing methanogens seem to be more sensitive than acetoclastic methanogens (Rinzema and Lettinga, 1988).

The inhibitory effect of sulfides may be due to sulfide *per se,* but it may also be caused by sulfate, which serves as a terminal electron acceptor for sulfate-reducing bacteria, possibly resulting in competition between this bacterial group and methanogens, since both groups utilize the same substrates (see Section 13.5.5). No inhibition was observed when the TOC/sulfate molar ratio was 1.3 (Karhadkar et al., 1987). In a thermophilic anaerobic fermenter, sulfide inhibition could be minimized by controlling the pH and by recycling the biomass to select for sulfide-tolerant microflora (McFarland and Jewell, 1990).

Tannins. Tannins are phenolic compounds that originate from grapes, bananas, apples, coffee, beans, and cereals. These compounds are generally toxic to methanogenic bacteria. While gallotannic acid is highly toxic to methanogens, its monomeric derivatives, gallic acid and pyrogallol, are much less toxic (Field and Lettinga, 1987). It is possible that tannins inhibit methanogens by reacting with accessible enzyme sites.

Salinity. Salinity is another toxicant encountered in anaerobic digestion of wastes. Since potassium antigonizes sodium toxicity, this type of toxicity can be countered by adding potassium salts to wastewater.

Feedback inhibition. Anaerobic systems may also be inhibited by several of the intermediates produced during the process. High concentrations of these intermediates (H_2, volatile fatty acids) are toxic by virtue of feedback inhibition (Barnes and Fitzgerald, 1987).

In order to avoid some of the problems discussed above, it has been suggested that two-phase anaerobic digestion systems be used to spatially separate acidogenic bacteria from methanogenic bacteria (Ghosh and Klass, 1978; Cohen et al., 1980; Pipyn et al., 1979). Some of the advantages of the two-phase system are enhanced stability and increased resistance to toxicants (toxicants are removed or reduced in the first stage). A long solids retention time also allows methanogens to acclimate to toxicants such as ammonia, sulfides, and formaldehyde. Thus, anaerobic digestion of industrial wastes containing toxic chemicals should be undertaken in reactors (e.g., anaerobic filter, anaerobic fluidized bed, anaerobic upflow sludge blanket) that allow a long SRT at relatively low hydraulic retention times (Bhattacharya and Parkin, 1988; Parkin et al., 1983).

13.6. ANAEROBIC TREATMENT OF WASTEWATER

Anaerobic digestion processes have been traditionally used for the treatment of wastewater sludges. There is now a growing interest in anaerobic treatment of wastewater as an energy-efficient approach to waste treatment. For the processing of low-strength wastes, the solids retention time must be controlled independently of the hydraulic retention time. A strategy for a successful treatment of wastewater by anaerobic systems is to find means of concentrating the slow-growing microorganisms.

We will now discuss the microbiological aspects of the main types of anaerobic systems used in waste treatment. The reactor configurations are displayed in Figure 13.6 (Speece, 1983).

13.6.1. Septic Tanks

The septic tank system, the oldest and most widely used anaerobic treatment system, was introduced at the end of the last century. Approximately 25% of the U.S. population is served by septic tanks.

A septic tank system is composed of a tank and an absorption field (Fig. 13.7):

Tank. The primary function of the tank is the

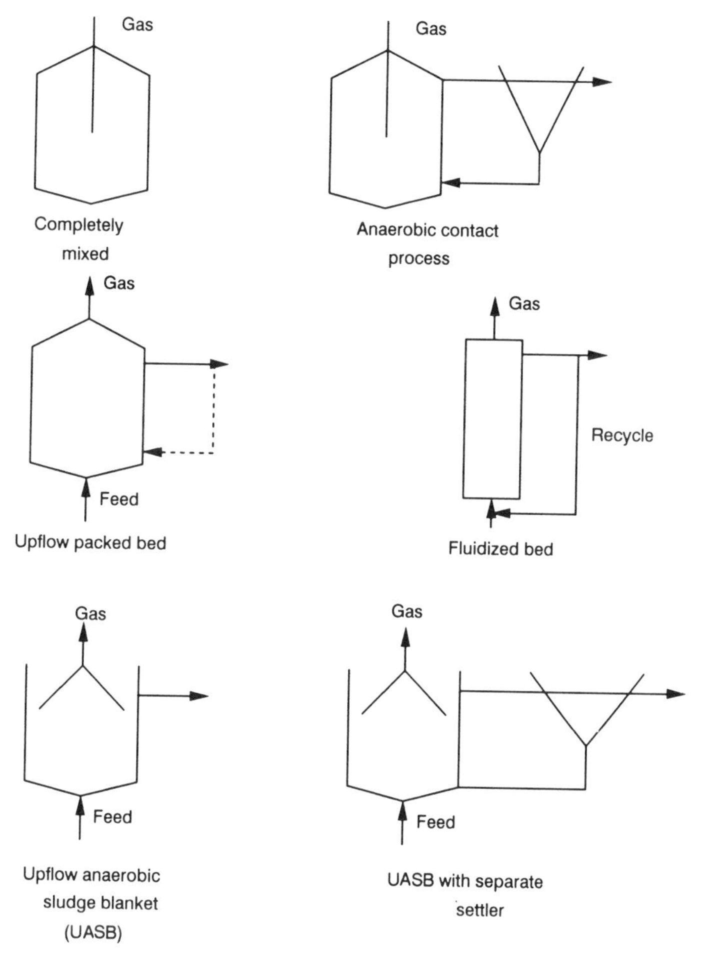

Fig. 13.6. Some bioreactor configurations used in anaerobic waste treatment. Adapted from Speece (1983).

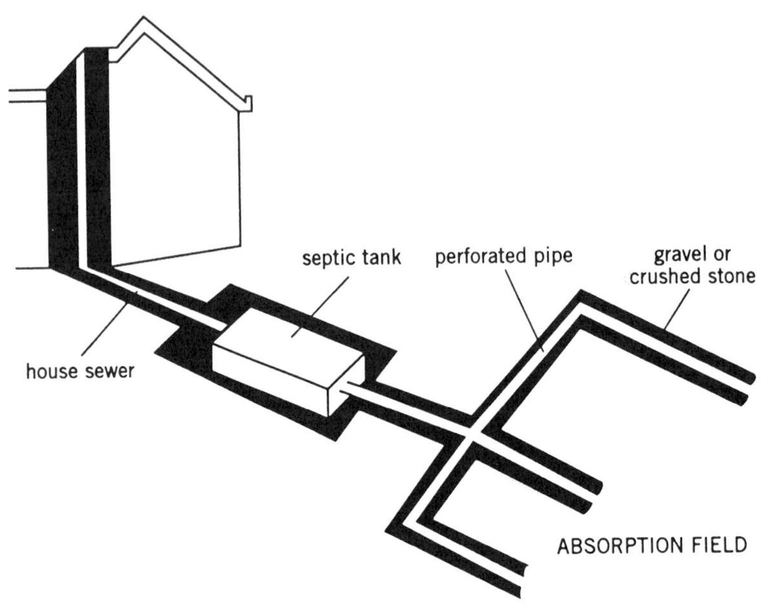

Fig. 13.7. Septic tank system. Adapted from U.S. DH/E/W (1969).

biological digestion of waste organic matter under anaerobic conditions followed by separation of solids from the incoming wastewater (Scalf, et al., 1977). The tank is made of concrete, metal, or fiberglass and is designed to remove the wastewater solids to avoid clogging of the absorption field. Wastewater undergoes anaerobic digestion, resulting in the production of sludge called septage and a floating layer of light solids (fats) called scum. The detention time of wastewater within a septic tank varies from 24 hr to 72 hr. A septic tank system generates a relatively small amount of septage (1,000–2,000 gallons per tank per 2–5 years). Septage is disposed of mainly by land application or is combined with municipal wastewater to be treated in wastewater treatment plants. Septage is approximately 50 times as concentrated as municipal wastewater. For a plant loaded with municipal wastewater to 75% of its design capacity, the recommended septage loading is approximately 1% (Segall and Ott, 1980). The tank should be inspected regularly and should be cleaned every 3–5 years to remove the accumulated septage. In hot climates, septic tanks provide efficient removal of suspended solids (approximately 80%) and BOD (> 90%). Otherwise, septic tanks achieve 65%–80% and 70%–80% for BOD and suspended solids, respectively. However, anaerobic digestion in the tank provides limited inactivation of pathogens (Hagedorn, 1984). Viruses are not significantly inactivated following wastewater digestion in septic tanks and may therefore be transported into groundwater (Hain and O'Brien, 1979; Sinton, 1986; Vaughn et al., 1983).

Absorption field. The effluents from septic tanks reach an absorption field through a system of perforated pipes that are surrounded by gravel or crushed stones. The septic tank effluent is treated by the soil as it percolates downward to the groundwater. There are several designs of soil absorption systems, including trenches, beds, mounds, and seepage pits (U.S. EPA, 1980). Proper functioning of the soil absorption field depends on many factors, including the wastewater characteristics, rate of wastewater loading, geology, and soil characteristics. The absorption field, under normal conditions, should operate under unsaturated flow conditions (Scalf et al., 1977).

Septic tank effluents are major contributors to groundwater pollution with pathogenic microorganisms (see Chapter 19). Wastewater volumes as high as (820–1,460) × 10^9 gallons per year are contributed by septic tanks in the United States (Office of Technology Assessment, 1984). These relatively large wastewater inputs to the subsurface environment are responsible for a significant percentage of the disease outbreaks that result from the consumption of untreated groundwater. Septic tanks are probably the major contributors of enteric viruses found in the subsurface environment. Drinking water wells should therefore be properly situated to avoid groundwater contamination and to allow for sufficient distance to bring about an efficient inactivation of microbial pathogens. Geostatistical techniques have been considered to estimate virus inactivation rates in groundwater in order to predict safe septic tank setback distances for installation of drinking water wells (Yates, 1985; Yates et al., 1986; Yates and Yates, 1987).

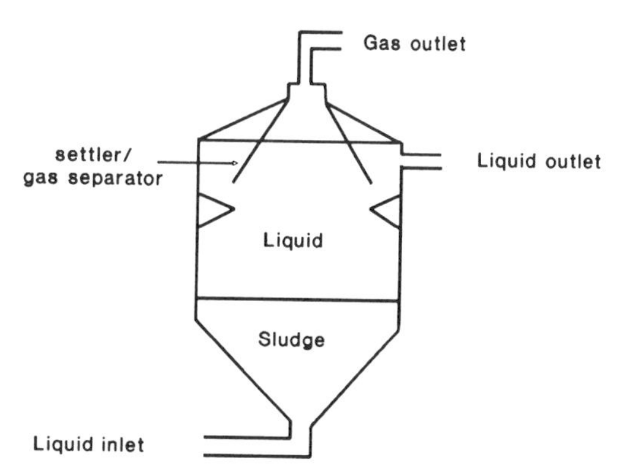

Fig. 13.8. Upflow anaerobic sludge blanket (UASB) process. Adapted from Lettinga et al. (1980).

13.6.2. Upflow Anaerobic Sludge Blanket (UASB)

The UASB digester was introduced at the beginning of the century, and after numerous modifications it was put into commercial use in the Netherlands for the treatment of industrial wastewater generated by the food industry (e.g., beet sugar, and corn and potato starch).

The UASB type of digester consists of a bottom layer of packed sludge, a sludge blanket, and an upper liquid layer (Fig. 13.8) (Lettinga et al., 1980). Wastewater flows upward through a sludge bed that is covered by a floating blanket of active bacterial flocs. Settler screens separate the sludge flocs from the treated water, and gas is collected at the top of the reactor (Schink, 1988). This process results in the formation of granular sludge that settles well. The sludge is immobilized by the formation of highly settleable microbial aggregates, which grow into distinct granules (< 1–5 mm) that have a high VSS content and specific activity. Immunological techniques, SEM examination, and energy-dispersive X-ray analysis of granular sludge have shown that the granules are composed of methanogens such as *Methanothrix soehngenii* (Brummeler et al., 1985; Hulshoff Pol et al., 1982, 1983), *Methanobacterium, Methanobrevibacter,* and *Methanosarcina,* as well as Ca precipitates (Vissier et al., 1991; Wu et al., 1987). Scanning and transmission electron microscopy showed that the granules are three-layered structures (Fig. 13.9) (McLeod et al., 1990). The inner layer consists of *Methanothrix*-like cells that may act as nucleation centers, which are necessary for the initiation of granule development. The middle layer consists of bacterial rods that include both H_2-producing acetogens and H_2-consuming organisms. The outermost layer consists of a mixture of rods, cocci, and filamentous microorganisms. This layer consists of a mixture of fermentative and H_2-producing organisms. Thus, a granule appears to harbor the necessary physiological groups to convert organic compounds to methane. The bacteriological composition of granules depends on the type of growth substrate (Grotenhuis et al., 1991). The factors that affect the rate of granulation include wastewater characteristics (higher rate when wastewater is composed of soluble carbohydrates), conditions of operation (e.g., sludge loading rate), temperature, pH, and availability of essential nutrients (Hulshoff et al., 1983; Wu et al., 1987).

The treatment of distillery wastewater by the UASB process has been found to yield 92% BOD removal (Pipyn et al., 1979).

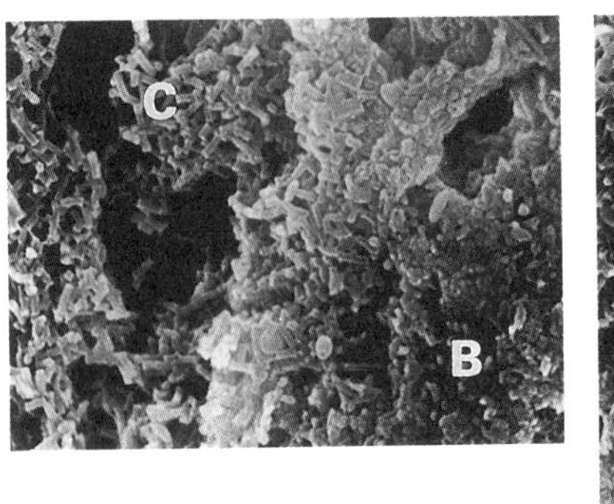

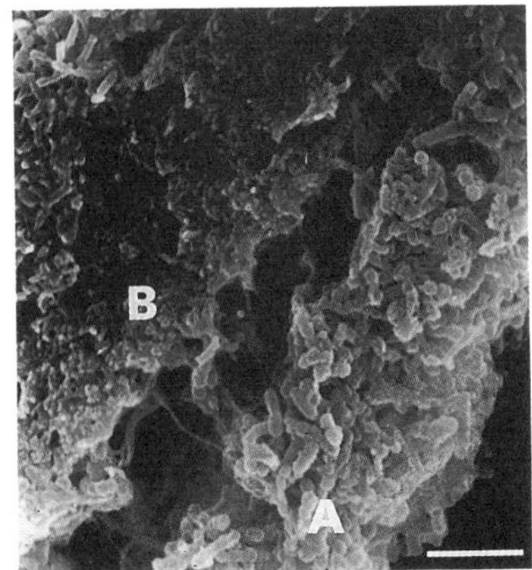

Fig. 13.9. Three-layered structure of a sludge anaerobic granule. (A). Exterior heterogenous layer with rods, filaments, and cocci. (B). Second layer containing predominantly rods and cocci. (C). Homogenous core containing a large number of cavities, surrounded by one bacterial morphology. Bar = 5μM. From MacLeod et al. (1990), with permission of the publisher, and courtesy of S.R. Guiot.

13.6.3. Anaerobic Filter

Anaerobic filters (Fig. 13.10), the anaerobic equivalent of trickling filters, were first introduced at the beginning of the century and further developed in 1969 by Young and McCarty. They contain support media (rock, gravel, plastic) with a void space of approximately 50% or more (Frostell, 1981; Jewell, 1987). The bulk of anaerobic bacteria grow attached to the filter medium, but some form flocs that become trapped inside the filter medium. The upflow of wastewater through the reactor helps retain suspended solids in the column. This process is particularly efficient for wastewaters rich in carbohydrates (Sahm, 1984). The loading rate varies with the type of waste and with the type of support medium. It generally falls within the range of 5–20 kg COD m^{-3} d^{-1} (Barnes and Fitzgerald, 1987). This system achieves modest BOD removal but higher removal of solids. Approximately 20% of the BOD is converted to methane.

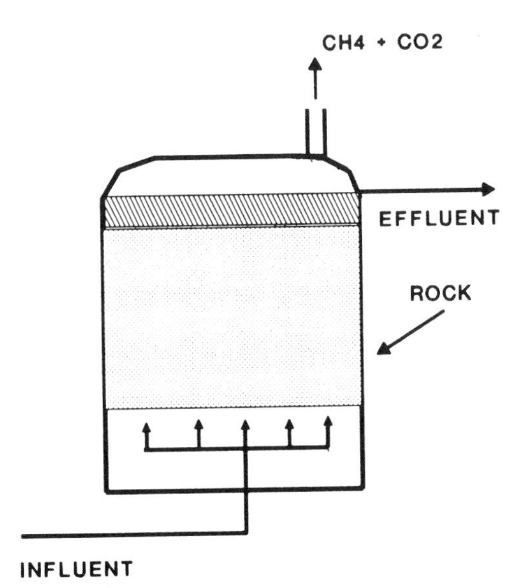

Fig. 13.10. Anaerobic filter. From Jewell (1987), with permission of the publisher.

Another version of the anaerobic filter is the thin-film reactor developed by van den Berg and collaborators (van den Berg and Kennedy, 1981). This reactor contains several clay tubes 5–10 cm in diameter. Incoming wastewater flows downwards and is treated by the 1- to 3-mm-thick anaerobic biofilm that develops on the surfaces of the clay tubes (Fig. 13.11).

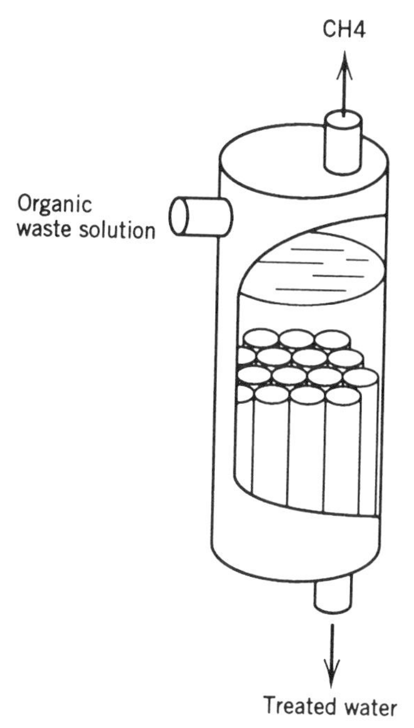

Fig. 13.11. Thin film bioreactor. Adapted from van den Berg & Kennedy, (1981).

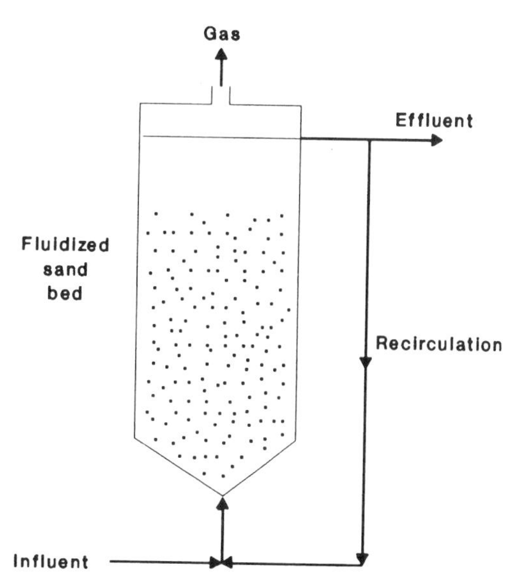

Fig. 13.12. Fluidized-bed reactor.

13.6.4. Anaerobic Attached-Film Expanded-Bed and Fluidized-Bed Reactors

These reactors were introduced in the 1970s. Some features distinguish the expanded beds from the fluidized ones (Fig. 13.12). Bed expansion caused by the upflow of wastewater is much greater in expanded than in fluidized beds. Wastewater flows upwards through a sand bed (diameter < 1 mm), which provides a surface area for the growth of bacterial biofilms. The flow rate is high enough to obtain an expanded or fluidized bed. This in turn necessitates the recirculation of the wastewater through the bed. This process is effective for the treatment of low-strength organic substrates (COD less than 600 mg/L) at short hydraulic retention times (several hours) and it allows high solids retention times (Speece, 1983; Switzenbaum and Jewell, 1980). This process offers the following advantages (Cooper and Wheeldon, 1981; Sahm, 1984; Switzenbaum and Jewell, 1980):

1. Good contact between wastewater and microorganisms.
2. Clogging and channeling are avoided.
3. High biomass concentrations can be achieved and this is associated with reduced reactor volume.
4. Biofilm thickness can be controlled.
5. Successful treatment of low-strength wastewater (≤ 600 mg/L COD) is possible at low temperature and at relatively short hydraulic retention times (< 6 hr.)

Some investigators have argued that fluidized-bed reactors can be applicable to aerobic treatment. An advantage of such application would be lower detention times, and thus, higher loading rates. Furthermore, nitrification would be favored as a result of the retention of the slow-growing nitrifiers in the biofilm (Rittmann, 1987).

Some of the disadvantages of this process are the high energy necessary for sufficient upflow velocity for bed expansion and the relatively large volume (30%–40%) occupied by the support material. Some have proposed an anaerobic expanded microcarrier bed (MCB) to solve these

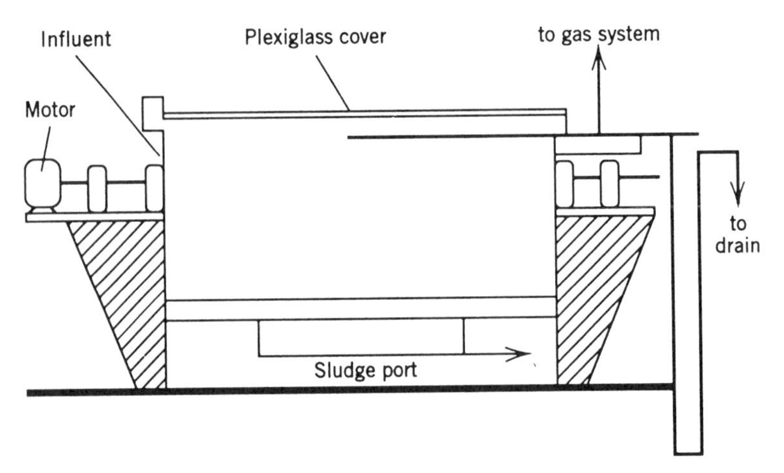

Fig. 13.13. Anaerobic rotating biological contactor. Adapted from Laquidara et al. (1986).

problems. Powdered zeolite is used as support material in the MCB process, which promotes the formation of granular sludge as in the USAB process (Yoda et al., 1989).

13.6.5. Anaerobic Rotating Biological Contactor

An anaerobic rotating biological contactor is similar to its aerobic counterpart (see Chapter 10) except that the reactor is sealed to create anoxic conditions (Fig. 13.13) (Laquidara et al., 1986). This process allows greater disk submergence because oxygen transfer is not considered. Development of the attached anaerobic biofilm is a function of applied organic loading and time. Approximately 85% COD removal is achievable even at loading rates as high as 90 $g/m^2 \cdot d$ COD (Fig. 13.14) (Laquidara et al., 1986). At this organic loading, methane is produced at a rate of 20 $L/m^2 \cdot d$. The following are some of the advantages of this fixed-film anaer-

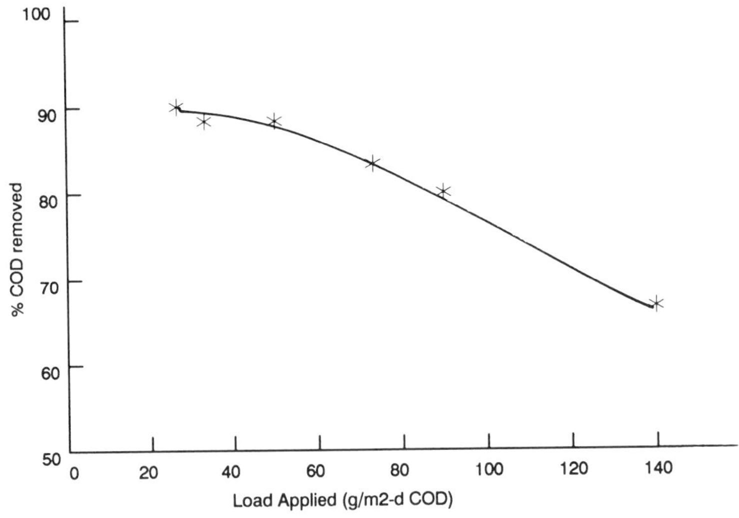

Fig. 13.14. Efficiency of an anaerobic biological contactor as a function of organic load. From Laquidara et al. (1986), with permission of the publisher.

obic system (Laquidara et al., 1986; Tait and Friedman, 1980):

1. Potential for higher organic loadings
2. Cell residence time independent of hydraulic detention time
3. Low production of waste solids
4. Ability to withstand toxic shock loads
5. Methane production

A close examination of the anaerobic processes for the treatment of wastewater shows that more research is needed concerning the microbiological aspects of these processes. A thorough examination of attached biofilms is also warranted. We also need to examine thoroughly the biodegradation potential of recalcitrant compounds, using these processes (Jewell, 1987).

13.7. FURTHER READING

Barnes, D., and P.A. Fitzgerald. 1987. Anaerobic wastewater treatment processes, pp. 57–113, in: *Environmental Biotechnology,* C.F. Forster and D.A.J. Wase, Eds. Ellis Horwood, Chichester, U.K.

Jewell, W.J. 1987. Anaerobic sewage treatment. Environ. Sci. Technol. 21: 14–20.

Kirsop, B.H. 1984. Methanogenesis. Crit. Rev. Biotechnol. 1: 109–159.

Koster, I.W. 1988. Microbial, chemical and technological aspects of the anaerobic degradation of organic pollutants, pp. 285–316, in: *Biotreatment Systems,* Vol. 1, D.L. Wise, Ed. CRC Press, Boca Raton, FL.

Lettinga, G., A.F.M. van Velsen, S.W. Hobma, W. de Zeeuw, and A. Klapwijk. 1980. Use of upflow sludge blanket (USB) reactor concept for biological wastewater treatment, especially for anaerobic treatment. Biotechnol. Bioeng. 22: 699–734.

Sahm, H. 1984. Anaerobic wastewater treatment. Adv. Biochem. Eng. Biotechnol. 29: 84–115.

Speece, R.E. 1983. Anaerobic biotechnology for industrial waste treatment. Environ. Sci. Technol. 17: 416A–427A.

Zehnder, A.J.B., Ed. 1988. *Biology of Anaerobic Microorganisms.* Wiley, New York.

14

BIOLOGICAL AEROSOLS AND BIOODORS FROM WASTEWATER TREATMENT PLANTS

14.1 INTRODUCTION

Biological aerosols are defined as biological contaminants present as suspended particles in the air. The size of aerosolized biological particles varies widely, from viruses to airborne algae or protozoa. Microorganisms may also attach to airborne dust particles that enter into the respiratory system. The sources of biological aerosols are of two kinds.

Natural sources include coughing, sneezing, shedding from human skin or animal hide, and disturbance of soil and aquatic environments by wind action.

Production sources include agricultural practices (cleaning of silos, chicken houses), textile mills (inhalation of *Bacillus anthracis*), processing of diseased animals in abattoirs and rendering plants, medical and dental facilities (breathing equipment, humidifying devices in nurseries, surgical pumps, rotary instruments used by dentists), aerosols from lab operations (blenders, sonication devices), and finally wastewater treatment operations.

Aerosols in indoor environments are produced by human and animal sources (droplets resulting from sneezing, coughing or talking, desquamation of skin or hide), plants, ventilation and air conditioning, home humidifiers, household toilets, and floor materials and draperies (Spendlove and Fannin, 1983).

Early laboratory studies, using simulated wastewaters contaminated with microorganisms, have suggested that aeration of wastewater may be responsible for the generation of biological aerosols, some of which could infect humans and cause diseases (Dart and Stretton, 1980).

This chapter also covers the microbiological aspects of bioodors production in wastewater treatment plants as well as microbiological approaches for treating the odors.

14.2 DEFENSE MECHANISMS OF THE RESPIRATORY SYSTEM AGAINST AIRBORNE PARTICLES

In the respiratory system, air is conveyed from the nose to the pharynx, larynx (voice box), trachea (windpipe), bronchi, bronchioles, terminal bronchioles, and finally the alveoli (Fig. 14.1) (Vander et al., 1985). Airborne particles are retained according to their size at the various levels of the respiratory tract. Particles of 10–20 μm are trapped in the nose. Particles ranging between 2 μm and 10μm are retained in nasal passages and/or in the bronchial tree (trachea, bronchi, and bronchioles), whereas those less than 2 μm are deposited in terminal bronchioles and alveoli.

Particle size determines the degree of penetration and retention in the respiratory system. Particles with a size range of 1–2 μm escape trapping by the upper respiratory tract and dis-

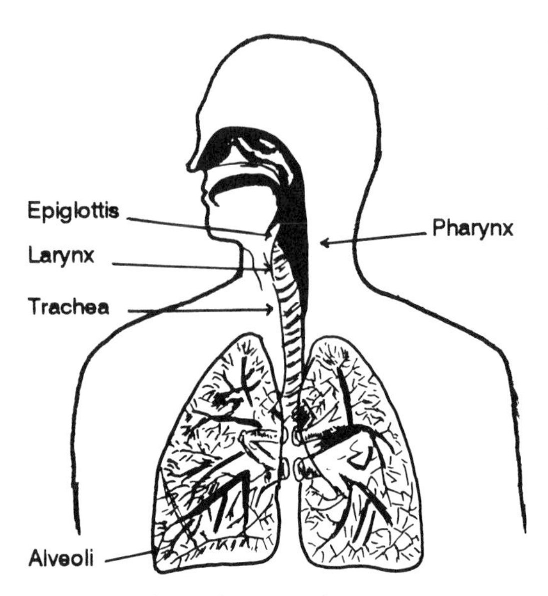

Fig. 14.1. Human respiratory system.

play the greatest deposition at the level of the alveoli.

The respiratory system provides defense mechanisms, however, against inhaled airborne particles. In the tracheobronchial region, inhaled particles are transported via the mucociliary escalator up to the mouth, where they are expelled or swallowed. In the pulmonary region, foreign particles are ingested by alveolar macrophages.

14.3 SAMPLING OF BIOLOGICAL AEROSOLS

Several samplers have been proposed for the collection of biological aerosols in various fluids or directly on solid growth media. An important consideration is the ability to gain information on the particle size distribution of biological aerosols. This information is important from a public health viewpoint since, as discussed above, the depth of penetration of biological aerosols into the respiratory system depends on the aerosol size. Selective growth media are needed for the recovery of specific categories of airborne microorganisms. Furthermore, airborne microorganisms are subjected to desiccation, sunlight, and other environmental stresses. Thus, growth media may have to be modified to recover the stressed airborne microorganisms. For example, the recovery of airborne bacteria is greatly enhanced when resuscitation agents such as betain and catalase are incorporated into the growth medium (Marthi and Lighthart, 1990; Marthi et al., 1991).

The various approaches for collecting biological aerosols are described below.

14.3.1. Sedimentation

Airborne microorganisms are allowed to deposit on a surface coated with a nutrient medium or on open Petri dishes containing a suitable growth medium. The composition of the growth medium depends on the type of airborne microorganism sought. However, this approach gives only qualitative information.

14.3.2. Membrane Filters

Air is passed through a membrane filter preferably made of polycarbonate, which can then be placed on a suitable growth medium. The filter may also be eluted, and the eluates can be plated on appropriate solid growth media to give an indication of viable airborne microorganisms or can be stained with acridine orange and examined by epifluoresence microscopy to obtain a total (viable plus nonviable) microbial count (Palmgren et al., 1986; Thorne et al., 1992).

14.3.3. Reuter Centrifugal Air Sampler

Air is collected by centrifugation followed by impaction on an agar medium surface. This portable sampler collects microorganisms on an agar medium-coated plastic strip lining the sampler drum. The maximum air sampling capacity is 11.3 ft^3 (Placencia et al., 1982).

14.3.4. Liquid Impingement Device (All-Glass Impinger)

This is one of the most popular means for collecting biological aerosols (Fig. 14.2). A vacuum

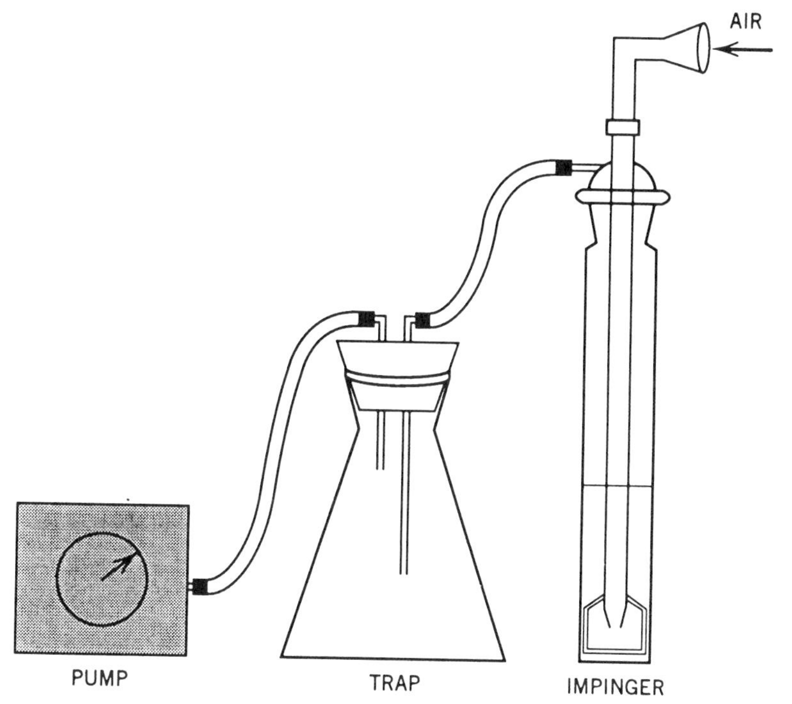

Fig. 14.2. Liquid impingement device for collecting biological aerosols.

pump is attached to the impinger and draws the air sample, resulting in the collection of biological aerosols in a suitable fluid (e.g., phosphate buffer, peptone water). Following aerosol collection, the fluid can be plated directly on a suitable growth medium. It can also be concentrated by membrane filtration and the filter placed on a solid nutrient medium. For viruses and bacteriophages, the collecting fluid should be preferably concentrated prior to assay on a suitable host cell. A drawback of this device is the loss of microbial viability because of impingement, sudden hydration, and osmotic shock (Thorne et al., 1992).

14.3.5. Sieve-Type Sampler

The most used sieve-type air sampler (Fig. 14.3) is the Andersen multistage sieve sampler. A measured volume of air is drawn and passed through a series of sieves of decreasing pore size. Growth media plates are included at each stage. This sampler, in contrast to the liquid impinger, gives information, useful from a public health viewpoint, about the particle size distribution of airborne microorganisms. Some drawbacks of the Andersen sampler include the possible overloading of the growth media plates, microbial stress due to impaction on agar surface, aggregation of microorganisms, and detection of only viable airborne microorganisms (Andersen, 1958; Lembke et al., 1981; Thorne et al., 1992).

14.3.6. May Three-Stage Glass Impinger

The multistage May sampler is a liquid impingement device with the advantage of particle size distribution. This three-stage glass impinger helps collect particles of sizes 6.0, 3.3, and 0.7 μm, which respectively simulate approximately the three major portions of the respiratory tract: nasopharyngeal, tracheobronchial,

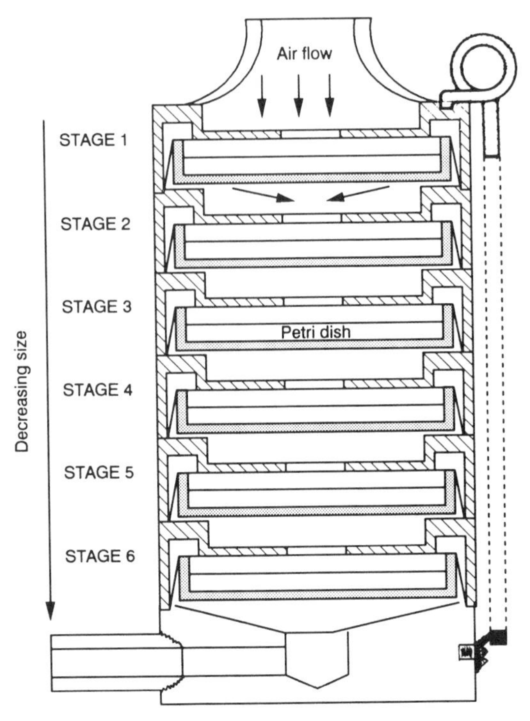

Fig. 14.3. Sieve-type sampler (Andersen sampler) for biological aerosols.

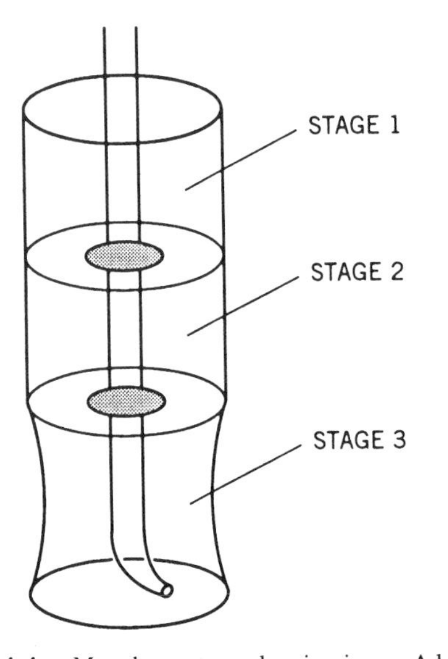

Fig. 14.4. May three-stage glass impinger. Adapted from Zimmerman et al. (1987).

and pulmonary regions (Fig. 14.4) (May, 1966; Zimmerman et al., 1987). The results obtained with this sampler correlate well with those obtained with the Andersen sampler (Zimmerman et al., 1987).

14.3.7. Large-Volume Electrostatic Precipitators

Since enteric viruses are found in relatively low concentrations in the air, it is necessary to use large-volume electrostatic precipitators for their detection. These costly air samplers can be operated at flow rates of approximately 1,000 L/min. The aerosols are collected in a liquid medium, and thousands of liters must be sampled (20,000–50,000 L) for virus detection in air samples (Shuval et al., 1989; Sorber et al., 1984).

14.4. PREDICTIVE MODELS FOR ESTIMATING DOWNWIND LEVELS OF AIRBORNE MICROORGANISMS

Biological aerosol monitoring is costly and labor-intensive, particularly with regard to airborne viruses which are found at relatively very low concentrations in the air we breathe. Dispersion models that have been borrowed from the field of aerosol mechanics and require the integration of biological and meteorological data have been developed to predict the downwind concentrations of aerosolized microorganisms from known sources (e.g., activated sludge units, spray irrigation sites) and to assess the possible health significance of microbial aerosols. These models assume a Gaussian (normal) distribution of pollutants at any specific downwind location (Mohr, 1991; Reed et al., 1988; U.S. EPA, 1982). Early dispersion models dealt with the dispersion and deposition of inert particles (Pasquill, 1961). Later models took into account the viability of microorganisms in the aerosolized state (Camann et al., 1978; Lighthart and Frisch, 1976; Lighthart and Mohr, 1987; U.S. EPA, 1982).

The downwind concentration of C_d of aerosolized microorganisms is given by the following equation (U.S. EPA, 1982):

$$C_d = Q\, D_d\, e^{xa} + B \qquad (14.1)$$

where Q = microbial concentration at the source (number/m^3); D_d = atmospheric dispersion factor (s/m^3); x = decay rate (s^{-1}); a = aerosol age (downwind distance/wind speed) (s); and B = background concentration of biological aerosol (number/m^3).

The microbial concentration at the source is given by the following equation:

$$Q = WFEI \qquad (14.2)$$

where W = microbial concentration in wastewater (number/L); F = flow rate of wastewater (L/s); E = aerosolization efficiency (s/m^3) (E = 0.3% for wastewater, and 0.04% for sludge applied with a spray gun); I = survival factor of aerosolized microorganisms; it is the fractional reduction of microorganisms during the aerosol formation process ($I > 0$). I varies with the pathogen under consideration.

Certain model components are site-specific and need to be determined at the site under study. These include meterological data (wind direction and speed, atmospheric stability), microbial concentration in wastewater, flow rate, and background aerosol concentration. Other model components are non-site-specific and include aerosolization efficiency, microbial decay rate (depends on the microorganism type and is generally obtained from laboratory studies), and survival factor, which represents the initial shock during the aerosolization process.

A similar dispersion model has been used to predict the fate of airborne microorganisms in a spray irrigation site, in Pleasanton, California. This model also incorporates essential microbiological parameters such as decay rates (Camann et al., 1978; Johnson et al., 1980).

14.5. FACTORS CONTROLLING THE SURVIVAL OF BIOLOGICAL AEROSOLS

Immediately after aerosolization of microorganisms they are subjected to an initial shock. A rapid die-off of microorganisms is observed following this initial shock due mainly to desiccation. Thereafter, microbial decay will continue at a slower rate as the aerosols are transported downwind from the source. The persistence of airborne microorganisms depends upon several environmental factors, the most important of which are relative humidity (RH), desiccation, solar radiation, and temperature (Cox, 1987; Mohr, 1991).

14.5.1. Relative Humidity

Relative humidity is probably the most crucial factor controlling aerosol stability (Cox, 1987). Conflicting results have been obtained regarding the effect of relative humidity (RH) on microbial stability; they are probably due to variations in experimental procedures (Mohr, 1991). Bacteria and viruses generally survive best at high relative humidity. Nonenveloped viruses (picornaviruses and adenoviruses) survive best at high relative humidity, whereas enveloped viruses (e.g., myxoviruses) survive best at low relative humidity. Figure 14.5 illustrates the effect of relative

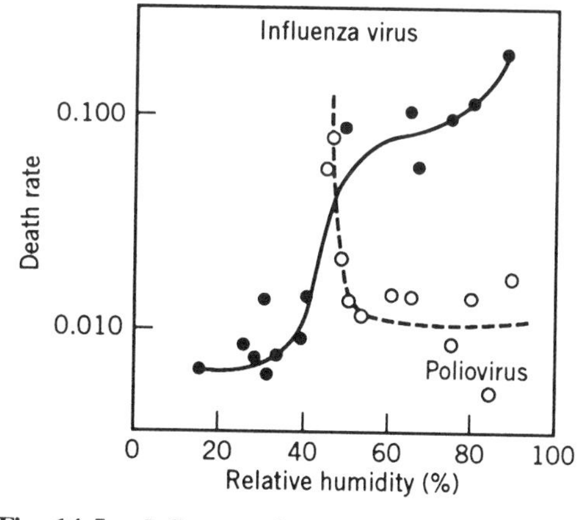

Fig. 14.5. Influence of relative humidity on the survival of influenza virus and poliovirus. From Hemmes et al. (1960), with permission of the publisher.

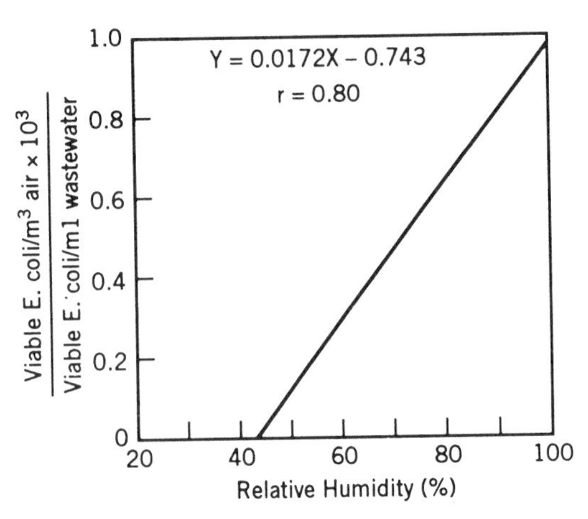

Fig. 14.6. Normalized *E. coli* aerosol concentration as a function of relative humidity. Adapted from Teltsch and Katzenelson (1978).

humidity on an enveloped virus (influenza virus) and a nonenveloped one (poliovirus) (Hemmes et al., 1960). However, rotaviruses are inactivated at high RH and they survive best at 50% relative humidity (Ijaz and Sattar, 1985; Sattar et al., 1984). Controlled field experiments with a nalidixic acid-resistant *E. coli* showed that survival was correlated with relative humidity (Fig. 14.6) (Teltsch and Katzenelson, 1978). However, in atmospheres containing inert gases such as nitrogen, argon, or helium, the survival of *E. coli* was best at low RH (Cox, 1987).

14.5.2. Temperature

Stability of microbial aerosols generally decreases as temperature increases (Marthi et al., 1990).

14.5.3. Solar Radiation

Aerosols survive longer at nightime, in the absence of sunlight (Shuval et al., 1989). Controlled field experiments with aerosolized *E. coli* showed that bacterial survival was negatively correlated with solar radiation intensity (Fig. 14.7) (Teltsch and Katzenelson, 1978).

14.5.4. Type of Microorganism

Aerosol persistence depends on the type of microorganisms. In aerosols, enteroviruses are generally hardier than bacteria or coliphages. Encapsu-

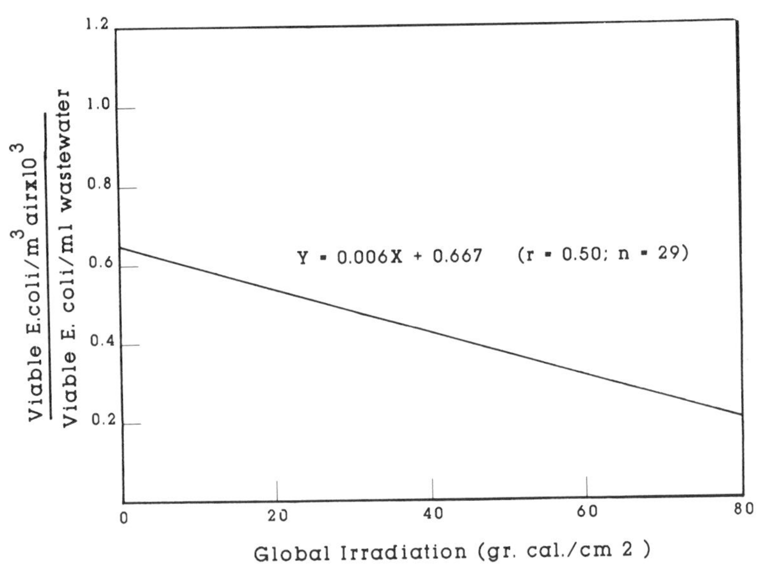

Fig. 14.7. Normalized *E. coli* aerosol concentration as a function of solar irradiation. From Teltsch and Katzenelson (1978), with permission of the publisher.

lated bacteria (e.g., *Klebsiella*) appear to survive better than nonencapsulated counterparts.

14.5.5. Other Factors

Fluctuations in air pressure, presence of free radicals of oxygen, and atmospheric pollutants (e.g., NO_2, SO_2, O_3) also affect microbial stability in aerosols (Mohr, 1991).

14.6. PRODUCTION OF BIOLOGICAL AEROSOLS BY WASTEWATER TREATMENT OPERATIONS

Biological aerosols are produced whenever aeration is used in treatment processes (e.g., activated sludge, aerated lagoons, aerobic digestion of sludge). Mechanical devices that assist in the aerosolization of raw wastewater or wastewater effluents may lead to the production of biological aerosols. For example, airborne bacterial densities above the aeration tank of an activated sludge system appear to be directly related to the aeration rate. Figure 14.8 shows that bacterial densities are related to aeration rate and they peak during night time (Fedorak and Westlake, 1980). Biological aerosols are also generated by routine activities indoors (e.g., sludge dispensing or floor mopping) (Fedorak and Westlake, 1980). These aerosols may be transported for relatively great distances from the wastewater treatment plant.

Spray irrigation with wastewater effluents also leads to the production of biological aerosols. Indicators and pathogenic bacteria as well as enteroviruses have been isolated at spray irrigation sites, sometimes hundreds of meters from the production source (Applebaum et al., 1984; Camann et al., 1988; Fannin et al., 1985; Moore et al., 1979; Shuval et al., 1989; Teltsch and Katzenelson, 1978; Teltsch et al., 1980). In Israel, spray irrigation with oxidation pond effluents produced bacterial (total bacteria, total and fecal coliforms, fecal streptococci), viral, and algal aerosols up to distance of 730 m from the source. At that distance, 79% of the air samples were positive for indicator bacteria during daytime and 100% were positive during nightime. Approximately 10% of the samples were positive for viruses. Virus concentration in positive samples ranged from 0.03 to 2 PFU/m^3 (Shuval et al., 1989). Other studies in Israel have demonstrated the presence of enteroviruses (poliovirus 2, echovirus types 1, 7, 17, and 25, coxsackievirus B1, and other unidentified viruses), *Salmonella*, and total coliforms in biological aerosols (Teltsch et al., 1980).

In Lubbock, Texas, a poor-quality trickling filter effluent was used for spray irrigation of agricultural crops. The fecal coliform levels in the wastewater effluent exceeded 10^6/100 ml and the enterovirus level during the summer season sometimes exceeded 1,000 PFU/L. Impoundment of the wastewater effluent in a reservoir reduced the fecal coliform level by 99% and the enterovirus level to less than 10 PFU/L (Moore et al., 1988). Spray irrigation with the effluent prior to reservoir impoundment significantly increased the downwind levels of indicator bacteria (fecal coliform, fecal streptococci, *Clostridium perfringens*, mycobacteria) and bacterial phage relative to background levels (Table 14.1) (Camann et al., 1988). Microbial levels decreased as the distance from the spray irrigation site increased. Enteroviruses, mostly

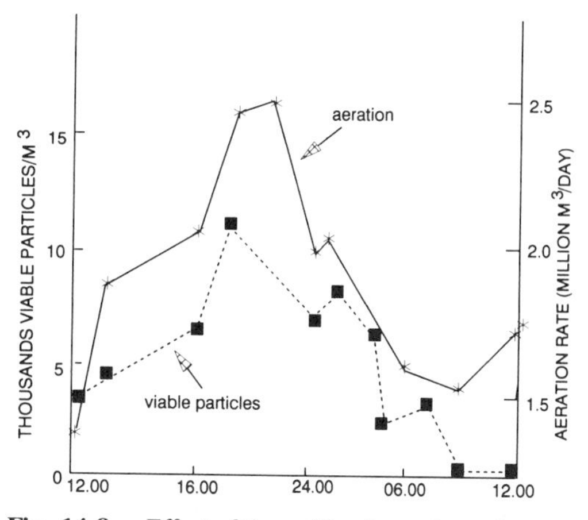

Fig. 14.8. Effect of time of the day and aeration rate on bacterial densities above an activated sludge unit. From Fedorak and Westlake (1980), with permission of the publisher.

TABLE 14.1. Biological Aerosols Downwind From a Site Spray-Irrigated With Wastewater Effluents[a]

Microorganism group/ wastewater source	Microorganism concentration in air (number/m³)				
	Ambient background		Downwind of irrigation line		
	Homes	Fields	30–89 m	90–149 m	250–409 m
Fecal coliforms, CFU	0.01	<0.006			
Pipeline			180	1.8	0.3
Reservoir			1.4	0.2	<0.08
Fecal streptococci, CFU	0.5	0.07			
Pipeline			140	16	0.5
Reservoir			0.5	0.2	0.3
Mycobacteria, CFU	0.05	0.1			
Pipeline			2.5	0.8	0.2
Reservoir			0.05	0.1	<0.03
Clostridium perfringens, CFU		0.08			
Pipeline			9	1.2	0.6
Coliphage, PFU	<0.005	<0.003			
Pipeline			9.9	1.8	0.1
Reservoir			0.06	0.07	0.07
Enteroviruses, PFU					
Pipeline			0.05		

[a]Adapted from Camann et al., with permission of the publisher.

TABLE 14.2. Viral Aerosols Downwind From Two Spray Irrigation Sites[a]

Distance from spray line, m	Enterovirus density		
	Host cell line	In wastewater, PFU/mL	In air, PFU/m³
Hancock farm, Texas			
60	HeLa	0.16	0.0029
	RD		0.0015
46	HeLa	0.10	0.011
	RD		0.018
44	HeLa	2.2	16.2
	RD		18.3
49	HeLa	0.066	0.010
	RD		0.013
Pleasanton, Calif.			
63	HeLa	0.036	0.0047
63	HeLa	0.18	0.0074

[a]Adapted from Camann (1988), with permission of the publisher.

polioviruses, were consistently detected in air samples at 44–60 m downwind from the spray irrigation site (Table 14.2) (Camann et al., 1988). Thus, storage impoundment of wastewater should be an integral part of an irrigation scheme since this practice significantly reduces indicator and pathogen levels in wastewater and aerosol samples (Camann et al., 1988; Moore et al., 1988).

Air samples were taken at the Muskegon County (Michigan) Wastewater Management System #1, with an army prototype XM2 Biological Sampler, which collects 1,050 liters of air/min (Brenner et al., 1988). Bacteria (86–7,143 CFU/m^3) and coliphage (0–9 PFU/m^3) were isolated but no animal virus was recovered even in air samples over 400 m^3.

In sludge application sites, microbial indicators (total coliforms, fecal coliforms, fecal streptococci, bacteriophages) and mycobacteria were detected downwind from the spray irrigation site. Enteroviruses were not detected in a 1,470–m^3 pooled air sample. It was concluded that sludge application to land by spray irrigation is not a serious threat to human health (Sorber et al., 1984).

14.7. HEALTH HAZARDS OF BIOLOGICAL AEROSOLS GENERATED BY WASTE TREATMENT OPERATIONS

An evaluation of the potential hazard due to biological aerosols must take into account several critical factors, particularly the pathogen concentration of the wastewater effluent being sprayed, the degree of aerosolization (which depends on meteorological conditions and type of equipment), and meteorological conditions (wind velocity, temperature, solar radiation, and relative humidity).

14.7.1. Spray Irrigation With Wastewater Effluents and Liquid Sludge

We have shown that microbial aerosols are produced by spray irrigation of agricultural land with wastewater effluents and, sometimes, liquid sludges. Studies have been undertaken in the United States, Canada, Israel, and other countries on the health aspect of aerosols generated by spray irrigation of wastewater (Bausum et al., 1983; Camann et al., 1986, 1988; Fattal et al., 1986; Sekla et al., 1980). High-pressure sprinklers can aerosolize from 0.1%–1% of the wastewater effluent, leading to the spreading of aerosols hundreds of meters from the production source. Furthermore, 66%–78% of particles generated as a result of spray irrigation are in the 1- to 5-μm range and are efficiently deposited in the lungs (Bausum et al., 1983). Others have reported that 20% of the bacterial aerosols collected with an Andersen sampler at a spray irrigation site 730 m from the source were in the respirable range (0.65–2.0 μm) (Shuval et al., 1989). However, in the United States, no significant increase in incidence of human disease was observed as a result of exposure to biological aerosols generated by sewage treatment plants or by spray irrigation with wastewater effluents (U.S., EPA, 1982). Most epidemiological studies have shown little evidence of increased risk of infection or increased occurrence of disease related to spray irrigation with wastewater. The Lubbock Infection Surveillance Study (LISS), conducted to monitor microbial infections in a rural community situated near a wastewater irrigation site, indicated that spray irrigation with wastewater effluents did not result in any significant increase in rotavirus infections at spray irrigation sites (rotavirus infection was defined as a twofold increase in rotavirus serum antibody between two blood collection periods) (Camann et al., 1986; Ward et al., 1989).

In Israel, a retrospective epidemiological study was conducted to record disease incidence resulting from irrigation with oxidation pond effluents. Clinical records for enteric diseases were examined for kibbutzim (i.e., agricultural settlements) practicing spray irrigation with wastewater; they were compared with those from kibbutzim that did not irrigate with wastewater. Although preliminary data showed increase in the incidence of some diseases (e.g., shigellosis, salmonellosis, hepatitis) in the kibbutzim that practiced spray irrigation (Katzenelson et al., 1976), there was no conclusive evidence of significant increased risk of disease resulting from exposure to aerosolized wastewater effluents. Except for echovirus type 4 (probably due to an echovirus 4 epidemic prior to the collection of blood samples for the study), no significant increase of antibodies to seven enteroviruses was observed in communities exposed to biological aerosols from sprinkler irrigation with partially treated wastewater (Fattal et al., 1987). However, within 11 kibbutzim, the practice of spray irrigation with wastewater effluents resulted in a small excess risk of enteric diseases during the irrigation season, particularly for the age group 0–4 years. No such excess was observed, however, on a year-round basis (Fattal et al., 1984, 1986).

Possible reasons for difficulties in drawing definite conclusions about human health hazards due to biological aerosols are the following.

1. Airborne microbes are not unique to wastewater.
2. There are confounding factors such as other sources and routes of infection.
3. Induced immunity by low aerosol levels.
4. Sensitivity of monitoring techniques is insufficient.
5. Minimal infectious doses for aerosolized microorganisms are not well known.

14.7.2. Health Risks Associated With Composting of Wastewater Sludge

Some composting operations may result in workers' exposure to pathogens from aerosols or direct contact with the composted material. The high temperatures (55–65°C) generated during composting are efficient for inactivation of pathogenic microorganisms normally found in wastewater sludge. However, these temperatures allow the proliferation of thermophilic actinomycetes as well as pathogenic fungi such as *Aspergillus fumigatus*. A study of workers in four composting facilities in the United States has revealed the following effects on compost workers (Clark et al., 1984): nasal, ear, and skin infections; burning eyes and skin irritation; and throat and anterior nares cultures positive for *A. fumigatus*. Infection is caused by inhalation of spores, which can reach the alveoli. Its course is determined by the immunological status of the host, since infection can be very severe in immunodeficient patients. Allergic bronchopulmonary aspergillosis also develops in persons with asthma (Clark et al., 1981, 1984; Rippon, 1974). Workers also display high levels of antibodies to endotoxin (lipolysaccharide) produced by gram-negative bacteria. Endotoxins may cause headache, fever, and nose and eye irritation.

14.7.3. Health Hazards to Wastewater Treatment Plant Workers

Opportunities exist for wastewater treatment operators to become exposed to pathogenic bacteria, viruses, and parasites from biological aerosols or from contact with contaminated materials and surfaces in the plant. In a wastewater treatment plant in Oslo, Norway, *Salmonella* spp. was isolated from wastewater, sludge, and floor surfaces in an eating area, but there was little correlation between the *Salmonella* serotypes isolated from the plant and those isolated directly from the population (Langeland, 1982). Endotoxin concentration in a wastewater treatment plant in Finland varied between 0.6 and 310 ng/m^3. A proposed occupational exposure limit (8-hr time-weighted average) for airborne endotoxin is 30 ng/m^3 (Palchak et al., 1988). Operators may display an increased level of antibodies to endotoxins from gram-negative bacteria and to some enteroviruses, particularly hepatitis A and coxsackie B3 viruses (Clark et al., 1977; Rylander et al., 1976; Skinhoj et al., 1981). In Copenhagen, Denmark, higher levels of antibody against HAV were observed among sewage workers (81%) than among gardeners (61%) or city clerks (48%) (Skinhoj et al., 1981). However, the slightly increased risk of infection resulting from exposure to wastewater does not appear to translate into increased overt disease incidence among wastewater treatment plant workers.

14.8. MICROBIOLOGICAL ASPECTS OF ODORS GENERATED BY WASTEWATER TREATMENT PLANTS

During wastewater treatment operations, odor-producing compounds are caused by the anaerobic decomposition of organic matter containing sulfur and nitrogen. These volatile compounds have relatively low molecular weight and are generally identified, by odor panels consisting of at least five people, or by gas chromatography. Odorous compounds are assigned a threshold odor number (TON) that is defined as the concentration below which the odor is no longer detectable by the human nose.

14.8.1. Types of Odor-Causing Compounds

The main odor-causing compounds (Table 14.3) (U.S. EPA, 1985) are the following (Henry and Gehr, 1980; Jenkins et al., 1980; U.S. EPA, 1985; van Langenhove et al., 1985).

14.8.1.1. Inorganic Gases

H_2S is results from sulfate reduction by sulfate-reducing bacteria, such as *Desulfovibrio desulfuricans*, which use sulfate as an electron acceptor. It can also be produced by the anaerobic decomposition of sulfur-containing amino acids such as methionine, cysteine, and cystine (see Chapter 3). However, sulfate reduction is the most important contributor of H_2S in wastewater. H_2S is a colorless toxic gas with a rotten-egg odor and has a low olfactory threshold (below 0.2 ppm) (National Research Council, 1979). H_2S is quite toxic to humans and is fatal at concentrations exceeding 500 ppm (National Research Council, 1979). This gas is also corrosive to materials (concrete, copper, lead, iron) commonly found in wastewater treatment plants.

The partitioning of H_2S between the liquid and gas phases depends mainly on pH, initial dissolved hydrogen sulfide concentration, and temperature. At pH = 7.0, H_2S represents 50% of the dissolved sulfides in wastewater. Its concentration increases as the pH decreases (Sawyer and McCarty, 1967; Pisarczyk and Rossi, 1982). H_2S solubility in wastewater also decreases as temperature increases. Figure 14.9 shows that, at one atmosphere pressure, the H_2S level in air increases as temperature and dissolved hydrogen sulfide increase (Bowker et al., 1989).

TABLE 14.3. Some Odorous Sulfur Compounds in Wastewater[a]

Substance	Formula	Characteristic odor	Odor threshold	Molecular weight
Allyl mercaptan	$CH_2=CH\text{-}CH_2\text{-}SH$	Strong garlic-coffee	0.00005	74.15
Amyl mercaptan	$CH_3\text{-}(CH_2)_3\text{-}CH_2\text{-}SH$	Unpleasant, putrid	0.0003	104.22
Benzyl mercaptan	$C_6H_5CH_2\text{-}SH$	Unpleasant, strong	0.00019	124.21
Crotyl mercaptan	$CH_3\text{-}CH=CH\text{-}CH_2\text{-}SH$	Skunk-like	0.000029	90.19
Dimethyl sulfide	$CH_3\text{-}S\text{-}CH_3$	Decayed vegetables	0.0001	62.13
Ethyl mercaptan	$CH_3CH_2\text{-}SH$	Decayed cabbage	0.00019	62.10
Hydrogen sulfide	H_2S	Rotten eggs	0.00047	34.10
Methyl mercaptan	CH_3SH	Decayed cabbage	0.0011	48.10
Propyl mercaptan	$CH_3\text{-}CH_2\text{-}CH_2\text{-}SH$	Unpleasant	0.000075	76.16
Sulfur dioxide	SO_2	Pungent, irritating	0.009	64.07
Tert-butyl mercaptan	$(CH_3)_3C\text{-}SH$	Skunk, unpleasant	0.00008	90.10
Thiocresol	$CH_3\text{-}C_6H_4\text{-}SH$	Skunk, rancid	0.000062	124.21
Thiophenol	C_6H_5SH	Putrid, garlic-like	0.000062	110.18

[a]From U.S. EPA (1985) with permission of the publisher.

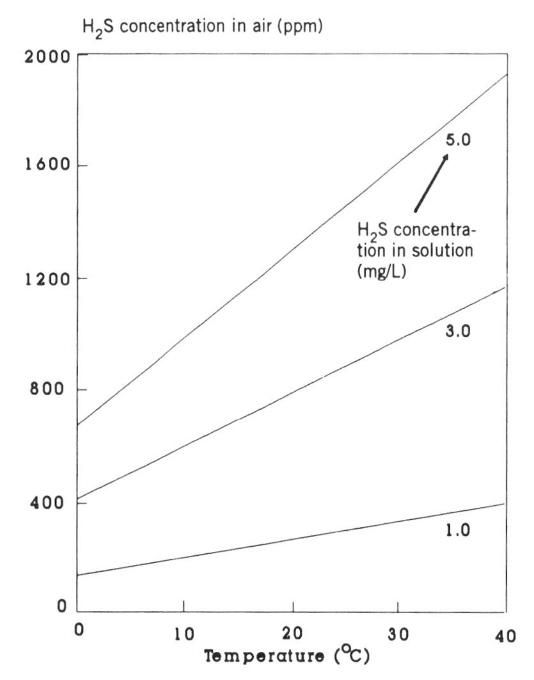

Fig. 14.9. Effect of temperature and dissolved hydrogen sulfide on H_2S concentration in air. Adapted from Bowker et al. (1989).

14.8.1.2. *Mercaptans*

The most important volatile organic compounds that contribute to odor problems in wastewater treatment plants are the mercaptans (e.g., ethyl mercaptan, methyl mercaptan, isopropyl mercaptan), organic sulfides (e.g., diallyl sulfide, dimethyl sulfide, methyl isopropyl sulfide, diethyl sulfide, methyl pentyl sulfide), polysulfides (e.g., dimethyl disulfide, methyl ethyl disulfide), and thiophenes (e.g., alkylthiophenes) (Forster and Wase, 1987; van Langenhove et al., 1985). Some characteristics of odorous sulfur compounds are displayed in Table 14.3 (U.S. EPA, 1985).

14.8.1.3. *Other Compounds*

Some of the other compounds that contribute to odor problems are organic acids, phenol, *p*-cresol (van Langenhove et al., 1985).

14.8.2. Preventive and Treatment Methods for Odor Control in Wastewater Treatment Plants

14.8.2.1. *Preventive Measures*

Several wastewater process units or operations (e.g., preliminary treatment, primary treatment, trickling filters, rotating biological contactors, activated sludge) can be the source of odors if improperly designed and maintained (e.g., poor housekeeping, insufficient aeration). Several operating practices can help prevent odors in wastewater treatment plants. These preventive measures include, for example, mechanical cleaning of collection systems, frequent washing to remove grit and organic debris, frequent scraping to remove scums and grease, keeping vents clear in trickling filters to keep biofilms under aerobic conditions, and reduce foam spray in aeration tanks. In activated sludge systems tank walls, air pipes, and diffusers should be kept clean (Bowker et al., 1989; Henry and Gehr, 1980; U.S. EPA, 1985).

Malodorous production of gas can also be prevented by addition of chlorine, hydrogen peroxide, potassium permanganate or sodium nitrate. Another preventive measure is the covering of odor-producing units and treatment of odorous air prior to release (Bowker et al., 1989).

14.8.2.2. *Treatment Methods*

Wastewater odors can be treated prior to their release into the atmosphere. The various treatment methods are the following.

Wet scrubbers. The odorous gases are contacted with a solution that may contain an oxidizing agent such as chlorine, potassium permanganate, hydrogen peroxide, or ozone. The treated air is then discharged to the atmosphere. Wet scrubbers remove more than 90% of gases.

Combustion of odors. This is done by a direct flame process at temperatures varying from approximately 500°C to more than 800°C or by catalytic oxidation (the catalyst may be platinum or palladium) at lower temperatures in the range of 300–500°C (Bowker et al., 1989; Henry and Gehr, 1980; Pope and Lauria, 1989).

Adsorption to activated carbon and other media. Activated carbon offers a large surface area (up to 950 m^2/g of carbon) for the nonselective adsorption of malodorous gases. It is regenerated by thermal or chemical treatment. A 1-kg amount of activated carbon can treat 276–735 m^3 of air (Huang et al., 1979). Activated carbon is specifically used in the rendering and food industries (Henry and Gehr, 1980). Other adsorbents are activated alumina impregnated with potassium permanganate, mixtures of wood chips with iron oxide, or a chelated iron adsorbent. The efficiency of the latter was demonstrated under field conditions in a wastewater treatment plant in Hawaii (Mansfield et al., 1992).

Ozone contactors. Odor-causing compounds are readily oxidized with ozone. Hydrogen sulfide, methyl mercaptan, and amines are oxidized to sulfur, methyl sulfonic acid, and amine oxides, respectively.

Microbiological methods. Soil/compost filters ("Bodenfilters") consist of a network of perforated PVC pipes buried 1–3 m below the soil or compost surface (Fig. 14.10) (U.S. EPA, 1985; Pomeroy, 1982). As gases flow upward through soil macropores, they are adsorbed by the soil and compost matrix and rapidly oxidized by microbiological or chemical means (Bohn and Bohn, 1988; Rands et al., 1981). Odorous gas oxidation by microbial action appears to be the primary means for odor reduction in soil/compost beds (Carlson and Leiser, 1966). Adsorption of gases to soil varies with soil type (sandy loams are typically used) and gas type. Gas retention is relatively high in soils with a high content of clay minerals and humic substances. The soil must be kept moist and warm to maintain microbial activity and thus achieve good performance. Anaerobic conditions developing as a result of excess water inside the bed should be avoided. This can be accomplished by covering the bed or by draining the excess water (U.S. EPA, 1985).

Some data are available on the identity of the microorganisms responsible for the degradation of methyl sulfides and hydrogen sulfide. Methyl sulfide, dimethyl sulfide, and dimethyl disulfide can be degraded by bacteria belonging to the genera *Thiobacillus* and *Hyphomicrobium* (de Bont et al., 1981; Kanagawa and Kelly, 1986; Sivela and Sundman, 1975). These compounds, along with methylmercaptans are efficiently removed from contaminated air by oxidation to sulfate with a culture of *Thiobacillus thioparus* (Kanagawa and Mikami, 1989).

Installation and operating costs for soil/compost beds are considerably lower than for incin-

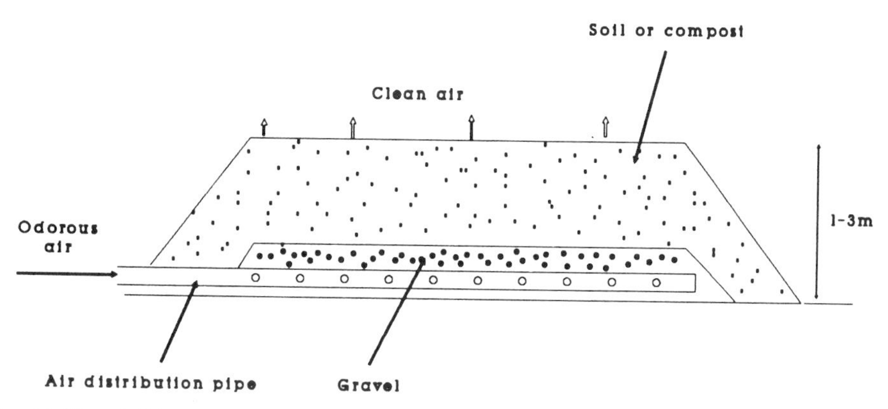

Fig. 14.10. Soil/compost filter for odor control. From U.S. EPA (1985), with permission of the publisher.

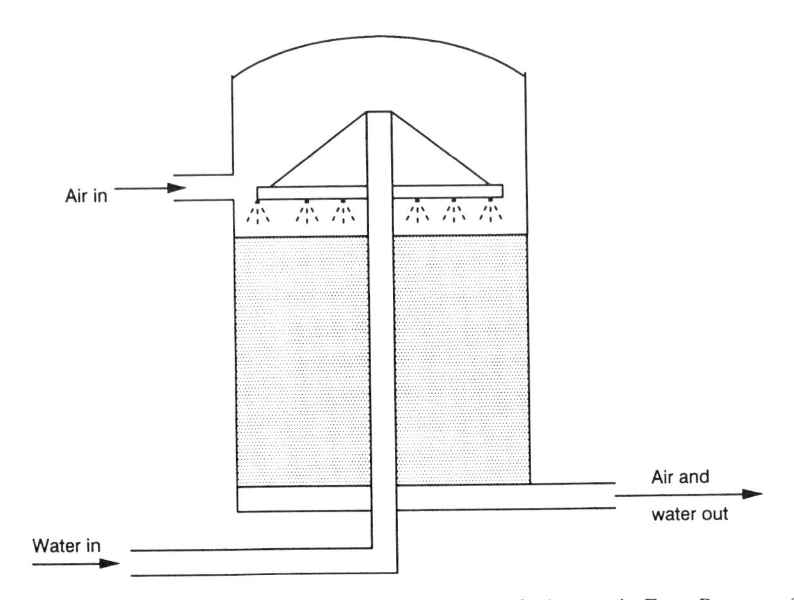

Fig. 14.11. Deodorizing towers for the biological treatment of odorous air. From Pomeroy (1982), with permission of the publisher.

eration or wet scrubbing. The practice of using soil beds for the treatment of gases is environmentally safe because soils do not adsorb more gas than they can handle, thus avoiding groundwater pollution.

Trickling filters are also efficient systems for odor removal from wastewater. Several designs are available for this purpose. Figure 14.11 shows deodorizing towers for treating odorous air (Pomeroy, 1982). Biological treatment of odorous air can also be accomplished by diffusion into an activated sludge aeration tank. However, some gases (e.g., hydrogen sulfide) may corrode the air compressors.

Other microbiological methods have also been suggested to remove H_2S, a foul-smelling and dangerous gas. Bioreactor columns loaded with attached photosynthetic bacteria (e.g., *Chlorobium*) remove up to 95% of H_2S from anaerobic digester effluent (Kobayashi et al., 1983). Photosynthetic bacteria, especially members of chromatiaceae (purple sulfur bacteria) and chlorobiaceae (green sulfur bacteria) are known to use H_2S as an electron donor in photosynthesis and oxidize it to elemental sulfur and sulfate (Pfenning, 1978).

14.9. FURTHER READING

Applebaum, J., N. Gutman-Bass, M. Lugten, B. Teltsch, B. Fattal, and H. I. Shuval. 1984. Dispersion of aerosolized enteric viruses and bacteria by sprinkler irrigation with wastewater. Monogr. Virol. 15: 193–210.

Cox, C.S. 1987. *The Aerobiological Pathway of Microorganisms*. Wiley, Chichester, U.K.

Mohr, A.J. 1991. Development of models to explain the survival of viruses and bacteria in aerosols, pp. 160–190, in: *Modeling the Environmental Fate of Microorganisms*, C.J. Hurst, Ed. American Society for Microbiology, Washington, DC.

Moore, B.E., D.E. Camman, C.A. Turk, and C.A. Sorber. 1988. Microbial characterization of municipal wastewater at a spray irrigation site: The Lubbock infection surveillance study. J. Water Pollut. Control Fed. 60: 1222–1230.

U.S. EPA. 1982. *Estimating Microorganisms Densities in Aerosols from Spray Irrigation of Wastewater*, EPA 600/9-82-003. Center for Environmental Research Information, Cincinnati, OH.

U.S. EPA. 1985. *Odor and Corrosion Control in Sanitary Systems and Treatment Plants*. Report # EPA/625/1-85-018.

15

MICROBIOLOGICAL ASPECTS OF DRINKING WATER TREATMENT AND DISTRIBUTION

15.1. INTRODUCTION

Drinking water safety is a worldwide concern. Contaminated drinking water has the greatest impact on human health worldwide (Geldreich, 1990). Drinking nontreated or improperly treated water is a major cause of illness in developing countries. For example, in 1980, 25,000 persons per day died worldwide as a result of consumption of contaminated water. Twenty five percent of hospital beds were occupied by people who had become ill after consuming contaminated water (U.S. EPA, 1989e; World Health Organization, 1979). In the United States a retrospective epidemiological study showed that waterborne disease outbreaks are due to the consumption of untreated and inadequately treated water or to subsequent contamination in the distribution network (Fig. 15.1) (Lippy and Waltrip, 1984). Ingestion of even low levels of pathogens, particularly viruses, may pose some risk of infection, clinical illness, or even mortality to susceptible populations (Gerba and Haas, 1986) (see Chapter 4). Advances in drinking water research, followed by the establishment of multiple barriers against microbial pathogens and parasites, have significantly increased the safety of the water we drink daily, particularly in industrialized nations.

Communities obtain their potable water from surface or underground sources. In the United States, surface waters are used as source water by an estimated 6,000 community water systems and serve a population of approximately 155 million people (Craun, 1988; Federal Register, 1987). Surface waters are often contaminated by domestic wastewater, storm water runoff, discharges from food-processing plants, and resuspension of microbial pathogens and parasitic cysts and ova that have accumulated in bottom sediments. In some areas, these surface waters are practically "diluted wastewaters," and problems arise when upstream communities discharge pathogen-laden wastewater effluents into surface waters that become drinking water supplies to downstream communities. Contamination of groundwaters is also well documented (see Chapter 19). Concern over these subsurface waters stems from the fact that they are often consumed without any treatment.

In this chapter, we will cover the microbiological and public health aspects of water treatment and distribution with particular emphasis on the formation of biofilms in distribution lines.

15.2. OVERVIEW OF PROCESSES INVOLVED IN WATER TREATMENT PLANTS

Water contains several chemical and biological contaminants that must be removed efficiently in order to produce drinking water that is safe and

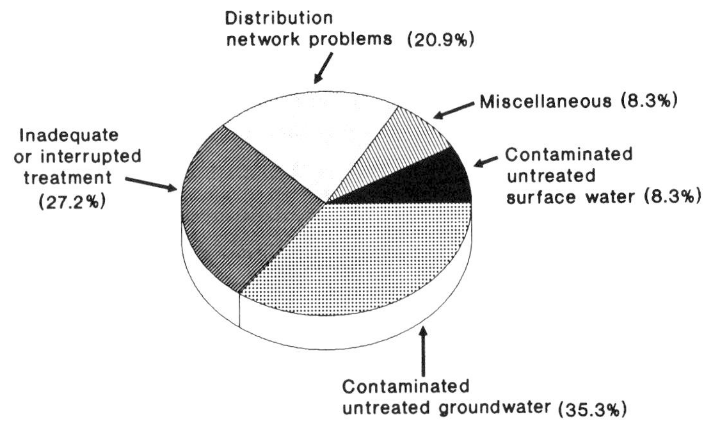

Fig. 15.1. Percentage distribution of waterborne disease outbreaks caused by deficiency in public water systems. From Lippy and Waltrip (1984), with permission of the publisher.

aesthetically pleasing to the consumer. The chemical contaminants include nitrates, heavy metals, radionuclides, pesticides, and other xenobiotics. The finished product must also be free of microbial pathogens and parasites, turbidity, color, taste, and odor. To achieve this goal, raw water (surface water or groundwater) is subjected to a series of physicochemical processes that will be described in detail in Section 15.3. Disinfection alone is sufficient if the raw water originates from a protected source. More commonly, several processes are used to treat water. Disinfection may be combined with coagulation, flocculation, and filtration. Additional treatments to remove specific compounds include preaeration and activated carbon treatment. The treatment train depends on the quality of the source water under consideration.

There are two main categories of water treatment plants.

Conventional filter plants. Conventional plants (Fig. 15.2) serve an estimated population of 108 million consumers in the United States (U.S. EPA, 1989e). The leading processes in this type of plant are coagulation and filtration. The raw water is rapidly mixed with a coagulant (aluminum sulfate, ferric chloride, ferric sulfate). Following coagulation, the flocs that are produced are allowed to settle in a clarifier. Clarified effluents are then passed through sand or diatomaceous earth filters. The water is finally disinfected prior to distribution.

Softening plants. The leading process in these plants is softening, which helps remove hardness due to the presence of calcium and magnesium in water and results in the formation of calcium and magnesium precipitates. Following settling of the precipitates, the water is filtered and disinfected.

15.3. PROCESS MICROBIOLOGY AND FATE OF PATHOGENS AND PARASITES IN WATER TREATMENT PLANTS

In water treatment plants, microbial pathogens and parasites can be physically removed by processes such as coagulation, precipitation, filtration, and adsorption, or they can be inactivated by disinfection or by the high pH that results from water softening (Engelbrecht, 1983).

The types of pathogens and parasites of most concern in drinking water are the following (AWWA, 1987) (Chapter 4 should be consulted for more details on these pathogens and parasites):

Viruses. Viruses are occasionally detected in drinking water from conventional water treatment plants that meet the standards currently used to judge treatment efficiency. For example,

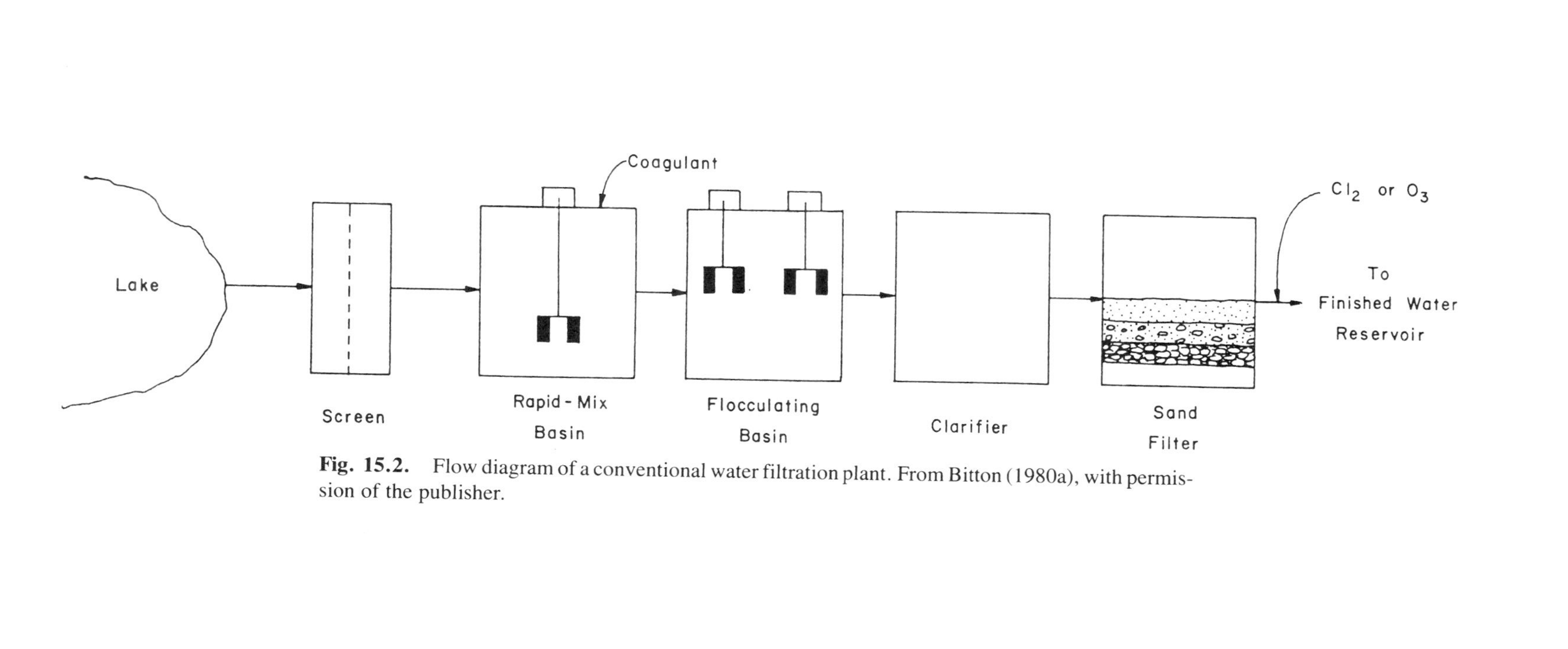

Fig. 15.2. Flow diagram of a conventional water filtration plant. From Bitton (1980a), with permission of the publisher.

enteroviruses have been detected at levels ranging from 3 to 20 viruses per 1,000 L in finished drinking water from water treatment plants that include prechlorination, flocculation, sedimentation, sand filtration, ozonation, and final chlorination (Payment, 1989a). However, at a water filtration plant in Laval, Canada, no virus was detected in 162 finished water samples (1,000–2,000 L per sample), although coliphage and *Clostridia* were detected in finished water (Payment, 1991). Virus removal also was investigated at a full-scale water treatment plant in Mery sur Oise, France. The treatment train in this plant included the following treatments: a preozonation step followed by coagulation–flocculation, sand filtration, ozonation, activated carbon, and a final ozonation step. No virus was detected in the final product (Table 15.1) (Joret et al., 1986).

Isolation of enteric viruses (polioviruses, coxsackie A and B viruses, echoviruses, rotaviruses, reoviruses, adenoviruses) has been documented worldwide in partially treated drinking water and in marginally operated and even apparently well-operated water treatment plants (Bitton et al., 1986). Their finding may be due to relatively drastic changes in source water quality or to equipment and process failure.

Giardia lamblia. The methodology for detecting this parasite is under development (see Chapter 4), but the skill to routinely monitor its presence in drinking water is not yet available in water treatment plants.

Opportunistic pathogens. These are waterborne pathogens (e.g., *Ps. putida, Alcaligenes, Acinetobacter,* and *Flavobacterium* species) that cause secondary infections in hospitals and, potentially, among consumers.

Legionella. This pathogen is an example of a nonenteric microorganism that can be transmitted by inhalation of drinking water aerosols from shower heads or humidifiers. Nosocomial Legionnaire's disease may be contracted by exposure to *Legionella* from the water distribution system in hospitals (Best et al., 1984b).

As mentioned previously, several unit processes and operations are used in water treatment plants to produce microbiologically and chemically safe drinking water. The extent of treatment depends on the source of raw water; surface waters generally require more treatment than groundwaters. The unit processes designed for water treatment, with the exception of the disinfection step, do not specifically address the destruction or removal of parasites or bacterial and viral pathogens.

15.3.1. Storage of Raw Water

Raw water can be stored in reservoirs to minimize fluctuations in water quality. Storage can affect the microbiological quality of the water. The reduction of pathogens, parasites, and indicator microorganisms during storage is variable and is influenced by a number of factors such as temperature, sunlight, sedimentation, and adverse biological phenomena such as predation, antagonism, and lytic action of bacterial phages. Temperature is a significant factor influencing pathogen survival in reservoirs. It appears that,

TABLE 15.1. Virus Concentrations (PFU/1,000 L) at Various Stages of a Water Treatment Plant[a]

Sampling event	Stored water	Sedimentation	Sand filtration	Ozonation
1	10.4	<25	9.1	<1
2	6.1	132	<1	<1
3	100	75	<2	<2
4	90	5	<1	<1
5	10	20	3	<1
6	30.7	10	5	<1

[a]Adapted from Joret et al. (1986).

under suitable conditions, storage can lead to approximately one-log reduction of organisms. However, higher reductions have been observed.

15.3.2. Prechlorination

A prechlorination step is sometimes included to improve the performance of unit processes (e.g., filtration, coagulation–flocculation) or to oxidize color-producing substances such as humic acids. Although prechlorination reduces somewhat the levels of pathogenic microorganisms, its use is being questioned because of increased chances of forming trihalomethanes (see Chapter 6).

15.3.3. Coagulation–Flocculation

Coagulation involves the destabilization of colloidal particles (e.g., mineral colloids, microbial cells, virus particles) by coagulants (Al and Fe salts) and sometimes by coagulant aids (e.g., activated silica, bentonite, polyelectrolytes, starch). The most common coagulants are alum, ferric chloride, and ferric sulfate. Following slow mixing, the colloidal particles form flocs that are large enough to allow rapid settling (Fig. 15.3) (Williams and Culp, 1986). The pH of the water is probably the most significant factor affecting coagulation. Other factors include turbidity, temperature, and mixing regime. Coagulation is the most important process used in water treatment plants for clarification of colored and turbid waters.

Removal of bacteria, although variable, may exceed 90%. Coagulation removes 74%–99.4% of *E. coli* and coliforms (Sobsey and Nagy, 1984). Under laboratory conditions, coagulation–flocculation is effective in removing 90%–99% of viruses from water (Table 15.2) (Bitton, 1980a). However, removal of virus under field conditions is generally lower than under controlled laboratory conditions. In a pilot water treatment plant treating water from the Seine River in France, indigenous virus removal by coagulation–flocculation varied between 31% and 90%, with a average removal of 61%. This shows that virus removal obtained with laboratory strains is much higher than with indigenous viruses (Joret et al., 1986). Virus removal is not influenced by turbidity or enhanced by the addition of a nonionic coagulant aid (Rao et al., 1988). Removal of protozoan cysts by coagulation–sedimentation may also exceed 90%.

Coagulation is sometimes improved by using coagulant aids such as polyelectrolytes, bentonite, or activated silica. Polyelectrolytes help form large flocs that settle out rapidly. Their concentration is an important parameter, since excessive dosage can inhibit flocculation.

Coagulation merely transfers pathogenic microorganisms from water to the flocculated material, which is incorporated into a sludge that must thus be disposed of properly.

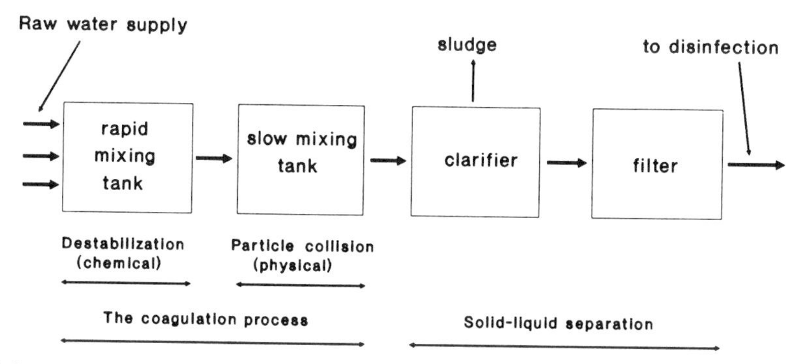

Fig. 15.3. Schematic diagram of a coagulation process. From Williams and Culp (1986), with permission of the publisher.

TABLE 15.2. Removal of Viruses by Coagulation–Sedimentation[a]

Coagulant	Concentration of coagulant (ml/L)	Clay (mg/L)	Virus	Removals	
				Virus (%)	Turbidity (%)
Alum	10	50	Poliovirus 1	86	96
	25.7	120	Phage T4	98	99
	25.7	120	Phage MS2	99.8	98
Ferric sulfate	40	—	Poliovirus 1	99.8	—
	40	—	Phage T4	99.8	—
Ferric chloride	60	50	Poliovirus 1	97.8	97.5
	60	100	Poliovirus 1	93.3	97.8
	60	500	Poliovirus 1	99.7	99.7

[a]From Bitton (1980a), with permission of the publisher.

15.3.4. Water Softening

Hardness is caused by the presence of calcium and magnesium compounds in the water. There are two categories of hardness: carbonate hardness, which is due to bicarbonates of Ca and Mg, and noncarbonate hardness, which is due to Ca and Mg chlorides. Hardness is responsible for increased consumption of soap and formation of scale in pipes. Water softening is the removal of Ca and Mg hardness by the lime-soda process or by ion-exchange resins (Bitton, 1980a). The lime–soda process consists of adding lime (calcium hydroxide) and soda ash (sodium carbonate) to the water. The carbonate hardness is removed by the following chemical process:

$$\underset{\text{Ca bicarbonate}}{Ca(HCO_3)_2} + \underset{\text{lime}}{Ca(OH)_2} \rightarrow \underset{\text{Ca carbonate}}{2CaCO_3} + 2H_2O \quad (15.1)$$

$$\underset{\text{Mg bicarbonate}}{Mg(HCO_3)_2} + \underset{\text{lime}}{2Ca(OH)_2} \rightarrow \underset{\text{Mg hydroxide}}{Mg(OH)_2} + CaCO_3 + 2H_2O \quad (15.2)$$

The high pH (>11) generated by water softening with lime leads to an effective inactivation of bacterial and viral pathogens. Poliovirus type 1, rotavirus, and hepatitis A virus are effectively removed (>95% removal) during water softening (pH = 11) (Table 15.3) (Rao et al., 1988). Bacterial pathogens are also efficiently reduced following liming to reach a pH that exceeds 11. The inactivation rate is temperature-dependent (Riehl et al., 1952; Wattie and Chambers, 1943).

Microbial removal during water softening is due (1) to microbial inactivation at detrimentally high pHs (pH ≥ 11) from loss of structural integrity or inactivation of essential enzymes and (2) to physical removal of microorganisms by adsorption to positively charged magnesium hydroxide flocs ($CaC0_3$ precipitates are negatively charged and do not adsorb microorganisms).

As regards ion-exchange resins, Ca and Mg are removed from water by exchange with Na present on the exchange sites. Viruses are well removed by anion-exchange resins but not as much by cation-exchange resins.

15.3.5. Filtration

Filtration is defined as the passage of fluids through porous media to remove turbidity (suspended solids such as clays, silt particles, microbial cells) and flocculated particles. This process depends on the filter medium, concentration and type of solids to be filtered out, and the operation of the filter.

Filtration is one of the oldest processes used for water treatment. In 1685, an Italian physi-

TABLE 15.3. Removal of Virus by Water Softening[a,b]

Virus[c]	Virus input PFU/500 mL	Percentage removal		
		Virus	Total hardness	Turbidity
PV	3.9×10^6	96	54	70
RV	7.8×10^5	>99	47	60
HAV	6.8×10^8	>97	76	62

[a]Adapted from Rao et al. (1988).
[b]pH—adjusted to 11.0.
[c]PV = poliovirus; RV = rotavirus; HAV = hepatitis A virus.

cian, Luc Antonio Porzio, conceived a filtration system for protecting soldiers' health in military installations. This process has been in use since and has contributed greatly to the reduction of waterborne diseases such as typhoid fever and cholera. An examination of waterborne disease outbreaks around the world clearly shows that filtration has been instrumental as a barrier against pathogenic microorganisms and has largely contributed to the reduction of waterborne diseases. Figure 15.4 shows the dramatic reduction of the typhoid fever death rate in Albany, New York, following adoption of sand filtration and then of chlorination about a decade later (Logsdon and Lippy, 1982; Willcomb, 1923).

15.3.5.1. *Slow Sand Filtration*

Although more popular in Europe than in the United States, slow sand filters mostly serve communities of less than 10,000 population because capital and operating costs are lower than for rapid sand filters (Slezak and Sims, 1984). In the United States the first slow sand filter was installed in Lawrence, Massachusetts, to remove *Salmonella typhi* from water.

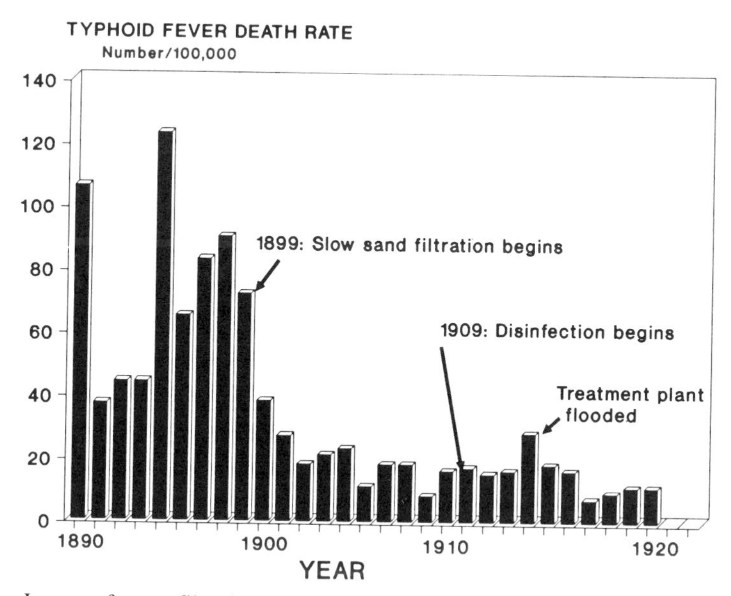

Fig. 15.4. Impact of water filtration and chlorination on typhoid fever death rate in Albany, N.Y. From Logsdon and Lippy (1982), with permission of the publisher.

Slow sand filters (Fig. 15.5) contain a layer of sand (60–120 cm deep) supported by a graded gravel layer (30–50 cm deep). The sand grain size varies between 0.15 and 0.35 mm and the hydraulic loading range is between 0.04 and 0.4 m/hr (Bellamy et al., 1985b).

Biological growth inside the filter includes a wide variety of organisms, including bacteria, algae, protozoa, rotifers, microtubellaria (flatworms), nematodes (round worms), annelids (segmented worms), and arthropods (Duncan, 1988). The buildup of a biologically active layer, called the schmutzdecke, occurs during the normal operation of a slow sand filter. The top layer is composed of biological growth and filtered particulate matter. This leads to a head loss across the filter, a problem corrected by removing or scraping the top layer of sand. The length of time between scrapings depends on the turbidity of the raw water and varies from 1–2 weeks to several months (Logsdon and Hoff, 1986). Scraping is followed by replenishing of the filter bed with clean sand, an operation called resanding. There is sometimes a deterioration of water quality for some days after scraping, but it later improves during the ripening period (Cullen and Letterman, 1985). However, some investigators (Logsdon and Lippy, 1982) have not observed a deterioration after scraping. Bacterial activity in the schmutzdecke helps remove assimilable organic compounds, some of which are precursors of chlorinated organics such as trihalomethanes (Collins et al., 1992; Eightmy et al., 1992; Fox et al., 1984).

The removal efficiency for coliform bacteria, but not of *Giardia* cysts, is influenced by temperature, sand grain size, and filter depth (Bellamy et al., 1985a, 1985b). Figure 15.6 (Bellamy et al., 1985a) displays the effect of filter depth and sand grain size on total coliform removal. Removal of *Giardia* and total coliforms exceeds 99% even at the highest loading rate of 0.4 m/hr. Pilot plant studies have shown that slow sand filtration achieves 4- to 5-log reductions of coliforms (Fox et al., 1984). A survey of slow sand filters in the United States showed that most of the plants reported coliform levels of 1 per 100 ml or less (Slezak and Sims, 1984). The microbial community developing on the sand filter greatly influences the removal of bacteria, *Giardia,* and turbidity by slow sand filters (Bellamy et al., 1985b; Cleasby et al., 1984; Huisman and Wood, 1974).

With regard to viruses, a removal exceeding

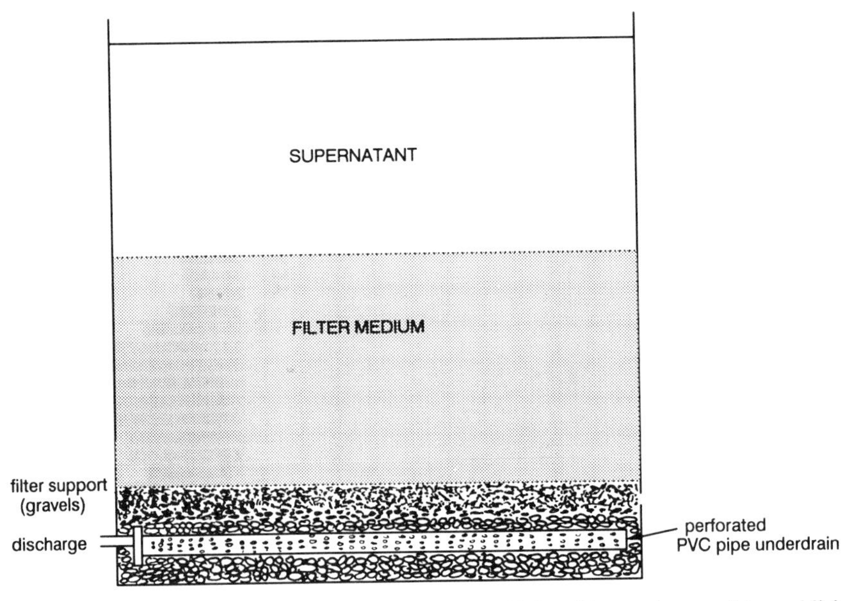

Fig. 15.5. Slow sand filter. From Slezak and Sims (1984), with permission of the publisher.

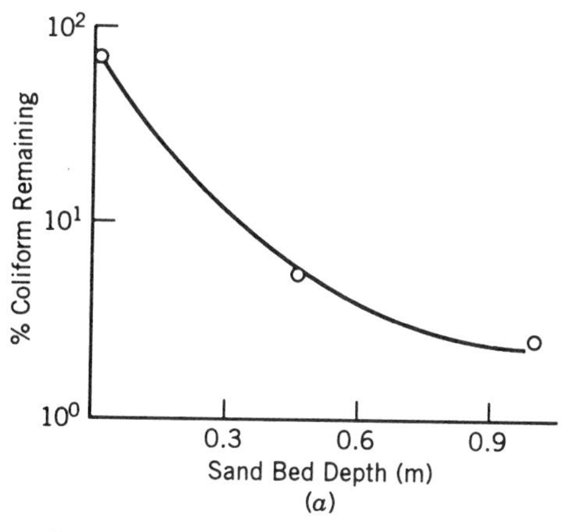

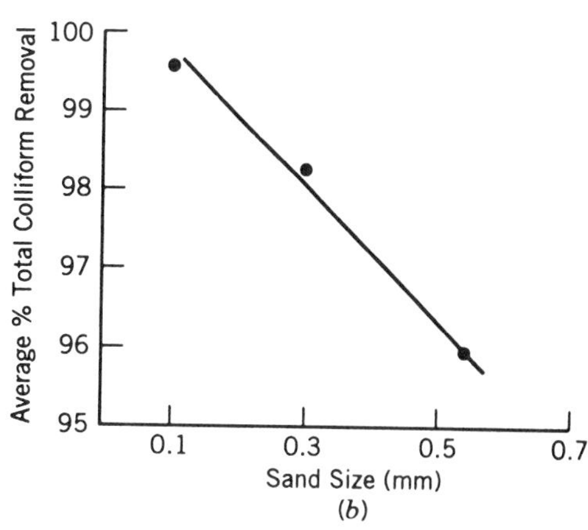

Fig. 15.6. Effect of sand bed depth **(a)** and sand grain size **(b)** on the removal of total coliforms by slow sand filtration. From Bellamy et al. (1985a), with permission of the publisher.

99.999% was observed in a slow sand filter operated at 11°C at a rate of 0.2 m/hr (Poynter and Slade, 1977). An established filter seems to remove virus better than a new filter (Wheeler et al., 1988). The factors controlling virus removal by slow sand filters include filter depth, flow rate, temperature, and the presence of a well-developed biofilm on the filter.

15.3.5.2. Rapid Sand Filtration

A rapid sand filter consists of a layer of sand supported by a layer of anthracite, gravel, or calcite (Bitton, 1980a). The rapid sand effluent is collected by an underdrain system. Rapid sand filters are operated at filtration rates of 5–24 $m^3/hr/m^2$ as compared with 0.1–1 $m^3/hr/m^2$ for slow sand filters (Huisman and Wood, 1974; Logsdon and Hoff, 1986). These filters are periodically cleaned by backwashing (i.e., reversing the flow) at a sufficient flow rate to allow a thorough cleaning of the sand.

This process appears to be less effective for the removal of bacteria, viruses, and protozoan cysts unless it is preceded by coagulation–flocculation. The removal of *Salmonella* and *Shigella* by filtration and coagulation is similar to that of coliforms. *Giardia* cysts pass through water treatment plants that use direct filtration without any chemical pretreatment. The breakthrough is more pronounced during winter months when temperature is below 5°C (Hibler and Hancock, 1990). The combination of filtration with coagulation–flocculation is particularly important in respect to removal of *Giardia* cysts from water (Logsdon et al., 1981). For low-turbidity water, proper chemical coagulation of the water prior to rapid sand filtration is necessary to achieve a good removal of turbidity and *Giardia* cysts. It has been proposed that removal of turbidity could serve as a surrogate indicator of removal of *Giardia* cysts in low-turbidity (<1 NTU) water (Al-Ani et al., 1986). However, *Giardia* and *Cryptosporidium* monitoring in filtered drinking water samples indicated that removal of turbidity was a good predictor of removal of *Cryptosporidium* oocysts but not *Giardia* cysts (LeChevallier et al., 1991a). Filtration without prior addition of coagulants removes approximately 90% of *Cryptosporium* oocysts. If the water source is a surface water, then a dual barrier is necessary to prevent contamination with *Giardia lamblia* cysts. This dual barrier consists of coagulation–sand filtration to retain cysts and turbidity, and disinfection (AWWA, 1987).

Entamoeba histolytica cysts are also effectively removed when rapid sand filtration is preceded by coagulation (Baylis *et al.*, 1936). Since sand particles are essentially poor adsorbents of viruses, pathogen removal by sand filtration is variable and often low. However, coagulation prior to sand filtration removes more than 99% of viruses (Table 15.4) (Bitton, 1980a). A coagulation, settling, and filtration process removes more than 1 log of hepatitis virus, simian rotavirus (SA-11), and poliovirus (Table 15.5) (Rao et al., 1988). A pilot plant using Seine River water that contained 190–1,420 PFU/1,000 L removed 1–2 logs of viruses after coagulation–flocculation followed by sand filtration (Joret et al., 1986). A preozonation step (0.8 ppm ozone level) greatly improved virus removal, which was increased to 2–3 logs. Helminths form relatively large eggs, which are effectively removed by sedimentation, coagulation, and filtration.

Thus, optimum coagulation in water treatment plants is essential for control of parasites and pathogens. Other parameters that adversely affect filter operation are sudden changes in water flow rates, interruptions in chemical feed, inadequate filter backwashing, and the use of clean sand that has not been allowed to undergo a ripening period (Logsdon and Hoff, 1986; Ongerth, 1990).

Dual-stage filtration (DSF) is an alternative to rapid sand filtration for small water treatment plants. This process consists of chemical coagulation followed by a filter assembly consisting of two tanks, a depth clarifier, and a depth filter. At a flow rate of 10 GPM/ft^2, this treatment removed more than 99% of *Giardia* cysts from water with an effluent turbidity less than 1 NTU (Horn et al., 1988).

15.3.5.3. Diatomaceous Earth Filtration

Diatomaceous earth (DE) or diatomite is made of remains of siliceous shells of diatoms. DE filtration includes two steps (Lange et al., 1986).

Precoating. A precoat of DE (3–5 mm) is applied on a porous filter septum that serves as a support for the buildup of a 1/8-in precoat of filter medium (Fig. 15.7) (Bitton, 1980a) to form a filter cake.

Filtration. Raw water, treated with additional DE (body feed), is passed through the filter cake to keep the filter running for longer periods of time.

At the end of a run, the filter cake is removed and replaced by fresh diatomite.

The performance of a DE filter is affected by DE grade, thickness of the DE cake, size of microorganisms, and chemical conditioning of DE. Chemical coating of DE with Al and Fe salts or cationic polyelectrolytes improves the microbe-removing efficiency of this material. A cationic polyelectrolyte, at a concentration of 0.15 ppm, brings about 100% removal of poliovirus, as compared with 62% without polyelectrolyte (Bitton, 1980a). Virus removal by DE is also enhanced by *in situ* precipitation of metallic salts (Al, Fe) (Farrah et al., 1991). The particle size of the diatomaceous earth used for precoating and body feed also influences the efficiency of this material for removing microorganisms (Logsdon and Lippy, 1982; Logsdon and Hoff, 1986). Removal of bacteria is influenced by DE grade, hydraulic loading rate, influent bacteria concentration, and filtration time. Chemical coating with alum and coagulants greatly improves their removal (Table 15.6) (Lange et al., 1986; Hunter et al., 1966; Schuler and Ghosh, 1990).

DE is generally effective in removing cysts of parasites. The removal of *Giardia lamblia* cysts and *Cryptosporidium* oocysts by DE exceeds 99% (Logsdon and Lippy, 1982; Schuler and

TABLE 15.4. Removal of Poliovirus by Rapid Sand Filters[a,b]

Treatment	% Removal
A. Sand filtration	1–50
B. Sand filtration + coagulation with alum	
1. Without settling	90–99
2. With settling	>99.7

[a]Adapted from Robeck et al. (1962).
[b]The flow rate was 2–6 gpm/ft^2.

TABLE 15.5. Removal of Virus by Coagulation–Settling–Sand Filtration[a]

Virus[b]	Viral assays, total PFU/200 L (percentage removed)		
	Input	Settled water	Filtered water
PV	5.2×10^7	1.0×10^6 (98)	8.7×10^4 (99.84)
RV	9.3×10^7	4.6×10^6 (95)	1.3×10^4 (99.987)
HAV	4.9×10^{10}	1.6×10^9 (97)	7.0×10^8 (98.6)

[a]Adapted from Rao et al. (1988).
[b]PV = poliovirus; RV = rotavirus; HAV = hepatitis A virus.

Ghosh, 1990). Furthermore, removal of *Cryptosporidium* is greatly improved upon addition of alum (Schuler and Ghosh, 1990). However, cyst removal is not influenced by DE grade, hydraulic loading rate, or temperature (Lange et al., 1986).

15.3.6. Activated Carbon

Activated carbon is an adsorbent, derived from wood, bituminous coal, or lignite, that is activated by a combustion process. It offers a large surface area (500–600 m^2/g) for the adsorption of taste and odor compounds, color, excess chlorine, toxic and mutagenic substances (e.g., chlorinated organic compounds, including trihalomethanes), trihalomethane precursors, pesticides, and substances that cause aftergrowth (Najm et al., 1991; van Puffelen, 1983).

Activated carbon may be used in the form of granular activated carbon (GAC), applied after sand filtration and prior to chlorination and in the form of powdered activated carbon (PAC), which has a smaller particle size than GAC and can be applied at various points in water treatment plants, mainly prior to filtration. It is less costly than GAC treatment and is applied only when needed.

Activated carbon has been used since antiquity but its microbiological aspects have been investigated only recently (LeChevallier and McFeters, 1990; Weber et al., 1978). The functional groups on the carbon surface help in adsorption of microorganisms, including bacteria

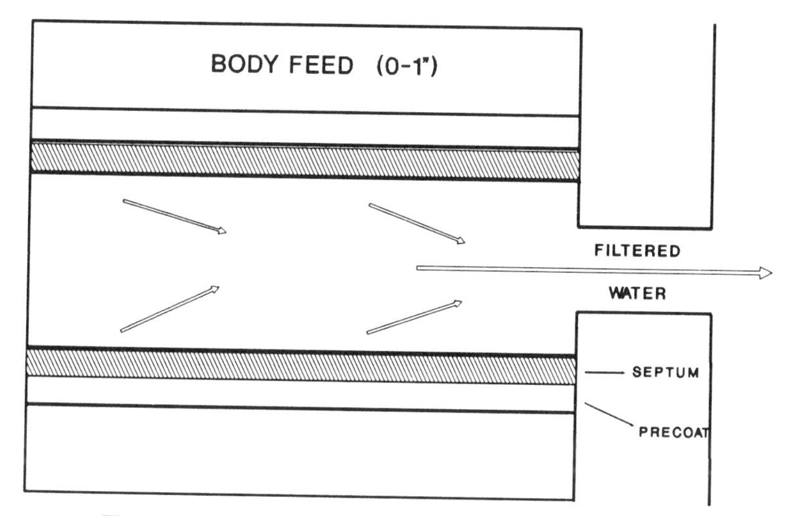

Fig. 15.7. Diatomite filter. Adapted from Baumann (1971).

TABLE 15.6. Alum Enhancement of Removal of Bacteria by Diatomaceous Earth[a]

Grade	Alum-to-DE ratio,[b] g alum / g DE	Body feed concentration, mg/L	Average percentage removals		
			Total coliform	Standard plate count	% Turbidity removal
B[c]	0.02	25			66.11
	0.04	25	99.02	95.02	86.41
	0.04	25	99.02	95.02	86.41
	0.05	25	99.86	98.56	98.38
D[d]	0.05	25[e]	98.01	79.31	79.06
	0.05	25	99.56	93.25	94.41
	0.05	50	96.33	99.57	98.61
	0.08	25	99.83	99.52	98.80

[a]Adapted from Lange et al. (1986).
[b]Alum coating ratio is grams of alum as $Al_2SO_4 \cdot 14.3\ H_2O$ slurry per gram of diatomaceous earth in slurry.
[c]C-545.
[d]C-503.
[e]Bodyfeed concentration was increased to 50 mg/L after 3 h of testing at 25 mg/L.

(rods, cocci, and filamentous bacteria), fungi, and protozoa. Scanning electron microscope observations have shown colonization of carbon surface by microorganisms (polysaccharide-producing bacteria, stalked protozoa) (Fig. 15.8). The dominant bacterial genera identified on GAC particles or in interstitial water were *Pseudomonas, Alcaligenes, Aeromonas, Acinetobacter, Arthrobacter, Flavobacterium, Chromobacterium, Bacillus, Corynebacterium, Micrococcus, Paracoccus,* and *Moraxella* (Camper et al., 1986; Wilcox et al., 1983). GAC may also harbor bacterial indicators and pathogens.

Activated carbon has the ability to remove organic materials that in turn serve as carbon substrates that support growth of bacteria in the filter matrix (De Laat et al., 1985; Servais et al., 1991). The magnitude of biological assimilation depends on various factors, including temperature, running time and intensity, and frequency of filter backwashing (van der Kooij, 1983). Removal of organics appears to result from microbial proliferation in the activated carbon column. Some of these bacteria, however, can produce endotoxins, which may enter the treated water. Treatment to enhance growth of bacteria on GAC produces biologically activated carbon (BAC), which has been recommended for increasing the removal of organics as well as extending the lifetime of GAC columns. Growth of bacteria in activated carbon columns can be enhanced by ozone, which makes organic matter more biodegradable. Micropollutants, such as phenol, can be degraded by biofilms grown on ozonated natural organic matter (DeWaters and DiGiano, 1990). The bacterial community in BAC is made mostly of aerobic, gram-negative, oxidase-negative, catalase-positive, motile rods collectively grouped as biotype 1. Other bacterial genera are *Pseudomonas, Acinetobacter, Enterobacter,* and *Moraxella*-like bacteria (Rollinger and Dott, 1987).

Pathogenic bacteria may be successful in colonizing mature GAC filters (Grabow and Kfir, 1990; LeChevallier and McFeters, 1985a). However, pathogenic and indicator bacteria can be inhibited by biofilm microorganisms on activated carbon, a phenomenon that is due to nutritional competitive inhibition or to the production of bacteriocin-like substances by the filter microbial community (Camper et al., 1985; LeChevallier and McFeters, 1985a; Rollinger and Dott, 1987).

Problems arise when bacteria or bacterial mi-

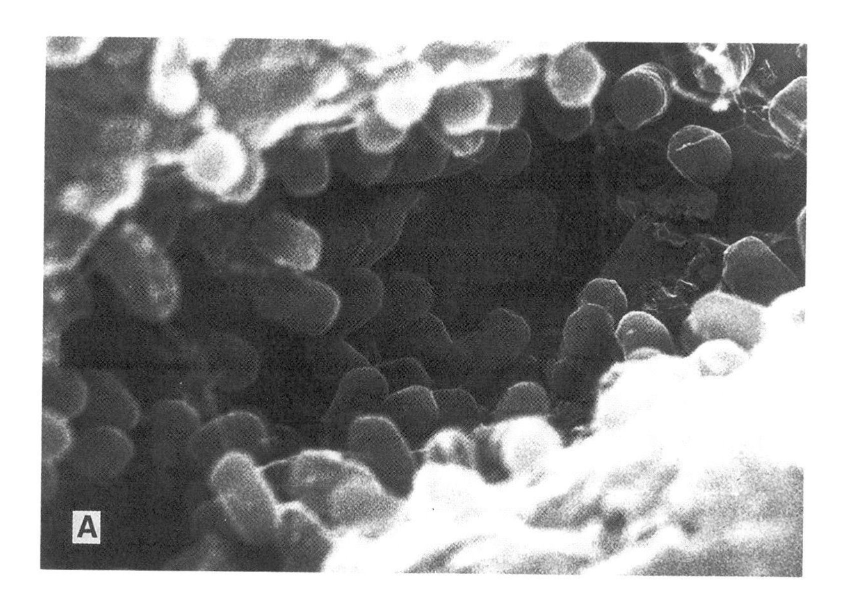

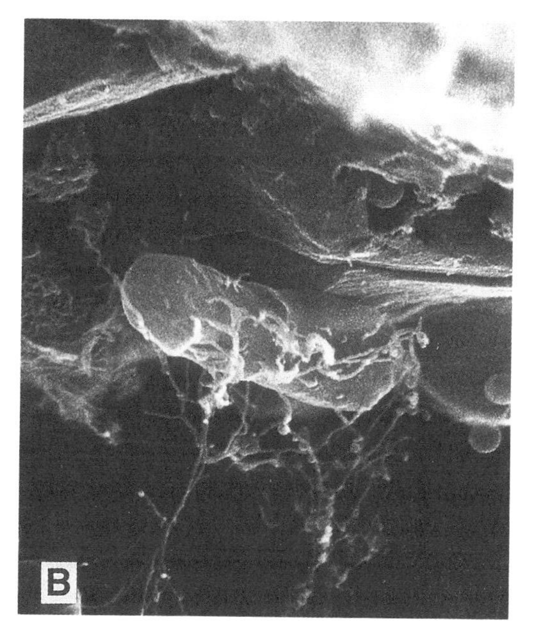

Fig. 15.8. Bacterial colonization of activated carbon. **A.** Bacteria growing in a crevice on activated carbon surface. **B.** Capsular material covering a bacterium attached to a carbon particle surface. Courtesy of M.W. LeChevallier.

crocolonies attached to carbon filters are sloughed off the filter or when bacteria-coated carbon particles penetrate the treatment system (Camper et al., 1986). An examination of 201 samples showed that heterotrophic bacteria and coliform bacteria are associated with the carbon particles. The attached bacteria display increased resistance to chlorination (Camper et al., 1986; LeChevallier et al., 1984a; Stewart et al., 1990). Figure 15.9 shows that *E. coli* and natu-

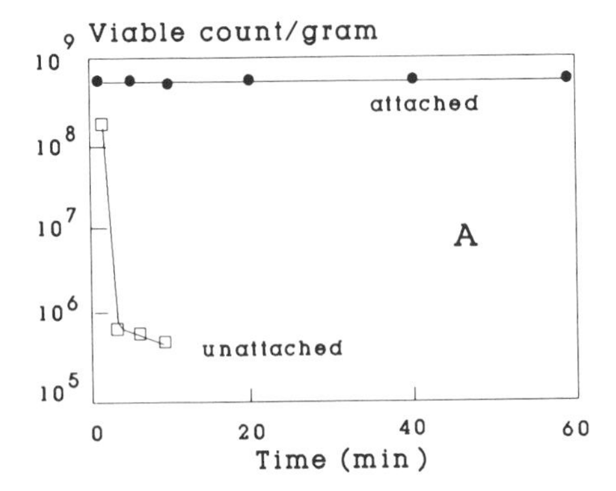

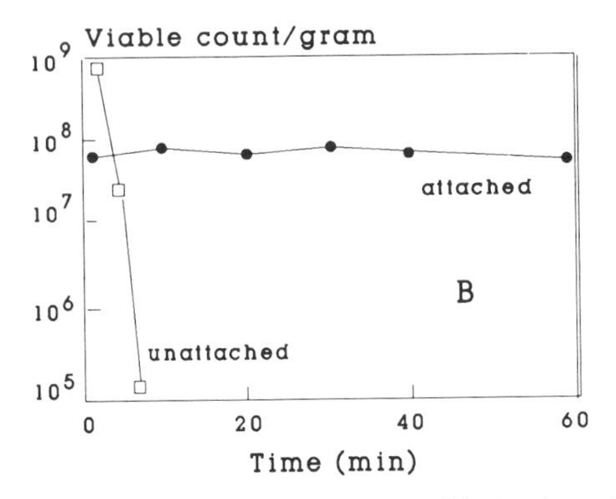

Fig. 15.9. Increased resistance to chlorination of bacteria attached to GAC. **A.** *E. coli.* **B.** Heterotrophic plate count. Adapted from LeChevallier et al. (1984a).

rally occurring heterotrophic bacteria attached to carbon particles are more resistant to chlorine than unattached bacteria (LeChevallier et al., 1984a). Monochloramine appears to control biofilm bacteria more efficiently than free chlorine or chlorine dioxide (LeChevallier et al., 1988). Operational variables that contribute to the release of particles from activated carbon beds include filter backwashing and increase in bed depth, applied water turbidity, and filtration rate (Camper et al., 1987). Thus, activated carbon particles entering the water distribution system can carry potential pathogens that are resistant to chlorination as well as nutrients that support microbial growth within the distribution system.

Viruses are adsorbed to activated carbon by electrostatic forces and the interaction is controlled by pH, ionic strength, and the organic matter content of the water. The competition of organics with viruses for the attachment sites on the carbon surface makes this material an unreliable sorbent for removing viruses from water (Bitton, 1980a).

The following are some possible disadvantages of carbon filters:

1. Production and release of endotoxins
2. Creation of anaerobic conditions inside filters with subsequent production of odorous compounds (e.g., H_2S)
3. Production of effluents with high colony counts
4. Occasional growth of zooplankton in carbon filters and their release in the filter effluents

15.3.7. Biological Treatment of Water

Water containing organic compounds as well as ammonia nitrogen is biologically unstable. The presence of organic carbon and nitrogen in water causes significant problems such as trihalomethane formation following disinfection with chlorine, taste and odor problems, regrowth of bacteria in distribution systems, and reduced bed life of granular activated carbon columns (Hozalski et al., 1992). Several European countries (e.g., France, Germany, Netherlands) and Japan include biological processes in the treatment train to remove biodegradable materials and thus obtain biologically stable water. The biological treatment (e.g., biofiltration involving biologically active GAC filters) is based on aerobic biofilm processes in regard to drinking water treatment. This type of treatment removes organic compounds (total and assimilable organic carbon), including trihalomethane precursors and xenobiotics (e.g., chlorinated phenols and benzenes), as well as NH_4 and trace elements such as iron and manganese (Bouwer and Crowe, 1992; LeChevallier et al., 1992; Manem and Rittmann, 1992; Rittmann, 1987, 1989; Rittmann and Snoeyink, 1984). Preozonation may enhance the biodegradability of organic compounds by biofilms (Hozalski et al., 1992).

15.3.8. Disinfection

This topic is covered in Chapter 6.

15.4. WATER DISTRIBUTION SYSTEMS

15.4.1. Introduction

Drinking water quality may deteriorate during storage and transport through water distribution pipes. This deterioration is due to the following (Sobsey and Olson, 1983):

1. Improperly built and operated storage reservoirs (which should be covered to prevent airborne contamination and to exclude animals)
2. Microbial regrowth in storage reservoirs
3. Taste and odor problems due to the growth of algae, actinomycetes, and fungi
4. Bacterial colonization of water distribution pipes

Drinking water may contain humic and fulvic acids as well as easily biodegradable natural organics such as carbohydrates, proteins and lipids. The presence of dissolved organic compounds in finished drinking water is responsible for several problems, among which are taste and odors, enhanced chlorine demand, trihalomethane formation, and bacterial colonization of water distribution lines (Allen et al., 1980; Bourbigot et al., 1984; Hoehn et al., 1980; Rook, 1974; Servais et al., 1991).

Biofilms develop at solids–water interfaces and are widespread in natural environments as well as in engineered systems. They are ubiquitous and are commonly found in trickling filters, rotating biological contractors, activated carbon beds, pipe surfaces, groundwater aquifers, aquatic weeds, tooth surfaces (i.e., dental plaque), and medical prostheses (Anwar and Costerton, 1992; Bryers and Characklis, 1981; Characklis, 1988; Trulear and Characklis, 1982; van der Wende and Characklis, 1990; Hamilton, 1987). Some of the concerns about biofilms in various fields are shown in Table 15.7 (Characklis et al., 1982; Trulear and Characklis, 1982).

15.4.2. Processes Contributing to Development of Biofilms in Distribution Systems

Biofilms are relatively thin layers (up to a few hundred microns thick) of microorganisms that attach to and grow on surfaces. Biofilms also include corrosion by-products, organic detritus, and inorganic particles such as silt and clay minerals. They contain heterogenous assemblages of microorganisms, depending on the chemical composition of pipe surface, the chemistry of finished water, and oxido-reduction potential in the biofilm. They take days to weeks to develop, depending on nutrient availability and environmental conditions. Growth of biofilms proceeds up to a critical thickness, when nutrient diffusion across the biofilm becomes limiting. The decreased diffusion of oxygen is conducive to the development of facultative and anaerobic microorganisms in the deeper layers of the biofilm.

Various processes contribute to the development of biofilms on surfaces exposed to water flow. The processes involved are the following (Bitton and Marshall, 1980; Marshall, 1976; Olson et al., 1991; Trulear and Characklis, 1982).

15.4.2.1. Surface Conditioning

Surface conditioning is the first step in biofilm formation. Minutes after exposure of inorganic surfaces to water flow, a surface conditioning layer, made of organic molecules, initially adsorb to the surfaces.

15.4.2.2. Transport of Microorganisms to the Conditioned Surface

Diffusion and turbulent eddy transport are involved in a turbulent flow regime. Chemotaxis may also enhance the rate of bacterial adsorption

TABLE 15.7. Some Concerns About Accumulation of Biofilms[a]

Effects	Specific process	Concerns
Heat transfer reduction	Biofilm formation on condenser tubes and cooling tower fill material. Energy losses.	Power industry Chemical process industry U.S. Navy Solar energy systems
Increase in fluid frictional resistance	Biofilm formation in water and wastewater conduits as well as condenser and heat exchange tubes. Causes increased power consumption for pumped systems or reduced capacity in gravity systems. Energy losses.	Municipal utilities Power industry Chemical process industry
	Biofilm formation on ship hulls causing increased fuel consumption. Energy losses.	U.S. Navy Shipping industry
Mass transfer and chemical transformations	Accelerated corrosion caused by processes in the lower layers of the biofilm. Results in material deterioration in metal condenser tubes, wastewater conduits, and cooling tower fill.	Power industry U.S. Navy Municipal utilities Chemical process industry
	Biofilm formation on remote sensors, submarine periscopes, sight glasses, and so on, causing reduced effectiveness.	U.S. Navy Water quality data collection
	Detachment of microorganisms from biofilms in cooling towers. Releases pathogenic organisms (for example, *Legionella* in aerosols).	Public health
	Biofilm formation and detachment in drinking water distribution systems. Changes water quality in distribution system.	Municipal utilities Public health
	Biofilm formation on teeth. Causes dental plaque and caries.	Dental health
	Attachment of microbial cells to animal tissue. Causes disease of lungs, intestinal tract, and urinary tract.	Human health
	Extraction and oxidation of organic and inorganic compounds from water and wastewater (for example, rotating biological contactors, biologically aided carbon adsorption, and benthal stream activity). Reduced pollutant load.	Wastewater treatment Water treatment Stream analysis

(continued)

TABLE 15.7. (*Continued*)

Effects	Specific process	Concerns
	Biofilm formation in industrial production processes reduces product quality.	Pulp and paper industry
	Immobilized organisms or community of organisms for conducting specific chemical transformation.	Chemical process industry

[a]Adapted from Trulear and Characklis (1982).

to surfaces under more quiescent flow (Young and Mitchell, 1973).

15.4.2.3. Adsorption of Microorganisms to the Surface

Adsorption of bacteria to surfaces is a two-step process. The first step is reversible sorption, mainly controlled by electrostatic interactions between the adsorbent and the cell. The second step consists of irreversible adsorption of cells, resulting from the production of extracellular exopolymers at the surface. Other attachment organelles include flagella, pili (fimbriae), stalks, and holdfasts (e.g., *Caulobacter*) (Bitton and Marshall, 1980; Olson et al., 1991).

15.4.2.4. Biofilm Accumulation

The presence of even low levels of organic matter in distribution lines allows the growth and accumulation of biofilm microorganisms. A number of factors (pH, redox potential, TOC, chlorine residual, temperature, hardness) control the growth of microorganisms on pipe surfaces. Two important rate-limiting factors for microbial succession are chlorine residual and TOC in the distribution line (Olson and Nagy, 1984).

Biofilm microorganisms are held together by an extracellular polymeric matrix (mannans, glucans, uronic acids, glycoproteins) called glycocalyx (Characklis and Cooksey, 1983; Costerton and Geesey, 1979). The glycocalyx helps protect microorganisms from predation and from chemical insult (Hamilton, 1987). Growth of bacteria on stainless steel and PVC surfaces, as measured by epifluorescence microscopy, was shown to proceed with a doubling time of 11 days up to 4 months. Afterwards, growth slowed down and the doubling time was 47 days. The mean number of bacteria on the surfaces was 4.9×10^6 cells per 1 cm^2 (Pedersen, 1990).

Several direct and indirect methods are available for determining biofilm accumulation on surfaces. The methods are summarized in Table 15.8 (Characklis et al., 1982). There are still some difficulties in estimating bacterial distribution, activity, and numbers in biofilms.

15.4.2.5. Biofilm Detachment from the Surface

The detachment of biofilms from surfaces is due to shear forces and to the sloughing that occurs under low nutrient or low oxygen levels. Detachment increases with flow velocity, biofilm thickness and shock chlorine treatment, and is decreased by surface roughness (Pedersen, 1990).

The net accumulation of biofilms on surfaces is described by the following equation (Trulear and Characklis, 1982):

Net rate of attached biofilm accumulation = Rate of biomass production − Rate of biomass detachment

$$Ap \frac{dh}{dt} = R_g YA - R_d A \qquad (15.3)$$

where h = biofilm thickness; A = reactor wetted surface; p = biofilm density; R_g = nutrient (glu-

TABLE 15.8. Measurement of Biofilm Accumulation[a]

Type	Analytical method
Direct measurement of biofilm quantity	Biofilm thickness Biofilm mass
Indirect measurement of biofilm quantity: specific biofilm constituent	Polysaccharide Total organic carbon Chemical oxygen demand Protein
Indirect measurement of biofilm quantity: microbial activity within the biofilm	Viable cell count Epifluorescence microscopy ATP Lipopolysaccharide Substrate removal rate
Indirect measurement of biofilm quantity: effects of biofilm on transport properties	Frictional resistance Heat transfer resistance

[a]Adapted from Characklis et al. (1982).

cose) removal rate by attached biomass; Y = yield coefficient of attached biomass; and R_d = biofilm detachment rate.

15.4.3. Problems Caused by Biofilms in Distribution Networks

The development of biofilms on surfaces can be beneficial or detrimental to processes in water and wastewater treatment plants. Trickling filters and rotating biological contractors are examples of processes that rely on microbial activity in biofilms. However, biofilms can cause problems in water distribution pipes (van der Wende and Characklis, 1990):

1. Biofilm accumulation increases the frictional resistance of fluids (Fig. 15.10) (Bryers and Characklis, 1981), leading to greater losses of pressure or to reduced water flow if the drop in pressure is held constant. The community structure of the biofilm also affects frictional resistance. A predominantly filamentous biofilm appears to increase frictional resistance (Trulear and Characklis, 1982).
2. Anaerobic conditions lead to the production of H_2S, a toxic gas with a rotten-egg smell.
3. Accumulation of biofilms also causes taste and odor problems.
4. Complaints arise about red and black waters, which result from the activity of iron- and manganese-oxidizing bacteria (e.g., *Hyphomicrobium*).
5. Resistance of biofilm microorganisms to disinfection by chlorine increases, contributing to the regrowth of indicator and pathogenic bacteria in distribution systems. It has been suggested that this increased resistance may be due to protective effect by extracellular polymeric materials, selection of biofilm microorganisms with enhanced resistance to the disinfectant, or to attachment of bacteria to biological (e.g., macroinvertebrates and algal surfaces) and nonbiological surfaces (turbidity particles, activated carbon). As shown for activated carbon, biofilm bacteria appear to be more resistant to residual

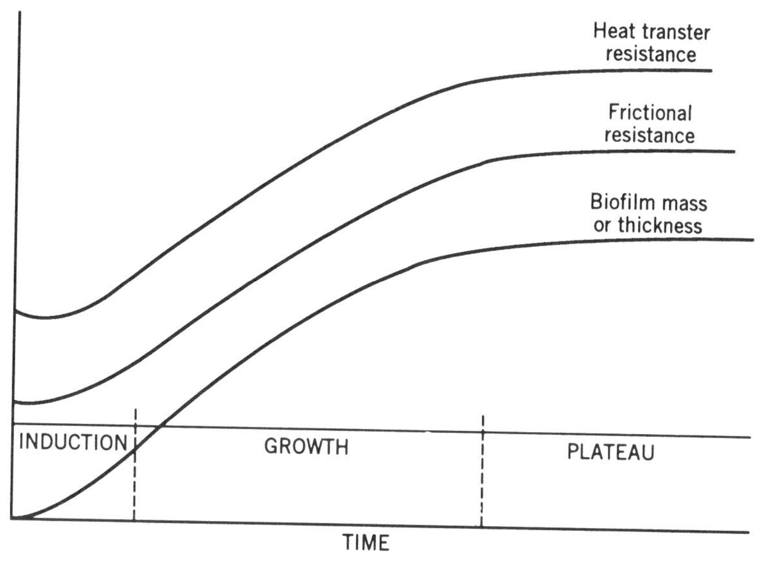

Fig. 15.10. Increase in heat transfer resistance and frictional resistance resulting from biofilm growth. From Bryers and Characklis (1981), with permission of the publisher.

chlorine than suspended bacteria (LeChevallier et al., 1988). Particulates, including particulate organic matter, protect pathogenic microorganisms from disinfectant action (Hoff, 1978). It was found that *Enterobacter cloacae,* in the presence of particulates from water distribution systems, is protected from chlorine action as a result of its attachment to the particles (Herson et al., 1987). Figure 15.11 shows a scanning electron micrograph of particle-associated bacteria in drinking water. Chloramines appear to be more efficient in biofilm control than free chlorine (HOCl or OCl^-). This may be due to the lower affinity of chloramines for bacterial polysaccharides (LeChevallier et al., 1990; van der Wende and Characklis, 1990). It has been proposed that, for biofilm control, free chlorine should be used as a primary disinfectant but the residual should be converted to chloramine (LeChevallier et al., 1990). The increased resistance of biofilm microorganisms to chlorine applies to other antibacterial agents such as antibiotics. Biofilms developing on implanted medical devices are generally resistant to antibiotic action (Anwar and Costerton, 1992).

The free chlorine residual decreases as the water flows through the distribution system. This, as well as the detachment of bacteria from pipe surfaces, leads to an increase in heterotrophic plate count in the distribution system.

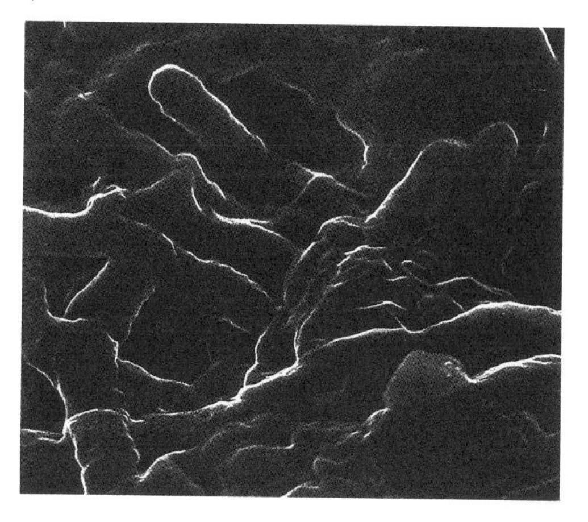

Fig. 15.11. Encapsulated bacteria attached to a particle in drinking water. Courtesy of M.W. LeChevallier.

15.4.4. Growth of Pathogenic and Indicator Bacteria in Water Distribution Systems

Bacteria grow in water distribution systems and colonize pipe connections, tubercles, and dead ends. The presence of a wide range of microorganisms (eubacteria, filamentous bacteria, actinomycetes, diatoms) has been demonstrated in distribution pipes by scanning electron microscopy. Some of the bacteria are seen attached to the surfaces by means of extracellular fribrillar materials (Ridgway and Olson, 1981; Ridgway et al., 1981). Turberculated cast iron pipe sections from the water distribution system in Columbus, Ohio, metropolitan area harbored high numbers of aerobic and anaerobic (e.g., sulfate-reducing) bacteria. Some samples contained up to 3.1×10^7 bacteria per gram of tubercle material (Allen et al., 1980; Geldreich, 1980; Tuovinen and Hsu, 1982) (Fig. 15.12). Several investigations have also dealt with the proliferation of iron and manganese bacteria, which grow attached to pipe surfaces and subsequently cause a deterioration of water quality (e.g., pipe clogging and color problems). Attached iron bacteria such as *Gallionella* have stalks that are partially covered with iron hydroxide. This was confirmed by X-ray energy-dispersive microanalysis (Ridgway et al., 1981). Several materials used in reservoirs and water distribution systems have been found to support the growth of *Pseudomonas, Aeromonas,* and coliform bacteria. Their growth is limited when the assimilable organic carbon (AOC) ranges between 10 μg and 15 μg of acetate equivalent per liter (van der Kooij and Hijnen, 1985a). Coliform (e.g., *K. pneumonae, E. coli, E. aerogenes, E. cloacae*) growth was demonstrated under very low nutrient conditions in water distribution systems. Environmental isolates appear to grow better (i.e., they have higher growth rates, higher yields and higher affinity for substrates) than clinical isolates and are thus better indicators of the conditions prevailing in distribution lines (Camper et al., 1991). However, bacteria belonging to the *Flavobacterium/Moraxella* group produce bacteriocin-like substances, which are inhibitory to coliform bacteria (Means and Olson, 1981). Growth of bacteria in water distribution systems is important from a public health viewpoint because isolation of pathogenic bacteria (e.g., *Salmonella, Klebsiella pneumonae, Yersinia, Legionella*) and the routine occurrence of opportunistic pathogens have been documented in the literature.

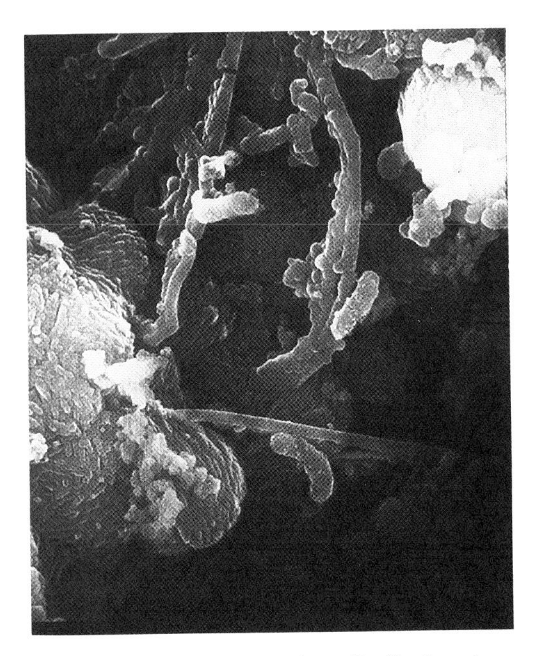

Fig. 15.12. Tubercle on an iron distribution pipe. Courtesy of M.W. LeChavallier.

Much work has been carried out on the survival and growth of Legionellae in potable water distribution systems and plumbing in hospitals and homes (Colbourne et al., 1988; Muraca et al., 1988; Stout et al., 1985). Legionellae appear to be more resistant to chlorine than *E. coli* (States et al., 1989), and small numbers may survive in distribution systems that have been judged to be microbiologically safe. These organisms may also grow to detectable levels inside hot water tanks in hospitals and homes and thus pose a health threat (Stout et al., 1985; Witherell et al., 1988). *Legionella* survives well at 50°C (Dennis et al., 1984), and environmental isolates are able to grow and multiply in tapwater at 32°C, 37°C, and 42°C (Yee and Wadowsky, 1982). The enhanced survival and growth in these systems have also been linked to stagnation (Ciesielski et al., 1984), stimulation by rubber fittings in the plumbing system (Colbourne et al., 1988), and trace concentrations of metals such as Fe, Zn, and K (States et al., 1985, 1989). It has also been found that sediments found in water distribution systems and tanks (i.e., scale and organic particulates) and the natural microflora significantly improved the survival of *Legionella pneumophila*. Sediments indirectly stimulate *Legionella* growth by promoting the growth of commensalistic microorganisms (Stout et al., 1985; Wadowsky and Yee, 1985).

Another condition that is favorable to the multiplication of *Legionella* in hot water tanks is the stagnation of the water within the tank (Ciesielski et al., 1984). *Legionella pneumophila* was detected in higher numbers in tanks that were not in use than in on-line tanks. Thus, the prevention of stagnation of water in these tanks may help control the multiplication of *Legionella*. Several methods have been considered for controlling *Legionella* in water: superchlorination (chlorine concentration of 2–6 mg/L), heat treatment at 60–80°C for many days, UV irradiation (Antopol and Ellner, 1979; Knudson, 1985; Muraca et al., 1987; Stout et al., 1986), biocides (Fliermans and Harvey, 1984; Grace et al., 1981; Skaliy et al., 1980; Soracco et al., 1983), and alkaline treatment (States et al., 1987). The control of Legionellae in potable water systems can also be achieved by the control of protozoa, which sometimes harbor these pathogens (States et al., 1990).

The heterotrophic plate count indicates bacterial levels in water distribution systems and should not exceed 500 organisms per 1 ml. A high HPC is indicative of deterioration of water quality in distribution pipes. HPC includes gram-negative bacteria belonging to the following genera: *Pseudomonas, Aeromonas, Klebsiella, Flavobacterium, Enterobacter, Citrobacter, Serratia, Acinetobacter, Proteus, Alcaligenes, Enterobacter,* and *Moraxella*. Gram-positive bacteria found are *Bacillus* and *Micrococcus* (Agoustinos et al., 1992). The growth of these bacteria can be promoted by trace organics in the distribution line (van der Kooij and Hijnen, 1988). HPC in chlorinated distribution water serving an Oregon coastal community were characterized (LeChevallier et al., 1980). It was shown that actinomycetes and *Aeromonas* spp. were the microorganisms most frequently detected in the distribution system. At the Pont-Viau water filtration plant in Canada, the heterotrophic plate count (20°C or 35°C) increased as the treated water traveled through the distribution system. *Pseudomonas aeruginosa, Aeromonas hydrophila,* and *Clostridium perfringens* were detected in the distribution system (Table 15.9) (Payment et al., 1989).

Many of the bacteria isolated in water distribution systems are opportunistic pathogens (Table 15.9) (Payment et al., 1989). The presence of high numbers of opportunistic pathogens in drinking water is of concern because these microorganisms can lead to infection of certain segments of the population (newborns, the sick, and the elderly). They can also cause secondary infections among patients in hospitals (see more details in Chapter 4).

15.4.5. Determination of Assimilable Organic Carbon in Drinking Water

Bacterial aftergrowth in distribution systems is influenced by several factors, among them the level of trace biodegradable organic materials found in finished water, water temperature, nature of pipe surfaces, disinfectant residual concentration, and detention time within the distribution system (Joret et al., 1988). Drinking water harbors bacteria (e.g., *Flavobacterium, Pseudomonas*) that can grow at extremely low substrate concentrations (van der Kooij and Hijnen, 1981). *Pseudomonas aeruginosa* (strain P1525) and *P. fluorescens* (strain P17) are able to grow in tapwater at relatively low concentrations (μg/L) of low-molecular-weight organic substrates such as acetate, lactate, succinate, and amino acids. A strain of *K. pneumonae* is able to utilize organic compounds at concentrations of a few micrograms per liter. This bacterium grows

TABLE 15.9. Bacteria Isolated on R2A or M-Endo Media From Water Distribution System Samples[a,b]

	Colonies detected on R2A medium (20° or 35°C)		Colonies from M-Endo medium
o	*Acinetobacter* spp	o	*Aeromonas hydrophila*
	Alcaligenes spp	o	*Citrobacter freundii*
	Arthrobacter spp	o	*Enterobacter aerogenes*
o	*Bacillus* spp (mainly *cereus* and *sphaericus*)	o	*Enterobacter agglomerans*
	Corynebacterium spp	o	*Enterobacter cloacae*
	Empedobacter spp	o	*Hafnia alvei*
o	*Flavobacterium* spp		*Klebsiella oxytoca*
	Flexibacter spp		*Klebsiella ozaenae*
	Micrococcus spp	o	*Klebsiella pneumoniae*
o	*Moraxella* spp	o	*Serratia fonticala*
o	*Pseudomonas* (non-*aeruginosa*)	o	*Serratia liquefaciens*
o	*Serratia marcescens*	p	*Vibrio fluvialis*
	Spirillum spp		
	Sporosarcina spp		
	Staphylococcus spp		
p	*Vibrio fluvialis*		

[a]From Payment et al. (1989), with permission of the publisher.
[b]o = opportunistic pathogen; p = primary pathogen.

on maltose with a yield $Y = 4.1 \times 10^6$ CFU per 1 μg C (van der Kooij and Hijnen, 1988).

It is estimated that the assimilable organic carbon in tapwater is between 0.1% and 9% of the total organic carbon (van der Kooij and Hijnen, 1985a; van der Kooij et al., 1982b; LeChevallier et al., 1991c), although this fraction may be higher if the treatment train includes an ozonation step, which makes refractory compounds more available to microorganisms (Servais et al., 1991). *Klebsiella pneumonae* is another example of microorganisms capable of regrowth in distribution systems as well as in drinking water stored in redwood reservoirs (Clark et al., 1982; Geldreich and Rice, 1987; Seidler et al., 1977; Talbot et al., 1979).

Bioassays have been developed to assess the potential of drinking water to support microbial growth, to exert a chlorine demand, or to lead to the formation of disinfection by-products. These bioassays include (Huck, 1990) (1) Biomass-based methods, the goal of which is to measure the production of bacterial biomass following consumption of assimilable organic carbon; and (2) dissolved organic carbon-based methods, the goal of which is to assess chlorine demand and disinfection by-products formation by measuring biodegradable dissolved organic carbon (BDOC).

15.4.5.1. *Biomass-Based Methods*

These methods are based on the measurement of the growth potential of pure bacterial cultures or indigenous microorganisms following incubation in the water samples for several days. Colony-forming units or ATP are determined to assess numbers of bacteria or biomass.

A pasteurized or filtered water sample is seeded with pure bacterial cultures of *Pseudomonas fluorescens* (strain P17), *Spirillum* (strain NOX), or *Flavobacterium* (strain S12). Their growth in the water sample is compared with that obtained in a sample spiked with substrates such as acetate or oxalate. The AOC is calculated by the following formula (van der Kooij, 1990):

$$AOC\ (\mu g\ C/L) = \frac{N_{max} \times 1{,}000}{Y} \qquad (15.4)$$

where N_{max} = maximum colony count (CFU/ml); and Y = yield coefficient (CFU/μg of carbon).

When *P. Fluorescens* P17 is used, AOC is calculated by applying the yield factor Y for acetate ($Y = 4.1 \times 10^6$ CFU/μg C) (van der Kooij and Hijnen, 1985; van der Kooij et al., 1982a). For *Spirillum* sp., strain NOX, $Y_{acetate} = 1.2 \times 10^7$ CFU/μg C (van der Kooij and Hijnen, 1984). AOC concentration in the sample is expressed as micrograms acetate-C equivalents per liter. This approach necessitates a relatively long incubation period (van der Kooij et al., 1982b; van der Kooij and Hijnen, 1984, 1985). P17 and NOX strains can also be used in combination, sequentially or simultaneously, to measure AOC in environmental samples. Efforts have been made to simplify and standardize this method (Kaplan et al., 1993).

Other bacterial mixtures (e.g., mixture of *P. fluorescens, Curtobacterium* sp, *Corynebacterium* sp.) have also been used to estimate AOC. Calibration curves help estimate the AOC from the bacterial colony count (Fig. 15.13) (Kemmy et al., 1989). Some observers criticized the use of pure bacterial cultures for measuring AOC in environmental samples and prefer the use of natural assemblages of bacteria derived from the sample being examined. Biomass production can also be determined by ATP measurement (Stanfield and Jago, 1987) or turbidimetry (Werner, 1982, 1985). Rapid determination of AOC can be achieved by increasing the incubation temperature and bacterial inoculum density, and enumerating the test organisms by ATP measurement (LeChevallier et al., 1993).

AOC levels in various environmental samples, using the method based on the growth potential of *P. fluorescens* P17, are displayed in Table 15.10 (van der Kooij et al., 1982b). AOC levels vary from less than 1 μg C per 1 L to more than 3 mg/L and account for 0.03% to 27% of the total dissolved organic carbon. AOC of drinking water from 20 water treatment plants in the Netherlands varied between 1.1 and 57 μg acetate-C equivalents per 1 L. A significant correlation was observed between AOC of water leaving the plants and counts of heterotrophic bacteria in distribution systems (van der Kooij, 1992). This technique is useful for determining the effect of water treatment processes on AOC and the release of biodegradable materials from pipe surfaces (van der Kooij et al., 1982b). Co-

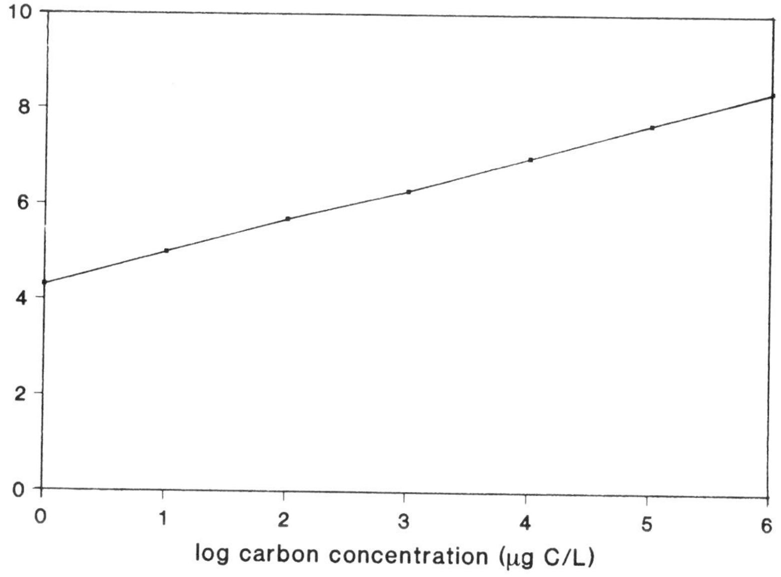

Fig. 15.13. Calibration curves for estimation of assimilable organic carbon (AOC) from bacterial colony count. From Kemmy et al. (1989), with permission of the publisher.

TABLE 15.10. Concentrations of Assimilable Organic Carbon (AOC) in Various Water Samples[a]

Source of water	Dissolved organic carbon (DOC), mg C/L	Assimilable organic carbon (AOC), mg C/L
Biologically treated wastewater	13.5	3.0–4.3
River Lek	6.8	0.062–0.085
River Meuse	4.7	0.118–0.128
Brabantse Diesbosch	4.0	0.08–0.103
Lake Yssel, after open storage	5.6	0.48–0.53
River Lek, after bank filtration	1.6	0.7–1.2
Aerobic groundwater	0.3	<0.15

[a]Adapted from van der Kooij et al. (1982b).

agulation, sedimentation, sand filtration, and activated carbon (GAC) treatment generally reduce the AOC concentration whereas ozonation, and to a lesser extent chlorination, increase the AOC levels as a result of production of low-molecular-weight compounds (van der Kooij and Hijnen, 1985; LeChevallier et al., 1992; Servais et al., 1987; Servais et al., 1991). It was recommended that the AOC concentration of drinking water in distribution systems be less than 10 μg of acetate-C equivalents to limit aftergrowth of heterotrophic bacteria (van der Kooij, 1990, 1992). AOC concentration is also an important parameter to consider in regard to the biological clogging of sand beds that occurs as a result of infiltration of pretreated surface water in recharge wells. AOC level should be less than 10 μg acetate-C equivalents per 1 L to prevent biological clogging (Hunen and van der Kooij, 1992).

Determination of AOC level, based on the growth potential of *P. fluorescens*, does not appear to be a good indicator of the growth potential of coliforms in water distribution systems; coliform monitoring appears to be more suitable (McFeters and Camper, 1988). In a water distribution system in New Jersey, coliform regrowth was associated with rainfall, water temperature higher than 15°C, and AOC levels higher than 50 μg acetate-carbon equivalents per 1 L (LeChevallier et al., 1991c). Coliform regrowth is limited at AOC levels lower than 100 μg/L (Fig. 15.14) (LeChevallier et al., 1992).

A coliform growth response test, using *Enterobacter cloacae* as the test organism, was developed by the U.S. EPA to measure specifically the growth potential of coliform bacteria in water. Some investigators recommend the use of environmental coliform isolates found in drinking water (e.g., *Enterobacter aerogenes, Klebsiella pneumonae, E. coli*) to determine assimilable organic carbon in distribution systems (Camper et al., 1991). These isolates appear to be more adapted to oligotrophic conditions than their clinical counterparts (McFeters and Camper, 1988; Rice et al., 1988).

15.4.5.2. DOC-Based Methods

These assays are based on the measurement of dissolved organic carbon before and after incubation of the sample in the presence of an indigenous bacterial inoculum (e.g., river water or sand filter bacteria). It is argued that indigenous bacterial populations are more suitable than pure cultures for testing the biodegradation of natural organic compounds (Block et al., 1992; Neilson et al., 1985). The biodegradable dissolved organic carbon is given by the following formula:

$$\text{BDOC (mg/L)} = \text{initial DOC} - \text{final DOC} \quad (15.5)$$

The general approach is as follows.

A water sample is sterilized by filtration

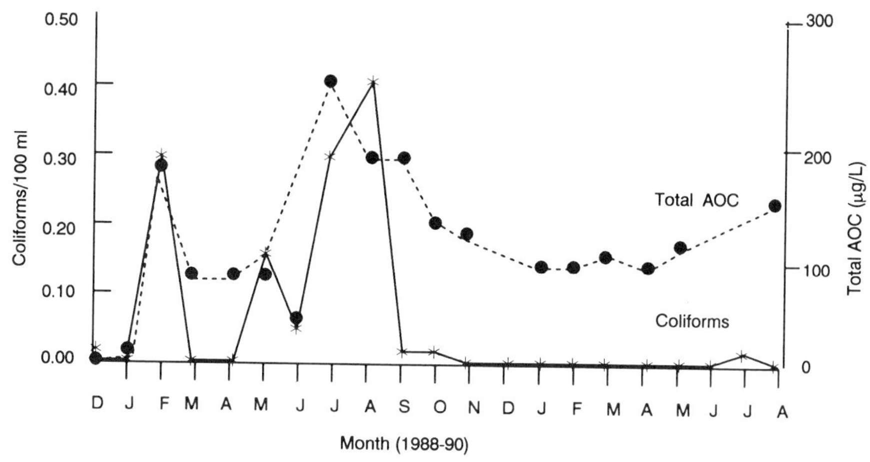

Fig. 15.14. Relationship between mean coliform densities and total assimilable organic carbon (AOC) levels. From LeChevallier et al. (1992), with permission of the publisher.

trough with a 0.2-µm pore size filter, inoculated with indigenous microorganisms, and incubated in the dark at 20°C for 10–30 days, until DOC reaches a constant level. AOC is the difference between the initial and final DOC values (Servais et al., 1987). This method showed that the BDOC in the Seine River in France was approximately 0.7 mg/L (Servais et al., 1989).

Another approach consists of seeding the water sample (300 ml) with prewashed biologically active sand with mixed populations of attached indigenous bacteria. A European collaborative study has demonstrated that the source of the indigenous bacterial inoculum is not a major source of variance in the BDOC procedure (Block et al., 1992). BDOC is estimated by monitoring the decrease in dissolved organic carbon (Joret et al., 1988). The experimental setup for this method is displayed in Figure 15.15 (Joret and Levi, 1986; Joret et al., 1988). The incubation period at 20°C is 3–5 days. One advantage of this method is the use of biofilm microorganisms as inoculum, thus simulating situations that occur in water distribution systems. The monitoring of water treatment plants in Paris suburbs showed that treated drinking water con-

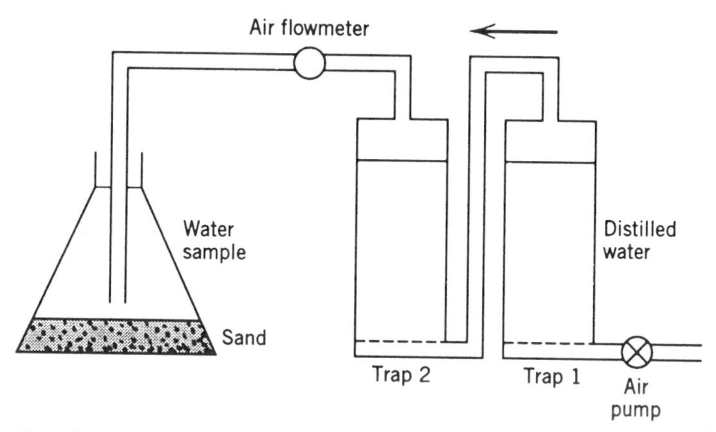

Fig. 15.15. Experimental setup for BDOC determination. From Joret et al. (1988), with permission of the publisher.

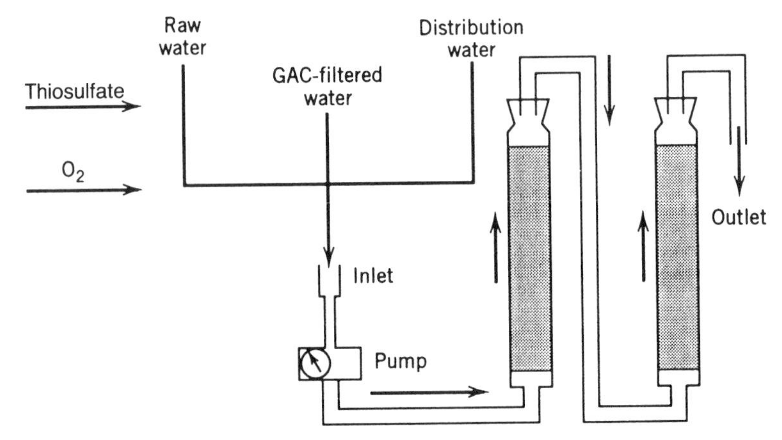

Fig. 15.16. Apparatus for BDOC determination. Adapted from Ribas et al. (1991).

tains enough BDOC to sustain bacterial growth in distribution systems (Joret et al., 1988). Nutrient availability, as measured by this technique, appears to correlate well with the regrowth potential of bacteria (Joret et al., 1990).

A novel proposed method consists of passing water continuously through one or two glass columns filled with sand or sintered porous glass and conditioned to obtain the development of a biofilm on the supports provided. BDOC is the difference between the inlet of the first column and the outlet of the second column (Fig. 15.16) (Ribas et al., 1991; Frias et al., 1992). BDOC is obtained in hours to days, depending of the type of water being assayed.

15.5. OTHER BIOLOGICAL PROBLEMS ASSOCIATED WITH WATER TREATMENT AND DISTRIBUTION

15.5.1. Taste and Odor Problems

Several surveys have shown that water treatment plants, especially those using surface waters, have taste and odor problems. The sources of the taste and odor in water may be anthropogenic or natural.

Anthropogenic sources include phenol, chlorinated phenols (2-chlorophenol, 2,4-dichlorophenol, 2,6-dichlorophenol; Burttschell et al., 1959), hydrocarbons, and halogenated compounds such as chloroform (Zoeteman et al., 1980).

Among *natural sources,* the major taste- and odor-causing compounds are geosmin and 2-methyl isoborneol (MIB), which are products of actinomycete metabolism (e.g., *Streptomyces* and *Nocardia* species) and cyanobacteria (e.g., *Oscillatoria, Anabaena, Lyngbia*) (Gerber, 1979; Izaguirre et al., 1982; Lalezary et al., 1986; Markovic and Kroeger, 1989; Medsker et al., 1968; Safferman et al., 1967). Geosmin, 2-methyl isoborneol, sesquiterpenes, β-cyclocitral, 3-methyl-1-butanol, and others have been isolated from water supplies and implicated as the cause of an earthy-musty odor in drinking water (Hayes and Burch, 1989; Izaguirre et al., 1982). Geosmin and 2-methyl isoborneol have very low threshold odor concentrations of 10 and 29 ng/L, respectively (Cees et al., 1974, Persson, 1979). Other microbial metabolites are 2-isopropyl-3-methyoxypyrazine and 2-isobutyl-3-methoxypyrazine (Fig. 15.17) (Lalezary et al., 1986).

Various approaches are used to control taste- and odor-causing compounds in water treatment plants (Namkung and Rittmann, 1987):

1. *Adsorption to solids.* Adsorption to activated carbon is the most popular treatment for removing these compounds. Concentrations as low as 5 mg/L can successfully reduce geosmin

Molecular structure	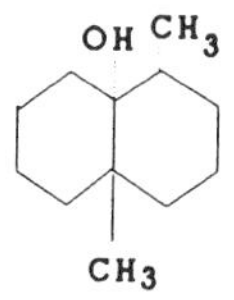	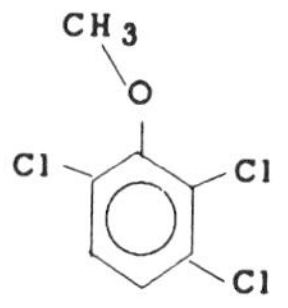	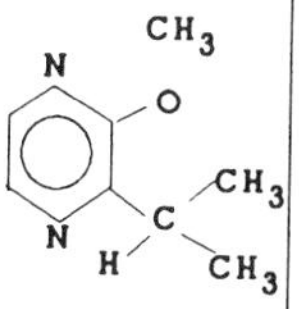	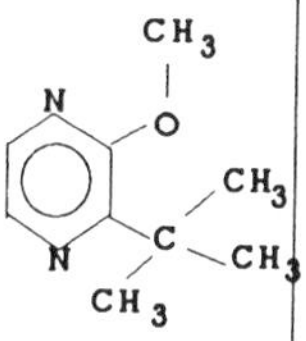	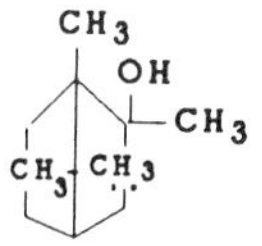
Symbol	Geosmin	TCA	IPMP	IBMP	MIB
Molecular weight	182	212	152	166	168
Molecular formula	$C_{12}H_{22}O$	$C_7H_5OCl_3$	$C_8H_{12}ON_2$	$C_9H_{14}ON_2$	$C_{11}H_{20}O$
Name	trans-1,10-dimethyl trans-9 decalol	2,3,6-trichloro anisole	2-isopropyl-3-methoxy pyrazine	2-isobutyl-3-methoxy pyrazine	2-methyl-isoborneol

Fig. 15.17. Molecular makeup of five taste and odor compounds. Adapted from Lalezary et al. (1986).

and MIB to acceptable levels (Lalezary-Craig et al., 1988). Zeolites have also been suggested for taste and odor removal (Ellis and Korth, 1993).

2. *Oxidation processes* (chlorination, ozonation, K permanganate). Chlorine dioxide is one of the most effective oxidant for removing taste- and odor-causing compounds, but it is not effective in eliminating odors caused by hydrocarbons (Walker et al., 1986). Chlorine dioxide is also effective against geosmin-producing actinomycetes such as *Streptomyces griseus* (Whitmore and Denny, 1992). Peroxone, a mixture of ozone and hydrogen peroxide, appears to be more effective for the oxidation of MIB and geosmin than ozone alone. The optimum $H_2O_2 : O_3$ ratio for oxidation of taste and odor compounds is 0.1 to more than 0.3, depending on the water being treated (Ferguson et al., 1990).

3. *Biodegradation*. Various laboratory bacterial cultures have been considered for the removal of these compounds. *Bacillus cereus* can degrade geosmin in water (Hoehn, 1965). The taste- and odor-causing compounds (e.g., geosmin and MIB) are also used cometabolically by biofilms grown on fulvic acid, which serves as a primary substrate (Namkung and Rittmann, 1987).

4. *Artificial groundwater recharge* (i.e., percolation of surface water through sand and gravel ridges leads to higher removal of odorous compounds, including geosmin and 2-methyl isoborneol, than alum coagulation (Savenhed et al., 1987).

15.5.2. Algae

Raw surface waters may contain high numbers of cyanobacteria (e.g., *Microcystis aeruginosa, Anabaena flos-aquae, Oscillatoria*), many of which produce allergenic, hepatoxic, neurotoxic, and possibly tumor-promoting chemicals (Carmichael, 1989; Falconer and Buckley, 1989). Several cyanobacterial hepatoxins, also called microcystins, have been isolated from freshwater environments (Sivonen et al., 1992). A waterborne outbreak of diarrheal illness (watery diarrhea) was associated with cyanobacteria-like organisms in Chicago (Center for Disease Control, 1991). The long-term chronic effects of these algal by-products on human health are not well known (Falconer, 1989).

Algal cells are removed by coagulation and filtration and effectively controlled by chlorination (Kay et al., 1980). However, blue-green algal toxins are not removed by conventional flocculation–filtration–chlorination treatment, although the addition of activated carbon filtration or ozonation to the treatment sequence leads to their efficient removal (Himberg et al., 1989).

In finished waters, the algae found include green algae (e.g., *Chlorella, Scenedesmus, Ankistrodesmus),* cyanobacteria (e.g., *Schizothrix*), diatoms (e.g., *Achnanthes*), and flagellated pyrrophytes (e.g., *Glenodinium*). Algae have a tendency to grow in water reservoirs especially during the warm season. Algal blooms in reservoirs can be prevented by covering the reservoir to block sunlight and are generally effectively controlled by copper or chlorine treatment. Chlorine, at a level of approximately 1 ppm can effectively control algae such as *Chlorella* (Fig. 15.18) (Kay et al., 1980).

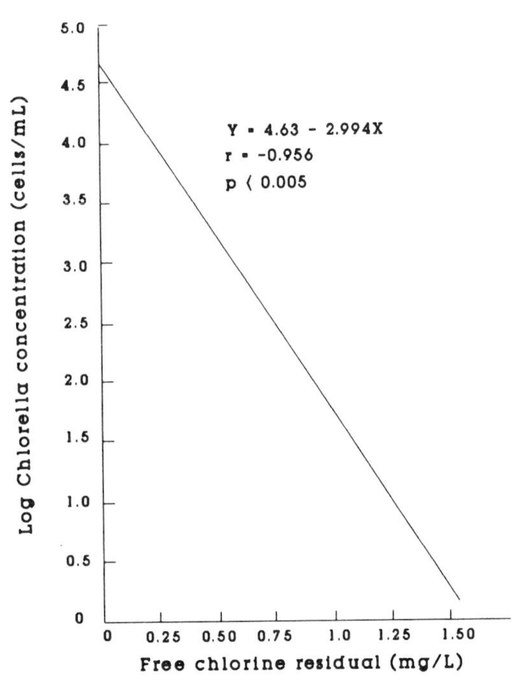

Fig. 15.18. Effect of free chlorine on *Chlorella vulgaris*. From Kay et al., (1980), with permission of the publisher.

Algae may cause the following problems in water treatment plants:

1. Clogging of sand filters, particularly during the warm season.

2. Taste and odor problems. Cyanobacteria (*Oscillatoria, Anabaena, Microcystis*) are responsible for taste and odor problems in drinking water (Medsker et al., 1968). The chrysophite *Synura petersenii* is responsible for the cucumber odor of some drinking waters. This odor is caused by a chemical identified as 2,6-nonadienal (Hayes and Burch, 1989). Another diatom species, *Sinura uvella,* gives a cod liver oil odor to water (Juttner, 1981).

3. Increase in trihalomethane precursors with a subsequent increase in trihalomethanes upon chlorination of drinking water (Karimi and Singer, 1991). Several blue-green algae (*Anabaena, Anacystis*), green algae (*Selenastrum, Scenedesmus*), and diatoms (*Navicula*) were implicated in the production of chloroform upon chlorination (Oliver and Schindler, 1980).

15.5.3. Fungi

Fungi are often isolated from water distribution systems (Hinzelin and Block, 1985; O'Connor et al., 1975; Rosenzweig and Pipes, 1989). Filamentous fungi are found in drinking water at levels up to approximately 10^2 CFU/100 ml (Olson and Nagy, 1984), while yeast cell levels may reach numbers up to 10^3 per 1 ml (O'Connor et al., 1975). Approximately 50% of samples taken from chlorinated groundwater or surface water contained fungi. Filamentous fungi account for most the isolates (89%), whereas yeasts account for 11% of the isolates. Genera such as *Penicillium, Verticillium, Fusarium, Alternaria, Trichoderma, Cephalosporium, Aspergillus, Cladosporium, Mucor, Epicoccum, Phialophora,* and *Rhodotorula* are encountered in drinking water because of their ability to survive chlorination and grow on the surface of drinking water reservoirs and distribution pipes (Nagy and Olson, 1982; Rosenzweig et al., 1989). Water treatment processes can remove up to 2 logs of fungal numbers. However, these microorganisms appear to be more resistant to chlorine and ozone than coliform bacteria (Haufele and Sprockhoff, 1973; Rosenzweig et al., 1983). Monitoring of mesophilic fungi in raw waters and tapwaters from several municipalities in Finland showed that they occur at concentrations up to approximately 100 per 1 L of tapwater (Niemi et al., 1982). Approximately 50% of potable water samples taken from small municipal water distribution systems contained fungi at levels of 1–6 fungal propagules per 50 ml. The predominant genera found were *Aspergillus, Alternaria, Cladosporium,* and *Penicillium* (Rosenzweig et al., 1986).

Fungi may cause the following problems in water distribution systems:

1. They exert a chlorine demand and may protect bacterial pathogens from inactivation by chlorine (Rosenzweig and Pipes, 1989; Rosenzweig et al., 1983; Seidler et al., 1977).
2. They may degrade some of the jointing compounds used in distribution systems.
3. Some of them form humic-like substances that may act as precursors of trihalomethanes (Day and Felbeck, 1974).
4. They may cause taste and odor problems.
5. Some fungi may be pathogenic (e.g., *A. flavus, A. fumigatus*) and some of them may cause allergic reactions in sensitive persons (Rosenzweig et al., 1986). The public health significance of their presence in water is practically unknown.

15.5.4. Actinomycetes

Soil runoff is the most likely source of actinomycetes in water. Chemical coagulation followed by slow sand filtration and disinfection removes actinomycetes from water. However, routine chlorination has little effect on actinomycetes and is not very effective in taste and odor reduction (Sykes and Skinner, 1973). Despite treatment, actinomycetes are found in drinking water (Niemi et al., 1982); the most widely found actinomycetes in water distribution systems belong to the genus *Streptomyces*. Other genera found are *Nocardia* and *Micromonospora*. These organisms are found at levels up to 10^3 CFU/100 ml. Actinomycetes were the most frequently detected microorganisms in chlorinated distribution water from a coastal community in Oregon (LeChevallier et al., 1980).

15.5.5. Protozoa

Protozoa may be found in treated drinking water (e.g., *Hartmanella*) and some are part of the microflora of biofilms on the surface of water reservoirs (e.g., *Bodo, Vorticella, Euplotes*). Although much effort has been focused on pathogenic protozoa (e.g., *Giardia lamblia, Cryptosporidium, Entamoeba histolytica, Acanthamoeba, Balantidium coli*), little is known about the ecology of these microorganisms in water treatment plants.

To address removal of protozoan cysts from water, recent amendments of the Safe Drinking Water Act (PL 93-523) mandates EPA to require filtration for all surface water supplies. However, exceptions (e.g., effective disinfection) to this requirement are being considered.

15.5.6. Invertebrates

This group includes nematodes, crustaceans, flatworms, water mites, and insect larvae. Their presence in water treatment plants and in distribution lines may be of public health significance because these organisms may harbor potential pathogens and protect them from disinfectant action. Invertebrates can be controlled in water distribution lines by chemical (copper sulfate, chlorine) and physical methods (pipe cleaning, hydrant flushing, filtration of source waters) (Levy, 1990; Levy et al., 1986).

15.5.7. Endotoxins

These heat-stable lipopolysaccharides are structural components of the outer membrane of gram-negative bacteria and cyanobacteria that can be detected in surface water and groundwater, as well as in drinking water produced from reclaimed wastewater (Burger et al., 1989; Carmichael, 1981b; Goyal and Gerba, 1982b). High concentrations of endotoxin-producing cyanobacteria (e.g., *Schizothrix calcicola*) have been found in uncovered finished water reservoirs (Sykora et al., 1980).

Although these chemicals have been associated with human diseases and allergies, the health implications of their presence in drinking water are unclear at the present time.

15.5.8. Iron and Manganese Bacteria

Iron and manganese bacteria attach to and grow on pipe surfaces and subsequently cause a deterioration of water quality (e.g., pipe clogging and color problems leading to laundry staining).

15.6. POINT-OF-USE HOME DEVICES FOR WATER TREATMENT

The public at large is interested in point-of-use home devices to remove toxic and carcinogenic chemicals and improve the aesthetic quality of drinking water (removal of taste and odor, turbidity, color). The home devices can be connected to a third faucet separate from the traditional cold and hot water faucets. Treatment is accomplished by filtration, adsorption, ion exchange, reverse osmosis, distillation, or UV irradiation (Geldreich and Reasoner, 1990; Reasoner et al., 1987). The most frequently used process is filtration through activated carbon.

The contaminants of concern are pathogenic bacteria, viruses, protozoan cysts (e.g., *Giardia*), toxic metals (e.g., cadmium, mercury, lead), iron, manganese, organic substances of potential health significance (e.g., trichloroethylene, hydrocarbons, benzene), particulates, color, odor and chlorine taste (Geldreich and Reasoner, 1990). As to microbial contaminants, the U.S. EPA requires a minimum removal capacity of 99.9%, 99.99% and 99.9999% for *Giardia* cysts, viruses, and bacteria, respectively.

These devices can be installed in a home as add-on units consisting of small activated carbon cartridges, in-line devices that are installed under the kitchen sink, or point-of-entry devices for treating the entire home water supply. However, there are some problems associated with use of these devices (Reasoner et al., 1987):

1. Heterotrophic bacteria and, possibly, pathogenic microorganisms may colonize the activated carbon surface, leading to the development of high levels of bacteria in the product water. An example of heterotrophic plate count bacteria found in the product water is shown in Figure 15.19 (Reasoner et al., 1987; Geldreich et al., 1985; Taylor et al., 1979; Wallis et al., 1974). Static conditions overnight or following vacation periods as well as favorable temperatures and nutritional conditions provide opportunity for bacterial growth in the treatment device. Pathogenic bacteria such as *Pseudomonas aeruginosa* or *Klebsiella pneumonae* are also able to colonize the filter surface (Geldreich et al., 1985; Reasoner et al., 1987; Tobin et al., 1981) and may pass into the product water. However, the epidemiological significance of this phenomenon is unknown. It is thus advisable to flush the unit for 1–3 min prior to use in the morning or after returning from vacation.

2. Silver-impregnated filters are sold by some manufacturers to control bacterial growth inside the filter. The antibacterial effect of silver is rather slow and significant bacterial reductions are obtained only after hours of contact (Bell, 1991). Furthermore, at the concentrations used in point-of-use devices, silver does not exert any significant detrimental effect on growth of heterotrophic bacteria (possibly owing to the selection of silver-resistant bacteria), and it has no significant antiviral effect (Gerba and Thurman, 1986; Tobin et al., 1981). Furthermore, because of concern over consumers' health, the silver released in the treated water cannot exceed 50 μg/L (Geldreich et al., 1985). Water filtration units based on reverse osmosis may produce water with bacterial counts of approximately 10^3 to 10^4 CFU/ml. Bacterial isolates that have been identified are *Pseudomonas, Flavobacterium, Alcaligenes, Acinetobacter, Chromobacterium*

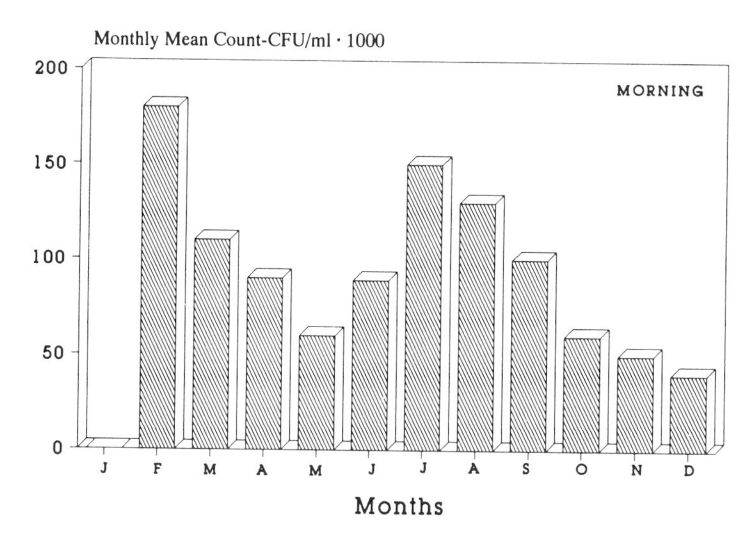

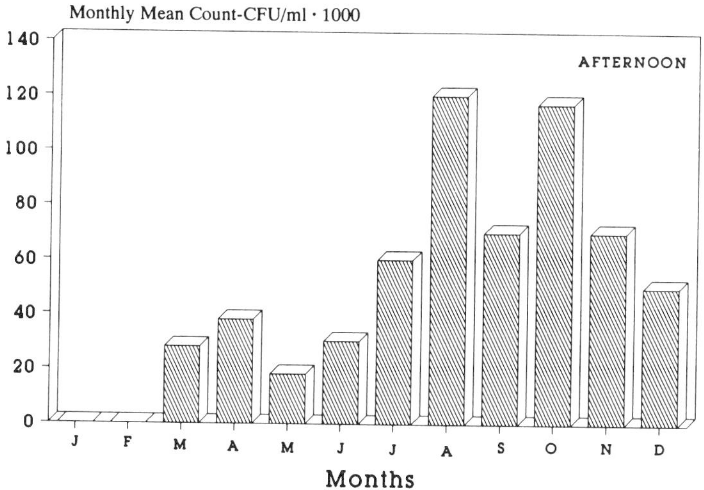

Fig. 15.19. Mean heterotrophic plate count found in product water from home point-of-use devices. From Reasoner et al. (1987), with permission of the publisher.

and *Moraxella* (Payment, 1989b). A prospective epidemiological study undertaken in a Montreal suburban area, showed a correlation between bacterial counts in drinking water at 35°C and the incidence of gastrointestinal symptoms in 600 households consuming water treated by reverse osmosis. The level of gastrointestinal illnesses in the group consuming regular tapwater was 30% higher than that in the group consuming water treated by reverse osmosis (Payment et al., 1991, 1993).

Thus, although reverse osmosis appears to reduce the incidence of gastrointestinal illnesses, treatment of drinking water by some home devices may have an adverse effect on the microbiological quality of drinking water. The use of point-of-use devices by consumers is considered by some experts as a potential health hazard and some strongly believe that water treatment should be carried out by trained professionals and not by poorly trained consumers (Geldreich et al., 1985).

15.7. FURTHER READING

Anwar, H., and J.W. Costerton. 1992. Effective use of antibiotics in the treatment of biofilm-associated infections. ASM News 58: 665–668.

AWWA Organisms in Water Committee. 1987. Committee report: Microbiological considerations for drinking water regulation revisions. J. Am. Water Works Assoc. 79: 81–88.

Bitton, G., S.R. Farrah, C. Montague, and E.W. Akin. 1986. Global survey of virus isolations from drinking water. Environ Sci. Technol. 20: 216–222.

Characklis, W.G., and K.E. Cooksey. 1983. Biofilms and microbial fouling. Adv. Appl. Microbiol. 29: 93–138.

Geldreich, E.E. 1990. Microbiological quality of source waters for water supply, pp. 3–31, in: *Drinking Water Microbiology,* G.A. McFeters, Ed. Springer-Verlag, New York.

van der Kooij, D. 1990. Assimilable organic carbon (AOC) in drinking water, pp. 57–87, in: *Drinking Water Microbiology,* G.A. McFeters, Ed. Springer-Verlag, New York.

LeChevallier, M.W., and G.A. McFeters. 1990. Microbiology of activated carbon, pp. 104–119, in: *Drinking Water Microbiology,* G.A. McFeters, Ed. Springer Verlag, New York.

McFeters, G.A., Ed. 1990. *Drinking Water Microbiology.* Springer-Verlag, New York.

Olson, B.H., and L.A. Nagy, 1984. Microbiology of potable water. Adv. Appl. Microbiol. 30: 73–132.

U.S. EPA. Drinking Water Health Effects Task Force. 1989e. *Health Effects of Drinking Water Treatment Technologies.* Lewis, Chelsea, MI, pp. 146.

16

POLLUTION CONTROL BIOTECHNOLOGY

16.1. INTRODUCTION

Advances in microbial genetics and genetic engineering have given great impetus to the field of pollution control biotechnology. In some countries, efforts are being focused on the application of biotechnology to wastewater treatment. In Japan, the government is interested in the application of genetic engineering methods and immobilization techniques for waste treatment, as well as the development and improvement of bioreactors for wastewater treatment (Matsui et al., 1991). Owing to the lack of information on the fate of genetically engineered microorganisms in the environment, only microorganisms isolated by traditional enrichment techniques are being marketed at the present time.

In this chapter, we will discuss the use of commercial enzyme and microbial blends, immobilized microorganisms, and recombinant DNA technology for enhancing biological wastewater treatment. We will also explore the role of microorganisms in removal of metals in wastewater treatment plants.

16.2. USE OF COMMERCIAL BLENDS OF MICROORGANISMS AND ENZYMES IN WASTEWATER TREATMENT

16.2.1. Use of Microbes in Pollution Control

The early attempts to use microorganisms in the pollution control field have focused on anaerobic digestion. Later on, microorganisms capable of degrading herbicides and other chemicals in industrial wastes were isolated and used in commercial preparations designed for pollution control. In the 1970s, microbial preparations were marketed for enhancing the operational efficiency of waste treatment processes. This approach has been called bioaugmentation (Johnson et al., 1985).

16.2.2. Production of Microbial Seeds

Microbial strains for enhancing biodegradation of specific chemicals are generally isolated from environmental samples (wastewater, sludge, compost, soil) and selected by conventional enrichment techniques. They are grown in nutrient media that contain a specific organic chemical as the sole source of carbon and energy or as a sole source of nitrogen. Furthermore, strains that can handle relatively high concentrations of the target chemical are selected. Some of the microbial strains may be subsequently irradiated to obtain a desirable mutation (Fig. 16.1) (Johnson et al., 1985).

Prior to using commercial preparations of microorganisms for pollution control, one must first gain information about the biodegradability of the chemical pollutant under investigation. This can be done by searching the literature or by undertaking biodegradation studies under laboratory conditions. Bioassays are also used for assessing the toxicity of the wastewater under consideration to the commercial preparation of microbial seeds (see Chapter 17). Enrichment

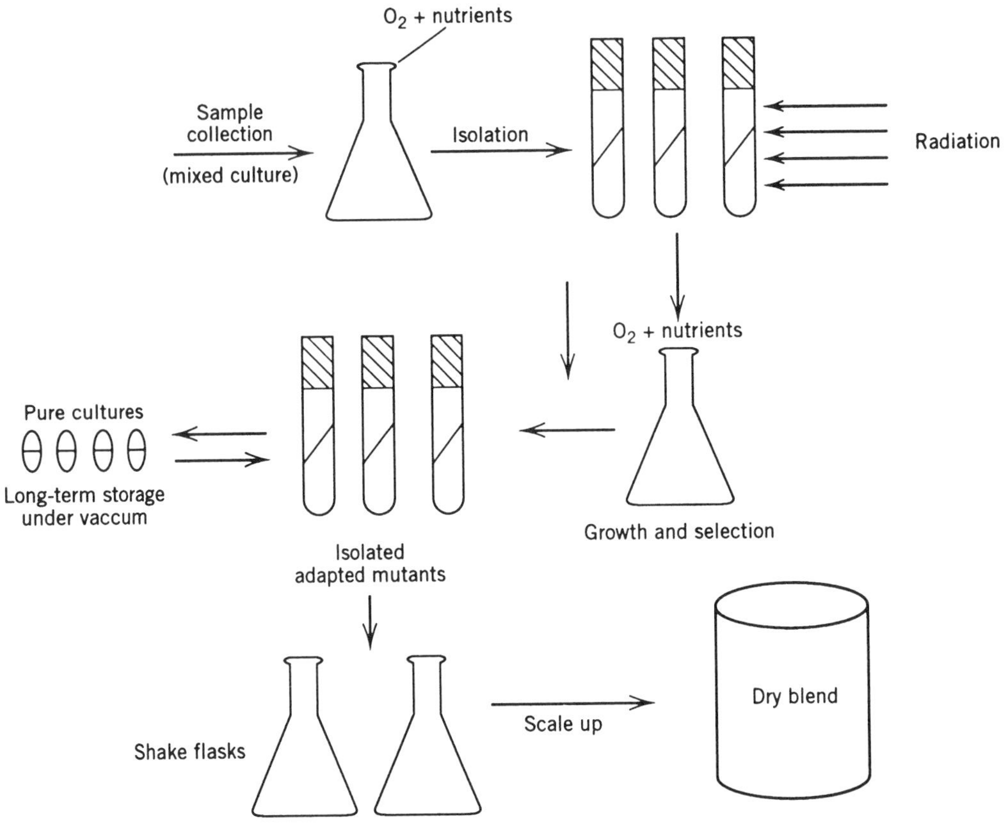

Fig. 16.1. Isolation and preparation of microbial blends for pollution control. Adapted from Johnson et al. (1985).

culture techniques have been used to isolate commercial strains of bacteria that degrade petroleum hydrocarbons. For example, mixtures of bacterial isolates belonging to the genus *Pseudomonas* are marketed for the *in situ* biorestoration of aquifers contaminated with aliphatic or aromatic hydrocarbons (von Wedel et al., 1988). The selected strains are grown in large fermenters and then concentrated by centrifugation or filtration. They are then preserved by lyophilization, drying, or freezing. For successful results, the applied microorganism must withstand the desired environmental conditions. These conditions include temperature, pH, dissolved oxygen, nutrient availability and ability to withstand potential toxicity of the wastewater.

16.2.3. Use of Bioaugmentation in Waste Treatment

Bioaugmentation is the use of selected strains of microbes isolated from the environment to improve some of the processes involved in traditional waste treatment. It may involve the addition of selected bacteria to a bioreactor to maintain or enhance the biodegradation potential in the reactor. This technology has been known for decades. Unfortunately, information on the formulation of the mixtures of the microbial cultures is scanty because of trade secrets. Some of the applications of bioaugmentation are the following (Grubbs, 1984; Rittmann et al., 1990):

Increased BOD removal. Microbial strains

may be used to increase BOD removal in wastewater treatment plants.

Reduction of sludge volume. Production of large amounts of sludge is a serious problem associated with aerobic waste treatment, and thus reduction of sludge volumes is highly desirable. The reduction is the result of increased organic removal following addition of a mixed culture of selected microorganisms. Reductions in volumes of generated sludge of 17% to nearly 30% have been documented.

Use of mixed cultures in sludge digestion. In aerobic digesters, the use of mixed cultures has led to significant savings in energy requirements. In anaerobic digesters, bioaugmentation has resulted in enhanced methane production.

Biotreatment of hydrocarbon wastes. Commercial bacterial formulations have been traditionally used for the treatment of hydrocarbon wastes. For example, the addition of cultures of mutant bacteria improved the effluent quality of a petrochemical wastewater treatment plant (Christiansen and Spraker, 1983).

Biotreatment of hazardous wastes. The use of added microorganisms for treating hazardous wastes (e.g., phenols, ethylene glycol, formaldehyde) has been attempted and has a promising future. Bioaugmentation with parachlorophenol-degrading bacteria achieved a 96% removal in 9 hr, as compared with a control that exhibited 57% removal after 58 hr (Fig. 16.2) (Kennedy et al., 1990). *Candida tropicalis* cells also have been used to remove high concentrations of phenol in wastewater (Kumaran and Shivaraman, 1988). *Delsulfomonile tiedjei,* when added to a methanogenic upflow anaerobic granular-sludge blanket, increased the ability of the bioreactor to dechlorinate 3-chlorobenzoate (Ahring et al., 1992).

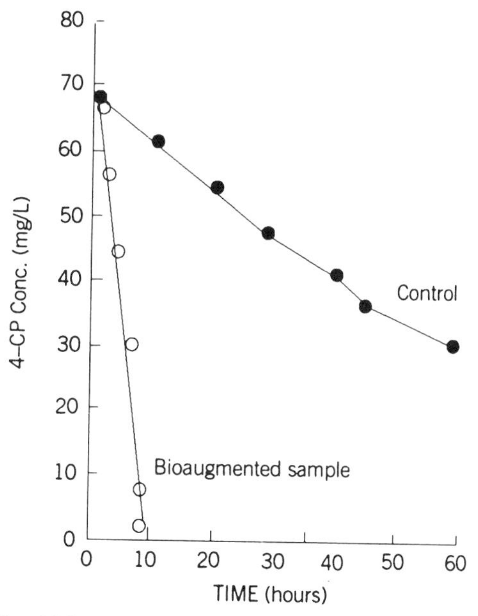

Fig. 16.2. Biodegradation of parachlorophenol by bioaugmentation. Adapted from Kennedy et al. (1990).

Some major drawbacks of bioaugmentation are the need for an acclimation period prior to onset of biodegradation, and short survival or lack of growth of microbial inocula in the seeded bioreactors. Evaluations of some of these commercial products are sometimes negative or inconclusive. Furthermore, addition of cell-free extracts does not significantly improve the performance of biological waste treatment (Jones and Schroeder, 1989). Some novel bioaugmentation schemes for the degradation of hazardous wastes in wastewater treatment plants are being investigated. For example, the enricher–reactor process, which involves the maintenance of a separate acclimated culture for addition to an activated sludge system, does improve the biodegradation of xenobiotics (Babcock et al., 1992). The enrichment cultures can be maintained in the enricher–reactor by adding a less hazardous inducing compound of structure similar to that of the target compound (Babcock and Stenstrom, 1993).

16.2.4. Use of Enzymes in Waste Treatment

Enzymes play a key role in the hydrolysis and biotransformation of organics in wastewater treatment plants, and several of them have been detected in wastewater samples (e.g., phosphatases, aminopeptidases, esterases) (Boczar et al., 1992). It has also been suggested that enzymes can be added to wastewaters to improve the treatability of xenobiotic compounds.

Microbial exoenzymes are used to detoxify pesticides in soils (Munnecke, 1981). For example, carbaryl is transformed to 1-naphtol, which is 920 times less toxic than the parent compound. Parathion hydrolases, produced by *Pseudomonas* sp. and *Flavobacterium* species, degrade parathion to diethylthiophosphoric acid and *p*-nitrophenol (Mulbry and Karns, 1989). These enzymes have been used for the cleanup of containers for parathion, detoxification of wastes containing high concentrations of organophosphates, and in soil cleanup operations (Karns et al., 1987). Less is known about the use of enzymes in wastewater treatment plants. To be useful, the added enzymes must be stable under conditions (e.g., temperature, pH, toxicant levels) that prevail in waste treatment processes. An application of this technology is illustrated by the use of specific enzymes to reduce the production of excessive amounts of extracellular polysaccharides during wastewater treatment. An overproduction of polysaccharides may lead to increased water retention, resulting in reduced sludge dewatering. A number of enzymes can degrade these exopolymers (Sutherland, 1977). Phage-induced depolymerases have been described in several phage–bacterial systems (Adams and Park, 1956; Bartel et al., 1968; Bessler et al., 1973; Nelson et al., 1988; Vandenbergh et al., 1985). A phage-induced depolymerase has been isolated from a sludge sample and found to readily increase the degradation, as shown by viscosity reduction, of exopolysaccharides of sludge bacteria (Fig. 16.3) (Nelson et al., 1988). The addition of this enzyme or mixtures of enzymes could eventually be used to improve sludge dewatering.

The polymerization and precipitation of aromatic compounds (e.g., substituted phenols and anilines) in drinking water and wastewater can be catalyzed by specific enzymes such as horseradish peroxidase. This enzyme catalyzes the oxidation of phenol and chlorophenols by hydrogen peroxide (Kilbanov et al., 1983; Maloney et al., 1986; Nakamoto and Machida, 1992; Nicell et al., 1992).

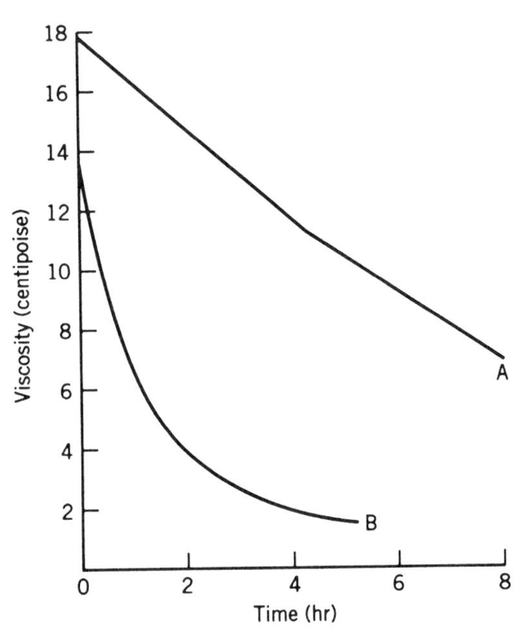

Fig. 16.3. Viscosity reduction of polysaccharides by depolymerase. **A.** 1.8×10^9 PFU/ml; **B.** 18×10^9 PFU/ml. Adapted from Nelson et al. (1988).

16.3. USE OF IMMOBILIZED CELLS IN WASTE TREATMENT

Some wastewater treatment processes are based on naturally immobilized microorganisms. Aggregated cells in activated sludge systems as well as cells attached to rock or plastic surfaces in trickling filters are good examples of exploitation of immobilized cells in waste treatment (Webb, 1987). Fluidized bed reactors, as used in water and wastewater treatment, are other examples of cell immobilization on surfaces such as sand or other particles. These fluidized bed reactors are used in the anaerobic treatment of wastewater (see Chapter 13). Addition of immobilized microorganisms for wastewater treatment is now under consideration in several laboratories.

16.3.1. Immobilization Techniques

Various approaches have been taken to the immobilization of microorganisms, animal and plant cells, and organelles. The various immobilization techniques used are summarized in Table 16.1 (Brodelius and Mosbach, 1987). The most popular approach to immobilization is entrapment of cells in polymeric materials such as alginate, carrageenan, or polyacrylamide. Owing to the toxicity and detrimental effect of polyacryalamide on cell viability, natural algal polysaccharides such as alginate and carrageenan have been the polymers of choice for microbial cell immobilization (Cheetham and Bucke, 1984). Carrageenan is an algal polysaccharide extracted from algae of the class Rhodophyceae; alginate is extracted from algae belonging to the class Phaeophyceae (e.g., *Laminaria hyperborea* or *Macrocystis pyrifera*). Alginate reacts with most divalent cations, particularly Ca^{2+}, to form gels. Briefly, the cells are mixed with a solution of sodium alginate and the mixture is poured dropwise over a solution of $CaCl_2$. Beads form instaneously and are left in the $CaCl_2$ solution for approximately 1 hr for complete gel formation (Bucke, 1987). However, calcium alginate beads can be destroyed if the surrounding medium contains phosphates and other calcium chelators. Loss of bead integrity may also result from excess cell growth or from gas production by the immobilized microorganisms.

Other techniques include the immobilization

TABLE 16.1. Cell Immobilization Techniques[a]

Species	Immobilization technique/support
Microoganisms	Entrapment
	Alginate
	Carrageenan
	Polyacrylamide
	Polyacrylamide-hydrazide
	Agarose
	Photo-cross-linkable resin, prepolymers, and urethane
	Epoxy carrier
	Chitosan
	Cellulose; cellulose acetate
	Gelatin
	Adsorption
	Sand
	Porous brick, porous silica
	Celite
	Wood chips
	Covalent binding
	Hydroxyethyl acrylate
Animal cells	Hollow fiber
	Entrapment (agarose and fibrin)
Plant cells	Entrapment
	Polyurethane
	Agarose, carrageenan, alginate
Plant protoplasts	Microcarriers
	Entrapment
	Alginate, agarose, and carrageenan
Organelles	Entrapment
	Alginate
	In cross-linked protein

[a]Adapted from Brodelius and Mosbach (1987).

of activated sludge microorganisms in polyvinyl alcohol (PVA) dropped in saturated boric acid solution or refrigerated for gel formation (Matsui et al., 1991). Cells can also be co-immobilized with magnetic particles in a polyacrylamide gel. Batch and continuous-flow experiments showed almost 100% phenol removal for at least 40 days by sludge microorganisms co-immobilized with ferromagnetic particles (Ozaki et al., 1991). A distinct advantage of this system is the ability to recover the immobilized microorganisms with a magnet.

Encapsulation of microorganisms within polymers generally produces relatively large beads (2–3 mm in diameter), in which immobilized microorganisms may be subjected to oxygen limitations. Additionally, these beads would not be suitable for certain environmental applications (e.g., groundwater restoration). An improved procedure for encapsulation of bacteria produces microspheres with a 2- to 50-μm diameter. The entrapped *Flavobacterium* cells are as active as free cells in pentachlorophenol degradation (Stormo and Crawford, 1992).

16.3.2. Some Examples of the Use of Immobilized Cells in Waste Treatment

The use of immobilized enzymes for the degradation of pesticides has been widely investigated. However, immobilized cells have also been considered for the treatment of various wastes and for decontaminating water or wastewater containing natural or xenobiotic compounds (Crawford and O'Reilly, 1989). The following are some examples of the use of this technology in pollution control.

Removal of brown lignin compounds. Brown lignin compounds found in paper mill effluents can be removed by immobilized white-rot fungus (*Coriolus versicolor*) (Livernoche et al., 1983).

Cyanide degradation by immobilized fungi. The fungi produce an enzyme, cyanide hydratase, that transforms cyanide into formamide (Nazaly and Knowles, 1981).

Biodegradation of phenolic compounds. Several studies addressed the degradation of phenolic compounds by using immobilized bacteria (Bettmann and Rehm, 1985; Bisping and Rehm, 1988; Hackel et al., 1975; Heitkamp et al., 1990). Although lower rates of phenol degradation were achieved by an immobilized consortium of methanogens, the immobilized microorganisms were able to tolerate higher phenol concentrations (Dwyer et al., 1986). Chlorinated phenols are degraded by bacteria (*Pseudomonas, Arthrobacter*) immobilized on chitin surfaces by covalent bonding (Portier, 1986), by *Alcaligenes* spp. entrapped in polyacrylamide-hydrazide (Bisping and Rhem, 1988), or by *Rhodococcus* immobilized on a polyurethane carrier (Valo et al., 1990). The biodegradation of 2-chlorophenol by activated sludge microorganisms immobilized in calcium alginate has also been demonstrated (Sofer et al., 1990). Bioreactors containing *Flavobacterium* immobilized in calcium alginate can degrade pentachlorophenol at a maximum degradation rate of 0.85 mg PCP per gram beads per hour (O'Reilly et al., 1988). Tyrosinase immobilized on magnetite can rapidly remove chlorophenols, methoxyphenols, and cresols from wastewater by oxidation. The *p*-substituted phenols are more readily removed than the *m*- or the *o*-substituted ones (Wada et al., 1992).

CH_4 production by immobilized methanogens (Karube et al., 1980). Anaerobic waste treatment may consist of two-stage bioreactors containing immobilized miroorganisms. The first stage contains acid formers; the second stage contains methanogens (Messing, 1988).

Dehalogenation of chloroaromatics (e.g., monochlorobenzoates, 2,4-dichlorophenoxyacetic acid) by immobilized *Pseudomonas* spp. cells (Sahasrabudhe et al., 1991).

Use of immobilized nitrifiers and denitrifiers to tackle nitrogen problems (Nilsson and Ohlson, 1982; Nitisoravut and Yang, 1992; van Ginkel et al., 1983). Nitrifying bacteria, immobilized in polyethylene glycol resin and added to activated sludge as suspended beads, enhance nitrification in activated sludge, thus reducing the time necessary for complete nitrification. The enhancement of nitrification by the immobilized bacteria is illustrated in Figure 16.4 (Tan-

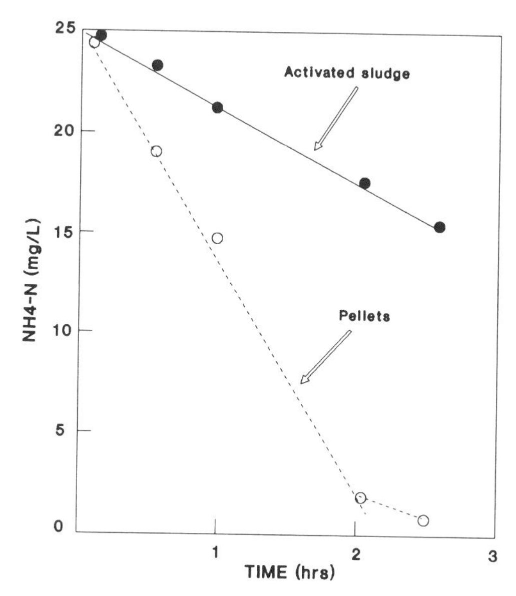

Fig. 16.4. Enhancement of nitrification by immobilization of nitrifiers. Adapted from Tanaka et al. (1991).

aka et al., 1991). Cell immobilization in polyvinyl alcohol has also been considered for enhancing nitrification in waste treatment (Ariga et al., 1987; Myoga et al., 1991). It appears that the immobilized nitrifiers are more stable than free cells (Asano et al., 1992). Bioreactors containing denitrifiers immobilized in cellulose triacetate carry out denitrification with more than 99% efficiency at nitrate-N loading rates lower than 420 mg/L·hr (Nitisoravut and Yang, 1992).

Immobilized activated sludge microorganisms. A two-step process, consisting of a reactor containing immobilized activated sludge microorganisms followed by a biofilm reactor, achieved a high treatment efficiency when the BOD load was 1.4 $kg \cdot m^{-3} \cdot day^{-1}$. Moreover, the operation of the first bioreactor under anaerobic conditions helped prevent the biofouling of the beads by filamentous fungi (Inamori et al., 1989).

Use of immobilized algae to remove micronutrients from wastewater effluents. Scenedesmus entrapped in carrageenan beads and *Phormidium* immobilized on the surface of chitosan flakes were shown to remove nitrogen and phosphorus from wastewater effluents (Chevalier and de la Noue, 1985; Proulx and de la Noue, 1988).

16.3.3. Use of Immobilized Cells and Enzymes in Biosensor Technology

A biosensor is composed of a biological sensing element (e.g., immobilized microorganism or enzyme) connected to a transducer. The signal given by the transducer is proportional to the chemical being analyzed. Biosensors are made of a wide range of biological elements and transducers (Table 16.2) (Turner et al., 1987). Several biosensors have been developed and applied to microbiological and biochemical processes in clinical, pharmaceutical, and food industries as well as wastewater treatment (Karube, 1987; Karube and Tamiya, 1987). A breakthrough in medical technology was the development of biosensors capable of rapidly measuring glucose and urea in body fluids. Examples of biosensors that can be applied to wastewater treatment are those designed for the detection of BOD, ammonia, organic acids, and methane (Karube and Tamiya, 1987; Matsui et al., 1991).

16.3.3.1. BOD Sensor

More rapid methods for determination of BOD are being sought. Biosensors using pure microbial cultures (e.g., *Bacillus subtilis, Clostridium butyricum, Trichosporon cutaneum*) or mixtures

TABLE 16.2. Biosensor Components[a]

Biological elements	Transducers
Organisms	Potentiometric
Tissues	Amperometric
Cells	Conductimetric
Organelles	Impedimetric
Membranes	Optical
Enzymes	Calorimetric
Receptors	Acoustic
Antibodies	Mechanical
Nucleic acids	"Molecular" electronic
Organic molecules	

[a]Adapted from Turner et al. (1987).

of activated sludge microorganisms have been considered. A biosensor consisting of immobilized yeast, *Trichosporon cutaneum,* and an oxygen probe was developed for BOD estimation (Karube, 1987; Karube et al., 1977; Karube and Tamiya, 1987). The BOD biosensor includes an oxygen electrode that consists of a platinum cathode and an aluminum anode bathed in saturated KCl solution, and a Teflon membrane. Yeast cells are immobilized on a porous membrane and are trapped between the porous and the Teflon membranes. Oxygen consumption by the immobilized microorganisms causes a decrease in current until a steady state is reached. A good correlation was observed between the current drop and BOD values as obtained by standard methods. The BOD biosensor measures BOD between 3 mg/L and 60 mg/L (Karube and Tamiya, 1987). A biosensor using *T. cutaneum* immobilized in polyvinyl alcohol displays a very short response time (<30 s) and is stable for 48 days. This sensor showed a good correlation with the 5-day BOD test (Fig. 16.5) (Riedel et al., 1990).

16.3.3.2. Methane Biosensor

This biosensor consists of immobilized methanotrophic bacteria (*Methylomonas flagellata*) in contact with an oxygen electrode. The immobilized bacteria utilize methane as well as oxygen according to the following reaction (Matsunaga et al., 1980):

$$CH_4 + NADH_2 + O_2 \rightarrow CH_3OH + NAD + H_2O \quad (16.1)$$

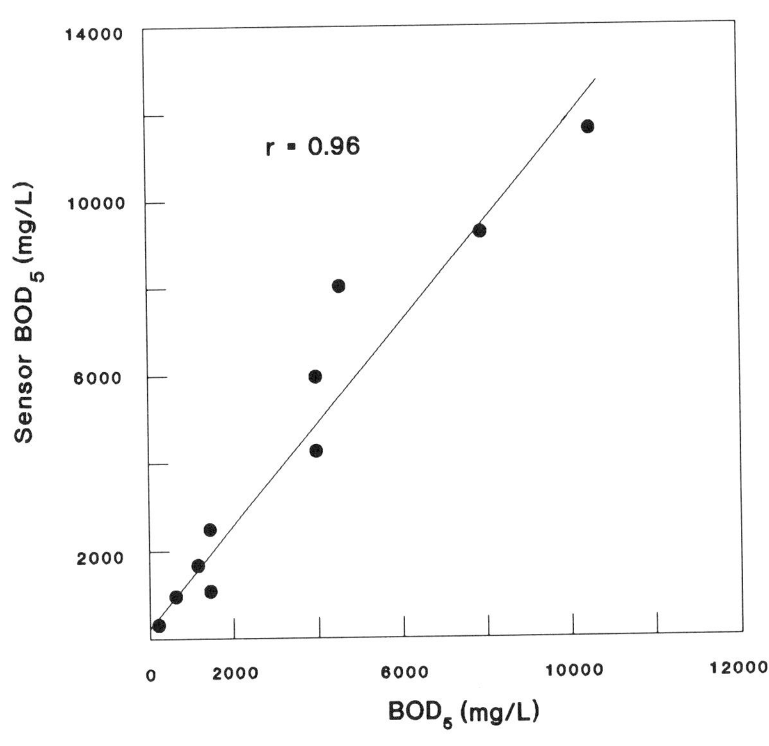

Fig. 16.5. Correlation between biosensor BOD value with the 5-day BOD value of wastewater. From Riedel et al. (1990), with permission of the publisher.

Oxygen consumption leads to a decrease in current, which is proportional to methane concentration in the sample.

16.3.3.3. Ammonia Biosensor

This biosensor, based on amperometry, consists of immobilized nitrifying bacteria and an oxygen electrode. This biosensor has been used for ammonia determination in wastewaters.

16.3.4. Advantages and Disadvantages of Immobilized Cells

Some advantages of cell immobilization are the following (Brodelius and Mosbach, 1987; Webb, 1987):

1. Continuous reactor operation without risk of cell washout
2. Ease of cell separation from reaction mixture
3. High cell density
4. Ability to reuse cells
5. Ability of different microbial species spacially separated to perform different functions
6. Enhanced overall productivity due to increased cell concentration in a given volume
7. Enhanced stability of immobilized microorganisms or enzymes
8. Decrease in volume of the bioreactors
9. Tolerance by immobilized microorganisms of higher levels of toxicants

However, cell immobilization has limitations:

1. Diffusion problems due to high cell density and to low solubility of oxygen in water
2. Changes in cell physiology that might affect productivity
3. Changes in composition of the microbial population, which can be a problem in wastewater that harbors mixed populations of microorganisms.
4. Cost of immobilization

16.4. ROLE OF MICROORGANISMS IN METAL REMOVAL IN WASTEWATER TREATMENT PLANTS

Physical/chemical processes are generally employed for removal of heavy metals from wastewater. These include ion exchange, oxidation/reduction, precipitation, ultrafiltration and many others. Much of the particulate-associated metals are removed by sedimentation during primary treatment of wastewater. The removal of soluble forms in the activated sludge system depends on the type of metal and is around 50%–60% for Cd, Hg, Cu, and Zn but may be lower for other metals (e.g., Ni, Co) (Sterrit and Lester, 1986). Microorganisms offer an alternative to physical/chemical methods for metal removal and recovery (Forster and Wase, 1987). Metals are removed (i.e., immobilized) by the activated sludge biomass. Their removal increases with sludge age and is partially due to increase in mixed-liquor concentration of solids.

16.4.1. Removal of Metal by Wastewater Microorganisms

Extracellular polymers produced by microorganisms commonly found in activated sludge display a great affinity for metals. Several types of bacteria (e.g., *Zooglea ramigera, Bacillus licheniformis*), some of which have been isolated from activated sludge, produce extracellular polymers that are able to complex and subsequently accumulate metals such as iron, copper, cadmium, nickel, or uranium. The accumulated metals can be easily released from the biomass by treatment with hydrochloric acid. For example, *Zooglea ramigera* can accumulate up to 0.17 g of Cu per gram of biomass (Norberg and Persson, 1984; Norberg and Rydin, 1984). This bacterium, when immobilized in alginate beads, is also able to accumulate Cd from solutions containing Cd concentrations as high as 250 mg/L (the alginate beads adsorb some of the cadmium) (Kuhn and Pfister, 1990). Some microorganisms may also synthesize siderophores, which chelate iron and facilitate its transport inside the cell

(Lundgren and Dean, 1979). Nonliving immobilized microbial systems are also able to remove metals from wastewaters. Patented processes involving immobilized bacteria, fungi, and algae have been developed to remove heavy metals from wastewater (Brierley et al., 1989). A proprietary product, called *Algasorb*, consists of algae cells embedded in a silican gel polymer material, and can remove heavy metals, including uranium (Anonymous, 1991). Fungal mycelia (e.g., *Aspergillus* and *Penicilium*) have also been considered for removal of metal from wastewater (Galun et al., 1982) and thus may offer a good alternative for detoxification of effluents. Metals are removed by fungi from solutions by sorption to the fungal surface or by a much slower, energy-dependent intracellular uptake. Biosorption column studies have shown that immobilized *Aspergillus oryzae* remove Cd efficiently from solution. It has been shown that detergent treatment of fungal biomass considerably improves removal of metal (Ross and Townsley, 1986). Cadmium appears to adsorb to the fungal biomass but active uptake of the metal by the fungus does not appear to be significant (Kiff and Little, 1986). Therefore, most of the removal of metals by fungi does not appear to be linked to a metabolic process. Table 16.3 (Eccles and Hunt, 1986) shows a list of some microorganisms used for removal and/or recovery of metals from industrial wastewaters.

16.4.2. Mechanisms of Metal Removal by Microorganisms

In the environment, including wastewater treatment plants, metals are removed by microorganisms by the following mechanisms (Brierley et al., 1989; Sterrit and Lester, 1986; Trevors, 1989; Trevors et al., 1985).

16.4.2.1. Adsorption to Cell Surfaces

Microorganisms bind metals as a result of interactions between metal ions and the negatively charged microbe surfaces. Gram-positive bacteria are particularly suitable for metal binding. Fungal and algal cells also display a high affinity for heavy metals (Darnall et al., 1986; Ross and Townsley, 1986). Sorption of metals to activated sludge solids has been found to conform to the Langmuir and Freundlich isotherms. Figure 16.6 (Mullen et al., 1989) displays the Freundlich isotherms for the adsorption of cadmium and copper to *Bacillus cereus* and *Pseudomonas aeruginosa*.

16.4.2.2. Complexation

Microorganisms can produce organic acids (e.g., citric acid), which may chelate toxic metals, resulting in the formation of metallorganic molecules. Metals may also be complexed by carboxyl groups found in microbial polysaccharides and other polymers. This phenomenon

TABLE 16.3. Some Microorganisms Involved in Metal Removal/Recovery From Industrial Wastewaters[a]

Microorganism removed/recovered	Metal
Zooglea ramigera	Copper
Saccharomyces cerevisieae	Uranium and other metals
Rhizopus arrhizus	Uranium
Chlorella vulgaris	Gold, zinc, copper, mercury
Aspergillus orhizae	Cadmium
Aspergillus niger	Copper, cadmium, zinc
Pecicillium spinulosum	Copper, cadmium, zinc
Trichoderma viride	Copper
AMT-Bioclaim(TM)	Biotechnology-based use of granulated product derived from biomass

[a]Adapted from Eccles and Hunt (1986).

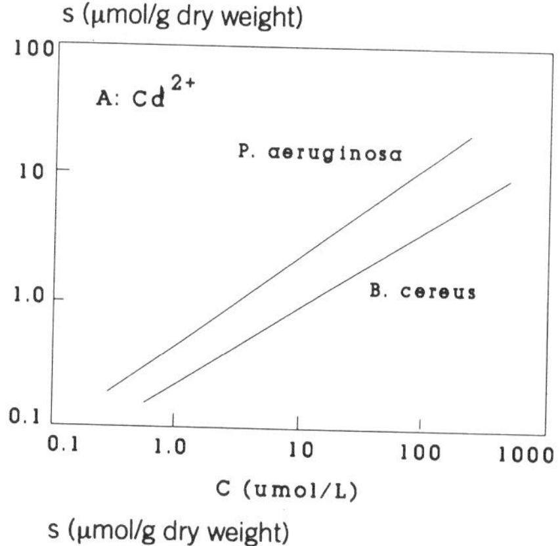

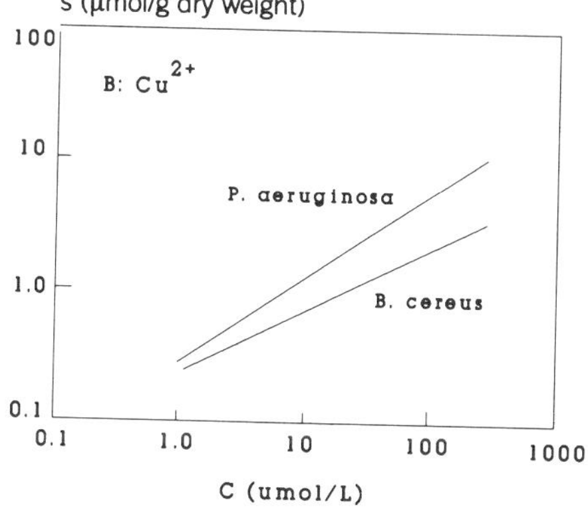

Fig. 16.6. Freundlich isotherms for sorption of cadmium **(A)** and copper **(B)** by *B. cereus* and *P. aeruginosa*. Adapted from Mullen et al. (1989).

is of great importance in wastewater treatment plants, particularly those using the activated sludge process, where industrial wastes are treated (Bitton and Freihoffer, 1978; Brown and Lester, 1979, 1982; McLean et al., 1990; Sterritt and Lester, 1986; Rudd et al., 1984). *Pseudomonas putida* has a cystein-rich protein that binds cadmium (Higham et al., 1984).

16.4.2.3. *Precipitation*

Some bacteria promote metal precipitation by producing ammonia, organic bases, or hydrogen sulfide, which precipitate metals as hydroxides or sulfides. Sulfate-reducing bacteria transform SO_4 to H_2S, which promotes the extracellular precipitation of metals from solution. *Klebsiella aerogenes* is able to detoxify cadmium to a cadmium sulfide (CdS) form, which precipitates as electron-dense granules at the cell surface. This process is induced by cadmium (Aiking et al., 1982).

16.4.2.4. *Volatilization*

Some metals are transformed to volatile species as a result of microbial action. For example, bacterially mediated methylation converts Hg^{2+} to dimethyl mercury, a volatile compound. Some bacteria have the ability to detoxify mercury by transforming Hg^{2+} into Hg^0, a volatile species. This detoxification process is plasmid-encoded and is regulated by an operon consisting of several genes. The most important gene is the *merA* gene, which is responsible for the production of mercuric reductase, the enzyme that catalyzes the transformation of Hg^{2+} to Hg^0.

16.4.2.5. *Intracellular Accumulation of Metals*

Microbial cells can accumulate metals, which gain entry into the cell by specific transport systems.

16.5. POTENTIAL APPLICATION OF RECOMBINANT DNA TECHNOLOGY IN WASTE TREATMENT

The application of recombinant DNA technology to domestic and industrial waste treatment is in its infancy. The relatively slow application of this technology in full-scale waste treatment is partially due to our lack of knowledge concerning the release of engineered microorganisms into the environment. In addition, the proposed new technology does not appear at the present time to be more economical than existing technologies.

Major applications of recombinant DNA technology include the enhancement of biodegradation of xenobiotics in wastewater treatment plants and the use of nucleic acid probes to detect pathogens and parasites in wastewater effluents and other environmental samples.

16.5.1. Some Tools in Recombinant DNA Technology

16.5.1.1. *Nucleic Acid Probes*

A gene probe is made of a piece of DNA controlling a desirable function in a cell (e.g., biodegradation of a given xenobiotic), and labeled with a radioactive element such as ^{32}P or with an enzyme (e.g., β-galactosidase, alkaline phosphatase). The probe can hybridize (i.e., become associated) with a complementary strand of target DNA isolated from a given environmental sample or a bacterial colony (Jain et al., 1988; Sayler and Blackburn, 1989). Nucleic acid probes are now available for the detection of a wide range of microorganisms. They can be used, for example, to detect microorganisms that are capable of degrading specific chemicals in wastewater samples. The use of the polymerase chain reaction (PCR) (see section 16.5.1.2) greatly enhances the sensitivity of nucleic acid probes. Some of these probes that can be combined with PCR technology are available for detecting bacterial, viral, and protozoan pathogens and parasites, and for tracking genetically engineered microorganisms (Sayler and Layton, 1990).

16.5.1.2. *Polymerase Chain Reaction (PCR)*

The PCR technique was developed in 1986 at Cetus Corporation by Mullis and collaborators (Mullis and Fallona, 1987). This technique essentially simulates *in vitro* the DNA replication process that occurs *in vivo;* it consists of amplifying discreet fragments of DNA by generating millions of copies of the target DNA (Atlas, 1991; Oste, 1988).

During cell division, two new copies of DNA are made and one set of genes is passed on to each daughter cell. Copies of genes increase exponentially as the number of generations increases. PCR simulates *in vitro* the DNA duplication process and can create millions of copies of the target DNA sequence. PCR consists of three steps, which constitute one cycle in DNA replication. The three steps are the following (Fig. 16.7):

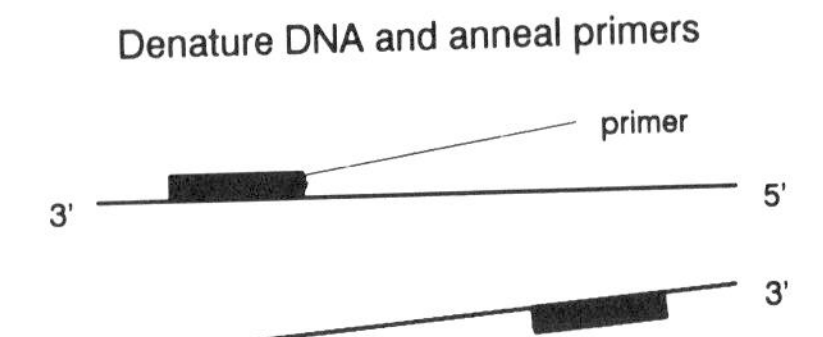

First cycle: Primer extension (complementary strand synthesis)

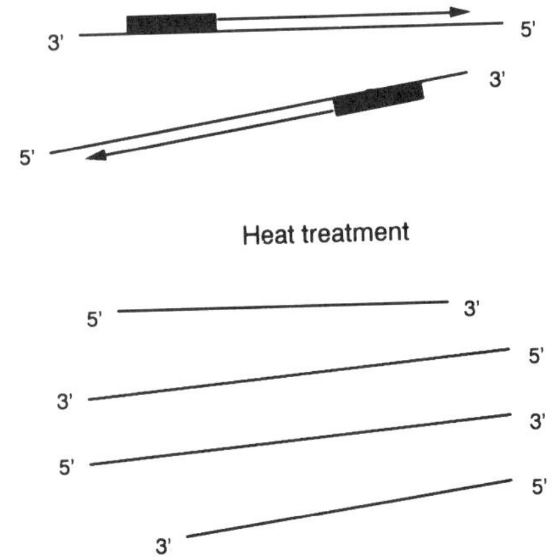

Second cycle: Primer extension (complementary strand synthesis)

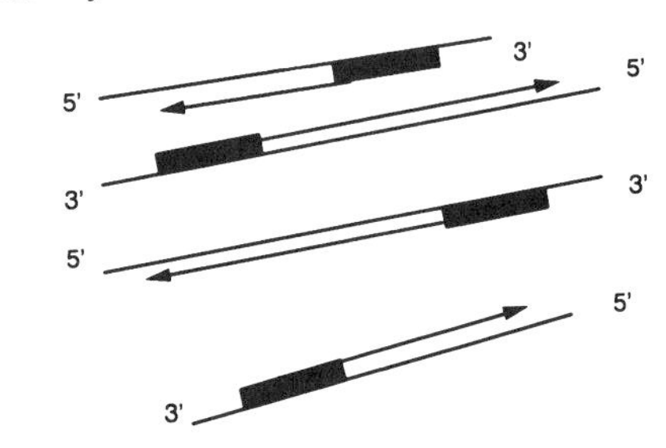

Fig. 16.7. Polymerase chain reaction (PCR). Adapted from Brown (1990).

1. *DNA denaturation.* When incubated at high temperature, the target double-stranded DNA fragment is denatured and dissociates into two strands.

2. *Annealing of primers.* When the temperature is lowered, the target DNA fragment anneals to synthetic nucleotide primers, which are made of 18 to 28 nucleotides and which flank the target DNA fragment.

3. *Primer extension or amplification step.*

The primers are extended with a thermostable DNA polymerase, the enzyme responsible for DNA replication in cells. This thermostable enzyme (*Taq* DNA polymerase) is extracted from *Thermus aquaticus,* a bacterium that is found in hot springs.

Following approximately 30 cycles that last approximately 3 hr, the target DNA fragment is amplified and accumulates exponentially. The PCR technique can be automated by using a *DNA thermal cycler,* which automatically controls the temperatures necessary for the denaturation and annealing steps.

This procedure is very useful in cloning, DNA sequencing, tracking genetic disorders, and forensic analysis and could be a powerful tool in diagnostic microbiology and virology. The tests presently used to detect exposure to the human immunodeficiency virus (HIV) detect only antibodies to the virus and not the virus itself. However, HIV was identified directly, with PCR, in the blood of AIDS patients from whom the virus was also isolated by more traditional tissue culture techniques. Virus identification by PCR is relatively rapid as compared to traditional culture techniques (Ou et al., 1988). The following are the environmental applications of PCR technology (Atlas, 1991):

1. Detection of specific bacteria. Specific bacteria in environmental samples, including wastewater, effluents, and sludges can be detected by PCR.

2. Environmental monitoring of GEMs. Genetically engineered microbes that perform certain useful functions (e.g., pesticide or hydrocarbon degradation) can be traced with PCR technology. The target DNA sequence is amplified *in vitro* and then hybridized to a constructed DNA probe.

3. Detection of indicator and pathogenic microorganisms. PCR technology has been considered for the detection of foodborne and waterborne pathogens and parasites in water, wastewater, and food. This sensitive technique can help detect as little as one organism in large volumes of water or wastewater. Examples of pathogens and parasites detected by PCR are invasive *Shigella flexneri,* enterotoxigenic *E. coli, Legionella pneumophila, Salmonella* (Bej et al., 1991b; Bej et al., 1992; Koide et al., 1993; Lampel et al., 1990; Olive, 1989; Tsai et al., 1993), hepatitis A virus (Divizia et al., 1993; Prevot et al., 1993), Norwalk virus (DeLeon et al., 1992), adenoviruses (Girones et al., 1993), enteroviruses (Abbaszadegan et al., 1993), human immunodeficiency virus (Ansari et al., 1992), *Giardia* (Mahbubani et al., 1991), and *Cryptosporidium* (Johnson et al., 1992). Multiplex PCR using several sets of primers allows the simultaneous detection of several pathogens or the selective detection of a specific pathogen (e.g., *Salmonella*) in an environmental sample (Way et al., 1993). PCR was also used to amplify *lacZ* and *uidA* genes for the detection of total coliforms (β-galactosidase producers) and *E. coli* (β-glucuronidase producer), respectively. An advantage of PCR is the detection of phenotypically negative *E. coli* in environmental samples (see more details in Chapter 4). Improvements in PCR technology for environmental monitoring are needed. Environmental samples as well as chemicals used for virus concentration contain substances (e.g., humic and fulvic acids, beef extract used in virus concentration) that interfere with pathogen detection by PCR. DNA or rRNA purification by gel filtration, using Sephadex or G-75, G-100, or G-200 helps remove the interference due to humic substances (Abbaszadegan et al., 1993; Tsai and Olson, 1992; Tsai et al., 1993). Inhibitors found in shellfish extracts can also be removed by treatment with cetyltrimethylammonium bromide prior to PCR (Atmar et al., 1993; Moran et al., 1993). Another problem is that PCR does not give an indication of the viability of the pathogens and parasites detected in the environment.

16.5.2. Potential Application of Recombinant DNA Technology to Waste Treatment

Some potential improvements in biological treatment of municipal wastewaters are presented in Table 16.4 (Rittmann, 1984). Some of these improvements can be accomplished by modifying existing technologies, by selecting

TABLE 16.4. Some Improvements in Biological Treatment of Municipal Wastewaters[a]

Improvement	Problem type[b]	Likely solution type[c]
Eliminate activated sludge bulking	2	a, d
Improve biofilm attachment	2	a, d
Stable nitrification	2	a
	3	d
Prevent sloughing in trickling filters	2	a
Reduce O_2 limitation in aerobic processes	3	d
Reduce energy consumption	2	a, b
Reduce sludge quantities produced	2	a, b
Enhance removal of phosphorus	2, 3	a, b, c, d
Biodegrade xenobiotic organics	2	a, b, c, d
Resist toxic upsets	2	a, d
Prevent generation of odors	2, 3	a, b, c
Make simple efficient processes for small communities	2	a, b

[a]From Rittman (1984), with permission of the publisher.
[b]Problem types: 1. Not feasible. 2. Not reliable or efficient. 3. Not economic.
[c]Solution types: a. Improve existing process. b. Use of new process. c. Use a novel microorganism. d. Apply genetic manipulation.

for novel microorganisms by traditional methods, or by using recombinant DNA technology. Successful use of these technologies can result, for example, in enhancement of denitrification in sewage treatment plants, increased growth rates of nitrifying bacteria (thus speeding up nitrification and reducing the sludge age in wastewater treatment plants), improved bacterial flocculation in activated sludge, improved performance of biofilm processes, enhanced biological removal of phosphorus, improved performance of methanogens, and better control of activated sludge bulking (Johnson and Robinson, 1982; Rittmann, 1984).

Genetically engineered microbes can be useful in several areas of waste treatment, among which are biomass production, biodegradation of recalcitrant molecules, fermentation (methane and organic acid production), enhancement of enzyme activity, and increased resistance to toxic inhibition. Waste treatment processes can be improved by selection of novel microorganisms that can thrive in domestic wastewater and perform a desirable function (Patterson, 1984; Rittmann, 1984). There are two steps involved in the application of recombinant DNA technology to waste treatment (Rittmann, 1984): The first step consists of finding a microorganism that has the desirable function (e.g., ability to degrade a pesticide). The second step consists of transferring this desirable function to a suitable host, preferably a microorganism with some relevance from an environmental viewpoint. Efforts are now being focused on understanding the genetic basis of biodegradation of xenobiotics in the environment in general and in biological treatment processes, in particular. Advances in recombinant DNA technology will help future development of engineered microbial strains with enhanced and broader biodegradative ability (Chaudhury and Chapalamadugu, 1991; Rittmann et al., 1990; Sayler and Blackburn, 1989). A useful application of genetic engineering would be in the treatment of industrial wastes. These wastes offer a harsh environment for the maintenance and growth of GEMs. Extremes in temperatures, pH, salinity, ionic composition, oxygen, and redox potential are often encountered in industrial wastes. Treatment of these wastes by conventional biological treatment technologies often necessitates several adjustments (temperature, pH, salinity) of wastewater for a successful treatment. Some (Kobayashi, 1984; Shilo, 1979) have proposed the use of GEMs originating from extreme environments (e.g., hypersaline waters, acid hot springs, alkaline lakes). These microorganisms would be able to degrade industrial wastes in special biological reactors. Biofilm reactors would constitute an acceptable option because they help minimize the potential release of GEMs into the environment.

Other potential applications include the use of genetically engineered multi-plasmid *Pseudomonas* strains for degrading several components found in crude oils (Friello et al., 1976). The enhancement of the level of several enzymes has been achieved by molecular biologists, using recombinant DNA technology. These enzymes include DNA ligase, tryptophan synthetase, α-amylase, benzylpenicillin acylase, etc. (Demain, 1984). Enhancing the production of enzymes involved in the degradation of recalcitrant organic molecules would be a useful undertaking.

However, the usefulness of DNA technology in pollution control is offset by the following limitations (Hardman, 1987):

1. The existence of multistep pathways in the degradation of xenobiotics raises the possibility that an engineered organism may not be able to completely mineralize the target xenobiotic.

2. The engineered microorganism may be capable of degrading only one or two compounds.

3. Knowledge about the degradative pathways of interest is limited or lacking. This knowledge would be useful for the identification of the responsible genes to be cloned.

4. The recombinant strain of interest may be unstable in the natural environment. Little is known about the competitive ability of recombinant strains with the more adapted indigenous population. Microcosms simulating an activated sludge system have been useful for studying the fate of GEMs that degrade substituted aromatic compounds. Recent studies showed that the GEMs survive well and degrade aromatic compounds in the complex activated sludge environment. There was no demonstrable negative effect of the GEMs on the indigenous microflora (Dwyer et al., 1988).

5. Finally public concern about the accidental or deliberate release of GEMs into the environment limit their application. The issues of concern include the persistence and reproduction of recombinant microorganisms in the natural environment, their interaction with indigenous organisms, and their impact on ecosystem function. Of specific concern is the persistence of added plasmids in wastewater treatment plants (Phillips et al., 1989). The use of suicide genes has been suggested for controlling the spread of GEMs in the environment. The GEMs would carry an inducible suicide gene that, once induced, would kill the GEM and thus halt its spread (Fox, 1989).

The benefits and risks of GEM use in the environment, and in wastewater treatment plants in particular, need to be addressed (Tiedje et al., 1989).

16.6. FURTHER READING

Atlas, R.M. 1991. Environmental applications of the polymerase chain reaction. ASM News 57: 630–632.

Brierley, C.L., J.A. Brierley, and M.S. Davidson. 1989. Applied microbial for metal recovery and removal from wastewater, pp. 359–381, in: *Metal Ions and Bacteria,* T.J. Beveridge and R.J. Doyle, Eds. Wiley, New York.

Cheetham, P.S.J., and C. Bucke. 1984. Immobilization of microbial cells and their use in waste water treatment, pp. 219–235, in: *Microbiological Methods for Environmental Biotechnology.*

Hardman, D.J. 1987. Microbial control of environmental pollution: The use of genetic techniques to engineer organisms with novel catabolic capabilities, pp. 295–317, in: *Environmental Biotechnology,* C.F. Forster and D.A.J. Wase, Eds. Ellis Horwood, Chichester, U.K.

Johnson, J.B., and S.G. Robinson. 1982. Opportunities for development of new detoxification processes through genetic engineering, pp. 301–314, in: *Detoxification of Hazardous Wastes,* J.H. Exner, Ed. Ann Arbor Science, Ann Arbor, MI.

Kobayashi, H.A. 1984. Application of genetic engineering to industrial waste/wastewater treatment, pp. 195–214, in: *Genetic Control of Environmental Pollutants,* G.S. Omenn and A. Hollaender, Eds. Plenum Press, New York.

Patterson, J.W. 1984. Perspectives on opportunities for genetic engineering applications in industrial pollution control, pp. 187–193, in: *Genetic Control of Environmental Pollutants,* G.S Omenn and A. Hollaender, Eds. Plenum Press, New York.

Rittmann, B.E. 1984. Needs and strategies for genetic control: Municipal wastes, pp. 215–228, in: *Genetic Control of Environmental Pollutants,* G.S. Omenn and A. Hollaender, Eds. Plenum Press, New York.

Sayler, G.S., and A.C. Layton. 1990. Environmental application of nucleic acid hybridization. Annu. Rev. Microbiol. 44: 625–648.

Tiedje, J.M., R. Colwell, Y.L. Grossman, R.E. Hodson, R.E. Lenski, R.N. Mack, and P.J. Regal. 1989. The planned introduction of genetically engineered organisms: Ecological considerations and recommendations. Soc. Ind. Microbiol. News 39: 149–165.

TRANSFORMATIONS AND TOXIC EFFECTS OF CHEMICALS IN WASTEWATER TREATMENT PLANTS

17

BIOTRANSFORMATIONS OF XENOBIOTICS AND METALS IN WASTEWATER TREATMENT PLANTS

17.1. INTRODUCTION

During the last few decades, an array of foreign compounds (i.e., foreign to biological systems) called xenobiotics has been introduced into the environment. These compounds are generally resistant or recalcitrant to biodegradation. Some chemicals such as the halogenated organic compounds (halogenated hydrocarbons, halogenated aromatics, pesticides, PCBs) are very resistant to microbial action. However, there are also refractory compounds of natural origin. These compounds include, among others, humic substances and lignin, as well as halogenated natural compounds generally found in the marine environment.

Industrial wastewaters contain relatively high concentrations of recalcitrant organic compounds that may be toxic, mutagenic, or carcinogenic (see Chapter 18) and may be bioaccumulated or biomagnified (i.e., their concentration increases at higher trophic levels) by the biota. These wastewaters, sometimes treated on site mostly by physicochemical processes, are discharged to surface waters or to conventional municipal wastewater treatment plants.

Microorganisms play a key role in biogeochemical cycles, particularly the carbon cycle. They degrade natural and anthropogenic compounds and release CO_2, CO, or CH_4. It is widely assumed that microorganisms can degrade any natural organic compounds although some (e.g., humic compounds, lignin) are quite resistant to biodegradation and have formed as a result of conditions unfavorable to their biodegradation.

Biotransformation is the alteration of organic compounds by microbial action, sometimes by *consortia* of microorganisms. The biological transformation of a xenobiotic may result in mineralization, accumulation, or polymerization of the compound with naturally occurring compounds (e.g., humic compounds) (Hardman, 1987; Bollag, 1979) (Fig. 17.1). Mineralization is the transformation of organic compounds to halide (e.g., chlorine, bromine), CO_2, and/or CH_4 (Rochkind-Dubinsky et al., 1987). Despite the abundant literature on biodegradability of toxic pollutants in wastewater treatment plants, little has been done to identify the microorganisms responsible for their biodegradation in these engineered systems.

17.2. BIODEGRADATION IN AQUATIC ENVIRONMENTS

Biological, chemical, and environmental factors affect the fate of chemicals in the environment (Giger and Roberts, 1978; Hardman, 1987; Johnston and Robinson, 1984; Lesinger et al., 1981; Sayler and Blackburn, 1989; Sayler et al., 1984). Table 17.1 gives more details concerning these factors.

Several investigators have isolated and identified microorganisms or microbial consortia that are capable of partially or completely degrading several classes of trace organics. More recently, much effort has been devoted to understanding the genetics of biodegradation with emphasis on

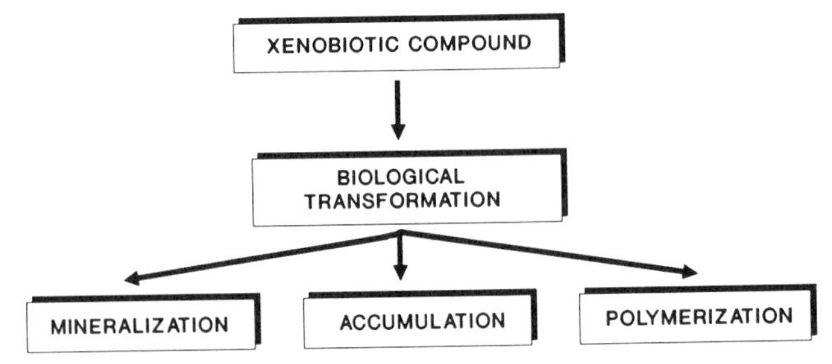

Fig. 17.1. The biological fate of xenobiotic compounds in the environment. From Hardman (1987), with permission of the author.

the study of catabolic plasmids, which code for the biodegration of xenobiotics. The ultimate goal is the construction of microbial strains that can be used for the bioremediation of hazardous waste sites. Several halogenated xenobiotics, however, are quite resistant to microbial action and several causes have been suggested to explain their recalcitrance.

17.2.1. Recalcitrance of Compounds

Resistance to biodegradation can result from the following (Alexander, 1985; Anderson, 1989; Hanstveit et al., 1988):

1. Molecular structure (e.g., substitutions with chlorine and other halogens)

TABLE 17.1. Factors Controlling the Fate of Xenobiotics in the Environment

Factors	Consequences
Chemical factors	
Molecular weight or size	Limited active transport
Polymeric nature	Extracellular metabolism required
Aromaticity	Oxygen-requiring enzymes (in aerobic environment)
Halogen substitution	Lack of dehalogenating enzymes
Solubility	Competitive partitioning
Toxicity	Enzyme inhibition, cell damage
Environmental factors	
Dissolved oxygen	O_2-sensitive and O_2-requiring enzymes
Temperature	Mesophilic temperature optimum
pH	Narrow pH optimum
Dissolved carbon	Concentration-dependence of organic/pollutant complexes for growth
Particulates, surfaces	Sorptive competition for substrate
Light	Photochemical enhancement
Nutrient and trace elements	Limitations on growth and enzyme synthesis
Biological factors	
Enzyme ubiquity	Low frequency of degradative species
Enzyme specificity	Analogous substrates not metabolized
Plasmid-encoded enzymes	Low frequency of degradative species
Enzyme regulation	Repression of catabolic enzyme synthesis Required acclimation or induction
Competition	Extinction of low density populations
Habitat selection	Lack of establishment of degradative populations

[a]Adapted from Sayler et al. (1984).

2. Failure of a compound to enter a cell owing to the absence of suitable permeases
3. Unavailability of the compound as a result of insolubility or adsorption, may make it inaccessible to microbial action
4. Unavailability of the proper electron acceptor
5. Unfavorable environmental factors such as temperature, light, pH, O_2, moisture, or redox potential
6. Unavailability of other nutrients (e.g., N, P) and growth factors necessary for microorganisms
7. Compound toxicity can affect the biodegradation potential by microorganisms, which have numerous ways to detoxify chemicals (e.g., catabolic plasmids). Some metabolites formed as a result of biodegradation may be more toxic than the parent compound.
8. Low substrate concentration can also affect biodegradation by microorganisms. Organisms growing at very low substrate concentrations have a high affinity for substrates (i.e., a very low half-saturation constant K_s). Some microorganisms in the environment may not be able to assimilate and grow on organic substrates below a threshold concentration. Most biodegradation studies have been carried out with relatively high substrate concentrations. This is not realistic because environmental concentrations (ppm or ppb level) are much lower than those that are sometimes utilized under laboratory conditions (Alexander, 1985).

On the basis of substrate concentration, there are two categories of microorganisms: (1) Copiotrophs, which grow rapidly in the presence of high substrate concentrations; and (2) oligotrophs, which grow slowly in the presence of low substrate concentrations. The latter have a high affinity for substrates (i.e., low K_s) and thus grow slowly but efficiently in the presence of low substrate concentrations. Microorganisms respond to low nutrient concentrations by adapting morphologically, leading to cells with high surface-to-volume ratios.

17.2.2. Dehalogenation of Organic Compounds

Research efforts have focused on the mechanisms of dehalogenation of synthetic halogenated organics because these chemicals are resistant to degradation (i.e., recalcitrant), are lipophilic, and tend to bioaccumulate in the food chain. Microorganisms must be capable of cleaving the carbon–halogen bond of halogenated compounds. However, this necessitates a lag period for the induction of dehalogenase enzymes, called hydrohalidases, which hydrolyze the carbon–halogen bond (Johnston and Robinson, 1984). Some mechanisms of dehalogenation are the following (Colwell and Sayler, 1978; Linkfield et al., 1989; Sims et al., 1991):

Reductive dehalogenation is the replacement of one Cl atom with one H atom. Reductive dehalogenation of halogenated organic compounds occurs under anaerobic conditions, particularly under methanogenic conditions, and requires the induction of dehalogenating enzymes.

Hydrolytic dehalogenation is the replacement of a halogen by a hydroxyl group.

17.2.3. Cometabolism

In cometabolism, an organic compound is converted to metabolic products but does not serve as a source of energy or nutrients to microorganisms (Alexander, 1981, 1985). The organisms derive their energy and carbon from a primary substrate but none from the xenobiotic, which acts as a secondary substrate. The reactions involved in cometabolism include dehalogenation, introduction of hydroxyl groups, ring cleavage, and oxidation of methyl groups (Rittmann et al., 1988).

Cometabolism has been demonstrated in pure and mixed cultures for a wide range of compounds (insecticides, herbicides, surfactants, aliphatic and aromatic hydrocarbons, and other industrial chemicals) and it has been suggested that it occurs in nature (Alexander, 1979; Horvath, 1972). There is some evidence of the occurrence of cometabolism in wastewater (Jacobson et al., 1980). Metabolic products of ^{14}C-labeled herbicides have been demonstrated in nonsterile but not in sterile primary sewage effluent after a relatively long period of incubation. However, the carbonaceous substrates were not incorporated into wastewater microorganisms, an indication that they did not serve as a source of carbon to microorganisms.

17.2.4. Genetic Regulation of Xenobiotic Biodegradation

The genes regulating the catabolism of many xenobiotics are plasmid-borne. Catabolic plasmids (or degradative plasmids) are extrachromosomal elements that control the transformation of xenobiotic compounds. They have been identified in a limited number of types of bacteria, mostly in the genus *Pseudomonas* (see Table 17.2) (Sayler and Blackburn, 1989). These catabolic plasmids may be lost when the microorganisms are not maintained on the substrate specific for the enzyme encoded by the plasmid. Catabolic plasmids may complement chromosome-encoded pathways. A well-described plasmid is the TOL plasmid (pWWO). This 117-kb plasmid encodes for the degradation of toluene and xylenes, and for its own replication and transfer (Saunders and Saunders, 1987; Sayler and Blackburn, 1989). Multiplasmid microbial strains have been constructed for the biodegradation of hydrocarbons in crude oil (Friello et al., 1976). These strains are capable of degrading toluene, xylenes, camphor, octane, and naphthalene (Saunders and Saunders, 1987). Degradative plasmids have also been constructed for the biodegradation of highly persistent and toxic xenobiotics, namely the chlorinated compounds. However, the use of these novel plasmid-bearing microorganisms under field conditions remains to be investigated.

Genetic ecologists work towards enhancing the capabilities of natural microbial communities at the gene level through gene amplification and increased expression (Olson and Goldstein, 1988; Olson, 1991). The increased activity of the natural microbial communities would lead to an enhanced degradation of toxic organic pollutants and to biotransformation of toxic metals to types that are less toxic as well as less available to the biota. This approach appears to be less troublesome than the one based on introduction of genetically engineered microorganisms into the environment.

There are certain advantages to a genetic ecological approach. In the first place, genes control specific functions, whereas whole cells carry out a multitude of functions many of which are useless to a particular goal. In addition, genes controlling a specific function (e.g., degradation of a xenobiotic compound) can be amplified through genetic manipulations (e.g., increasing the copy number of a desirable plasmid). Since several catabolic plasmids have been reported, the biodegradation of a given xenobiotic can be increased by encouraging gene transfer by manipulation of the environment (Olson, 1991).

TABLE 17.2. Some Degradative Plasmids[a]

Plasmid	Substrate	Conjugative ability
TOL	Toluene, meta-xylene, para-xylene	+
CAM	Camphor	+
OCT	Octane, hexane, decane	−
SAL	Salicylate	+
NAH	Naphthalene	+
NIC	Nicotine/nicotinate	+
pJP1	2,4-Dichlorophenoxyacetic acid	+
pAC21	4-Chlorobinphenyl	+
pAC25	3-Chlorobenzoate	+
pAC27	3- and 4-Chlorobenzoate	+

[a]Adapted from Hooper (1987) and Sayler and Blackburn (1989).

17.3. FATE OF XENOBIOTICS IN WASTEWATER TREATMENT PLANTS

There is great concern over the impact of toxic industrial pollutants on wastewater treatment processes and upon receiving waters such as lakes and rivers. Hazardous organic and inorganic pollutants are removed by biological and advanced wastewater treatment processes via the following mechanisms: sorption to sludge biomass and to activated carbon, volatilization, chemical oxidation, chemical flocculation and biodegradation (Grady, 1986; Metcalf and Eddy, 1991).

17.3.1. Physicochemical Processes

Sorption to sludge biomass and activated carbon. Nonpolar trace organics tend to adsorb to wastewater solids and some are removed following sedimentation, resulting in their transfer into sludge solids. It is well known that bacterial, algal, and fungal cells are capable of adsorbing and accumulating organic pollutants (Baughman and Paris, 1981; Lal and Saxena, 1982). The activated sludge biomass is able to adsorb organic pollutants such as lindane, diazinon, pentaclorophenol, and PCBs. Adsorption of these compounds generally fits the Freundlich isotherm. There is a good correlation between compound adsorption and the octanol/water partition coefficient. Since the adsorption of organic compounds is generally reversible, concern has arisen over their leaching following land application of sludges (Bell and Tsezos, 1987).

Activated carbon is suitable for removing residual refractory organic compounds and heavy metals. The carbon columns can be operated in the downflow or upflow modes. A process, called PACT, that combines the use of activated carbon and activated sludge removes ammonia, toxic metals, and trace organics.

Volatilization. Air stripping removes volatile organic compounds (VOC) and ammonia. Since air stripping transfers VOC from the water phase to the air phase, concern has arisen over the release of potential toxicants in the air. This has mainly to do with chlorinated compounds, 31%–

TABLE 17.3. Some Microorganisms Involved in the Biodegradation of Xenobiotics[a]

Organic pollutants	Organism
Phenolic compounds	*Achromobacter, Alcaligenes, Acinetobacter, Arthrobacter, Azotobacter, Bacillus cereus, Flavobacterium, Pseudomonas putida, P. aeruginosa,* and *Nocardia* *Candida tropicalis, Debaromyces subglobosus,* and *Trichosporon cutaneoum* *Aspergillus, Penicillium,* and *Neurospora*
Benzoates and related compounds	*Arthrobacter, Bacillus* spp., *Micrococcus, Moraxella, Mycobacterium, P. putida,* and *P. fluorescens*
Hydrocarbons	*Escherichia coli, P. putida, P. aeruginosa,* and *Candida*
Surfactants	*Alcaligenes, Achromobacter, Bacillus, Citrobacter, Clostridium resinae, Corynebacterium, Flavobacterium, Nocardia, Pseudomonas, Candida,* and *Cladosporium*
Pesticides	
DDT	*P. aeruginosa*
Linurin	*B. sphaericus*
2,4-D	*Arthrobacter* and *P. cepacia*
2,4,5-T	*P. cepacia*
Parathion	*Pseudomonas* spp. and *E. coli; P. stutzeri* and *P. aeruginosa*

[a]From Kumaran and Shivaraman (1988), with permission of the publisher.

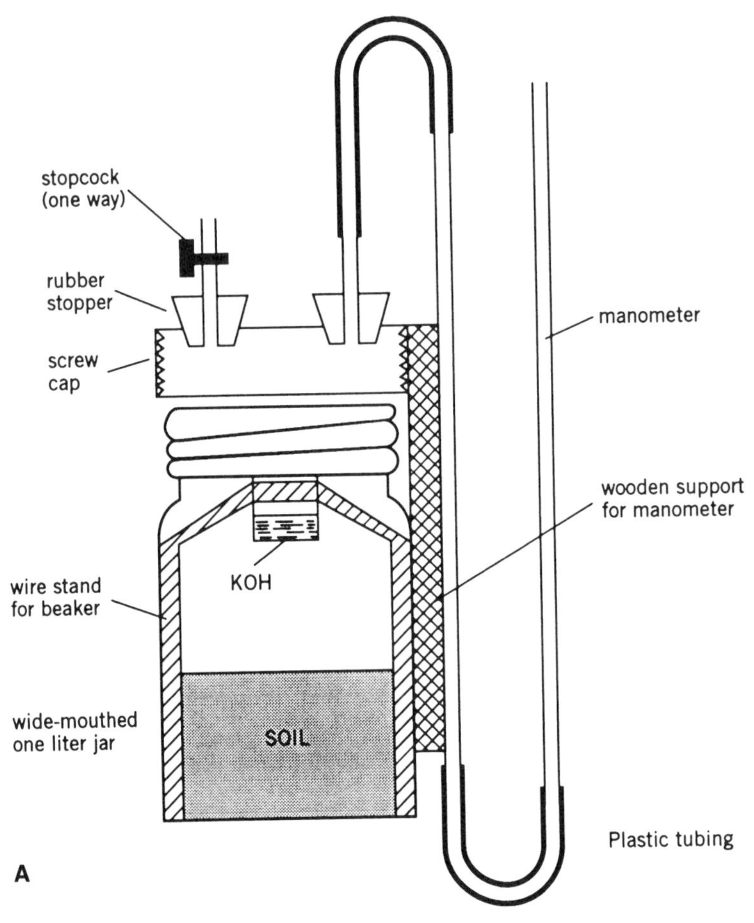

Fig. 17.2. Some biodegradability tests. **A:** Measurement of oxygen consumption. **B.** Measurement of CO_2 evolution. **C.** Measurement of gas CH_4/CO_2 production under anaerobic conditions. From Anderson (1989), with permission of the publisher.

71% of which air stripping transfers to the atmosphere (Parker et al., 1993).

Chemical oxidation. In addition to their disinfecting property, disinfectants such as chlorine dioxide or ozone are capable of oxidizing trace organics.

Photocatalytic degradation of organics in wastewater. This treatment, which consists of exposing wastewater to sunlight in the presence of a suitable catalyst (e.g., titanium dioxide), achieves complete mineralization of toxic organics (e.g., phenol, chlorinated compounds, surfactants) (Koramann et al., 1991; Okamoto et al., 1985; Ollis, 1985). The large-scale application of this technology to waste treatment is yet to be demonstrated.

Chemical coagulation. Chemical coagulation is suitable for removing toxic metals and trace organics.

17.3.2. Microbial Processes: Biodegradation

Biodegradation of organics by wastewater microorganisms is variable and is generally low for chlorinated compounds. Exposure of wastewater microbial communities to toxic xenobiotics results in the selection of resistant microor-

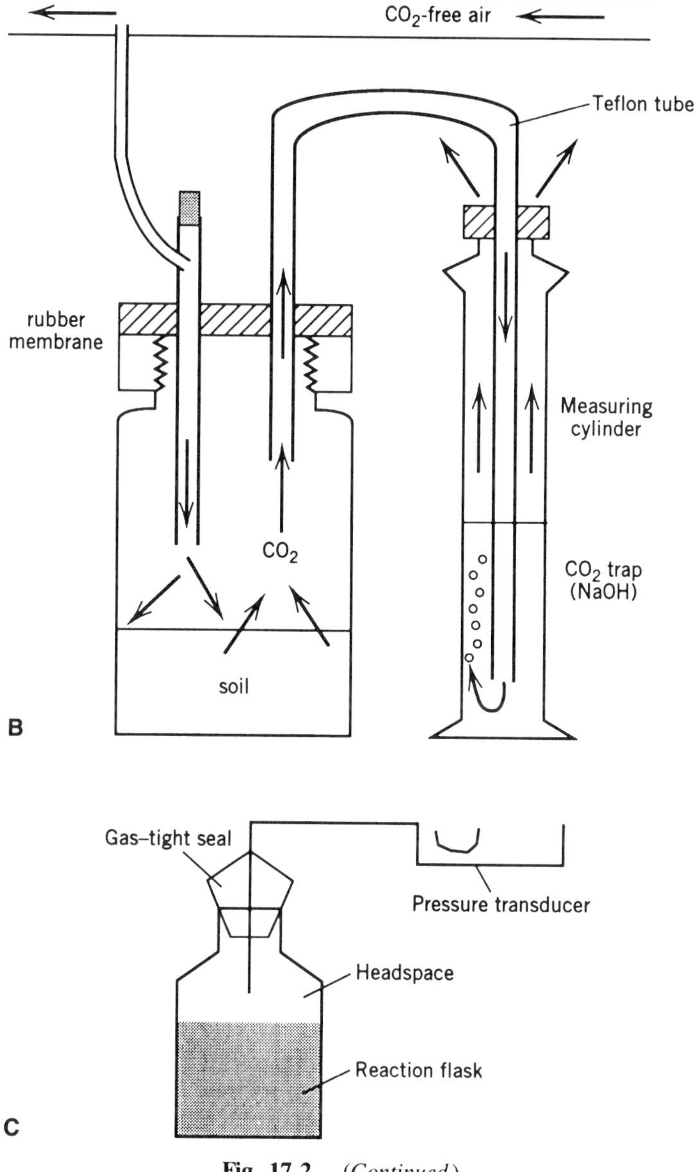

Fig. 17.2. (*Continued*)

ganisms that have the appropriate enzymes to use the xenobiotic as the sole source of carbon and energy. This process is called acclimation or adaptation. Most microorganisms need a period of acclimation prior to the onset of metabolism. Prior exposure to the xenobiotic helps reduce the acclimation period. Microbiologists have isolated a vast array of microorganisms that have the ability to degrade organic toxicants. A list of some of these microbial species is shown in Table 17.3 (Kumaran and Shivaraman, 1988). Since several xenobiotics are aromatic molecules, the microorganisms must be capable of cleaving the aromatic ring in these compounds. The biochemical pathways have been elucidated for many of the compounds. The fission of the aromatic ring is mediated by enzymes called monooxygenases, which introduce molecular oxygen into the ring prior to cleavage.

The biodegradability of organic compounds is often assessed by the disappearance of substrate or by removal of BOD or COD. More accurate approaches include recovery of radiolabeled parent substrate and/or metabolic products and mineralization products (measuring CO_2 or CH_4 production) (Rochkind-Dubinsky et al., 1987).

The assay systems for biodegradability test include the following (Fig. 17.2) (Anderson, 1989):

1. Measurement of oxygen consumption by manometric and electrolytic systems (Fig. 17.2A).
2. Measurement of CO_2 evolution by infrared or chemical methods (Fig. 17.2B).
3. Use of radiolabeled substrates (^{14}C labeled substrate).
4. Measurement of the disappearance of the chemical by gas chromatography or HPLC.
5. Determination of the reduction of dissolved organic carbon (DOC). For example, in Europe, the Association Française de Normalisation (AFNOR) test measures the reduction of DOC in aquatic samples. The pass level is 70% DOC reduction. In Japan, the Ministry of International Trade and Industry (MITI) test measures BOD reduction. The pass level is 60% BOD reduction.
6. Chemical biodegradability under anaerobic conditions (Fig. 17.2C). The test essentially consists of spiking a diluted sludge sample with the test chemical in a sealed bottle and measuring gas production ($CH_4 + CO_2$) by gas chromatography or by other means (Owen et al., 1979; Shelton and Tiedje, 1984).

17.4. REMOVAL OF TOXIC ORGANIC POLLUTANTS BY AEROBIC BIOLOGICAL PROCESSES

Several organic compounds have been detected in municipal wastewaters that are contributed by households, industry, commercial establishments, schools, universities, hospitals, and storm waters. These compounds belong to several categories which include phenols; aliphatic, polycyclic, and chlorinated aromatic hydrocarbons; phthalates; phosphate esters; ethers; terpenes; sterols; aldehydes; and acids and their esters. Hydrophobic organic compounds are generally effectively removed by wastewater treatment (Paxeus et al., 1992).

The removal of organic compounds by biological waste treatment depends mainly on waste composition, type of treatment, and solid retention time (SRT). Conventional aerobic biological treatment removes up to 85% of dissolved organic carbon. Table 17.4 shows that about half of the dissolved organic compounds remaining after biological treatment (activated sludge, trickling filter, or stabilization ponds) is composed of humic, fulvic, and hymathomelanic acids. The easily degradable compounds such as carbohydrates and proteins represent approximately 25% of the DOC in biologically treated wastewater (Giger and Roberts, 1978; Manka et al., 1974).

The removal of several organic toxicants by activated and trickling filter processes has been investigated (Hannah et al., 1986, 1988). In general, the activated sludge process is quite efficient in decreasing the concentration of many priority pollutants and other xenobiotics to concentrations below detection limits (Grady, 1986). The average removal for seven volatile organic pollutants exceeded 91% for both activated sludge and trickling filter processes (Table 17.5) (Hannah et al., 1988). However, removal is variable for semivolatile compounds and varies from 41% to 91% for trickling filters, and from 57% to 96% for activated sludge (Table 17.5). In general, the activated sludge process provides the best removal for both volatile and semivolatile organic pollutants. Faculative lagoons with a long detention time (25.6 days) are the best alternative to activated sludge for the removal of organics (Hannah et al., 1986). An activated rotating biological contactor removed 75%–96% of chlordane from contaminated wastewater. It was suggested that biodegradation is the main mechanism responsible for

TABLE 17.4. Distribution of Categories of Organic Compounds in Biologically Treated Effluents[a]

	Total chemical oxygen demand, %		
Chemical class	Trickling filter	Stabilization pond	Activated sludge (extended aeration)
Proteins	21.6	21.1	23.1
Carbohydrates	5.9	7.8	4.6
Tannins and lignins	1.3	2.1	1.0
Anionic detergents	16.6	12.2	16.0
Ether extractables	13.4	11.9	16.3
Fulvic acid	25.4	26.6	24.0
Humic acid	12.5	14.7	6.1
Hymathomelanic acid	7.7	6.7	4.8

[a]From Manka et al. (1974), with permission of the publisher.

TABLE 17.5. Removal of Volatile and Semivolatile Compounds by Biological Treatment[a]

	% Removal	
Pollutants	Trickling filter	Activated sludge
A. Semivolatile compounds		
Bis(2-ethylhexyl)phthalate	76	71
Dibutylphthalate	81	71
Naphthalene	89	95
Phenanthrene	90	93
Pyrene	83	91
Fluoranthene	85	92
Isophorone	65	96
Bis(2-chloroethyl)ether	31	64
p-Dichlorobenzene	86	94
Phenol	91	91
2-4-Dichlorophenol	55	86
Pentachlorophenol	41	60
Lindane	46	57
Heptachlor	64	67
B. Volatile compounds		
Carbon tetrachloride	90	81
1,1-Dichloroethane	92	97
1,1-Dichloroethylene	>97	>97
Chloroform	89	98
1,2-Dichloroethane	87	95
Bromoform	82	68
Ethylbenzene	>98	>98

[a]Adapted from Hannah et al. (1988).

chlordane removal (Sabatini et al., 1990). Anionic surfactants such as the linear alkylbenzenesulfonates (LAS) are thoroughly removed (average removal about 99%) following activated sludge treatment (Brunner et al., 1988). Cationic surfactants are removed in activated sludge processes by adsorption/precipitation as well as biodegradation. Phthalate esters, commercially used as plasticizers, some of which are listed as priority pollutants by the U.S. EPA, are rapidly biodegraded by activated sludge microorganisms. Their biodegradability decreases as their molecular weight increases (O'Grady et al., 1985; Shugatt et al., 1984). The majority (i.e., >76%) of *para*-dichlorobenzene is biodegraded during activated sludge treatment. Some of it is volatilized, and the extent of volatilization increases with increasing aeration rate (Topping, 1987).

Pentachlorophenol (PCP) is a biocide widely used as a wood preservative. It is also used as a fungicide/bactericide in many other products (Guthrie et al., 1984). Strains of *Flavobacterium* and *Pseudomonas* are known to degrade PCP under aerobic conditions (Saber and Crawford, 1985; Watanabe, 1973). *Flavobacterium* and *Pseudomonas* strains, isolated from PCP-contaminated soils, are capable of mineralizing up to 200 ppm and 160 ppm of PCP, respectively (Radehaus and Schmidt, 1992; Saber and Crawford, 1989). *Phanerochaete* spp., the white-rot fungus, is also capable of mineralizing pentachlorophenol as well as other xenobiotics such as PCBs, chlorinated anilines, and polycyclic aromatic hydrocarbons (Brodkorb and Legge, 1992; Field et al., 1992; Lamar and Dietrich, 1990; Lamar et al., 1990). A U.S. EPA survey has shown, however, that PCP removal by activated sludge is generally low and erratic. Much of the removal is due to PCP adsorption to sludge solids (Feiler, 1980). In laboratory-scale activated sludge units, PCP biodegradation increases with increasing solids retention time (SRT). Sorption is an important removal mechanism at low SRTs (Jacobsen et al., 1991), particularly for nonacclimated biomass (Blackburn et al., 1984; Edgehill and Finn, 1983a; Moos et al., 1983; Rochkind-Dubinsky et al., 1987). PCP is biotransformed (and sometimes completely mineralized) in reactors that contain acclimated biomass or are modified with PCP-degrading bacteria.

Because they are thermally stable and have good dielectric properties, polychlorinated biphenyls (PCB) are used as dielectric fluids and as fire retardants. Bacteria (e.g., *Pseudomonas, Achromobacter*) and fungi (e.g., *Aspergillus, Phanerochaete chrysosporium*) carry out the biodegradation of PCBs in the environment (Ahmed and Focht, 1973; Dmochewitz and Ballschmiter, 1988; Takase et al., 1986). The genes that regulate PCB biodegradation are either plasmid- or chromosome-encoded (Chaudhry and Chapalamadugu, 1991). The few studies that deal with the fate of PCBs in the activated sludge system have shown the involvement of sorption and stripping in the removal of these compounds but definite proof of biotransformation or mineralization is still lacking (Herbst et al., 1977; Kaneko et al., 1976; Rochlind-Dubinsky et al., 1987; Tucker et al., 1975).

The biotransformation of chloroaromatics by microorganisms is summarized in Figure 17.3 (Rochkind-Dubinsky et al., 1987). Chlorobenzenes and chlorophenoxy herbicides are metabolized to chlorophenols which in turn can be metabolized to chlorocatechols or chloroanisoles. Chloroaromatics that contain nitrogen are metabolized to chloroanilines, which form chlorocatechols, among other products. Chlorocathechols can be further metabolized to nonchlorinated by-products such as pyruvate, succinate, and acetyl-CoA. There are few data that demonstrate the biotransformation or mineralization of chloroaromatics in full-scale treatment plants (Rochkind-Dubinsky et al., 1987). Table 17.6 (Jordan, 1982) shows the removal of four chloroaromatics by an activated sludge unit. Their removal varies from 30% to 80%.

Aromatic and polynuclear aromatic hydrocarbons (PAHs) include compounds such as benzene, xylene, toluene, naphthalene, anthracene, styrene, phenantrene, and benzo(a)pyrene. Several of these chemicals or their oxidation products are carcinogenic. These chemicals are degraded under aerobic conditions (Crawford and

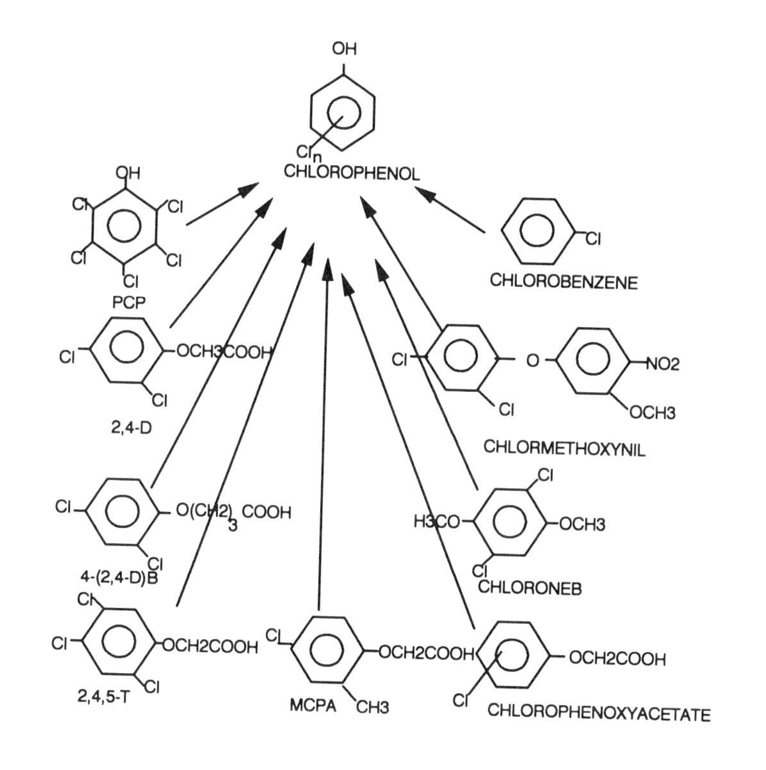

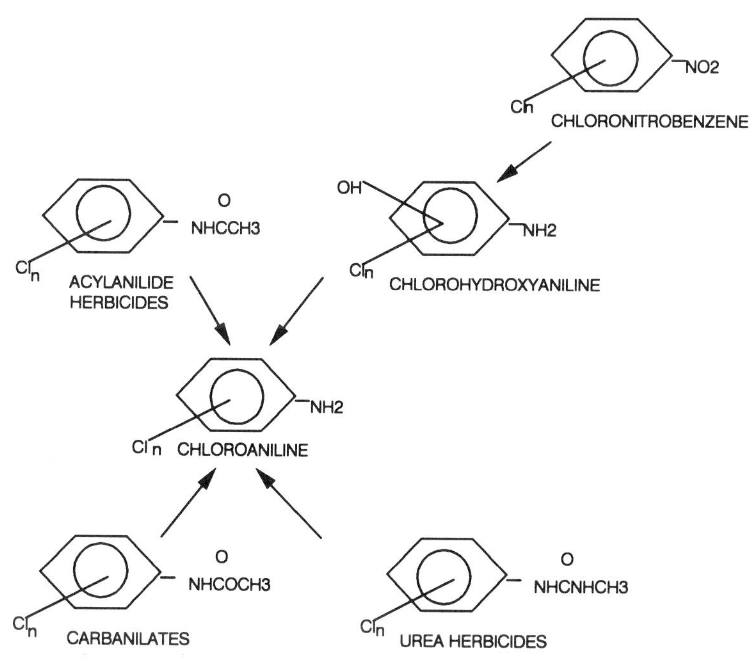

Fig. 17.3. Biodegradation of some chloroaromatic compounds. Adapted from Rochkind-Dubinsky et al. (1987).

TABLE 17.6. Removal of Chloroaromatic Compounds by an Activated Sludge Unit[a]

Compound	Percentage removal
2,4-Dichlorophenol	46
1,3-Dichlorobenzene	30
1,4-Dichlorobenzene	80
1,2,4-Trichlorobenzene	79

[a]Adapted from Jordan (1982).

O'Reilly, 1989). Under these conditions, benzene is converted to catechol, followed by ring fission ultimately to give CO_2 and H_2O (Fig. 17.4) (Gibson and Subramanian, 1984). With regard to hydrocarbons with side chains, biodegradation may proceed either by ring fission or oxidation of the side chain (Gibson and Subramanian, 1988).

The aerobic degradation of trichloroethylene (TCE) by mixed microbial consortia has been demonstrated in experimental expanded-bed bioreactors. More than 90% of TCE is degraded when propane or methane plus propane are used as primary substrates. The microbial consortia use propane more efficiently than methane, and no TCE intermediates (e.g., dichloroethylene, vinyl chloride) are detected in the bioreactors (Fig. 17.5) (Phelps et al., 1990). Fixed-film bioreactors, using methane or natural gas as the primary substrate, are able to remove TCE at levels less than 1,000 μg/L (TCE is toxic to methanotrophic bacteria at concentrations higher than 1,000 μg/L). Methanotrophic biofilms are also capable of removing more than 90% of TCE from vapor streams generated by air stripping of polluted groundwater and soil venting (Canter et al., 1990). In expanded-bed bioreactors, methane- or propane-fed microbial consortia can degrade mixed organic wastes containing benzene, toluene, xylene, TCE, vinyl chloride, and nine other chlorinated hydrocarbons. These consortia are able to degrade 80%–95% of TCE and 99% of the other organics (Phelps et al., 1991). Removal of nitrogen and phosphorus from wastewater can also be achieved by using methanotrophic attached-film bioreactors. Phosphorus is removed to less than 0.1 mg P/L and the ammonia removal efficiency varies between 94% and 99% (Jewell et al., 1992).

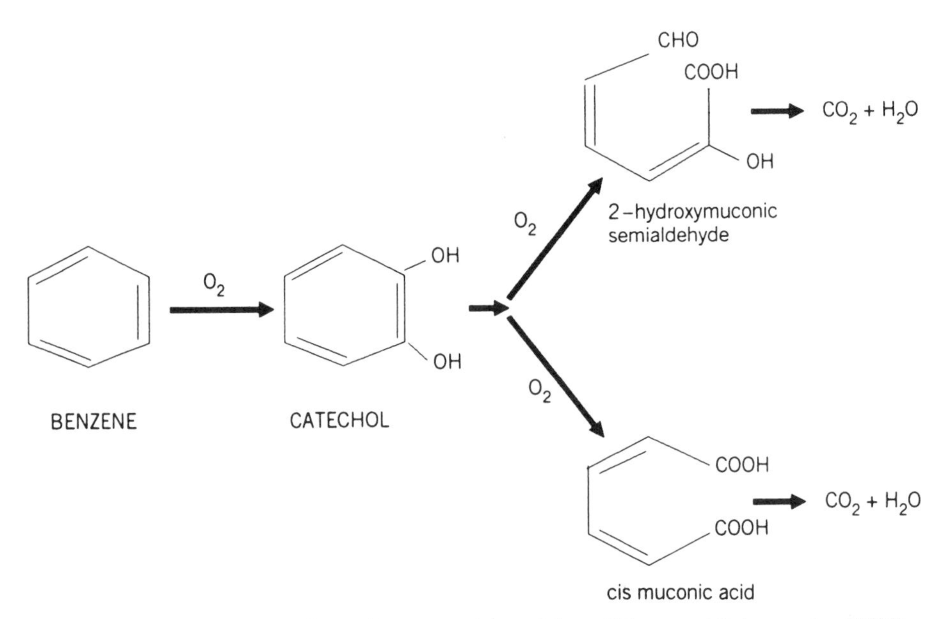

Fig. 17.4. Bacterial catabolism of benzene. Adapted from Gibson and Subramanian (1988).

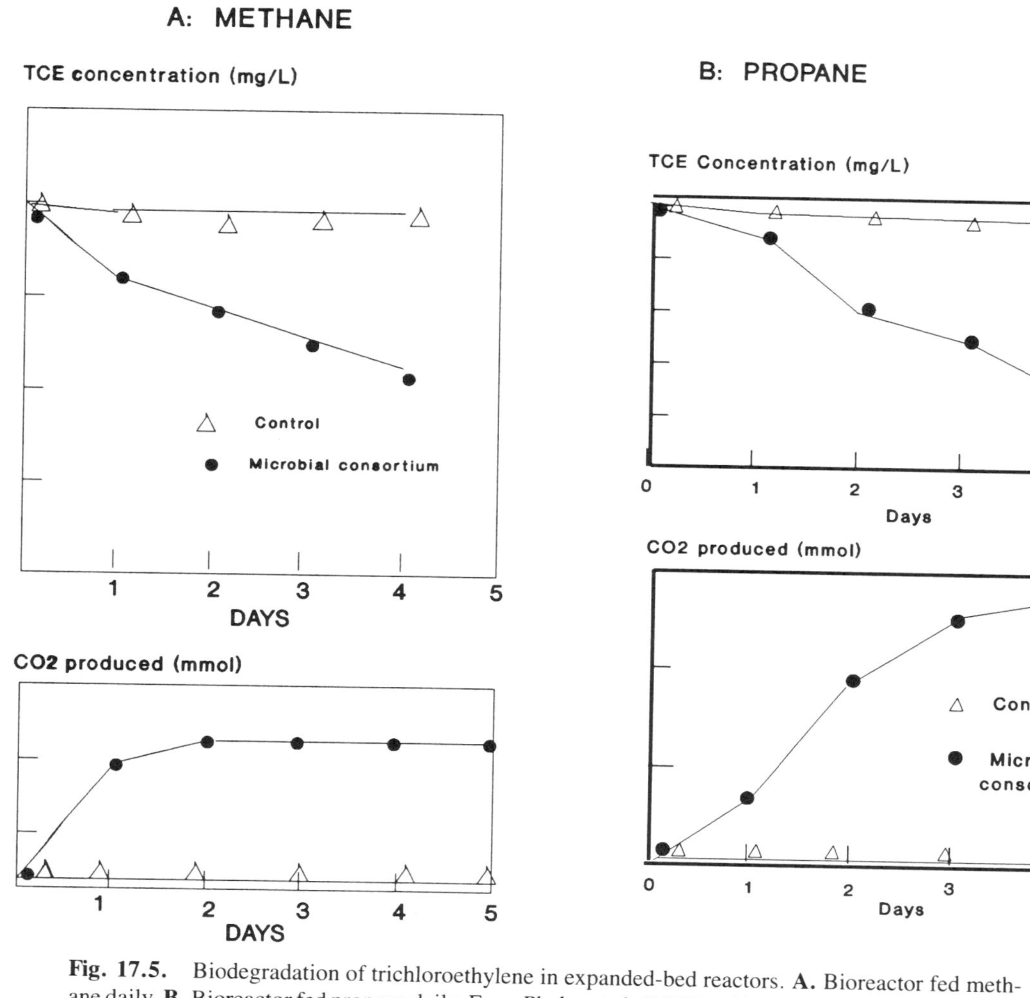

Fig. 17.5. Biodegradation of trichloroethylene in expanded-bed reactors. **A.** Bioreactor fed methane daily. **B.** Bioreactor fed propane daily. From Phelps et al. (1990), with permission of the publisher.

17.5. REMOVAL OF TOXIC ORGANIC POLLUTANTS BY ANAEROBIC AND ANOXIC BIOLOGICAL PROCESSES

In anaerobic respiration the oxidation of organic matter is coupled with the reduction of alternate electron acceptors such as nitrate (denitrification), sulfate (sulfate reduction), ferric iron (iron reduction), and CO_2 (methanogenesis). During the past two decades, significant advances have been made in our understanding of anaerobic processes (Suflita and Sewell, 1991). New high-rate anaerobic treatment processes have been developed and show great potential for the treatment of industrial wastewaters. These processes include suspended biomass and attached growth systems (Sterrit and Lester, 1988). The advantages of anaerobic treatment of wastewaters and sludges have been discussed in Chapter 13. We will now discuss the anaerobic biodegradation of selected categories of xenobiotics.

Phenolic compounds (e.g., phenol, catechol, resorcinol, *p*-cresol) make up 60%–80% of coal conversion wastewaters. They are degraded to methane and carbon dioxide under anaerobic conditions (Blum et al., 1986; Fedorak and Hrudey, 1984; Khan et al., 1981; Suidan et al., 1981; Young and Häggblom, 1989). Culture acclimation improves the metabolic and gas production rates for phenolic compounds. These compounds may also serve as a carbon source and electron donors to denitrifiers and thus are degraded under anoxic conditions (Fedorak and Hrudey, 1988).

Chlorinated phenols are used as biocides (e.g., wood preservation) and their formation under environmental conditions results from the use of chlorine as an oxidant in the paper industry or as a disinfectant in water and wastewater containing phenols (Ahlborg and Thunberg, 1980). Concern has arisen over the release of chlorophenols in the environment because some of these compounds are toxic, some are carcinogenic, and some may act as precursors of dioxins (Boyd and Shelton, 1984). Some chlorophenols (e.g., pentachlorophenol, 2,4-dichlorophenol, 2,4,5-trichlorophenol) have been placed on the U.S. EPA list of priority pollutants. Chlorophenols are biodegraded under both aerobic (Tokuz, 1989) and anaerobic conditions. They are degraded in wastewater sludge under methanogenic conditions via reductive dechlorination (chlorine is removed from the aromatic ring) as the initial step, followed by mineralization to CO_2 and CH_4 (Krumme and Boyd, 1988; Nicholson et al., 1992; Tiedje et al., 1987; Woods et al., 1989). In fresh sludge, the relative rate of biodegradation is ortho > meta > para (Boyd and Shelton, 1984; Boyd et al., 1983). *o*-Chlorophenol is degraded to phenol, thus indicating reductive dechlorination of the molecule (Boyd et al., 1983). In acclimated sludge, more than 90% of 2-chlorophenol, 4-chlorophenol, and 2,4-dichlorophenol was shown to be mineralized to CO_2 and CH_4 (Boyd and Shelton, 1984). Chlorophenol, used as the sole source of carbon and energy in an anaerobic upflow bioreactor, was dechlorinated by reductive dechlorination and mineralized to CO_2 and CH_4. Most of the methanogenic activity was located at the bottom of the bioreactor (Krumme and Boyd, 1988). There is evidence that chlorophenol biodegradation may also occur under sulfate-reduction conditions (Haggblom and Young, 1990; Kohring et al., 1989).

Pentachlorophenol is also biodegraded during anaerobic digestion of sludge and is transformed by reductive dechlorination to 3,5-dichlorophenol and other chlorinated phenols. Two thirds of labeled PCP was mineralized to CO_2 and CH_4 (Guthrie et al., 1984; Mikesell and Boyd, 1985, 1986).

The biodegradation of phenolic compounds under anaerobic conditions has been summarized by Fedorak and Hrudey (1988). In general, alkyl phenolics (e.g., cresols, dimethyl phenols, ethyl phenols) are the most resistant to biodegradation by methanogenic cultures.

Benzene and toluene are degraded under methanogenic conditions to CO_2 and CH_4 by a methanogenic consortium originally isolated from sewage sludge. Most of the CO_2 is derived from the methyl group of toluene (Grbic-Galic and Vogel, 1987). Chlorinated benzenes are widely used as fungicides, as industrial solvents, and in the manufacturing of various chemicals. They are widely detected in environmental samples, including wastewater, surface waters, groundwater, and sediments (Fathepure et al., 1988). Chlorobenzene and dichlorobenzene are utilized by bacteria (e.g., *Alcaligenes* spp. and *Pseudomonas* spp.) as a sole source of carbon (Chaudhury and Chapalamadugu, 1991). Available evidence shows that chlorobenzenes are degraded to chlorophenols and chlorocatechols (Rochkind-Dubinsky et al., 1987). Hexachlorobenzene is biodegraded to tri- and dichlorobenzene in anaerobically digested sludge by reductive dechlorination at a rate of 13.6 μM/L/day (Fathepure et al., 1988).

Chlorinated hydrocarbons (pentachloroethylene, trichloroethylene, and carbon tetrachloride) are also degraded under anaerobic conditions. Pentachloroethylene (PCE) is transformed by reductive dehalogenation to trichloroethylene (TCE), dichloroethylene (DCE), and vinyl chloride (VC) (Fathepure et al., 1987; Vogel and McCarty, 1985). Almost complete dechlorination of PCE to ethene is possible under anaerobic conditions in the presence of methanol. The microorganisms responsible for the dechlorination under those specific conditions have not been identified (DiStefano et al., 1991). TCE can be degraded under both aerobic or anaerobic conditions. Under methanogenic conditions, TCE biodegradation results in the formation of vinyl chloride, which is more genotoxic than the parent compound (Fig. 17.6) (Vogel and McCarty, 1985). Furthermore, vinyl chloride can be mineralized to CO_2, under aerobic conditions, by a wide range of microorganisms (Davis and Carpenter, 1990; Hartmans et al., 1985). Carbon tetrachloride is transformed to chloroform, dichloromethane, and CO_2 (Egli et al., 1988) but can be mineralized directly to CO_2, under denitrification conditions, without simultaneous production of chloroform, a dangerous by-product (Criddle et al., 1990).

Several other compounds are biodegraded under anaerobic conditions. These include phthalic acid esters, anionic surfactants, and polyethylene glycol (PEG). Phthalic acid esters can be biodegraded aerobically in activated sludge (O'Grady et al., 1985; Hannah et al., 1986), as well as anaerobically during anaerobic sludge digestion. However, some of these compounds

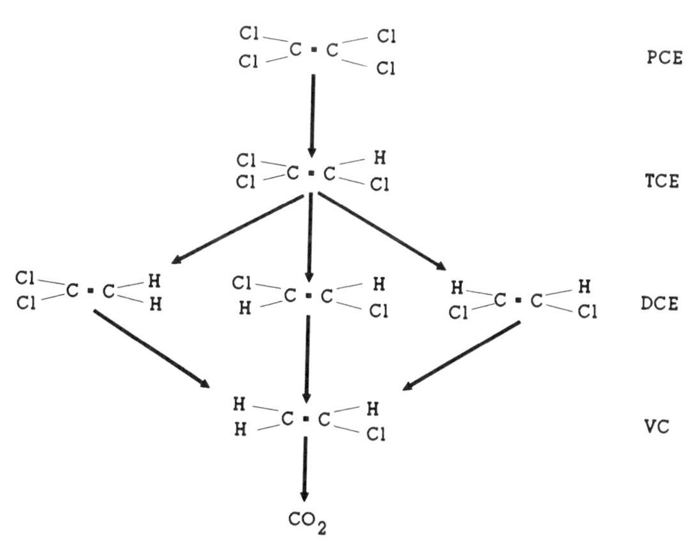

Fig. 17.6. Pathway for the conversion of pentachloroethylene to vinyl chloride through reductive dehalogenation. From Vogel and McCarty (1985), with permission of the publisher.

(e.g., di-2-ethylhexyl and di-*n*-octyl phthalates) are resistant to biodegradation (Ziogou et al., 1989). Anionic surfactants such as the linear alkylbenzenesulfonates (LAS) are partially transferred to sludge. They are slightly affected by anaerobic digestion and thus accumulate at levels in the range of approximately 4g kg^{-1} of sludge (Brunner et al., 1988). PEG is relatively resistant to microbial degradation in aerobic environments (Cox, 1978), whereas under anaerobic conditions methanogenic consortia biodegrade PEG to ethanol, acetate, methane, and ethylene glycol (Dwyer and Tiedje, 1983).

Several anthropogenic compounds can be metabolized by denitrifying organisms. Substrates metabolized by denitrifiers include benzoate (Garcia et al., 1981), phthalates (Aftring and Taylor, 1981), nonionic detergents (Dodgson et al., 1984), toluene and *m*-xylene (Zeyer et al., 1989), and chlorinated compounds (Bouwer and McCarty, 1983; Criddle et al., 1990).

17.6. BIODEGRADATION IN BIOFILMS

Adsorption of microorganisms to surfaces leads to the formation of biofilms in diverse environments such as trickling filters, water distribution pipes, and aquifer materials. We have discussed the mechanisms involved in biofilm formation as well as the microbial ecology of biofilms in Chapter 15.

Trace organics can be removed from water and wastewater by physical, chemical, and biological processes. Fixed-film or biofilm processes are simple and inexpensive means for biological removal of trace organics from water and wastewater (Namkung et al., 1983; Rittmann and McCarty, 1980a, 1980b; Williamson and McCarty, 1976). Models have been proposed to describe the kinetics the uptake of trace organics and subsequent growth and decay of biofilm microorganisms. The parameters of biofilm kinetics are generally estimated through conventional batch and continuous-culture techniques but can also be determined *in situ* by using biofilm reactors (Rittmann et al., 1986).

17.6.1. Steady-State Model

The steady-state model (microbial biomass growth balances biomass loss) describes the growth of biofilm bacteria in the presence of a single growth-limiting substrate (Rittmann and McCarty, 1980a, 1980b; McCarty et al., 1986). The model considers the diffusion of the substrate from the bulk solution into the biofilm and its subsequent utilization by biofilm bacteria, as well as bacterial growth and decay (Fig. 17.7) (Rittmann and McCarty, 1980a).

$$L_f = \frac{YJ}{b\,X_f} \quad (17.1)$$

where L_f = steady-state biofilm thickness (cm); Y = true yield of cell biomass per unit substrate used (mg VSS/mg substrate); X_f = biofilm cell density (mg VSS/cm^3); b = bacterial decay rate (day^{-1}); and J = substrate flux (mg/cm^2/day). J is the rate at which a substrate is transported from the bulk solution into the biofilm (Rittmann, 1987).

J is given by Fick's first law:

$$J = D\,\frac{S - S_s}{L} \quad (17.2)$$

where S = substrate concentration in the bulk solution (mg/L); S_s = substrate concentration at the biofilm surface (mg/L); L= effective diffusion layer (cm); and D= molecular diffusivity of substrate in water (cm^3/day).

There is a minimum substrate flux to support a deep biofilm. For heterotrophic microorganisms, this minimum was reported as 0.1 mg BOD_L/cm^2/day (BOD_L is the ultimate BOD as measured by oxygen demand). Substantial growth of the biofilm is obtained at J values three times higher than the minimum substrate flux. Nitrifiers are repressed at these high J values and are able to compete favorably with heterotrophs only at J values lower than the minimum of 0.1 mg BOD_L/cm^2/day (Rittmann, 1987).

There is a minimum steady-state substrate concentration, S_{min}, below which there is no biofilm growth. Figure 17.8 (Rittmann and McCarty, 1980b) shows that no further substrate degra-

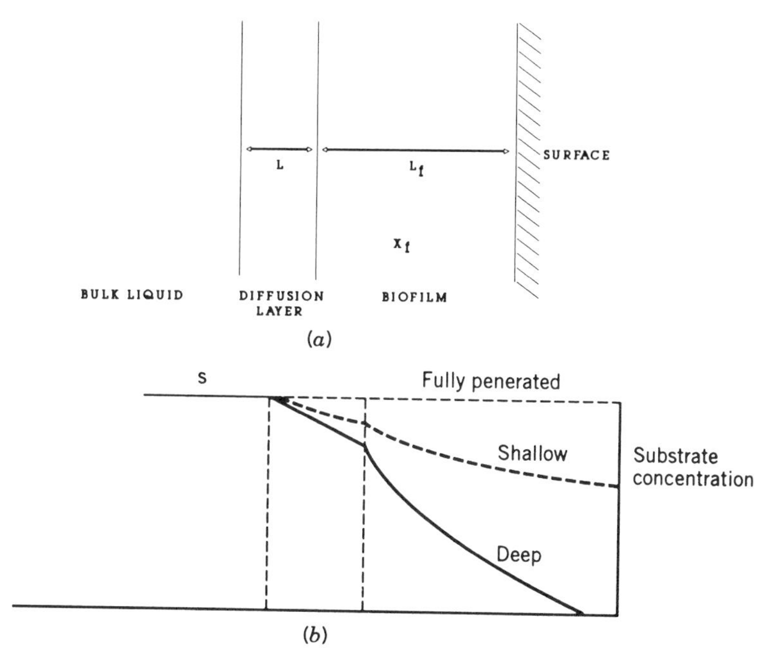

Fig. 17.7. Conceptual basis for biofilm model. **a.** Physical concepts. **b.** Substrate concentration profiles. Adapted from Rittmann and McCarty (1980a).

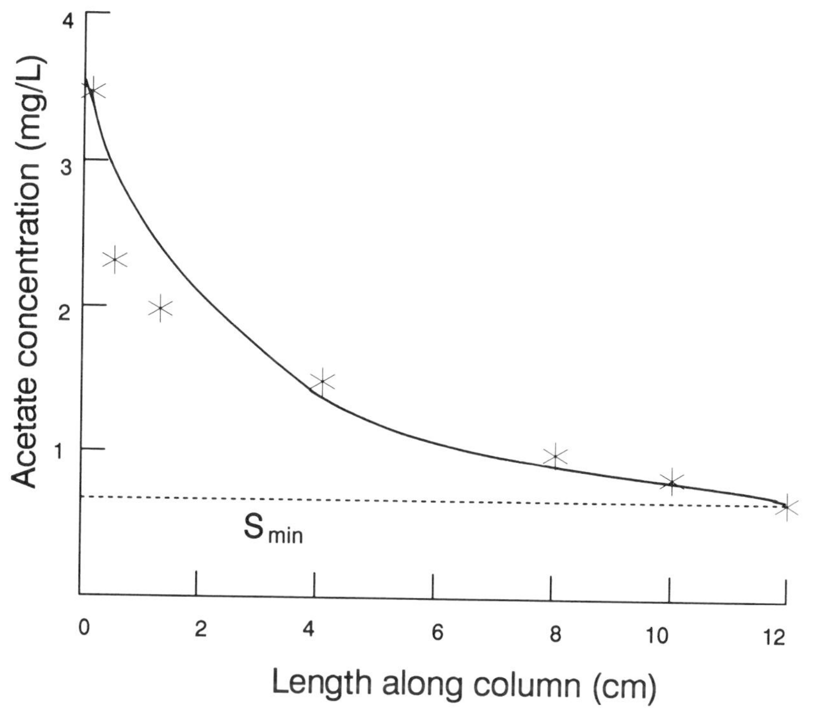

Fig. 17.8. Steady-state results for substrate (acetate) biodegradation in biofilms. Adapted from Rittmann and McCarty (1980b).

dation occurs when the substrate concentration approaches S_{min}. S_{min} is given by the following formula (Rittmann, 1987; Rittmann and McCarty, 1980a, 1980b; 1981; Rittmann et al., 1986, Rittmann et al., 1988):

$$S_{min} = \frac{K_s b}{Yk - b} = \frac{K_s b}{\mu_{max} - b} \quad (17.3)$$

where K_s = half-saturation constant (mg/L); b = bacterial decay rate (day^{-1}); Y = yield of cell biomass per unit substrate used (mg VSS/mg substrate); k = maximum specific substrate utilization rate (mg/mg VSS); and μ_{max} = maximum specific growth rate (day^{-1}).

In oligotrophic biofilms, bacteria have a high affinity for substrates (i.e., low K_s) and therefore a low S_{min} (Rittmann et al., 1986). Oligotrophic microorganisms (or K-strategists) also have a low μ_{max} and their K_s values vary from 0.0005 to 0.33 mg/L for suspended and biofilm oligotrophs (Table 17.7) (Rittmann et al., 1986)

TABLE 17.7. K_s Values of Suspended and Biofilm Oligotrophs[a]

Microorganisms and substrate	K_s mg/L
Pseudomonas aeruginosa	
Acetate	0.075
Aspartate	0.23
Flavobacterium spp.	
Maltose	0.065
Glucose	0.33
Starch	0.037
Amylose	0.083
Amylopectin	0.029
Spirillum spp.	
Oxalate	0.005
Salmonella typhimurium	
Glucose	0.25
Biofilm cultures	
Acetate	0.018
Phenol	0.012
Salicylate	0.0005–0.11

[a]Adapted from Rittmann et al. (1986).

whereas copiotrophs display K_s values between 1 and 20 mg/L (van der Kooij et al., 1982a; van der Kooij and Hijnen, 1984, 1985a; Rittmann et al., 1986; Schmidt and Alexander, 1985). Acetate-fed biofilm oligotrophs growing on activated carbon have a K_s value lower than 10 μg/L. The μ_{max} values for these oligotrophs vary between 0.14 and 0.20 hr^{-1} (Nakamura et al., 1989), whereas their maximum substrate uptake rate varies between 0.15 and 1.7 g COD per gram biomass-d (Rittmann et al., 1986).

For simple substrates, under aerobic conditions, S_{min} varies between 0.1 and 1 mg/L (Stratton et al., 1983). The S_{min} under methanogenic conditions is much higher and ranges from 3.3 to 67 mg/L (Rittmann et al., 1988).

17.6.2. Non-Steady-State Model

Substrate removal by biofilms at concentrations below S_{min} can be achieved by a non-steady-state biofilm (Rittmann and McCarty, 1981). Under laboratory conditions, this type of biofilm sustained more than 85% removal of galactose for a year, at substrate concentrations below S_{min} (galactose S_{min} = 0.41 mg/L) (Rittmann and Brunner, 1984). Cometabolism can also explain the utilization of trace organics by biofilms at concentrations below S_{min}.

17.7. METAL BIOTRANSFORMATIONS

Heavy metal sources in wastewater treatment plants include mainly industrial discharges and urban stormwater runoff. Biological treatment processes (activated sludge, trickling filter, oxidation ponds) remove from 24% (e.g., Cd) to 82% (e.g., Cu, Cr) of metals (Table 17.8) (Hannah et al., 1986). Toxic metals may adversely affect biological treatment processes as well as the quality of receiving waters. They are inhibitory to both anaerobic and aerobic processes in wastewater treatment (Barth et al., 1965).

The impact of metals on wastewater microorganisms, as well as bioassays for assessing their toxic action, will be discussed in Chapter 18.

TABLE 17.8. Percentage Metal Removal by Wastewater Treatment Plants[a]

Metal	Primary clarification	Trickling filter	Activated sludge	Aerated lagoon	Facultative lagoon
Cr	7	52	82	71	79
Cu	19	60	82	74	79
Ni	4	30	43	35	43
Pb	30	48	65	58	50
Cd	12	28	24	—	32

[a]From Hannah et al. (1986), with permission of the publisher.

17.7.1. Metabolic Activity and Metal Biotransformations

The metabolic activity of microorganisms can result in the solubilization, precipitation, chelation, biomethylation, or volatilization of heavy metals (Iverson and Brinckman, 1978). Microbial activity may result in the following:

1. Production of strong acids such as H_2SO_4 by chemoautotrophic bacteria (e.g., *Thiobacillus*), which dissolve minerals (see Chapter 3 for more details).
2. Production of organic acids (e.g., citric acid), which dissolve but also chelate metals to form metallorganic molecules.
3. Production of ammonia or organic bases, which precipitate heavy metals as hydroxides.
4. Production of H_2S by sulfate-reducing bacteria (see Chapter 3), which precipitates heavy metals as insoluble sulfides.
5. Production of extracellular polysaccharides, which can chelate heavy metals and thus reduce their toxicity (Bitton and Freihoffer, 1978).
6. Ability of certain bacteria (e.g., sheathed filamentous bacteria) to fix Fe and Mn on their surface in the form of hydroxides or some other insoluble metal salts.
7. Biotransformation by certain bacteria that have the ability to biomethylate or volatilize heavy metals (see Section 17.7.2).

17.7.2. Biotransformation of Specific Metals

Biomethylation of mercury was first reported in 1969. Since then, many bacterial (e.g., *Pseudomonas fluorescens, E. coli, Clostridium* sp.) and fungal (*Aspergillus niger, Saccharomyces cerevisiae*) isolates have been shown to methylate Hg(II) to methyl mercury. Vitamin B_{12} stimulates the production of methyl mercury by microorganisms.

$$Hg^{2+} \rightarrow \underset{\text{methyl mercury}}{Hg^{+}\text{—}CH_3} \rightarrow \underset{\text{dimethyl mercury}}{Hg(CH_3)_2} \qquad (17.4)$$

Methyl mercury accumulates under anaerobic conditions and is degraded under aerobic conditions. Sulfate-reducing bacteria are also capable of biomethylating mercury (Compeau and Bartha, 1985). Molybdate, an inhibitor of sulfate reducers, decreases mercury methylation in anoxic environments by more than 95% (Compeau and Bartha, 1987). Microorganisms can also transform $Hg^{(2+)}$ or organic mercury compounds (methyl mercury, phenylmercuric acetate, ethyl mercuric phosphate) to the volatile form Hg^0. Mercury volatilization is a detoxification mechanism, the genetic control of which has been elucidated. Plasmid genes code for resistance to metals such as mercury or cadmium. Resistance to mercury is carried out by the *mer* operon, which consists of a series of genes (*mer*A, *mer*C, *mer*D, *mer*R, *mer*T). The *mer*A gene is responsible for the production of mercuric reductase enzyme which transforms Hg^{2+} to Hg^0 (Silver and Misra, 1988).

Cadmium, Cd^{2+}, can be accumulated by bacteria (e.g., *E. coli, B. cereus*) and fungi (e.g., *Aspergillus niger*). This metal can also be volatilized in the presence of vitamin B_{12}. Like mercury, lead can be methylated by bacteria (e.g., *Pseudomonas, Alcaligenes, Flavobacterium*) to tetramethyl lead $(CH_3)_4Pb$.

Fungi (e.g., *Aspergillus, Fusarium*) are capable of transforming arsenic to trimethyl arsine, a volatile form with a garlic-like odor.

Selenium is another metal that is metylated through the metabolic activity of bacteria (*Aeromonas, Flavobacterium*) and fungi (*Penicillium, Aspergillus*).

Chromium-rich wastewaters are generated by industrial processes such as leather tanning, metal plating, and cleaning. An *Enterobacter cloacae* strain, isolated from municipal wastewater, was able to reduce hexavalent chromium (Cr^{6+}) to trivalent chromium (Cr^{3+}), which precipitates as a metal hydroxide at neutral pH, thus reducing the bioavailability and toxicity of this metal (Ohtake and Hardoyo, 1992).

17.8. MOBILE WASTEWATER PROCESSING SYSTEMS FOR BIODEGRADATION OF HAZARDOUS WASTES

Mobile waste treatment systems consist of modular equipment units that can be transported to a contaminated site by truck or railcar. They may also consist of larger units that require assembly on the site. Mobile waste treatment technology is not new, since it has been used for many years by the military. More recently, the U.S. EPA has used this technology for emergency response and remedial action for hazardous waste sites and contaminated aquifers. Mobile waste treatment systems are being investigated as possible alternatives to land treatment. Moreover, the use of the transportable units avoids the risks to humans and the environment that are associated with the transport of hazardous wastes to conventional treatment facilities. However, because of the limited data available on the performance and reliability of these systems, communities are sometimes reluctant to adopt them to tackle their pollution problems.

Mobile systems treat wastes by physical, chemical, and biological processes. These processes are listed in Table 17.9. We will concentrate on the biological processes used in mobile waste treatment systems (Glynn et al., 1987).

17.8.1. Aerobic Treatment Processes

Processes such as activated sludge or fixed-film bioreactors are used in mobile treatment units. In some instances powdered activated carbon is added to wastewater to help remove halogenated organics. Removal efficiency for organics is sometimes improved by addition of commercial acclimated bacterial strains. The removal efficiency for some organics can be as high as 99%. These systems can also be used for on-site treatment of groundwater from contaminated aquifers. Sludges generated by the mobile treatment units must be dewatered and sent off to a treatment and disposal facility.

TABLE 17.9. Processes Involved in Mobile Waste Treatment Systems[a]

Category	Process
Physical processes	Thermal treatments (incineration, wet air oxidation, etc. . .)
	Air stripping
	Adsorption to activated carbon
	Evaporation/dewatering
	Filtration
	Ion exchange
	Membrane separation
Chemical processes	Reduction/oxidation (redox)
	Precipitation
	Neutralization
	Dechlorination
Biological processes	Aerobic digestion
	Anaerobic digestion
	In situ biodegradation

[a]From Glynn et al. (1987), with permission of the publisher.

17.8.2. Anaerobic Digestion Processes

The processes involved in anaerobic digestion have been described in Chapter 13. Anaerobic digestion units are useful for the handling of high-strength wastewaters and leachates from hazardous waste sites. However, manufacturers rarely recommend them over aerobic systems, because the anaerobic biodegradation process can be inhibited by several factors, including sensitivity to heavy metal toxicity, variable temperature and pH, and sudden changes in wastewater characteristics.

17.9. FURTHER READING

Alexander, M. 1985. Biodegradation of organic chemicals. Environ. Sci. Technol. 18: 106–111.

Chaudhury, G.R., and S. Chapalamadugu. 1991. Biodegradation of halogenated organic compounds. Microbiol. Rev. 55: 59–79.

Grady, C.P.L. 1986. Biodegradation of hazardous wastes by conventional biological treatments. Hazard Wastes and Hazard. Materials 3: 333–365.

Hardman, D.J. 1987. Microbial control of environmental pollution: The use of genetic techniques to engineer organisms with novel catabolic capabilities, pp. 295–317, in: *Environmental Biotechnology,* C.F. Forster and D.A.J. Wase, Eds. Ellis Horwood, Chichester, U.K.

Leisinger, T., R. Hutter, A.M. Cook, and J. Nuesch, Eds. 1981. *Microbial Degradation of Xenobiotics and Recalcitrant Compounds.* Academic Press, New York.

Rittmann, B.E., D.E. Jackson, and S.L. Storck. 1988. Potential for treatment of hazardous organic chemicals with biological processes, pp. 15–64, in: *Biotreatment Systems,* Vol. 3, D.L. Wise, Ed. CRC Press, Boca Raton, FL.

Rochkind-Dubinsky, M.L., G.S. Sayler, and J.W. Blackburn. 1987. *Microbiological Decomposition of Chlorinated Aromatic Compounds.* Marcel Dekker, New York, pp. 315.

Sayler, G.S., and J.W. Blackburn. 1989. Modern biological methods: The role of biotechnology, pp. 53–71, in: *Biotreatment of Agricultural Wastewater,* M.E. Huntley, Ed. CRC Press, Boca Raton, FL.

18

TOXICITY TESTING IN WASTEWATER TREATMENT PLANTS USING MICROORGANISMS

18.1. INTRODUCTION

Serious concern has arisen over the release of more than 50,000 xenobiotics into the environment. Their impact on aquatic environments, including wastewaters, is generally determined by acute and chronic toxicity tests, consisting mostly of fish and invertebrate bioassays (Peltier and Weber, 1985). However, because of the large inventory of chemicals, short-term bioassays are now being considered for handling this task. These tests are mostly based on inhibition of the activity of enzymes, bacteria, fungi, algae, and protozoa (Bitton and Dutka 1986; Blaise, 1991; Dutka and Bitton 1986; Liu and Dutka 1984). These enzymatic and microbial assays, also called microbiotests, are simple, rapid, and cost-effective, and they can be miniaturized. The advantages of microbiotests are summarized in Table 18.1 (Blaise, 1991).

18.2. IMPACT OF TOXICANTS ON WASTEWATER TREATMENT

Toxic inhibition by organic (e.g., chlorinated organics, phenolic compounds, surfactants, pesticides) and inorganic (e.g., heavy metals, sulfides, ammonia) chemicals is a major problem encountered during the biological treatment of industrial wastewaters. Some of the chemicals that enter wastewater treatment plants, particularly the volatile compounds, are a potential health threat to plant operators. Many of the toxic chemicals or their metabolites, however, are transferred to wastewater sludges. The application of these sludges to agricultural soils may result in the uptake and accumulation of toxic and genotoxic chemicals by crops and grazing animals, eventually posing a threat to humans (see Chapter 19).

Chemical toxicants may also adversely affect biological treatment processes. Toxic inhibition is sometimes a serious problem in plants that treat industrial effluents. Activated sludge is the process most studied with regard to toxic inhibition (see anaerobic digestion in Chapter 13). The major effects of toxicants on activated sludge are reduced BOD and COD removal, reduced efficiency in solids separation, and modification of sludge compaction properties.

Chemical toxicants can also diminish the quality of receiving waters. Toxic wastewater effluents may threaten aquatic organisms in receiving waters, the use of which may be restricted. Guidelines are available for the levels of several heavy metals in receiving waters but less is known as to the levels of organic toxicants.

18.2.1. Heavy Metals

Heavy metals are major toxicants found in industrial wastewaters that may adversely affect the biological treatment of wastewater. The sources

TABLE 18.1. Attractive Features of Microbiotests[a]

Feature	Explanatory remark
Inexpensive or cost-efficient	Cost is test-dependent and can vary from a few dollars to several hundred dollars.
Generally not labor-intensive	As opposed to steps involved in undertaking fish bioassays, for example.
High sample throughput potential	When automation technology can be applied.
Cultures easily maintained or maintenance-free	Freeze-drying technology can be applied.
Modest laboratory and incubation space requirement	As opposed to a specialized laboratory essential for fish bioassays, for example.
Insignificant postexperimental chores	Owing to disposable plastic ware, which is recycled instead of having to be washed for reuse, as in the case of large experimental vessels.
Low sample volume requirements	Often, a few milliliters suffice to initiate tests instead of liters.
Sensitive/rapid responses to toxicants	Short life cycles of (micro)organisms enable end-point measurements after just minutes or several hours of exposure to toxic chemicals.
Precise/reproducible responses	High number of assayed organisms, increased number of replicates, and error-free robotic technology are contributors to this feature.
Surrogate testing potential	Microbiotests are adequate substitutes for macrobiotests in some cases.
Portability	For cases in which microbiotests are amenable to being applied in the field.

[a]From Blaise (1991), with permission of the publisher.

of heavy metals in wastewater treatment plants are mainly industrial discharges and urban stormwater runoffs. Some of the heavy metals are priority pollutants (e.g., Cd, Cr, Pb, Hg, Ag). Heavy metal toxicity is mainly due to soluble metals. It is controlled by various factors such as pH, type and concentration of complexing agents in wastewater, antagonistic effects by toxicant mixtures, oxidation state of the metal, and redox potential (Babich and Stotzky, 1986; Jenkins et al., 1964; Kao et al., 1982; Sujarittanonta and Sherrard, 1981). Metals may be complexed with natural (e.g., humic substances) or anthropogenic substances (e.g., nitrilotriacetic acid, which is used as a builder in detergents). Microorganisms can also affect the complexation of metals and modify their solubility (see Chapter 17).

Anaerobic wastewater treatment processes appear to be more sensitive to heavy metals than aerobic processes (Barth, 1975). Heavy metals inhibit two important processes that occur during aerobic treatment of wastewater, namely COD removal and nitrification. Shock loads of heavy metals (e.g., Hg, Cd, Zn, Cr, Cu) may lead to deflocculation of activated sludge (Henney et al., 1980; Lamb and Tollefson, 1973; Neufield, 1976). The compaction behavior of activated sludge may be affected as a result of differential effects of heavy metals on filamentous and floc-forming microorganisms. Activated sludge bulking appears to occur less in systems receiving heavy metals (Barth et al., 1965; Neufield, 1976). This phenomenon is also observed in the presence of organic toxicants (Monsen and Davis, 1984).

Methanogenesis is affected by heavy metals, although some metals (e.g., Ni, Co), at trace levels, may be stimulatory to methanogenic bacteria. Heavy metal toxicity depends on sulfide and phosphate levels as well as other ligands present in wastewater. Other inhibitors of methanogenesis include ammonia, phenols, chlorinated hydrocarbons, benzene ring compounds, formaldehyde, detergents, and sulfides (this topic is covered in more detail in Chapter 13).

Activated sludge and facultative lagoons provide the best removal for toxic metals such as Cd, Cr, Cu, Zn, Ni, and Pb. Metals are generally concentrated in sludges (Hannah et al., 1986). Metal removal by activated sludge is due to sorption of the metals to flocs. Removal by biological solids depends on pH, solubility and concentration of metals, concentration of organic matter, amount of biomass, and biological solid retention time (Cheng et al., 1975; Nelson et al., 1981; Sujarittanonta and Sherrard, 1981). The affinity of biological solids for heavy metals was found to follow the order Pb > Cd > Hg > Cr^{3+} > Cr^{6+} > Zn > Ni (Neufield and Hermann, 1975). Other means of removal of heavy metals by microorganisms are complexation by carboxyl groups of microbial polysaccharides and other polymers, precipitation (e.g., precipitation of Cd by *Klebsiella aerogenes*), volatilization (e.g., mercury), and intracellular accumulation. The role of microorganisms in metal removal in waste treatment was explored in Chapter 16.

18.2.2. Organic Toxicants

There are two concerns over the fate of organic toxicants in wastewater treatment plants.

1. *Biodegradation of organic toxicants in wastewater treatment plants* (see Chapter 17). It is desirable that xenobiotics be mineralized to CO_2 or, at least, to less toxic metabolites.

2. *Toxicity of xenobiotics to waste treatment organisms,* with subsequent reduction in the removal of biogenic organic compounds (i.e., lower BOD reduction) or inhibition of nitrification (Boethling, 1984) and methane production (Fedorak and Hrudey, 1984).

The toxicity of phenolic compounds has been widely studied in wastewater treatment plants. Phenol and *p*-cresol inhibit methane production when their concentration is above 2,000 and 1,000 mg/L, respectively (Fedorak and Hrudey, 1984). Phenol toxicity is also observed in fixed-film biological reactors at concentrations between 1,000 and 3,000 mg/L (Sayama and Itokawa, 1980). Pentachlorophenol, a widely used pesticide in the United States, is inhibitory to ATP synthesis and methane production in methanogenic bacteria (Roberton and Wolfe, 1970). PCP is inhibitory to unacclimated methanogens at a threshold concentration of 200 μg/L. Acclimation raises the threshold to 600 μg/L (Guthrie et al., 1984). The toxic effect of organic and inorganic toxicants on methanogenesis is discussed in Chapter 13.

Inhibitors affect substrate removal by microorganisms in a manner similar to the way they affect enzyme activity (Hartmann and Laubenberger, 1968; Volskay and Grady, 1988). There are four major types of reversible inhibitors (competitive, noncompetitive, uncompetitive, and mixed-type) that affect biodegradation, based on their effect on the kinetic parameters μ_{max} and K_s (see Chapter 2 for more details). Respiration-based toxicity studies in pilot plant bioreactors have shown that 13 of 33 RCRA-listed (RCRA = Resource Conservation and Recovery Act) organic compounds display on EC_{50} above 100 mg/L and therefore will not have a significant impact on activated sludge operation at the lower concentrations actually found. Characterization of the inhibition caused by RCRA compounds showed that none of the compounds displayed competitive inhibition (Table 18.2) (Volskay et al, 1990).

TABLE 18.2. Characterization of Inhibition Caused by RCRA-Listed Organic Compounds[a]

Inhibitor	Type of inhibition
Carbon tetrachloride	Mixed
Chlorobenzene	Mixed
Chloroform	Mixed
1,2-Dichloroethane	Mixed
1,2-Dichloropropane	Mixed
2-4-Dimethylphenol	Uncompetitive
Methylene chloride	Mixed
Nitrobenzene	Uncompetitive
Phenol	Uncompetitive
Tetrachloroethylene	Noncompetitive
Toluene	Mixed
1,1,1-Trichloroethane	Mixed
1,1,2-Trichloroethane	Mixed
Trichloroethylene	Mixed

[a]From Volskay et al. (1990), with permission of the publisher.

18.3. TOXICITY ASSAYS USING ENZYMES AND MICROORGANISMS

Applications of microbial toxicity assays in wastewater treatment plants fall into four categories.

1. The first category involves the use of these assays to monitor the toxicity of wastewaters at various points in the collection system, the major goal being the protection of biological treatment processes from toxicant action. These screening tests should be useful for pinpointing the source of the toxicants entering the wastewater treatment plant.

2. The second category involves the use of these toxicity assays in process control to evaluate pretreatment options for detoxifying incoming industrial wastes.

3. The third category concerns the application of short-term microbial and enzymatic assays to detect toxic inhibition of biological processes used in the treatment of wastewaters and sludges.

4. The last category deals with the use of these rapid assays in toxicity reduction evaluation (TRE) to characterize the problem toxic chemical(s) (See section 18.4).

We will now review the most frequently used enzymatic and microbial assays for toxicity assessment in wastewater treatment plants.

18.3.1. Enzymatic Assays

Enzymes are proteins that serve as catalysts of biological reactions in animal, plant, and microbial cells. The use of enzymes for indicating the adverse effect of toxic chemicals on microbial populations in soil is well known (Burns, 1978). In aquatic environments, some enzymes (e.g., dehydrogenases) are well correlated with microbial activity (Dermer et al., 1980). Chemical toxicity can be conveniently and rapidly determined in water and wastewater by using simple enzymatic assays that can be miniaturized and automated. Several enzymes (dehydrogenases, ATPase, phosphatase, esterase, urease, luciferase, β-galactosidase, -glucosidase) were explored for assessing toxicity in aquatic environments, including wastewater treatment plants. Table 18.3 lists the short-term toxicity assays based on enzyme activity or biosynthesis (Bitton and Koopman, 1986, 1992; Christensen et al., 1982; Obst et al., 1988).

Dehydrogenases are the enzymes most utilized in toxicity testing in wastewater treatment plants. Dehydrogenase activity is assayed by measuring the reduction of oxidoreduction dyes such as triphenyl tetrazolium chloride (TTC), nitroblue tetrazolium (NBT), 2-(*p*-iodophenyl)-3-(*p*-nitrophenyl)-5-phenyltetrazolium chloride (INT), or resazurin. Several toxicity tests based on inhibition of dehydrogenase activity in wastewater have been developed (Bitton and Koopman, 1986).

Recent developments include tests based on the inhibitory effect of chemicals on enzyme biosynthesis instead of enzyme activity. For example, the *de novo* biosynthesis of β-galactosidase in *Escherichia coli* is more sensitive to toxicants than is enzyme activity (Dutton et al., 1988; Reinhartz et al., 1987). Toxi-Chromotest is a commercial toxicity assay that is based on the inhibitory effect of chemicals on β-galactosidase biosynthesis in *E. coli* (Reinhartz et al., 1987). However, enzyme biosynthesis is less sensitive than other assays (e.g., Microtox, *Ceriodaphnia dubia*) used for assessing the toxicity of wastewater effluents or sediment extracts (Koopman et al., 1988; Koopman et al., 1989; Kwan and Dutka 1990). A higher sensitivity is obtained with a toxicity test based on inhibitory effect of chemicals on the biosynthesis of α-glucosidase in *Bacillus licheniformis* (Campbell et al., 1993; Dutton et al., 1990).

18.3.2. Microbial Bioassays

Several bacterial assays are available for determining the toxicity of environmental samples (Bitton and Dutka, 1986; Dutka and Bitton, 1986; Liu and Dutka, 1984). A selection of bacterial assays is displayed in Table 18.4.

Toxic chemicals may adversely affect the light output of bioluminescent bacteria. A toxicity assay, commercialized under the name of Microtox, utilizes freeze-dried cultures of the bioluminescent marine bacterium *Photobacterium phosphoreum* (Bulich, 1979, 1986) (Fig.

TABLE 18.3. Short-Term Toxicity Assays Based on Enzyme Activity or Biosynthesis[a]

Enzyme	End point measured	Comments
Dehydrogenases	Measure reduction of oxidoreduction dyes such as INT or TTC.	Widely tested in water, wastewater, soils, sediments.
ATPase	Measure phosphate concentration using ATP as a substrate.	*In vivo* and *in vitro* tests have been used.
Esterases	Nonfluorescent substrates degraded to fluorescent products.	Acetylcholinesterase sensitive to organophosphates and carbamates.
Phosphatases	Measure organic portion of substrate (e.g., phenol) or inorganic phosphate.	Heavy metals toxicity in soils.
Uease	Measure ammonia production from urea.	Studied mostly in soils.
Luciferase	Measure light production using ATP as a substrate.	Used in ATP-TOX bioassay in conjunction with inhibition of ATP levels in a bacterial culture.
β-galactosidase	Measure hydrolysis of *o*-nitrophenyl-β-D-galactoside.	Toxicant effect on both enzyme activity and biosynthesis was tested.
α-glucosidase	*p*-Nitrophenyl-α-*D*-glucoside.	Toxicant effect on enzyme biosynthesis has been tested.
Tryptophanase	Add Ehrlich's reagent and measure absorbance at 568 nm.	Toxicant effect on enzyme biosynthesis has been tested.

[a]From Bitton and Koopman (1992), with permission of the publisher.

TABLE 18.4. Short-Term Bacterial Toxicity Assays[a]

Assay	Basis for the test
Microtox	Inhibition of bioluminescence of *Photobacterium phosphoreum*.
Spirillum volutans	Toxicants cause loss of coordination of rotating fascicles of flagella with concomitant loss of motility.
Growth inhibition	Measure growth inhibition of pure (e.g., *Aeromonas, Pseudomonas*) or mixed cultures by absorbance determination for microbial suspensions or by measurement of zones of inhibition on solid growth media.
Viability assays	Measure effect of toxicants on the viability of bacterial cultures on agar plates.
ATP assay	Inhibitory effect of toxic chemicals on ATP levels in microorganisms.
ATP-TOX assay	Test based on both the growth inhibition, by ATP measurement, of bacterial culture and inhibition of luciferase activity.
Respirometry	Measures effect of toxicants on microbial respiration in environmental samples.
Toxi-Chromotest	Based on inhibition of biosynthesis of β-galactosidase in *E. coli*.
α-Glucosidase biosynthesis assay	Based on inhibition of biosynthesis of α-glucosidase in *Bacillus licheniformis*.
Nitrobacter bioassay	Measures inhibition of nitrite oxidation to nitrate.
Microcalorimetry	Measures decreases in heat production by microbial communities.

[a]From Bitton and Koopman (1992), with permission of the publisher.

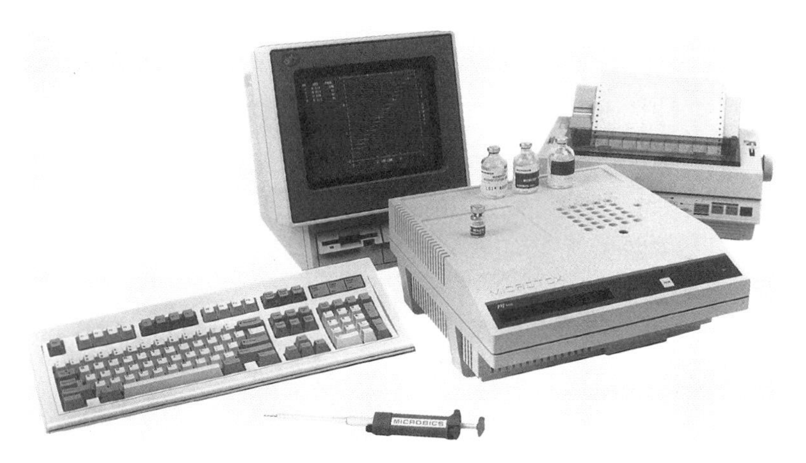

Fig. 18.1. Microtox model 500 toxicity analyzer. Courtesy of A. Bulich, Microbics Corp., Carlsbad, CA.

18.1). This test is based on the inhibition of *P. phosphoreum* by toxic chemicals (Fig. 18.2). It has been used for determining the toxicity of wastewater effluents, complex industrial wastes (oil refineries, pulp and paper), fossil fuel process water, sediment extracts, sanitary landfill, and hazardous waste leachates (Munkittrick et al., 1991). This assay shows good correlation with fish, *Daphnia,* and algal bioassays (Blaise et al., 1987; Curtis et al., 1982; Giesy et al., 1988a; Logue et al., 1989). However, it appears that it is not as sensitive to toxic metals. Microtox could be useful for on-line monitoring of toxicants entering wastewater or drinking water treatment plants (Levi et al., 1989).

Toxicity testing can also be based on growth

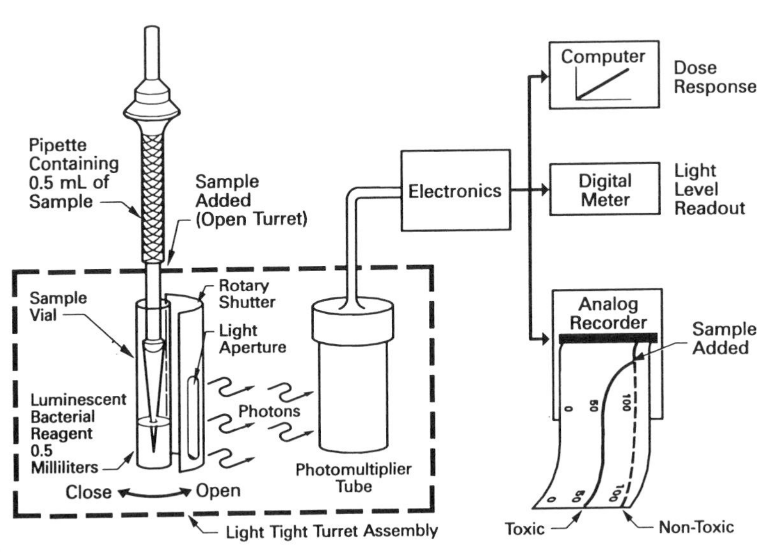

Fig. 18.2. Microtox toxicity test system. Courtesy of A. Bulich, Microbics Corp., Carlsbad, CA.

inhibition of pure or mixed bacterial cultures (Alsop et al., 1980; Trevors, 1986). The assays consist of determining changes in bacterial densities by measurement of the optical density of the bacterial supensions, determination of inhibition zones on solid growth media, or ATP measurement. The ATP-TOX assay is based on both the growth inhibition, by ATP measurement, of *E. coli* and inhibition of luciferase activity (Xu and Dutka, 1987). In tests based on inhibition of ^{14}C-glucose uptake by activated sludge microorganisms, activated sludge is spiked with labeled glucose; substrate uptake is measured following a 15-min incubation (Larson and Schaeffer, 1982).

Toxic chemicals may also exert an adverse effect on nutrient cycling (C, N, P, and S cycles) by microorganisms. As to the nitrogen cycle, nitrification is probably the step most sensitive to environmental toxicants. A nitrifying process is affected more rapidly than a carbonaceous BOD removal process. Toxicity assays based on the inhibition of both *Nitrosomonas* and *Nitrobacter* have been developed for determining the toxicity of wastewater samples (Alleman, 1988; Williamson and Johnson, 1981). However, *Nitrosomonas* appears to be much more sensitive to toxicants than *Nitrobacter* (Blum and Speece, 1992). The impact of toxic chemicals on the carbon cycle is conveniently determined by measuring the inhibition of microbial respiration. Microbial respiration can be measured, with oxygen electrodes, manometers, or electrolytic respirometers (King and Dutka, 1986).

Respiration inhibition kinetics analysis (RIKA) involves the measurement of the effect of toxicants on the kinetics of biogenic substrate (e.g., butyric acid) removal by activated sludge microorganisms. The kinetic parameters studied are q_{max}, the maximum specific substrate removal rate (determined indirectly by measuring the V_{max}, the maximum respiration rate), and K_s, the half-saturation coefficient. The procedure consists of measuring with a respirometer the Monod kinetic parameters, V_{max} and K_s, in the absence and in the presence of various concentrations of the inhibitory compound (Volskay and Grady, 1990; Volskay et al., 1990). Figure 18.3 (Volskay et al., 1990) illustrates the results obtained by using the RIKA procedure with 1,2-dichloropropane.

Exposure of microbial cells to toxicants also results in the induction of stress proteins, which can be detected by gel electrophoresis. Some of these proteins overlap with heat shock (Neidhardt et al., 1984) and starvation proteins (Matin et al., 1989), but others are produced in response to specific inorganic and organic toxicants (Blom et al., 1992). The pattern of stress proteins may possibly be used as an index of exposure to toxic chemicals but a rapid test is not yet available.

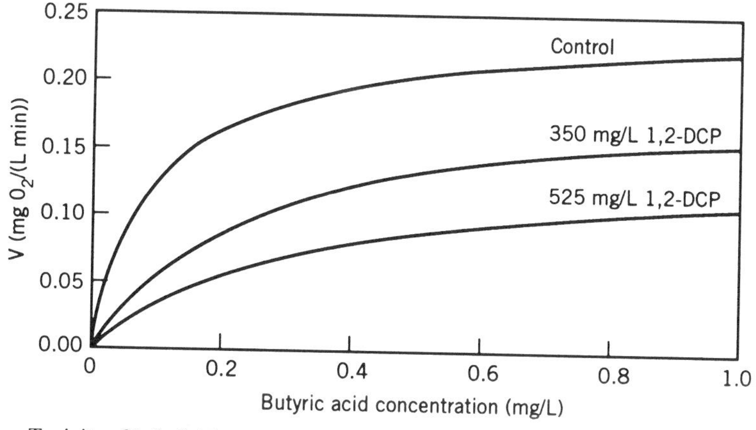

Fig. 18.3. Toxicity of 1,2-dichloropropane (1,2-DCP), measured by the RIKA procedure (1,2-DCP is an inhibitor which increases K_s). Adapted from Volskay et al. (1990).

Algae may also serve as test organisms in toxicity testing. The algal bottle test proposed by the U.S. EPA (U.S. EPA, 1989c) is based on growth inhibition of the green alga *Selenastrum capricornutum*. This assay was miniaturized by means of 96-well microplates. The toxicity end points are growth inhibition as determined by cell counts after 96-hr or ATP inhibition after 4-hr exposure. This alga is mostly sensitive to heavy metals and herbicides (Blaise, 1991; Hickey et al., 1991; St. Laurent et al., 1992).

18.3.3. Commercial Rapid Assays for Toxicity Testing in Wastewater Treatment Plants

A few microbial or enzymatic assays are now marketed and some could be useful for toxicity testing in wastewater treatment plants (Bitton and Koopman, 1992).

Microtox. This test, discussed in Section 18.3.2, is probably the most popular commercial test for assessing toxicity in wastewater treatment plants. An on-line model is now available.

Polytox. The assay microorganisms in Polytox are a blend of bacterial strains originally isolated from wastewater. The Polytox kit, specifically designed to assess the effect of toxic chemicals on biological waste treatment, is based on the reduction of the respiratory activity of the rehydrated cultures in the presence of toxicants.

Toxi-Chromotest. The Toxi-Chromotest is based on inhibition of β-galactosidase biosynthesis in *E. coli* by toxic chemicals. This test is not as sensitive as Microtox to aquatic toxicants.

MetPAD. MetPAD is the first microbial assay for the direct assessment of the toxicity of specific categories of chemicals. It is a bioassay designed for the specific determination of heavy metal toxicity in environmental samples and is based on inhibition of β-galacosidase activity in a mutant strain of *E. coli* by bioavailable heavy metals. The MetPAD methodology is displayed in Figure 18.4. This bioassay was useful for the determination of the toxicity of bioavailable heavy metals in wastewaters and in sediment elu-

MetPAD™ Methodology

Add 3 ml of DILUENT to bottle containing BACTERIAL REAGENT. Mix well.

Add 0.1 ml of bacterial suspension to 0.9 ml of sample in assay tubes. Shake and incubate for 90 min at 35C.

Add 0.1 ml of BUFFER to tubes and mix. Dispense 10μL drops on ASSAY PADS.

Incubation of PADS for 30 min at 35C. → Observe PURPLE color intensity.

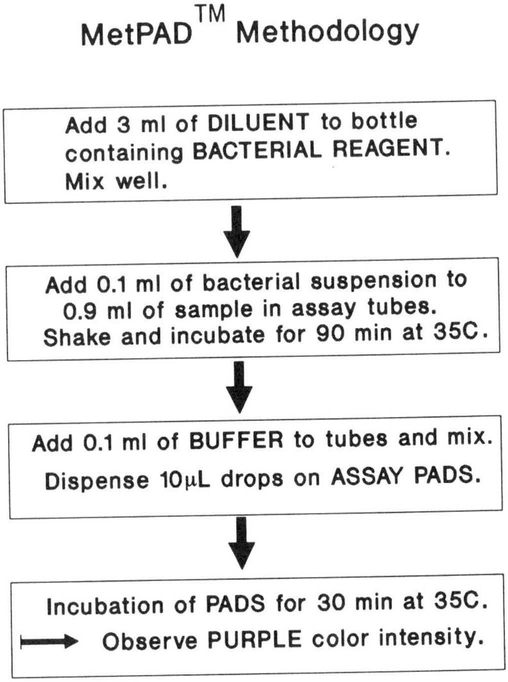

Fig. 18.4. MetPAD methodology for assessment of heavy metal toxicity. From Bitton et al. (1992b), with permission of the publisher.

triates (Bitton et al., 1992a,b; Campbell et al., 1992). This kit is also available in the 96-well microplate format and marketed as MetPLATE.

18.4. APPLICATION OF MICROBIAL AND ENZYMATIC TESTS TO TOXICITY ASSESSMENT IN WASTEWATER TREATMENT PLANTS

Microbial toxicity tests could be convenient tools in toxicant characterization by wastewater fractionation in Phase 1 of toxicity reduction evaluation (TRE) as proposed by the U.S. EPA (U.S. EPA, 1988, 1989a, 1989b). Toxicity reduction evaluation consists of a series of tests that are carried out systematically to determine the sources of effluent toxicity, the specific causative toxicant(s), and the effectiveness of pollution control measures to reduce effluent toxicity.

Phase 1 of TRE consists of a series of fractionation steps that provide information about the physicochemical properties of the toxicants in a given wastewater effluent. The physicochemical treatments are followed by acute and/or chronic toxicity assays, using daphnids, fish, or possibly microbial and enzymatic assays. In acute tests, the endpoints are LC_{50}'s (concentration that leads to 50% mortality during the testing period) or ET_{50}'s (elapsed time for 50% mortality at one effluent concentration). These fractionation steps include the following tests (Fig. 18.5) (U.S. EPA, 1988):

1. Baseline effluent toxicity test. The original wastewater effluent must be tested and LC_{50} and ET_{50} determined.

2. Degradation test. These tests provide information concerning the biodegradation, photodegradation, and/or chemical oxidation of the effluent. This information can be useful in treatability studies during Phase II of TRE.

3. Filtration test. This test gives an indication of whether the causative toxicant is associated with solids.

4. Air stripping test. Air stripping provides information about whether the effluent toxicity is caused by volatile or oxidizable compounds. This test is carried out at ambient, acid (pH = 3.0), and basic pH (pH = 11.0).

5. Oxidant reduction test. This test determines the presence of oxidants (e.g., chlorine) that may be responsible for effluent toxicity.

6. EDTA chelation test. Absence or reduction of toxicity following treatment of the sample with the sodium salt of ethylenediamine tetraacetic acid indicates the presence of cationic toxicants.

7. Solid phase extraction (SPE) test. It consists of passing the filtered effluent though small columns containing a C_{18} hydrocarbon adsorbent resin. Absence or reduction of toxicity following the SPE test indicates the presence of nonpolar compounds.

8. Other tests. Cationic and anionic exchange resins are also used to test for the presence of cationic and anionic toxicants, respectively. Chelating resins have also been considered for removing heavy metal toxicity. XAD resins are sometimes employed for the removal or reduction of toxicity caused by polar organic compounds (Walsh and Garnas, 1983; DiGiano et al., 1992; Mazidji et al., 1992; Mirenda and Hall, 1992).

Since these fractionation schemes generate more than 50 fractions for each sample tested, it would appear that short-term microbial assays would be useful in toxicant characterization for toxicity reduction evaluation. Microtox, in conjunction with *Ceriodaphnia* bioassay, was used for toxicant characterization in wastewater samples from Jacksonville, Florida (Mazidji et al., 1990). The wastewater fractionation procedures give valuable information about the chemical properties of toxicants in wastewater, thus aiding

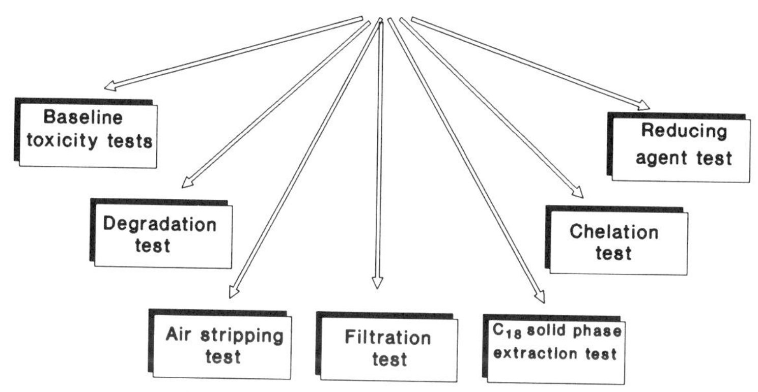

Fig. 18.5. Phase 1 effluent characterization tests. Adapted from U.S. EPA (1988).

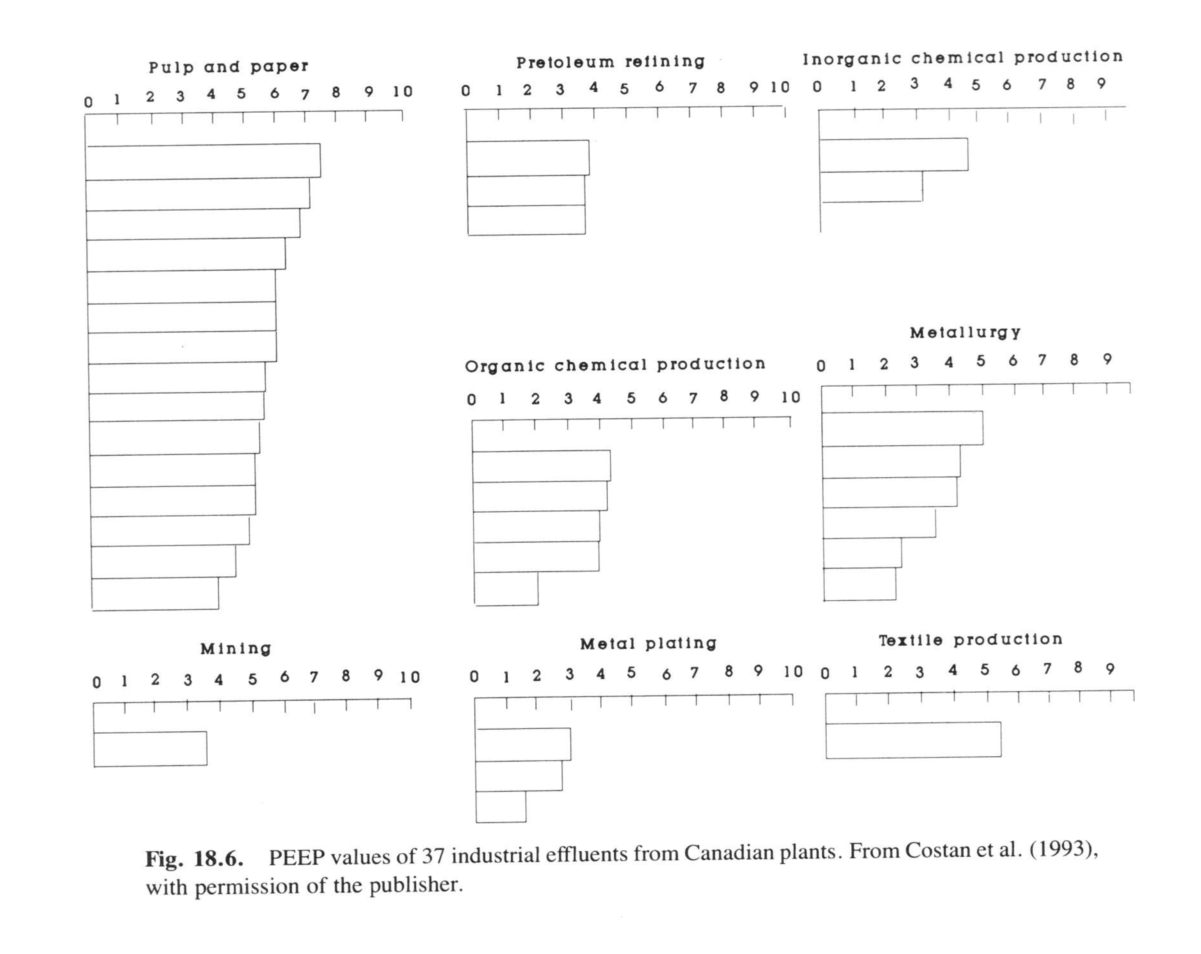

Fig. 18.6. PEEP values of 37 industrial effluents from Canadian plants. From Costan et al. (1993), with permission of the publisher.

both the evaluation of control techniques and specific identification of the problem chemicals.

There is no microbial bioassay that can detect all the categories of environmental toxicants with equal sensitivity. Therefore, a battery-of-tests approach has been suggested that consists of using concurrently some of the short-term assays discussed previously. The battery-of-tests approach has been found to be useful in the formulation of a new index for assessing the toxic potential of wastewater effluents (Costan et al., 1993). This toxicity index, called PEEP (potential ecotoxic effects probe), is expressed as a $\log_{10}$ value that varies from 0 to 10. This index takes into account the toxic/genotoxic response of several short-term bioassays with test organisms belonging to different trophic levels (bacteria, algae, crustaceans), the persistence of the toxic/genotoxic effect, and the flow rate of the wastewater effluent. PEEP is given by the following formula:

$$P = \log_{10}\left[1 + n\left(\frac{\sum_{i=1}^{N} T_i}{N}\right) Q\right] \quad (18.1)$$

where P = potential ecotoxic effects probe (numerical value of 0 to 10); n = number of tests that exhibit toxic/genotoxic responses; N = maximum number of obtainable toxic/genotoxic responses; T_i = toxic response in toxicity units exhibited by a given bioassay before or after biodegradation; and Q = wastewater effluent flow rate (m^3/h).

Figure 18.6 (Costan et al., 1993) shows the PEEP values of 37 industrial effluents from canadian plants. The PEEP ranged from 0 (nontoxic effluents) to 7.5 (very toxic effluents) and was highest for pulp and paper mill effluents.

18.5. FURTHER READING

Bitton, G., and B.J. Dutka, Eds. 1986. *Toxicity Testing Using Microorganisms,* Vol. 1. CRC Press, Boca Raton, FL.

Bitton, G., and B. Koopman, 1992. Bacterial and enzymatic bioassays for toxicity testing in the environment. Rev. Environ. Contam. Toxicol. 125: 1–22.

Dutka, B.J., and G. Bitton, Eds. 1986. *Toxicity Testing Using Microorganisms,* Vol. 2. CRC Press, Boca Raton, FL.

Liu, D., and B.J. Dutka, Eds. 1984. *Toxicity Screening Procedures Using Bacterial Systems.* Marcel Dekker, New York.

MICROBIOLOGY AND PUBLIC HEALTH ASPECTS OF WASTEWATER DISPOSAL AND REUSE

19

PUBLIC HEALTH ASPECTS OF WASTEWATER AND SLUDGE DISPOSAL

Wastewater treatment plants generate effluents and sludges that must be disposed of safely and economically. In this chapter, we will discuss the public health aspects regarding the disposal of wastewater effluents and sludges into the environment. We have limited our discussion to two popular and most studied approaches to waste disposal, land application and ocean outfalls.

A. LAND DISPOSAL OF WASTEWATER EFFLUENTS AND SLUDGES

Land application of wastewater effluents and sludges is one of the most popular options for disposing of these cumbersome materials. Following is a discussion of the public health aspects of disposal of wastewater effluents and sludges.

19.1. LAND TREATMENT SYSTEMS FOR WASTEWATER EFFLUENTS AND SLUDGES

19.1.1. Wastewater Effluents

The main objectives of land application of wastewater are further effluent treatment, groundwater recharge, and the provision of nutrients for agricultural crops. During land treatment of wastewater effluents, biological and chemical pollutants are removed by physical (settling, filtration), chemical (adsorption, precipitation), and biological (e.g,. plant uptake, microbial transformations, biological decay) processes. Land treatment systems for wastewater effluents are capable of removing microbial pathogens and parasites, BOD, suspended solids, nutrients (nitrogen and phosphorus), toxic metals, and trace organics (Fig. 19.1). Suspended solids are removed by filtration and sedimentation. Soluble organic compounds are removed by microbial action, particularly by biofilms developing on soil particles. Nitrogen is removed by sedimentation–filtration (e.g., particulate-associated organic nitrogen), adsorption to soil and volatilization (e.g., NH_4), uptake by crops, and biological denitrification (Fig. 19.2) (Lance, 1972). Phosphorus is removed by adsorption to soil particles, chemical precipitation, and uptake by vegetation. The effluent chemical quality expected to result from land treatment is shown in Table 19.1 (Metcalf and Eddy, 1991). The capacity of soils for retention of metals is generally high, particularly in alkaline soils. The removal of trace organics by soils and aquifers is generally carried out by adsorption (organic soils have a high retention capacity), volatilization, and biodegradation (see Section 19.7). There is concern over the uptake of

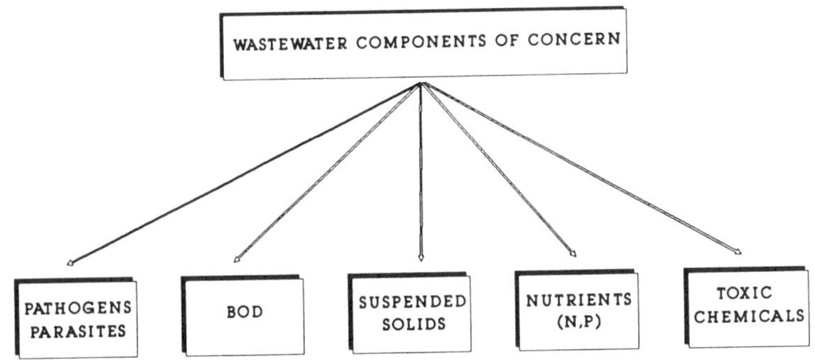

Fig. 19.1. Wastewater components of concern in land application of wastewater effluents.

trace organics found in wastewaters and sludges by agricultural crops and animals (Majeti and Clark, 1981).

The three types of land treatment systems are the following (Polprasert, 1989; Reed et al., 1988; U.S. EPA, 1981):

19.1.1.1. Slow-Rate Irrigation System

Slow-rate irrigation is the most frequently used land treatment system. It provides essential nutrients to satisfy the growth requirements of agricultural crops. Pretreated wastewater is applied

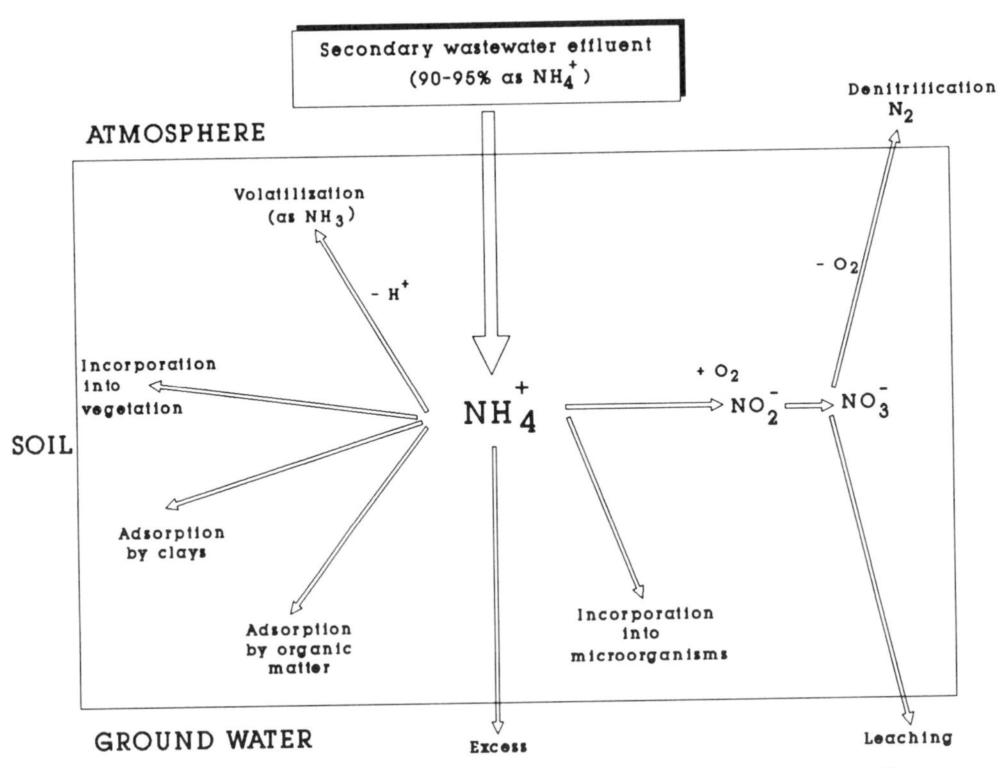

Fig. 19.2. A summary of nitrogen transformations during land disposal of wastewater effluents. Adapted from Lance (1972).

TABLE 19.1. Expected Chemical Quality of Wastewater Effluents Following Land Application[a]

Constituent	Value, mg/L					
	Slow-rate[b]		Rapid infiltration[c]		Overland-flow[d]	
	Average	Maximum	Average	Maximum	Average	Maximum
BOD	<2	<5	2	<5	10	<15
Suspended solids	<1	<5	2	<5	15	<25
Ammonia nitrogen as N	<0.5	<2	0.5	<2	1	<3
Total nitrogen as N	3	<8	10	<20	5	<8
Total phosphorus as P	<0.1	<0.3	1	<5	4	<6

[a]From Metcalf and Eddy (1991), with permission of the publisher.
[b]Percolation of primary or secondary effluent through 5 ft (1.5 m) of soil.
[c]Percolation of primary or secondary effluent through 15 ft (4.5 m) of soil.
[d]Runoff of continued municipal wastewater over about 150 ft (45 m) of slope.

to land (soil texture ranges from sandy loams to clay loams), by sprinkler or surface distribution, at a relatively slow rate (0.5–6 m/year) (weekly loading rate of 1.3–10 cm) and serves as a source of nutrients for forage (e.g., alfalfa, bermuda grass, ryegrass) or field crops (e.g., corn, cotton, barley) (Fig. 19.3); (U.S. EPA, 1981). Physical, chemical, and biological processes contribute to the treatment of the incoming wastewater. This system provides the highest treatment potential and removes approximately 99% of BOD, suspended solids, and coliforms. Some limitations of slow-rate irrigation systems are land cost, and high operating cost of transport of wastewater to the treatment site.

19.1.1.2. Overland Flow System

Wastewater is applied at a rate of 3–20 m/year or more and flows down a grass-covered slope (2%–8%) with a length of 30–60 m (Fig. 19.4) (U.S. EPA, 1981). The most suitable soils are clay or clay-loamy soils with low permeability (≤ 0.5 cm/hr) to limit wastewater percolation through the soil profile. The treated effluent is captured in a collection channel. Nutrients (N, P,

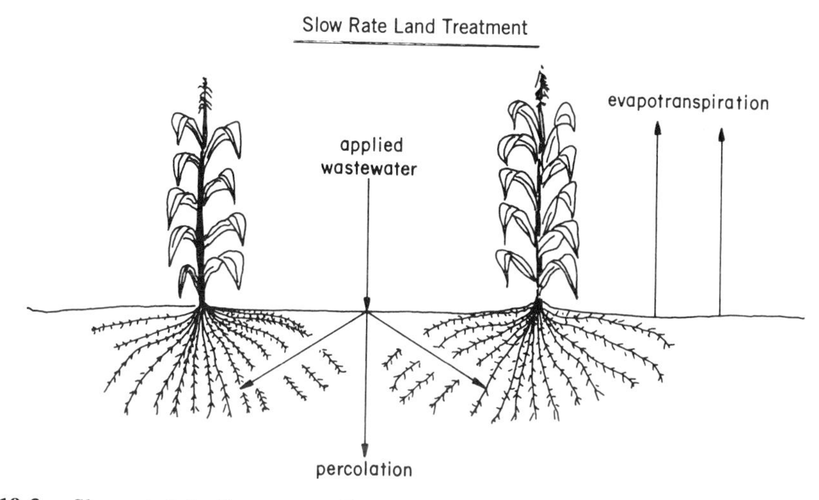

Fig. 19.3. Slow-rate irrigation system. From U.S. EPA (1981), with permission of the publisher.

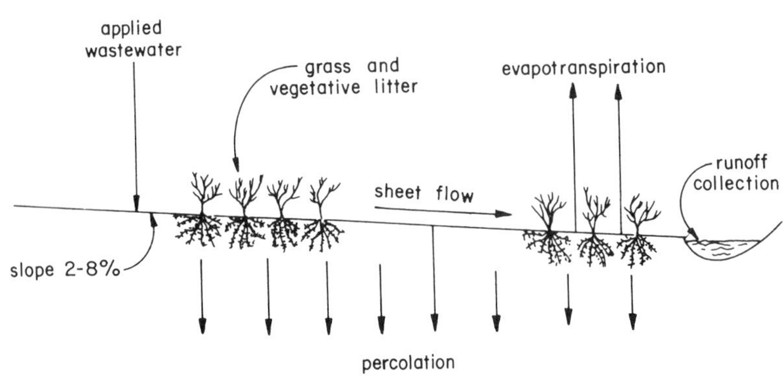

Fig. 19.4. Overland flow system. From U.S. EPA (1981), with permission of the publisher.

and BOD), suspended solids, and pathogens are removed as wastewater flows down the slope. This system achieves 95%–99% removal of BOD and suspended solids. Removal of nitrogen in overland flow systems is due to nitrification followed by denitrification and crop uptake. Phosphorus is removed by adsorption and precipitation.

19.1.1.3. *Rapid Infiltration Systems*

Wastewater is applied intermittently at high loading rates (6–125 m/year) onto a permeable soil (e.g., sands or loamy sands (Fig. 19.5) (U.S. EPA, 1981). The hydraulic pathway displayed in Figure 19.5 shows that most of the applied wastewater flows to groundwater aquifers. The treated wastewater can be collected by recovery wells. The minimum depth to groundwater is 1 m during flooding periods and 1.5–3 m during drying periods. The treatment potential of rapid infiltration systems is lower than in slow-rate systems. Removal of nitrogen is generally low but can be increased by encouraging denitrification. Denitrification requires adequate carbon levels (as found in primary effluents) and

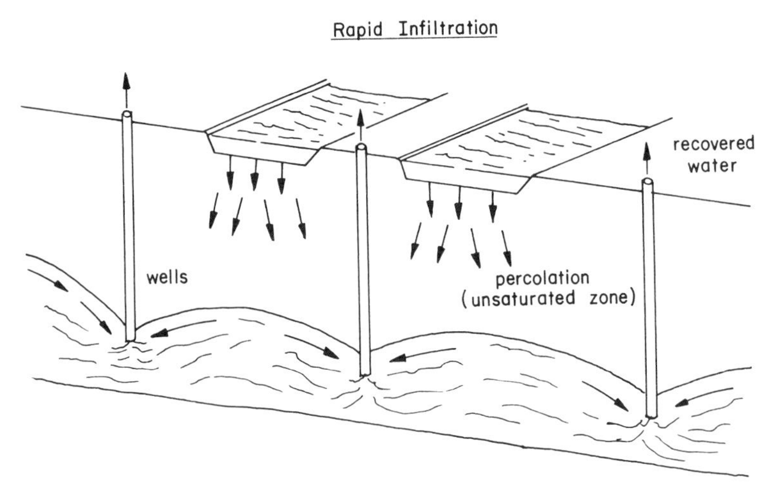

Fig. 19.5. Rapid infiltration system. From U.S. EPA (1981), with permission of the publisher.

low oxygen levels, necessitating flooding periods as long as 9 days, followed by drying periods of about 2 weeks.

19.1.2. Wastewater Sludges

Approximately 7 million dry tons of sludge are produced annually in the United States, a number that is expected to increase in the future and to double by the year 2000. Sludge is disposed of by land application (21%–39%, depending on the size of the plant), landfills (12%–35%), incineration (1%–32%, mostly by large plants) distribution and marketing (13%–19%), and ocean dumping (1%–4%) (U.S. EPA, 1984). The various sludge disposal methods are summarized in Figure 19.6 (Metcalf and Eddy 1991). Sludge is also used in horticulture and by home gardeners, generally after composting. It is applied to forests to increase productivity and to strip-mined land for reclamation. Recent U.S. legislation has banned sludge disposal in the ocean. Of the options available for sludge disposal, it seems that the agricultural option is the most convenient and least costly.

The European Community generates approx-

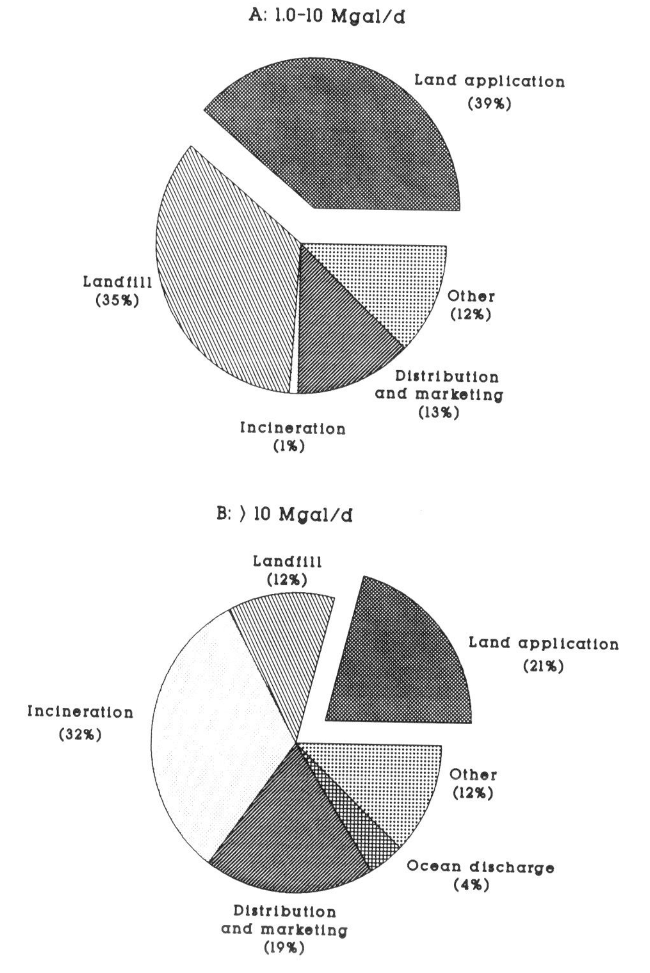

Fig. 19.6. Comparative utilization of sludge disposal options. **A.** Plant capacity 1.0–10 Mgal/d. **B.** Plant capacity > 10 Mgal/d. Adapted from Metcalf and Eddy (1991).

imately 6 million dry tons of sludge, 30% of which is used by agriculture. In Great Britain, approximately two thirds of the sludge produced is applied to land and one third is dumped into the ocean. About 4% of the sludge is incinerated (Bruce and Davis, 1989; Forster and Senior, 1987; Kofoed, 1984; Wallis and Lehmann, 1983).

Sludge contains useful nutrients such as nitrogen (5.1% dw basis), phosphorus (1.6% dw basis), and potassium (0.4% dw basis) as well as micronutrients. Sludge organic matter helps improve soil structure by increasing its water-holding capacity and by aerating the soil. Design sludge-loading rates are based on nitrogen-loading rates necessary for plant growth (Metcalf and Eddy, 1991).

Wastewater sludge also contains heavy metals and trace organics that can present potential health risks to humans, animals, and agricultural crops. Much is known about phytotoxicity of heavy metals but fewer data are available on the impact of trace organics on soils and crops. Cadmium is the metal of greatest concern because it can accumulate in plants and can pose a threat to humans and grazing animals. Design sludge-loading rates are based on cadmium levels in the applied sludge.

The methods of applying sludge to land are the following (Bruce and Davis, 1989; Wallis and Lehmann, 1983):

1. *Surface spreading by tanker or by rain gun.* This practice suffers from uneven sludge spreading. The sludge is worked into the soil to control odors and avoid nitrogen losses.

2. *Sludge injection into the soil.* Although more expensive than surface spreading, this practice has the advantage of minimizing odors, nitrogen loss, and surface runoff.

19.2. PUBLIC HEALTH ASPECTS OF WASTEWATER AND SLUDGE APPLICATION TO LAND

The problems that are associated with land application of wastewater and sludges are contamination of groundwater, soils and crops contamination with pathogens, heavy metals, nitrate, and toxic and carcinogenic organic compounds (Bitton et al., 1980).

19.2.1. Pathogenic Microorganisms and Parasites

Pathogenic microorganisms and parasites may survive on crops, particularly leafy vegetables (e.g., spinach, lettuce), irrigated with wastewater or wastewater effluents. Market vegetables can also become contaminated with pathogens and parasites (e.g., *Salmonella, Giardia lamblia, Entamoeba histolytica*) during irrigation, transportation, and subsequent handling (Ercolani, 1976; Kowal, 1982; Pude et al., 1984). Survival of pathogens on crops depends on the type of pathogen, the type of crop, and environmental conditions (e.g., sunlight, temperature, wind, rain) and may vary from days to months (Rose, 1986). The concentration of parasite eggs applied to land can be high and can reach levels of 6,000–12,000 viable eggs per $1 m^2$/year. Parasite eggs, especially those of *Ascaris,* can persist in soils for years (5–7 years or more) (Little, 1986). Therefore, the irrigation of processed food crops with wastewater effluents should be stopped some weeks (4–6 weeks) prior to harvesting to allow a sufficient inactivation of potential pathogens and parasites. Furthermore, irrigation of food crops that are eaten raw should be restricted.

19.2.2. Chemical Contaminants

Crops and grazing animals may become contaminated with heavy metals (e.g., Cd, Zn, Cu, Ni) present in sludge. Nitrate from sludge and wastewater effluents can also contaminate groundwater used for drinking water supply (Bouchard et al., 1992). The health effects of nitrates in water are discussed in Chapter 3. Lipophilic trace organic compounds are recalcitrant to biodegradation and may accumulate in the fatty tissues of grazing animals.

19.3. TRANSPORT OF PATHOGENS THROUGH SOILS

The detection of pathogenic microorganisms in groundwater has triggered research on their fate (transport and persistence) in the soil matrix.

19.3.1. Bacterial Transport Through Soils

Owing to their size, bacterial pathogens can be filtered out during their transport through the soil matrix. Removal of bacteria by soils is inversely proportional to the particle size of the soils. Also, bacteria are charged biocolloids, which can adsorb to soils, providing there are optimal conditions that favor their attachment to soil particles. These conditions are the presence of cations, presence of clay minerals that provide adsorption sites, low concentrations of soluble organics, and low pH conditions (Gerba and Bitton, 1986). Heavy rainfall promotes bacterial transport through soils (Lamka et al., 1980; Zyman and Sorber, 1988). Laboratory experiments with sludge–soil mixtures (equivalent to an average loading rate of 0.05 tons/ha) challenged with bacterial indicators have shown that only heavy rainfall (12.3 cm/day) promotes significant downward transport of the bacteria to the bottom of an 8-in-deep column. Lower rainfall did not cause significant migration of the bacterial cells (Zyman and Sorber, 1988).

Under field conditions, indicator bacteria are efficiently retained by soils and are detected at low levels in groundwater. They are not suitable indicators of virus transport into groundwater (Alhajjar et al., 1988). Following sludge application to land, most of the sludge-associated bacteria are retained at the soil surface and their transport to groundwater is unlikely (Liu, 1982).

19.3.2. Virus Transport Through Soils

The major factors controlling virus transport through soils are soil type, virus serotype, ionic strength of soil solution, soluble organic compounds present in wastewater effluents, and hydraulic flow rate. Retention of viruses by the soil matrix is primarily governed by the adsorption to surfaces, particularly those provided by clays and other minerals such as hematite and magnetite. Both electrostatic and hydrophobic interactions are involved in virus adsorption to soils (Bitton, 1980b; Bitton and Harvey, 1992; Gerba, 1984; Lipson and Stotzky, 1987).

Adsorption of virus to soils is affected by soil texture. Clay soils generally have a greater virus retaining capacity than sandy ones. Muck soils display a low affinity for viruses (Scheuerman et al., 1979; Sobsey et al., 1980). Adsorption varies with the type and strain of virus. Some viruses (e.g., poliovirus 1) notoriously adsorb well to soils, whereas others (e.g., echovirus 1 and 11) display a low adsorption capacity (Gerba et al., 1980; Jansons et al., 1989b; Sobsey et al., 1986). Several enteroviruses (e.g., poliovirus, coxsackie virus, and echoviruses) and bacterial phages have been used as models to study virus transport in column experiments. These model viruses do not always simulate the survival and distribution of hepatitis A virus and rotaviruses (Dizier et al., 1984; Sobsey et al., 1986). Viruses adsorb poorly to soils in low-ionic-strength solutions. This explains why rainwater tends to desorb soil-bound viruses and redistribute them within the soil profile (Alhajjar et al., 1988; Duboise et al., 1976; Lance and Gerba, 1980). Adsorption is inhibited and thus virus transport is promoted by soluble organic materials found in wastewater effluents and sludges and by humic and fulvic acids (Dizier et al., 1984; Schaub and Sorber, 1977; Scheuerman et al., 1979). Virus transport is also promoted by increasing hydraulic flow rate (Lance and Gerba, 1980; Vaughn et al., 1981).

Both column experiments and field studies have shown that sludge application to land does not result in transport of virus to aquifers. Viruses have not been detected in groundwater beneath sludge application sites (Farrah et al., 1981). Indeed, sludge-associated virus particles often become trapped at the soil surface and their migration through the soil matrix is thus limited (Bitton et al., 1984; Damgaard-Larsen et al., 1977; Moore et al., 1978; Pancorbo et al., 1988). The limited leaching of viruses from sludge par-

TABLE 19.2. Summary of the Main Factors Governing the Transport of Microbial Pathogens Through Soils[a]

Factor	Comments
Soil type	Fine-textured soils retain microorganisms more effectively than light-textured soils. Iron oxides increase the adsorptive capacity of soils. Muck soils are generally poor virus adsorbents.
Filtration	Straining of bacteria at soil surface limits their movement.
pH	Generally, adsorption increases when pH decreases.
Cations	Adsorption increases in the presence of cations (cations help reduce repulsive forces on both microorganisms and soils particles). Rainwater may desorb viruses from soil owing to its low conductivity.
Soluble organics	Generally compete with microorganisms for adsorption sites. Humic and fulvic acid reduce virus adsorption to soils.
Microbial type	Adsorption to soils varies with microbial type and strain.
Flow rate	The higher the flow rate, the lower the microbial adsorption to soils.
Saturated versus unsaturated flow	Virus movement is less under unsaturated flow conditions.

[a]Adapted from Gerba and Bitton (1984).

ticles has been confirmed for both viruses and bacteria (Hurst and Brashear, 1987; Liu, 1982). Table 19.2 summarizes the main factors governing the transport of microbial pathogens through soils. Several models have been proposed for predicting viral and bacterial transport through soils and aquifers; the models are based on the information on microbial transport and survival in these environments that has been generated during the past 20 years (e.g., Harvey and Garabedian, 1991; Park et al., 1990; Tim and Mostaghimi, 1991; Yates and Ouyang, 1992; Yates and Yates, 1989). These models vary in complexity and require environmental (e.g., temperature), soil/hydrogeologic (e.g., texture, adsorption coefficient), and microbiological (e.g., microbial type, inactivation rate) input parameters. Most of these models simulate microbial transport under saturated flow conditions, only a few of them addressing transport under unsaturated conditions (e.g., VIRTUS model; Yates and Ouyang, 1992). Some of these models, after some fine-tuning, could be useful in estimating microbial pathogen numbers in groundwater following transport through soils.

19.4. PERSISTENCE OF PATHOGENS IN SOILS

19.4.1. Persistence of Bacterial Pathogens

The main factors that affect the survival of pathogenic bacteria in soils are temperature, moisture content, sunlight, pH, organic matter, type of bacteria, and antagonistic microflora, which include indigenous soil bacteria and predatory protozoa (Bitton and Harvey, 1992; Foster and Engelbrecht, 1973; Gerba and Bitton, 1984; Gerba et al., 1975b). Soil desiccation is important in the control of survival of bacteria in soils. The decay rate of total and fecal coliforms in soil–sludge mixtures increases as the mixture is allowed to dry naturally (Zyman and Sorber, 1988).

19.4.2. Persistence of Viruses in Soil

The two decisive factors that control the persistence of virus in soils treated with wastewater effluents or sludge are soil temperature and moisture (Bitton, 1980a). Recent studies with

TABLE 19.3. Summary of the Main Factors Governing the Persistence of Microbial Pathogens in Soils[a]

Factor	Comments
Physical factors	
Temperature	Longer survival at low temperatures; longer survival in winter than in summer.
Water-holding capacity	Survival is lower in sandy soils with lower water-holding capacity.
Light	Lower survival at soil surface.
Soil texture	Clays and humic materials increase water retention by soils and thus affect microbial survival.
Chemical factors	
pH	May indirectly control survival by controlling adsorption to soils, particularly for viruses.
Cations	Some (e.g., Mg^{2+}) may thermally stabilize viruses.
Organic matter	May influence bacterial survival and regrowth.
Biological factors	
Antagonism from soil microflora	Increased survival in sterile soils. No clear trend as regards the effect of soil microflora on viruses.

[a]Adapted from Bitton et al. (1987) and Gerba and Bitton (1984).

sludge-modified desert soils confirmed the importance of these two environmental factors (Straub et al., 1992). Soil microorganisms can also produce antiviral substances that increase the rate of viral inactivation (Hurst et al., 1980; Sobsey et al., 1980). Other factors include soil and virus type (e.g., HAV has a longer persistence than other enteric viruses in soil; Sobsey et al., 1986). Viruses in sludge applied to soil persist for 23 weeks during the winter season in Denmark (Damgaard-Larsen et al., 1977) but for only 2–4 weeks during the summer or fall in Florida (Bitton et al., 1984).

Table 19.3 summarizes the main factors affecting the persistence of microbial pathogens in soils.

19.5. DISPOSAL OF SEPTIC TANK EFFLUENTS ON LAND

The description and microbiology of septic tank systems are addressed in Chapter 13. The number of septic tanks in the United States has been estimated at 22 million units. These on-site treatment systems serve approximately one fourth to one third of the U.S. population (U.S. EPA, 1986). They generate 800 billion gallons of sewage per year (Canter and Knox, 1984). In Florida, over 1.3 million families (more than 27% of the state housing units) are served by on-site sewage disposal systems (Bicki et al., 1984). Septic tanks are major contributors to the contamination of subsurface environments. The contaminants are household chemicals (nitrate, heavy metals, organic toxicants), pathogenic microorganisms, and parasitic cysts.

Septic tank effluents contribute significantly to groundwater contamination. The use of untreated groundwater was responsible for more than one third of disease outbreaks in the United States between 1971 and 1980 (Craun, 1986a). There are several documented instances of groundwater pollution by septic tank effluents (Bicki et al., 1984; Hagedorn, 1984). Groundwater contamination is often due to the use of unsuitable soils (e.g., coarse-textured soils) for receiving septic tanks effluents, high water table, saturated flow in the soil, and high loading rates. The extent of groundwater contamination depends on the climatic, soil, and biological factors that influence the transport and persistence of bacterial pathogens in soils (Gerba and Bitton, 1984). Bacterial transport through the absorption field is also controlled by the degree of soil saturation with water. In unsaturated soils movement of bacteria can be restricted within 3 ft, whereas under saturated conditions bacteria may be transported over much greater distances. The operation of an absorption field sometimes leads to the formation of a biological clogging mat or crust, which appears to be an effective barrier to bacterial breakthrough (Bouma et al., 1972). The formation of this mat is predominantly controlled by biological factors such as the accumulation of microbial polysaccharides (Mitchell and Nevo, 1964; Nevo and Mitchell, 1967; Vandevivere and Baveye, 1992). Figure 19.7 shows scanning electron micrographs of sand grains colonized by slime-producing bacteria (Vandevivere and Baveye, 1992).

Virus transport from septic tanks to groundwater has been documented in several studies. In a study by Hain and O'Brien (1979) poliovirus type 1 was found not to have been substantially inactivated in a tank and was detected in groundwater. This virus, when used as a tracer, was detected in monitoring wells and in lake water in the vicinity of a septic tank (Stramer and Cliver, 1981). Other studies have documented the breakthrough of both enteroviruses and coliform bacteria (Vaughn et al., 1983).

Septic tanks must be cleaned periodically (every 2–5 years) to remove the sludge, called septage, that has accumulated in the tank (Canter and Knox, 1985). Septage must be disposed of properly because it may contain high levels of inorganic nutrients (nitrogen, phosphorus), toxic heavy metals, hazardous organic compounds resulting from the use of household chemicals, pathogens, and parasites (Ridgley and Calvin, 1982; Stramer and Cliver, 1984; U.S. EPA, 1980; Ziebell et al., 1975). While much is known about the fate of bacterial pathogens, less is known about viruses in septage, particularly those serving multiple housing units. Improper disposal of septage may lead to subsurface pollution. Septage may be disposed of by land application (surface spreading, subsurface incorporation, trenching, and landfilling), handled in

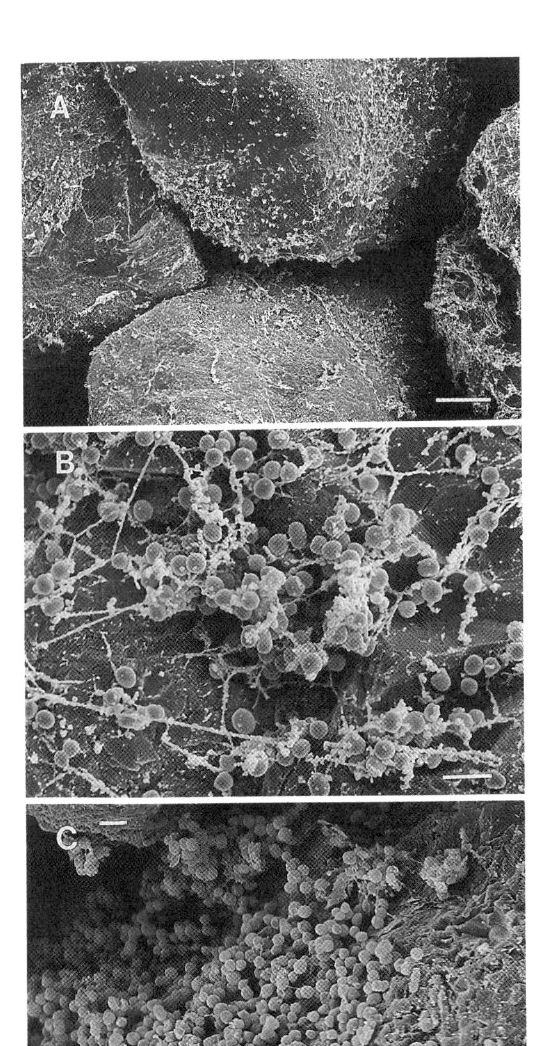

Fig. 19.7. SEM micrographs of sand grains colonized by slime-producing bacteria. **A, B.** Depth 16 mm: Bacteria entangled in a fibrillar material. Bars are 20 μm(**A**) and 2 μm (**B**). **C.** Depth 23 mm: No exopolymers were observed. Bar is 2 μm. From Vandevivere and Baveye (1992), with permission of the publisher.

special facilities (e.g., lagooning, lime stabilization, chlorination, or treated in municipal wastewater treatment plants (U.S. EPA, 1980). Unfortunately, septage is also dumped illegally onto land and into waterways without any treatment. Following a survey in Gainesville, Florida, it was estimated that as much as 60% of septage is illegally disposed of (Gainesville Regional Utilities, 1985).

Problems of groundwater contamination with septic tank effluents prompted State and local government agencies to require minimum setback distances between septic tanks and drinking water wells. Recently, geostatistical techniques were proposed to predict safe setback distances (Yates et al., 1985, 1986; Yates and Yates, 1987, 1989; see Section 19.6).

19.6. SURVIVAL OF PATHOGENS IN GROUNDWATER

Gastroenteritis and hepatitis A outbreaks due to the consumption of untreated groundwater have been documented. As discussed previously, contamination can be caused by groundwater recharge or land application of wastewater effluents or septic tank effluents. Viruses (e.g., polioviruses, coxsackieviruses, echoviruses, hepatitis A virus) have been isolated from groundwaters around the globe (Bitton and Farrah, 1986; Farrah and Bitton, 1990). A draft of the Ground Water Disinfection Rule was released recently by the U.S. EPA to address the problem of groundwater contamination by viruses and other microorganisms such as *Legionella*. This piece of legislation requires disinfection (e.g., chlorination, UV irradiation) for all community and noncommunity public water systems using groundwater, and it sets a maximum contaminant level goal (MCLG) of zero for viruses (Grubbs and Pontius, 1992).

Persistence of viruses is generally higher in groundwater than in surface waters. The decay rates of several enteroviruses in groundwater vary between 0.0004 and 0.0037 hr^{-1} (Bitton et al., 1983; Jansons et al., 1989a). Temperature is the most decisive factor that controls virus survival in groundwater (Jansons et al., 1989a; Yates and Gerba, 1984; Yates et al., 1985). The relationship between virus inactivation rate and temperature is described by the following equation:

$$I = 0.018\,(T) - 0.144 \qquad (19.1)$$

where I = virus inactivation rate (hr^{-1}); and T = groundwater temperature (°C).

Safe setback distances for drinking water wells in the vicinity of septic tanks can be predicted by geostatistical techniques that help estimate virus inactivation in groundwater, by means of a regression equation that describes the relationship between virus inactivation rates and groundwater temperature (Yates et al., 1985; Yates and Yates, 1987; Yates et al., 1986).

19.7. BIODEGRADATION IN SOILS AND AQUIFERS: AN INTRODUCTION TO BIOREMEDIATION

There is growing concern over soil and groundwater contamination with hazardous wastes, sometimes caused by effluents from wastewater treatment plants. It has been estimated that groundwater contamination with hazardous chemicals occurs or is suspected in 70%–80% of land disposal facilities in the United States (Ouellette, 1991). Environmental biotechnologists are working at finding ways to reduce environmental contamination by using modern tools in microbiology, molecular ecology, chemistry, and environmental and engineering sciences (Sayler et al., 1991). The transformation of xenobiotics by microbial action has already been discussed in Chapter 17. In this section, we will concentrate on biotreatment strategies for soils and aquifers.

19.7.1. Bioremediation of Soils

There are three categories of microbiological soil decontamination techniques (Compeau et al., 1991; Hanstveit et al., 1988): (1) in situ techniques; (2) addition of microorganisms or enzymes; and (3) bioreactors for treatment of excavated soil.

19.7.1.1. *In Situ Decontamination Techniques*

The contaminated soil is amended with nutrients (N and P) and plowed to provide oxygen. For example, the biodegradation of diesel oil in soil can be stimulated by remediation measures consisting of liming, addition of nitrogen and phosphorus, and tilling. This approach was found to reduce total hydrocarbons by 95%, eliminate polycyclic aromatic hydrocarbons, and result in complete detoxification in 20 weeks, as measured with Microtox (a microbial toxicity test) and the Ames test (a mutagenicity test assay) (Wang et al., 1990).

In situ biorestoration that made use of above-ground biodegradation cells resulted in 95% removal of oil and grease in a clay soil (Vance, 1991). Some have proposed the stimulation of indigenous soil microflora by treating the soil with a structural analog of the chemical to be removed. For example, removal of 3,4-dichloroaniline can be enhanced by stimulation of aniline-degrading microorganisms (You and Bartha, 1982).

19.7.1.2. *Addition of Microorganisms or Enzymes to Soils*

Addition of microorganisms and enzymes stimulates the biodegradation of xenobiotics in soils (Crawford and O'Reilly, 1989). Fungal laccases are polyphenol oxidases that catalyze the binding, by oxidative cross-coupling, of phenolic compounds to the humic fraction of soils and their subsequent immobilization and detoxification (Bollag, 1992). Most efforts have been concentrated on the use of bacterial inocula grown in large fermenters (e.g., *Arthrobacter; Rhodococcus chlorophenolicus, Flavobacterium* spp.) for the bioremediation of pentachlorophenol-contaminated soils. Under laboratory conditions at 30°C, soil inoculated with 10^6 PCP-degrading *Arthrobacter* per g dry soil reduced the half-life of PCP from 2 weeks to less than 1 day (Edgehill and Finn, 1983b). Microbial immobilization on bark chips or their encapsulation in polyurethane or alginate enhances their PCP-degrading ability as well as their resistance to PCP toxicity (Crawford et al., 1989; Salkinoja-Salonen et al., 1989). The use of commercial bacterial inocula, however, did not enhance the biodegradation of hydrocarbons in a site contaminated with bunker C fuel (Compeau et al., 1991).

19.7.1.3. *Bioreactors for Treatment of Excavated Soil*

These bioreactors include soil slurry reactors, composting, and land treatment units. The removal of PCP and hydrocarbons by means of land treatment units and soil slurries has been demonstrated. PCP biodegradation in soil slurries was enhanced by the addition of a PCP microbial consortium (Compeau et al., 1991). Composting was also explored as a technology for bioremediation of soils contaminated with explosives such as trinitrotoluene (Myler and Sisk, 1991). More than 90% of the explosives were removed from contaminated soils within 80 days (Williams et al., 1989).

19.7.2. Treatment Strategies for Aquifers

There are three basic approaches to the treatment of contaminated aquifers (Bouwer et al., 1988): physical containment, above-ground treatment, and *in situ* bioremediation.

19.7.2.1. *Physical Containment*

Physical containment includes the use of temporary physical barriers to slow down or halt movement of contaminants. This approach has been adopted with some success in hazardous waste sites.

19.7.2.2. *Above-Ground Treatment ("Pump-and-Treat Technology")*

The contaminated water is pumped out from the aquifer by extraction wells and treated above ground by one of several treatment processes. However, it is difficult to extract chemical contaminants adsorbed to the aquifer matrix. Since the removal of organic contaminants by this technology is relatively slow, this approach can be regarded as a means for preventing further migration of the contaminant in the aquifer (Mackay and Cherry, 1989). The major available treatment technologies are air stripping to remove volatile organic compounds, adsorption to granular activated carbon, ultrafiltration, oxidation with ozone/UV or ozone /H_2O_2, activated sludge, and fixed-film biological reactors (Bouwer et al., 1988). For example, fixed-film bioreactors, using sand as the matrix and methane or natural gas as the primary substrate, are capable of removing up to at least 60% of trichloroethylene from polluted water. They also remove more than 90% of TCE and TCA from vapor streams generated by air stripping of polluted groundwater (Canter et al., 1990).

19.7.2.3. *In Situ Bioremediation*

In situ bioremediation is the enhancement of the catabolic activity of indigenous microorganisms by adding nutrients and, if necessary, oxygen (added as air, pure oxygen, or hydrogen peroxide). *In situ* treatment depends on aquifer characteristics (e.g., permeability as measured by hydraulic conductivity), contaminant characteristics, oxygen level, pH, availability of nutrients, redox conditions, and the presence of microorganisms able to degrade the contaminant under consideration (Alexander, 1985; Bedient and Rifai, 1992; McCarty et al., 1984; Rittmann, 1987; Thomas and Ward, 1989; Wilson et al., 1986). This approach has been used mostly for gasoline spills. Indigenous subsurface bacteria are able to grow on aromatic compounds such as naphthalene, toluene, benzene, ethylbenzene, *p*-cresol, xylene, phenol, and cresol, which are used as a sole source of carbon and energy (Brockman et al., 1989; Frederickson et al., 1991; Glynn et al., 1987). *In situ* bioremediation of aquifers contaminated with pentachlorophenol and creosote can be enhanced by injection of hydrogen peroxide (100 mg/L) as well as inorganic nutrients such as nitrogen and phosphorus (Piotrowski, 1989).

Much effort has been concentrated on the fate of chlorinated aliphatic hydrocarbons (e.g., trichloroethylene, dichloroethylene) in aquifers. These chemicals undergo reductive dehalogenation under anaerobic conditions in aquifers (Vogel and McCarty, 1985). Under aerobic conditions, these compounds are degraded by methane-utilizing bacteria (Fogel et al., 1986; Moore et al., 1989). Methanotrophic (i.e., methane-utilizing) bacteria use methane as a sole source of energy and as a major source of carbon (Haber et al., 1983). They can transform more than 50% of trichloroethene (TCE) into CO_2 and bacterial biomass (Fogel et al., 1986). High conversion rates of TCE are obtained with methanotrophs with V_{max} up to 290 nmol/min/mg of cells of *Methylosinus trichosporium* (Oldenhuis et al., 1991). A methanotrophic biofilm reactor was shown to be capable of degrading TCE and TCA in a continuous-flow operation for a period of 6 months. The maximum degradation rate for TCE was 400 μg/L·h (Strand et al., 1991). TCE was ultimately converted to CO_2 and CO by a microbial consortium composed of a methanotroph (*Methylocystis* spp.) and heterotrophic bacteria. The methanotroph converts TCE into glycoxylic, dichloroacetic, and trichloroacetic acids while heterotrophic bacteria carry out the biotransformation to CO_2 and CO (Uchiyama et al., 1992). A field demonstration of *in situ* biorestoration of an aquifer was undertaken at Moffett Naval Air Station (Roberts et al., 1989). Aquifer indigenous methanotrophic bacteria, stimulated by addition of oxygen and methane, were able to metabolize chlorinated aliphatic solvents such as TCE, *cis*- and *trans*-1,2-dichloroethene (DCE), and vinyl chloride (VC) under *in situ* conditions. The extent of biotransformation was 20% for TCE, 40% for *cis*-DCE, 85% for *trans*-DCE, and 95% for VC. A methanotrophic bacterium isolated from groundwater cometabolically degraded TCE in the presence of methane or methanol used as primary substrates. TCE can also be transformed to TCE epoxide by methane monooxygenase produced by bacteria. The epoxide then breaks down spontaneously to dichloroacetic acid and glyoxilic acid (Little et al., 1988). Methane monooxygenase activity is associated with TCE biodegradation by the methanotrophs *Methylosinus trichosporium* and *Methylomonas methanica* (Koh et al., 1993; Tsien et al., 1989). The synthesis of this enzyme is stimulated only under copper stress (Oldenhuis et al., 1991; Stanley et al., 1983) and is suppressed at copper concentrations as low as 0.25 μM (Brusseau et al., 1990). Mutants capable of producing methane monooxygenase in the presence of high levels of copper (up to 12 μM copper) have been recently isolated (Phelps et al., 1992).

Prior to field trials, laboratory work should be undertaken to assess the solubility, toxicity, and biodegradability of a contaminant and to determine any limiting nutrients in the aquifer. The site should also be evaluated for its suitability for *in situ* bioreclamation. The criteria for a suitable site are the following (Glynn et al., 1987):

1. Hydrogeological features must allow extraction of contaminated groundwater (i.e., recovery of the free contaminant phase) and reinjection and circulation of the treated water. Other hydreogeological characteristics that must be evaluated include depth and specific yield of the aquifer and permeability and direction of groundwater flow.

2. The groundwater parameters that are favorable to contaminant biodegradation (dissolved oxygen, pH, alkalinity and available nutrients) are the following.

Electron acceptor. Oxygen can be added to the aquifer by injecting pure oxygen, air, or hydrogen peroxide. Hydrogen peroxide, at a concentration of up to 750 mg/L, was used as a source of oxygen for the bioremediation of a shallow sandy aquifer contaminated with fuel spills in Michigan (Wilson et al., 1989). Since H_2O_2 may be toxic to aquifer microflora, some recommend gradually increasing its concentration to avoid toxicity. Solid peroxides (e.g., Na percarbonate, Ca peroxide) are also being studied as sources of oxygen for *in situ* subsurface bioremediation. Toxicity is reduced when these solid peroxides are encapsulated in polyvinylidene chloride (Vesper et al., 1992). Nitrate, and perhaps nitrous oxide, can also serve as electron acceptors for groundwater bacteria. Aromatic hydrocarbons (e.g., toluene, xylene, ethylbenzene) are biodegraded under denitrifying conditions in aquifer materials (Evans et al., 1991; Hutchins, 1989, 1991; Hutchins et al., 1991). Sulfate is another useful electron acceptor for the biotransformation of halogenated organic compounds (Cobb and Bouwer, 1991). The use of oxygen in conjunction with other electron acceptors (e.g., nitrate, nitrous oxide) for bioremediation of aquifers is being investigated.

Addition of inorganic nutrients. Sometimes, it is necessary to add nitrogen and phosphorus to enhance biodegradation. They are generally added through injection wells or infiltration galleries (Thomas and Ward, 1989).

Addition of methane and air. Methane and air can enhance the biodegradation of chlorinated hydrocarbons such as TCE.

19.7.2.4. Bioremediation by Addition of Microorganisms

Mixed microbial cultures can be added to contaminated aquifers to enhance biodegradation (Thomas and Ward, 1989). These cultures are obtained by traditional enrichment techniques or recombinant DNA technology. However, the latter approach has not been approved yet (see Chapter 16). The added microorganisms must reach, and grow in, the contaminated zone. In California, commercial blends of hydrocarbon-degrading bacteria were utilized for the bioreclamation of soils and aquifers contaminated with petroleum hydrocarbons (von Wedel et al., 1988). The process consists of the following:

1. Contaminated groundwater is treated above ground in chemostat bioreactors in the presence of the commercial bacterial cultures.

2. The treated groundwater is further modified with additional bacteria and is reinfiltrated into the contaminated soil for *in situ* treatment of both contaminated soil and aquifer.

In the study by von Wedel et al. (1988), following several months of operation, this augmented-biorestoration scheme resulted in 93% removal of total hydrocarbons. For benzene, toluene, and xylene, the rates of removal were 85%, 98%, and 33%, respectively (Fig. 19.8) (von Wedel et al., 1988).

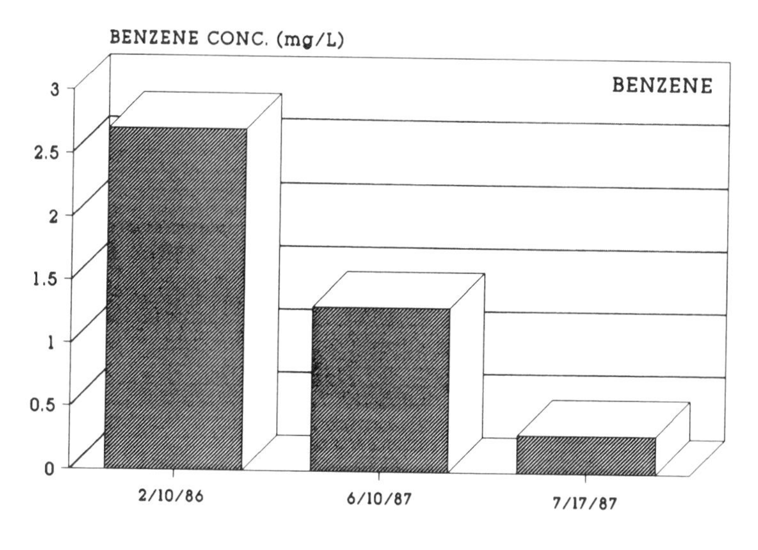

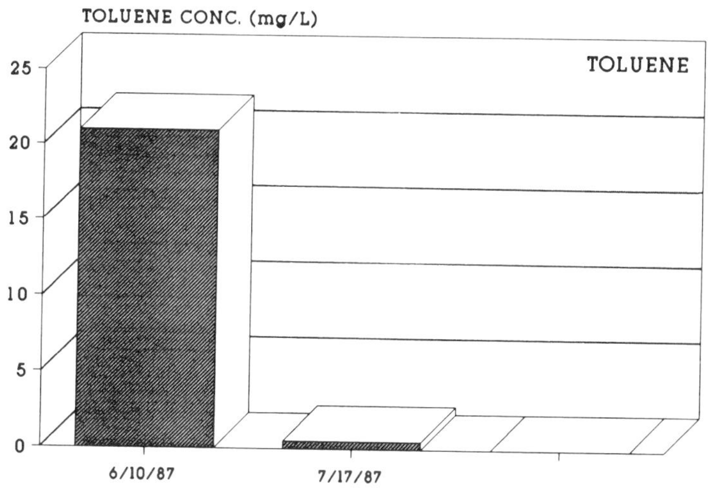

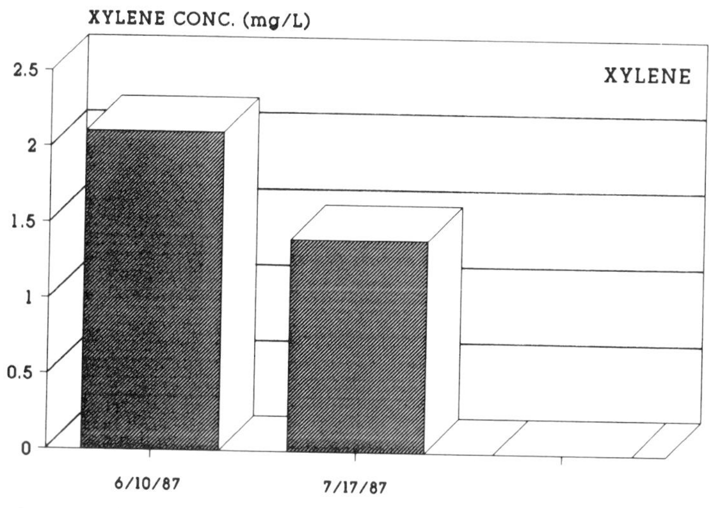

Fig. 19.8. Augmented biorestoration of groundwater for reducing petroleum hydrocarbon levels. From von Wedel et al. 1988, with permission of the publisher.

B. DISPOSAL OF WASTEWATER EFFLUENTS AND SLUDGE IN THE MARINE ENVIRONMENT

19.8. INTRODUCTION

Billions of gallons of wastewater effluents, sometimes receiving only primary treatment or less, are disposed of daily by the ever increasing populations along the coastlines around the world, sometimes not far away from public bathing beaches. The effluents include domestic wastewater containing fecal materials and industrial wastes harboring metals and xenobiotic compounds. Millions of tons of digested sludge are also disposed of at sea by pumping or transport by barges to the disposal site. Ocean outfalls are generally an inexpensive alternative to tertiary treatment. The outfall pipes may extend 1–4 miles offshore. Disposal of wastewater effluents and sludges into the ocean adversely impacts on marine life, contaminates shellfish beds, and may have public health implications for swimmers in public beaches.

In the 1960s the realization that oceans are not the "infinite sink" for dumping human wastes triggered extensive research on the fate of pathogenic microorganisms and parasites in seawater. We will now examine the main factors that control the fate of enteric pathogens in water and sediments in the marine environment.

19.9. GLOBAL SURVEYS OF ENTERIC PATHOGENS IN CONTAMINATED SEAWATER

Contamination of the marine environment by pathogenic microorganisms is mainly due to the disposal of wastewater or wastewater effluents into estuarine waters, to offshore disposal by sewage outfalls, and to rivers contaminated with wastewater effluents. Microbiological examination of coastal waters near sewage outfalls shows the presence of pathogenic bacteria such as *Salmonella* and *Vibrio chloerae* (Grimes et al., 1984; Morinigo et al., 1992a). The City of Miami, Florida, discharges domestic wastewater from a sewage outfall located 3.6 km offshore. Fecal coliforms (9,000–55,000/100 ml), fecal streptococci (0.5–19/100 ml) and enteroviruses (21–59 PFU/400 L) were detected within 200 m from the outfall. However, viruses were detected in the sediments at recreational bathing beaches situated up to 3.6 km from the outfall (Table 19.4) (Schaiberger et al., 1982).

Monitoring of bacterial indicators and enteroviruses in the vicinity of a sewage outfall off the Israeli coast showed that bacteria were reduced more rapidly than viruses. Only fecal streptococci displayed an inactivation rate similar to that of enteroviruses (Fig. 19.9) (Fattal et al., 1983).

The detection of enteric viruses in seawater was also documented off the coast of several countries, including Brazil (Marques and Martins, 1983), France (Hugues et al., 1980, 1981), Israel (Fattal et al., 1983), Italy (Petrilli et al., 1980), Spain (Bravo and de Vicente, 1992; Finance et al., 1982; Lucena et al., 1982; Morinigo et al., 1992a), and the United States (Goyal et al., 1979; Rao et al., 1984; Vaughn et al., 1979). The types of viruses found were identified as polioviruses, coxsackie viruses A and B, echoviruses, adenoviruses, and rotaviruses; their levels varied between 0.007 and 100 PFU/L (Bitton et al., 1985). In the vicinity of sewage outfalls, enteric viruses are also often detected in marine sediments, where they may persist for long time periods (Bitton et al., 1981b; Goyal et al., 1984; Rao et al., 1984; Schaiberger et al., 1982). More recently, coliphages, *B. fragilis* phages, entero-

TABLE 19.4. Detection of Enteroviruses in Sediments Along a Bathing Beach[a,b]

Station	Depth (m)	Enteroviruses (PFU/L)
1	1.2	0
2	0.9	15
3	1.5	30
4	1.5	0
5	1.5	0

[a]Adapted fom Schaiberger et al. (1982).
[b]Sample collected at 400 m apart on beach 3.6 km from outfall pipe's discharge.

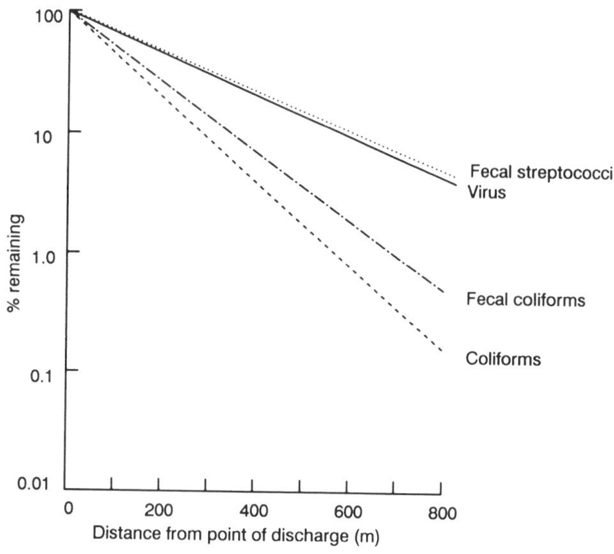

Fig. 19.9. Inactivation of enteric microorganisms at various distances from a sewage outfall in Israel. From Fattal et al. (1983), with permission of the publisher.

viruses, and rotaviruses were detected in marine sediments at 300 m to 12 km off the coast of Barcelona, an area impacted by fecal pollution. There was a significant correlation between the presence of enteric viruses and *B. fragilis* phages (Jofre et al., 1989).

19.10. SURVIVAL OF PATHOGENIC AND INDICATOR MICROORGANISMS IN SEAWATER

Laboratory experiments and *in situ* survival studies (e.g., use of dialysis bags and flow-through systems) have shown that a number of environmental and biological factors control the fate of enteric microorganisms in the marine environment.

19.10.1. Temperature

Temperature is a decisive factor controlling the survival of pathogenic microorganisms in seawater. The die-off of wastewater microorganisms increases at higher temperatures (El-Sharkawi et al., 1989). Decay rates of *E. coli* and enterococci in diffusion chambers were correlated with temperature in the 0–20°C range. *E. coli* survival was generally lower than that of enterococci (Lessard and Sieburth, 1983). Figure 19.10 (Won and Ross, 1973) also shows that the survival of an enterovirus, echovirus 6, in seawater is much lower at 22°C than at 3–5°C. Similarly, hepatitis A virus and phage indicators (F^+ and *B. fragilis* phages) are more persistent in seawater at 5°C than at 25°C (Chung and Sobsey, 1993).

19.10.2. Solar Radiation

Solar radiation also plays a key role in the decline of indicator and pathogenic bacteria in seawater (Bellair et al., 1977; Chamberlin and Mitchell, 1978; El-Sharkawi et al., 1989; Fujioka et al., 1981; Gameson and Gould, 1975; Kapuschinski and Mitchell, 1982). Figure 19.11 (McCambridge and McMeekin, 1981) shows that both solar radiation and biological factors have a detrimental effect on the survival of *E. coli* in seawater. Moreover, *E. coli* was more sensitive to sunlight than *Salmonella typhimurium*. Phages or fecal streptococci survive longer than fecal coliforms (Fujioka et al., 1981;

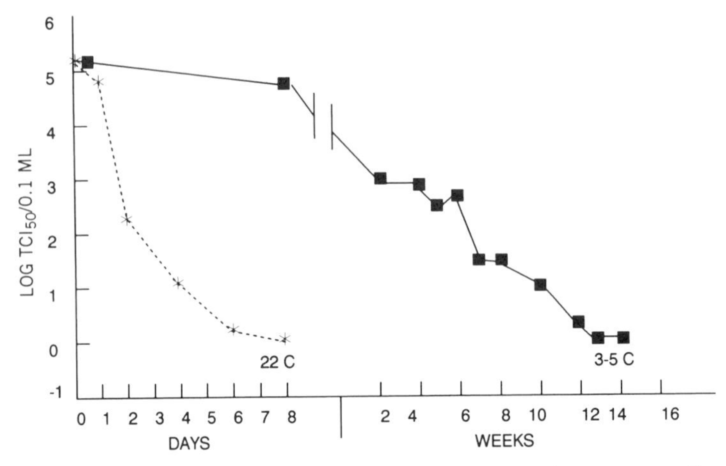

Fig. 19.10. Inactivation of echovirus 6 in seawater as a function of temperature. From Won and Ross (1973), with permission of the publisher.

Kapuschinski and Mitchell, 1982). Thus, the FC/FS ratio generally used to indicate the source (human vs. animal origin) of fecal pollution would be less valid in the marine environment (Fujioka et al., 1981). Analysis of hundreds of seawater samples showed a negative correlation between the mean monthly log coliform count and the mean monthly duration of sunshine ($r = 0.93$) (Fig. 19.12) (Fattal et al., 1983). Thus coliform counts should be higher in the winter than in the summer season. Exposure to sunlight in coastal seawater can also cause sublethal injury in microorganisms and may reduce the activity of several enzymes as well as bacterial culturability, because of the production of highly reactive types of oxygen (singlet oxygen, superoxide, hydrogen peroxide). Catalase may be the site of sunlight-induced damage in *E. coli*, and addition of this enzyme to minimal growth media improves the recovery of sunlight-injured cells (Arana et al., 1992; Kapuschinski and Mitchell, 1981).

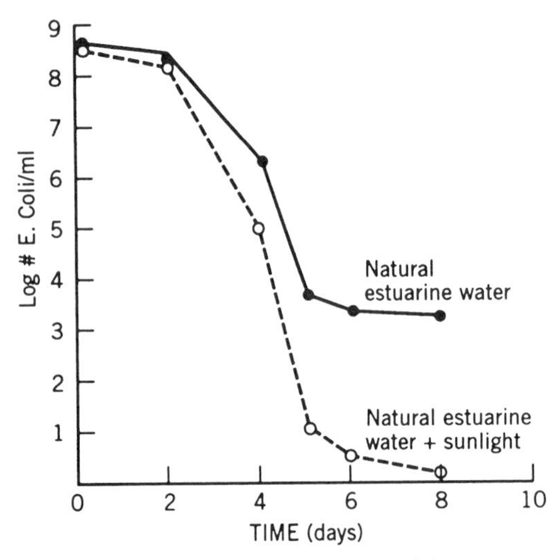

Fig. 19.11. Effect of sunlight on *E. coli* in seawater. Adapted from McCambridge and McMeekin (1981).

The killing effect of ultraviolet on viruses is well known (see Chapter 6). Solar radiation is an important contributor to the loss of infectivity of marine bacteriophages in seawater. In full sunlight, the decay rates may be 0.4–0.8 hr^{-1}. The decay rate was estimated at 0.033 hr^{-1} when averaged over 24 hr and integrated over the upper 30 m of the water column (Suttle and Chen, 1992). However, less is known about the effect of natural solar radiation on *enteric* viruses. Exposure of poliovirus 1 for 3 hr to solar radiation in Florida (light intensity was 0.646 cal cm^{-2} min^{-1} and the mean temperature was 26°C) resulted in approximately 1-log inactivation of virus (Bitton et al., 1979). A similar inactivation rate was observed for phage T_4 in seawater (Attree-Pietri and Breittmayer, 1970).

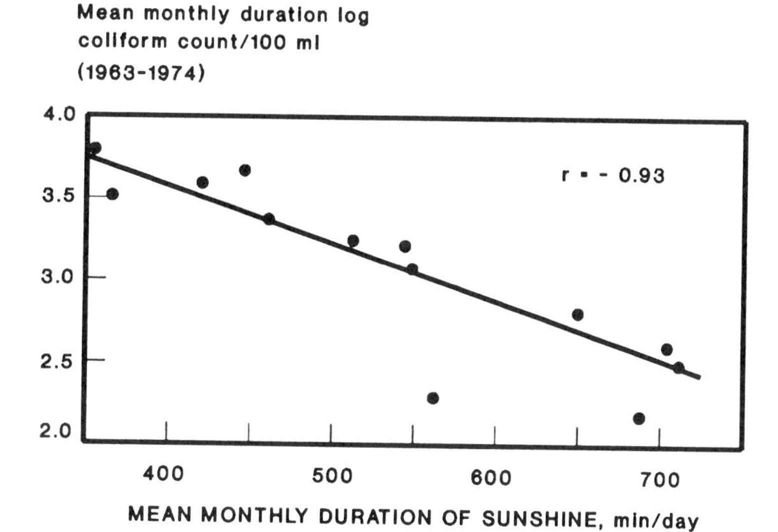

Fig. 19.12. Relationship between duration of sunshine and mean coliform count (1963–1974). From Fattal et al. (1983), with permission of the publisher.

19.10.3. Osmotic Stress

Genetically controlled osmoregulatory processes, induced by salts, help enteric bacteria survive osmotic stress in the marine environment (Munro et al., 1987). Osmoregulatory processes involve K^+ uptake as well as accumulation of compatible organic osmolytes such as glycine-betaine, trehalose, and glutamate (Gauthier et al., 1991; Larsen et al., 1987; Strom et al., 1986). Glycine-betaine has been found in marine sediments (King, 1988) and may protect enteric bacteria from osmotic stress in this environment (Munro et al., 1989).

19.10.4. Adsorption to Particulates

Adsorption of bacteria or viruses to particulates (silts, clay minerals, cell debris, or particulate organic matter) appears to provide protection to microorganisms from environmental insults (Bitton and Mitchell, 1974; Gerba and Schaiberger, 1975). Figure 19.13 illustrates the effect of a clay mineral, montmorillonite, in protecting *E. coli* and bacteriophage T_7 (Bitton and Mitchell, 1974a, 1974b). This phenomenon was also observed in lake water (Babich and Stotzky,

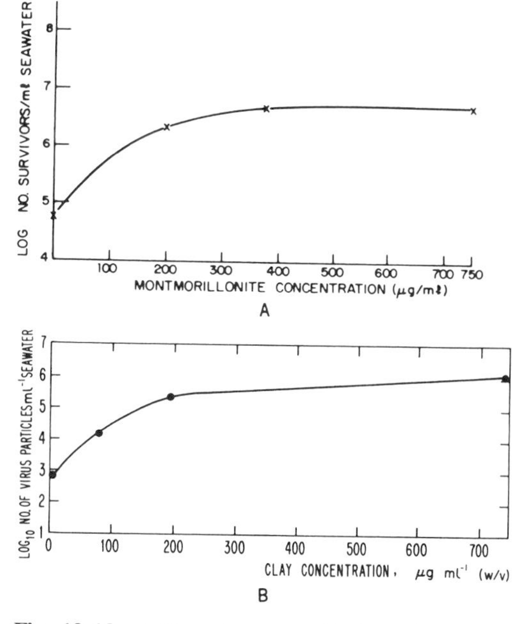

Fig. 19.13. Effect of montmorillonite on the survival of *E. coli* (**A**) and bacteriophage T7 (**B**) in seawater. Adapted from Bitton and Mitchell (1974a) and Bitton and Mitchell (1974b).

1980). Solid-associated microorganisms may settle and accumulate in aquatic sediments (see Section 19.11) and may remain infective to their host cells.

19.10.5. Biological Factors

Biological factors are also implicated in the decline of enteric pathogens in the marine environment. Some 30 years ago, it was demonstrated that a heat-labile substance was involved in virus inactivation in seawater (Plissier and Therre, 1961). Since then, this phenomenon has repeatedly been demonstrated in the marine environment around the globe. The contribution of small protozoan flagellates to the decay of marine bacteriophages in seawater has been documented. These nanoflagellates can ingest approximately three viruses per flagellate per hour (Suttle and Chen, 1992). Figure 19.14 shows that both enteric bacteria and viruses are readily inactivated in natural seawater but to a lesser extent in heat-treated seawater (Bitton and Mitchell, 1974b; Mitchell, 1971; Pietri and Breittmayer, 1976; Shuval et al., 1971). Thus, lytic and antagonistic microorganisms (e.g., *Vibrio marinus; Bdellovibrio bacteriovorus;* predacious protozoa such as *Vexillifera*) and other unknown biological factors are implicated in inactivation of enteric microorganisms in seawater (Borrego and Romero, 1985; Magnusson et al., 1967; Patti et al., 1987). Some contend that protozoa are the most important predators of *E. coli* in the marine environment (Enzinger and Cooper, 1976). The major

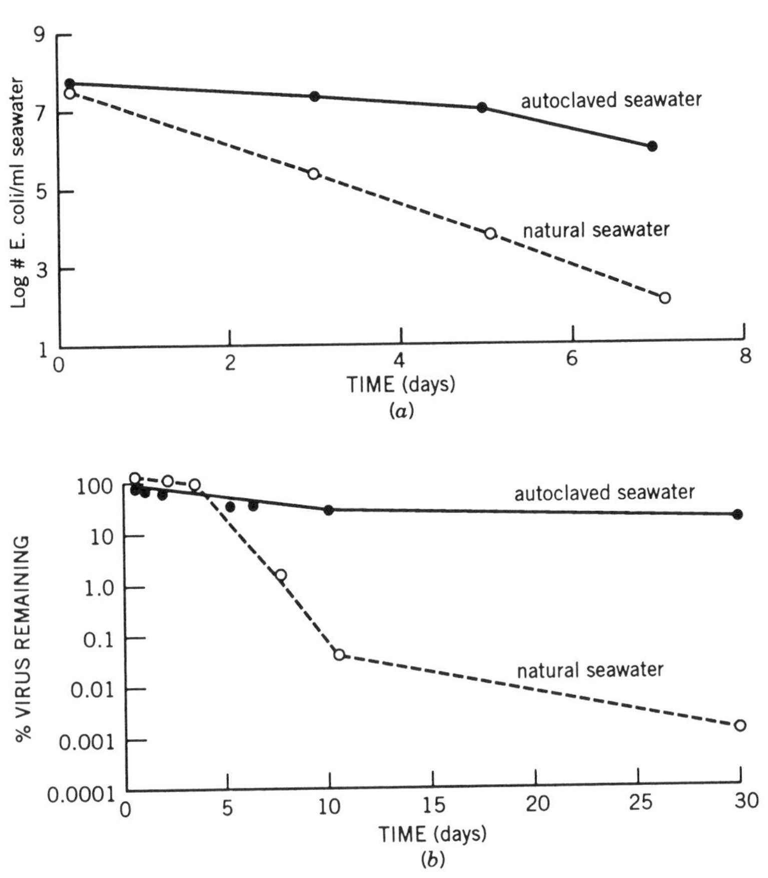

Fig. 19.14. Inactivation of enteric microorganisms in natural seawater. **a.** *E. coli*. **b.** Poliovirus 1. From Bitton and Mitchell (1974b) and Shuval et al. (1971), with permission of the publisher.

influence of protozoa is exerted during the first two days of incubation (McCambridge and McMeekin, 1980).

Other factors that control the survival of enteric pathogens in aquatic environments are microorganism type (e.g., viruses survive better than bacteria), growth stage, aggregation, pH, dissolved oxygen, and heavy metals (Bitton, 1978; Block, 1981; Gauthier et al., 1992). Fecal streptococci generally persist longer in seawater than total or fecal coliforms (Bravo and de Vicente, 1992). As regards the persistence of enteric pathogens in the marine environment, we can draw some general conclusions (Akin et al., 1976; Hetrick, 1978):

1. Enteric pathogens, especially viruses, are more labile in seawater than in fresh water.
2. The results of several studies show that microbial inactivation in seawater is variable and unpredictable.
3. There are probably other, yet unknown, factors contributing to the decline of enteric pathogens in seawater.

TABLE 19.5. Accumulation of Indicator Bacteria (MPN/100 ml) in Marine Sediments[a]

	Total coliform	Fecal coliform
Overlying water		
Station #1	6,886	2,382
2	5,320	1,528
3	64	19
4	92	10
Sediment		
Station #1	382,143	9,731
2	192,857	16,806
3	16,791	152
4	14,279	151

[a]Adapted from Goyal et al. (1977).

19.11. SURVIVAL OF PATHOGENIC AND INDICATOR MICROORGANISMS IN SEDIMENTS

Sewage and sludge disposal off the coast from sewage outfalls or barges affect marine sediments, which can be important reservoirs of pathogenic microorganisms and parasites. Microbial pathogens and indicattors associated with organic and inorganic particulates settle into the bottom sediments of freshwater and marine environments, where they accumulate and reach higher concentrations than in the water column. A buildup of indicator bacteria in sediments has been observed around sewage outfalls (Table 19.5) (Goyal et al., 1977). The accumulation of enteric pathogens is due to their longer survival in sediments. Figure 19.15 displays the longer persistence of *E. coli* in marine sediments than in seawater (Gerba and McLeod, 1976). Furthermore, in *E. coli*, expression of the genes responsible for osmoregulation has been found to be enhanced in marine sediments containing organic matter. This may explain, at least partially, why sediments act as reservoirs for pathogens (Gauthier and Breittmayer, 1990). Enteric viruses also find their way into sediments, where they accumulate and survive longer than in the water column (Bitton, 1980a). For example, hepatitis A viruses, poliovirus 1, and phage indicators (F^+ and *B. fragilis* phages) persist longer in sediments than in the water column (Chung and Sobsey, 1993). As a result of sludge dumping off the Delaware–Maryland coast, bacterial indicators (TC, FC, FS) and amaoebas were detected as far as 40 km from the sludge dumpsite. Amoebas, total coliforms, and fecal streptococci persisted longer in sediments than did fecal coliforms (O'Malley et al., 1982). Several species of *Acanthamoeba* were isolated from the sediments of a Philadelphia–Camden sludge disposal site (Sawyer et al., 1982). Some of these amoeba species are of potential health significance (e.g., *A. culbertsoni, A. hatchetti*). The levels of pathogenic and indicator bacteria in intertidal sediments in Boston Harbor were followed up to 460 m from a storm and wastewater discharge point. However, despite their accumulation in the sediments, all bacterial indicators (total bacteria, fecal coliforms, fecal streptococci) declined with increasing distance from the sewage outfall (Shiaris et al., 1987).

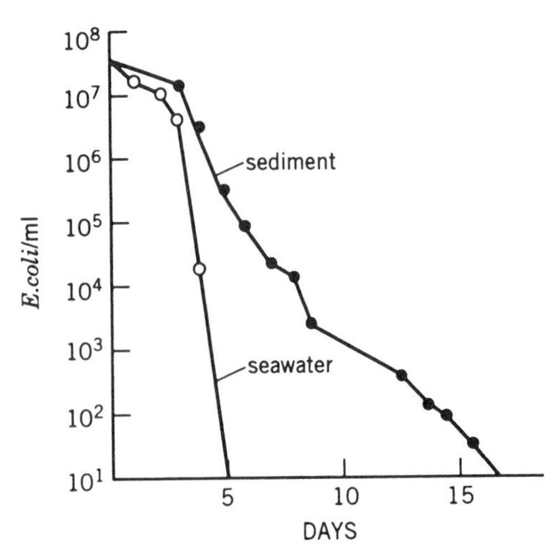

Fig. 19.15. Comparison of *E. coli* persistence in seawater and in sediments. Adapted from Gerba and McLeod (1976).

Thus, sediments can serve as reservoirs of enteric pathogens. Sediment resuspension through motor boat activity, currents, swimming, and changes in water quality, would increase pathogen levels in the water column.

19.12. HEALTH ASPECTS OF SWIMMING IN FECALLY CONTAMINATED RECREATIONAL WATERS

The health risks associated with water sports (swimming, water skiing, scuba diving, surfing) fall into two categories according to the source of exposure (Dufour, 1986).

19.12.1. Exposure to Water Containing Pathogenic Microorganisms of Wastewater Origin

Several diseases have been associated with swimming in wastewater-contaminated recreational waters. These include typhoid fever, salmonellosis, shigellosis, hepatitis, and gastroenteritis. The sources of the infectious agents are wastewater treatment plant effluents, septic tank effluents, stormwater runoff, and discharges from lower animals (Cabelli, 1989). A waterborne shigellosis disease outbreak was reported in Iowa and was linked to swimming in a fecally contaminated portion of the Mississippi River (Rosenberg et al., 1976). Outbreaks of hepatitis and viral gastroenteritis have been associated with swimming in contaminated aquatic environments (Bryan et al., 1974; Denis et al., 1974; Koopman et al., 1982). Prospective epidemiological studies have also shown an association between gastroenteritis and swimming (Cabelli, 1981). Enterococcus levels in recreational waters appear to be a good indicator of the risk of swimming-associated gastroenteritis. Thus, a recent U.S. EPA guideline sets the enterococcus level in bathing waters at 35 CFU/100 ml (Cabelli, 1989). A prospective epidemiological study in South Africa has shown that the relative risk (i.e., incidence rate among swimmers divided by the incidence rate among nonswimmers) for developing gastrointestinal symptoms was higher at a moderately polluted beach than at a control beach (von Schirnding et al., 1993).

19.12.2. Exposure to Authochthonous (i.e., Indigenous) Microorganisms

Water-associated activities may also be the source of infections caused by autochthonous microorganisms in marine and freshwater environments. The main agents are *Aeromonas, Vibrio,* and *Pseudomonas aeruginosa.* The etiologic agents in ear infections ("swimmer's ear") are *Pseudomonas aeruginosa* and *Vibrio* species (e.g., *V. parahaemolyticus, V. vulnificus, V. alginolyticus*). *Pseudomonas aeruginosa* is ubiquitous in nature but it is also recovered from human stools and finds its way into wastewater. It is responsible for ear and urinary tract infections as well as dermatitis and folliculitis (Fox and Hambricks, 1984; Havelaar et al., 1983; Salmen et al., 1983). This pathogen was found in 45% of coastal water samples in Israel and correlated well with the presence of total and fecal coliforms (Yoshpe-Purer and Golderman, 1987).

Swimming in fecally polluted waters may not cause only enteric disturbances but also ailments of the upper respiratory tract. Risk of pneumonia can be high in near-drowning situations. Pneumonia cases attributed to *Pseudomonas putrefaciens, Staphylococcus aureus, Aeromonas hydrophila,* and *Legionella pneumophila* have been documented (Reines and Cook, 1981; Rosenthal et al., 1975; Sekal et al., 1982). Swimming in recreational waters can also be the cause of skin infections due to presence of opportunistic microorganisms (*Aeromonas, Mycobacterium, Staphylococcus,* and *Vibrio* species). These infections are associated with skin abrasion and laceration. Recreational waters may serve as a vehicle for skin infections caused by *Staphylococcus aureus,* and some observers have recommended that this organism be used as an additional indicator of the sanitary quality of recreational waters, since its presence is associated with human activity in recreational waters (Charoenca and Fujioka, 1993; Yoshpe-Purer and Golderman, 1987). A prospective epidemiological study in South Africa indicated that the relative risk for developing skin symptoms was significantly higher among whites at a moderately polluted beach than at a control beach (von Schirnding et al., 1993). Swimming in areas contaminated by urine from infected domestic or wild animals was related to outbreaks of leptospirosis, also named Weil's disease, and hemorrhagic jaundice, which causes headaches, fever, chills, and nausea. The infectious agents are *Leptospira* species (Dufour, 1986).

Swimming in warm recreational lakes can be associated with primary meningoencepahlitis (PAM), which is caused by a free-living amoeba *Naegleria fowleri.* This protozoan gains access to the brain through the nose and can cause death within 3 days.

19.13. FURTHER READING

Bitton, G. 1980b. Adsorption of viruses to surfaces: Technological and ecological implications, pp. 331–374, in: *Adsorption of Microorganisms to Surfaces,* G. Bitton and K.C. Marshall, Eds. Wiley Interscience, New York.

Bitton, G., and R.W. Harvey. 1992. Transport of pathogens through soils and aquifers, Chapter 7, in: *New Concepts in Environmental Microbiology,* R. Mitchell, Ed. Wiley-Liss, New York.

Canter, L.W., and R.C. Knox. 1985. *Septic Tank Systems: Effects on Groundwater Quality.* Lewis Publishing, Chelsea, MI.

Dufour, A.P. 1986. Diseases caused by water contact, pp. 23–41, in: *Waterborne Diseases in the United States,* G.F. Craun, Ed. CRC Press, Boca Raton, FL.

Gerba, C.P., and G. Bitton. 1984. Microbial pollutants: Their survival and transport pattern to groundwater, in: *Groundwater Pollution Microbiology,* G. Bitton and C.P. Gerba, Eds. Wiley, New York.

Gerba, C.P., C. Wallis, and J.L. Melnick. 1975b. Fate of wastewater bacteria and viruses in soils. J. Irrig. Drainage Div. ASCE 3: 157–168.

Lipson, S.M., and G. Stotzky. 1987. Interactions between viruses and clay minerals, pp. 197–230, in: *Human Viruses in Sediments, Sludges, and Soils,* V.C. Rao and J.L. Melnick, Eds. CRC Press, Boca Raton, FL.

McCarty, P.L., B.E. Rittmann, and E.J. Bouwer. 1984. Microbiological processes affecting chemical transformations in groundwater, pp. 89–115, in: *Groundwater Pollution Microbiology,* G. Bitton and C.P. Gerba, Eds. Wiley, New York.

Rose, J.B. 1986. Microbial aspects of wastewater reuse for irrigation. CRC Crit. Rev. Environ. Control. 16: 231–256.

U.S. EPA. 1981. *Process Design Manual for Land Treatment of Municipal Wastewater.* EPA-625/1--81-013. U.S. EPA, Cincinnati, OH.

U.S. EPA. 1983. *Process Design Manual for Land Application of Municipal Sludge.* EPA 625/1--83-016.

Wallis, P.M., and D.L. Lehmann, Eds. 1983. *Biological Health Risks of Sludge Disposal to Land in Cold Climates.* Univ. of Calgary Press, Calgary, Canada.

20 WASTEWATER REUSE

20.1. INTRODUCTION

Indirect reuse of wastewater has been practiced for centuries around the globe. Planned reuse of this resource has been documented as early as the sixteenth century in Europe. In the United States this practice was initiated around the beginning of this century in Arizona and California for irrigation of lawns and gardens or for use as cooling water. Some distinguish between "reuse," "recycling," and "reclamation." Recycling is the internal reuse of a wastewater by a given industry prior to its ultimate disposal. Wastewater reuse is the use of treated wastewater for a beneficial goal such as crop irrigation. Wastewater reclamation is the treatment of wastewater to make it reusable. This practice has gained importance worldwide because of water shortages, particularly in arid areas (e.g., California, Arizona, Texas, Colorado) and wastewater disposal regulations (DeBoer and Linstedt, 1985).

The health effects related to wastewater reuse are classified into two categories.

1. *Health effects due to parasites as well as bacterial and viral pathogens.* The exposure routes for the infectious agents are direct contact from contaminated surfaces, accidental ingestion of contaminated water, consumption of raw vegetables that have been irrigated with reclaimed water, and long-term exposure to biological aerosols in the vicinity of spray irrigation sites or cooling towers (see Chapters 4 and 14). The risk of transmission of infectious disease is mainly associated with the use of untreated sewage or wastewater effluents of very poor quality (Cooper, 1991; Rose, 1986).

2. *Chemicals.* The chemicals of concern are heavy metals, pesticides, chlorinated compounds, and other xenobiotics. The adverse effects of these chemicals, many of which are mutagenic or carcinogenic, are of special concern when the reclaimed wastewater is used for crop irrigation or groundwater recharge (Bitton and Gerba, 1984; Cooper, 1991; Nellor et al., 1985).

20.2. CATEGORIES OF WASTEWATER REUSE

The various categories of wastewater reuse are agricultural use (land application), landscape irrigation, groundwater recharge, recreational use, nonpotable urban use, potable reuse, and industrial use (Table 20.1) (Asano and Tchobanoglous, 1991).

20.2.1. Agricultural Reuse: Land Application

Reclaimed wastewater is most commonly used for irrigation of agricultural crops. While the planned effluent reuse for agricultural irrigation in the United States is less than 1%, other countries such as India, Israel, and South Africa are using 20%–25% of wastewater effluents for agricultural purposes (Rose, 1986). Wastewater reuse for agricultural purposes is also practiced in North Africa (Morocco, Tunisia, Lybia), the Middle East (Jordan, Egypt, Saudi Arabia), Latin America (Chile, Peru, Mexico), and Asia (India, China) (Bartone, 1991).

The advantages of land application of waste-

TABLE 20.1. Categories of Municipal Wastewater Reuse[a,b]

Wastewater reuse categories	Potential constraints
Agricultural irrigation Crop irrigation Commercial nurseries	Effect of water quality, particularly salts, on soils and crops Public health concerns related to pathogens (bacteria, viruses, and parasites)
Landscape irrigation Park Schoolyard Freeway median Golf course Cemetery Greenbelt Residential	Surface and groundwater pollution if not properly managed Marketability of crops and public acceptance
Industrial reuse Cooling Boiler feed Process water Heavy construction	Reclaimed wastewater constituents related to scaling, corrosion, biological growth, and fouling Public health concerns, particularly aerosol transmission of organics, and pathogens in cooling and boiler feed water
Groundwater recharge Groundwater replenishment Salt water intrusion Subsidence control	Organic chemicals in reclaimed wastewater and their toxicological effects Total dissolved solids, metals, and pathogens in reclaimed wastewater
Recreational/environmental uses Lakes and ponds Marsh enhancement Streamflow augmentation Fisheries Snowmaking	Health concerns over bacteria and viruses Eutrophication due to N and P
Nonpotable urban uses Fire protection Air conditioning Toilet flushing	Public health concerns about pathogens transmitted by aerosols Effects of water quality on scaling, corrosion, biological growth, and fouling
Potable reuse Blending in water supply Pipe-to-pipe water supply	Organic chemicals in reclaimed wastewater and their toxicological efects Esthetics and public acceptance Public health concerns on pathogen transmission including viruses

[a]From Asano and Tchobanoglous (1991), with permission of the publisher.
[b]Arranged in descending order of volume of use.

water effluents are the provision of a supply of water and valuable nutrients to crops, and its function as an additional treatment for the effluents prior they reach groundwater. The main disadvantage is the potential contamination of groundwater resources and agricultural crops with parasites, bacterial and viral pathogens, toxic metals, and mutagenic/carcinogenic trace organics. Toxicity to crops is caused by excess salinity and toxic ions such as sodium, boron, chloride, cadmium, copper, zinc, nickel, beryllium, and cobalt (see Chapter 19 for further details).

There is obviously a risk of the transmission of disease through the use of untreated wastewater for vegetable irrigation. In Mexico, study of the irrigation of vegetable crops with domestic wastewater has shown that the highest bacterial contamination is observed in leafy vegetables such as lettuce (37,000 total coliforms per 100 g and 3,600 fecal coliforms per 100 g) and spinach (8,700 total coliforms per 100 g; 2,400 fecal coliforms per 100 g). The common rinsing of vegetables with tapwater does not reduce the indicator organisms to safe levels (Table 20.2) (Rosas et al., 1984). Outbreaks of diseases such as cholera have been associated with wastewater irrigation of vegetables. Outbreaks of disease caused by parasites can also be linked to this practice. Figure 20.1 shows the association between the consumption of wastewater-irrigated vegetables in Israel and the percentage of stool samples positive for *Ascaris*. After climbing to 35%, it decreased to less than 1% after the government banned the use of wastewater for vegetable irrigation (Gunnerson et al., 1984).

TABLE 20.2. Effect of Vegetable Rinsing on Coliform Levels[a]

	Geometric mean[b] of the following samples of the following types of bacteria:			
	Rinsed		Unrinsed	
Crop	TC	FC	TC	FC
Celery	300	30	1,300	300
Spinach	2,400	1,700	8,700	2,400
Lettuce	700	570	37,000	3,600
Parsley	370	300	3,100	660
Radish	650	300	2,600	360

[a]From Rosas et al. (1984), with permission of the publisher.
[b]Most probable number per 100 g.

A coliform level of 2.2 per 100 ml (7-day median) is allowed for food crops in California. The U.S. standard for irrigation of nonedible crops (e.g., seed and fiber crops) is a coliform level of 5,000 per 100 ml. Access to the public is restricted by posting of warning signs, and a buffer zone is allowed if spray irrigation is conducted at the site. A level of 23 coliforms per 100 ml has been adopted for irrigation of pastures used for milk animals and for recreational use (e.g., golf courses). In Florida, a level of 23 total coliforms per 100 ml is allowed for most agricultural uses.

20.2.2. Groundwater Recharge

Reclaimed water can be used for groundwater recharge to augment groundwater supplies and to prevent saltwater intrusion in coastal areas. Rechage is carried out by surface spreading or by direct injection. The reclaimed water must be of drinking water quality and it requires biological, chemical, and physical treatments. An example is Water Factory 21 reclamation facility in Orange County, California. The advanced treatment train for wastewater reclamation includes lime treatment, recarbonation, filtration, activated carbon adsorption, reverse osmosis, and final chlorination (Asano, 1985; Metcalf and Eddy, 1991). The treated effluent is directly injected into groundwater. A final chlorine residual is necessary to avoid bioclogging of the recharge well and aquifer (Bouwer, 1992). Inactivation and removal of pathogens in soils and groundwater are discussed in Chapter 19. There are no U.S. federal regulations concerning groundwater recharge.

20.2.3. Recreational Use

In arid areas, treated wastewater is used to fill recreational lakes (e.g., boating, fishing, and water sports). One of the best examples of use of

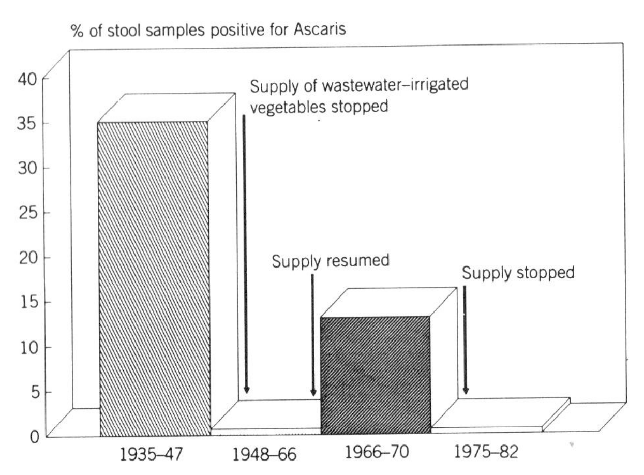

Fig. 20.1. Relationship between *Ascaris*-positive stool samples and supply of wastewater-irrigated vegetables in Jerusalem (1935–1982). From Gunnerson et al. (1984), with permission of the publisher.

reclaimed wastewater for recreational purposes is Santee, California, where lakes receiving treated wastewater are used for boating, fishing, and even swimming. Another example of recreational use of reclaimed wastewater is South Lake Tahoe, California.

People may come in contact with potential pathogens when engaging in water sports. The potential public health aspects of these activities have been discussed in Chapter 19.

20.2.4. Urban Nonpotable Use

The category of urban nonpotable use includes the use of reclaimed wastewater for private lawns and parks irrigation, fire protection, air conditioning, and toilet flushing. Some states such as California have guidelines for irrigation of parks, playgrounds, schoolyards, and other public access areas. Some states (e.g., Georgia) have less stringent guidelines, while others do not allow such uses.

The City of Colorado Springs, Colorado, uses a tertiary effluent to spray-irrigate city parks. The tertiary treatment consists of activated sludge treatment followed by dual-media (sand and anthracite) gravity filtration and chlorination to maintain a residual chlorine of 4–6 mg/L (Schwebach et al., 1988). Fecal coliform densities are 91% of the time below 23 per 100 ml, and 99% of the time below 500 per 100 ml. A two-year prospective epidemiological study did not show any significant difference in reported gastrointestinal illness rates between visitors of parks irrigated with wastewater effluent and those of parks irrigated with potable water. However, a significant increase in rates was observed when the fecal coliform level of the reclaimed wastewater was above 500 per 100 ml. This study essentially showed that the city standard of 200 fecal coliforms per 100 ml was adequate for protecting public health.

A dual distribution system consists of providing two water supply systems: One system provides water of very good microbiological and chemical quality for drinking, cooking, and washing (this category represents only 2% of the total amount of water used in households); the second system provides water of lower quality (reclaimed wastewater) for other household uses such as lawn irrigation. The systems are color-coded in order to avoid errors. A dual water supply system went in operation in 1977 in St. Petersburg, Florida. Little is known about any adverse health effect on urban populations. It was reported that sensitive ornamental plants may be adversely affected if the reclaimed wastewater contains high chloride concentrations (>600 mg/L) (Berger, 1982; Johnson, 1991).

20.2.5. Direct Potable Reuse of Reclaimed Wastewater as a Domestic Water Supply

This category includes the intentional reuse of wastewater as part of the potable water supply of a given community. Direct potable reuse is being contemplated in arid zones in response to severe water shortages. Indirect potable reuse of wastewater has been often practiced by many communities the world over. This occurs when the treated wastewater effluent of one community becomes the drinking water supply for another community downstream (e.g., Cincinnati, Ohio, which uses water from communities upstream along the Ohio River) (Donovan et al., 1980; Dean and Lund, 1981; Hammer, 1986). Some examples of direct potable reuse of reclaimed wastewater are the following.

20.2.5.1. Windhoek, Namibia

Treated wastewater is intentionally added to the drinking water supply in Windhoek, Namibia, in southwest Africa. Biologically treated effluents are subjected to tertiary treatment that comprises lime treatment, ammonia stripping, sand filtration, breakpoint chlorination, activated carbon, and final chlorination.

A long-term epidemiological study showed no adverse health effects associated with the consumption of directly reclaimed drinking water. Heterotrophic plate count (<100 per ml), total coliforms (0 per 100 ml), and coliphage (0 per 100 ml) were proposed for the routine monitoring of the quality of reclaimed wastewater (Grabow, 1990; Isaacson et al., 1986).

20.2.5.2. The Denver Potable Water Reuse Demonstration Project

The city of Denver, Colorado, situated in a semiarid area, expects an increased demand for potable water in the next decades. The city's water department built a 1-MGD (3.8 million L/day) water reuse plant to examine the feasibility of treating wastewater plant effluents to potable quality on a continuous basis. The process proposed provides multiple barriers against contaminants such as bacteria, viruses, protozoa, heavy metals, and trace organics (Lauer, 1991; Rogers and Lauer, 1992). The treatment train, illustrated in Figure 20.2 (Lauer et al., 1985), consists essentially of the following steps: lime clarification, recarbonation (use of CO_2 as a neutralizing agent), filtration, selective ion exchange for ammonia removal, two carbon adsorption steps

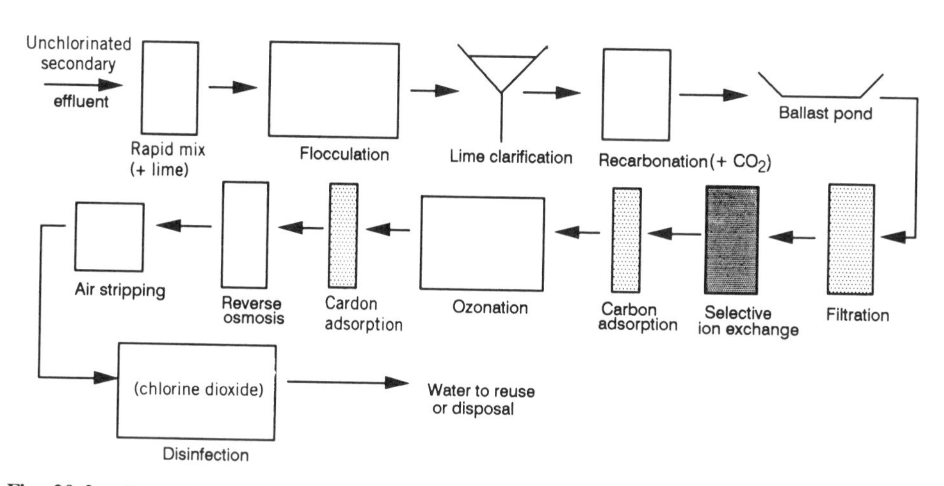

Fig. 20.2. Denver's potable water reuse project: Water treatment train. Adapted from Lauer et al. (1985).

separated by ozonation, reverse osmosis, air stripping, and final disinfection with chlorine dioxide (Lauer et al., 1985; Rogers and Lauer, 1986). The plant effectively removes total organic carbon (Fig. 20.3) and total coliforms (Fig. 20.4) (Lauer et al., 1985). As to other indicators (fecal coliforms, fecal streptococci, heterotrophic plate count bacteria, and coliphage), the product water was of quality equal to or better than that of the Denver drinking water, and no coliphage, fecal coliform, or fecal streptococci were detected in the finished water (Arber, 1983; Rogers and Lauer, 1986). The treatment train results in an approximately 7- to 8-log decrease in total coliform counts (Fig. 20.4; Lauer et al., 1985).

20.2.5.3. *San Diego Project*

The City of San Diego, California is studying the potential use of reclaimed wastewater to supplement the raw water source of potable supply (Cooper, 1991). The plan is to build a reclamation plant with a capacity of 450,000 m^3/day (120 MGD) to reclaim and beneficially reuse 86 million m^3 of reclaimed water per year by the year 2010 (Bayley et al., 1992). The treatment train includes primary and biological treatments followed by tertiary treatment that consists of coagulation, sand filtration, reverse osmosis, and activated carbon. The chemical and microbiological quality of the reclaimed wastewater was found to be equal to or better than that of the city's present raw water supply (Cooper, 1991).

22.2.6. Industrial Use

Reclaimed wastewater is used by industry mainly as cooling water for power plants. Other uses include boiler feed, washing, and processing. For example, in Israel a municipal wastewater effluent, subjected to lime treatment followed by ammonia stripping and pH adjustment, is used as make-up water for a cooling tower serving a refinery and petrochemical complex (Rebuhn and Engel, 1988).

20.2.7. Wetlands and Aquaculture for Wastewater Renovation

Wetlands constitute a low-cost and low-energy alternative to traditional tertiary treatment. They remove BOD and nutrients (N, P) through biological uptake by plants and microbial action. The treatment efficiency of wetlands is controlled by hydraulic loading, water depth, and extent of coverage by aquatic plants (DeBoer and Lindstedt, 1985).

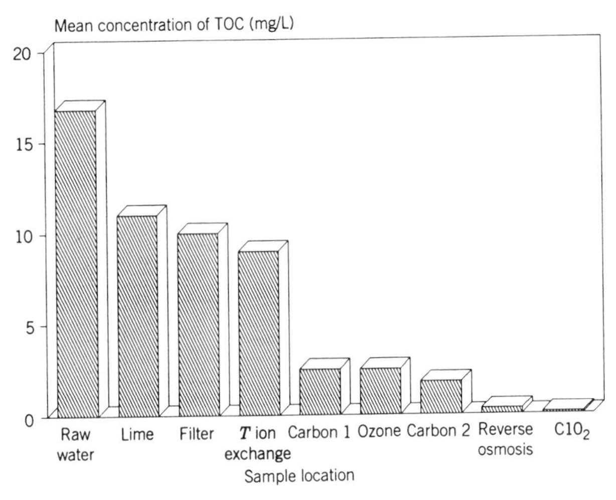

Fig. 20.3. Denver's potable water reuse project: TOC removal. From Lauer et al. (1985), with permission of the publisher.

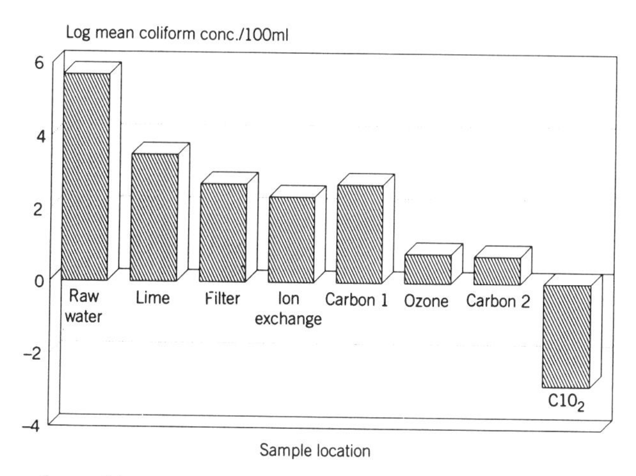

Fig. 20.4. Denver's potable water reuse project: Removal of total coliforms. From Lauer et al. (1985), with permission of the publisher.

Aquaculture also provides a means for renovating wastewater while allowing the growth of aquatic plants such as water hyacynths (Reddy, 1984) or raising fish for human consumption (Buras, 1984).

20.3. THE U.S. EXPERIENCE IN WASTEWATER REUSE

Most wastewater reuse projects are located in the arid and semiarid areas of western and southwestern United States. The State of California uses approximately 1 million m^3 of reclaimed wastewater per day (Asano and Tchobanoglous, 1991; Ongerth and Jopling, 1977). This state early recognized wastewater as a valuable resource, actively promoted the use of reclaimed wastewater for irrigation and other uses (Table 20.3) (Crook, 1985; Metcalf and Eddy, 1991), and has the highest number of water reuse projects. Southern California imports water from Northern California and the Colorado River and has thus promoted several water reclamation projects (Nichols, 1988). The leading projects are Water Factory 21 and Irvine Ranch Water District. Water Factory 21 produces a highly polished tertiary effluent that meets California potable water standards. This effluent is mixed with deep-well groundwater and is used for groundwater recharge. The Irvine Ranch Water District uses reclaimed water in a dual distribution system for office buildings.

In Southern California, approximately 27% of the effluents from reclamation plants are reused. About two thirds of the reused water is intended for groundwater recharge by surface spreading or deep-well injection. The public health aspects of the use of reclaimed water for groundwater recharge at Whittier Narrows, California, have been evaluated. Epidemiological studies did not show evidence of any measurable health effects associated with the consumption of the water in the study area (Nellor et al., 1985).

The California Department of Health Services has established criteria to address public health concerns over wastewater reuse. These criteria specify the level of wastewater treatment and coliform levels for various types of uses of reclaimed wastewater (crop and landscape irrigation, impoundments, and groundwater recharge) (Table 20.3). Tertiary treatment (oxidation, coagulation, clarification, filtration and disinfection) is required for water used for spray or surface irrigation of food crops that are eaten raw. The tertiary effluent must be essentially pathogen-free. The guidelines establishing the treatment and bacteriological quality of reclaimed water call for 2.2 total coliforms per 100 ml (with an upper limit of 23 TC per 100 ml in

TABLE 20.3. State of California Wastewater Reclamation Criteria[a]

Use of reclaimed wastewater	Description of minimum treatment requirements: Primary	Secondary and disinfected	Secondary coagulated, filtered, and disinfected	Coliform MPN/100 ml median (daily sampling)
Irrigation				
Fodder crops	X			No requirement
Fiber	X			No requirement
Seed crops	X			No requirement
Produce eaten raw, surface-irrigated		X		2.2
Produce eaten raw, spray-irrigated			X	2.2
Processed produce, surface-irrigated	X			No requirement
Processed produce, spray-irrigated		X		23
Landscapes: golf course, cemeteries, freeways		X		23
Landscapes: parks, playgrounds, schoolyards			X	2.2
Recreational impoundments				
No public contact		X		23
Boating and fishing only		X		2.2
Body access (bathing)			X	2.2

[a]Adapted from Metcalf and Eddy (1991).

10% of the samples) and a turbidity standard of 2 NTU. These guidelines apply to uses such as irrigation of food crops, recreational uses permitting unrestricted body access, and irrigation of parks, playgrounds, and schoolyards.

According to some observers, the California standards are too strict and are not based on sound epidemiological evidence (Shuval, 1991). It has been argued that these standards could hardly be achieved by most of the U.S. wastewater treatment plants. More liberal guidelines have been proposed by public health experts meeting in Engelberg, Switzerland, in 1985 (Table 20.4) (International Reference Centre for Waste Disposal, 1985; World Health Organization, 1989). For example, the new proposed fecal coliform guideline for wastewater used for irrigation of crops likely to be eaten uncooked is ≧1,000 fecal coliforms per 100 ml instead of ≧2.2 coliforms per 100 ml. The wastewater effluent should also contain less than 1 nematode egg per liter.

Water reuse in Arizona is employed at 180 plants treating approximately 200 MGD (Rose, 1986; Rose et al., 1989b). The state has established a compliance program for monitoring viruses, *Giardia,* and fecal coliforms in reused water (Anonymous, 1984). Arizona is the only state in the United States that has adopted standards for enteric viruses. The standards specify that virus levels should not exceed 1 PFU/40 L for reclaimed water used for spray-irrigation of food eaten raw or for unrestricted-access water sports. For irrigated landscape areas and golf courses with full access to the public, the virus level should not exceed 125 PFU/40 L. With regard to

TABLE 20.4. WHO-Recommended Microbiological Quality Guidelines for Wastewater Use in Agriculture[a]

Category	Reuse conditions	Exposed group	Intestinal nematodes (arithmetic mean no. of eggs per liter)	Fecal coliforms (geometric mean no. per 100 ml)	Wastewater treatment expected to achieve the required microbiological quality
A	Irrigation of crops likely to be eaten uncooked, sports fields, public parks	Workers, consumers, public	≤ 1	$\leq 1{,}000$	A series of stabilization ponds designed to achieve the microbiological quality indicated, or equivalent treatment
B	Irrigation of cereal crops, industrial crops, fodder crops, pasture, and trees	Workers	≤ 1	No standard recommended	Retention in stabilization ponds for 8–10 days or equivalent helminth and fecal coliform removal
C	Localized irrigation crops in category B if exposure of workers and the public does not occur	None	Not applicable	Not applicable	Pretreatment as required by the irrigation technology, but not less than primary sedimentation

[a]From World Health Organization (1989), with permission of the publisher.

Giardia, none should be detected in 40 L of water. Virus monitoring for activated sludge and oxidation pond effluents showed that about 60% of the samples met the compliance standard of 1 PFU/40 L. Furthermore, 97% of sand-filtered activated sludge effluents met the virus standard and two thirds of these samples met the *Giardia* standard (Arizona Department of State, 1984; Rose and Gerba, 1991; Rose et al., 1989).

The State of Florida requires treatment steps such as filtration to improve the efficiency of the disinfection step, which requires a maintenance of a chlorine residual of 1 mg/L after 30-min contact time. The Florida Department of Environmental Protection has established a standard of no detectable fecal coliforms per 100 ml for reclaimed water.

20.4. WATER REUSE IN SPACE

Drinking water is an expensive commodity in space. It was estimated that it would cost more than US $8,000 to send one gallon of water in space (Nicks, 1986). Therefore, wastewater recycling is a necessity under these conditions. NASA engineers and scientists are investigating means of recycling water from wastewater generated by astronauts working in space stations. The wastewater contains human fecal wastes, urine, wash water, and humidity condensate (Wachinski, 1988). However, this is no easy task, since processes for waste collection and wastewater and water treatment must be designed to function in a zero-gravity environment. Some processes considered for waste treatment in space include dry incineration at 600°C, wet oxidation at 230–290°C at pressures between 70 and 150 bars, and supercritical water oxidation at 374°C and 215 bars (Wachinski, 1988).

20.5. FURTHER READING

Asano, T., and G. Tchobanoglous. 1991. The role of wastewater reclamation and reuse in the USA. Water Sci. Technol. 23: 2049–2059.

Cooper, R.C. 1991. Public health concerns in wastewater reuse. Water Sci. Technol. 24: 55–65.

Grabow, W.O.K. 1990. Microbiology of drinking water treatment: Reclaimed wastewater, pp. 185–203, in: *Drinking Water Microbiology*, G.A McFeters, Ed. Springer-Verlag, New York.

Rose, J.B. 1986. Microbial aspects of wastewater reuse for irrigation. Crit. Rev. Environ. Control 16: 231–256.

Shuval, H.I., Ed. 1977. *Water Renovation and Reuse*. Academic Press, New York.

REFERENCES

Abbaszadegan, M., C.P. Gerba, and J.B. Rose. 1991. Detection of *Giardia* cysts with a cDNA probe and applications to water samples. Appl. Environ. Microbiol. 57: 927–931.

Abbaszadegan, M., M.S. Huber, C.P. Gerba, and I.L. Pepper. 1993. Detection of enteroviruses in groundwater with the polymerase chain reaction. Appl. Environ. Microbiol. 59: 1318–1324.

Abram, J.W., and D.B. Nedwell. 1978. Hydrogen as a substrate for methanogenesis and sulphate reduction in anaerobic salt march sediments. Arch. Microbiol. 117: 93–97.

Abu-Ghararah, Z.H., and C.W. Randall. 1990. The effect of organic compounds on biological phosphorus removal. Water Sci. Technol. 23: 585–594.

Acher, A., and B.I. Juven. 1977. Destruction of fecal coliforms in sewage water by dye-sensitized photooxidation. Appl. Environ. Microbiol. 33: 1019–1023.

Acher, A., E. Fischer, R. Zellingher, and Y. Manor. 1990. Photochemical disinfection of effluents: Pilot plant studies. Water Res. 24: 837–843.

Adams, M.H. 1959. *Bacteriophages*, Interscience, New York.

Adams, M.H. and B.H. Park. 1956. An enzyme produced by a phage-bacterial system. Virology 2: 719–736.

Aftring, R.P. and B.F. Taylor. 1981. Aerobic and anaerobic catabolism of phthalate acid by a nitrate-respiring bacterium. Arch. Microbiol. 130: 101–104.

Agoustinos, M.T., S.N. Venter, and R. Kfir. 1992. Assessment of water quality problems due to microbial growth in drinking water distribution systems. Paper presented at the International Water Pollution Research Conference International Symposium, Washington, DC, May 26–29, 1992.

Ahlborg, U.G., and T.M. Thunberg. 1980. Chlorinated phenols: Occurrence, toxicity, metabolism, and environmental impact. Crit. Rev. Toxicol. 7: 1–35.

Ahlstrom, S.B. and T. Lessel. 1986. Irradiation of municipal sludge for pathogen control. In: *Control of Sludge Pathogens*, C.A. Sorber, Ed. Water Pollution Control Federation, Washington, DC.

Ahmed, M., and D.D. Focht. 1973. Degradation of polychlorinated biphenyl by two species of *Achromobacter*. Can. J. Microbiol. 19: 47–52.

Aieta, E.M., and J.D. Berg. 1986. A review of chlorine dioxide in drinking water treatment. J. Am. Water Works Assoc. 78: 62–72.

Aiking, H., K. Kok, H. van Heerikhuizen, and J. van't Riet. 1982. Adaptation to cadmium by *Klebsiella aerogenes* growing in continuous culture proceeds mainly via formation of cadmium sulfate. Appl. Environ. Microbiol. 44: 938–944.

Akin, E.W., W.F. Hill, G.B. Cline, and W.H. Benton. 1976. The loss of poliovirus 1 infectivity in marine waters. Water Res. 10: 59–63.

Al-Ani, M.Y., D.W. Hendricks, G.S. Logsdon, and C.P. Hibler. 1986. Removing *Giardia* cysts from low turbidity waters by rapid rate filtration. J. Am. Water Works Assoc. 78: 66–73.

Albert, M., E. Biziagos, J.M. Crance, R. Deloince, and L. Schwartzbrod. 1990. Detection des virus enteriques cultivables in vitro et de l'antigene du virus de l'hepatite A dans les boues primaires de stations d'epuration. J. Fr. Hydrol. 2: 275–283.

Alberts, B., D. Bray, J. Lewis, M. Raff, K. Roberts and J.D. Watson. 1989. *Molecular Biology of the Cell.* Garland, New York.

Albertson, G.E., and P. Hendricks. 1992. Bulking and foaming organism control at Phoenix, AZ WWTP. Water Sci. Technol. 26:461–472.

Alcaid, E., and E. Garay. 1984. R-plasmid transfer in *Salmonella* spp. isolated from wastewater and sewage-contaminated surface waters. Appl. Environ. Microbiol. 48: 435–438.

Alexander, M. 1977. *Introduction to Soil Microbiology* (2nd ed). Wiley, New York.

Alexander, M. 1979. Role of cometabolism, pp. 67–75, in: *Microbial Degradation of Pollutants in Marine Environments*, A.L. Bourquin and P.H. Pritchard, Eds. U.S. EPA, Gulf Breeze, FL.

Alexander, M. 1981. Biodegradation of chemicals of environmental concern. Science 211: 132–138.

Alexander, M. 1985. Biodegradation of organic chemicals. Environ. Sci. Technol. 19: 106–111.

Alhajjar, B.J., S.L. Stramer, D.O. Cliver, and J.M. Harkin. 1988. Transport modelling of biological tracers from septic systems. Water Res. 22: 907–915.

Alleman, J.E. 1988. Respiration-based evaluation of nitrification inhibition using enriched *Nitrosomonas* cultures, pp. 642–650, in: Scholze, R.J, E.D. Smith, J.T. Bandy, Y.C. Yu, and J.V. Basilico, Eds. *Biotechnology for Degradation of Toxic Chemicals in Hazardous Wastes Sites*. Noyes Data Corp., Park Ridge, NJ.

Alleman, J.E., J.A. Veil, and J.T. Canaday. 1982. Scanning electron microscope evaluation of rotating biological contactor biofilm. Water Res. 16: 543–550.

Allen, M.J., R.H. Taylor, and E.E. Geldreich. 1980. The occurrence of microorganisms in water main encrustations. J. Am. Water Works Assoc. 72: 614–626.

Alsop, G.M., G.T. Waggy, and R.A. Conway. 1980. Bacterial growth inhibition test. J. Water Pollut. Control Fed. 52:2452–2456.

Alvarez, M.E., and R.T. O'Brien. 1982. Mechanism of inactivation of poliovirus by chlorine dioxide and iodine. Appl. Environ. Microbiol. 44: 1064–1071.

Andersen, A.A. 1958. New sampler for the collection, sizing and enumeration of viable airborne particles. J. Bacteriol. 76: 471–484.

Anderson, G.K., T. Donnelly, and K.J. McKeown. 1982. Identification and control of inhibition in the anaerobic treatment of industrial wastewater. Process Biochem. 17: 28–32.

Anderson, J.P.E. 1989. Principles of and assay systems for biodegradation, pp. 129–145, in: *Biotechnology and Biodegradation*, D. Kamely, A. Chakrabarty, and G.S. Omenn, Eds. Gulf Publishing Co., Houston, TX.

Andrin, C., and J. Schwartzbrod. 1992. Isolating *Campylobacter* from wastewater. Paper presented at the International Water Pollution Research Conference International Symposium , Washington, DC, May 26–29, 1992.

Anonymous. 1984. *Water Quality Standards for Wastewater Reuse*. Arizona Department of State, Phoenix, AZ.

Anonymous, 1991. Microbes help clean heavy metals. ASM News 57: 296.

Ansari, S.A., S.R. Farrah, and G.R. Chaudhry. 1992. Presence of human immunodefiency virus in wastewater and their detection by the polymerase chain reaction. Appl. Environ. Microbiol. 58: 3984–3990.

Antonie, R.L. 1976. *Fixed Biological Surfaces-Wastewater Treatment: The Rotating Biological Contactor*. CRC Press, New York.

Antopol, S.C., and P.D. Ellner. 1979. Susceptibility of *Legionella pneumophila* to ultraviolet radiation. Appl. Environ. Microbiol. 38: 347–348.

Anwar, H., and J.W. Costerton. 1992. Effective use of antibiotics in the treatment of biofilm-associated infections. ASM News 58: 665–668.

APHA. 1985. *Standard Methods for the Examination of Water and Wastewater* (16th ed.), American Public Health Association. Washington, DC.

APHA. 1989. *Standard Methods for the Examination of Water and Wastewater* (17th ed.), American Public Health Association, Washington, DC.

Applebaum, J., N. Gutman-Bass, M. Lugten, B. Teltsch, B. Fattal, and H.I. Shuval. 1984. Dispersion of aerosolized enteric viruses and bacteria by sprinkler irrigation with wastewater. Monogr. Virol. 15: 193–210.

Appleton, A.R., Jr., C.J. Leong, and A.D. Venosa. 1986. Pathogen and indicator organism destruction by the dual digestion system. J. Water Pollut. Control Fed. 58: 992–999.

Appleton, A.R., Jr., and A.D. Venosa. 1986. Technology evaluation of the dual digestion system. J. Water Pollut. Control Fed. 58: 764–769.

Arana, I., A. Muela, J. Iriberri, L. Egeas, and I. Barcina. 1992. Role of hydrogen peroxide in loss of culturability mediated by visible light in *Escherichia coli* in a freshwater ecosystem. Appl. Environ. Microbiol. 58: 3903–3907.

Arber, R.P. 1983. From wastewater to drinking water. Civil Eng. (ASCE) Feb. 1983: 46–49.

Archer, D.B. 1984. Detection and quantitation of

methanogens by enzyme-linked immunosorbent assay. Appl. Environ. Microbiol. 48: 797–801.

Archer, D.B., and B.H. Kirsop. 1991. The microbiology and control of anaerobic digestion, pp. 43–91, in: *Anaerobic Digestion: A Waste Treatment Technology*, A. Wheatly, Ed. Elsevier Applied Science, London, U.K.

Arhing, B.K., N. Christiansen, I. Mathrani, H.V. Hendriksen, A.J.L. Macario, and E.V. de Macario. 1992. Introduction of *de novo* bioremediation ability, aryl reductive dechlorination, into anaerobic granular sludge by inoculation of sludge with *Desulfomonile tiedjei*. Appl. Environ. Microbiol. 58: 3677–3682.

Ariga, O., H. Takagi, H. Nishizawa, and Y. Sano. 1987. Immobilization of microoorganisms with PVA hardened by iterative freezing and thawing. J. Ferment. Technol. 65: 651–658.

Arizona Department of State. 1984. *Water Quality Standards of Wastewater Reuse*. Phoenix, AZ, author.

Armstrong, G.L., J.J. Calomiris, and R.J. Seidler. 1982. Selection of antibiotic-resistant standard plate count bacteria during water treatment. Appl. Environ. Microbiol. 44: 308–316.

Armstrong, G.L., D.S. Shigeno, J.J. Calomiris, and R.J. Seidler. 1981. Antibiotic-resistant bacteria in drinking water. Appl. Environ. Microbiol. 42: 277–283.

Arther, R.G., P.R. Fitzgerald, and J.C. Fox. 1981. Parasitic ova in anaerobically digested sludge. J. Water Pollut. Control Fed. 53: 1334–1338.

Arun, V., T. Mino, and T. Matsuo. 1988. Biological mechanism of acetate uptake mediated by carbohydrate consumption in excess phophorus removal systems. Water Res. 22: 565–570.

Arvin, E. 1985. Observations supporting phosphate removal by biologically mediated chemical precipitation: A review. Water Sci. Technol. 15: 43–63.

Arvin, E., and G.H. Kristensen. 1983. Phosphate precipitation in biofilms and flocs. Water Sci. Technol. 15: 65–85.

Asano, H., H. Myoga, M. Asano, and M. Toyao. 1992. A study of nitrification utilizing whole microorganisms immobilized by the PVA-freezing method. Water Sci. Technol. 26: 1037–1046.

Asano, T., Ed. 1985. *Artificial Recharge of Groundwater*, Butterworth, Boston.

Asano, T., and G. Tchobanoglous. 1991. The role of wastewater reclamation and reuse in the USA. Water Sci. Technol. 23: 2049–2059.

Ashgari, A., S.R. Farrah, and G. Bitton. 1992. Use of hydrogen peroxide treatment and crystal violet agar plates for selective recovery of bacteriophages from natural environments. Appl. Environ. Microbiol. 58: 1158–1162.

Ashley, N.V., and T.J. Hurst. 1981. Acid and alkaline phosphatase activity in anaerobic digested sludge: A biochemical predictor of digester failure. Water Res. 15: 633–638.

Atlas, R.M. 1986. *Basic and Practical Microbiology*. Macmillan, New York.

Atlas, R.M. 1991. Environmental applications of the polymerase chain reaction. ASM News 57: 630–632.

Atlas, R.M., and R. Bartha. 1987. *Microbial Ecology: Fundamentals and Applications*. Addison Wesley, Reading MA.

Atlas, R., A. Bej, R. Steffan, J. Dicesare, and L. Haff. 1989. Detection of coliforms in water by polymerase chain reaction (PCR) and gene probe methods. Abstr. Annu. Meet. Am. Soc. Microbiol., Vol. 89.

Atmar, R.L, T.G. Metcalf, F.H. Neill, and M.K. Estes. 1993. Detection of enteric viruses in oysters by using the polymerase chain reaction. Appl. Environ. Microbiol. 59: 631–635.

Attree-Pietri, C., and J.P. Breittmayer. 1970. Etude comparee de germes-tests de contamination fecale et de bacteriophages en eau de mer. J. Fr. Hydrol. 10: 103–106.

Auling, G., F. Pilz, H-J Busse, S. Karrasch, M. Streichan, and G. Schon. 1991. Analysis of polyphosphate accumulating microflora in phosphorus-eliminating, anaerobic-aerobic activated sludge systems by using diaminopropane as a biomarker for rapid estimation of *Acinetobacter* spp. Appl. Environ. Microbiol. 57: 3585–3592.

AWWA, 1985a. Waterborne *Giardia*: It's enough to make you sick (Roundtable). J. Am. Water Works Assoc. 77: 14.

AWWA, 1985b. Trends in ozonation. Roundtable held in Washington, D.C., June 25, 1985. J. Am. Water Works Assoc. 77: 19–30.

AWWA "Organisms in Water" Committee. 1987. Committee report: Microbiological consideration for drinking water regulations revisions. J. Am. Water Works Assoc. 79: 81–88.

AWWA, 1988. *Cryptosporidium* roundtable (Nov. 18,

1987). Baltimore, MD. J. Am. Water Works Assoc. 80: 14–27.

Ayres, R.M., G.P. Alabaster, D.D. Mara, and D.L. Lee. 1992. A design equation for human intestinal nematode egg removal in waste stabilization ponds. Water Res. 26: 863–865.

Babcock, R.W., Jr., K.S. Ro, C.-C. Hsieh, and M.K. Stenstrom. 1992. Development of an off-line enricher-reactor process for activated sludge degradation of hazardous wastes. Water Environ. Res. 64: 782–791.

Babcock, R.W., Jr., and M.K. Stenstrom. 1993. Use of inducer compounds in the enricher-reactor process for degradation of 1-naphthylamine wastes. Water Environ. Res. 65: 26–33.

Babich, H., and G. Stotzky. 1980. Reductions in inactivation rates of bacteriophages by clay minerals in lake water. Water Res. 14: 185–187.

Babich, H., and G. Stotzky. 1986. Environmental factors that affect the utility of microbial assays for the toxicty and mutagenicity of chemical pollutants, pp. 9–42, in: *Toxicity Testing Using Microorganisms*, Vol. 2, B.J. Dutka and G. Bitton, Eds. CRC Press, Boca Raton, FL.

Bach, P.D., M. Shoda, and H. Kubota. 1984. Rate of composting of dewatered sewage sludge in continuously mixed isothermal reactor. J. Ferment. Technol. 62: 285–292.

Balakrishnan, S., and W.W. Eckenfelder. 1969. Nitrogen relationships in biological treatment processes: II. Nitrification trickling filters. Water Res. 3: 167–174.

Balch, W.E., G.E. Fox, L.J. Magnum, C.R. Woese, and R.S. Wolfe. 1979. Methanogenesis: Reevaluation of a unique biological group. Microbiol. Rev. 143: 260–296.

Bancroft, K., P. Chrostowski, R.L. Wright, and I.H. Suffet. 1984. Ozonation and oxidation competition values. Relationship to disinfection and microorganisms regrowth. Water Res. 18: 473–478.

Barbaree, J.M., B.S. Fields, J.C. Feeley, G.W. Gorman, and W.T. Martin. 1986. Isolation of protozoa from water associated with a legionellosis outbreak and demonstration of intracellular multiplication of *Legionellla pneumophila*. Appl. Environ. Microbiol. 51: 422–424.

Barbier, D., D. Ferrine, C. Duhamel, R. Doublet, and P. Georges. 1990. Parasitic hazard with sewage sludge applied to land. Appl. Environ. Microbiol. 56: 1420–1422.

Barker, J., M.R.W. Brown, P.J. Collier, I. Farrell, and P. Gilbert. 1992. Relationship between *Legionella pneumophila* and *Acanthamoeba polyphaga*: Physiological status and susceptibility to chemical inactivation. Appl. Environ. Microbiol. 58: 2420–2425.

Barnard, J.L. 1973. Biological denitrification. J. Water Pollut. Control Fed. 72: 705–709.

Barnard, J.L. 1975. Biological nutrient removal without the addition of chemicals. Water Res. 9: 485–490.

Barnes, D., and P.J. Bliss. 1983. *Biological Control of Nitrogen in Wastewater Treatment*. E. & F.N. Spon, London.

Barnes, R., J.I. Curry, L.M. Elliott, C.R. Peter, B.R. Tamplin, and B.W. Wilcke, Jr. 1989. Evaluation of the 7-hr membrane filter test for quantitation of fecal coliforms in water. Appl. Environ. Microbiol. 55: 1504–1506.

Barnes, D., and P.A. Fitzgerald. 1987. Anaerobic wastewater treatment processes, pp. 57–113, in: *Environmental Biotechnology*, C.F. Forster and D.A.J. Wase, Eds. Ellis Horwood, Chichester, U.K.

Bartel, P.F., G.K. Lam, and T.E. Orr. 1968. Purification and properties of polysaccharide depolymerase associated with phage-infected *Pseudomonas aeruginosa*. J. Biol. Chem. 243: 2077–2088.

Barth, E.F. 1975. The effects and removal of heavy metals in biological treatment: Discussion. in: *Heavy Metals in Aquatic Environment*, P.A. Krenkel, Ed. Pergamon Press, Oxford.

Barth, E.F., M.B. Ettinger, B.V. Salotto, and G.N. McDermott. 1965. Summary report on the effects of heavy metals on biological treatment processes. J. Water Pollut. Control Fed. 37: 86–96.

Bartley, T.D., T.J. Quan, M.T. Collins, and S.M. Morrison. 1982. Membrane filter technique for the isolation of *Yersinia enterocolitica*. Appl. Environ. Microbiol. 43: 829–834.

Bartone, C.R. 1991. International perspective on water resources management and wastewater reuse: Appropriate technologies. Water Sci. Technol. 23: 2039–2047.

Bates, R.C., P.T.B. Shaffer, and S.M. Sutherland. 1977. Development of poliovirus having increased resistance to chlorine inactivation. Appl. Environ. Microbiol. 33: 849–853.

Battigelli, D.A., D. Lobe, and M.D. Sobsey. 1993. Inactivation of hepatitis A virus and other enteric viruses in water by ultraviolet. Water Sci. Technol. 27: 339–342.

Baumann, E.R. 1971. Diatomite filtration of potable water, pp. 280–294, in: *Water Quality and Treatment. A Handbook of Public Water Supplies.* McGraw-Hill, New York.

Baughman, G.L., and D.F. Paris. 1981. Microbial bioconcentration of organic pollutants from aquatic systems: A critical review. Crit. Rev. Microbiol. 8: 205–228.

Baumann, M., H. Lemmer, and H. Ries. 1988. Scum actinomycetes in sewage treatment plants. Part 1: Growth kinetics of *Nocardia amarae* in chemostat culture. Water Res. 22: 755–759.

Bausum, H.T., S.A. Schaub, R.E. Bates, H.L. McKim, P.W. Schumacher, and B.E. Brockett. 1983. Microbiological aerosols from a field-source wastewater irrigation system. J. Water Pollut. Control Fed. 55: 65–80.

Bayley, H.E., J.G. Moutes, and F.D. Schlesinger. 1992. An analysis of changing constraints and planning for flexibility in a water reclamation program. Water Sci. Technol. 26: 1525–1535.

Baylis, J.R., O. Gullans, and B.K. Spector. 1936. The efficiency of rapid sand filters in removing the cysts of amoebic dysentery organisms from water. Public Health Rep. 50: 1567–1571.

Bedient, P.B., and H.S. Rifai. 1992. Bioremediation, pp. 117–41, in: *Groundwater Remediation*, R.J. Charbeneau, P.B. Bedient, and R.C. Loehr, Eds. Technomic Publishing Co., Lancaster, PA.

van Beelen, P., A.C. Dijkstra, and G.D. Vogels. 1983. Quantitation of coenzyme F_{420} in methanogenic sludge by the use of reversed-phase high-performance liquid chromatography and a fluorescence detector. Eur. J. Microbiol. Biotechnol. 18: 67–69.

Bej, A.K., J. DiCesare, L. Haff, and R.M. Atlas. 1991a. Detection of *Escherichia coli* and *Shigella* spp. in water by using the polymerase chain reaction and gene probes for *uid*. Appl. Environ. Microbiol. 57: 1013–1017.

Bej, A.K., M.H. Mahbubani, and R.M. Atlas. 1991b. Detection of viable *Legionella pneumophila* in water by polymerase chain reaction and gene probe methods. Appl. Environ. Microbiol. 57: 597–600.

Bej, A.K., S.C. McCarty, and R.M. Atlas. 1991c. Detection of coliform bacteria and *Escherichia coli* by multiplex polymerase chain reaction: Comparison with defined substrate and plating methods for water quality monitoring. Appl. Environ. Microbiol. 57: 2429–2432.

Bej, A.K., M.H. Mahbubani, and R.M. Atlas. 1992. Detection of genus *Salmonella* using polymerase chain reaction (PCR) method. Abstr. Annu. Meet. Am. Soc. Microbiol. 92: Q-251.

Bej, A.K., R. Steffan, J. DiCesare, L. Haff, and R.M. Atlas. 1990. Detection of coliform bacteria in water by polymerase chain reaction and gene probes. Appl. Environ. Microbiol. 56: 307–314.

Bell, F.A. 1991. Review of effects of silver-impregnated carbon filters on microbial water quality. J. Am. Water Works Assoc. 83: 7476.

Bell, J.P., and M. Tsezos. 1987. Removal of hazardous organic pollutants by biomass adsorption. J. Water Pollut. Control Fed. 59: 191–198.

Bell, R.B. 1978. Antibiotic resistance patterns of fecal coliforms isolated from domestic sewage before and after treatment in an aerobic lagoon. Can. J. Microbiol. 24: 886–888.

Bell, R.B., W.R. Macrae, and G.E. Elliott. 1981. R factors in coliform–fecal coliform sewage flora of the prairies and northwest territories of Canada. Appl. Environ. Microbiol. 42: 204–210.

Bellair, J.T., G.A.P. Smith, and I.G. Wallis. 1977. Significance of diurnal variations in fecal coliform die-off in the design of ocean outfalls. J. Water Pollut. Control Fed. 49: 2022–2030.

Bellamy, W.D., D.W. Hendricks and G.S. Logsdon. 1985a. Slow sand filtration: Influences of selected process variables. J. Am. Water Works Assoc. 77: 62–66.

Bellamy, W.D., G.P. Silverman, D.W. Hendricks, and G.S. Logsdon. 1985b. Removing *Giardia* cysts with slow sand filtration. J. Am. Water Works Assoc. 77: 52–60.

Benedict, A.H., E. Epstein, and J. Alpert. 1988. *Composting Municipal Sludge: A Technology Evaluation*. Noyes Data Corp., Park Ridge, NJ.

Benson, H.J. 1973. Microbiological Applications: A Laboratory Manual in General Microbiology. W.C. Brown, Dubuque, IA.

Bercovier, H., M. Deral-Cochin, J. Sherman, B. Fattal, and H.I. Shuval. 1984. Seroepidemiological survey of irrigation workers in Israel and isolation of *Legionella* spp. in the environment. in: *Legionella*, C. Thornsberry, A. Balows, J.C. Feeley, and W. Jakubowski, Eds. American Society for Microbiology, Washington, DC.

Berendt, R.F. 1980. Influence of blue-green algae (cyanobacteria) on survival of *Legionella pneumophila* in aerosols. Infect. Immun. 32: 690–692.

Berg, G., Ed. 1978. *Indicators of Viruses in Water and Food*. Ann Arbor Science, Ann Arbor, MI.

Berg, G., and D. Berman. 1980. Destruction by anaerobic mesophilic and thermophilic digestion of viruses and indicator bacteria indigenous to domestic sludges. Appl. Environ. Microbiol. 39: 361–368.

Berg, G., H. Sanjahsaz, and S. Wangwongwatana. 1989. Potentiation of the virucidal effectiveness of free chlorine by substances in drinking water. Appl. Environ. Microbiol. 55: 390–393.

Berg, G., H. Sanjahsaz, and S. Wangwongwatana. 1990. KCl potentiation of the virucidal effectiveness of free chlorine at pH 9. Appl. Environ. Microbiol. 56: 1571–1575.

Berg, J.D., and L. Fiksdal. 1988. Rapid detection of total and fecal coliforms in water by enzymic hydrolysis of 4-methylumbelliferone-ß-D-galactoside. In: *Proceedings of the International Conference on Water and Wastewater Microbiology*, Newport Beach, CA, Feb. 8–11, 1988, Vol. 1.

Berg, J.D., P.V. Roberts, and A. Matin. 1986. Effect of chlorine dioxide on selected membrane functions of *Escherichia coli*. J. Appl. Bacteriol. 60: 213–220.

van den Berg, L., and K.J. Kennedy. 1981. Support materials for stationary fixed films reactors for high rate methonogen fermentations. Biotechnol. Lett. 3: 165–170

van den Berg, L., C.P. Lentz, R.J. Athey, and E.A. Rook. 1974. Assessment of methanogenic activity in anaerobic digestion. Apparatus and methods. Biotechnol. Bioeng. 16: 1459–1465.

Berger, B.B. 1982. Water and wastewater quality control and the public health. Annu. Rev. Public Health 3: 359–392.

Berman, D., and J.C. Hoff. 1984. Inactivation of simian rotavirus SA11 by chlorine, chlorine dioxide and monochloramine. Appl. Environ. Microbiol. 48: 317–323.

Berman, D., E.W. Rice, and J.C. Hoff. 1988. Inactivation of particle-associated coliforms by chlorine and monochloramine. Appl. Environ. Microbiol. 54: 507–512.

Bermudez, M., and T.C. Hazen. 1988. Phenotypic and genotypic comparison of *Escherichia coli* from pristine tropical waters. Appl. Environ. Microbiol. 54: 979–983.

Bernarde, M.A., N.B. Snow, V.P. Olivieri, and B. Davidson. 1967. Kinetics and mechanism of bacterial disinfection by chlorine dioxide. Appl. Microbiol. 15: 257–265.

Bessler, W., E. Freund-Molbert, H. Knufermann, R.C. Thurow, and S. Stirm. 1973. A bacteriophage induced depolymerase active on *Klebsiella KII* capsular polysaccharide. Virology 56: 134–151.

Best, D.J., J. Jones, and D. Strafford. 1985. The environment and biotechnology, pp. 213–256, in: *Biotechnology: Principles and Applications*, I.J. Higgins, D.J. Best, and J. Jones, Eds. Blackwell Scientific Publishers, Oxford, U.K.

Best, M.G., A. Goetz, and V.L. Yu. 1984. Heat eradication measures for control of nosocomial Legionnaire's disease: Implementation, education and cost analysis. Am. J. Infect. Control 12: 26–30.

Bettmann, H., and H.J. Rehm. 1985. Continuous degradation of phenol(s) by *Pseudomonas putida* P8 entrapped in polyacrylamide-hydrazide. Appl. Microbiol. Biotechnol. 22: 389–393.

Bhattacharya, S.K., and G.F. Parkin. 1988. Fate and effect of methylene chloride and formaldehyde in methane fermentation systems. J. Water Pollut. Control Fed. 60: 531–536.

Bhattacharya, S.K., and G.F. Parkin. 1989. The effect of ammonia on methane fermentation processes. J. Water Pollut. Control Fed. 61: 56–59.

Bicki, T.J., R.B. Brown, M.E. Collins, R.S. Mansell, and D.F. Rothwell. 1984. *Impact of On-Site Sewage Disposal Systems on Surface and Groundwater Quality*. Report to Florida Department of Health Rehabilitation Service, Tallahassee, FL.

Bifulco, J.M., and F.W. Schaefer, III. 1993. Antibody–magnetite method for selective concentration of *Giardia lamblia* cysts from water samples. Appl. Environ. Microbiol. 59: 772–776.

Bisping, B., and H.J. Rehm. 1988. Multistep reactions with immobilized microorganisms. Biotechnol. Appl. Biochem. 10: 87–98.

Bissonette, G.K., J.J. Jezeski, G.A. McFeters, and D.G. Stuart. 1975. Influence of environmental stress on enumeration of indicator bacteria from natural waters. Appl. Microbiol. 29: 186–194.

Bissonette, G.K., J.J. Jezeski, G.A. McFeters, and D.G. Stuart. 1977. Evaluation of recovery methods to detect coliforms in water. Appl. Environ. Microbiol. 33: 590–595.

Bitton, G. 1978. Survival of enteric viruses, pp. 273–

299, in: *Water Pollution Microbiology*, Vol. 2, R. Mitchell, Ed. Wiley, New York.

Bitton, G. 1980a. *Introduction to Environmental Virology*. Wiley, New York.

Bitton, G. 1980b. Adsorption of viruses to surfaces: Technological and ecological implications, pp. 331–374, in: *Adsorption of Microorganisms to Surfaces*, G. Bitton and K.C. Marshall, Eds. Wiley Interscience, New York.

Bitton, G. 1983. Bacterial and biochemical tests for assessing chemical toxicity in the aquatic environment: A review. Crit. Rev. Environ. Control 13: 51-67.

Bitton, G. 1987. Fate of bacteriophages in water and wastewater treatment plants, pp. 181–195, in: *Phage Ecology*, S.M. Goyal, C.P. Gerba, and G. Bitton, Eds. Wiley Interscience, New York.

Bitton, G., M. Campbell, and B. Koopman. 1992a. MetPAD: A bioassay kit for the specific determination of heavy metal toxicity in sediments from hazardous waste sites. Environ. Toxicol. Water Qual. 7: 323–328.

Bitton, G., L.T. Chang, S.R. Farrah, and K. Clifford. 1981a. Recovery of coliphages from wastewater and polluted lake water by magnetite–organic flocculation. Appl. Environ. Microbiol. 41: 93–97.

Bitton, G., Y.J. Chou, and S.R. Farrah. 1981b. Techniques for virus detection in aquatic sediments. J. Virol. Methods 4: 1–8.

Bitton, G., B.L. Damron, G.T. Edds, and J.M. Davidson, Eds. 1980. *Sludge-Health Risks of Land Application*. Ann Arbor Science, Ann Arbor, MI.

Bitton, G., and B.J. Dutka, Eds. 1986. *Toxicity Testing Using Microorganisms*, Vol. 1. CRC Press, Boca Raton, FL.

Bitton, G., B.J. Dutka, and C.W. Hendricks. 1989. Microbial toxicity tests, p. 6–44 to p. 6–66, in: *Ecological Assessment of Hazardous Waste Sites*, W. Warren-Hicks, B.R. Parkhurst, and S.S. Baker, Jr., Eds. EPA 600/3–89/013. U.S. EPA, Corvallis, OR.

Bitton, G., and S.R. Farrah. 1986. Contamination des eaux souterraines par les virus. Rev. Int. Sci. Eau 2: 31–37.

Bitton, G., S.R. Farrah, C. Montague, M.W. Binford, P.R. Scheuerman, and A. Watson. 1985. *Survey of Virus Isolation Data From Environmental Samples*. Research Report (contract # 68–03–3196) submitted to U.S. EPA, Health Effect Research Lab, Cincinnati, Ohio.

Bitton, G., S.R. Farrah, C. Montague, and E.W. Akin. 1986. Global survey of virus isolations from drinking water. Environ. Sci. Technol. 20: 216–222.

Bitton, G., S.R. Farrah, R.H. Ruskin, J. Butner, and Y.J. Chou. 1983. Survival of pathogenic and indicator organisms in groundwater. Ground Water 213: 405–410.

Bitton, G., R. Fraxedas, and G. Gifford. 1979. Effect of solar radiation on poliovirus: Preliminary experiments. Water Res. 13: 225–228.

Bitton, G., and V. Freihoffer. 1978. Influence of extracellular polysaccharides on the toxicity of copper and cadmium towards *Klebsiella aerogenes*. Microb. Ecol. 4: 119–125.

Bitton, G., and C.P. Gerba, Eds. 1984. *Groundwater Pollution Microbiology*. Wiley, New York.

Bitton, G., and R.W. Harvey. 1992. Transport of pathogens through soils and aquifers. Chapter 7, in: *New Concepts in Environmental Microbiology*, R. Mitchell, Ed. Wiley-Liss, New York.

Bitton, G., Y. Henis, and N. Lahav. 1972. Effect of several clay minerals and humic acid on the survival of *Klebsiella aerogenes* exposed to ultraviolet irradiation. Appl. Microbiol. 23: 870–874.

Bitton G., and B. Koopman. 1986. Biochemical tests for toxicity screening, pp. 27–55, in: *Toxicity Testing Using Microorganisms*, G. Bitton and B.J. Dutka, Eds. CRC Press, Boca Raton, FL.

Bitton, G., and B. Koopman. 1992. Bacterial and enzymatic bioassays for toxicity testing in the environment. Rev. Environ. Contam. Toxicol. 125: 1–22.

Bitton, G., B. Koopman, and O. Agami. 1992b. MetPAD™: A bioassay for rapid assessment of heavy metal toxicity in wastewater. Water Environ. Res. 64: 834–836.

Bitton, G., and K.C. Marshall, Eds. 1980. *Adsorption of Microorganisms to Surfaces*, Wiley Interscience, New York.

Bitton, G., J.E. Maruniak, and F.W. Zettler. 1987. Virus survival in natural ecosystems, pp. 301–332, in: *Survival and Dormancy of Microorganisms*, Y. Henis, Ed. Wiley-Interscience, New York.

Bitton, G., and R. Mitchell. 1974a. Effect of colloids

on the survival of bacteriophages in seawater. Water Res. 8: 227–229.

Bitton, G., and R. Mitchell. 1974b. Protection of *E. coli* by montmorillonite in seawater. J. Sanit. Engin. Div. Am. Soc. Civ Eng. 100: 1310–1313.

Bitton, G., O.C. Pancorbo, and S.R. Farrah. 1984. Virus transport and survival after land application of sewage sludge. Appl. Environ. Microbiol. 47: 905–909.

Blackall, L. L., A.E. Harbers, P.F. Greenfield, and A.C. Hayward. 1988. Actynomycete scum problems in Australian activated sludge plants. In: *Proceedings of the International Conference on Water and Wastewater Microbiology*, Newport Beach, CA, Feb. 8–11, 1988, Vol. 2.

Blackall, L. L., A.E. Harbers, P.F. Greenfield, and A.C. Hayward. 1991. Foaming in activated sludge plants: A survey in Queensland, Australia and an evaluation of some control strategies. Water Res. 25: 313–317.

Blackall, L.L., and K.C. Marshall. 1989. The mechanism of stabilization of actinomycete foams and the prevention of foaming under laboratory conditions. J. Ind. Microbiol. 4: 181–188.

Blackbeard, J.R., G.A. Elkana, and G.R. Marais. 1986. A survey of filamentous bulking and foaming in activated sludge plants in South Africa. Water Pollut. Control 1: 90–100.

Blackburn, J.W., W.L. Troxler, and G.S. Sayler. 1984. Prediction of the fate of organic chemicals in a biological treatment process—An overview. Environ. Prog. 3: 163–176.

Blaise, C. 1991. Microbiotests in aquatic ecotoxicology: Characteristics, utility, and prospects. Environ. Toxicol. Water Qual. 6: 145–155.

Blaise C., R. van Coillie, N. Bermingham, and G. Coulombe. 1987. Comparaison des reponses toxiques de trois indicateurs biologiques (bacteries, algues, poissons) exposes a des effluents de fabriques de pates et papiers. Rev. Int. Sci. Eau 3: 9–17.

Blaser, M.J., and L.B. Peller. 1981. *Campylobacter* enteritis. N. Engl. J. Med. 305: 1444–1452.

Blaser, M.J., P.F. Smith, W.L.L. Wang, and J.C. Hoff. 1986. Inactivation of *Campylobacter* by chlorine and monochloramine. Appl. Environ. Microbiol. 51: 307–311.

Blaser, M.J., D.N. Taylor, and R.A. Feldman. 1983. Epidemiology of *Campylobacter jejuni* infections. Epidemiol. Rev. 5: 157–176.

Blewett, D.A., S.E. Wright, D.P. Casemore, N.E. Booth, and C.E. Jones. 1993. Infective dose size studies on *Cryptosporidium parvum* using gnotobiotic lambs. Water Sci. Technol. 27: 61–64.

Block, J.C. 1981. Viruses in environmental waters, pp. 117–145, in: *Viral Pollution of the Environment*, G. Berg, Ed. CRC Press, Boca Raton, FL.

Block, J.-C. 1983. A review of some problems related to epidemiological studies, pp. 33–46, in: *Biological Health Risks of Sludge Disposal to Land in Cold Climates*, P.M. Wallis and D.L. Lehmann, Eds. University of Calgary Press, Calgary, Canada.

Block, J.C., A.H. Havelaar, and P. l'Hermite, Eds. 1986. Epidemiological Studies of Risks Associated with the Agricultural Use of Sewage Sludge: Knowledge and Needs. Elsevier Applied Science, London.

Block, J.C., L. Mathieu, P. Servais, D. Fontvielle, and P. Werner. 1992. Indigenous bacterial inocula for measuring the biodegradable dissolved organic carbon (BDOC) in waters. Water Res. 26: 481–486.

Blom, A., W. Harder, and A. Matin. 1992. Unique and overlapping pollutant stress proteins of *Escherichia coli*. Appl. Environ. Microbiol. 58: 331–334.

Blum, D.J.W., R. Hergenroeder, G.F. Parkin, and R.E. Speece. 1986. Anaerobic treatment of coal conversion wastewater constituents: Biodegradability and toxicity. J. Water Pollut. Control Fed. 58: 122–131.

Blum, D.J.W., and R.E. Speece. 1992. The toxicity of organic chemicals to treatment processes. Water Sci. Technol. 25: 23–31.

Blyth, W. 1973. Farmer's lung disease, pp. 261–276, in: *Actinomycetales: Characteristics and Practical Importance*, G. Sykes and F.A. Skinner, Eds. Academic, New York.

Bock, E., P.A. Wilderer, and A. Freitag. 1988. Growth of *Nitrobacter* in the absence of dissolved oxygen. Water Res. 22: 243–250.

Boczar, B.A., W.M. Begley, and R.J. Larson. 1992. Characterization of enzyme activity in activated sludge using rapid analyses for specific hydrolases. Water Environ. Res. 64: 792–797.

Boethling, R.S. 1984. Environmental fate and toxicity in wastewater treatment of quaternary ammonium surfactants. Water Res. 18: 1061–1076.

Bohn, H., and R. Bohn. 1988. Soil beds weed out air pollutants. Chem. Eng. April 25, 1988, pp. 73–76.

Bollag, J.M. 1979. Transformation of xenobiotics by microbial activity, pp. 19–27, in: *Microbial Degradation of Pollutant in Marine Environments*, A.W. Bourquin and P.H. Pritchard, Eds. EPA-600/9–79–012.

Bollag, J-M. 1992. Decontaminating soils with enzymes. Environ. Sci. Technol. 26: 1876–1881.

de Bont, J.A., J.P. van Dijken, and W. Harder. 1981. Dimethylsulphoxide and dimethyl sulphide as a carbon, sulfur and energy source for growth of *Hyphomicrobium*. J. Gen. Microbiol. 127: 315–323.

Borrego, J.J., M.A. Morinigo, A. de Vicente, R. Cornax, and P. Romero. 1987. Coliphages as an indicator of faecal pollution in water. Its relationship with indicator and pathogenic microorganisms. Water Res. 21: 1473–1480.

Borrego, J.J., and P. Romero. 1985. Coliphage survival in seawater. Water Res. 19: 557–562.

Bosch, A., J.M. Diez, and F.X. Abad. 1993. Disinfection of human enteric viruses in water by copper:silver and reduced levels of chlorine. Water Sci. Technol. 27: 351–356.

Bosch, A., C. Tartera, R. Gajardo, J.M. Diez, and J. Jofre. 1989. Comparative resistance of bacteriophages active against *Bacteroides fragilis* to inactivation by chlorination or ultraviolet radiation. Water Sci. Technol. 21: 221–226.

Bouchard, D.C., M.K. Williams, and R.Y. Surampalli. 1992. Nitrate contamination of groundwater: Sources and potential health effects. J. Am. Water Works Assoc. 84(9): 85–90.

Bouma, J., W.A. Ziebell, W.G. Walker, P.G. Olcott, E. McCoy, and F.D. Hole. 1972. Soil absorption of septic tank effluent: A field study of some major soils in Wisconsin. Information circular #20, University of Wisconsin Extension and Geological–Natural History Survey. University of Wisconsin, Madison.

Bourbigot, M.M., A. Dodin, and R. Lheritier. 1984. Bacteria in distribution systems. Water Res. 18: 585–591.

Bourbigot, M.M., M.C. Hascoet, Y. Levi, F. Erb, and N. Pommery. 1986. Role of ozone and granulated activated carbon in the removal of mutagenic compounds. Environ. Health Perspect. 69: 159–163.

Bouwer, E.J., and P.B. Crowe. 1992. Assessment of biological processes in drinking water treatment. J. Am. Works Assoc. 80: 82–93.

Bouwer, E.J., and P.L. McCarty. 1983. Transformations of halogenated organic compounds under denitrification conditions. Appl. Environ. Microbiol. 45: 1295–1299.

Bouwer, E., J. Mercer, M. Kavanaugh, and F. DiGiano. 1988. Coping with groundwater contamination. J. Water Pollut. Control Fed. 60: 1415–1421.

Bouwer, H. 1992. Agricultural and municipal use of wastewater. Water Sci. Technol. 26: 1583–1591.

Bowden, A.V. 1987. Survey of European sludge treatment and disposal practices. Water Research Centre Report #1656-M.

Bowker, R.P.G., J.M. Smith, and N.A. Webster. 1989. *Odor and Corrosion Control in Sanitary Sewerage Systems and Treatment Plants*. Hemisphere Publishing Corp., New York.

Boyce, D.S., O.J. Sproul, and C.E. Buck. 1981. The effect of bentonite clay on ozone disinfection of bacteria and viruses in water. Water Res. 15: 759–767.

Boyd, R.F. 1988. *General Microbiology* (2nd ed). Times Mirror/Mosby Cool. Publishers, St. Louis.

Boyd, S.A., and D.R. Shelton. 1984. Anaerobic biodegradation of chlorophenols in fresh and acclimated sludge. Appl. Environ. Microbiol. 47: 272–277.

Boyd, S.A., D.R. Shelton, D. Berry, and J.M. Tiedje. 1983. Anaerobic biodegradation of phenolic compounds in digested sludge. Appl. Environ. Microbiol. 46: 50–54.

Brandon, J.R. 1978. *Parasites in Soil/Sludge Systems*. SAND 77–1970 Report. Sandia Laboratory, Alburquerque, NM.

Brandon, J.R., W.D. Burge, and N.K. Enkiri. 1977. Inactivation by ionizing radiation of *Salmonella enteriditis* serotype montevideo grown in composted sewage sludge. Appl. Environ. Microbiol. 33: 1011–1012.

Bravo, J.M., and A. de Vicente. 1992. Bacterial die-off from sewage discharged through submarine outfalls. Water Sci. Technol. 25: 9–16.

Brayton, P.R., M.L. Tamplin, A. Huk, and R.R. Colwell. 1987. Enumeration of *Vibrio cholerae* 01 in Bangladesh waters by fluorescent-antibody direct viable count. Appl. Environ. Microbiol. 53: 2862–2865.

Brenner, K.P., P.V. Scarpino, and C.S. Clark. 1988. Animal viruses, coliphages, and bacteria in aerosols and wastewater at a spray irrigation site. Appl. Environ. Microbiol. 54: 409–415.

Brierley, C.L., J.A. Brierley, and M.S. Davidson. 1989. Applied microbial processes for metal recovery and removal from wastewater, pp. 359–381, in: *Metal Ions and Bacteria*, T.J. Beveridge and R.J. Doyle, Eds. Wiley, New York.

Brigano, F.A.O., P.V. Scarpino, S. Cronier, and M.L. Zink. 1979. Effect of particulates on inactivation of enteroviruses in water by chlorine dioxide. In: *Progress in Water Disinfection Technology*, A.D. Venosa, Ed. EPA-600/9-79-018, Cincinnati, Ohio.

Brock, T.D., and M.T. Madigan. 1991. *Biology of Microorganisms* (6th ed.). Prentice-Hall, Englewood Cliffs, NJ.

Brockman, F.J., B.A. Denovan, R.J. Hicks, and J.K. Frederickson. 1989. Isolation and characterization of quinoline-degrading bacteria from subsurface sediments. Appl. Environ. Microbiol. 55: 1029–1032.

Brodelius, P., and K. Mosbach. 1987. Overview. Methods Enzymol. 135: 173–175.

Brodisch, K.E.U., and S.J. Joyner. 1983. The role of microorganisms other than *Acinetobacter* in biological phosphate removal in activated sludge processes. Water Sci. Technol. 15: 117–122.

Brodkorb, T.S., and R.L. Legge. 1992. Enhanced biodegradation of phenanthrene in oil tar-contaminated soils supplemented with *Phanerochaete chrysosporium*. Appl. Environ. Microbiol. 58: 3117–3121.

Brown T.A. 1990. *Gene Cloning: An Introduction*. Chapman and Hall, London.

Brown, M.J., and J.N. Lester. 1979. Metal removal in activated sludge: The role of bacterial extracellular polymers. Water Res. 13: 817–837.

Brown, M.J., and J.N. Lester. 1982. Role of bacterial extracellular polymers in metal uptake in pure bacterial culture and activated sludge. I. Effect of metal concentration. Water Res. 16: 1539–1548.

Bruce, A.M., and R.D. Davis. 1989. Sewage sludge disposal: Current and future options. Water Sci. Technol. 21: 1113–1128.

Brummeler, E., L.W. Hulshoff Pol, J. Dolfing, G. Lettinga, and A.J.B. Zehnder. 1985. Methanogenesis in an upflow anaerobic sludge blanket reactor at pH 6 on an acetate–propionate mixture. Appl. Environ. Microbiol. 49: 1472–1477.

Brunner, P.H., S. Capri, A. Marcomini, and W. Giger. 1988. Occurrence and behaviour of linear alkylbenzenesulfonates, nonylphenol, nonylphenol mono- and nonylphenol diethoxylates in sewage and sewage sludge treatment. Water Res. 22: 1465–1472.

Brusseau, G.A., H.C. Tsien, R.S. Hanson, and S.P. Wackett. 1990. Optimization of trichloroethylene oxidation by methanotrophs and the use of a colorimetric assay to detect soluble methane monooxygenase activity. Biodegradation 1: 19–29.

Bryan, F.L. 1977. Diseases transmitted by foods contaminated by wastewater. J. Food Protection 40: 45–56.

Bryan, J.A., J.D. Lehmann, I.F. Setiady, and M.H. Hatch. 1974. An outbreak of hepatitis A associated with recreational lake water. Am. J. Epidemiol. 99: 145.

Bryant, R.D., and E.J. Laishley. 1990. The role of hydrogenase in anaerobic biocorrosion. Can. J. Microbiol. 36: 259–264.

Bryant, R.D., W. Jansen, J. Boivin, E.J. Laishley, and J.W. Costerton. 1991. Effect of hydrogenase and mixed sulfate-reducing bacterial populations on the corrosion of steel. Appl. Environ. Microbiol. 57: 2804–2809.

Bryers, J., and W. Characklis. 1981. Early fouling biofilm formation in a turbulent flow system: Overall kinetics. Water Res. 15: 483–491.

Bucke, C. 1987. Cell immobilization on calcium alginate. Methods Enzymol. 135: 175–189.

Bulich A.A. 1979. Use of luminescent bacteria for determining toxicity in aquatic environments. In: *Aquatic Toxicology*, L.L. Markings, and R.A. Kimerle, Eds. American Society for Testing and Materials, Philadelphia, PA.

Bulich, A.A. 1986. Bioluminescent assays, pp. 57–74, in: *Toxicity Testing Using Microorganisms*, Vol. 1, G. Bitton and B.J. Dutka, Eds. CRC Press, Boca Raton, FL.

Buras, N.S. 1984. Water reuse for aquaculture: Public Health aspects. in: *Water Reuse Symposium III*, San Diego, CA, August 26–31, 1984. AWWA, Denver, CO.

Burger, J.S., W.O.K. Grabow, and R. Kfir. 1989. Detection of endotoxins in reclaimed and conventionally treated drinking water. Water Res. 23: 733–738.

Burkhardt, W., III, and W.D. Watkins. 1992. *Clostridium perfringens* provides the only reliable measure of human contamination in the marine environment. Abstr. Annu. Meet. Am. Soc. Microbiol. 92: Q-257.

Burkhardt, W. III, W.D. Watkins, and S.R. Rippey. 1992. Survival and replication of male-specific bacteriophages in molluscan shellfish. Appl. Environ. Microbiol. 58: 1371–1373.

Burns, R.G, Ed. 1978. *Soil Enzymes*. Academic, London.

Burttschell, R.H., A.A. Rosen, F.M. Middleton, and M.B. Ettinger. 1959. Chlorine derivatives of phenol causing taste and odor. J. Am. Water Works Assoc. 51: 205–214.

Busta, F.F. 1976. Practical implications of injured microorganisms in foods. J. Milk Food Technol. 39: 138–145.

Cabelli, V.J. 1981. *Health Effects Criteria for Marine Recreational Waters*. EPA-600/1–80–031. U.S. Environmental Protection Agency, Cincinnati, Ohio.

Cabelli, V.J. 1989. Swimming-associated illness and recreational water quality criteria. Water Sci. Technol. 21: 13–21.

Calabrese, J.P., and G.K. Bissonnette. 1989. Modification of standard recovery media for enhanced detection of chlorine-stressed coliform and heterotrophic bacteria. Abstr. Annu. Meet. Am. Soc. Microbiol., Vol. 89.

Calabrese, J.P., and G.K. Bissonnette. 1990. Improved membrane filtration method incorporating catalase and sodium pyruvate for detection of chlorine-stressed coliform bacteria. Appl. Environ. Microbiol. 56: 3558–3564.

Callaway, W.T. 1968. The metazoa of waste treatment. J. Water Pollut. Control Fed. 40: R412-R422.

Calomiris, J.J., J.L. Armstrong, and R.J. Seidler. 1984. Association of metal tolerance with multiple antibiotic resistance of bacteria isolated from drinking water. Appl. Environ. Microbiol. 47: 1238–1242.

Camann, D.E., C.A. Sorber, B.P. Sagik, J.P. Glennon, and D.E. Johnson. 1978. In: *Risk Assessment and Health Effects of Land Application of Municipal Wastewater and Sludges*, (pp. 240–266), B.P. Sagik and C.A. Sorber, Eds. Center for Applied Research and Technology, University of Texas, San Antonio.

Camann, D.E., P.J. Graham, M.N. Guentzel, H.J. Harding, K.T. Kimball, B.E. Moore, R.L. Northrop, N.L. Altman, R.B. Harrist, A.H. Holguin, R.L. Mason, C.B. Popescu, and C.A. Sorber. 1986. *The Lubbock Land Treatment System, Research and Demonstration Project, Vol. 4, Lubbock Infection Surveillance Study (LISS)*. EPA-600/S2–86/027d, U.S. EPA, Research Triangle Park, NC.

Camann, D.E., B.E. Moore, H.J. Harding, and C.A. Sorber. 1988. Microorganism levels in air near spray irrigation of municipal wastewater: The Lubbock infection surveillance study. J. Water Pollut. Control Fed. 60: 1960–1970.

Campbell, M., G. Bitton, and B. Koopman. 1993. Toxicity testing of sediment elutriates based on inhibition of α-glucosidase biosynthesis in *Bacillus licheniformis*. Arch. Environ. Contam. Toxicol. 24: 469–472.

Camper, A.K., M.W. Lechevallier, S.C. Broadaway, and G.A. McFeters. 1985. Growth and persistence of pathogens on granular activated carbon filters. Appl. Environ. Microbiol. 50: 1378–1382.

Camper, A.K., M.W. Lechevallier, S.C. Broadaway, and G.A. McFeters. 1986. Bacteria associated with granular activated carbon particles in drinking water. Appl. Environ. Microbiol. 52: 434–438.

Camper, A.K., S.C. Broadaway, M.W. LeChevallier, and G.A. McFeters. 1987. Operational variables and the release of colonized granular activated carbon particles in drinking water. J. Am. Water Works Assoc. 79: 70–74.

Camper, A.K., and G.A. McFeters. 1979. Chlorine injury and the enumeration of waterborne coliform bacteria. Appl. Environ. Microbiol. 37: 633–641

Camper, A.K., G.A. McFeters, W.G. Characklis, and W.L. Jones. 1991. Growth kinetics of coliform bacteria under conditions relevant to drinking water distribution systems. Appl. Environ. Microbiol. 57: 2233–2239.

Campbell, A.T., L.J. Robertson, and H.V. Smith. 1992. Viability of *Cryptosporidium parvum* oocysts: Correlation of in vitro excystation with inclusion or exclusion of fluorogenic vital dyes. Appl. Environ. Microbiol. 58: 3488–3493.

Campbell, I., S. Tzipori, G. Hutchinson, and K.W. Angus. 1982. Effect of disinfectants on survival of *Cryptosporidium* oocysts. Vet. Rec. 111:414–415.

Canter, L.W., and R.C. Knox. 1984. *Evaluation of Septic Tank System Effects on Groundwater Quality*. EPA-600/S2–84–107, R.S. Kerr Research Laboratories, Ada, OK.

Canter, L.W., and R.C. Knox. 1985. Septic Tank System: Effects on Groundwater Quality, Lewis Pubs., Chelsea, MI.

Canter, L.W., L.E. Streebin, M.C. Arquiaga, F.E. Carranza, D.E. Miller, and B.H. Wilson. 1990. Innovative processes for reclamation of contaminated subsurface environments. EPA-600/S2–90/017, U.S. EPA, Ada, OK.

Carlson, D.A., and C.P. Leiser. 1966. Soil beds for the control of sewage odors. J. Water Pollut. Control Fed. 38: 829–840.

Carmichael, W.W., Ed. 1981a. *The Aquatic Environment: Algal Toxins and Health*. Plenum Press, New York.

Carmichael, W.W. 1981b. Freshwater blue-green algae toxins, pp. 7–13, in: *The Aquatic Environment: Algal Toxins and Health*, W.W. Carmichael, Ed., Plenum, New York.

Carmichael, W.W. 1989. Freshwater cyanobacteria (blue-green algae) toxins, pp. 3–16, in: *Natural Toxins: Characterization, Pharmacology and Therapeutics*, C.L. Ownby and G.V. Odell, Eds. Pergamon Press, Oxford, U.K.

Carrington, B.G. 1985. Pasteurization: Effects on *Ascaris* eggs, pp. 121–125, in: *Inactivation of Microorganisms in Sewage Sludge by Stabilization Processes*, D. Strauch, A.H. Havelaar, and P. L'Hermite, Eds. Elsevier Applied Science, London.

Carson, L.A., and N.J. Petersen. 1975. Photoreactivation of *Pseudomonas cepacia* after ultraviolet exposure: A potential source of contamination in ultraviolet-treated waters. J. Clin. Microbiol. 1:462–464.

Carter, A.M., R.E. Pacha, G.W. Clark, and E.A. Williams. 1987. Seasonal occurrence of *Campylobacter* spp. in surface waters and their correlation with standard indicator bacteria. Appl. Environ. Microbiol. 53: 523–526.

Casey, T.G., M.C. Wentzel, R.E. Lowenthal, G.A. Ekama, and G.v.R. Marais. 1992. A hypothesis for the cause of low F/M filament bulking in nutrient removal activated sludge systems. Water Res. 26: 867–869.

Casson, L.W., C.A. Sorber, J.L. Sykora, P.D. Cavaghan, M.A. Shapiro, and W. Jakubowski. 1990. *Giardia* in wastewater—Effect of treatment. J. Water Pollut. Control Fed. 62: 670–675.

Cees, B., J. Zoeteman, and G.J. Piet. 1974. Cause and identification of taste and odour compounds in water. Sci. Total Environ. 3: 103–115.

Center for Disease Control. 1991. Outbreaks of diarrheal illness associated with cyanobacteria (blue-green algae)-like bodies: Chicago and Nepal. Morb. Mort. Rep. 40: 325.

Cha, D.K., D. Jenkins, W.P. Lewis, and W.H. Kido. 1992. Process control factors influencing *Nocardia* populations in activated sludge. Water Environ. Res. 64: 37–43.

Chakrabarti, T., and P.H. Jones. 1983. Effect of molybdenum and selenium addition on the denitrification of wastewater. Water Res. 17: 931–936.

Chamberlin, C.E., and R. Mitchell. 1978. A decay model for enteric bacteria in natural waters, pp. 325–348, in: *Water Pollution Microbiology*, Vol. 2, R. Mitchell, Ed. Wiley, New York.

Chambers, B. 1982. Effect of longitudinal mixing and anoxic zones on settleability of activated sludge. in: *Bulking of Activated Sludge: Preventive and Remedial Methods*, B. Chambers and E.J. Tomlinson, Eds. Ellis Horwood, Chichester, U.K.

Chang, G.W., J. Brill, and R. Lum. 1989. Proportion of ß-glucuronidase-negative *Escherichia coli* in human fecal samples. Appl. Environ. Microbiol. 55: 335–339.

Chang, J.C.H., S.F. Ossof, D.C. Lobe, M.H. Dorfman, C.M. Dumais, R.G. Qualls, and J.D. Johnson. 1985. UV inactivation of pathogenic and indicator microorganisms. Appl. Environ. Microbiol. 49: 1361–1365.

Chang, S.D., and P.C. Singer. 1991. The impact of ozonation on particle stability and the removal of TOC and THM precursors. J. Am. Water Works Assoc. 83: 71–79.

Chang, S.-L. 1982. The safety of water disinfection. Annu. Rev. Public Health 3: 393–418.

Chang, S.L., R.L. Woodward, and P.W. Kabler. 1960. Survey of free living nematodes and amoebas in municipal supplies. J. Am. Water Works Assoc. 52: 613–618.

Chang, Y., J.T. Pfeiffer, and E.S.K. Chian. 1979. Comparative study of different iron compounds in inhibition of *Sphaerotilus* growth. Appl. Environ. Microbiol. 38: 385–389.

Chao, A.C., and T.M. Keinath. 1979. Influence of process loading intensity on sludge clarification and thickening characteristics. Water Res. 13: 1213–1217.

Characklis, W.G., Ed. 1988. *Bacterial Regrowth in Distribution Systems*. Research Report, AWWA Research Foundation, Denver, CO.

Characklis, W.G., and K.E. Cooksey. 1983. Biofilms and microbial fouling. Adv. Appl. Microbiol. 29: 93–138.

Characklis, W.G., M.G. Trulear, J.D. Bryers, and N.

Zelver. 1982. Dynamic of biofilm processes: Methods. Water Res. 16: 1207–1216.

Charoenca, N., and R.S. Fujioka. 1993. Assessment of *Staphylococcus* bacteria in Hawaii recreational waters. Water Sci. Technol. 27: 283–289.

Chaudhry, G.R,. and S. Chapalamadugu. 1991. Biodegradation of halogenated organic compounds. Microbiol. Rev. 55: 59–79.

Cheeseman, P., A. Toms-Wood, and R.S. Wolfe. 1972. Isolation and properties of a fluorescent compound, factor F_{420}, from *Methanobacterium* strain M.O.H. J. Bacteriol. 112: 527–531.

Cheetham, P.S.J., and C. Bucke. 1984. Immobilization of microbial cells and their use in waste water treatment, pp. 219–235, in: *Microbiological Methods for Environmental Biotechnology*, J.M. Grainger and J.M. Lynch, Eds. Academic, London.

Chen, Y.S.R., O.J. Sproul, and A. Rubin. 1985. Inactivation of *Naegleria gruberi* cysts by chlorine dioxide. Water Res. 19: 783–789.

Chen, Y.-S., and J. Vaughn. 1990. Inactivation of human and simian rotaviruses by chlorine dioxide. Appl. Environ. Microbiol. 56: 1363–1366.

Chen, Y.S., J.M. Vaughn, and R.M. Niles. 1987. Rotavirus RNA and protein alterations resulting from ozone treatment. Abstr. Annu. Meet. Am. Soc. Microbiol. 87: Q-22.

Cheng, M.H., J.W. Patterson, and R.A. Minear. 1975. Heavy metal uptake by activated sludge. J. Water Pollut. Control Fed. 47: 362–376.

Chet, I., and R. Mitchell. 1976. Ecological aspects of microbial chemotactic behavior. Annu. Rev. Microbiol. 30: 221–239.

Chevalier, P., and J. de la Noue. 1985. Wastewater nutrient removal with microalgae immobilized in carrageenan. Enzyme Microbiol. Technol. 7: 621–624.

Chiesa, S.C., and R.L. Irvine. 1985. Growth and control of filamentous microbes in activated sludge: An integrated hypothesis. Water Res. 19: 471–479.

Choi, E., and J.M. Rim. 1991. Competition and inhibition of sulfate reducers and methane producers in anaerobic treatment. Water Sci. Technol. 23: 1259–1264.

Christensen, G.M., D. Olson, and B. Reidel. 1982. Chemical effects on the activity of eight enzymes: A review and a discussion relevant to environmental monitoring. Environ. Res. 29: 247–255.

Christensen, M.H., and P. Harremoes. 1977. Biological denitrification of sewage: A literature review. Prog. Water Technol. 8: 509–555.

Christensen, M.H., and P. Harremoes. 1978. Nitrification and denitrification in wastewater treatment, pp. 391–414, in: *Water Pollution Microbiology*, Vol. 2, R. Mitchell, Ed. Wiley, New York.

Christiansen, J.A., and P.W. Spraker. 1983. Improving effluent quality of petrochemical wastewaters with mutant bacterial cultures. In: *Proceedings of the 37th Industrial Waste Conference*, Purdue University, Lafayette, Indiana, pp. 567–576.

Chudoba, J. 1985. Control of activated sludge filamentous bulking. VI: Formulation of basic principles. Water Res. 19: 1017–1022.

Chudoba, J. 1989. Activated sludge-Bulking control, pp. 171–202, in: *Encyclopedia of Environmental Control Technology*, Vol. 3, *Wastewater Treatment Technology*, P.N. Cheremisinoff, Ed. Gulf Publishing Co., Houston, TX.

Chudoba, J., J.S. Cesh, J. Farkac, and P. Grau. 1985. Control of activated sludge filamentous bulking. Experimental verification of kinetic selection theory. Water Res. 19: 191–196.

Chudoba, J., P. Grau, and V. Ottova. 1973. Control of activated sludge filamentous bulking. II: Selection of microorganisms by means of a selector. Water Res. 7: 1389–1406.

Chung, H., and M.D. Sobsey. 1993. Comparative survival of indicator viruses and enteric viruses in seawater and sediment. Water Sci. Technol. 27: 425–428.

Chung, Y.C., and J.B. Neethling. 1988. ATP as a measure of anaerobic sludge digester activity. J. Water Pollut. Control Fed. 60: 107–112.

Chung, Y.C., and J.B. Neethling. 1989. Microbial activity measurements for anaerobic sludge digestion. J. Water Pollut. Control Fed. 61: 343–349.

Ciesielski, C.A., M.J. Blaser, and W.L. Wang. 1984. Role of stagnation and obstruction of water flow in isolation of *Legionella pneumophila* from hospital plumbing. Appl. Environ. Microbiol. 48: 984–987.

Clark, C.S., A.B. Bjornson, G.M. Schiff, J.P. Phair, G.L. van Meer, and P.S. Gartside. 1977. Sewage worker's syndrome. Lancet 1: 1009.

Clark, C.S., H.S. Bjornson, J.W. Holland, V.J. Elia, V.A. Majeti, C.R. Meyer, W.F. Balistreri, G.L. van Meer, P.S. Gartside, B.L. Specker, C.C. Linnemann, Jr., R. Jaffa, P.V. Scarpino, K. Brenner,

W.J. Davis-Hoover, G.W. Barrett, T.S. Anderson,and D.L. Alexander. 1981. *Evaluation of the Health Risks Associated with the Treatment and Disposal of Municipal Wastewater and Sludge*. EPA-600/S1–81–030, U.S. EPA, Cincinnati, Ohio.

Clark, C.S., H.S. Bjornson, J. Schwartz-Fulton, J.W. Holland, and P.S. Gartside. 1984. Biological health risks associated with the composting of wastewater treatment plant sludge. J. Water Pollut. Control Fed. 56: 1269–1276.

Clark, D.L., B.B. Milner, M.H. Stewart, R.L. Wolfe, and B.H. Olson. 1991. Comparative study of commercial 4-methylumbelliferyl-ß-D-glucuronide preparations with the Standard Methods membrane filtration fecal coliform test for the detection of *Escherichia coli* in water samples. Appl. Environ. Microbiol. 57: 1528–1534.

Clark, J.A., C.A. Burger, and L.E. Sabatinos. 1982. Characterization of indicator bacteria in municipal raw water, drinking water, and new main water. Can. J. Microbiol. 28: 1002–1013.

Clark, R.M., E.J. Read, and J.C. Hoff. 1989. Analysis of inactivation of *Giardia lamblia* by chlorine. J. Environ. Eng. Div. Am. Soc. Civ. Eng. 115: 80–90.

Clarke, N.A., and S.L. Chang. 1975. Removal of enteroviruses from sewage by bench-scale rotary-tube trickling filters. Appl. Microbiol. 30: 223–228.

Cleasby, J.L., D.J. Hilmoe, and C. J. Dimitracopoulos. 1984. Slow sand and direct in-line filtration of a surface water. J. Am. Water Works Assoc. 76: 44–55.

Cleuziat, P., and J. Robert-Baudouy. 1990. Specific detection of *Escherichia coli* and *Shigella* species using fragments of genes coding for ß-glucuronidase. FEMS Microbiol. Lett. 72: 315–322.

Cliver, D.O. 1984. Significance of water and environment in the transmission of virus disease. Monogr. Virol. 15: 30–42.

Cliver, D.O. 1985. Vehicular transmission of hepatitis A. Public Health Rev. 13: 235–292.

Cloete, T.E., and P.L. Steyn. 1988. The role of *Acinetobacter* as a phosphorus removing agent in activated sludge. Water Res. 22: 971–976.

Cobb, G.D., and E.J. Bouwer. 1991. Effect of electron acceptors on halogenated organic compounds biotransformations in a biofilm column. Environ. Sci. Technol. 25: 1068–1074.

Coetzee, J.N. 1987. Bacteriophage taxonomy, pp. 45–86, in: *Phage Ecology*, S.M. Goyal, C.P. Gerba, and G. Bitton, Eds. Wiley, New York.

Cohen, A., A.M. Breure, J.G. van Andel, and A. van Deursen. 1980. Influence of phase separation on the anaerobic digestion of glucose. I. Maximum COD-turnover rate during continuous operation. Water Res. 14: 1439–1448.

Cohen, M.L. 1992. Epidemiology of drug resistance: Implications for a post-antimicrobial era. Science 257: 1050–1055.

Colbourne, J.S., P.J. Dennis, R.M. Trew, C. Berry, and G. Vesey. 1988. *Legionella* and public water supplies. In: *Proceedings of the International Conference on Water and Wastewater Microbiology*, Newport Beach, CA, Feb. 8–11, 1988, Vol. 1.

Cole, C.A., J.B. Stamberg, and D.F. Bishop. 1973. Hydrogen peroxide cures filamentous growth in activated sludge. J. Water Pollut. Control Fed. 45: 829–836.

Colignon, A., M.N. Fortin, and G. Martin. 1986. Action de l'ozone sur le foisonnement filamenteux. Sci. Eau 5: 137–142.

Collins, M.R., T.T. Eightmy, J. Fenstermacher, and S.K. Spanos. 1992. Removing natural organic matter by conventional slow sand filtration. J. Am. Water Works Assoc. 84: 80–90.

Colwell, R.R., and D.J. Grimes. 1986. Evidence of genetic modification of microorganisms occurring in natural aquatic environments, pp. 222–230, in: *Aquatic Toxicology and Environmental Fate*, Vol. 9, T.M. Poston and R. Purdy, Eds. ASTM Pub. No. 921. ASTM, Philadelphia.

Colwell, R.R., and G.S. Sayler. 1978. Microbial degradation of industrial chemicals, pp. 111–134, in: *Water Pollution Microbiology*, Vol. 2, R. Mitchell, Ed. Wiley, New York.

Commeau, Y., K.J. Hall, R.E.W. Hancock, and W.K. Oldham. 1986. Biochemical model for enhanced biological phosphorus removal. Water Res. 20: 1511–1521.

Comeau, Y., B. Rabinowitz, K.J. Hall, and W.K. Oldham. 1987. Phosphate release and uptake in enhanced biological phosphorus removal from wastewater. J. Water Pollut. Control Fed. 59: 707–715.

Compeau, G., and R. Bartha. 1985. Sulfate reducing bacteria: Principal methylators of mercury in anoxic estuarine sediments. Appl. Environ. Microbiol. 50: 498–502.

Compeau, G., and R. Bartha. 1987. Effect of salinity on mercury-methylating activity of sulfate-reducing bacteria in estuarine sediments. Appl. Environ. Microbiol. 53: 261–265.

Compeau, G.C., W.D. Mahaffey, and L. Patras. 1991. Full-scale bioremediation of contaminated soil and water, pp. 91–109, in: *Environmental Biotechnology for Waste Treatment*, G.S. Sayler, R. Fox and J.W. Blackburn, Eds. Plenum, New York.

Condie, L.W. 1986. Toxicological problems associated with chlorine dioxide. J. Am. Water Works Assoc. 78: 73–78.

Cooper, P.F., and D.H.V. Wheeldon. 1981. Fluidized- and expanded-bed reactors for wastewater treatment. Water Pollut. Control 79: 286–306.

Cooper, R.C. 1991. Public health concerns in wastewater reuse. Water Sci. Technol. 24: 55–65.

Cordes, L.G., D.W. Fraser, P. Skaliy, C.A. Perlino, W.R. Elsea, G.F. Mallison, and P.S. Hayes. 1980. Legionaire's disease outbreak at an Atlanta, Georgia, country club: Evidence for spread from an evaporative condenser. Am. J. Epidemiol. 111: 425–431.

Cornax, R., M.A. Morinigo, I.G. Paez, M.A. Munoz, and J.J. Borrego. 1990. Application of direct plaque assay for detection and enumeration of bacteriophages of *Bacteroides fragilis* from contaminated water samples. Appl. Environ. Microbiol. 56: 3170–3173.

Correa, I.E., N. Harb, and M. Molina. 1989. Incidence and prevalence of *Giardia* spp. in Puerto rican waters: Removal of cysts by conventional sewage treatment plants. Abstr. Annu. Meet. Am. Soc. Microbiol., New Orleans.

Costan, G., N. Bermingham, C. Blaise, and J.F. Ferard. 1993. Potential ecotoxic effects probe (PEEP): A novel index to assess and compare the toxic potential of industrial effluents. Environ. Toxicol. Water Qual. 8: 115–140.

Costerton, J.W. 1980. Some techniques involved in study of adsorption of microorganisms to surfaces, pp. 403–423, in: *Adsorption of Microorganisms to Surfaces*, G. Bitton and K.C. Marshall, Eds. Wiley, New York.

Costerton, J.W., J. Boivin, E.J. Laishley, and R.D. Bryant. 1989. A new test for microbial corrosion, pp. 20–25, in: *Sixth Asian-Pacific Corrosion Control Conference*. Singapore Asian-Pacific Materials and Corrosion Association, Singapore.

Costerton, J.W., and G.C. Geesey. 1979. Microbial contamination of surfaces, pp. 211–221, in: *Surface Contamination*, K.L. Mittal, Ed. Plenum, New York.

Covert, T.C., E.W. Rice, S.A. Johnson, D. Berman, C.H. Johnson, and P.J. Mason. 1992. Comparing defined-substrate tests for the detection of *Escherichia coli* in wastewater. J. Am. Water Works Assoc. 84: 98–105.

Covert, T.C., L.C. Shadix, E.W. Rice, J.R. Haines, and R.W. Freyberg. 1989. Evaluation of the Autoanalysis Colilert test for detection and enumeration of total coliforms. Appl. Environ. Microbiol. 55: 2443–2447.

Cox, C.S. 1987. *The Aerobiological Pathway of Microorganisms*. Wiley, Chichester, U.K.

Cox, D.P. 1978. The biodegradation of polyethylene glycols. Adv. Appl. Microbiol. 23:173–194.

Craun, G.F. 1979. Waterborne giardiasis in the United States. Am. J. Public Health 69: 817–820.

Craun, G.F. 1984a. Health aspects of groundwater pollution, pp. 135–179, in: *Groundwater Pollution Microbiology*, G. Bitton and C.P. Gerba, Eds. Wiley, New York.

Craun, G.F. 1984b. Waterborne outbreaks of giardiasis: Current status, pp. 243–261, in: *Giadia and Giardiasis*, S.L. Erlandsen and E.A. Meyers, Eds. Plenum, New York.

Craun, G.F. 1986a. Statistics of waterborne outbreaks in the U.S. (1920–1980), pp. 73–159, in: *Water Diseases in the United States*, G.F. Craun, Ed. CRC Press, Boca Raton, FL.

Craun, G.F., Ed. 1986b. *Waterborne Diseases in the United States*. CRC Press, Boca Raton, FL.

Craun, G.F. 1988. Surface water supplies and health. J. Am. Water Works Assoc. 80: 40–52.

Crawford, R.L., and K.T. O'Reilly. 1989. Bacterial decontamination of agricultural wastewaters, pp. 73–89, in: *Biotreatment of Agricultural Wastewater*, M.E. Huntley, Ed. CRC Press, Boca Raton, FL.

Crawford, R.L., K.T. O'Reilly, and H.-L. Tao. 1989. Microorganism stabilization for *in situ* degradation of toxic chemicals, pp. 203–211, in: *Biotechnology and Biodegradation*, D. Kamely, A. Chakrabarty, and G.S. Omenn, Eds. Gulf Publishing Co., Houston, TX.

Criddle, C.S., J.T. DeWitt, D. Grbic-Galic, and P.L. McCarty. 1990. Transformation of carbon tetrachloride by *Pseudomonas* sp., strain KC under denitrification conditions. Appl. Environ. Microbiol. 56: 3240–3246.

Crook, J. 1985. Water reuse in California. J. Am. Water Works Assoc. 77: 60–71.

Cross, T., and M. Goodfellow. 1973. Taxonomy and classification of the actinomycetes, pp. 11–112, in: *Acinomycetales*, G. Sykes and F.A. Skinner, Eds. Academic, London.

Cullen, T.R., and R.D. Letterman. 1985. The effect of slow sand filter maintenance on water quality. J. Am. Water Works Assoc. 77: 48–55.

Curds, C.R. 1975. Protozoa, pp. 203–268, in: *Ecological Aspects of Used-Water Treatment*, Vol. 1, C.R. Curds and H.A. Hawkes, Eds. Academic, London.

Curds, C.R. 1982. The ecology and role of protozoa in aerobic sewage treatment processes. Annu. Rev. Microbiol. 36: 27–46

Curds, C.R., and H.A. Hawkes, Eds. 1975. *Ecological Aspects of Used-Water Treatment*, Vol. 1, Academic, London.

Curds, C.R., and H.A. Hawkes, Eds. 1983. *Ecological Aspects of Used-Water Treatment*, Vol. 2, Academic, London.

Current, W.L. 1987. *Cryptosporidium*: Its biology and potential for environmental transmission. Crit. Rev. Environ. Control 17: 21.

Current, W.L. 1988. The biology of *Cryptosporidium*. ASM News 54: 605–611.

Curtis C., A. Lima, S.J. Lorano, and G.D. Veith. 1982. Evaluation of a bacterial bioluminescence bioassay as a method for predicting acute toxicity of organic chemicals to fish, pp. 170–178, in: *Aquatic Toxicity and Hazard Assessment*, J.G. Pearson, R.B. Foster and W.E. Bishop, Eds. STP #766. ASTM, Philadelphia.

Curtis, T.P., D.D. Mara, and S.A. Silva. 1992a. The effect of sunlight on faecal coliforms: Implications for research and design. Water Sci. Technol. 26: 1729–1738.

Curtis, T.P., D.D. Mara, and S.A. Silva. 1992b. Influence of pH, oxygen, and humic substances on ability of sunlight to damage fecal coliforms in waste stabilization pond water. Appl. Environ. Microbiol. 58: 1335–1343.

Dagues, R.E. 1981. Inhibition of nitrogenous BOD and treatment plant performance evaluation. J. Water Pollut. Control Fed. 53: 1738–1741.

Daigger, G.T., M.H. Robbins, Jr., and B.R. Marshall. 1985. The design of a selector to control low-F/M filamentous bulking. J. Water Pollut. Control Fed. 57: 220–226.

Damgaard-Larsen, S., K.O. Jensen, E. Lund, and B. Nisser. 1977. Survival and movement of enterovirus in connection with land disposal of sludges. Water Res. 11: 503–508.

Darnall, D.W., B. Greene, M.T. Henzl, J.M. Hosea, R.A. McPherson, J. Sneddon, and M.D. Alexander. 1986. Selective recovery of gold and other metal ions from an algal biomass. Environ. Sci. Technol. 20: 206–210.

Dart, R.K., and R.J. Stretton. 1980. *Microbiological Aspects of Pollution Control*. Elsevier, Amsterdam.

Davis, J.W., and C.L. Carpenter. 1990. Aerobic biodegradation of vinyl chloride in groundwater sample. Appl. Environ. Microbiol. 56: 3878–3880.

Davis, M.L., and D.A. Cornwell. 1985. *Introduction to Environmental Engineering*. PWS Engineering, Boston.

Day, H.R., and G.T. Felbeck, Jr. 1974. Production and analysis of a humic-like exudate from the aquatic fungus *Aureobasidium pullulans*. J. Am. Water Works Assoc. 66: 484–489.

Deakyne, C.W., M.A. Patel, and D.J. Krichten. 1984. Pilot plant demonstration of biological phosphorus removal. J. Water Pollut. Control Fed. 56: 867–873.

Dean, R.B., and E. Lund. 1981. *Water Reuse: Problems and Solutions*. Academic, London, U.K.

Debartolomeis, J., and V.J. Cabelli. 1991. Evaluation of an *Escherichia coli* host strain for enumeration of F male-specific bacteriophages. Appl. Environ. Microbiol. 57: 1301–1305.

DeBoer, J., and K.D. Linstedt. 1985. Advances in water reuse applications. Water Res. 19: 1455–1461.

De Laat, J., F. Bouanga, M. Dore, and J. Mallevialle. 1985. Influence du developpement bacterien au sein des filtres de charbon actif en grains sur l'elimination de composes organiques biodegradables et non biodegradables. Water Res. 19: 1565–1578.

DeLeon, R., M.D. Sobsey, R.M. Matsui, and R.S. Baric. 1992. Detection of Norwalk virus by reverse transcriptase–polymerase chain reaction and non-reactive oligoprobes (RT-PCR-OP). Paper presented at the International Water Pollution Research Conference International Symposium, Washington, DC, May 26–29, 1992.

Delwiche, C.C. 1970. The nitrogen cycle. Sci. Am. 223: 137–146.

Demain, A.L. 1984. Capabilities of microorganisms (and microbiologists), pp. 277–299, in: *Genetic*

Control of Environmental Pollutants, G.S. Omenn and A. Hollaender, Eds. Plenum, New York.

De Mik, G., and I. De Groot. 1977. Mechanisms of inactivation of bacteriophage φX174 and its DNA in aerosols by ozone and ozonized cyclohexene. J. Hyg. 78: 199–211.

Denis, F.A., E. Blanchovin, A. DeLigneres, and P. Flamen. 1974. Coxsackie A16 infection from lake water. J. Am. Med. Assoc. 228: 1370.

Dennis, P.J., D. Green, and B.P. Jones. 1984. A note on the temperature tolerance of *Legionella*. J. Appl. Bacteriol. 56: 349–350.

Dennis, W.H., V.P. Olivieri, and C.W. Kruse. 1979. Mechanism of disinfection: Incorporation of Cl-36 into f2 virus. Water Res. 13: 363–369.

Dermer, O.C, V.S. Curtis, and F.R Leach. 1980. *Biochemical Indicators of Subsurface Pollution*. Ann Arbor Science Publishers, Ann Arbor, MI.

DeWaters, J.E., and F.A. DiGiano. 1990. The influence of ozonated natural organic matter on the biodegradation of a micropollutant in a GAC bed. J. Am. Water Works Assoc. 82: 69–75.

Dice, J.C. 1985. Denver's seven decades of experience with chloramination. J. Am. Water Works Assoc. 77: 34–37.

Diahl, J.D., Jr. 1991. Improved method for coliform verification. Appl. Environ. Microbiol. 57: 604–605.

Diekert, G., U. Konheiser, K. Piechulla, and R.K. Thauer. 1981. Nickel requirement and factor F_{430} content of methanogenic bacteria. J. Bacteriol. 148: 459–465.

DiGiano, F.A., C. Clarkin, M.J. Charles, M.J. Maerker, D.E. Francisco, and C. LaRocca. 1992. Testing of the EPA toxicity identification evaluation protocol in the textile dye manufacturing industry. Water Sci. Technol. 25: 55–63.

DiStefano, T.D., J.M. Gossett, and S.H. Zinder. 1991. Reductive dechlorination of high concentrations of pentachloroethene to ethene by an anaerobic enrichment culture in the absence of methanogenesis. Appl. Environ. Microbiol. 57: 2287–2292.

Divizia, M., C. Gnesivo, R.A. Bonapasta, G. Morace, G. Pisani, and A. Pana. 1993. Virus isolation and identification by PCR in an outbreak of hepatitis A: Epidemiological investigation. Water Sci. Technol. 27: 199–205.

Dizer, H., W. Bartocha, H. Bartel, K. Seidel, J.M. Lopez-Pila, and A. Grohmann. 1993. Use of ultraviolet radiation for inactivation of bacteria and coliphages in pretreated wastewater. Water Res. 27: 397–403.

Dizer, H., A. Nasser, and J.M. Lopez. 1984. Penetration of different human pathogenic viruses into sand columns percolated with distilled water, groundwater, or wastewater. Appl. Environ. Microbiol. 47: 409–415.

Dmochewitz, S., and K. Ballschmiter. 1988. Microbial transformation of technical mixtures of polychlorinated biphenyls (PCB) by the fungus *Aspergillus niger*. Chemosphere 17: 111–121.

Dodgson, K.S., G.F. White, J.A. Massey, J. Shapleigh, and W.J. Payne. 1984. Utilization of sodium dodecyl sulphate by denitrifying bacteria under anaerobic conditions. FEMS Microbiol. Lett. 24: 53–56.

Dolfing, J., and J.W. Mulder. 1985. Comparison of methane production rate and coenzyme F_{420} content of methanogenic consortia in anaerobic granular sludge. Appl. Environ. Microbiol. 49: 1142–1145.

Domek, M.J., M.W. LeChevallier, S.C. Cameron, and G.A. McFeters. 1984. Evidence for the role of copper in the injury process of coliform bacteria in drinking water. Appl. Environ. Microbiol. 48: 289–293.

Dondero, T.J., R.C. Rendtorff, G.F. Mallison, R.M. Weeks, J.S. Levy, E.W. Wong, and W. Schaffner. 1980. An outbreak of Legionnaire's disease associated with contaminated air-conditioned cooling tower. N. Engl. J. Med. 302: 362–370.

Donovan, J.F., J.E. Bates, and C.H. Rowell. 1980. *Guidelines for Water Reuse*. Camp, Dresser and McKee, Boston.

Doohan, M. 1975. Rotifera, pp. 289–304, in: *Ecological Aspects of Used-Water Treatment*, Vol. 1: *The Organisms and Their Ecology*, C.R. Curds and H.A. Hawkes, Eds. Academic, London.

Dott, W., and P. Kampfer. 1988. Biochemical methods for automated bacterial identification and testing metabolic activities in water and wastewater. In: *Proceedings of the International Conference on Water and Wastewater Microbiology*, Newport Beach, CA, Feb. 8–11, 1988, Vol. 1.

Drakides, C. 1980. La microfaune des boues activees. Etude d'une methode d'observation et application au suivi d'un pilote en phase de demarrage. Water Res. 14: 1199–1207.

Duboise, S.M., B.E. Moore, and B.P. Sagik. 1976.

Poliovirus survival and movement in a sandy forest soil. Appl. Environ. Microbiol. 31: 536–543.

Dufour, A.P. 1986. Diseases caused by water contact, pp. 23–41, in: *Waterborne Diseases in the United States*, G.F. Craun, Ed. CRC Press, Boca Raton, FL.

Dugan, P.R. 1987a. Prevention of formation of acid drainage from high-sulfur coal refuse by inhibition of iron- and sulfur-oxidizing microorganisms. I. Preliminary experiments in controlled shaken flasks. Biotechnol. Bioeng. 29: 41–48.

Dugan, P.R. 1987b. Prevention of formation of acid drainage from high-sulfur coal refuse by inhibition of iron- and sulfur- oxidizing microorganisms. II. Inhibition in "run of mine" refuse under simulated field conditions. Biotechnol. Bioeng. 29: 49–54.

duMoulin, G.C., I. Sherman, and K.D. Stottmeier. 1981. *Mycobacterium intracellulare*: An emerging pathogen. Abstr. Annu. Meet. Am. Soc. Microbiol., Washington, DC.

Duncan, A. 1988. The ecology of slow sand filters, pp. 163–180, in: *Slow Sand Filtration: Recent Development in Water Treatment Technology*, N.J.D. Graham, Ed. Ellis Horwood, Chichester, U.K.

Dutka B.J., and G. Bitton, Eds. 1986. *Toxicity Testing Using Microorganisms*, Vol 2. CRC Press, Boca Raton, FL.

Dutka, B.J., A. El Shaarawi, M.T. Martins, and P.S. Sanchez. 1987. North and South American studies on the potential of coliphage as a water quality indicator. Water Res. 21: 1127–1134.

Dutton, R.J., G. Bitton, and B. Koopman. 1988. Enzyme biosynthesis versus enzyme activity as a basis for microbial toxicity testing. Toxicity Assess. 3: 245–253.

Dutton, R.J, G. Bitton, B. Koopman, and O. Agami. 1990. Inhibition of ß-galactosidase biosynthesis in *Escherichia coli*: Effect of alterations of the outer membrane permeability to environmental toxicants. Toxicity Assess. 5: 253–264.

Duvoort-van Engers, L.E., and S. Coppola. 1986. State of the art on sludge composting, pp. 59–75, in: *Processing and Use of Organic Sludge and Liquid Agricultural Wastes*, P. l'Hermite, Ed. D. Reidel Publishing Co., Dordrecht, Netherlands.

Dwyer, D.F., M.L. Krumme, S.A. Boyd, and J.M. Tiedje. 1986. Kinetics of phenol biodegradation by an immobilized methanogenic consortium. Appl. Environ. Microbiol. 52: 345–351.

Dwyer, D.F., F. Rojo, and K.N. Timmis. 1988. Fate and behaviour in an activated sludge of a genetically-engineered microorganism designed to degrade substituted aromatic compounds, pp. 77–88, in: *The Release of Genetically-Engineered Microorganisms*, M. Sussman, C.H. Collins, F.A. Skinner, and D.E. Stewart-Tull, Eds. Academic, London.

Dwyer, D.F., and J.M. Tiedje. 1983. Degradation of ethylene glycol and polyethylene glycols by methanogenic consortia. Appl. Environ. Microbiol. 46: 185–190.

Eaton, J.W., C.F. Kolpin, and H.S. Swofford. 1973. Chlorinated urban water: A cause of dyalysis-induced hemolytic anemia. Science 181: 463–464.

Eccles, H., and S. Hunt, Eds. 1986. *Immobilization of Ions by Bio-Sorption*. Ellis Horwood, Chichester, U.K.

Edberg, S.C., M.J. Allen, D.B. Smith, and the National Collaborative Study. 1988. National field evaluation of a defined substrate method for the simultaneous enumeration of total coliforms and *Escherichia coli* from drinking water: Comparison with the standard multiple-tube fermentation method. Appl. Environ. Microbiol. 54: 1595–1601.

Edberg, S.C., M.J. Allen, and D.B. Smith. 1989. Rapid, specific autoanalytical method for the simultaneous detection of total coliforms and *E. coli* from drinking water. Water Sci. Technol. 21: 173–177.

Edberg, S.C., M.J. Allen, D.B. Smith, and N.J. Kriz. 1990. Enumeration of total coliforms and *Escherichia coili* from source water by the defined substrate technology. Appl. Environ. Microbiol. 56: 366–369.

Edeline, F. 1988. *L'Epuration Biologique des Eaux Residuaires: Theorie et Technologie*. Editions CEBEDOC, Liege, Belgium.

Edgehill, R.U., and R.K. Finn. 1983a. Activated sludge treatment of synthetic wastewater containing pentachlorophenol. Biotechnol. Bioeng. 25: 2165–2176.

Edgehill, R.U., and R.K. Finn. 1983b. Microbial treatment of soil to remove pentachlorophenol. Appl. Environ. Microbiol. 45: 1122–1125.

Edmonds, P. 1978. *Microbiology: An Environmental Perspective*, Macmillan, New York.

Edwards, T., and B.C. McBride. 1975. New method for the isolation and identification of methanogenic bacteria. Appl. Microbiol. 29: 540–545.

Egli, C., T. Tschan, R. Schlotz, A.M. Cook, and

T. Leisinger. 1988. Transformation of tetrachoromethane to dichloromethane and carbon dioxide by *Acetobacterium woodii*. Appl. Environ. Microbiol. 54: 2819–2823.

Ehrlich, H.L. 1981. *Geomicrobiology*. Marcel Dekker, New York.

Eighmy, T.T., M.R. Collins, S.K. Spanos, and J. Fenstermacher. 1992. Microbial populations, activities and carbon metabolism in slow sand filters. Water Res. 26: 1319–1328.

Eighmy, T.T., D. Maratea, and P.L. Bishop. 1983. Electron microscopic examination of wastewater biofilm formation and structural components. Appl. Environ. Microbiol. 45: 1921–1931.

Eikelboom, D.H. 1975. Filamentous organisms observed in activated sludge. Water Res. 9: 365–388.

Eikelboom, D.H., and H.J.J. van Buijsen. 1981. *Microscopic Sludge Investigation Manual*. Report No. A94a, TNO Research Institute, Delft, The Netherlands.

Eisenberg, T.N., E.J Middlebrooks, and V.D. Adams. 1987. Sensitizer photooxidation for wastewater disinfection and detoxification. Water Sci. Technol. 19: 1225–1258.

Eisenhardt, A., E. Lund, and B. Nissen. 1977. The effect of sludge digestion on virus infectivity. Water Res. 11: 579–581.

Ellis, J., and W. Korth. 1993. Removal of geosmin and methylisoborneol from drinking water by adsorption on ultrastable zeolite-Y. Water Res. 27: 535–539.

El-Sharkawi, F., L. El-Attar, A. Abdel Gawad, and S. Molazem. 1989. Some environmental factors affecting survival of fecal pathogens and indicator organisms in seawater. Water Sci. Technol. 21: 115–120.

Engelbrecht, R.S. 1983. Source, testing and distribution. In: *Assessment of Microbiology and Turbidity Standards for Drinking Water*, P.S. Berger and Y. Argaman, Eds. EPA-570–9-83001.

Engelbrecht, R.S., D.H. Foster, E.O. Greening, and S.H. Lee. 1974. *New Microbial Indicators of Wastewater Chlorination Efficiency*. EPA-670/2–73/082, Cincinnati, Ohio.

Enzinger, R.M., and R.C. Cooper. 1976. Role of bacteria and protozoa in the removal of *E. coli* from estuarine waters. Appl. Environ. Microbiol. 31: 758–763.

Epstein, E. 1979. In: *Workshop on the Health and Legal Implications of Sewage Sludge Composting*, Vol. 2. Energy Resources Co., Cambridge, MA.

Ercolani, G.L. 1976. Bacteriological quality assessment of fresh marketed lettuce and fennel. Appl. Environ. Microbiol. 31: 847–852.

Ericksen, T.H., and A.P. Dufour. 1986. Methods to identify water pathogens and indicator organisms, pp. 195–214, in: *Waterborne Diseases in the United States*, G.F. Craun, Ed. CRC Press, Boca Raton, FL.

van Etten, J.L., L.C. Lane, and R.H. Meints. 1991. Viruses and viruslike particles of eucaryotic algae. Microbiol. Rev. 55: 586–620.

Evans, P.J., D.T. Mang, K.S. Kim, and L.Y. Young. 1991. Anaerobic degradation of toluene by a denitrifying bacterium. Appl. Environ. Microbiol. 57: 1139–1145.

Evans, T.M., J.E. Schillinger, and D.G. Stuart. 1978. Rapid determination of bacteriological water quality by using *limulus* lysate. Appl. Environ. Microbiol. 35: 376–382.

Falconer, I.R. 1989. Effect on human health of some toxic cyanobacteria (blue-green algae) in reservoirs, lakes and rivers. Toxicity Assess. 4: 175–184.

Falconer, I.R., and T.H. Buckley. 1989. Tumour promotion by *Microcystis* sp., a blue-green alga occurring in water supplies. Med. J. Aust. 150: 351.

Fannin, K.F., S.C. Vana, and W. Jakubowski. 1985. Effect of an activated sludge wastewater treatment plant on airborne air densities of aerosols containing bacteria and viruses. Appl. Environ. Microbiol. 49: 1191–1196.

Farooq, S., and S. Akhlaque. 1983. Comparative response of mixed cultures of bacteria and virus to ozonation. Water Res. 17: 809–812.

Farooq, S., C.S. Kurucz, T.D. Waite, W.J. Cooper, S.R. Mane, and J.H. Greenfield. 1992. Treatment of wastewater with high enegy electron beam irradiation. Water Sci. Technol. 26: 1265–1274.

Farquhar, G.J., and W.C. Boyle. 1972. Control of *Thiothrix* in activated sludge. J. Water Pollut. Control Fed. 44: 14–19.

Farrah, S.R., and G. Bitton. 1982. Methods (other than microporous filters) for concentration of viruses from water, pp. 117–149, in: *Methods in Environmental Virology*, C.P. Gerba and S.M. Goyal, Eds. Marcel Dekker, New York.

Farrah, S.R., and G. Bitton. 1984. Enteric bacteria in aerobically digested sludge. Appl. Environ. Microbiol. 47: 831–834.

Farrah, S.R., and G. Bitton. 1990. Viruses in the soil

environment, pp. 529–556, in: *Soil Biochemistry*, Vol. 6, J-M Bollag and G. Stotzky, Eds. Marcel Dekker, New York.

Farrah, S.R., G. Bitton, E.M. Hoffmann, O. Lanni, O.C. Pancorbo, M.C. Lutrick, and J.E. Bertrand. 1981. Survival of enteroviruses and coliform bacteria in a sludge lagoon. Appl. Environ. Microbiol. 41: 459–465.

Farrah, S.R., D.R. Preston, G.A. Toranzos, M. Girard, G.A. Erdos, and V. Vasuhdivan. 1991. Use of modified diatomaceous earth for removal and recovery of viruses in water. Appl. Environ. Microbiol. 57: 2502–2506.

Farrel, J.B., A.E. Erlap, J. Rickabaugh, D. Freedman, and S. Hayes. 1988. Influence of feeding procedure on microbial reductions and performance on anaerobic digestion. J. Water Pollut. Control Fed. 60: 635–644.

Farrel, J.B., J. Smith, S. Hathaway, and R. Dean. 1974. Lime stabilization of primary sludges. J. Water Pollut. Control Fed. 46: 113–122.

Fathepure, B.Z., J.P. Nengu, and S.A. Boyd. 1987. Anaerobic bacteria that dechlorinate perchloroethene. Appl. Environ. Microbiol. 53: 2671–2674.

Fathepure, B.Z., J.M. Tiedje, and S.A. Boyd. 1988. Reductive dechlorination of hexachlorobenzene to tri- and dichlorobenzenes in anaerobic sewage sludge. Appl. Environ. Microbiol. 54: 327–330.

Fattal, B., M. Margalith, H.I. Shuval, and A. Morag. 1984. Community exposure to wastewater and antibody prevalence to several enteroviruses. In: *Proceedings of the Third Water Reuse Symposium*. American Water Works Association, Denver.

Fattal, B., M. Margalith, H.I. Shuval, Y. Wax, and A. Morag. 1987. Viral antibodies in agricultural populations exposed to aerosols from wastewater irrigation during a viral disease outbreak. Am. J. Epidemiol. 125: 899–906.

Fattal, B., R.J. Vasl, E. Katzenelson, and H.I. Shuval. 1983. Survival of bacterial indicator organisms and enteric viruses in the Mediterranean coastal waters off Tel-Aviv. Water Res. 17: 397–402.

Fattal, B., Y. Wax, M. Davies, and H.I. Shuval. 1986. Health risks associated with wastewater irrigation: An epidemiological study. Am. J. Public Health 76: 977–979.

Fayer, R., and B.L.P. Ungar. 1986. *Cryptosporidium* spp. and cryptosporidiosis. Microbiol. Rev. 50: 458–483.

Feachem, R.G., D.J. Bradley, H. Garelick, and D.D. Mara. 1983. *Sanitation and Disease: Health Aspect of Excreta and Wastewater Management*. Wiley, Chichester, U.K.

Federal Register. 1987. *National Primary Drinking Water Regulations: Filtration, Disinfection, Turbidity, Giardia lamblia, Viruses, Legionella and Heterotrophic Bacteria*. Proposed rule, 40 CFR parts 141 and 142. Fed. Reg. 52:212: 42718 (Nov. 3, 1987).

Fedorak, P.M., and S.E. Hrudey. 1984. The effects of phenols and some alkyl phenolics on batch anaerobic methanogenesis. Water Res. 18: 361–367.

Fedorak, P.M., and S.E. Hrudey. 1988. Anaerobic degradation of phenolic compounds with applications to treatment of industrial wastewaters, pp. 169–225, in: *Biotreatment Systems*, Vol. 1, D.L. Wise, Ed. CRC Press, Boca Raton, FL.

Fedorak, P.M., D.J. Roberts, and S.E. Hrudey. 1986. The effects of cyanide on the methanogenic degradation of phenolic compounds. Water Res. 20: 1315–1320.

Fedorak, P.M., and D.W.S. Westlake. 1980. Airborne bacterial densities at an activated sludge treatment plant. J. Water Pollut. Control Fed. 52: 2185–2192.

Feiler, H. 1980. *Fate of Priority Pollutants in Publicly Owned Treatment Works*. EPA-440/1-80-301, U.S. EPA, Washington, DC.

Fenchel, T.M., and B.B. Jorgensen. 1977. Detritus food chain in aquatic ecosystems. Adv. Microb. Ecol. 1: 1–58.

Feng, P.C.S., and P.A. Hartman. 1982. Fluorogenic assay for immediate confirmation of *Escherichia coli*. Appl. Environ. Microbiol. 43: 1320–1329.

Ferguson, D.W., M.J. McGuire, B. Koch, R.L. Wolfe, and E.M. Aieta. 1990. Comparing PEROXONE and ozone for controlling taste and odor compounds, disinfection by-products and microorganisms. J. Am. Water Works Assoc. 82: 181–191.

Fernandez, A., C. Tejedor, and A. Chordi. 1992. Effect of different factors on the die-off of fecal bacteria in a stabilization pond purification plant. Water Res. 26: 1093–1098.

Field, J.A., E. de Jong, G.F. Costa, and J.A.M. de Bont. 1992. Biodegradation of polycyclic aromatic hydrocarbons by new isolates of white rot fungi. Appl. Environ. Microbiol. 58: 2219–2226.

Field, J.A., and G. Lettinga. 1987. The methanogenic

toxicity and anaerobic degradability of a hydrolyzable tannin. Water Res. 21: 367–374.

Fields, B.S., E.B. Shotts, Jr., J.C. Feeley, G.W. Gorman, and W.T. Martin. 1984. Proliferation of *Legionella pneumophila* as an intracellular parasite of the ciliated protozoan *Tetrahymena pyriformis*. Appl. Environ. Microbiol. 47: 467–471.

Fiksdal, L., J.S. Maki, S.J. LaCroix, and J.T. Staley. 1985. Survival and detection of *Bacteroides* spp., prospective indicator bacteria. Appl. Environ. Microbiol. 49: 148–150.

Finance, C., F. Lucena, M. Briguaud, M. Aymard, R. Pares, and L. Schwartzbrod. 1982. Etude quantitative et qualitative de la pollution virale de l'eau de mer a Barcelone. Rev. Fr. Sci. Eau 1: 139–149.

Finch, G.R., and N. Fairbairn. 1991. Comparative inactivation of poliovirus type 3 and MS2 coliphage in demand-free phosphate buffer by using ozone. Appl. Environ. Microbiol. 57: 3121–3126.

Finlay, B.B., and S. Falkow. 1989. Common themes in microbial pathogenicity. Microbiol. Rev. 53: 210–230.

Finstein, M.S., J. Cirello, D.J. Suler, M.L. Morris, and P.F. Strom. 1980. Microbial ecosystems responsible for anaerobic digestion and composting. J. Water Pollut. Control Fed. 52: 2675–2685.

Fitzgerald, P.R. 1982. *Proceedings of the Symposium on Microbial Health Considerations of Soil Disposal of Wastewaters*, University of Oklahoma, Norman, Oklahoma, pp. 101–120. EPA-600/9–83–017.

Fitzmaurice, G.D., and N.F. Gray. 1989. Evaluation of manufactured inocula for use in the BOD test. Water Res. 23: 655–657.

Flewett, T.H. 1982. Clinical features of rotavirus infections, pp. 125–137, in: *Virus Infections of the Gastrointestinal Tract*, D.A. Tyrell and A.Z. Kapikian, Eds. Marcel Dekker, New York.

Fliermans, C.B., W.B. Cherry, L.H. Orrison, S.J. Smith, and L. Thacker. 1979. Isolation of *Legionella pneumophila* from nonepidemic aquatic habitats. Appl. Environ. Microbiol. 37: 1239–1242.

Fliermans, C.B., W.B. Cherry, L.H. Orrison, S.J. Smith, and D.H. Pope. 1981. Ecological distribution of *Legionella pneumophila*. Appl. Environ. Microbiol. 41: 9–16.

Fliermans, C.B., and R.S. Harvey. 1984. Effectiveness of 1-bromo-3-chloro-5,5-dimethylhydantoin against *Legionella pneumophila* in a cooling tower. Appl. Environ. Microbiol. 47: 1307–1310.

Florentz, M., and P. Granger. 1983. Phosphorus-31 nuclear magnetic resonance of activated sludge: Use for the study of the biological removal of phosphates from wastewater. Environ. Technol. Lett. 4: 9–12.

Focht, D.D., and W. Verstraete. 1977. Biochemical ecology of nitrification and denitrification. Adv. Microb. Ecol. 1:135–214.

Fogarty, A.M., and O.H. Tuovinen. 1991. Microbiological degradation of pesticides in yard waste composting. Microbiol. Rev. 55: 225–233.

Fogel, M.M., A.R. Taddeo, and S. Fogel. 1986. Biodegradation of chlorinated ethenes by a methane-utilizing mixed culture. Appl. Environ. Microbiol. 50: 720–724.

Ford, T., and R. Mitchell. 1990. The ecology of microbial corrosion. Adv. Microb. Ecol. 11: 231–262.

Forster, C.F., and J. Dallas-Newton. 1980. Activated sludge settlement—Some suppositions and suggestions. Water Pollut. Control 79: 338–351.

Forster, C.F., and D.W.M. Johnston. 1987. Aerobic processes, pp. 15–56, in: *Environmental Biotechnology*, C.F. Forster and D.A.J. Wase, Eds. Ellis Horwood, Chichester, U.K.

Forster, C.F., and E. Senior. 1987. Solid waste, pp. 176–233, in: *Environmental Biotechnology*, C.F. Forster and D.A.J. Wase, Eds. Ellis Horwood, Chichester, U.K.

Forster, C.F., and D.A.J. Wase, Eds. 1987. *Environmental Biotechnology*, Ellis Horwood, Chichester, U.K.

Foster, D.H., and R.S. Engelbrecht. 1973. Microbial hazards in disposing of wastewater on soil, pp. 247–259, in: *Recycling Treated Municipal Wastewater and Sludge Through Forest and Cropland*, W.E. Sopper and L.T. Kardos, Eds. The Pennsylvania State University Press, University Park, PA.

Foster, D.M., D.S. Walsh, and O.J. Sproul. 1980. Ozone inactivation of cell-and fecal-associated viruses and bacteria. J. Water Pollut. Control Fed. 52: 2174–2184.

Foster, S.O., E.L. Palmer, G.W. Gary, M.L. Martin, K.L. Herrmann, P. Beasley, and J. Sampson. 1980. Gastroenteritis due to rotavirus in an isolated Pacific Island group: An epidemic of 3,439 cases. J. Infect. Dis. 141: 32–35.

Fox, A.B., and G.W. Hambricks. 1984. Recreationally associated *Pseudomonas aeruginosa* folliculitis: Report of an epidemic. Arch. Dermatol. 120: 1304–1307.

Fox, J.L. 1989. Contemplating suicide genes in the environment. ASM News 55: 259–261.

Fox, K.R., R.J. Miltner, G.S. Logsdon, D.L. Dicks, and L.F. Drolet. 1984. Pilot-plant studies of slow-rate filtration. J. Am. Water Works Assoc. 76: 62–68.

Frederickson, J.K., F.J. Brockman, D.J. Workman, S.W. Li, and T.O. Stevens. 1991. Isolation and characterization of a subsurface bacterium capable of growth on toluene, naphthalene, and other aromatic compounds. Appl. Environ. Microbiol. 55: 796–803.

Frias, J., F. Ribas, and F. Lucena. 1992. A method for the measurement of biodegradable organic carbon in waters. Water Res. 26: 255–258.

Friedman, B.A., P.R. Dugan, R.A. Pfister, and C.C. Remsen. 1969. Structure of exocellular polymers and their relationship to bacterial flocculation. J. Bacteriol. 98: 1328–1334.

Friello, D.A., J.R. Mylroie, and A.M. Chakrabarty. 1976. Use of genetically engineered microorganisms for rapid degradation of fuel hydrocarbons, pp. 205–213, in: *Proceedings of the International Biodegradation Symposium*, J.M. Sharpley and A.M. Kaplan, Eds. Applied Science Publishers, London.

Frostell, B. 1981. Anaerobic treatment in a sludge bed system compared with a filter system. J. Water Pollut. Control Fed. 53: 216–222.

Fuhs, G.W., and M. Chen. 1975. Microbiological basis of phosphate removal in the activated sludge process for the treatment of wastewater. Microb. Ecol. 2: 119–138.

Fujioka, R.S., H.H. Hashimoto, E.B. Siwak, and R.H.F. Young. 1981. Effect of sunlight on survival of indicator bacteria in seawater. Appl. Environ. Microbiol. 41: 690–696.

Funderburg, S.W., and C.A. Sorber. 1985. Coliphages as indicators of enteric viruses in activated sludge. Water Res. 19: 547–555.

Funderburg, S.W., B.E. Moore, C.A. Sorber, and B.P. Sagik. 1978. Survival of poliovirus in model wastewater holding pond. Prog. Water Technol. 10: 619–629.

Gainesville Regional Utilities. 1985. *On-site Systems for Wastewater Treatment in the Gainesville Urban Area*. Report prepared for the Water Management Advisory Commission to the Gainesville City Commission.

Galun, M., P. Keller, D. Malki, H. Feldstein, E. Galun, S.M. Siegel, and B.Z. Siegel. 1982. Removal of uranium (VI) from solution by fungal biomass and fungal wall-related biopolymers. Science 219: 285–286.

Gameson, A.L.H., and D.J. Gould. 1975. Effects of solar radiation on the mortality of some terrestrial bacteria in seawater, pp. 209–219, in: *Discharge of Sewage from Sea Outfalls*, A.L.H. Gameson, Ed. Pergamon Press, London.

Garcia, J.L., S. Roussos, and M. Bensoussan. 1981. Etudes taxonomiques des bacteries denitrifiantes isolees sur benzoate dans les sols de rizieres du Senegal. Cahiers ORSTOM Ser. Biol. 43: 13–25.

Gaudy, A.F., Jr. 1972. Biochemical oxygen demand, pp. 305–332, in: *Water Pollution Microbiology*, R. Mitchell, Ed. Wiley, New York.

Gaudy, A.F., Jr. and E.T. Gaudy. 1988. *Elements of Bioenvironmental Engineering*. Engineering Press, San Jose, CA.

Gauthier, M.J., and V.A. Briettmayer. 1990. Regulation of gene expression in *E. coli* cells starved in seawater: Influence on their survival in marine environments. Paper presented at the International Symposium on Health-Related Water Microbiology, Tubingen, West Germany, April 1–6, 1990.

Gauthier, M.J., G.N. Flatau, D. Le Rudulier, R.L. Clement, and M.-P. Combarro. 1991. Intracellular accumulation of potassium and glutamate specifically enhances survival of *Escherichia coli* in seawater. Appl. Environ. Microbiol. 57: 272–276.

Gauthier, M.J., G.N. Flatau, R.L. Clement, and P.M. Munro. 1992. Sensitivity of *Escherichia coli* cells to seawater closely depends on their growth stage. J. Appl. Bacteriol. 73: 257–262.

Gavaghan, P.D., J.L. Sykora, W. Jakubowski, C.A. Sorber, L.W. Casson, A.M. Sninsky, M.D. Lichte, and G. Keleti. 1992. Viability and infectivity of *Giardia* and *Cryptosporidium* in digested sludge. Paper presented at the International Water Pollution Research Conference International Symposium, Washington, DC, May 26–29, 1992.

Gealt, M.A., M.D. Chai, K.B. Alpert, and J.C. Boyer. 1985. Transfer of plasmid pBR322 and pBR325 in wastewater from laboratory strains of *Escherichia coli* to bacteria indigenous to the waste disposal system. Appl. Environ. Microbiol. 49: 836–841.

Geldenhuys, J.C., and P.D. Pretorius. 1989. The occurrence of enteric viruses in polluted water, correlation to indicators organisms and factors influencing their numbers. Water Sci. Technol. 21: 105–109.

Geldreich, E.E. 1980. Microbiological processes in water supply distribution (seminar). Annu. Meet. Am. Soc. Microbiol., Miami Beach, FL.

Geldreich, E.E. 1990. Microbiological quality of source waters for water supply, pp. 3–31, in: *Drinking Water Microbiology*, G.A. McFeters, Ed. Springer-Verlag, New York.

Geldreich, E.E., M.J. Allen, and R.H. Taylor. 1978. Interferences to coliform detection in potable water supplies, pp. 13–20, in: *Evaluation of the Microbiology Standards for Drinking Water*, C.W. Hendricks, Ed. U.S. EPA, Washington, DC.

Geldreich, E.E., K.R. Fox, J.A. Goodrich, E.W. Rice, R.M. Clark, and D.L. Swerdlow. 1992. Searching for a water supply connection in the Cabool, Missouri disease outbreak of *Escherichia coli* 0157:H7. Water Res. 26: 1127–1137.

Geldreich E.E., and B.A. Kenner. 1969. Concepts of fecal streptococci in stream pollution. J. Water Pollut. Control Fed. 41: R336-R341.

Geldreich, E.E., and D.J. Reasoner. 1990. Home treatment devices and water quality, pp. 147–167, in: *Drinking Water Microbiology*, G.A. McFeters, Ed. Springer-Verlag, New York.

Geldreich, E.E., and E.W. Rice. 1987. Occurrence, significance, and detection of *Klebsiella* in water systems. J. Am. Water Works Assoc. 79: 74–80.

Geldreich, E.E., R.H. Taylor, J.C. Blannon, and D.J. Reasoner. 1985. Bacterial colonization of point-of-use water treatment devices. J. Am. Water Works Assoc. 77: 72–80.

Gerba, C.P. 1984. Applied and theoretical aspects of virus adsorption to surfaces. Adv. Appl. Microbiol. 30: 133–168.

Gerba, C.P. 1987a. Phage as indicators of fecal pollution, pp. 197–209, in: *Phage Ecology*, S.M. Goyal, C.P. Gerba and G. Bitton, Eds., Wiley Interscience, New York.

Gerba, C.P. 1987b. Recovering viruses from sewage, effluents, and water, pp. 1–23, in: *Methods for Recovering Viruses from the Environment*, G. Berg, Ed. CRC Press, Boca Raton, FL.

Gerba, C.P., and G. Bitton. 1984. Microbial pollutants: Their survival and transport pattern to groundwater, in: *Groundwater Pollution Microbiology*, G. Bitton and C.P. Gerba, Eds. Wiley, New York.

Gerba, C.P., S.M. Goyal, C.J. Hurst, and R.L. LaBelle. 1980. Type and strain dependence of enterovirus adsorption to activated sludge, soils and estuarine sediments. Water Res. 14: 1197–1198.

Gerba, C.P., and C.N. Haas. 1986. Risks associated with enteric viruses in drinking water, pp. 460–468, in: *Progress in Chemical Disinfection*, G.E. Janauer, Ed. State University of New York, Binghamton, N.Y.

Gerba, C.P., and J.S. McLeod. 1976. Effect of sediments on the survival of *Escherichia coli* in marine waters. Appl. Environ. Microbiol. 32: 114–120.

Gerba, C.P., and G. Schaiberger. 1975. Effect of particulates on virus survival in seawater. J. Water Pollut. Control Fed. 47: 93–103.

Gerba, C.P., S.N. Singh, and J.B. Rose. 1985. Waterborne gastroenteritis and viral hepatitis. CRC Crit. Rev. Environ. Control 15: 213–236.

Gerba, C.P., Stagg, C.H., and M.G. Abadie. 1978. Characterization of sewage solids-associated viruses and behavior in natural waters. Water Res. 12: 805–812

Gerba, C.P., and R.B. Thurman. 1986. *Evaluation of the Efficacy of Microdyn Against Waterborne Bacteria and Viruses*. Monograph, University of Arizona, Tucson, AZ.

Gerba, C.P., C. Wallis, and J.L. Melnick. 1975a. Viruses in water: The problem, some solutions. Environ. Sci. Technol. 9: 1122–1126.

Gerba, C.P., C. Wallis, and J.L. Melnick. 1975b. Fate of wastewater bacteria and viruses in soils. J. Irrig. Drainage Div., ASCE 3: 157–168.

Gerba, C.P., C. Wallis, and J.L. Melnick. 1977. Disinfection of wastewater by photodynamic action. J. Water Pollut. Control Fed. 49: 578–583.

Gerber, N.N. 1979. Volatile substances from actinomycetes: Their role in the odor pollution of water. Crit. Rev. Microbiol. 7: 191–214.

Ghosh, S., and D.L. Klass. 1978. Two-phase anaerobic digestion. Proc. Biochem. 13: 15–24.

Ghosh, S., J.P. Ombregt, and P. Pipyn. 1985. Methane production from industrial wastes by two-phase anaerobic digestion. Water Res. 19: 1083–1088.

Gibson, D.T., and V. Subramanian. 1988. Microbial degradation of aromatic hydrocarbons, pp. 181–193, in: *Microbial Degradation of Organic Com-*

pounds, D.T. Gibson, Ed. Marcel Dekker, New York.

Giesy, J.P., R.L. Craney, J.L. Newsted, C.J. Rosiu, A. Benda, R.G. Kreis, and F.J. Horvath. 1988. Comparison of three sediments bioassay methods using Detroit River sediments. Environ. Toxicol. Chem. 7:483–498.

Giger, W., and P.V. Roberts. 1978. Characterization of persistent organic carbon, pp. 135–175, in: *Water Pollution Microbiology*, Vol. 2, R. Mitchell, Ed. Wiley, New York.

Gilpin, R.W. 1984. Laboratory and field applications of U.V light disinfection on six species of *Legionella* and other bacteria in water. Paper presented at the Second International Symposium of the American Society for Microbiology, Washington, DC.

van Ginkel, C.G., J. Tramler, K. Ch. A.M. Luyben, and A. Klapwijk. 1983. Characterization of *Nitrosomonas europaea* immobilized in calcium alginate. Enzyme Microb. Technol. 5: 297–303.

Girones, R., A. Allard, G. Wadell, and J. Jofre. 1993. Application of PCR for the detection of adenovirus in polluted waters. Water Sci. Technol. 27: 235–241.

Glass, J.S., and R.T. O'Brien. 1980. Enterovirus and coliphage inactivation during activated sludge treatment. Water Res. 14: 877–882.

Gloyna, E.F. 1971. *Waste Stabilization Ponds*. WHO Monograph Series #60. WHO, Geneva.

Glynn, W., C. Baker, A. LoRe, and A. Quaglieri. 1987. *Mobile Waste Processing Systems and Treatment Technologies*. Noyes Data Corp., Park Ridge, NJ.

Goddard, A.J., and C.F. Forster. 1987a. Stable foams in activated sludge plants. Enzyme Microb. Technol. 9: 164–168.

Goddard, A.J., and C.F. Forster. 1987b. A further examination into the problem of stable foams in activated sludge plants. Microbios 50: 29–42.

Goddard, M.R., J. Bates, and M. Butler. 1981. Recovery of indigenous enteroviruses from raw and digested sludges. Appl. Environ. Microbiol. 42: 1023–1028.

Godfrey, A.J., and L.E. Bryan. 1984. Intrinsic resistance and whole cell factors contributing to antibiotic resistance, pp. 113–145, in: *Antimicrobial Drug Resistance*, L.E. Bryan, Ed. Academic, Orlando, FL.

Goldstein, N. 1988. Steady growth for sludge composting. Biocycle 29(10): 29–43.

Gotaas, H.B. 1956. *Composting: Sanitary Disposal and Reclamation of Organic Wastes*. WHO Monograph Series #31. WHO, Geneva.

Goyal, S.M. 1987. Methods in phage ecology, pp. 267–287, in: *Phage Ecology*, S.M. Goyal, C.P. Gerba and G. Bitton, Eds. Wiley Interscience, New York.

Goyal, S.M., W.N. Adams, M.L. O'Malley, and D.W. Lear. 1984. Human pathogenic viruses at sewage disposal sites in the middle Atlantic region. Appl. Environ. Microbiol. 48: 758–763.

Goyal, S.M., and C.P. Gerba. 1982a. Concentration of viruses from water by membrane filters, pp. 59–116, in: *Methods in Environmental Virology*, C.P. Gerba and S.M. Goyal, Eds. Marcel Dekker, New York.

Goyal, S.M., and C.P. Gerba. 1982b. Occurrence of endotoxins in groundwater during land application of wastewater. J. Environ. Sci. Health A17: 187–196.

Goyal, S.M., and C.P. Gerba. 1983. Viradel method for detection of rotavirus from seawater. J. Virol. Methods 7: 279–285.

Goyal, S.M., C.P. Gerba, and G. Bitton, Eds. 1987. *Phage Ecology*, Wiley Interscience, New York.

Goyal, S.M., C.P. Gerba, and J.L. Melnick. 1977. Occurrence and distribution of bacterial indicators and pathogens in canal communities along the Texas coast. Appl. Environ. Microbiol. 34: 139–149.

Goyal, S.M., C.P. Gerba, and J.L. Melnick. 1979. Human enteroviruses in oysters and their overlying water. Appl. Environ. Microbiol. 37: 572–581.

Goyal, S.M., K.S. Zerda, and C.P. Gerba. 1980. Concentration of coliphage from large volumes of water and wastewater. Appl. Environ. Microbiol. 39: 85–91.

Grabow, W.O.K. 1986. Indicator systems for assessment of the virological safety of treated drinking water. Water Sci. Technol. 18: 159–165.

Grabow, W.O.K. 1990. Microbiology of drinking water treatment: Reclaimed wastewater, pp. 185–203, in: *Drinking Water Microbiology*, G.A. McFeters, Ed. Springer-Verlag, New York.

Grabow, W.O.K., J.S. Burger, and E.M. Nupen. 1980. Evaluation of acid-fast bacteria, *Candida albicans*, enteric viruses and conventional indicators for monitoring wastewater reclamation systems. Prog. Water Technol. 12: 803–817.

Grabow, W.O.K, and P. Coubrough. 1986. Practical direct plaque assay for coliphages in 100 ml samples of drinking water. Appl. Environ. Microbiol. 52: 430–433.

Grabow, N.A., and R. Kfir. 1990. Growth of *Legionella* bacteria in activated carbon filters. Paper presented at the International Symposium on Health-Related Water Microbiology, Tubingen, West Germany, April 1–6, 1990.

Grabow, W.O.K., and O.W. Prozesky. 1973. Drug resistance of coliform bacteria in hospital and city sewage. Antimicrob. Agents Chemother. 3: 175–180.

Grace, R.D., N.E. Dewar, W.G. Barnes, and G.R. Hodges. 1981. Susceptibility of *Legionella pneumophila* to three cooling towers microbicides. Appl. Environ. Microbiol. 41: 233–236.

Grady, C.P.L. 1986. Biodegradation of hazardous wastes by conventional biological treatments. Haz. Wastes Haz. Mater. 3: 333–365.

Grady, C.P.L., Jr., and H.C. Lim. 1980. *Biological Waste Treatment*. Marcel Dekker, New York.

Grant, W.D., and P.E. Long. 1981. *Environmental Microbiology*. Halsted Press, New York.

Grbic-Galic, D., and T.M. Vogel. 1987. Transformation of toluene and benzene by mixed methanogenic cultures. Appl. Environ. Microbiol. 53: 254–260.

Grimes, D.J., F.L. Singleton, J. Stemmler, L.M. Palmer, P. Brayton, and R.R. Colwell. 1984. Microbiological effects of wastewater effluent discharge into coastal waters of Puerto Rico. Water Res. 18: 613–619.

van Groenestijn, J.W., G.J.F.M. Vlekke, D.M.E. Anink, M.H. Deinema, and A.J.B. Zehnder. 1988a. Role of cations in accumulation and release of phosphate by *Acinetobacter* strain 210A. Appl. Environ. Microbiol. 54: 2894–2901.

van Groenestijn, J.W., M.M.A. Bentvelsen, M.H. Deinema, and A.J.B. Zehnder. 1988b. Polyphosphate-degrading enzymes in *Acinetobacter* spp. and activated sludge. Appl. Environ. Microbiol. 55: 219–223.

Groethe, D.R., and J.G. Eaton. 1975. Chlorine-induced mortality in fish. Trans. Am. Fish Soc. 104: 800–805.

Grotenhuis, J.T.C., M. Smit, C.M. Plugge, X. Yuansheng, A.A.M. van Lammeren, A.J.M. Stams, and A.J.B. Zehnder. 1991. Bacteriological composition and structure of granular sludge adapted to different substrates. Appl. Environ. Microbiol. 57: 1942–1949.

Grubbs, R.B. 1984. Panel discussion: Emerging industrial applications, pp. 331–349, in: *Genetic Control of Environmental Pollutants*, G.S. Omenn and A. Hollaender, Eds. Plenum, New York.

Grubbs, T.R., and F.W. Pontius. 1992. USEPA releases draft ground water disinfection rule. J. Am. Water Works Assoc. 84: 25–31.

Guest, H. 1987. *The world of Microbes*. Science Tech. Pubs., Madison, WI.

Gujer, W., and A.J.B. Zehnder. 1983. Conversion processes in anaerobic digestion. Water Sci. Technol. 15: 127–134.

Gunnerson, C.G., H.I. Shuval, and S. Arlosoroff. 1984. Health effects of wastewater irrigation and their control in developing countries. In: *Proceedings of the Third Water Reuse Symposium*. American Water Works Association, Denver, CO.

Guthrie, M.A., E.J. Kirsch, R.F. Wukasch, and C.P.L. Grady, Jr. 1984. Pentachlorophenol biodegradation. II. Anaerobic. Water Res. 18: 451–461.

Haas, C.N., and R.S. Engelbrecht. 1980. Physiological alterations of vegetative microorganisms resulting from chlorination. J. Water Pollut. Control Fed. 52: 1976–1989.

Haas, C.N., M.G. Kerallus, D.M. Brncich, and M.A. Zapkin. 1986. Alteration of chemical and disinfectant properties of hypochlorite by sodium, potassium and lithium. Environ. Sci. Technol. 20: 822–826.

Haas, C.N., M.A. Meyer, M.S. Paller, and M.A. Zapkin. 1983. The utility of endotoxins as surrogate indicator in potable water microbiology. Water Res. 17: 803–807.

Haas, C.N., B.F. Severin, D. Roy, R.S. Engelbrecht, A. Lalchandani, and S. Farooq. 1985. Field observations on the occurrence of new indicators of disinfection efficiency. Water Res. 19: 323–329.

de Haas, D.W. 1989. Fractionation of bioaccumulated phosphorus compounds in activated sludge. Water Sci. Technol. 21: 1721–1725.

Haber, C.L, L.N. Allen, S. Zhao, and R.S. Hanson. 1983. Methylotrophic bacteria: Biochemical diversity and genetics. Science 221: 1147–1153.

Hackel, U., J. Klein, R. Megnet, and F. Wagner. 1975. Immobilization of cells in polymeric matrices. Eur. J. Appl. Microbiol. 1: 291–294.

Hagedorn, C. 1984. Microbiological aspects of groundwater pollution due to septic tanks, pp. 181–195, in: *Groundwater Pollution Microbiology*, G. Bitton and C.P. Gerba, Eds. Wiley, New York.

Haggblom, M.M., and L.Y. Young. 1990. Chlorophenol degradation coupled to sulfate reduction. Appl. Environ. Microbiol. 56: 3255–3260.

Hain, K.E., and R.T. O'Brien. 1979. *The Survival of Enteric Viruses in Septic Tanks and Septic Tank Drain Fields*. Water Resources Research Institute Report No. 108. New Mexico Water Resources Research Institute, New Mexico State University, Las Cruces, NM.

Hall, J.C., and R.J. Foxen. 1983. Nitrification in BOD_5 test increases POTW noncompliance. J. Water Pollut. Control Fed. 55: 1461–1469.

Hall, R.M., and M.D. Sobsey. 1993. Inactivation of hepatitis A virus (HAV) and MS-2 by ozone and ozone-hydrogen peroxide in buffered water. Water Sci. Technol. 27: 371–378.

Halvorson, H.O., D. Pramer, and M. Rogul, Eds. 1985. *Engineered Organisms in the Environment: Scientific Issues*. American Society for Microbiology, Washington, DC.

Hamelin, C., F. Sarhan, and Y.S. Chung. 1978. Induction of deoxyribonucleic acid degradation in *Escherichia coli* by ozone. Experientia 34: 1578–1579.

Hamilton, W.A. 1985. Sulphate anaerobic bacteria and anaerobic corrosion. Annu. Rev. Microbiol. 39: 195–217.

Hamilton, W.A. 1987. Biofilms: Microbial interactions and metabolic activities, pp. 361–385, in: *Ecology of Microbial Communities*, M. Fletcher, T.R.G. Gray, and J.G. Jones, Eds. Cambridge University Press, Cambridge, U.K.

Hammer, M.J. 1986. *Water and Wastewater Technology*. Wiley, New York.

Hanaki, K., Z. Hong, and T. Matsuo. 1992. Production of nitrous oxide gas during denitrification of wastewater. Water Sci. Technol. 26: 1027–1036.

Hanaki, K., and C. Polprasert. 1989. Contribution of methanogenesis to denitrification in an upflow filter. J. Water Pollut. Control Fed. 61: 1604–1611.

Hancock, R.E.W. 1984. Alterations in outer membrane permeability. Annu. Rev. Microbiol. 38: 237–264.

Handzel, T.R., R.M. Green, C. Sanchez, H. Chung, and M.D. Sobsey. 1993. Improved specifity in detecting F-specific coliphages in environmental samples by suppression of somatic phages. Water Sci. Technol. 27: 123–131.

Hanel, L. 1988. *Biological Treatment of Sewage by the Activated Sludge Process*. Ellis Horwood, Chichester, U.K.

Hannah, S.A., B.M. Austern, A.E. Eralp, and R.H Wise. 1986. Comparative removal of toxic pollutants by six wastewater treatment processes. J. Water Pollut. Control Fed. 58: 27–34.

Hannah, S.A., B.M. Austern, A.E. Eralp, and R.A. Dobbs. 1988. Removal of organic toxic pollutants by trickling filter and activated sludge. J. Water Pollut. Control Fed. 60: 1281–1283.

Hanstveit, A.O., W.J. Th. van Gemert, D.B. Janssen, W.H. Rulkens, and H.J. van Veen. 1988. Literature study on the feasibility of microbiological decontamination of polluted soils, pp. 63–128, in: *Biotreatment Systems*, Vol. 1, D.L. Wise, Ed. CRC Press, Boca Raton, FL.

Harakeh, M. 1985. Factors influencing chlorine disinfection of wastewater effluent contaminated by rotaviruses, enteroviruses, and bacteriophages, pp. 681–690, in: *Water Chlorination: Chemistry, Environmental Impact and Health Effects*, Vol. 5, R.L. Joley, Ed. Lewis Pub., Chelsea, MI.

Hardman, D.J. 1987. Microbial control of environmental pollution: The use of genetic techniques to engineer organisms with novel catabolic capabilities, pp. 295–317, in: *Environmental Biotechnology*, C.F. Forster and D.A.J. Wase, Eds. Ellis Horwood, Chichester, U.K.

Harremoes, P. 1978. Biofilm kinetics, pp. 71–109, in: *Water Pollution Microbiology*, Vol. 2, R. Mitchell, Ed. Wiley, New York.

Harremoes, P., and M.H. Christensen. 1971. Denitrifikation med methan. Vand 1: 7–11.

Harris, G.D., V.D. Adams, D.L. Sorensen, and R.R. Dupont. 1987. The influence of photoreactivation and water quality on ultraviolet disinfection of secondary municipal wastewater. J. Water Pollut. Control Fed. 59: 781–787.

Harris, J.R. 1986. Clinical and epidemiological characteristics of common infectious diseases and chemical poisonings caused by ingestion of contaminated drinking water, pp. 11–21, in: *Waterborne Diseases in the United States*, G.F. Craun, Ed. CRC Press, Boca Raton, FL.

Harris, R.H., and R. Mitchell. 1973. The role of polymers in microbial aggregation. Annu. Rev. Microbiol. 27: 27–50.

Harrison, J.R., and G.T. Daigger. 1987. A comparison of trickling filter media. J. Water Pollut. Control Fed. 59: 679–685.

Hart, C.A., D. Baxby, and N. Blundell. 1984. Gastroenteritis due to *Cryptosporidium*: A prospective survey in a children's hospital. J. Infect. 9: 264–270.

Hartmann, L., and G. Laubenberger. 1968. Toxicity measurements in activated sludge. J. Sanit. Eng. Div. Am. Soc. Civ. Eng. 94: 247–252.

Hartmans, S., J.A.M. de Bont, J. Tramper, and K.C.A.M. Luyben. 1985. Bacterial degradation of vinyl chloride. Biotechnol. Lett. 7: 383–388.

Harvey, R.W., and S.P. Garabedian. 1991. Use of colloid filtration theory in modeling movement of bacteria though a contaminated sandy aquifer. Environ. Sci. Technol. 25: 178–185.

Haufele, A., and H.V. Sprockhoff. 1973. Ozone for disinfection of water contaminated with vegetative and spore forms of bacteria, fungi and viruses. Zentralbl. Bakteriol. Hyg. Abt. 1, Orig. Reihe B 175: 53–70.

Havelaar, A. H., M. Bosman, and J. Borst. 1983. Otitis externa by *Pseudomonas aeruginosa* associated with whirlpools. J. Hyg. 90: 489–498.

Havelaar, A.H., and W.M. Hogeboom 1983. Factors affecting the enumeration of coliphage in sewage and sewage-polluted waters. Antonie van Leeuwenhoek J. Microbiol. 49: 387–397.

Havelaar, A.H., W.M. Hogeboom, K. Furuse, R. Pot, and M.P. Hormann. 1990a. F-specific RNA bacteriophages and sensitive host strains in faeces and wastewater of human and animal origin. J. Appl. Bacteriol. 69: 30–37.

Havelaar, A.H., C.C.E. Meulemans, W.M. Pot-Hogeboom, and J. Koster. 1990b. Inactivation of bacteriophage MS2 in wastewater effluent with monochromatic and polychromatic ultraviolet light. Water Res. 24: 1387–1391.

Hawkes, H.A. 1983a. Activated sludge. In: *Ecological Aspects of Used Water Treatment*, Vol. 1, C.R. Curds and H.A. Hawkes, Eds. Academic, London.

Hawkes, H.A. 1983b. Stabilization ponds. In: *Ecological Aspects of Used Water Treatment*, Vol. 2, C.R. Curds and H.A. Hawkes, Eds. Academic, London.

Hayes, E.B., T.D. Matte, T.R. O'Brien, T.W. McKinley, G.S. Logsdon, J.B. Rose, B.L.P. Ungar, D.M. Word, P.F. Pinsky, M.L. Cummings, M.A. Wilson, E.G. Long, E.S. Hurwittz, and D.D. Juranek. 1989. Large community outbreak of cryptosporidiosis due to contamination of a filtered public water supply. N. Engl. J. Med. 320: 1372–1376.

Hayes, K.P., and M.D. Burch. 1989. Odorous compounds associated with algal blooms in south australian waters. Water Res. 23: 115–121.

Hazen, T.C. 1988. Fecal coliforms as indicators in tropical waters. Toxicity Assess. 3: 461–477.

Heijnen, J.J., and J.A. Roels. 1981. A macroscopic model describing yield and maintenance relationships in aerobic fermentation. Biotechnol. Bioeng. 23: 739–741.

Heitkamp, M.A., V. Camel, T.J. Reuter, and W.J. Adams. 1990. Biodegradation of *p*-nitrophenol in an aqueous waste stream by immobilized bacteria. Appl. Environ. Microbiol. 56: 2967–2973.

Hejkal, T.W., F.M. Wellings, P.A. LaRock, and A.L. Lewis. 1979. Survival of poliovirus within organic solids during chlorination. Appl. Environ. Microbiol. 38: 114–118.

Hejzlar, J., and J. Chudoba. 1986. Microbial polymers in the aquatic environment: I. Production by activated sludge microorganisms under different conditions. Water Res. 20: 1209–1216.

Hemmes, J.H., K.C. Winkler, and S.M. Kool. 1960. Virus survival as a seasonal factor in influenza and poliomyelitis. Nature 188: 430–431.

Henney, R.C., M.C. Fralish, and W.V. Lacina. 1980. Shock load of chromium (VI). J. Water Pollut. Control Fed. 52: 2755–2760.

Henry, J.G., and R. Gehr. 1980. Odor control an operator's guide. J. Water Pollut. Control Fed. 52: 2523–2537.

Henson, J.M., P.H. Smith, and D.C. White. 1989. Examination of thermophilic methane-producing digesters by analysis of bacterial lipids. Appl. Environ. Microbiol. 50: 1428–1433.

Herbst, E., I. Scheunert, W. Klein, and F. Korte. 1977. Fate of PCB-^{14}C in sewage treatment-laboratory experiments with activated sludge. Chemosphere 6: 725–730.

Hernandez, J.F., J.M. Guibert, J.M. Delattre, C. Oger, C. Charriere, B. Hugues, R. Serceau, and F. Sinegre. 1990. A miniaturized fluorogenic assay for enumeration of *E. coli* and enterococci in marine waters (Abstract) In: International Symposium on Health-Related Water Microbiology, Tubingen, West Germany, April 1–6, 1990.

Hernandez, J.F., J.M. Guibert, J.M. Delattre, C.

Oger, C. Charriere, B. Hugues, R. Serceau, and F. Sinegre. 1991. Evaluation d'une methode miniaturisee de denombrement des *Escherichia coli* en eau de mer, fondee sur l'hydrolyse du 4-methylumbelliferyl ß-D-glucuronide. Water Res. 25: 1073–1078.

Hernandez, J.F., A.M. Pourcher, J.M. Delattre, C. Oger, and J.L. Loeuillard. 1993. MPN miniaturized procedure for the enumeration of faecal streptococci in fresh and marine waters: The MUST procedure. Water Res. 27: 597–606.

Herson, D.S., B. McGonigle, M.A. Payer, and K.H. Baker. 1987. Attachment as a factor in the protection of *Enterobacter cloacae* from chlorination. Appl. Environ. Microbiol. 53: 1178–1180.

Herwaldt, B.L., G.F. Craun, S.L. Stokes, and D.D. Juranek. 1992. Outbreaks of waterborne disease in the United States: 1989–90. J. Am. Water Works Assoc. 84: 129–135.

Hetrick, F.M. 1978. Survival of human pathogenic viruses in estuarine and marine waters. ASM News 44: 300–303.

Hibler, C.P., C.M. Hancock, L.M. Perger, J.G. Wegrzyn, and K.D. Swabby. 1987. Inactivation of *Giardia* cysts with chlorine at 0.5°C to 5.0°C. Report to the American Waterworks Association Research Foundation, Denver.

Hibler, C.P., and C.M. Hancock. 1990. Waterborne giardiasis, pp. 271–293, in: *Drinking Water Microbiology*, G.A. McFeters, Ed. Springer-Verlag, New York.

Hickey, C.W., C. Blaise, and G. Costan. 1991. Microtesting appraisal of ATP and cell recovery toxicity end points after acute exposure of *Selenastrum capricornutum* to selected chemicals. Environ. Toxicol. Water Qual. 6: 383–403.

Hickey, R.F., J. Vanderwielen, and M.S. Switzenbaum. 1987. The effect of organic toxicants on methane production and hydrogen gas levels during the anaerobic digestion of waste activated sludge. Water Res. 21: 1417–1427.

Higham, D.P., P.J. Sadler, and M.D. Schawen. 1984. Cadmium resistant *Pseudomonas putida* synthesizes novel cadmium binding proteins. Science 225: 1043–1046.

Hill, R.T., I.T. Knight, M.S. Anikis, and R.R. Col well. 1993. Benthic distribution of sewage sludge-indicated by *Clostridium perfringens* at a deep-ocean dump site. Appl. Environ. Microbiol. 59: 47–51.

Himberg, K., A.M. Keijola, L. Hiisvirta, H. Pyysalo, and K. Sivonen. 1989. The effect of water treatment processes on the removal of hepatoxins from *Microcystis* and *Oscillatoria* cyanobacteria: A laboratory study. Water Res. 23: 979–984.

Hinzelin, F., and J.C. Block. 1985. Yeast and filamentous fungi in drinking water. Environ. Lett. 6: 101–106.

Hiraishi, A., K. Masamune, and H. Kitamura. 1989. Characterization of the bacterial population structure in an anaerobic–aerobic activated sludge system on the basis of respiratory quinone profiles. Appl. Environ. Microbiol. 55: 897–901.

Hitdlebaugh, J.A., and R.D. Miller. 1981. Operational problems with rotating biological contactors. J. Water Pollut. Control Fed. 53: 1283–1293.

Ho, C.-F., and D. Jenkins. 1991. The effect of surfactants on *Nocardia* foaming in activated sludge. Water Sci. Technol. 23: 879–887.

Hoehn, R.C. 1965. Biological methods for the control of tastes and odors. Southwest Water Works J. 47: 26.

Hoehn, R.C., D.B. Barnes, B.C. Thompson, C.W. Randall, T.J. Grizzard, and P.T.B. Shaffer. 1980. Algae as sources of trihalomethane precursors. J. Am. Water Works Assoc. 72: 344–350.

Hoff, J.C. 1978. The relationship of turbidity to disinfection of potable water, pp. 103–117, in: *Evaluation of the Microbiology Standards for Drinking Water*, C.W. Hendricks, Ed. EPA-570/9–78/00C. U.S. EPA, Washington, DC.

Hoff, J.C. 1986. *Inactivation of Microbial Agents by Chemical Disinfectants*. EPA-600/2–86/067. U.S. EPA, Water Engineering Research Laboratory, Cincinnati, OH.

Hoff, J.C., and E.W. Akin. 1986. Microbial resistance to disinfectants: Mechanisms and significance. Environ. Health Perspect. 69: 7–13.

Hoffman, R.L., and R.M. Atlas. 1977. Measurement of the effect of cadmium stress on protozoan grazing of bacteria (bacterivory) in activated sludge by fluorescence microscopy. Appl. Environ. Microbiol. 53: 2440–2444.

Holdeman, L.V., I.J. Good, and W.E.C. Moore. 1976. Human fecal flora: Variation in human fecal composition within individuals and a possible effect of emotional stress. Appl. Environ. Microbiol. 31: 359–375.

Holms, H.W., and J.W. Vennes. 1970. Occurrence of

sulfur purple bacteria in a sewage treatment lagoon. Appl. Microbiol. 19: 988–996.

van Hoof, F., J.G. Janssens, and H. van Dyck. 1985. Formation of mutagenic activity during surface water preozonation and their behaviour in water treatment. Chemosphere 14: 501–510.

Hooper, S.W. 1987. Characterization of pSS50, the *4-Chlorobiphenyl Mineralization Plasmid*, Doctoral dissertation, University of Tennessee, Knoxville.

Horan, N.J., and C.R. Eccles. 1986. Purification and characterization of extracellular polysaccharide from activated sludges. Water Res. 20: 1427–1432.

Horn, J.B., D.W. Hendricks, J.M. Scanlan, L.T. Rozelle, and W.C. Trnka. 1988. Removing *Giardia* cysts and other particles from low-turbidity waters using dual-stage filtration. J. Am. Water Works Assoc. 80: 68–77.

Horvath, R.S. 1972. Microbial co-metabolism and the degradation of organic compounds in nature. Bacteriol. Rev. 36: 146–155.

Houghton, S.R., and D.D. Mara. 1992. The effect of sulphide generation in waste stabilization ponds on photosynthetic populations and effluent quality. Water Sci. Technol. 26: 1759–1768.

Houston, J., B.N. Dancer, and M.A. Learner. 1989a. Control of sewage filter flies using *Bacillus thuringiensis* var *israelensis*. I. Acute toxicity tests and pilot scale trial. Water Res. 23: 369–378.

Houston, J., B.N. Dancer, and M.A. Learner. 1989b. Control of sewage filter flies using *Bacillus thuringiensis* var *israelensis*. II. Full scale trials. Water Res. 23: 379–385.

Howgrave-Graham, A.R., and P.L. Steyn. 1988. Application of the fluorescent-antibody technique for the detection of *Sphaerotilus natans* in activated sludge. Appl. Environ. Microbiol. 54: 799–802.

Hoyana, Y., V. Bacon, R.E. Summons, W.E. Pereira, B. Helpern, and A.M. Duffield. 1973. Chlorination studies: IV. Reactions of aqueous hypochlorous acid with pyrimidine and purine bases. Biochem. Biophys. Res. Commun. 53: 1195–2001.

Hozalski, R.M., S. Goel, and E.J. Bouwer. 1992. Use of biofiltration for removal of natural organic matter to achieve biologically stable drinking water. Water Sci. Technol. 26: 2011–2014.

Hsu, S.C., R. Martin, and B.B. Wentworth. 1984. Isolation of *Legionella* species from drinking water. Appl. Environ. Microbiol. 48: 830–832.

Hu, M., L. Kang, and G. Yao. 1989. An outbreak of viral hepatitis A in Shangai, pp. 361–372, in: *Infectious Diseases of the Liver*. Proceedings of the 54th Falk Symposium, Oct. 12–14, 1989, Basel, Swizerland.

Hu, P., and P.F. Strom. 1991. Effect of pH on fungal growth and bulking in laboratory activated sludges. Res. J. Water Pollut. Control Fed. 63: 276–277.

Huang, J.C., and V.T. Bates. 1980. Comparative performance of rotating biological contactors using air and pure oxygen. J. Water Pollut. Control Fed. 52: 2686–2703.

Huang, J.Y.C., G.E. Wilson, and T.W. Schroepfer. 1979. Evaluation of activated carbon adsorption for sewer odor control. J. Water Pollut. Control Fed. 51: 1054–1062.

Huck, P.M. 1990. Measurement of biodegradable organic matter and bacterial growth potential in drinking water. J. Am. Water Works Assoc. 82: 78–86.

Hugues, B., A. Cini, M. Plissier, and J.R. Lefebvre. 1980. Recherche des virus dans le milieu marin a partir d'echantillon de volumes differents. Eau Quebec 13: 199–203.

Hugues, B., J.R. Lefebvre, M. Plissier, and A. Cini. 1981. Distribution of viral and bacterial densities in seawater near a coastal discharge of treated domestic sewage. Zentralbl. Bakt. Hyg. I. Abt. Orig. B 173: 509–516.

Huisman, L., and W.E. Wood. 1974. *Slow Sand Filtration*. World Health Organization, Geneva.

Huk, A., R.R. Colwell, R. Rahman, A. Ali, M.A.R. Chowdhury, S. Parveen, D.A. Sack, and E. Russek-Cohen. 1990. Detection of *Vibrio cholerae* 01 in the aquatic environment by fluorescent-monoclonal antibody and culture methods. Appl. Environ. Microbiol. 56: 2370–2373.

Hulshoff Pol, L.W., W.J. de Zeeuw, C.T.M. Velzeboer, and G. Lettinga. 1982. Granulation in UASB reactors. Water Sci. Technol. 15: 291–305.

Hulshoff Pol, L.W., J. Dolfing, W.J. de Zeeuw, and G. Lettinga. 1983. Cultivation of well adapted pelletized methanogenic sludge. Biotechnol. Lett. 5: 329–332.

Hunen, W.A.M, and D. van der Kooij. 1992. The effect of low concentrations of assimilable organic carbon (AOC) in water on biological clogging of sand beds. Water Res. 26: 963–972.

Hunter, G.V., G.R. Bell, and C.N. Henderson. 1966. Coliform organism removal by diatomite filtration. J. Am. Works Assoc. 58: 1160–1169.

Hurst, C.J., and D.A. Brashear. 1987. Use of a vacuum filtration technique to study leaching of indigenous viruses from raw wastewater sludge. Water Res. 21: 809–812.

Hurst, C.J., C.P. Gerba, and I. Cech. 1980. Effects of environmental variables and soil characteristics on virus survival in soil. Appl. Environ. Microbiol. 40: 1067–1079.

Huser, B.A., K. Wuhrmann, and A.J.B. Zehnder. 1982. *Methanothrix soehngenii* gen. nov. sp. nov., a new acetotrophic non-hydrogen-oxidizing bacterium. Arch. Microbiol. 132: 1–9.

Hussong, D., W.G. Burge, and N.K. Enkiri. 1985. Occurrence, growth and suppression of salmonellae in composted sewage sludge. Appl. Environ. Microbiol. 50: 887–893.

Hutchins, S.R. 1989. Biorestoration of fuel-contaminated aquifer using nitrate: Laboratory studies. Paper presented at the Tenth Annual Meeting of the Society of Environmental Toxicology and Chemistry, Toronto, Canada, Oct. 28–Nov. 2, 1989.

Hutchins, S.R. 1991. Biodegradation of monoaromatic hydrocarbons by aquifer microoorganisms using oxygen, nitrate, or nitrous oxide as the terminal electron acceptor. Appl. Environ. Microbiol. 57: 2403–2407.

Hutchins, S.R., G.W. Sewell, D.A. Kovacs, and G.A. Smith. 1991. Biodegradation of aromatic hydrocarbons by aquifer microorganisms under denitrifying conditions. Environ. Sci. Technol. 25: 68–76.

Ida, S., and M. Alexander. 1965. Permeability of *Nitrobacter agilis* to organic compounds. J. Bacteriol. 90: 151–155.

Ijaz, M.K., and S.A Sattar. 1985. Comparison of airborne survival of calf rotavirus and poliovirus type 1 (Sabin) aerosolized as a mixture. Appl. Environ. Microbiol. 49: 289–295.

Inamori, Y., Y. Kuniyasu, R. Sudo, and M. Koga. 1991. Control of the growth of filamentous microorganisms using predacious ciliated protozoa. Water Sci. Technol. 23: 963–971.

Inamori, Y., K. Matsusige, R. Sudo, K. Chiba, H. Kikuchi, and T. Ebisuno. 1989. Advanced wastewater treatment using an immobilized microorganism/biofilm two-step process. Water Sci. Technol. 21: 1755–1758.

International Reference Centre for Waste Disposal (IRCWD). 1985. Health aspects of wastewater and excreta use in agriculture and aquaculture: The Engelberg report. IRCWD News 23: 11–18, Dubendorf, Switzerland.

Isaac-Renton, J.L., C.P.J. Fung, and A. Lochan. 1986. Evaluation of tangential-flow multiple-filter technique for detection of *Giardia lamblia* cysts in water. Appl. Environ. Microbiol. 52: 400–402.

Isaacson, M., A.R. Sayed, and W.H.J. Hattingh. 1986. *Studies on Health Aspects of Water Reclamation during 1974 to 1983 in Windhoek, South West Africa/Namibia*. WRC Report No. 38/1/86. Water Research Commission, Pretoria.

Ishizaki, K., K. Sawadaishi, K. Miura, and N. Shinriki. 1987. Effect of ozone on plasmid DNA of *Escherichia coli in situ*. Water Res. 21: 823–827.

Ishizaki, K., N. Shinriki, and T. Ueda. 1984. Degradation of nucleic acids with ozone. V. Mechanism of action of ozone on deoxyribonucleoside 5'-monophosphates. Chem. Pharm. Bull. 32: 3601–3606.

Iverson, W.P., and F.E. Brinckman. 1978. Microbial metabolism of heavy metals, pp. 201–232, in: *Water Pollution Microbiology*, Vol. 2, R. Mitchell, Ed. Wiley, New York.

Izaguirre, G., C.J. Hwang, S.W. Krasner, and M.J. McGuire. 1982. Geosmin and 2-methylisoborneol from cyanobacteria in three water supply systems. Appl. Environ. Microbiol. 43: 708–714.

Jacobsen, B.N., N. Nyholm, B.M. Pedersen, O. Poulsen, and P. Ostfeldt. 1991. Microbial degradation of pentachlorophenol and lindane in laboratory-scale activated sludge reactors. Water Sci. Technol. 23: 349–356.

Jacobson, S.N., N.L. O'Mara, and M. Alexander. 1980. Evidence of cometabolism in sewage. Appl. Environ. Microbiol. 40: 917–921.

Jagger, J. 1958. Photoreactivation. Bacteriol. Rev. 22: 99–114.

Jahn, T.L., and F.L. John. 1949. *The Protozoa*. Wm. C. Brown, Dubuque, IA.

Jain, R.K., R.S. Burlage, and G.S. Sayler. 1988. Methods for detecting recombinant DNA in the environment. Crit. Rev. Biotechnol. 8: 33–47.

Jakubowski, W. 1986. USEPA-sponsored epidemiological studies of health risks associated with the treatment and disposal of wastewater and sewage sludge, pp. 140–153, in: *Epidemiological Studies of Risks Associated With Agricultural Use of Sewage Sludge: Knowledge and Needs*, J.C. Block, A.H. Havelaar, and P. l'Hermite, Eds. Elsevier Applied Science, London.

Jakubowski, W., and T.H. Ericksen. 1979. Methods of detection of *Giardia* cysts in water supplies. In: *Waterborne Transmission of Giardiasis*. EPA-600/9–79–001.

Jakubowski, W., and J.C. Hoff, Eds. 1979. Waterborne transmission of giardiasis. Proceedings of a symposium. EPA-600/9–79–001.

Jakubowski, W., J.L. Sykora, C.A. Sorber, L.W. Casson, and P.D. Cavaghan. 1990. Determining giardiasis prevalence by examination of sewage. International Symposium on Health-Related Water Microbiology, Tubingen, West Germany, April 1–6, 1990.

Jansons, J., L.W. Edmonds, B. Speight, and M.R. Bucens. 1989a. Movement of viruses after artificial recharge. Water Res. 23: 293–299.

Jansons, J., L.W. Edmonds, B. Speight, and M.R. Bucens. 1989b. Survival of viruses in groundwater. Water Res. 23: 301–306.

Jarroll, E.F., J.C. Hoff, and E.A. Meyer. 1984. Resistance of cysts to disinfection agents, pp. 311–328, in: *Giardia and Giardiasis*, S.L. Erlandsen and E.A. Meyer, Eds. Plenum, New York.

Jawetz, E., J.L. Melnick, and E.A. Adelberg. 1984. *Review of Medical Microbiology* (16th ed.). Lange Medical Publishers, Los Altos, CA.

Jeffrey, H.C., and R.M. Leach. 1972. *Atlas of Medical Helminthology and Protozoology*. Churchill Livingstone, Edinburgh.

Jehl-Pietri, C. 1992. Detection des virus enteriques dans le milieu hydrique: Cas du virus de l'hepatite dans l'environnement marin et les coquillages. Doctoral dissertation, Universite de Nancy 1, Nancy, France.

Jenkins, D. 1992. Towards a comprehensive model of activated sludge bulking and foaming. Water Sci. Technol. 25: 215–230.

Jenkins, D., M.G. Richard, and G.T. Daigger. 1984. *Manual on the Causes and Control of Activated Sludge Bulking and Foaming*. Water Research Commission, Pretoria, South Africa.

Jenkins, D., and M.G. Richard. 1985. The causes and control of activated-sludge bulking. Tappi 68: 73–76.

Jenkins, R.L., J.P. Gute, S.W. Krasner, and R.B. Baird. 1980. The analysis and fate of odorous sulfur compounds in wastewaters. Water Res. 14: 441–448.

Jenkins, S.H., D.G. Keight, and A. Ewins. 1964. The solubility of heavy metal hydroxides in water, sewage and sewage sludge. II. The precipitation of metals by sewage. Int. J. Air Water Pollut. 8: 679–693.

Jewell, W.J. 1987. Anaerobic sewage treatment. Environ. Sci. Technol. 21: 14–20.

Jewell, W.J., and R.M. Kabrick. 1980. Autoheated aerobic thermophilic digestion with aeration. J. Water Pollut. Control Fed. 52: 512–523.

Jewell, W.J., Y.M. Nelson, and M.S. Wilson. 1992. Methanotrophic bacteria for nutrient removal from wastewater: Attached film system. Water Environ. Res. 64:756–765.

Jofre, J. 1991. Les bacteriophages dans les milieux hydriques, pp. 253–275, in: *Virologie des Milieux Hydriques,* L. Schwartzbrod, Ed. TEC & DOC Lavoisier, Paris, France.

Jofre, J., M. Blasi, A. Bosch, and F. Lucena. 1989. Occurrence of bacteriophages infecting *Bacteroides fragilis* and other viruses in polluted marine sediments. Water Sci. Technol. 21: 15–19.

Johnson, D.E. Camann, J.W. Register, R.E. Thomas, C.A. Sorber, M.N. Guentzel, J.M. Taylor, and H.J. Harding. 1980. The evaluation of microbiological aerosols associated with the application of wastewater to land: Pleasanton, CA. EPA-600/1-80-015. U.S. EPA, Cincinnati, OH.

Johnson, D.W., N.J. Pienazek, and J.B. Rose. 1992. DNA probe hybridization and PCR detection of *Cryptosporidium* compared to immunofluorescence assay. Water Sci. Technol. 27: 77–84.

Johnson, D.W., N.J. Pienazek, and J.B. Rose. 1993. DNA probe hybridization and PCR detection of *Cryptosporidium* compared to immunofluorescence assay. Water Sci. Technol. 27: 77–84.

Johnston, J.B., and S.G. Robinson. 1982. Opportunities for development of new detoxification processes through genetic engineering, pp. 301–314, in: *Detoxification of Hazardous Wastes*, J.H. Exner, Ed. Ann Arbor Science, Ann Arbor, MI.

Johnston, J.B., and S.G. Robinson. 1984. *Genetic Engineering and the Development of New Pollution Control Technologies*. EPA 600/2-84-037.

Johnson, L.M., C.S. McDowell, and M. Krupta. 1985. Microbiology in pollution control: From bugs to biotechnology. Dev. Ind. Microbiol. 26: 365–376.

Johnson, W.D. 1991. Dual distribution systems: The public utility perspective. Water Sci. Technol. 24: 343–352.

Jolley, R.L., W.A. Brungs, and R.B. Cummings, Eds. 1985. *Water Chlorination: Chemistry, Environmental Impact, and Health Effects*. Lewis Pubs., Chelsea, MI.

Jones, B.R. 1956. Studies of pigmented non-sulfur purple bacteria in relation to cannery waste lagoon odors. Sewage Ind. Wastes 28: 883–893.

Jones, P.H., A.D. Tadwalkar, and C.L. Hsu. 1987. Enhanced uptake of phosphorus by activated sludge: Effect of substrate addition. Water Res. 21: 301–308.

Jones, W.L., and E.D. Schroeder. 1989. Use of cell-free extracts for the enhancement of biological wastewater treatment. J. Water Pollut. Control Fed. 61: 60–65.

Jordan, E.C. 1982. Fate of priority pollutants in publicly owned treatment works. 30-day study. EPA-440/1–82/302, Washington, DC.

Joret, J.C., P. Cervantes, Y. Levi, N. Dumoutier, L. Cognet, C. Hasley, M.O. Husson, and H. Leclerc. 1989. Rapid detection of *E. coli* in water using monoclonal antibodies. Water Sci. Technol. 21: 161–167.

Joret, J.-.C, A. Hassen, M.M. Bourbigot, F. Agbalika, P. Hartmann, and J.M Foliguet. 1986. Inactivation des virus dans l'eau sur une filiere de production à ozonation etagée. Water Res. 20: 871–876.

Joret, J.C., and Y. Levi. 1986. Method rapide d'évaluation du carbone éliminable des eaux par voie biologique. Tribune CEBEDEAU 510: 3–9.

Joret, J.C., Y. Levi, T. Dupin, and M. Gibert. 1988. Rapid method for estimating bioeliminable organic carbon in water. Paper presented at the American Water Works Association Conference, Orlando, FL, June 19–23, 1988.

Joret, J.C., Y. Levi, and C. Volk. 1990. Biodegradable dissolved organic carbon (BDOC) content of drinking water and potential regrowth of bacteria. Paper presented at the International Symposium on Health-Related Water Microbiology, Tubingen, West Germany, April 1–6, 1990.

Joret, J.-C., D. Perrine, and B. Langlais. 1992. Effect of temperature on the inactivation of *Cryptosporidium* oocysts by ozone. Paper presented at the International Water Pollution Research Conference International Symposium, Washington, DC, May 26–29, 1992.

Jorgensen, J.H., J.C. Lee, G.A. Alexander, and H.W. Wolf. 1979. Comparison of *Limulus* assay, standard plate count and total coliform count for microbiological assessment of renovated wastewater. Appl. Environ. Microbiol. 37: 928–931.

Juttner, F. 1981. Detection of lipid degradation products in the water of a reservoir during a bloom of *Synura uvella*. Appl. Environ. Microbiol. 41: 100–106.

Kabrick, R.M., and W.J. Jewell. 1982. Fate of pathogens in thermophilic aerobic sludge digestion. Water Res. 16: 1051–1060.

Kabrick, R.M., W.J. Jewell, B.V. Salotto, and D. Berman. 1979. Inactivation of viruses, pathogenic bacteria and parasites in the autoheated thermophilic digestion of sewage sludges. Proc. Ind. Waste Conf. Purdue Univ. 34:771–789.

Kanagawa, T., and D.P. Kelly. 1986. Breakdown of dimethylsulphide by mixed cultures and by *Thiobacillus thioparus*. FEMS Microbiol. Lett. 34: 13–19.

Kanagawa, T., and E. Mikami. 1989. Removal of methanethiol, dimethyl sulfide, dimethyl disulfide and hydrogen sulfide from contaminated air by *Thiobacillus thioparus* TK-m. Appl. Environ. Microbiol. 55: 555–558.

Kaneko, M. 1989. Effect of suspended solids on inactivation of poliovirus and T2-phage by ozone. Water Sci. Technol. 21: 215–219.

Kaneko, M., K. Morimoto, and S. Nambu. 1976. The response of activated sludge to a polychlorinated biphenyl (KC-500). Water Res. 10: 157–163.

Kao, J.F., L.P. Hsieh, S.S. Cheng, and C.P. Huang. 1982. Effect of EDTA on cadmium in activated sludge systems. J. Water Pollut. Control Fed. 54: 1118–1126.

Kaplan, J.E., G.W. Gary, and R.C. Baron. 1982. Epidemiology of Norwalk gastroenteritis and the role of Norwalk virus in outbreaks of acute nonbacterial gastroenteritis. Ann. Intern. Med. 96: 756–761.

Kaplan, L.A., T.L. Bott, and D.J. Reasoner. 1993. Evaluation and simplification of the assimilable organic carbon nutrient bioassay for bacterial growth in drinking water. Appl. Environ. Microbiol. 59: 1532–1539.

Kapuschinski, R.B., and R. Mitchell. 1981. Solar radiation induces sublethal injury in *Escherichia coli* in seawater. Appl. Environ. Microbiol. 41: 670–674.

Kapuschinski, R.B., and R. Mitchell, 1982. Sunlight-induced mortality of viruses and *Escherichia coli* in coastal seawater. Environ. Sci. Technol. 17: 1–6.

Karhadkar, P.P., J.M. Audic, G.M. Faup, and P. Khanna. 1987. Sulfide and sulfate inhibition of methanogenesis. Water Res. 21: 1061–1066.

Karimi, A.A., and P.C. Singer. 1991. Trihalomethane formation in open reservoirs. J. Am. Water Works Assoc. 83: 84–88.

Karns, J.S., M.T. Muldoon, W.W. Mulbry, M.K. Derbyshire, and P.C. Kearney. 1987. Use of microorganisms and microbial systems in the degradation of pesticides. in: *Biotechnology in Agricultural Chemistry*, H.M. LeBaron, R.O. Mumma, R.C. Honeycutt, and J.H. Duesing, Eds. ACS Symposium Series 334. American Chemical Society, Washington, DC.

Karube. I. 1987. Microorganism based sensors, pp. 13–29, in: *Biosensors: Fundamentals and Applications*, A.P.F. Turner, I. Karube, and G.S. Wilson, Eds. Oxford University Press, Oxford.

Karube, I., S. Kuriyama, T. Matsunaga, and S. Suzuki. 1980. Methane production from wastewaters by immobilized methanogenic bacteria. Biotechnol. Bioeng. 22: 847–857.

Karube, I., S. Mitsuda, T. Matsunaga, and S. Suzuki. 1977. A rapid method for estimation of BOD by using immobilized microbial cells. J. Ferment. Technol. 55: 243–246.

Karube, I., and E. Tamiya. 1987. Biosensors for environmental control. Pure Appl. Chem. 59: 545–554.

Katamay, M.M. 1990. Assessing defined-substrate technology for meeting monitoring requirements of the total coliform rule. J. Am. Water Works Assoc. 82: 83–87.

Kataoka, N., Y. Tokiwa, and K. Takeda. 1991. Improved technique for identification and enumeration of methanogenic bacterial colonies on roll tubes by epifluoresecnce microscopy. Appl. Environ. Microbiol. 57: 3671–3673.

Kato, K., and F. Kazama. 1991. Respiratory inhibition of *Sphaerotilus* by iron compounds and the distribution of the sorbed iron. Water Sci. Technol. 23: 947–954.

Katzenelson, E., I. Buium, and H.I. Shuval. 1976. Risk of communicable disease infection associated with wastewater irrigation in agricultural settlements. Science 194: 944–946.

Kay, G.P., G.L. Sykora, and R.A. Budgess. 1980. Algal concentration as a quality parameter of finished drinking waters in and around Pittsburgh, PA. J. Am. Water Works Assoc. 72: 170–176.

Kelly, S.M., and W.W. Sanderson. 1958. The effect of chlorine in water on enteric viruses. Am. J. Public Health. 48: 1323–1327.

Kemmy, F.A., J.C. Fry, and R.A. Breach. 1989. Development and operational implementation of a modified and simplified method for determination of assimilable organic carbon (AOC) in drinking water. Water Sci. Technol. 21: 155–159.

Kemp, H.A., D.B. Archer, and M.R.A. Morgan. 1988. Enzyme-linked immunosorbent assays for the specific and sensitive quantification of *Methanosarcina mazei* and *Methanobacterium bryantii*, Appl. Environ. Microbiol. 54: 1003–1008.

Kennedy, J.E., Jr., G. Bitton, and J.L. Oblinger. 1985. Comparison of selective media for assay of coliphages in sewage effluent and lake water. Appl. Environ. Microbiol. 49: 33–36.

Kennedy, M.S., J. Grammas, and W.B. Arbuckle. 1990. Parachlorophenol degradation using bioaugmentation. J. Water Pollut. Control Fed. 62: 227–233.

Keswick, B.H., N.R. Blacklow, G.C. Cukor, H.L. DuPont, and J.L. Vollet. 1982. Norwalk virus and rotavirus in travellers' diarrhea in Mexico. Lancet i: 110–111.

Ketratanakul, A., and S. Ohgaki. 1989. Indigenous coliphages and RNA-F-specific coliphages associated with suspended solids in the activated sludge process. Water Sci. Technol. 21: 73–78.

Kfir, R., M. du Preez, and B. Genthe. 1993. The use of monoclonal antibodies for the detection of faecal bacteria in water. Water Sci. Technol. 27: 257–260.

Khan, K.A., M.T. Suidan, and W.H. Cross. 1981. Anaerobic activated carbon filter for the treatment of phenol-bearing wastewater. J. Water Pollut. Control Fed. 53: 1519–1532.

Kiff, R.J., and D.R. Little. 1986. Biosorption of heavy metals by immobilized fungal biomass, pp. 71–80, in: *Immobilization of Ions by Bio-Sorption*, H. Eccles and S. Hunt, Eds. Ellis Horwood, Chichester, U.K.

Kilbanov, A.M., T.M. Tu, and K.P. Scott. 1983. Peroxidase catalyzed removal of phenols from coal-conversion wastewaters. Science 221: 259–261.

Kilvington, S., and D.G. White. 1985. Rapid identification of thermophilic *Naegleria* including *Naegleria fowleri* using API ZYM system. J. Clin. Pathol. 38: 1289–1292.

King, C.H., E.B. Shotts, Jr., R.E. Wooley, and K.G. Poter. 1988. Survival of coliform and bacterial pathogens within protozoa during chlorination. Appl. Environ. Microbiol. 54: 3023–3033.

King, E.F., and B.J. Dutka. 1986. Respirometric techniques, pp. 75–113, in: *Toxicity Testing Using Microorganisms*, Vol. 1, G. Bitton and B.J. Dutka, Eds. CRC Press, Boca Raton, FL.

King, G.M. 1988. Distribution and metabolism of quaternary amines in marine sediments, pp. 143–173, in: *Nitrogen Cycling in Costal Marine Environments*, T.H. Blackburn and J. Sorensen, Eds. Wiley, New York.

Kinner, N., D.L. Blackwill, and P.L. Bishop. 1983. Light and electron microscopic studies of microorganisms growing in rotating biological contactor biofilms. Appl. Environ. Microbiol. 45: 1659–1669.

Kinner, N.E., and C.R. Curds. 1989. Development of protozoan and metazoan communities in rotating biological contactor biofilms. Water Res. 23: 481–490.

Kirsop, B.H. 1984. Methanogenesis. Crit. Rev. Biotechnol. 1: 109–159.

Knudson, G.B. 1985. Photorectivation of UV-irradiated *Legionella pneumphila* and other *Legionella* species. Appl. Environ. Microbiol. 49: 975–980.

Kobayashi, H.A. 1984. Application of genetic engineering to industrial waste/wastewater treatment, pp. 195–214, in: *Genetic Control of Environmental Pollutants*, G.S. Omenn and A. Hollaender, Eds. Plenum, New York.

Kobayashi, H.A., M. Stenstrom, and R.A. Mah. 1983. Use of photosymthetic bacteria for hydrogen sulfide removal from anaerobic waste treatment effluent. Water Res. 17: 579–587.

Kodikara, C.P., H.H. Crew, and G.S.A.B. Stewart. 1991. Near on-line detection of enteric bacteria using *lux* recombinant bacteriophage. FEMS Microbiol. Lett. 83: 261–266.

Kofoed, A.D. 1984. Optimum use of sludge in agriculture, pp. 2–21, in: *Utilization of Sewage Sludge on Land: Rates of Application and Long-Term Effects of Metals*, S. Berglund, R.D. Davis, and P. L'Hermite, Eds. Reidel, Dordrecht.

Koh, S.-C., J.P. Bowman, and G.S. Sayler. 1993. Soluble methane monooxygenase production and trichloroethylene degradation by a type I methanotroph, *Methylomonas methanica* 68–1. Appl. Environ. Microbiol. 59: 960–967.

Kohring, G.W., X. Zhang, and J. Wiegel. 1989. Anaerobic dechlorination of 2,4-dichlorophenol in freshwater sediments in the presence of sulfate. Appl. Environ. Microbiol. 55: 2735–2737.

Koide, M., A. Saito, N. Kusano, and F. Higa. 1993. Detection of *Legionella* spp. in cooling tower water by the polymerase chain reaction method. Appl. Environ. Microbiol. 59: 1943–1946.

van der Kooij, D. 1983. Biological processes in carbon filters, pp. 119–152, in: *Activated Carbon in Drinking Water Technology*. Research Report, AWWA Research Foundation, Denver, CO.

van der Kooij, D. 1990. Assimilable organic carbon (AOC) in drinking water, pp. 57–87, in: *Drinking Water Microbiology*, G.A. McFeters, Ed. Springer-Verlag, New York.

van der Kooij, D. 1992. Assimilable organic carbon as an indicator of bacterial regrowth. J. Am. Water Works Assoc. 84:57–65.

van der Kooij, D., and W.A.M. Hijnen. 1981. Utilization of low concentrations of starch by a *Flavobacterium* species isolated from tap water. Appl. Environ. Microbiol. 41: 216–221.

van der Kooij, D., and W.A.M. Hijnen. 1984. Substrate utilization by an oxalate-consuming *Spirillum* species in relation to its growth in ozonated water. Appl. Environ. Microbiol. 47: 551–559.

van der Kooij, D., and W.A.M. Hijnen. 1985a. Determination of the concentration of maltose and starch-like compounds in drinking water by growth measurements with a well defined strain of *Flavobacterium* species. Appl. Environ. Microbiol. 49: 765–771.

van der Kooij, D., and W.A.M. Hijnen. 1985b. Measuring the concentration of easily assimilable organic carbon in water treatment as a tool for limiting regrowth of bacteria in distribution systems. Proceedings of the American Water Works Association Technological Conference.

van der Kooij, D., and W.A.M. Hijnen. 1988. Multiplication of a *Klebsiella pneumonae* strain in water at low concentration of substrate. In: *Proceedings of the International Conference on Water and Wastewater Microbiology*, Newport Beach, CA, Feb. 8–11, 1988, Vol. 1.

van der Kooij, D., J.P. Oranje, and W.A.M. Hijnen. 1982a. Growth of *Pseudomonas aeruginosa* in tap water in relation to utilization of substrates at concentrations of a few micrograms per liter. Appl. Environ. Microbiol. 44: 1086–1095.

van der Kooij, D., A. Visser, and W.A.M. Hijnen. 1982b. Determining the concentration of easily assimilable organic carbon in drinking water. J. Am. Water Works Assoc. 74: 540–545.

Koopman, B., G. Bitton, R.J. Dutton, and C.L. Logue. 1988. Toxicity testing in wastewater systems: Application of a short-term assay based on induction of the *lac* operon in *E. coli*. Water Sci. Technol. 20: 137–143.

Koopman, B., G. Bitton, J.J. Delfino, C. Mazidji, G. Voiland, and D. Neita. 1989. *Toxicity Screening in Wastewater Systems*. Final report (contract no. WM-222) to the Florida Department of Environmental Regulation, Tallahassee, FL.

Koopman, J.S., E.A. Eckert, H.B. Greenberg, B.C. Strohm, R.E. Isaacson, and A.S. Monto. 1982. Norwalk virus enteric illness aquired by swimming exposure. Am. J. Epidemiol. 115: 173.

Koramann, C., Bahnemann, D.W., and M.R. Hoffmann. 1991. Photolysis of chloroform and other molecules in aqueous TiO_2 suspensions. Environ. Sci. Technol. 25: 494–500.

Korich, D.G., J.R. Mead, M.S. Madore, and N.A. Sinclair. 1989. Effets of chlorine and ozone on *Cryptosporidium* oocyst viability. Abstr. Annu. Meet., Am. Soc. Microbiol., New Orleans.

Korich, D.G., J.R. Mead, M.S. Madore, N.A. Sinclair, and C.R. Sterling. 1990. Effect of ozone, chlorine dioxide, chlorine, and monochloramine on *Cryptosporidium parvum* oocyst viability. Appl. Environ. Microbiol. 56: 1423–1428.

Kornberg, S.R. 1957. Adenosine triphosphate synthesis from polyphosphate by an enzyme from *E. coli*. Biochim. Biophys. Acta 26:294–300.

Koster, I.W. 1988. Microbial, chemical and technological aspects of the anaerobic degradation of organic pollutants, pp. 285–316, in: *Biotreatment Systems*, Vol. 1, D.L. Wise, Ed. CRC Press, Boca Raton, FL.

Koster, I.W., and A. Cramer. 1987. Inhibition of methanogenesis from acetate in granular sludge by long-chain fatty acids. Appl. Environ. Microbiol. 53: 403–409

Koster, L.W., A. Rinzema, A.L. Vegt, and G. Lettinga. 1986. Sulfide inhibition of the methanogenic activity of granular sludge at various pH-levels. Water Res. 20: 1561–1567.

Kott, Y. 1966. Estimation of low numbers of *Escherichia coli* bacteriophage by use of the most probable number method. Appl. Microbiol. 14: 141–144.

Kott, Y., and N. Betzer. 1972. The fate of *Vibrio cholerae* (El Tor) in oxidation pond effluents. Israel J. Med. Sci. 8: 1912.

Kott, Y., N. Roze, S. Sperber, and N. Betzer. 1974. Bacteriophages as viral pollution indicators. Water Res. 8: 165–171.

Kowal, N.E. 1982. *Health Effects of Land Treatment: Microbiological Report*. EPA-600/1–82–007, U.S. EPA, Cincinnati, OH.

Kreft, P., M. Umphres, J.-M., Hand, C. Tate, M.J. McGuire, and R.R. Trussel. 1985. Converting from chlorine to chloramines: A case study. J. Am. Water Works Assoc. 77: 38–45.

Krumme, M.L., and S.A. Boyd. 1988. Reductive dechlorination of chlorinated phenols in anaerobic upflow bioreactor. Water Res. 22: 171–177.

Kuchenrither, R.D., and L.D. Benefield. 1983. Mortality patterns of indicator organisms during aerobic digestion. J. Water Pollut. Control Fed. 55: 76–80.

Kuchta, J.M., S.J. States, A.M. McNamara, R.M. Wadowsky, and R.B. Yee. 1983. Susceptibility of *Legionella pneumophila* to chlorine in tap water. Appl. Environ. Microbiol. 46: 1134–1139.

Kuhn, S.P., and R.M. Pfister. 1990. Accumulation of cadmium by immobilized *Zooglea ramigera* 115. J. Ind. Microbiol. 6: 123–128.

Kulikovsky, A., H.S. Pankratz, and S.L. Sadoff. 1975. Ultrastructural and chemical changes in spores of *Bacillus cereus* after action of disinfectants. J. Appl. Bacteriol. 38: 39–46.

Kumaran, P., and N. Shivaraman. 1988. Biological treatment of toxic industrial wastes, pp. 227–283, in: *Biotreament Systems*, Vol. 1, D.L. Wise Ed. CRC Press, Boca Raton, FL.

Kuter, G.A., H.A.J. Hoitink, and L.A. Rossman. 1985. Effect of aeration and temperature on composting of municipal sludge in a full-scale vessel system. J. Water Pollut. Control Fed. 57: 309–315.

Kwa, B.H., M. Moyad, M.A. Pentella, and J.B. Rose. 1993. A nude mouse model as an in vivo infectivity assay for cryptosporidiosis. Water Sci. Technol. 27: 65–68.

Kwan, K.K., and B.J. Dutka. 1990. Simple two-step sediment extraction procedure for use in genotoxicity and toxicity bioassays. Toxicity Assess. 5: 395–404.

Labatiuk, C.W., F.W. Schaefer III, G.R. Finch, and M. Belosevic. 1991. Comparison of animal infectivity, excystation, and fluorogenic dye as measures of *Giardia muris* cyst inactivation by ozone. Appl. Environ. Microbiol. 57: 3187–3192.

Laitinen, S., A. Nevalainen, M. Kotimaa, J. Liesivuori, and P.J. Marikainen. 1992. Relationship between bacterial counts and endotoxin concentration in the air of wastewater treatment plants. Appl. Environ. Microbiol. 58: 3474–3476.

Lal, R., and D.M. Saxena. 1982. Accumulation, metabolism, and effects of organochlorine insecticides on microorganisms. Microbiol. Rev. 46: 95–127.

Lalezary, S., M. Pirbazari, and M.J. McGuire. 1986. Oxidation of five earthy–musty taste and odor compounds. J. Am. Water Works Assoc. 78: 62–69.

Lalezary-Craig, S., M. Pirbazari, M.S. Dale, T.S. Tanaka, and M.J. McGuire. 1988. Optimizing the removal of geosmin and 2-methylisoborneol by powdered activated carbon. J. Am. Water Works Assoc. 80: 73–80.

Lamar, R.T., and D.M. Dietrich. 1990. *In situ* depletion of pentachlorophenol from contaminated soil by *Phanerochaete* spp. Appl. Environ. Microbiol. 56: 3093–3100.

Lamar, R.T., M.J. Larsen, and T.K. Kirk. 1990. Sensitivity to and degradation of pentachlorophenol by *Phanerochaete* spp. Appl. Environ. Microbiol. 56: 3519–3526.

Lamb, A., and E.L. Tollefson. 1973. Toxic effects of cupric, chromate and chromic ions on biological oxidation. Water Res. 7: 599–613.

Lamka, K.G., M.W. Lechevallier, and R.J. Seidler. 1980. Bacterial contamination of drinking water supplies in a modern rural neighborhood. Appl. Environ. Microbiol. 39: 734–738.

La Motta, E.J. 1976. Internal diffusion and reaction in biological films. Environ. Sci. Technol. 19: 765–769.

Lampel, K.A., J.A. Jagow, M. Trucksess, and W.E. Hill. 1990. Polymerase chain reaction for detection of invasive *Shigella flexneri* in food. Appl. Environ. Microbiol. 56: 1536–1540.

Lance, J.C. 1972. Nitrogen removal by soil mechanisms. Water Pollut. Control Fed. 44: 1352–1361.

Lance, J.C., and C.P. Gerba. 1980. Poliovirus movement during high rate land filtration of sewage water. J. Environ. Qual. 9: 31–34.

Landeen, L.K., M.T. Yahya, and C.P. Gerba. 1989. Efficacy of copper and silver ions and reduced levels of free chlorine in inactivation of *Legionella pneumophila*. Appl. Environ. Microbiol. 55: 3045–3050.

Lange, K.P., W.D. Bellamy, D.W. Hendricks, and G.S. Logsdon. 1986. Diatomaceous earth filtration of *Giardia* cysts and other substances. J. Am. Water Works Assoc. 78: 76–82.

Langeland, G. 1982. *Salmonella* spp. in the working environment of sewage treatment plants in Oslo, Norway. Appl. Environ. Microbiol. 43: 1111–1115.

van Langenhove, H., K. Roelstraete, N. Schamp, and J. Houtmeyers. 1985. GC-MS identification of odorous volatiles in wastewater. Water Res. 19: 597–603.

Laquidara, M.J., F.C. Blanc, and J.C. O'Shaughnessy. 1986. Development of biofilm, operating characteristics and operational control in the anaerobic rotating biological contactor process. J. Water Pollut. Control Fed. 58: 107–114.

LaRiviere, J.W.M. 1977. Microbial ecology of liquid waste treatment. Adv. Microb. Ecology 1: 215–259.

Larkin, J.M. 1980. Isolation of *Thiothrix* in pure culture and observation of a filamentous epiphyte on *Thiothrix*. Curr. Microbiol. 41: 155–158.

Larsen, P.I., L.K. Sydnes, B. Landfald, and A.R. Strom. 1987. Osmoregulation in *Escherichia coli* by accumulation of organic osmolytes: Betaines, glutamic acid and trehalose. Arch. Microbiol. 147: 1–7.

Larson, R.A., Ed. 1989. *Biohazards of Drinking Water Treatment*. Lewis Pubs., Chelsea, MI.

Larson, R.J., and S.L. Schaeffer. 1982. A rapid method for determining the toxicity of chemicals to activated sludge. Water Res. 16: 675–680.

Lau, A.O., P.F. Strom, and D. Jenkins. 1984a. Growth kinetics of *Sphaerotilus natans* and a floc former in pure and dual continuous culture. J. Water Pollut. Control Fed. 56: 41–51.

Lau, A.O., P.F. Strom, and D. Jenkins. 1984b. The competitive growth of floc-forming and filamentous bacteria: A model for activated sludge bulking. J. Water Pollut. Control Fed. 56: 52–61.

Lauer, W.C. 1991. Water quality for potable reuse. Water Sci. Technol. 23: 2171–2180.

Lauer, W.C., S.E. Rogers, and J.M. Ray. 1985. The current status of Denver's potable water reuse project. J. Am. Water Works Assoc. 77: 52–59.

Lawrence, A.W., and P.L. McCarty. 1969. Kinetics of methane fermentation in anaerobic treatment. J. Water Pollut. Control Fed. 41: R1-R17.

Lawrence, A.W., P.L. McCarty, and F.J.A. Guerin. 1966. The effects of sulfides on anaerobic treatment. Air Water Int. J. 110: 2207–2210.

LeChevalier, H.A., and D. Pramer. 1971. *The Microbes*. Lippincott, Philadelphia.

Lechevalier, H.A., and M.P. Lechevalier. 1975. *Actinomycetes of Sewage Treatment Plants*. EPA-600/2–75–031.

Lechevalier, H.A., M.P. Lechevalier, and P.E. Wyszkowski. 1977. *Actinomycetes of Sewage Treatment Plants*. EPA Report 3 600/2–77–145.

Lechevalier, M.P., and H.A. Lechevalier. 1974. *Nocardia amarae* sp. nov., an actinomycete common in foaming activated sludge. Int. J. Syst. Bacteriol. 24: 278–288.

LeChevallier, M.W., W.C. Becker, P. Schorr, and R.G. Lee. 1992. Evaluating the performance of biologically active rapid filters. J. Am. water Works Assoc. 84: 136–146.

LeChevallier, M.W., S.C. Cameron, and G.A. McFeters. 1983. New medium for improved recovery of coliform bacteria from drinking water. Appl. Environ. Microbiol. 45: 484–492.

LeChevallier, M.W., C.D. Cawthon, and R.G. Lee. 1988. Inactivation of biofilm bacteria. Appl. Environ. Microbiol. 54: 2492–2499.

LeChevallier, M.W., T.M. Evans, and R.J. Seidler. 1981. Effect of turbidity on chlorination efficiency and bacterial persistence in drinking water. Appl. Environ. Microbiol. 42: 159–167.

LeChevallier, M.W., T.S. Hassenauer, A.K. Camper, and G.A. McFeters. 1984a. Disinfection of bacteria attached to granular activated carbon. Appl. Environ. Microbiol. 48: 918–923.

Lechevallier, M.W., P.E. Jakanoski, A.K. Camper, and G.A. McFeters. 1984b. Evaluation of m-T7 as a fecal coliform medium. Appl. Environ. Microbiol. 48: 371–375.

LeChevallier, M.W., C.H. Lowry, and R.G. Lee. 1990. Disinfecting biofilm in a model distribution system. J. Am. Water Works Assoc. 82: 85–99.

Lechevallier, M.W., and G.A. McFeters. 1985a. Interactions between heterotrophic plate count bacteria and coliform organisms. Appl. Environ. Microbiol. 49: 1138–1141.

LeChevallier, M.W., and G.A. McFeters. 1985b. Enumerating injured coliforms in drinking water. J. Am. Water Works Assoc. 77: 81–87.

LeChevallier, M.W., and G.A. McFeters. 1990. Microbiology of activated carbon, pp. 104–119, in: *Drinking Water Microbiology*, G.A. McFeters, Ed. Springer-Verlag, New York.

LeChevallier, M.W., W.D. Norton, and R.G. Lee. 1991a. *Giardia* and *Cryptosporidium* spp. in filtered drinking water supplies. Appl. Environ. Microbiol. 57: 2617–2621.

LeChevallier, M.W., W.D. Norton, and R.G. Lee. 1991b. Occurrence of *Giardia* and *Cryotosporidium* spp. in surface water supplies. Appl. Environ. Microbiol. 57: 2610–2616.

LeChevallier, M.W., W. Schulz, and R.G. Lee. 1991c. Bacterial nutrients in drinking water. Appl. Environ. Microbiol. 57: 857–862.

LeChevallier, M.W., R.J. Seidler, and T.M. Evans. 1980. Enumeration and characterization of standard plate count bacteria in chlorinated and raw water supplies. Appl. Env. Microbiol. 40: 922–930.

LeChevallier, M.W., N.E. Shaw, L.A. Kaplan, and T.L. Bott. 1993. Development of a rapid assimilable organic carbon method in water. Appl. Environ. Microbiol. 59: 1526–1531.

LeChevallier, M.W., A. Singh, D.A. Schiemann, and G.A. McFeters. 1985. Changes in virulence of water enteropathogens with chlorine injury. Appl. Environ. Microbiol. 50: 412–419.

Lee, K.M., C.A. Brunner, J.B. Farrel, and A.E. Eralp. 1989. Destruction of enteric bacteria and viruses during two-phase digestion. J. Water Pollut. Control Fed. 61: 1421–1429.

Leisinger, T. 1983. Microorganisms and xenobiotic compounds. Experientia 39: 1183–1191.

Leisinger, T., R. Hutter, A.M. Cook, and J. Nuesch, Eds. 1981. *Microbial Degradation of Xenobiotics and Recalcitrant Compounds*. Academic, New York.

Lembke, L.L., R.N. Kniseley, R.C. van Nostrand, and M.D. Hale. 1981. Precision of the all-glass impinger and the Andersen microbial impactor for air sampling in solid waste handling facilities. Appl. Environ. Microbiol. 42: 222–225.

Lemmer, H. 1986. The ecology of scum-causing actinomycetes in sewage treatment plants. Water Res. 531–535.

Lemmer, H., and M. Baumann. 1988a. Scum actinomycetes in sewage treatment plants. Part 2: The effect of hydrophobic substrate. Water Res. 22: 761–763.

Lemmer, H., and M. Baumann. 1988b. Scum actinomycetes in sewage treatemnt plants. Part 3: Synergism with other sludge bacteria. Water Res. 22: 765–767.

Lemmer, H., and R.M. Kroppenstedt. 1984. Chemotaxonomy and physiology of some actinomycetes isolated from scumming activated sludge. System. Appl. Microbiol. 5: 124–135.

Lessard, E.J., and J. McN. Sieburth. 1983. Survival of natural sewage populations of enteric bacteria in diffusion and batch chambers in the marine environment. Appl. Environ. Microbiol. 45: 950–959.

Lettinga, G., A.F.M. van Velsen, S.W. Hobma, W. de Zeeuw, and A. Klapwijk. 1980. Use of upflow sludge blanket (USB) reactor concept for biological wastewater treatment, especially for anaerobic treatment. Biotechnol. Bioeng. 22: 699–734.

Levi, Y., C. Henriet, J.P. Coutant, M. Lucas, and G. Leger. 1989. Monitoring acute toxicity in rivers with the help of the Microtox test. Water Supply 7:25–31.

Levine, M.M. 1987. *Escherichia coli* that cause diarrhea: Enterotoxigenic, enteropathogenic, enteroinvasive, enterohemorrhagic, and enteroadherent. J. Infect. Dis. 155: 377–389.

Levy, R.V. 1990. Invertebrates and associated bacteria in drinking water distribution lines, pp. 224–248, in: *Drinking Water Microbiology*, G.A. McFeters, Ed. Springer-Verlag, New York.

Levy, R.V., R.D. Cheetham, J. Davis, G. Winer, and F.L. Hart. 1984. Novel method for studying the public health significance of macroinvertebrates in potable water. Appl. Environ. Microbiol. 47: 889–894.

Levy, R.V., F.L. Hart, and R.D. Cheetham. 1986. Occurrence and public health significance of invertebrates in drinking water systems. J. Am. Water Works Assoc. 78: 105–110.

Lewis, C.M., and J.L. Mak. 1989. Comparison of membrane filtration and Autoanalysis Colilert presence–absence techniques for analysis of total coliforms and *Escherichia coli* in drinking water samples. Appl. Environ. Microbiol. 55: 3091–3094.

Lewis, G.D., F.J. Austin, M.W Loutit, and K. Sharples. 1986. Enterovirus removal from sewage: The effectiveness of four different treatment plants. Water Res. 20: 1291–1297.

Leyval, C., C. Arz, J.C. Block, and M. Rizet. 1984. *Escherichia coli* resistance to chlorine after successive chlorinations. Environ. Technol. Lett. 5: 359–364.

Lighthart, B., and A.S. Frisch. 1976. Estimation of viable airborne microbes downwind from a point source. Appl. Environ. Microbiol. 31: 700–704.

Lighthart, B., and A.J. Mohr. 1987. Estimating downwind concentrations of viable airborne microorganisms in dynamic atmospheric conditions. Appl. Environ. Microbiol. 53: 1580–1583.

Lin, C.-Y. 1992. Effect of heavy metals on volatile fatty acid degradation in anaerobic digestion. Water Res. 26: 177–183.

Lin, S.D. 1985. *Giardia lamblia* and water supply. J. Am. Water Works Assoc. 77: 40–47.

Linkfield, T.G., J.M. Suflita, and J.M. Tiedje. 1989. Characterization of the acclimation period prior to the anaerobic biodegradation of haloaromatic compounds. Appl. Environ. Microbiol. 55: 2773–2778.

Lippy, E.C. 1986. Chlorination to prevent and control waterborne diseases. J. Am. Water Works Assoc. 78:49–52.

Lippy, E.C., and S.C. Waltrip. 1984. Waterborne disease outbreaks-1946–1980: A thirty-five-year perspective. J. Am. Water Works Assoc. 76: 60–67

Lipson, S.M., and G. Stotzky. 1987. Interactions between viruses and clay minerals, pp. 197–230, in: *Human Viruses in Sediments, Sludges, and Soils*, V.C. Rao and J.L. Melnick, Eds. CRC Press, Boca Raton, FL.

Little, C.D., A.V. Palumbo, S.E. Herbes, M.E. Lidstrom, R.L. Thyndall, and P.J. Gilmer. 1988. Trichloroethylene biodegradation by a methane-oxidizing bacterium. Appl. Environ. Microbiol. 54: 951–956.

Little, M.D. 1986. The detection and significance of pathogens in sludge: Parasites. In: *Control of Sludge Pathogens*, C.A. Sorber, Ed. Water Pollution Control Federation, Washington, DC.

Liu, D. 1982. Effect of sewage sludge land disposal on the microbiological quality of groundwater. Water Res. 16: 957–961.

Liu, D., and B.J. Dutka, Eds. 1984. *Toxicity Screening Procedures Using Bacterial Systems*. Marcel Dekker, New York.

Livernoche, D., L.Jurasek, M. Desrochers, J. Dorica, and I.A. Veliky. 1983. Removal of colour from kraft mill waste waters with cultures of white-rot fungi and with immobilized mycelia of *Coriolus versicolor*. Biotechnol. Bioeng. 25: 2055–2065.

Logan, K.B., G.E. Rees, N.D. Seeley, and S.B. Primrose. 1980. Rapid concentration of bacteriophages from large volumes of freshwater: Evaluation of positively-charged microporous filters. J. Virol. Methods 1: 87–97.

Logsdon, G.S., and J.C. Hoff. 1986. Barriers to the transmission of waterborne disease, pp. 255–274, in: *Waterborne Diseases in the United States*, G.F. Craun, Ed. CRC Press, Boca Raton, FL.

Logsdon, G.S., and E.C. Lippy. 1982. The role of filtration in preventing waterborne disease. J. Am. Water Works Assoc. 74: 649–655.

Logsdon, G.S., J.M. Symons, R.L. Hoye, Jr., and M.M. Arozarena. 1981. Alternative filtration methods for removal of *Giardia* cysts and cysts models. J. Am. Water Works Assoc. 73: 111–118.

Logue, C., B. Koopman, and G. Bitton. 1983. INT-reduction assays and control of sludge bulking. J. Environ. Eng. Sci. 109: 915–923.

Logue, C.L., B. Koopman, G.K. Brown, and G. Bitton. 1989. Toxicity screening in a large, municipal wastewater system. J. Water Pollut. Control Fed. 61: 632–640.

Longley, K.E., B.E. Moore, and C.A. Sorber. 1980. Comparison of chlorine and chlorine dioxide as disinfectants. J. Water Pollut. Control Fed. 52: 2098–2105.

Lucena, F., C. Finance, J. Jofre, J. Sancho, and L. Schwartzbrod. 1982. Viral pollution determination of superficial waters (river water and sea water) from the urban area of Barcelona (Spain). Water Res. 16: 173–177.

Luckiesh, M., and L.L Holladay. 1944. Disinfecting water by means of germicidal lamps. Gen. Electric Rev. 47: 45–54.

Lundgren, D.G., and W. Dean. 1979. Biogeochemistry of iron, pp. 211–223, in: *Biogeochemical Cycling of Mineral-Forming Elements*, P.A. Trudinger and D.J. Swaine, Eds. Elsevier, Amsterdam.

Lydholm, B., and A.L. Nielsen. 1981. The use of soluble polyelectrolytes for the isolation of virus from sludge, pp. 85–90, in: *Viruses and Wastewater Treatment*, M. Goddard and M. Butler, Eds. Pergamon Press, Oxford.

Macario, A.J.L., and E.C. de Macario. 1988. Quantitative immunologic analysis of the methanogenic flora of digestors reveals a considerable diversity. Appl. Environ. Microbiol. 54: 79–86.

Mach, P.A., and D.J. Grimes. 1982. R-plasmid transfer in a wastewater treatment plant. Appl. Environ. Microbiol. 44: 1395–1403.

Mackay, D.M., and J.A. Cherry. 1989. Groundwater contamination: Pump-and-treat remediation. Environ. Sci. Technol. 23: 630–636.

Mackie, R.I., and M.P. Bryant. 1981. Metabolic activity of fatty acid-oxidizing bacteria and the contribution of acetate, propionate, butyrate and CO_2 to methanogenesis in cattle waste at 40 and 60°C. Appl. Environ. Microbiol. 41: 1363–1373.

MacLeod, F.A., S.R. Guiot, and J.W. Costerton. 1990. Layered structure of bacterial aggregates in an upflow anaerobic sludge bed and filter reactor. Appl. Environ. Microbiol. 56: 1598–1607.

MacRae, J.D., and J. Smit. 1991. Characterization of Caulobacters isolated from wastewater treatment systems. Appl. Environ. Microbiol. 55: 751–758.

Madigan, M.T. 1988. Microbiology, physiology, and ecology of phototrophic bacteria, pp. 39–111, in: *Biology of Anaerobic Microorganisms*, A.J.B. Zehnder, Ed. Wiley-Interscience, New York.

Madore, M.S., J.B. Rose, C.P. Gerba, M.J. Arrowood, and C.R. Sterling. 1987. Occurrence of *Cryptosporidium* oocysts in sewage effluents and select surface waters. J. Parasitol. 73: 702–705.

Magnusson, S., K. Gundersen, A. Brandberg, and E. Lycke. 1967. Marine bacteria and their possible relation to the virus inactivation capacity of seawater. Acta Pathol. Microbiol. Scand. 71: 274–280.

Mahbubani, M.H., A.K. Bej, M. Perlin, F.W. Schaefer III, W. Jakubowski, and R.M. Atlas. 1991. Detection of *Giardia* cysts by using the polymerase chain reaction and distinguishing live from dead cysts. Appl. Environ. Microbiol. 57: 3456–3461.

Majeti, V.A., and C.S. Clark. 1981. Health risk of organics in land application. J. Environ. Eng. Div. Am. Soc. Civ. Eng. 107: 339–357.

Maloney, S.W., J. Maneim, J. Mallevialle, and F. Fiesssinger. 1986. Transformation of trace organic compounds in drinking water by enzymatic oxidative coupling. Environ. Sci. Technol. 20: 249–253.

Manafi, M., and R. Sommer. 1993. Rapid identification of enterococci with a new fluorogenic-chromogenic assay. Water Sci. Technol. 27: 271–274.

Manem, J.A., and B.E. Rittmann. 1992. Removing trace-level organic pollutants in a biological filter. J. Am. Water Works Assoc. 84: 152–157.

Manka, J., M. Rebhun, A. Mandelbaum, and A. Bortinger. 1974. Characterization of organics in secondary effluents. Environ. Sci. Technol. 8: 1017–1020.

Manning, J.F., and R.L. Irvine. 1985. The biological removal of phosphorus in a sequencing batch reactor. J. Water Pollut. Control Fed. 57: 87–94.

Mansfield, L.A., P.B. Melnyk, and G.C. Richardson. 1992. Selection and full-scale use of a chelated iron adsorbent for odor control. Water Environ. Res. 64: 120–127.

Marais, G.V.R. 1974. Fecal bacterial kinetics in stabilization ponds. J. Sanit. Eng. Div. Am. Soc. Civ. Eng. 100: 119–139.

March, D., L. Benefield, E. Bennett, D. Linstedt, and R. Hartman. 1981. Coupled trickling filter-rotating biological contactor nitrification process. J. Water Pollut. Control Fed. 53: 1469–1480.

Marciano-Cabral, F. 1988. Biology of *Naegleria* spp. Microbiol. Rev. 52: 114–133.

Marison, L.W. 1988a. Growth kinetics, pp. 184–217, in: *Biotechnology for Engineers: Biological Systems in Technological Processes*, A. Scragg, Ed. Ellis Horwood, Chichester, U.K.

Marison, L.W. 1988b. Enzyme kinetics, pp. 96–119, in: *Biotechnology for Engineers: Biological Systems in Technological Processes*, A. Scragg, Ed. Ellis Horwood, Chichester, U.K.

Markovic, M.J., and A.L. Kroeger. 1989. Saprophitic actinomycetes as a causative agent of musty odor in drinking water. Annu. Meet. Am. Soc. Microbiol., New Orleans.

Marques, E., and M.T. Martins. 1983. Enterovirus isolation from seawater from beaches of Baixada Santista [in Portugese]. Paper presented at the Ninth Latin American Congress for Microbiology, São Paulo, Brazil.

Marshall, K.C. 1976. *Interfaces in Microbial Ecology*. Harvard Univ. Press, Cambridge, Mass.

Marthi, B., V.P. Fieland, M. Walter, and R.J. Seidler. 1990. Survival of bacteria during aerosolization. Appl. Environ. Microbiol. 56: 3463–3467.

Marthi, B., and B. Lighthart. 1990. Effect of betaine on enumeration of airborne bacteria. Appl. Environ. Microbiol. 56: 1286–1289.

Marthi, B., B.T. Shaffer, B. Lighthart, and L. Ganio. 1991. Resuscitaion effects of catalase on airborne bacteria. Appl. Environ. Microbiol. 57: 2775–2776.

Martin, J.H., Jr., H.E. Bostian, and G. Stern. 1990. Reductions of enteric microorganisms during aerobic sludge digestion. Water Res. 24: 1377–1385.

Martins, M.T., I.G. Rivera, D.L. Clark, and B.H. Olson. 1992. Detection of virulence factors in culturable *Escherichia coli* isolates from water samples by DNA probes and recovery of toxin-bearing strains in minimal *o*-nitrophenol-β-D-galactopyranoside-4-methylumbelliferyl-β-D-glucuronide media. Appl. Environ. Microbiol. 58: 3095–3100.

Martz, R.F., D.I. Sebacher, and D.C. White. 1983. Biomass measurement of methane forming bacteria in environmental samples. J. Microbiol. Methods 1: 53–61.

Matin, A., E. Auger, P. Blum, and J. Schultz. 1989. Genetic basis of starvation survival in non-differentiating bacteria. Annu. Rev. Microbiol. 43: 293–316.

Matin, A., and S. Harakeh. 1990. Effect of starvation on bacterial resistance to disinfectants, pp. 88–103, in: *Drinking Water Microbiology*, G.A. McFeters, Ed. Springer-Verlag, New York.

Matsuda, H., H. Yamamori, T. Sato, Y. Ose, H. Nagase, H. Kito, and K. Sumida. 1992. Mutagenicity of ozonation products from humic substances and their components. Water Sci. Technol. 25: 363–370.

Matsui, T., S. Kyosai, and M. Takahashi. 1991. Application of biotechnology to municipal wastewater treatment. Water Sci. Technol. 23: 1723–1732.

Matsunaga, T., I. Karube, and S. Suzuki. 1980. A specific microbial sensor for formic acid. Eur. J. Appl. Microbiol. Biotechnol. 10: 235–243.

May, K.R. 1966. Multistage liquid impinger. Bacteriol. Rev. 30: 559–570.

Mazidji, C.N., B. Koopman, G. Bitton, and G. Voiland. 1990. Use of Microtox and *Ceriodaphnia* bioassays in wastewater fractionation. Toxicity Assess. 5: 265–277.

Mazidji, C.N., B. Koopman, and G. Bitton. 1992. Chelating resin versus ion-exchange resin for heavy metal removal in toxicity fractionation. Water Sci. Technol. 26: 189–196.

McCambridge, J., and T.A. McMeekin. 1980. Relative effects of bacterial and protozoan predators on survival of *Escherichia coli* in estuarine water samples. Appl. Environ. Microbiol. 40: 907–911.

McCambridge, J., and T.A. McMeekin. 1981. Effect of solar radiation and predacious microorganisms on survival of fecal and other bacteria. Appl. Environ. Microbiol. 41: 1083–1087.

McCarty, P.L., B.E. Rittmann, and E.J. Bouwer. 1984. Microbiological processes affecting chemical transformations in groundwater, pp. 89–115, in: *Groundwater Pollution Microbiology*, G. Bitton and C.P. Gerba, Eds. Wiley, New York.

McCarty, S.C., J.H. Standridge, and M.C. Stasiak. 1992. Evaluating a commercially available defined-substrate test for recovery of *E. coli*. J. Am. Water Works Assoc. 84: 91–97.

McClure, N.C., J.C. Fry, and A.J. Weightman. 1990. Gene transfer in activated sludge, pp. 111–129, in: *Bacterial Genetics in Natural Environments*, J.C. Fry and M.J. Day, Eds. Chapman and Hall, London.

McDonald, L.C., C.R. Hackney, and B. Ray. 1983. Enhanced recovery of injured *Escherichia coli* by compounds that degrade hydrogen peroxide or block its formation. Appl. Environ. Microbiol. 45: 360–365.

McFarland, M.J., and W.J. Jewell. 1990. The effect of sulfate reduction on the thermophilic (55°C) methane fermentation process. J. Ind. Microbiol. 5: 247–258.

McFeters, G.A., Ed. 1990. *Drinking Water Microbiology*, Springer-Verlag, New York.

McFeters, G.A., S.C. Cameron, and M.W. LeChevallier. 1982. Influence of diluents, media, and membrane filters on detection of injured waterborne coliform bacteria. Appl. Environ. Microbiol. 43: 97–103.

McFeters, G.A., and A.K. Camper. 1988. Microbiological analysis and testing, pp. 73–95, in: *Bacterial Regrowth in Distribution Systems*, Characklis, W.G., Ed. Research report, AWWA Research Foundation, Denver, CO.

McGowan, K.L., E. Wickersham, and N.A. Strockbine. 1989. *Escherichia coli* O157:H7 from water (Letter). Lancet i: 967–968.

McInernay, M.J., M.P. Bryant, R.B. Hespell, and J.W. Costerton. 1981. *Syntrophomonas wolfei*, gen. nov. sp. nov., an anaerobic syntrophic, fatty acid-oxidizing bacterium. Appl. Environ. Microbiol. 41: 1029–1039.

McKinley, V.L., and J.R. Vestal. 1985. Physical and chemical correlates of microbial activity and biomass in composting municipal sewage sludge. Appl. Environ. Microbiol. 50: 1395–1403.

McKinley, V.L., J.R. Vestal, and A.E. Eralp. 1985. Microbial activity in composting. Biocycle 26: 47–50.

McLean, R.J.C., D. Beauchemin, L. Clapham, and T.J. Beveridge. 1990. Metal-binding characteristics of the gamma-glutamyl capsular polymer of *Bacillus licheniformis* ATTC 9945. Appl. Environ. Microbiol. 56: 3671–3677.

McPherson, P., and M.A. Gealt. 1986. Isolation of indigenous wastewater bacterial strains capable of mobilizing plasmid pBR325. Appl. Environ. Microbiol. 51: 904–909.

Meadows, C.A., and B.H. Snudden. 1982. Prevalence of *Yersinia enterocolitica* in waters at the lower Chippewa River basin, Wisconsin. Appl. Environ. Microbiol. 43: 953–954.

Means, E.G., and B.H. Olson. 1981. Coliform inhibition by bacteriocin-like substances in drinking water distribution systems. Appl. Environ. Microbiol. 42: 506–512.

Meschsner, K., and G. Hamer. 1985. Denitrification by methanotrophic/methylotrophic bacterial associations in aquatic environments, pp. 257–271, in: *Denitrification in the Nitrogen Cycle*. Plenum, New York.

Meschner, K., T. Fleischmann, C.A. Mason, and G. Hamer. 1990. UV disinfection: Short-term inactivation and revival (Abstract). International Symposium on Health-Related Water Microbiology, Tubingen, West Germany, April 1–6, 1990.

Medsker, L.L., D. Jenkins, and J.F. Thomas. 1968. Odorous compounds in natural waters: An earthy-smelling compound associated with blue-green algae and actinomycetes. Environ. Sci. Technol. 2: 461–464.

Meganck, M.T.J., and G.M. Faup. 1988. Enhanced biological phosphorus removal from waste waters, pp. 111–203, in: *Biotreatment Systems*, Vol. 3, D.L. Wise, Ed. CRC Press, Boca Raton, FL.

Meganck, M.T.J., D. Malnou, P. Le Flohic, G.M. Faup, and J.M. Rovel. 1984. The importance of the acidogenic microflora in biological phosphorus removal, p. 254, in: *Enhanced Biological Phosphorus Removal from Wastewater*, Vol. 1. Proceedings of the IAWPRC Conference, Paris, September 1984.

Melnick, J.L. 1976. Taxonomy of viruses. Prod. Med. Virol. 22: 211–221.

Menard, A.B., and D. Jenkins. 1970. Fate of phosphorus in wastewater treatment processes: Enhanced removal of phosphate by activated sludge. Environ. Sci. Technol. 4: 1115–1119.

Mentzing, L.O. 1981. Waterborne outbreaks of *Campylobacter* enteritis in central Sweden. Lancet ii: 352–354.

Messing, R. 1988. Immobilized cells in anaerobic waste treatment, pp. 311–316, in: *Bioreactor Immobilized Enzymes and Cells: Fundamentals and Applications*, M. Moo-Young, Ed. Elsevier Applied Science, New York.

Metcalf and Eddy, Inc. 1991. *Wastewater Engineering: Treatment, Disposal and Reuse* (3rd ed.). McGraw-Hill, New York.

Michal, G. 1978. Determination of Michaelis constant and inhibitor constants, pp. 29–42, in: *Principles of Enzymatic Analysis*, H.U. Bergmeyer, Ed. Verlag Chemie, Wienheim.

Mikesell, M.D., and S.A. Boyd. 1985. Reductive dechlorination of the herbicides 2,4-D, 2,4,5-T and pentachlorophenol in anaerobic sewage sludges. J. Environ. Qual. 14: 337–340.

Mikesell, M.D., and S.A. Boyd. 1986. Complete reductive dechlorination and mineralization of pentachlorophenol by anaerobic microorganisms. Appl. Environ. Microbiol. 52: 861–865.

Mill, S.W., G.P. Alabaster, D.D. Mara, H.W. Pearson, and W.N. Thitai. 1992. Efficiency of faecal bacterial removal in waste stabilization ponds in Kenya. Water Sci. Technol. 26: 1739–1748.

Miller, R.A., M.A. Bronsdon, and W.R. Morton. 1986. Determination of the infectious dose of *Cryptosporidium* and the influence of inoculum size on disease severity in a primate model. Abstr. Annu. Meet. Am. Soc. Microbiol., Washington, DC.

Mirenda, R.J., and W.S. Hall. 1992. The application of effluent characterization procedures in toxicity identification evaluations. Water Sci. Technol. 25: 39–44.

Mitchell, R. 1971. Destruction of bacteria and viruses in seawater. J. Sanit. Eng. Div. Am. Soc. Civ. Eng. 97: 425–432.

Mitchell, R., and Z. Nevo. 1964. Effect of bacterial polysaccharide accumulation on infiltration of water through sand. Appl. Microbiol. 12: 219–223.

Moeller, J.R., and J. Calkins. 1980. Bactericidal agents in wastewater lagoons and lagoon design. J. Water Pollut. Control Fed. 52: 2442–2450.

Mohr, A.J. 1991. Development of models to explain the survival of viruses and bacteria in aerosols, pp. 160–190, in: *Modeling the Environmental Fate of Microorganisms*, C.J. Hurst, Ed. American Society Microbiology, Washington, DC.

Monsen, R.M., and E.M. Davis. 1984. Microbial responses to selected organic chemicals in industrial waste treatment units, pp. 233–249, in: *Toxicity Screening Procedures Using Bacterial Systems*, D. Liu and B.J. Dutka, Eds. Marcel Dekker, New York.

Moore, A.T., A. Vira, and S. Fogel. 1989. Biodegradation of *trans*-1,2-dichloroethylene by methane-utilizing bacteria in an aquifer simulator. Environ. Sci. Technol. 23: 403–406.

Moore, B.E., D.E. Camman, C.A. Turk, and C.A. Sorber. 1988. Microbial characterization of municipal wastewater at a spray irrigation site: The Lubbock infection surveillance study. J. Water Pollut. Control Fed. 60: 1222–1230.

Moore B.E., B.P. Sagik, and C.A. Sorber. 1976. An assessment of potention health risks associated with land disposal of residual sludges. Paper presented at the Third National Conference on Sludge Management, Disposal and Reuse, Miami, FL.

Moore, B.E., B.P. Sagik, and C.A. Sorber. 1978. Land application of sludge: Minimizing the impact of viruses on water resources, pp. 154–167, in: *Risk Assessment and Health Effects of Land Application of Municipal Wastewater and Sludges*, B.P. Sagik and C.A. Sorber, Eds. University of Texas, San Antonio.

Moore, B.E., B.P. Sagik, and C.A. Sorber. 1979. Procedure for recovery of airborne human enteric viruses during spray irrigation of treated wastewater. Appl. Environ. Microbiol. 38: 688–693.

Moore, B.E., B.P. Sagik, and C.A. Sorber. 1981. Viral transport to groundwater at a wastewater land application site. J. Water Pollut. Control Fed. 53: 1492–1502.

Moos, L.P., E.J. Kirsch, R.F. Wukasch, and C.P.L Grady, Jr. 1983. Pentachlorophenol biodegradation. I. Aerobic. Water Res. 17: 1575–1584.

Moran, M.A., V.L. Torsvik, T. Torsvik, and R.E. Hodson. 1993. Direst extraction and purification of rRNA for ecological studies. Appl. Environ. Microbiol. 59: 915–918.

Mori, T., K. Itokazu, Y. Ishikura, F. Mishina, Y. Sakai, and M. Koga. 1992. Evaluation of control strategies for actinomycete scum in full-scale treatment plants. Water Sci. Technol. 25: 231–237.

Morinigo, M.A., M.A. Munoz, R. Cornax, E. Martinez-Manzanares, and J.J. Borrego. 1992a. Presence of indicators and *Salmonella* in natural waters affected by outfall wastewater discharges. Water Sci. Technol. 25: 1–8.

Morinigo, M.A., D. Wheeler, C. Berry, C. Jones, M.A. Munoz, R. Cornax, and J.J. Borrego. 1992b. Evaluation of different bacteriophage groups as faecal indicators in contaminated natural waters in southern England. Water Res. 26: 267–271.

Morris, J.C. 1975. Aspects of the quantitative assessment of germicidal efficiency. in: *Disinfection of Water and Wastewater*, J.D. Johnson, Ed. Ann Arbor Science, Ann Arbor, MI.

Mueller, R.F., and A. Steiner. 1992. Inhibition of anaerobic digestion by heavy metals. Water Sci. Technol. 26: 835–846.

Mulbry, W.W., and J.S. Karns. 1989. Purification and characterization of three parathion hydrolases from gram-negative bacterial strains. Appl. Environ. Microbiol. 55: 289–293.

Mullen, M.D., D.C. Wolf, F.G. Ferris, T.J. Beveridge, C.A. Flemming, and G.W. Bayley. 1989. Bacterial sorption of heavy metals. Appl. Environ. Microbiol. 55: 3143–3149.

Mullis, K.B., and F.A. Fallona. 1987. Specific synthesis of DNA *in vitro* via a polymerase catalyzed chain reaction. Methods Enzymol. 155: 335–350.

Munkittrick, K.R, E.A. Power, and G.A. Sergy. 1991. The relative sensitivity of Microtox, daphnid, rainbow trout and fathead minnow acute lethality tests. Environ. Toxicol. Water Qual. 6:35–62.

Munnecke, D.M. 1981. The use of microbial enzymes for pesticides detoxification. In: *Microbial Degradation of Xenobiotics and Recalcitrant Compounds*, T. Leisinger et al., Eds. Academic, New York.

Munro, P.M., F. Laumond, and M.J. Gauthier. 1987. Previous growth of enteric bacteria on a salted medium increases their survival in seawater. Lett. Appl. Microbiol. 4: 121–124.

Munro, P.M., M.J. Gauthier, V.A. Breittmayer, and J. Bongiovanni. 1989. Influence of osmoregulation on starvation survival of *Escherichia coli* in seawater. Appl. Environ. Microbiol. 55: 2017–2024.

Muraca, P., J.E. Stout, and V.L. Yu. 1987. Comparative assessment of chlorine, heat, ozone, and UV light for killing *Legionella pneumophila* within a model plumbing system. Appl. Environ. Microbiol. 53: 447–453.

Muraca, P., V.L. Yu, and J.E. Stout. 1988. Environmental aspects of Legionaires' Disease. J. Am. Water Works Assoc. 80: 78–86.

Murray, G.E., R.S. Tobin, B. Junkins, and D.J. Kushner. 1984. Effect of chlorination on antibiotic resistance profiles of sewage-related bacteria. Appl. Environ. Microbiol. 48: 73–77.

Murray, W.D., and L. van den Berg. 1981. Effect of nickel, cobalt, and molybdenum on performance of methanogenic fixed-film reactors. Appl. Environ. Microbiol. 42: 502–505.

Musial, C.E., M.J. Arrowood, C.R. Sterling, and C.P. Gerba. 1987. Detection of *Cryptosporidium* in water by using propylene cartridge filters. Appl. Environ. Microbiol. 53: 687–692.

Myler, C.A., and W. Sisk. 1991. Bioremediation of explosive contaminated soils, pp. 137–146, in: *Environmental Biotechnology for Waste Treatment*, G.S. Sayler, R. Fox, and J.W. Blackburn, Eds. Plenum, New York.

Myoga, H., H. Asano, Y. Nomura, and H. Yoshida. 1991. Effect of immobilization on the nitrification treatability of entrapped cell reactors using the PVA freezing method. Water Sci. Technol. 23: 1117–1124.

Nagy L.A., and B.H. Olson. 1982. The occurrence of filamentous fungi in drinking water distribution systems. Can. J. Microbiol. 28: 667–671.

Najm, I.M., V.M. Snoeyink, B.W. Lykins, Jr., and J.Q. Adams. 1991. Using powdered activated carbon: A critical review. J. Am. Water Works Assoc. 83: 65–76.

Nakae, T. 1986. Outer membrane permeability of bacteria. Crit. Rev. Microbiol. 13: 1–62.

Nakamoto, S., and N. Machida. 1992. Phenol removal from aqueous solutions by peroxidase-catalyzed reaction using additives. Water Res. 26: 49–54.

Nakamura, K., M. Shibata, and Y. Miyaji. 1989. Substrate affinity of oligotrophic bacteria in biofilm reactors. Water Sci. Technol. 21: 779–790.

Nakasaki, K., M. Shoda, and H. Kubota. 1985a. Effect of temperature on composting of sewage sludge. Appl. Environ. Microbiol. 50: 1526–1530.

Nakhforoosh, N., and J.B. Rose. 1989. Detection of *Giardia* with a gene probe. Abstr. Annu. Meet. Am. Soc. Microbiol., New Orleans.

Namkung, E., and B.E. Rittmann. 1987. Removal of taste- and odor-causing compounds by biofims grown on humic substances. J. Am. Water Works Assoc. 79: 107–112.

Namkung, E., R.G. Stratton, and B.E. Rittmann. 1983. Predicting removal of trace organic compounds by biofilms. J. Water Pollut. Control Fed. 55: 1366–1372.

Narkis, N., and Y. Kott. 1992. Comparison between chlorine dioxide and chlorine for use as a disinfectant of wastewater effluents. Water Sci. Technol. 26: 1483–1492.

Nasser, A.M., Y. Tchorch, and B. Fattal. 1993. Comparative survival of *E. coli*, F^+ bacteriophages, HAV and poliovirus 1 in wastewaters and groundwaters. Water Sci. Technol. 27: 401–407.

Nathanson, J.A. 1986. *Basic Environmental Technology: Water Supply, Waste Disposal and Pollution Control*. Wiley, New York.

National Research Council. 1979. *Hydrogen Sulfide*. Report by the Committee on Medical and Biologic Effects of Environmental Pollutants, Division of Medical Science, National Research Council, Washington, DC.

Nazaly, N., and C.J. Knowles. 1981. Cyanide degradation by immobilized fungi. Biotechnol. Lett. 3: 363–368.

Negulescu, M. 1985. *Municipal Wastewater Treatment*. Elsevier, Amsterdam.

Neidhardt, F., R. Van Bogelen, and V. Vaughn. 1984. The genetics and regulation of heat-shock proteins. Annu. Rev. Genet. 18: 295–329.

Neilson, A.H., A.S. Allard, and M. Remberger. 1985. Biodegradation and transformation of recalcitrant compounds. in: *Handbook of Environmental Chemistry*, O. Hutzinger, Ed. Springer, New York.

Nellor, M.H., R.B. Baird, and J.R. Smyth. 1985. Health effects of indirect potable water reuse. J. Am. Water Works Assoc. 77: 88–96.

Nelson, P.O., A.K. Chung, and M.C. Hudson. 1981. Factors affecting the fate of heavy metals in the activated sludge process. J. Water Pollut. Control Fed. 53: 1323–1333.

Nelson, P.O., and A.W. Lawrence. 1980. Microbial viability measurements and activated sludge kinetics. Water Res. 14: 217–225.

Nelson, T.C., J.Y.C. Huang, and D. Ramaswami. 1988. Decomposition of exopolysaccharide slime by a bacteriophage enzyme. Water Res. 22: 1185–1188.

Neu, H.C. 1992. The crisis of antibiotic resistance. Science 257: 1064–1072.

Neufield, R.D. 1976. Heavy metal induced deflocculation of activated sludge. J. Water Pollut. Control Fed. 48: 1940–1947.

Neufield, R.D., and E.R. Hermann. 1975. Heavy metal removal by activated sludge. J. Water Pollut. Control Fed. 47: 310–329.

Nevo, Z., and R. Mitchell. 1967. Factors affecting biological clogging of sand associated with ground water recharge. Water Res. 1: 231–236.

Newsome, A.L., R.L. Baker, R.D. Miller, and R.R. Arnold. 1985. Interactions between *Naegleria fowleri* and *Legionella pneumophila*. Infect. Immun. 50: 449–452.

Nicell, J.A., J.K. Bewtra, K.E. Taylor, N. Biswas, and C. St. Pierre. 1992. Enzyme catalyzed polymerization and precipitation of aromatic compounds from wastewater. Water Sci. Technol. 25: 157–164.

Nichols, A.B. 1988. Water reuse closes water-wastewater loop. J. Water Pollut. Control Fed. 60: 1931–1937.

Nicholson, D.K., S.L. Woods, J.D. Istok, and D.C. Peek. 1992. Reductive dechlorination of chlorophenols by a pentachlorophenol-acclimated methanogenic consortium. Appl. Environ. Microbiol. 58: 2280–2286.

Nicks, O.W. 1986. *Conceptual Design for a Food Production, Water, and Waste Processing, and Gas Regeneration Module*. Progress Report, Contract No. NAG-9–161. Regenerative Concept Group, NASA, Johnson Space Center, Houston.

Niederwohrmeier, B., R. Bohm, and D. Strauch. 1985. Microwave treatment as an alternative pasteurization process for the disinfection of sewage sludge: Experiments with the treatment of liquid manure, pp. 135–147, in: *Inactivation of Microorganisms in Sewage Sludge by Stabilization Processes*, D. Strauch, A.H. Havelaar, and P. L'Hermite, Eds. Elsevier Applied Science, London.

van Niekerk, A.M., D. Jenkins, and M.G. Richard. 1987. The competitive growth of *Zooglea ramigera* and Type 021N in activated sludge and pure culture. A model for low F/M bulking. J. Water Pollut. Control Fed. 59: 262–273.

Niemi, R.M., S. Knuth, and K. Lundstrom. 1982. Actinomycetes and fungi in surface waters and in potable water. Appl. Environ. Microbiol. 43: 378–388.

Nilsson, I., and S. Ohlson. 1982. Columnar denitrification of water by immobilized *Pseudomonas denitrificans* cells. Eur. J. Appl. Microbiol. Biotechnol. 14: 86–90.

Nitisoravut, S., and P.Y. Yang. 1992. Denitrification of nitrate-rich water using entrapped-mixed-microbial cells immobilization technique. Water Sci. Technol. 26: 923–931.

Norberg, A.B., and S.-O. Enfors. 1982. Production of extracellular polysaccharide by *Zooglea ramigera*. Appl. Environ. Microbiol. 44: 1231–1237.

Norberg, A.B., and H. Persson. 1984. Accumulation of heavy metals ions by *Zooglea ramigera*. Biotechnol. Bioeng. 26: 239–246.

Norberg, A.B., and S. Rydin. 1984. Development of a continuous process for metal accumulation by *Zooglea ramigera*. Biotechnol. Bioeng. 26: 265–268.

Noss, C.I., F.S. Hauchman, and V.P. Olivieri. 1986. Chlorine dioxide reactivity with proteins. Water Res. 20: 351–356.

Noss, C.I., and V.P. Olivieri. 1985. Disinfecting capabilities of oxychlorine compounds. Appl. Environ. Microbiol. 50:1162–1164.

Nowak, G., and G.D. Brown. 1990. Characteristics of *Nostocoida lumicola* and its activity in activated sludge suspension. J. Water Pollut. Control Fed. 62: 137–142.

Nowak, G., G. Brown, and A. Yee. 1986. Effect of feed pattern and dissolved oxygen on growth of filamentous bacteria. J. Water Pollut. Control Fed. 58: 978–984.

O'Brien, R.T., and J. Newman. 1979. Structural and compositional changes associated with chlorine inactivation of polioviruses. Appl. Environ. Microbiol. 38: 1034–1039.

Obst, U., A. Holzapfel-Pschorn, and M. Wiegand-Rosinus 1988. Application of enzyme assays for toxicological water testing. Toxicity Assess. 3: 81–91.

O'Connor, J.T., L. Hash, and A.B. Edwards. 1975. Deterioration of water quality in distribution systems. J. Am. Water Works Assoc. 67: 113–116.

Odeymi, O. 1990. Use of solar radiation for drinking water disinfection in West Africa (Abstract). International Symposium on Health-Related Water Microbiology, Tubingen, West Germany, April 1–6, 1990.

Odom, J.M. 1990. Industrial and environmental concerns with sulfate reducing bacteria. ASM News 56: 473–476.

Office of Technology Asessment. 1984. *Protecting the Nation's Groundwater from Contamination*. OTA-0-233. U.S. Congress, Washington, D.C.

O'Grady, D.P., P.H. Howard, and A.F. Werner. 1985. Activated sludge biodegradation of 12 commercial phthalate esters. Appl. Environ. Microbiol. 49: 443–445.

Ohtake, H., and Hardoyo, 1992. New biological method for detoxification and removal of hexavalent chromium. Water Sci. Technol. 25: 395–402.

Ohtake, H., K. Takahashi, Y. Tsuzuki, and K. Toda. 1985. Uptake and release of phosphate by a pure culture of *Acinetobacter calcoaceticus*. Water Res. 19: 1587–1594.

Okamoto, K., Y. Yamamoto, H. Tanaka, M. Tanaka, and A. Itaya. 1985. Heterogenous photocatalytic decomposition of phenol over TiO_2 powder. Bull. Chem. Soc. Jpn. 58: 2015–2022.

O'Keefe, B., and J. Green. 1989. Coliphages as indicators of fecal pollution at three recreational beaches on the Firth of Forth. Water Res. 1027–1030.

Oldenhuis, R., J.Y. Oedzes, J.J. van der Waarde, and D.B. Janssen. 1991. Kinetics of chlorinated hydrocarbon degradation by *Methylosinus trichosporium* OB3b and toxicity of trichloroethylene. Appl. Environ. Microbiol. 57: 7–14.

Olive, D.M. 1989. Detection of enterotoxigenic *Escherichia coli* after polymerase chain reaction amplification with a thermostable DNA polymerase. J. Clin. Microbiol. 27: 261–265.

Oliver, B.G., and J.H. Carey. 1976. Ultraviolet disinfection, an alternative to chlorination. J. Water Pollut. Control Fed. 48: 2619–2627.

Oliver, B.G., and E.G. Cosgrove. 1977. The disinfection of sewage treatment plant effluents using ultraviolet light. Can J. Chem. Eng. 53: 170–174.

Oliver, B.G., and D.B. Shindler. 1980. Trihalomethanes from the chlorination of aquatic algae. Environ. Sci. Technol. 14: 1502–1505.

Olivieri, V.P. 1983. Measurement of microbial quality. in: *Assessment of Microbiology and Turbidity Standards For Drinking Water*, P.S. Berger and Y. Argaman, Eds. EPA-570–9-83–001, Office of Drinking Water, Washington, DC.

Olivieri, V.P., W.H. Dennis, M.C. Snead, D.R. Richfield, and C.W. Kruse. 1980. Reaction of chlorine and chloramines with nucleic acid under disinfection conditions, in: *Water Chlorination: Environmental Impacts and Health Effects*, Vol. 3, R.L. Jolley, W.A. Brungs, and R.B. Cumming, Eds. Ann Arbor Science, Ann Arbor, MI.

Olivieri, V.P., et al. 1985. Mode of action of chlorine dioxide on selected viruses. in: *Water Chlorina-*

tion: Environmental Implications and Health Effects, Vol. 5, R.L. Jolley, W.A. Brung, and R.B. Cumming, Eds. Ann Arbor Science, Ann Arbor, MI.

Ollis, D.F. 1985. Contaminant degradation in water. Environ. Sci. Technol. 19: 480–484.

O'Malley, M.L., D.W. Lear, W.N. Adams, J. Gaines, T.K. Sawyer, and E.J. Lewis. 1982. Microbial contamination of continental shelf sediments by wastewater. J. Water Pollut. Control Fed. 54: 1311–1317.

Olson, B.H. 1991. Tracking and using genes in the environment. Environ. Sci. Technol. 25: 604–611.

Olson, B.H., and R.A. Goldstein. 1988. Applying genetic ecology to environmental management. Environ. Sci. Technol. 22: 370–372.

Olson, B.H., R. McCleary, and J. Meeker. 1991. Background and models for bacterial biofilm formation and function in water distribution systems, pp. 255–285, in: *Modeling the Environmental Fate of Microorganisms*, C.J. Hurst, Ed. American Society for Microbiology, Washington, DC.

Olson, B.H., and L.A. Nagy. 1984. Microbiology of potable water. Adv. Appl. Microbiol. 30: 73–132.

Omura, T., M. Onuma, J. Aizawa, T. Umita, and T. Yagi. 1989. Removal efficiencies of indicator microorganisms in sewage treatment plants. Water Sci. Technol. 21: 119–124.

Ongerth, J.E., and H.H. Stibbs. 1987. Identification of *Cryptosporidium* oocysts in river water. Appl. Environ. Microbiol. 53: 672–676.

Ongerth, H.J., and W.F. Jopling. 1977. Water reuse in California, pp. 219–256, in: *Water Renovation and Reuse*, H.I. Shuval, Ed. Academic, New York.

Ongerth, J.E. 1990. Evaluation of treatment for removing *Giardia* cysts. J. Am. Water Works Assoc. 82: 85–96.

O'Reilly, K.T., R. Kadakia, R.A. Korus, and R.L Crawford. 1988. Utilization of immobilized bacteria to degrade aromatic compounds common to wood-treatment wastewaters. In: *Proceedings of the International Conference on Water and Wastewater Microbiology*, Newport Beach, CA, Feb 8–11, 1988, Vol. 1.

Oremland, R.S. 1988. Biogeochemistry of methanogenic bacteria, pp. 641–705, in: *Biology of Anaerobic Microorganisms*, A.J.B. Zehnder, Ed. Wiley, New York.

Oremland, R.S., and S. Polcin. 1982. Methanogenesis and sulfate reduction: Competitive and noncompetitive substrates in estuarine sediments. Appl. Environ. Microbiol. 44: 1270–1276.

Ortiz-Roque, C.M., and T.C. Hazen. 1987. Abundance and distribution of Legionellaceae in Puerto Rican water. Appl. Environ. Microbiol. 53: 2231–2236.

Oste, C. 1988. Polymerase chain reaction. Biotechniques 6: 162–167.

Ottolenghi, A.C., and V.V. Hamparian. 1987. Multiyear study of sludge application to farmland: Prevalence of bacterial enteric pathogens and antibody status of farm families. Appl. Environ. Microbiol. 53: 1118–1124.

Ou, C.Y., S. Kwok, S.W. Mitchell, D.H. Mack, J.J. Sninsky, J.W. Krebs, P. Feorino, D. Warfield, and G. Schochetman. 1988. DNA amplification for direct detection of HIV-1 in DNA of peripheral blood mononuclear cells. Science 239: 295–297.

Ouellette, R.P. 1991. A perspective on water pollution. Nat. Environ. J. 1: 20–24.

Owen, W.F., D.C. Stuckey, J.B. Healy, Jr., L.Y. Young, and P.L. McCarty. 1979. Bioassay for monitoring biochemical potential and anaerobic toxicity. Water Res. 13: 485–492.

Ozaki, H., Z. Liu, and Y. Terashima. 1991. Utilization of microorganisms immobilized with magnetic particles for sewage and wastewater treatment. Water Sci. Technol. 23: 1125–1136.

Painter, H.A. 1970. A review of literature on inorganic nitrogen metabolism in microorganisms. Water Res. 4:393–450.

Painter, H.A., and J.E. Loveless. 1983. Effect of temperature and pH value on the growth-rate constants of nitrifying bacteria in the activated sludge process. Water Res. 17: 237–248.

Painter H.A., and M. Viney. 1959. Composition of domestic sewage. J. Biochem. Microbiol. Technol. 1: 143–162.

Palchak, R.B., R. Cohen, M. Ainslie, and C. Lax Hoerner. 1988. Airborne endotoxin associated with industrial-scale production of protein products in gram-negative bacteria. Am. Ind. Hyg. Assoc. 49: 420–421.

Palm, J.C., D. Jenkins, and D.S. Parker. 1980. Relationship between organic loading, dissolved oxygen concentration and sludge settleability in the completely-mixed activated sludge process. J. Water Pollut. Control Fed. 52: 2484–2506.

Palmer, C.J., Y-L Tsai, A.L. Lang, and L.R. Sangermano. 1993. Evaluation of Colilert-Marine Water for detection of total coliforms and *Escherichia coli* in the marine environment. Appl. Environ. Microbiol. 59: 786–790.

Palmer, S.R., P.R. Gully, J.M. White, A.D. Pearson, W.G. Suckling, D.M. Jones, J.C.L. Rawes, and J.L. Penner. 1983. Waterborne outbreak of *Campylobacter* gastroenteritis. Lancet i: 287–290.

Palmgren, U., G. Strom, G. Blomquist, and P. Malmberg. 1986. The Nucleopore filter method: A technique for enumeration of viable and nonviable airborne microorganisms. Am. J. Ind. Med. 10: 325–327.

Pancorbo, O.C., G. Bitton, S.R. Farrah, G.E. Gifford, and A.R. Overman. 1988. Poliovirus retention in soil columns after application of chemical- and polyelectrolyte-conditioned dewatered sludges. Appl. Environ. Microbiol. 54: 118–123.

Panicker, P.V., and K.P. Krishnamoorthi. 1978. Elimination of enteric parasites during sewage treatement. Indian Assoc. Water Pollut. Control Tech. Annu. 5: 130–138. Cited by Feachem et al., 1983.

Panicker, P.V., and K.P. Krishnamoorthi. 1981. Parasite egg and cyst reduction in oxidation ditches and aerated lagoons. J. Water Pollut. Control Fed. 53: 1413–1419.

Parhad, N.M., and N.U. Rao. 1974. Effect of pH on survival of *Escherichia coli*. J. Water Pollut. Control Fed. 46: 980–986.

Park, N., T.N. Blandford, M.Y. Corapcioglu, and P.S. Huyakorn. 1990. VIRALT: A modular semi-analytical and numerical model for simulating viral transport in ground water. Office of Drinking Water, U.S. Environmental Protection Agency, Washington, DC.

Parker, D.S., D. Jenkins, and W.J. Kaufman. 1971. Physical conditioning of the activated sludge floc. J. Water Pollut. Control Fed. 43: 1897

Parker, D.S., and T. Richards. 1986. Nitrification in trickling filters. J. Water Pollut. Control Fed. 58: 896–902.

Parker, W.J., D.J. Thompson, J.P. Bell, and H. Melcer. 1993. Fate of volatile organic compounds in municipal activated sludge plants. Water Environ. Res. 65: 58–65.

Parkin, G.F., and R.E. Speece. 1982. Modeling toxicity in methane fermentation systems. J. Environ. Eng. Div. Am. Soc. Civ. Eng. 108: 515–531.

Parkin, G.F., R.E. Speece, C.H.J. Yang, and W.M. Kocher. 1983. Response of methane fermentation systems to industrial toxicants. J. Water Pollut. Control Fed. 55: 44–53.

Pasquill, F. 1961. The estimation of the dispersion of windborne material. Meteorol. Mag. 90: 33–49.

Patel, G.B., B.J. Agnew, and C.J. Dicaire. 1991. Inhibition of pure culture of methanogens by benzene ring compounds. Appl. Environ. Microbiol. 57: 2969–2974.

Patterson, J.W. 1984. Perspectives on opportunities for genetic engineering applications in industrial pollution control, pp. 187–193, in: *Genetic Control of Environmental Pollutants*, G.S. Ommen and A. Hollaender, Eds. Plenum, New York.

Patti, A.M., A.L. Santi, R. Gabrielli, S. Fiamma, M. Cauletti, and A. Pana. 1987. Hepatitis A virus and poliovirus 1 inactivation in estuarine water. Water Res. 21: 1335–1338.

Paul, E.A., and F.E. Clark. 1989. *Soil Microbiology and Biochemistry*. Academic, San Diego, CA.

Pavoni, J.L., M.W. Tenney, and W.F. Echelberger. 1972. Bacterial exocellular polymers and biological flocculation. J. Water Pollut. Control Fed. 44: 414–431.

Payment, P. 1989a. Elimination of viruses and bacteria during drinking water treatment: Review of 10 years of data from the Montreal metropolitan area, pp. 59–65, in: *Biohazards of Drinking Water Treatment*, R.A. Larson, Ed. Lewis Pubs., Chelsea, MI.

Payment, P. 1989b. Bacterial colonization of domestic reverse-osmosis filtration units. Can. J. Microbiol. 35: 1065–1067.

Payment, P. 1991. Fate of human enteric viruses, coliphages, and *Clostridium perfringens* during drinking-water treatment. Can. J. Microbiol. 37: 154–157.

Payment, P., E. Franco, L. Richardson, and J. Siemiatycki. 1991. Gastrointestinal health effects associated with the consumption of drinking water produced by point-of-use domestic reverse-osmosis filtration units. Appl. Environ. Microbiol. 57: 945–948.

Payment, P., E. Franco, and J. Siemiatycki. 1993. Absence of relationship between health effects due to tapwater consumption and drinking water quality parameters. Water Sci. Technol. 27: 137–143.

Payment, P., F. Gamache, and G. Paquette. 1989. Comparison of microbiological data from two water filtration plants and their distribution system. Water Sci. Technol. 21: 287–289.

Paxeus, N., P. Robinson, and P. Balmer. 1992. Study of organic pollutants in municipal wastewater in Goteborg, Sweden. Water Sci. Technol. 25: 249–256.

Pearson, H.W., D.D. Mara, S.W. Mills, and D.J. Sallman. 1987. Physicochemical parameters influencing faecal bacterial survival in waste stabilization ponds. Water Sci. Technol. 19: 145–152.

Peck, M.W. 1989. Changes in concentration of coenzyme F_{420} analogs during batch growth of *Metanosarcina barkeri* and *Methanosarcina mazei*. Appl. Environ. Microbiol. 55: 940–945.

Pedersen, D.C. 1981. *Density Levels of Pathogenic Organisms in Municipal Wastewater Sludge: A Literature Review*. EPA-600/2–81–170. U.S. EPA, Cincinnati, OH.

Pedersen, D.C. 1983. Effectiveness of sludge treatment processes in reducing levels of bacteria, viruses, and parasites, pp. 9–31, in: *Biological Health Risks of Sludge Disposal to Land in Cold Climates*, P.M. Wallis and D.L. Lehmann, Eds. University of Calgary Press, Calgary, Canada.

Pedersen, K. 1990. Biofilm development on stainless steel and PVC surfaces in drinking water. Water Res. 24: 239–243.

Peeters, J.E., E.A. Mazas, W.J. Masschelein, I.V. Martinez de Maturana, and E. Debacker. 1989. Effect of disinfection of drinking water with ozone or chlorine dioxide on survival of *Cryptosporidium parvum* oocysts. Appl. Environ. Microbiol. 55: 1519–1522.

Pelletier, P.A., and G.C. DuMoulin. 1988. Comparative resistance of mycobacteria to chloramine. Abstr. Annu. Meet. Am. Soc. Microbiol., Washington, DC.

Peltier, W.H., and C.I. Weber. 1985. *Methods for Measuring the Acute Toxicity of Effluents to Freshwater and Marine Organisms* (3rd ed.). EPA--600/4–85/013. U.S. EPA, Cincinnati, OH.

Persson, P.E. 1979. Notes on muddy odour. III. Variability of sensory response to 2-methylisoborneol. Aqua Fenn. 9: 48–52. Cited by Izaguirre et al., 1982.

Pescod, M.B., and J.V. Nair. 1972. Biological disk filtration for tropical waste treatment: Experimental studies. Water Res. 6: 1509–1523.

Petrilli, F.L., G.P. DeRenzi, P. Orlando, and S. DeFlora, 1980. Microbiological evaluation of coastal water quality in the Tyrrhenian Sea. Prog. Water Technol. 12: 129–136.

Pfenning, N. 1978. General physiology and ecology of photosynthetic bacteria, pp. 3–18, in: *The Photosynthetic Bacteria*, R.K. Clayton and W.R. Sistrom, Eds. Plenum, New York.

Pfuderer, G. 1985. Influence of lime treatment of raw sludge on the survival of pathogens, on the digestability of the sludge and on the production of methane: Hygienic investigations, pp. 85–97, in: *Inactivation of Microorganisms in Sewage Sludge by Stabilization Processes*, Strauch, D., A.H. Havelaar, and P. L'Hermite, Eds. Elsevier Applied Science, London.

Phelps, P.A., S.K. Agarwal, G.E. Speitel, Jr., and G. Georgiou. 1992. *Metlylosinus trichosporium* OB3b mutants having constitutive expression of soluble methane monooxygenase in the presence of high levels of copper. Appl. Environ. Microbiol. 58: 3701–3708.

Phelps, T.J., J.J. Niedzielski, K. Malachowski, R.M. Schram, S.E. Herbes, and D.C. White. 1991. Biodegradation of mixed-organic wastes by microbial consortia in continuous-recycle expanded-bed bioreactors. Environ. Sci. Technol. 25: 1461–1465.

Phelps, T.J., J.J. Niedzielski, R.M. Schram, S.E. Herbes, and D.C. White. 1990. Biodegradation of trichloroethylene in continuous-recycle expanded-bed biorectors. Appl. Environ. Microbiol. 56: 1701–1709.

Phillips, S.J., D.S. Dalgarn, and S.K. Young. 1989. Recombinant DNA in wastewater: pBR322 degradation kinetics. J. Water Pollut. Control Fed. 61: 1588–1595.

Pietri, C., and J.-P. Breittmayer. 1976. Etude de la survie d'un enterovirus en eau de mer. Rev. Int. Oceanogr. Med. 42: 77–86.

Pike, E.B., E.G. Carrington, and S.A. Harman. 1988. Destruction of Salmonellae, enteroviruses and ova of parasites by pasteurization and anaerobic digestion. In: *Proceedings of the International Conference on Water and Wastewater Microbiology*, Newport Beach, CA, Feb. 8–11, 1988, Vol. 1.

Pilly, E. 1990. *Maladies Infectieuses* (11th ed.). Editions C&R, La Madeleine, France.

Piotrowski, M.R. 1989. Bioremediation: Testing the waters. Civ. Eng. 59(8): 51–53.

Pipes, W.O. 1974. Control bulking with chemicals. Water Waste Eng. (Nov. 1974).

Pipes, W.O. 1978. Actinomycetes scum formation in activated sludge processes. J. Water Pollut. Control Fed. 5: 628–634.

Pipes, W.O., and W.B. Cooke. 1969. Proc. Ind. Waste Conf. Purdue Univ. 53: 170–182.

Pipyn, P., W. Verstraete, and J.P. Ombregt. 1979. A pilot scale anaerobic upflow reactor treating distillery wastewaters. Biotechnol. Lett. 1: 495–500.

Pisarczyk, K.S., and L.A. Rossi. 1982. Sludge odor control and improved dewatering with potassium permanganate. Paper presented at the 55th Annual Conference of the Water Pollution Control Federation, St. Louis, Oct. 5, 1982.

Pitt, P., and D. Jenkins. 1990. Causes and control of *Nocardia* in activated sludge. J. Water Pollut. Control Fed. 62: 143–150.

Placencia, A.M., J.T. Peeler, G.S. Oxborrow, and J.W. Danielson. 1982. Comparison of bacterial recovery by Reuter centrifugal air sampler and Slit-to-Agar sampler. Appl. Environ. Microbiol. 44: 512–513.

Plissier, M., and P. Therre, 1961. Recherches sur l'inactivation *in vitro* du poliovirus dans l'eau de mer. Ann. Inst. Pasteur 101:840–844.

Poggi, R. 1990. Impacts sanitaires des contaminations microbiologiques. In: *La Mer et les Rejets Urbains*, IFREMER Proc. 11: 115–132.

Polprasert, C. 1989. *Organic Wastes Recycling*. Wiley, Chichester, U.K.

Polprasert, C., M.G. Dissanayake, and N.C. Thanh. 1983. Bacterial die-off kinetics in waste stabilization ponds. J. Water Pollut. Control Fed. 55: 285–296.

Pomeroy, R.D. 1982. Biological treatment of odorous air. J. Water Pollut. Control Fed. 54: 1541–1545.

Pontius, F.W. 1992. A current look at the federal drinking water regulations. J. Am. Water Works Assoc. 84: 36–50.

Pope, R.J., and J.M. Lauria. 1989. Odors: The other effluent. Civ. Eng. 59(8): 42–44.

Popes, D.H., R.J. Soracco, H.K. Gill, and C.B. Fliermans. 1982. Growth of *Legionella pneumophila* in two-membered cultures with green algae and cyanobacteria. Curr. Microbiol. 7: 319–322.

Portier, R.J. 1986. Chitin immobilization systems for hazardous waste detoxification and biodegradation, pp. 229–244, in: *Immobilisation of Ions by Bio-Sorption*, H. Eccles and S. Hunt, Eds. Ellis Horwood, Chichester, U.K.

Pourcher, A.-M., L.A. Devriese, J.F. Hernandez, and J.M. Delattre. 1991. Enumeration by a miniaturized method of *Escherichia coli*, *Streptococcus bovis* and enterococci as indicators of the origin of faecal pollution of waters. J. Appl. Bacteriol. 70: 525–530.

Poynter, S.F.B., and J.S. Slade. 1977. The removal of viruses by slow sand filtration. Prog. Water Technol. 9: 75–78.

van Praagh, A.D., P.D. Gavaghan, and J.L. Sykora. 1993. *Giardia muris* cyst inactivation in anaerobic digester sludge. Water Sci. Technol. 27: 105–109.

Pretorius, W.A. 1971. Some operational characteristic of a bacterial disk unit. Water Res. 5: 1141–1146.

Prevot, J., S. Dubrou, and J. Marechal. 1993. Detection of human hepatitis A virus in environmental waters by an antigen-capture polymerase chain reaction method. Water Sci. Technol. 27: 227–233.

Price, G.J. 1982. Use of an anoxic zone to improve activated sludge settleability. in: *Bulking of Activated Sludge: Preventive and Remedial Methods*, B. Chambers and E.J. Tomlinson, Eds. Ellis Horwood, Chichester, U.K.

Proulx, D., and J. de la Noue. 1988. Removal of macronutrients from wastewater by immobilized microalgae, pp. 301–310, in: *Bioreactor Immobilized Enzymes and Cells: Fundamentals and Applications*, M. Moo-Young, Ed. Elsevier Applied Science, New York.

Pude, R.A., G.J. Jackson, J.W. Bier, T.K. Sawyer, and N.G. Risty. 1984. Survey of fresh vegetables for nematodes, amoebae and *Salmonella*. J. Assoc. Off. Anal. Chem. 67: 613–617.

van Puffelen, J. 1983. The importance of activated carbon, pp. 1–8, in: *Activated Carbon in Drinking Water Technology*. Research Report, AWWA Research Foundation, Denver, CO.

Pujol, R., P. Duchene, S. Schetrite, and J.P. Canler. 1991. Biological foams in activated sludge plants: Characterization and situation. Water Res. 25: 1399–1404.

Qin, D., P.J. Bliss, D. Barnes, and P.A. Fitzgerald. 1991. Bacterial (total coliform) die off in maturation ponds. Water Sci. Technol. 23: 1525–1534.

Qualls, R.G., M.H. Dorfman, and J.D. Johnson. 1989. Evaluation of the efficiency of ultraviolet disinfection systems. Water Res. 23: 317–325.

Qualls, R.G., M.P. Flynn, and J.D. Johnson. 1983. The role of suspended particles in ultraviolet irradiation. J. Water Pollut. Control Fed. 55: 1280–1285.

Qualls, R.G., and J.D. Johnson. 1983. Bioassay and dose measurement in U.V. disinfection. Appl. Environ. Microbiol. 45: 872–877.

Qualls, R.G., S.F. Ossoff, J.C.H. Chang, M.H. Dorfman, C.M. Dumais, D.C. Lobe, and J.D. Johnson. 1985. Factors controlling sensitivity in ultraviolet disinfection of secondary effluents. J. Water Pollut. Control Fed. 57: 1006–1011.

Radehaus, P.M., and S.K. Schmidt. 1992. Characterization of a novel *Pseudomonas* sp. that mineralizes high concentrations of pentachlorophenol. Appl. Environ. Microbiol. 58: 2879–2885.

Rands, M.B., D.E. Cooper, C.P. Woo, G.C. Fletcher, and K.A. Rolfe. 1981. Compost filters for H_2S removal from anaerobic digestion and rendering exhausts. J. Water Pollut. Control Fed. 53: 185–189.

Rao, V.C., S.B. Lakhe, S.V. Waghmare, and P. Dube, 1977. Virus removal in activated sludge sewage treatment. Prog. Water Technol. 9: 113–127.

Rao, V.C., T.G. Metcalf, and J.L. Melnick. 1986. Removal of pathogens during wastewater treatment, pp. 531–554, in: *Biotechnology*, Vol. 8, H.J. Rehm and G. Reed, Eds. VCH, Germany.

Rao, V.C., K.M. Seidel, S.M. Goyal, T.G. Metcalf, and J.L. Melnick. 1984. Isolation of enteroviruses from water, suspended solids and sediments from Galveston Bay: Survival of poliovirus and rotavirus adsorbed to sediments. Appl. Environ. Microbiol. 48: 404–409.

Rao, V.C., J.M. Symons, A. Ling, P. Wang, T.G. Metcalf, J.C. Hoff, and J.L. Melnick. 1988. Removal of hepatitis A virus and rotavirus by drinking water treatment. J. Am. Water Works Assoc. 80: 59–67.

Ratto, A., B.J. Dutka, C. Vega, C. Lopez, and A. El-Shaarawi. 1989. Potable water safety assessed by coliphage and bacterial tests. Water Res. 23: 253–255.

Reasoner, D.J. 1990. Monitoring heterotrophic bacteria in potable water, pp. 452–477, in: *Drinking Water Microbiology*, G.A. McFeters, Ed. Springer-Verlag, New York.

Reasoner, D.J., J.C. Blannon, and E.E. Geldreich. 1979. Rapid seven-hour fecal coliform test. Appl. Environ. Microbiol. 38:229–236

Reasoner, D.J., J.C. Blannon, and E.E. Geldreich. 1987. Microbiological characteristics of third-faucet point-of-use devices. J. Am. Water Works Assoc. 79: 60–66.

Reasoner, D.J., and E.E. Geldreich. 1985. A new medium for the enumeration and subculture of bacteria from potable water. Appl. Environ. Microbiol. 49: 1–7.

Rebuhn, M., and G. Engel. 1988. Reuse of wastewater for industrial cooling systems. J. Water Pollut. Control Fed. 60: 237–241.

Reddy, R.K. 1984. Use of aquatic macrophyte filters for water purification, pp. 660–678, in: *Proceedings of the Third AWWA Water Reuse Symposium*, San Diego, CA, Aug. 26–31, 1984, Vol. 2.

Reed, S.C., E.J. Middlebrooks, and R.W. Crites. 1988. *Natural Systems for Wastewater Management and Treatment*. McGraw-Hill, New York.

Reimers, R.S., D.B. McDonell, M.D. Little, T.G. Ackers, and W.D. Henriques. 1986. Chemical inactivation of pathogens in municipal sludges. In: *Control of Sludge Pathogens*, C.A. Sorber, Ed. Water Pollution Control Federation, Washington, DC.

Reines, H.D., and F.V. Cook. 1981. Pneumonia and bacteremia due to *Aeromonas hydrophila*. Chest 80: 264–168.

Reinhartz, A., I. Lampert, M. Herzberg, and F. Fish 1987. A new short-term, sensitive bacterial assay kit for the detection of toxicants. Toxicity Assess. 2: 193–206.

Rendtorff, R.C. 1954. The experimental transmission of human intestinal protozoan parasites: *Giardia lamblia* cysts given in capsules. Am. J. Hyg. 59: 209–212.

Ribas, F., J. Frias, and F. Lucena. 1991. A new dynamic method for the rapid determination of the biodegradable dissolved organic carbon in drinking water. J. Appl. Bacteriol. 71: 371–378.

Rice, E.W., M.J. Allen, D.J. Brenner, and S.C. Edberg. 1991. Assay for ß-glucuronidase in species of the genus *Escherichia* and its application for drinking-water analysis. Appl. Environ. Microbiol. 57: 592–593.

Rice, E.W., M.J. Allen, and S.C. Edberg. 1990. Efficacy of ß-glucuronidase assay for identification of *Escherichia coli* by the defined-substrate technology. Appl. Environ. Microbiol. 56: 1203–1205.

Rice, E.W., and J.C. Hoff. 1981. Inactivation of *Giardia lamblia* cysts by ultraviolet irradiation. Appl. Environ. Microbiol. 42: 546–547.

Rice, E.W., D.J. Reasoner, and P.V. Scarpino. 1988. Determining biodegradable organic matter in drinking water: A progress report. Paper presented at the Water Quality Technology Conference of the American Water Works Association, St. Louis, Nov. 13–17.

Rice, R.G. 1989. Ozone oxidation products—Implications for drinking water treatment, pp. 153–170, in: *Biohazards of Drinking Water Treatent*, R.A. Larson, Ed. Lewis Pubs., Chelsea, MI.

Richard, M.G., D. Jenkins, O. Hao, and G. Shimizu. 1982. *The Isolation and Characterization of Filamentous Microorganisms from Activated Sludge Bulking*. Report No. 81–2. Sanitary Engineering and Environmental Health Research Laboratory, University of California, Berkeley.

Richard, M.G., O. Hao, and D. Jenkins. 1985a. Growth kinetics of *Sphaerotilus* species and their significance in activated sludge bulking. J. Water Pollut. Control Fed. 57: 68–81.

Richard, M.G., G.P. Shimizu, and D. Jenkins. 1985b. The growth physiology of the filamentous organism type 021N and its significance to activated sludge bulking. J. Water Pollut. Control Fed. 57: 1152–1162.

Rickert, D.A., and J.V. Hunter. 1971. General nature of soluble and particulate organics in sewage and secondary effluent. Water Res. 5: 421–436.

Ridgley, S.M., and D.V. Calvin. 1982. Household hazardous waste disposal project. Metro toxicant Program #1. Toxicant Control Planning Section, Seattle, WA.

Ridgway, H.F., E.G. Means, and B.H. Olson. 1981. Iron bacteria in drinking-water distribution systems: Elemental analysis of *Gallionella* stalks, using x-ray energy-dispersive microanalysis. Appl. Environ. Microbiol. 41: 288–297.

Ridgway, H.F., and B.H. Olson. 1981. Scanning electron microscope evidence for bacterial colonization of a drinking-water distribution system. Appl. Environ. Microbiol. 41: 274–287.

Ridgway, H.F., and B.H. Olson. 1982. Chlorine resistance patterns of bacteria from two drinking water distribution systems. Appl. Environ. Microbiol. 44: 972–987.

Riedel, K., K.-P. Lange, H.J. Stein, M. Khun, P. Ott, and F. Scheller. 1990. A microbial sensor for BOD. Water Res. 24: 883–887.

Riehl, M.L., H.H. Wieser, and B.T. Rheins. 1952. Effect of lime-treated water upon survival of bacteria. J. Am. Water Works Assoc. 44: 466–470.

Riesser, V.W., J.R. Perrich, B.B. Silver, and J.R. McCammon. 1977. Possible mechanisms of poliovirus inactivation by ozone, pp. 186–192, in: *Forum on Ozone Disinfection*, E.G. Fochtman, R.G. Rice, and M.E. Browning, Eds. International Ozone Institute, New York.

Rinzema, A., and G. Lettinga. 1988. Anaerobic treatment of sulfate-containing waste water, pp. 65–109, in: *Biotreatment Systems*, Vol. 3, D.L. Wise, Ed. CRC Press, Boca Raton, FL.

Rippey, S.R., and W.D. Watkins. 1992. Comparative rates of disinfection of microbial indicator organisms in chlorinated sewage effluents. Water Sci. Technol. 26: 2185–2189.

Rippon, J.W. 1974. *Medical Mycology: The Pathogenic Fungi and Pathogenic Actinomycetes*. W.B. Saunders, Philadelphia.

Rittmann, B.E. 1984. Needs and stategies for genetic control: Municipal wastes, pp. 215–228, in: *Genetic Control of Environmental Pollutants*, G.S. Omenn and A. Hollaender, Eds. Plenum, New York.

Rittmann, B.E. 1987. Aerobic biological treatment. Environ. Sci. Technol. 21: 128–136.

Rittmann, B.E. 1989. Biodegradation processes to make drinking water biologically stable, pp. 257–263, in: *Biohazards of Drinking Water Treatment*, R.A. Larson, Ed. Lewis Pubs., Chelsea, MI.

Rittmann, B.E., and C.W. Brunner. 1984. The nonsteady-state biofilm process for advanced organic removal. J. Water Pollut. Control Fed. 56: 874–880.

Rittmann, B.E., L.A. Crawford, C.K. Tuck, and E. Namkung. 1986. *In situ* determination of kinetic parameters for biofilms: Isolation and characterization of oligotrophic biofilms. Biotechnol. Bioeng. 28: 1753–1760.

Rittmann, B.E., D.E. Jackson, and S.L Storck. 1988. Potential for treatment of hazardous organic chemicals with biological processes, pp. 15–64, in: *Biotreatment Systems*, Vol. 3, D.L. Wise, Ed. CRC Press, Boca Raton, FL.

Rittmann, B.E., and P.L. McCarty. 1980a. Model of steady-state biofilm kinetics. Biotechnol. Bioeng. 22: 2243–2357.

Rittmann, B.E., and P.L. McCarty. 1980b. Evaluation of steady-state biofilm kinetics. Biotechnol. Bioeng. 22: 2359–2373.

Rittmann, B.E., and P.L. McCarty. 1981. Substrate flux into biofilms of any thickness. J. Environ. Eng. Div. Am. Soc. Civ. Eng. 107: 831–849.

Rittmann, B.E., B.F. Smets, and D.A. Stahl. 1990.

The role of genes in biological treatment processes. Environ. Sci. Technol. 24: 23–29.

Rittmann, B.E., and V.L. Snoeyink. 1984. Achieving biologically stable drinking water. J. Am. Water Works Assoc. 76: 106–114.

Roberton, A.M., and R.S. Wolfe. 1970. ATP pools in *Methanobacterium*. J. Bacteriol. 102: 43–51.

Roberts, P.V., L. Semprini, G.D. Hopkins, D. Grbic-Galic, P.L. McCarty, and M. Reinhard. 1989. *In situ* aquifer restoration of chlorinated aliphatics by methanotrophic bacteria. EPA-600/S2–89/033, U.S. EPA, Ada, OK.

Robertson, L.J., A.T. Campbell, and H.V. Smith. 1992. Survival of *Cryptosporidium parvum* oocysts under various environmental pressures. Appl. Environ. Microbiol. 58: 3494–3500.

Robison, B.J. 1984. Evaluation of a fluorogenic assay for detection of *E. coli* in foods. Appl. Environ. Microbiol. 48: 285–288.

Rochkind-Dubinsky, M.L., G.S. Sayler, and J.W. Blackburn. 1987. *Microbiological Decomposition of Chlorinated Aromatic Compounds*, Marcel Dekker, New York.

Rodgers, F.G., P. Hufton, E. Kurzawska, C. Molloy, and S. Morgan. 1985. Morphological response of human rotavirus to ultraviolet radiation, heat and disinfectants. J. Med. Microbiol. 20: 123–130.

Rodgers, M.R., C.M. Bernardino, and W. Jakubowski. 1992. A comparison of methods for extracting amplifiable *Giardia* DNA from various environmental samples. Paper presented at the International Water Pollution Research Conference International Symposium, Washington, DC, May 26–29, 1992.

Rodriguez, R.L., and R.C. Tait. 1983. *Recombinant DNA Techniques: An Introduction*. Addison-Wesley, Reading, MA.

Rogers, S.E., and W.C. Lauer. 1986. Disinfection for potable reuse. J. Water Pollut. Control Fed. 58: 193–198.

Rogers, S.E., and W.C. Lauer. 1992. Denver's demonstration of potable water reuse: Water quality and health effects testing. Water Sci. Technol. 26: 1555–1564.

Rollinger, Y., and W. Dott. 1987. Survival of selected bacterial species in sterilized activated carbon filters and biological activated carbon filters. Appl. Environ. Microbiol. 53: 77–781.

Rook, J.J. 1974. Formation of haloforms during chlorination of natural waters. Water Treat. Exam. 23: 234–243.

Rosas, I., A. Baez, and M. Coutino. 1984. Bacteriological quality of crops irrigated with wastewater in the Xochimilco plots, Mexico City, Mexico. Appl. Environ. Microbiol. 47: 1074–1079.

Rose, J.B. 1986. Microbial aspects of wastewater reuse for irrigation. CRC Crit. Rev. Environ. Control 16: 231–256.

Rose, J.B. 1988. Occurrence and significance of *Cryptosporidium* in water. J. Am. Water Works Assoc. 80: 53–58.

Rose, J.B. 1990. Occurrence and control of *Cryptosporidium* in drinking water, pp. 294–321, in: *Drinking Water Microbiology*, G.A. McFeters, Ed. Springer-Verlag, New York.

Rose, J.B., A. Cifrino, M.S. Madore, C.P. Gerba, C.R. Sterling, and M.J. Arrowood. 1986. Detection of *Cryptosporidium* from wastewater and freshwater environments. Water Sci. Technol. 18: 233–237

Rose, J.B., R. De Leon, and C.P. Gerba. 1989a. *Giardia* and virus monitoring of sewage effluents in the state of Arizona. Water Sci. Technol. 21: 43–47.

Rose, J.B., R. de Leon, and C.P. Gerba. 1989b. *Giardia* and virus monitoring of sewage effluents in the State of Arizona. Water Sci. Technol. 21: 43–47.

Rose, J.B., and C.P. Gerba. 1991. Assessing potential health risks from viruses and parasites in reclaimed water in Arizona and Florida, USA. Water Sci. Technol. 23:2091–2098.

Rose, J.B., C.P. Gerba, and W. Jakubowski. 1991. Survey of potable water supplies for *Cryptosporidium* and *Giardia*. Environ. Sci. Technol. 25: 1393–1399.

Rose, J.B., L.K. Landeen, K.R. Riley, and C.P. Gerba. 1989c. Evaluation of immunofluoresecence techniques for detection of *Cryptosporidium* oocysts and *Giardia* cysts from environmental samples. Appl. Environ. Microbiol. 55: 3189–3196.

Rose, J.B., C.E. Musial, M.J. Arrowood, C.R. Sterling, and C.P. Gerba. 1985. Development of a method for the detection of *Cryptosporidium* in drinking water. Paper presented at the Water Technology Conference, AWWA, Houston, TX, Dec. 8–11, 1985.

Rosenberg, M.L., K.K. Hazlet, J. Schaefer, J.G. Wells, and R.C. Pruneda. 1976. Shigellosis from swimming. J. Am. Med. Assoc. 236: 1849–52.

Rosenszweig, W.D., H.A. Minnigh, and W.O. Pipes.

1983. Chlorine demand and inactivation of fungal propagules. Appl. Environ. Microbiol. 45: 182–186.

Rosenszweig, W.D., H.A. Minnigh, and W.O. Pipes. 1986. Fungi in potable distribution systems. J. Am. Water Works Assoc. 78: 53–55.

Rosenszweig, W.D., and W.O. Pipes. 1989. Presence of fungi in drinking water, pp. 85–93, in: *Biohazards of Drinking Water Treatment*, R.A. Larson, Ed. Lewis Pubs., Chelsea, MI.

Rosenszweig, W.D., G. Ramirez-Toro, H. Minnigh, and W.O. Pipes. 1989. Fungi in potable water. Abstr. Annu. Meet. Am. Soc. Microbiol.,New Orleans.

Rosenthal, S.L., J.H. Zuger, and E. Apollo. 1975. Respiratory colonization with *Pseudomonas putrefaciens* after near-drowning in salt water. J. Clin. Pathol. 64: 382.

Ross, I.S., and C.C. Townsley. 1986. The uptake of heavy metals by filamentous fungi, pp. 49–58, in: *Immobilisation of Ions by Bio-Sorption*, H. Eccles and S. Hunt, Eds. Ellis Horwood, Chichester, U.K.

Rouf, M.A., and J.L. Stokes. 1962. Isolation and identification of the sudanophilic granules of *Sphaerotilus natans*. J. Bacteriol. 83: 343–347.

Roy, D., P.K.Y. Wong, R.S. Engelbrecht, and E.S.K. Chian. 1981. Mechanism of enteroviral inactivation by ozone. Appl. Environ. Microbiol. 41: 718–723.

Rozich, A.F., R.M. Sykes, G.B. Walkenshaw, and D.R. Rodgers. 1982. Control of algal filamentous bulking at the Southerly wastewater treatment Plant. J. Water Pollut. Control Fed. 54: 231–237.

Rubin, A.J. 1988. Factors affecting the inactivation of *Giardia* cysts by monochloramine and comparison with other disinfectants, pp. 224–229, in: *Proceedings: Conference on Current Research in Drinking Water Treatment*. EPA-600/9–88/004, Cincinnati, OH.

Rubin, A.J., J.P. Engel, and O.J. Sproul. 1983. Disinfection of amoebic cysts in water with free chlorine. J. Water Pollut. Control Fed. 55: 1174–1182.

Rudd, T., R.M. Sterritt, and J.N. Lester. 1984. Complexation of heavy metals by extracellular polymers in the activated sludge process. J. Water Pollut. Control Fed. 56: 1260–1268.

Rush, B.A., R.A. Chapman, and R.W. Ineson. 1990. A probable waterborne outbreak of cryptosporidiosis in the Sheffield area. J. Med. Microbiol. 32: 239–242.

Russ, C.F., and W.A. Yanko. 1981. Factors affecting Salmonellae repopulation in composted sludge. Appl. Environ. Microbiol. 41: 597–602.

Rylander, R., K. Andersson, L. Belin, G. Berglund, R. Bergstrom, L.-A. Hanson, M. Lundholm, and I. Mattsby. 1976. Sewage worker's syndrome. Lancet ii: 478–479.

Sabatini, D.A., J.W, Smith, and L.W. Moore. 1990. Treatment of chlordane-contaminated water by the activated rotating biological contactor. J. Environ. Qual. 19: 334–338.

Saber, D.L., and R.L. Crawford. 1985. Isolation and characterization of *Flavobacterium* strains that degrade pentachlorophenol. Appl. Environ. Microbiol. 50: 1512–1518.

Saber, D.L., and R.L. Crawford. 1989. Isolation and characterization of *Flavobacterium* strains that degrade pentachlorophenol. Appl. Environ. Microbiol. 55: 1512–1518.

Saeger, V.W., and E.S. Tucker. 1976. Biodegradation of phthalic acid esters in river water and activated sludge. Appl. Environ. Microbiol. 31: 29–34.

Safferman, R.S., and M.E. Morris. 1976. Assessment of virus removal by a multi-stage activated sludge process. Water Res. 10: 413–420.

Safferman, R.S., A.A. Rosen, C.I. Mashni, and M.E. Morris. 1967. Earthy-smelly substance from a blue-green alga. Environ. Sci. Technol. 1: 429–430.

Sagy, M., and Y. Kott. 1990. Efficiency of rotating biological contactors in removing pathogenic bacteria from domestic sewage. Water Res. 24: 1125–1128.

Sahasrabudhe, A., A. Pande, and V. Modi. 1991. Dehalogenation of a mixture of chloroaromatics by immobilized *Pseudomonas* sp. US1 excels. Appl. Microbiol. Biotech. 35: 830–832.

Sahm, H. 1984. Anaerobic wastewater treatment. Adv. Biochem. Eng. Biotechnol. 29: 84–115.

Saier, M., K. Koch, and J. Wekerle. 1985. Influence of thermophilic anaerobic digestion (55°C) and subsequent mesophilic digestion of sludge on the survival of viruses with and without pasteurization of the digested sludge, pp. 28–37, in: *Inactivation of Microorganisms in Sewage Sludge by Stabilization Processes*, D. Strauch, A.H. Havelaar, and P. L'Hermite, Eds. Elsevier Applied Science, London.

Salkinoja-Salonen, M., P. Middeldorp, M. Briglia, R. Valo, M. Haggblom, and A. McBain. 1989.

Cleanup of old industrial sites, pp. 203–211, in: *Biotechnology and Biodegradation*, D. Kamely, A. Chakrabarty, and G.S. Omenn, Eds. Gulf Publishing Co., Houston.

Salmen, P., D.M. Dwyes, H. Vorse, and W. Kruse. 1983. Whirlpool-associated *Pseudomonas aeruginosa* urinary tract infections. J. Am. Med. Assoc. 250: 2025–2026.

Sandt, C., and D. Herson. 1989. Plasmid mobilization from a genetically engineered microorganism to an environmentally isolated strain of *Enterobacter cloacae* in sterile drinking water. Abstr. Annu. Meet. Am. Soc. Microbiol., New Orleans.

Saqqar, M.M., and M.B. Wood. 1991. Microbiological performance of multistage stabilization ponds for effluent use in agriculture. Water Sci. Technol. 23: 1517–1524.

Saqqar, M.M., and M.B. Pescod. 1992. Modelling coliform reduction in wastewater stabilization ponds. Water Sci. Technol. 26: 1667–1677.

Sarhan, H.R., and H.A. Foster. 1991. A rapid fluorogenic method for the detection of *Escherichia coli* by the production of ß-glucuronidase. J. Appl. Bacteriol. 70: 394–400.

Sarikaya, H.Z., and A.M. Saarci. 1987. Bacterial die-off in waste stabilization ponds. J. Environ. Eng. Div. Am. Soc. Civ. Eng. 113: 366–382.

Sarikaya, H.Z., A.M. Saarci, and A.F. Abdulfattah. 1987. Effect of pond depth on bacterial die-off. J. Environ. Eng. Div. Am. Soc. Civ. Eng. 113: 1350–1362.

Sarner, E. 1986. Removal of particulate and dissolved organics in aerobic fixed-film biological processes. J. Water Pollut. Control Fed. 58: 165–172.

Sato, C., S.W. Leung, and J.L. Schnoor. 1988. Toxic response of *Nitrosomonas europea* to copper in inorganic medium and wastewater. Water Res. 22: 1117–1127.

Sattar, S.A., M.K. Ijaz, C.M. Johnson-Lussenburg, and V.S. Springthorpe. 1984. Effect of relative humidity on the airborne survival of rotavirus SA-11. Appl. Environ. Microbiol. 47: 879–881.

Sattar, S.A., S. Ramia, and J.C.N. Westwood. 1976. Calcium hydroxide (lime) and the elimination of human pathogenic viruses from sewage: Studies with experimentally contaminated (poliovirus type I, Sabin) and pilot plant samples. Can. J. Public Health 67: 221–225.

Sauch, J.F. 1989. Use of immunofluorescence and phase-contrast microscopy for detection and identification of *Giardia* cysts in water samples. Appl. Environ. Microbiol. 50: 1434–1438.

Sauch, J.F., and D. Berman. 1991. Immunofluorescence and morphology of *Giardia lamblia* cysts exposed to chlorine. Appl. Environ. Microbiol. 57: 1573–1575.

Sauch, J.F., D. Flanigan, M.L. Galvin, D. Berman, and W. Jakubowski. 1991. Propidium iodide as an indicator of *Giardia* cyst viability. Appl. Environ. Microbiol. 57: 3243–3247.

Saunders, V.A., and J.R. Saunders. 1987. *Microbial Genetics Applied to Biotechnology: Principles and Techniques of Gene Transfer and Manipulation*. Macmillan, New York.

Savenhed, R., H. Boren, A. Grimwall, B.V. Lundgren, P. Balmer, and T. Hedberg. 1987. Removal of individual off-flavour compounds in water during artificial groundwater recharge and during treatment by alum coagulation/sand filtration. Water Res. 21: 277–283.

Sawyer, C.N., and P.L. McCarty. 1967. *Chemistry for Sanitary Engineers* (2nd ed.). McGraw-Hill, New York.

Sawyer, T.K., E.J. Lewis, M. Galassa, D.W. Lear, M.L. O'Malley, W.N. Adams, and J. Gaines. 1982. Pathogenic amoebas in ocean sediments near wastewater sludge disposal sites. J. Water Pollut. Control Fed. 54: 1318–1323.

Sayama, N., and Y. Itokawa. 1980. Treatment of cresol, phenol and formalin using fixed-film reactors. J. Appl. Bacteriol. 9: 395–403.

Sayler, G.S., and J.W. Blackburn. 1989. Modern biological methods: The role of biolotechnology, pp. 53–71, in: *Biotreatment of Agricultural Wastewater*, M.E. Huntley, Ed. CRC Press, Boca Raton, FL.

Sayler, G.S., A. Breen, J.W. Blackburn, and O. Yagi. 1984. Predictive assessment of priority pollutant bio-oxidation kinetics in activated sludge. Environ. Prog. 3: 153–163.

Sayler, G.S., R. Fox, and J.W. Blackburn, Eds. 1991. *Environmental Biotechnology for Waste Treatment*. Plenum, New York.

Sayler, G.S., and A.C. Layton. 1990. Environmental applications of nucleic acid hybridization. Annu. Rev. Microbiol. 44: 625–648.

Scalf, M.R., W.J. Dunlap, and J.F. Kreissel. 1977. *Environmental Effects of Septic Tank Systems*. EPA-600/3–77–096.

Schaefer, F.W. III, C.H. Johnson, C.H. Hsu, and E.W. Rice. 1991. Determination of *Giardia lamblia* infective dose for the mongolian gerbil (*Meriones unguiculatus*). Appl. Environ. Microbiol. 57: 2408–2409.

Schaiberger, G.E., T.D. Edmond, and C.P. Gerba. 1982. Distribution of enteroviruses in sediments contiguous with a deep marine sewage outfall. Water Res. 16: 1425–1428.

Schaub, S.A., and C.A. Sorber. 1977. Virus and bacteria removal from wastewater by rapid infiltration through soil. Appl. Environ. Microbiol. 33: 609–619.

Scheuerman, P.R. 1984. Fate of viruses during aerobic digestion of wastewater sludges. Doctoral dissertation, University of Florida, Gainesville, FL.

Scheuerman, P.R., G. Bitton, A.R. Overman, and G.E. Gifford. 1979. Transport of viruses through organic soils and sediments. J. Environ. Eng. Div. Am. Soc. Civ. Eng. 105: 629–640.

Scheuerman, P.R., S.R. Farrah, and G. Bitton. 1991. Laboratory studies of virus survival during aerobic and anaerobic digestion of sewage sludge. Water Res. 25: 241–245.

Schiemann, D.A. 1990. *Yersinia enterocolitica* in drinking water, pp. 322–339, in: *Drinking Water Microbiology*, G.A. McFeters, Ed. Springer-Verlag, New York.

Schiff, G.M., G.M. Stefanovic, E.C. Young, G.S. Sander, J.K. Pennekamp, and R.L. Ward. 1984a. Studies of echovirus 12 in volunteers: Determination of minimal infectious dose and the effect of previous infection on infectious dose. J. Infect. Dis. 150: 858–866.

Schiff, G.M., G.M. Stefanovic, E.C. Young, and J.K. Pennekamp. 1984b. Minimum human infective dose of enteric virus (echovirus 12) in drinking water. Monogr. Virol. 15: 222–228.

Schink, B. 1988. Principles and limits of anaerobic degradation: Environmental and technological aspects, pp. 771–846, in: *Biology of Anaerobic Microorganisms*, A.J.B. Zehnder, Ed. Wiley, New York.

von Schirnding, Y.E.R., N. Strauss, P. Robertson, R. Kfir, B. Fattal, A. Mathee, M. Franck, and V.J. Cabelli. 1993. Bather's morbidity from recreational exposure to sea water. Water Sci. Technol. 27: 183–186.

Schmidt, S.K., and M. Alexander. 1985. Effect of dissolved organic carbon and second substrates on the biodegradation of organic compounds at low concentrations. Appl. Environ. Microbiol. 49: 822–827.

Schuler, P.F., and M.M. Ghosh. 1990. Diatomaceous earth filtration of cysts and other particulates using chemical additives. J. Am. Water Works Assoc. 82: 67–75.

Schupp, D.G., and S.L. Erlandsen. 1987. A new method to determine *Giardia* cyst viability: Correlation of fluorescein diacetate and propidium iodide staining with animal infectivity. Appl. Environ. Microbiol. 53: 704–707.

Schwartzbrod, L., Ed. 1991. *Virologie des Milieux Hydriques*. TEC & DOC Lavoisier, Paris.

Schwartzbrod, L., C. Jehl-Pietri, S. Boher, B. Hugues, M. Albert, and C. Beril. 1990. Les contaminations par les virus. in: *La Mer et les Rejets Urbains*. IFREMER Proc. 11: 101–114.

Schwartzbrod, L., and C. Mattieu. 1986. Virus recovery from wastewater treatment plant sludges. Water Res. 20: 1011–1013.

Schwartzbrod, J., J.L. Stien, K. Bouhoum, and B. Baleux. 1989. Impact of wastewater treatment on helminth eggs. Water Sci. Technol. 21: 295–297.

Schwartzbrod, L., P. Vilagines, J. Schwartzbrod, B. Sarrette, R. Vilagines, and J. Collomb. 1985. Evaluation of the viral population in two wastewater treatment plants: Study of different sampling techniques. Water Res. 19: 1353–1356.

Schwebach, G.H., D. Cafaro, J. Egan, M. Grimes, and G. Michael. 1988. Overhauling health effects perspectives. J. Water Pollut. Control Fed. 60: 473–479.

Scragg, A., Ed. (1988). *Biotechnology for Engineers: Biological Systems in Technological Processes*. Ellis Horwood, Chichester, U.K.

Segall, B.A., and C.R. Ott. 1980. Septage treatment at a municipal plant. J. Water Pollut. Control Fed. 52: 2145–2157.

Segel, I.H. 1975. *Enzyme Kinetics*. Wiley, New York.

Seidler, R.J., and T.M. Evans. 1983. Analytical methods for microbial water quality. in: *Assessment of Microbiology and Turbidity Standards for Drinking Water*, P.S. Berger and Y. Argaman, Eds. EPA-570–9-83–001.

Seidler, R.J., J.E. Morrow, and S.T. Bagley. 1977. Klebsielleae in drinking water emanating from redwood tanks. Appl. Environ. Microbiol. 33: 893–900.

Sekla, L.H., W. Stackiw, A.G. Buchanan, and S.E. Parker. 1982. *Legionella pneumophila* pneumonia. Can. Med. Assoc. J. 126: 116–119.

Sekla, L., D. Gemmill, J. Manfreda, M. Lysyk, W. Stackiw, C. Kay, C. Hooper, L. van Buckenhout, and G. Eibisch. 1980. Sewage treatment plant workers and their environment: A health study, pp. 281–293, in: *Proceedings of the Symposium on Wastewater Aerosols and Disease*, H. Pahren and W. Jakubowski, Eds. EPA-600/9–80–028, U.S. EPA, Cincinnati, OH.

Servais, P., A. Anzil, and C. Ventresque. 1989. Simple method for determination of biodegradable dissolved organic carbon in water. Appl. Environ. Microbiol. 55: 2732–2734.

Servais, P., G. Billen, and M.-C. Hascoet. 1987. Determination of the biodegradable fraction of dissolved organic matter in waters. Water Res. 21: 445–452.

Servais, P., G. Billen, C. Ventresque, and G.P. Bablon. 1991. Microbial activity in GAC filters at the Choisy-le-Roi treatment plant. J. Am. Water Works Assoc. 83: 62–68.

Severin, B.F. 1980. Disinfection of municipal wastewater effluents with ultraviolet light. J. Water Pollut. Control Fed. 52: 2007–2018.

Seviour, E.M., C.J. Williams, R.J. Seviour, J.A. Soddell, and K.C. Lindrea. 1990. A survey of filamentous bacterial populations from foaming activated sludge plants in eastern states of Australia. Water Res. 24: 493–498.

Sezgin, M. 1982. Variation of sludge volume index with activated sludge characteristics. Water Res. 16: 83–88.

Sezgin, M., D. Jenkins, and J.C. Palm. 1980. Floc size, filament length and settling properties of prototype activated sludge plants. Prog. Water Technol. 12: 171–182.

Sezgin, M., D. Jenkins, and D.S. Parker. 1978. A unified theory of activated sludge bulking. J. Water Pollut. Control Fed. 50: 362–381.

Sezgin, M., and P.R. Karr. 1986. Control of actinomycete scum on aeration basins and clarifiers. J. Water Pollut. Control Fed. 58: 972–977.

Sezgin, M., M.P. Lechevalier, and P.R. Karr. 1988. Isolation and identification of actinomycetes present in activated sludge scum. In: *Proceedings of the International Conference on Water and Wastewater Microbiology*, Newport Beach, CA, Feb. 8–11, 1988, Vol. 1.

Shamat, N.A., and W.J. Maier. 1980. Kinetics of biodegradation of chlorinated organics. J. Water Pollut. Control Fed. 52: 2158–2166.

Shands, K.N., J.L. Ho, R.D. Meyer, G.W. Gorman, P.H. Edelstein, G.F. Mallison, S.M. Finegold, and D.W. Fraser. 1985. Potable water as a source of Legionnaire's disease. J. Am. Med. Assoc. 253: 1412–1416.

Shapiro, J. 1967. Induced rapid release and uptake of phosphate by microorganisms. Science 155: 1269–1271.

Shapiro, J., G.V. Levin, and H.G. Zea. 1967. Anoxically induced release of phosphate in wastewater treatment. J. Water Pollut. Control Fed. 39: 1810–1818.

Sharp, D.G., R. Floyd, and J.D. Johnson. 1976. Initial fast reaction of bromine on reovirus in turbulent flowing water. Appl. Environ. Microbiol. 31: 171–181.

Sharp, D.G., D.C. Young, R. Floyd, and J.D. Johnson. 1980. Effect of ionic environment on the inactivation of poliovirus in water by chlorine. Appl. Environ. Microbiol. 39: 530–534.

Shelton, D.R., and J.M. Tiedje. 1984. General method for determining anaerobic biodegradation potential. Appl. Environ. Microbiol. 47: 850–857.

Sherr, B.F., E.B. Sherr, and R.D. Fallon. 1987. Use of monodispersed, fluorescently labeled bacteria to estimate *in situ* protozoan bacterivory. Appl. Environ. Microbiol. 53: 958–965.

Shiaris, M.P., A.C. Rex, G.W. Pettibone, K. Keay, P. McNamus, M.A. Rex, J. Ebersole, and E. Gallagher. 1987. Distribution of indicator bacteria and *Vibrio parahaemolyticus* in sewage-polluted intertidal sediments. Appl. Environ. Microbiol. 53: 1756–1761.

Shih, J.L., and J. Lederberg. 1974. Effect of chloramine on *Bacillus subtilis* deoxyribonucleic acid. J. Bacteriol. 125: 934–945.

Shilo, M. 1979. *Strategies of Microbial Life in Extreme Environments*. Verlag Chemie, New York.

Shonheit, P., J.K. Kristjansson, and R.K. Thauer. 1982. Kinetic mechanism for the ability of sulfate reducers to outcompete methanogens for acetate. Arch. Microbiol. 132: 285–288.

Shonheit, P., J. Moll, and R.K. Thauer. 1979. Nickel, cobalt, and molybdenum requirement for growth of *Methanobacterium thermoautotrophicum*. Arch. Microbiol. 123: 105–107.

Shoop, D.S., L.L. Myers, and J.B. LeFever. 1990. Enumeration of enterotoxigenic *Bacteroides fragilis* in municipal sewage. Appl. Environ. Microbiol. 56: 2243–2244.

Shugatt, R.H., D.P. O'Grady, S. Banerjee, P.H. Howard, and W.E. Gledhill. 1984. Shake flask biodegradation of 14 commercial phthalate esters. Appl. Environ. Microbiol. 47: 601–606.

Shuval, H.I., Ed. 1977. *Water Renovation and Reuse*. Academic, New York.

Shuval, H. 1991. Health guidelines and standards for wastewater reuse in agriculture: Historical perpectives. Water Sci. Technol. 23: 2073–2080.

Shuval, H.I. 1992. Investigation of cholera and typhoid fever transmission by raw wastewater irrigated in Santiago, Chile. Paper presented at the International Water Pollution Research Conference International Symposium, Washington, DC, May 26–29, 1992.

Shuval, H.I., N. Guttman-Bass, J. Applebaum, and B. Fattal. 1989. Aerosolized enteric bacteria and viruses generated by spray irrigation of wastewater. Water Sci. Technol. 21: 131–135.

Shuval, H.I., A. Thompson, B. Fattal, S. Cymbalista, and Y. Wiener. 1971. Natural virus inactivation processes in seawater. J. Sanit. Eng. Div. Am. Soc. Civ. Eng. 97: 587–600.

Sias, S.R., A.S. Stouthamer, and J.L. Ingraham. 1980. The assimilatory and dissimilatory nitrate reductases of *Pseudomonas aeruginosa* are encoded by different genes. J. Gen. Microbiol. 118: 229–234.

Siefert, E., R.L Irgens, and N. Pfenning. 1978. Phototrophic purple and green bacteria in a sewage treatment plant. Appl. Environ. Microbiol. 35: 38–44.

Silver, S., and T.K. Misra. 1988. Plasmid-mediated heavy metal resistance. Annu. Rev. Microbiol. 42: 717–743.

Sims, J.L., J.M. Suflita, and H.H. Russell. 1991. *Reductive Dehalogenation of Organic Contaminants in Soils and Ground Water*, EPA-540/4–90/054. R.S. Kerr Environmental Research Laboratory, Ada, OK.

Singh, A., M.W. LeChevallier, and G.A. McFeters. 1985. Reduced virulence of *Yersinia enterocolitica* by copper-induced injury. Appl. Environ. Microbiol. 50: 406–411.

Singh, A., and G.A. McFeters. 1986. Recovery, growth, and production of heat-stable enterotoxin by *Escherichia coli* after copper-induced injury. Appl. Environ. Microbiol. 51:738–742.

Singh, A., and G.A. McFeters. 1987. Survival and virulence of copper- and chlorine-stressed *Yersinia enterocolitica* in experimentally infected mice. Appl. Environ. Microbiol. 53: 1768–1774.

Singh, A., and G.A. McFeters. 1990. Injury of enteropathogenic bacteria in drinking water, pp. 368–379, in: *Drinking Water Microbiology*, G.A. McFeters, Ed. Springer-Verlag, New York.

Singh, A., R. Yeager, and G.A. McFeters. 1986a. Assessment of *in vivo* revival, growth and pathogenicity of *E. coli* strains after copper- and chlorine-induced injury. Appl. Environ. Microbiol. 52: 832–837.

Singh, S.N., M. Bassous, C.P. Gerba, and L.M. Kelley. 1986b. Use of dyes and proteins as indicators of virus adsorption to soils. Water Res. 20: 267–272.

Sinton, L.W. 1986. Microbial contamination of alluvial gravel aquifers by septic tank effluents. Water Air Soil Pollut. 28: 407–425.

Sivela, S., and V. Sundman. 1975. Demonstration of *Thiobacillus* type bacteria which utilize methyl sulfides. Arch. Microbiol. 103: 303–304.

Sivonen, K., M. Namikoshi, W.R. Evans, W.W. Carmichael, F. Sun, L. Rouhiainen, R. Luukkainen, and K.L. Rinehart. 1992. Isolation and characterization of a variety of microcystins from seven strains of the cyanobacterial genus *Anabaena*. Appl. Environ. Microbiol. 58: 2495–2500.

Skaliy P, T.A. Thompson, G.W. Gorman, G.K. Morris, H.V. McEachern, and D.C. Mackel. 1980. Laboratory studies of disinfectants against *Legionella pneumophila*. Appl. Environ. Microbiol. 40: 697–700.

Skinhoj, P., F.B. Hollinger, K. Hovind-Hougen, and P. Lous. 1981. Infectious liver diseases in three groups of Copenhagen workers: Correlations of hepatitis A infections to sewage exposure. Arch. Environ. Health 36: 139–143.

Slezak, L.A., and R.C. Sims. 1984. The application and effectiveness of slow sand filtration in the United States. J. Am. Water Works Assoc. 76: 38–43.

Slijkhuis, H., and M.H. Deinema. 1988. Effect of environmental conditions on the occurrence of *Microthrix parvicella* in activated sludge. Water Res. 22: 825–828.

Small, I.C., and G.F. Greaves. 1968. A survey of

animals in distribution systems. Water Treat. Exam. 19: 150–183.

Smith, A.J., and D.S. Hoare. 1968. Acetate assimilation by *Nitrobacter agilis* in relation to it's "obligate autotrophy." J. Bacteriol. 95: 844–855.

Smith, M.R., and R.A. Mah. 1978. Growth and methanogenesis by *Methanosarcina* strain 227 on acetate and methanol. Appl. Environ. Microbiol. 36: 870–879.

Smith M.S., M.K. Firestone, and J.M. Tiedje. 1978. Acetylene inhibition method for short-term measurement of soil denitrification and its evaluation using nitrogen-13. Soil Sci. Soc. Am. J. 42:611–615.

Smith-Somerville, H.E., V.B. Huryn, C. Walker, and A.L. Winters. 1991. Survival of *Legionella pneumophila* in the cold-water ciliate *Tetrahymena vorax*. Appl. Environ. Microbiol. 57: 2742–2749.

Snoeyink, V.L., and D. Jenkins. 1980. *Water Chemistry*. Wiley, New York.

Sobsey, M.D. 1989. Inactivation of health-related microorganisms in water by disinfection processes. Water Sci. Technol. 21: 179–195.

Sobsey, M.D., C.W. Dean, M.E. Knuckles, and R.A. Wagner. 1980. Interactions and survival of enteric viruses in soil materials. Appl. Environ. Microbiol. 40: 92–101.

Sobsey, M.D., T. Fuji, and P.A. Shields. 1988. Inactivation of hepatitis A virus and model viruses in water by free chlorine and monochloramine. In: *Proceedings of the International Conference on Water and Wastewater Microbiology*, Vol. 1. Pergamon Press for IAWPRC.

Sobsey, M.D., and B. Olson. 1983. Microbial agents of waterborne disease. in: *Assessment of Microbiology and Turbidity Standards For Drinking Water*, P.S. Berger and Y. Argaman, Eds. EPA-570–9-83–001, Office of Drinking Water, Washington, D.C.

Sobsey, M.D., P.A. Shields, F.H. Hauchman, R.L. Hazard, and Canton, L.W. III. 1986. Survival and transport of hepatitis A virus in soils, groundwater and wastewater. Water Sci. Technol. 18: 97–106.

Soddell, J.A., and R.J. Seviour. 1990. Microbiology of foaming in activated sludge foams. J. Appl. Bacteriol. 69: 145–176.

Sofer, S.S., G.A. Lewandowski, M.P. Lodaya, F.S. Lakhwala, K.C. Yang, and M. Singh. 1990. Biodegradation of 2-chlorophenol using immobilized activated sludge. J. Water Pollut. Control Fed. 62: 73–80.

Soracco, R.J., H.K. Gill, C.B. Fliermans, and D.H. Pope. 1983. Susceptibility of algae and *Legionella pneumophila* to cooling tower biocides. Appl. Environ. Microbiol. 45: 1254–1260.

Sorber, C.A., B.E. Moore, D.E. Johnson, H.J. Harding, and R.E. Thomas. 1984. Microbiological aerosols from the application of liquid sludge to land. J. Water Pollut. Control Fed. 56: 830–836.

Sorensen, J., D. Christensen, and B.B. Jorgensen. 1981. Volatile fatty acids and hydrogen as substrates for sulfate-reducing bacteria in anaerobic marine sediments. Appl. Environ. Microbiol. 42: 5–11.

Speece, R.E. 1983. Anaerobic biotechnology for industrial wastewater treatment. Environ. Sci. Technol. 17: 416A-427A.

Speece, R.E., G.F. Parkin, and D. Gallagher. 1983. Nickel stimulation of anaerobic digestion. Water Res. 17: 677–683.

Spendlove, J.C., and K.F. Fannin. 1983. Source, significance and control of indoor microbial aerosols: Human health effects. Public Health Rep. 98: 229–244.

Sproul, O.J., R.M. Pfister, and C.K. Kim. 1982. The mechanism of ozone inactivation of water borne viruses. Water Sci. Technol. 14: 303–314.

Staley, J.T., J. Crosa, F. DeWalle, and D. Carlson. 1988. *Effect of Wastewater Disinfectants on Survival of R-factor Coliform Bacteria*. EPA-600/S2–87/092.

Stanfield, G., and P.H. Jago. 1987. The development and use of a method for measuring the concentration of assimilable organic carbon in water. Report PRU 1628-M. Water Research Centre, Medmenham, U.K. (Cited by Huck, 1990.)

Stanier, R.Y., E.A. Adelberg, J.L. Ingraham, and M.L. Wheelis. 1979. *Introduction to the Microbial World*. Prentice-Hall, Englewood Cliffs, NJ.

Stanley, S., S.D. Prior, D.J. Leak, and H. Dalton. 1983. Copper stress underlies the fundamental change in intracellular location of methane monooxygenase in methane-oxidizing organisms: Studies in batch and continuous cultures. Biotechnol. Lett. 5: 487–492.

States, S.J., L.F. Conley, M. Ceraso, T.E. Stephenson, R.S. Wolford, R.M. Wadosky, A.M. McNamara, and R.B. Yee. 1985. Effect of metals on

Legionella pneumophila growth in drinking water plumbing systems. Appl. Environ. Microbiol. 50: 1149–1154.

States, S.J., L.F. Conley, S.G. Towner, R.S. Wolford, T.E. Stephenson, A.M. McNamara, R.M. Wadowsky, and R.B. Yee. 1987. An alkaline approach to treating cooling waters for control of *Legionella pneumophila*. Appl. Environ. Microbiol. 53: 1775–1779.

States, S.J., J.M. Kuchta, L.F. Conley, R.S. Wolford, R.M. Wadowsky, and R.B. Yee. 1989. Factors affecting the occurrence of the legionnaires' disease bacterium in public water supplies, pp. 67–83, in: *Biohazards of Drinking Water Treatment*, R.A. Larson, Ed. Lewis Pubs., Chelsea, MI.

States, S.J., R.M. Wadowsky, J.M. Kuchta, R.S. Wolford, L.F. Conley, and R.B. Yee. 1990. *Legionella* in drinking water, pp. 340–367, in: *Drinking Water Microbiology*, G.A. McFeters, Ed. Springer-Verlag, New York.

Stathopoulos, G.A., and T. Vayonas-Arvanitidou. 1990. Detection of *Campylobacter* and *Yersinia* species in waters and their relationship to indicator microorganisms. Int. Symp. Health-Related Water Microbiol., Tubingen, W. Germany, April 1–6, 1990.

Stelzer, W. 1990. Detection and spread of *Campylobacter* in water (Abstract). International Symposium on Health-Related Water Microbiology, Tubingen, West Germany, April 1–6, 1990.

Stenquist, R.J., D.S. Parker, and T.J. Dosh. 1974. Carbon oxidation-nitrification in synthetic media trickling filters. J. Water Pollut. Control Fed. 46: 2327–2339.

Stenstrom, M.K., and S.S. Song. 1991. Effect of oxygen transport limitation on nitrification in the activated sludge process. Res. J. Water Pollut. Control Fed. 63: 208–219.

Sterritt, R.M., and J.N. Lester. 1986. Heavy metals immobilisation by bacterial extracellular polymers, pp. 121–134, in: *Immobilisation of Ions by Bio-Sorption*, H. Eccles and S. Hunt, Eds. Ellis Horwood, Chichester, U.K.

Sterritt, R.M., and J.N. Lester. 1988. *Microbiology for Environmental and Public Health Engineers*. E. & F.N. Spon, London.

Stewart, M.H., and B.H. Olson. 1992a. Impact of growth conditions on resistance of *Klebsiella pneumoniae* to chloramines. Appl. Environ. Microbiol. 58: 2649–2653.

Stewart, M.H., and B.H. Olson. 1992b. Physiological studies of chloramine resistance developed by *Klebsiella pneumonae* under low-nutrient growth conditions. Appl. Environ. Microbiol. 58: 2918–2927.

Stewart, M.H., R.L. Wolfe, and E.G. Means. 1990. Assessment of the bacteriological activity associated with granular activated carbon treatment of drinking water. Appl. Environ. Microbiol. 56: 3822–3829.

St. Laurent, D., C. Blaise, P. Macquarrie, R. Scroggins, and R. Trottier. 1992. Comparative assessment of herbicide phytotoxicity to *Selenastrum capricornutum* using microplate and flask bioassay procedures. Environ. Toxicol. Water Qual. 7: 35–48.

Stormo, K.E., and R.L. Crawford. 1992. Preparation of encapsulated microbial cells for environmental applications. Appl. Environ. Microbiol. 58: 727–730.

Stout, J.E., M.G. Best, and V.L. Yu. 1986. Susceptibility of members of the family Legionellaceae to thermal stress: Implications for heat eradication methods in water distribution systems. Appl. Environ. Microbiol. 52: 396–399.

Stout, J.E., V.L. Yu, and M.G. Best. 1985. Ecology of *Legionella pneumophila* within water distribution systems. Appl. Environ. Microbiol. 49: 221–228.

Stramer, S.L., and D.O. Cliver. 1981. Poliovirus removal from septic tank systems under conditions of saturated flow. Abstr. Annu. Meet. Am. Soc. Microbiol., Dallas, TX.

Stramer, S.L., and D.O. Cliver. 1984. Septage treatments to reduce the numbers of bacteria and polioviruses. Appl. Environ. Microbiol. 48: 566–572.

Strand, S.E., J.V. Wodrich, and H.D. Stensel. 1991. Biodegradation of chlorinated solvents in a sparged, methanotrophic biofilm reactor. Res. J. Water Pollut. Control Fed. 63: 859–867.

Stratton, R.G., E. Namkung, and B.E. Rittmann. 1983. Secondary utilization of trace organics by biofilms on porous media. J. Am. Water Works Assoc. 75: 463–469.

Straub, T.M., I.L. Pepper, and C.P. Gerba. 1992. Persistence of viruses in desert soils amended with anaerobically digested sewage sludge. Appl. Environ. Microbiol. 58: 636–641.

Strauch, D., A.H. Havelaar, and P. L'Hermite, Eds. 1985. *Inactivation of Microorganisms in Sewage Sludge by Stabilization Processes*, Elsevier Applied Science, London.

Streichan, M., J.R. Golecki, and G. Schon. 1990. Polyphosphate-accumulating bacteria from sewage plants with different processes for biological phosphorus removal. FEMS Microbiol. Ecol. 73: 113–124.

Stringer, R., and C.W. Kruse. 1970. Amoebic cystidal properties of halogens in water. Proceedings of the National Specialty Conference on Disinfection, American Society of Civil Engineers.

Strom, A.R., P. Falkenberg, and B. Landfald. 1986. Genetics of osmoregulation in *Escherichia coli*: Uptake and biosynthesis of organic solutes. FEMS Microbiol. Rev. 39: 79–86.

Strom, P.F. 1985a. Effect of temperature on bacterial species diversity in thermophilic solid-waste composting. Appl. Environ. Microbiol. 50: 899–905.

Strom, P.F. 1985b. Identification of thermophilic bacteria in solid-waste composting. Appl. Environ. Microbiol. 50: 906–913.

Strom, P.F., and D. Jenkins. 1984. Identification and significance of filamentous microorganisms in activated sludge. J. Water Pollut. Control Fed. 49: 584–589.

Suflita, J.M., and G.W. Sewell. 1991. Anaerobic biotransformation of contaminants in the subsurface. EPA-600-M-90/024. R.S. Kerr Environmental Research Laboratory, Ada, OK.

Suidan, M.T., W.H. Gross, M. Fong, and J.W. Calvert. 1981. Anaerobic carbon filter for degradation of phenols. J. Environ. Eng. Div. Am. Soc. Civ. Eng. 107: 563–579.

Sujarittanonta, S., and J.H. Sherrard. 1981. Activated sludge nickel toxicity studies. J. Water Pollut. Control Fed. 53: 1314–1322.

Suresh, N., R. Warburg, M. Timmerman, J. Wells, M. Coccia, M.F. Roberts, and H.O Halvorson. 1984. New strategies for the isolation of microorganisms responsible for phosphate accumulation, p.132, in: *Enhanced Biological Phosphorus Removal from Wastewater*, Vol. 1. Proceedings of the IAWPRC Post Conference, Paris.

Sutherland, I.W. 1977. Enzyme action on bacterial surface carbohydrates, pp. 209–245, in: *Surface Carbohydrates of the Prokaryotic Cell*, I.W. Sutherland, Ed. Academic, New York.

Suttle, C.A., and F. Chen. 1992. Mechanisms and rates of decay of marine viruses in seawater. Appl. Environ. Microbiol. 58: 3721–3729.

Sutton, R. 1992. Use of biosurfactants produced by *Nocardia amarae* for removal and recovery of non-ionic organics from aqueous solutions. Water Sci. Technol. 26: 2393–2396.

Swerdlow, D.L., B.A. Woodruff, R.C. Brady, P.M. Griffin, S. Tippen, H.D. Donnell, Jr., E.E. Geldreich, B.J. Payne, A. Meyer, Jr., J.S. Wells, K.D. Greene, M. Bright, N.H. Bean, and P.A. Blake. 1992. A waterborne outbreak in Missouri of *Escherichia coli* 0157:H7 associated with bloody diarrhea and death. Ann. Intern. Med. 117: 812–819.

Switzenbaum, M.S. 1983. Anaerobic treatment of wastewater: Recent developments. ASM News 49: 532–536.

Sykes, G., and F.A Skinner, Eds. 1973. *Actinomycetales: Characteristics and Practical Importance*, Academic, London.

Sykes, J.C. 1989. The use of biological selector technology to minimize sludge bulking. In: *Biological Nitrogen and Phosphorus Removal: The Florida Experience*. TREEO Center, University of Florida, Gainesville, FL.

Sykora, J.L., G. Keleti, R. Roche, D.R. Volk, G.P. Kay, R.A. Burgess, M.A. Shapiro, and E.C. Lippy. 1980. Endotoxins, algae and *Limulus* amoebocyte lysate test in drinking water. Water Res. 14: 829–839.

Sykora, J.L., C.A. Sorber, W. Jakubowski, L.W. Casson, P.D. Cavaghan, and M.A. Shapiro. 1990. Distribution of *Giardia* cysts in wastewater. International Symposium on Health-Related Water Microbiology, Tubingen, West Germany, April 1–6, 1990.

Szymona, M., and W. Ostrwoski. 1964. Inorganic polyphosphate glucokinase of *Mycobacterium phlei*. Biochim. Biophys. Acta 85: 283–295.

Tago, Y., and K. Aida. 1977. Exocellular micropolysaccharide closely related to bacterial floc formation. Appl. Environ. Microbiol. 34: 308–314.

Takase, L., T. Omori, and Y. Minoda. 1986. Microbial degradation products from biphenyl-related compounds. Agric. Biol. Chem. 50: 681–686.

Talbot, H.W., J.E. Morrow, and R.J. Seidler. 1979. Control of coliform bacteria in finished drinking water stored in redwood tanks. J. Am. Water Works Assoc. 71: 349–353.

Tanaka, K., M. Tada, T. Kimata, S. Harada, Y. Fujii, T. Mizugushi, N. Mori, and H. Emori. 1991. Development of new nitrogen system using nitrifying bacteria in synthetic resin pellets. Water Sci. Technol. 23: 681–690.

Tartera, C., and J. Jofre. 1987. Bacteriophage active against *Bacteroides fragilis* in sewage-polluted waters. Appl. Environ. Microbiol. 53: 1632–1637.

Tartera, C., F. Lucena, and J. Jofre. 1989. Human origin of *Bacteroides fragilis* bacteriophages present in the environment. Appl. Environ. Microbiol. 55: 2696–2701.

Taylor, D.N., K.T. McDermott, J.R. Little, J.G. Wells, and M.J. Blaser. 1983. *Campylobacter* enteritis from untreated water in the Rocky Mountains. Ann. Intern. Med. 99: 38–40.

Taylor, G.R., and M. Butler. 1982. A comparison of the virucidal properties of chlorine, chlorine dioxide, bromine chloride and iodine. J. Hyg. 89: 321–328.

Taylor, R.H, M.J. Allen, and E.E. Geldreich. 1979. Testing of home carbon filters. J. Am. Water Works Assoc. 71: 577–581.

Teltsch, B., and E. Katzenelson. 1978. Airborne enteric bacteria and viruses from spray irrigation with wastewater. Appl. Environ. Microbiol. 35: 290–296.

Teltsch, B., S. Kedmi, Y. Bonnet, Borenzstajn-Rotem, and E. Katzenelson. 1980. Isolation and identification of pathogenic microorganisms at wastewater irrigation fields: Ratios in air and wastewater. Appl. Environ. Microbiol. 39: 1183–1190.

Tenney, M.W., and W. Stumm. 1965. Chemical flocculation of microorganisms in biological waste treatment. J. Water Pollut. Control Fed. 37: 1370–1388.

Tetreault, M.J., A.H. Benedict, C. Kaempfer, and E.F. Barth. 1986. Biological phosphorus removal: A technology evaluation. J. Water Pollut. Control Fed. 58: 823–837.

Thomas, J.M., and C.H. Ward. 1989. *In situ* biorestoration of organic contaminants in the subsurface. Environ. Sci. Technol. 23: 760–766.

Thorne, P.S., M.S. Kiekhaefer, P. Whitten, and K.J. Donham. 1992. Comparison of bioaerosol sampling methods in barns housing swine. Appl. Environ. Microbiol. 58: 2543–2551.

Thyndall, R.L., and E.L. Domingue, 1982. Cocultivation of *Legionella pneumophila* and free-living amoebae. Appl. Environ. Microbiol. 44: 954–959.

Tiedje, J.M. 1988. Ecology of denitrification and dissimilatory nitrate reduction to ammonium, pp. 179–244, in: *Biology of Anaerobic Microorganisms*, A.J.B. Zehnder, Ed. Wiley, New York.

Tiedje, J.M., S.A. Boyd, and B.Z. Fathepure. 1987. Anaerobic biodegradation of aromatic hydrocarbons. Dev. Ind. Microbiol. 27: 117–127.

Tiedje, J.M., R. Colwell, Y.L. Grossman, R.E. Hodson, R.E. Lenski, R.N. Mack, and P.J. Regal. 1989. The planned introduction of genetically engineered organisms: Ecological considerations and recommendations. Soc. Ind. Microbiol. News 39: 149–165.

Tillet, D.M., and A.S. Myerson. 1987. The removal of pyritic sulfur from coal employing *Thiobacillus ferrooxidans* in a packed column recator. Biotechnol. Bioeng. 29: 146–150.

Tim, U.S., and S. Mostaghimi. 1991. Model for predicting virus movement through soils. Ground Water 29: 251–259.

Tison, D.L, D.H. Pope, W.B. Cherry, and C.B. Fliermans. 1980. Growth of *Legionella pneumophila* in association with blue-green algae. Appl. Environ. Microbiol. 39: 456–459.

Tobin, R.S., P. Ewan, K. Walsh, and B. Dutka. 1986. A survey of *Legionella pneumophila* in water in 12 canadian cities. Water Res. 20: 495–501.

Tobin, R.S., D.K. Smith, and J.A. Lindsay. 1981. Effects of activated carbon and bacteriostatic filters on microbiological quality of drinking water. Appl. Environ. Microbiol. 41: 646–651.

Toerien, D.F., A. Gerber, L.H. Lotter, and T.E. Cloete. 1990. Enhanced biological phosphorus removal in activated sludge systems. Adv. Microb. Ecol. 11: 173–230.

Tokuz, Y. 1989. Biodegradation and removal of phenols in rotating biological contactors. Water Sci. Technol. 21: 1751–1754.

Tomlinson, T.G., and I.L. Williams. 1975. Fungi, pp. 93–152, in: *Ecological Aspects of Used Water Treatment*, Vol. 1, C.R. Curds and H.A. Hawkes, Eds. Academic, London.

Topping, B. 1987. The biodegradability of *para-*

dichlorobenzene and its behaviour in model activated sludge plants. Water Res. 21: 295–300.

Torpey, W.N., H. Heukelekian, A.J. Kaplovski, and R. Epstein. 1971. Rotating disks with biological growths prepare wastewater for disposal or reuse. J. Water Pollut. Control Fed. 43: 2181–2188.

Tortora, G.J., B.R. Funke, and C.L. Case. 1989. *Microbiology: An Introduction*. Benjamin/Cummings Publishing Co., Redwood City, CA.

Toze, S., L.I. Sly, I.C. MacRae, and J.A. Fuerst. 1990. Inhibition of growth of *Legionella* species by heterotrophic plate count bacteria isolated from chlorinated drinking water. Curr. Microbiol. 21: 139–143.

Travers, S.M., and D.A. Lovett. 1984. Activated sludge treatment of abattoir wastewater. II. Influence of dissolved oxygen concentration. Water Res. 18: 435–439.

Trepeta, R.W., and S.C. Edberg. 1984. Methylumbelliferyl-ß-D-glucuronide-based medium for rapid isolation and identification of *E. coli*. J. Clin. Microbiol. 19: 172–174.

Trevors, J.T. 1986. Bacterial growth and activity as indicators of toxicity, pp. 9–25, in: *Toxicity Testing Using Microorganisms*, Vol. 1, G. Bitton and B.J. Dutka, Eds. CRC Press, Boca Raton, FL.

Trevors, J.T. 1989. The role of microbial metal resistance and detoxification mechanisms in environmental bioassay research. Hydrobiology 188: 143–147.

Trevors, J.T., K.M. Oddie, and B.H. Beliveau. 1985. Metal resistance in bacteria. FEMS Microbiol. Rev. 32: 39–54.

Trulear, M.G., and W.G. Characklis. 1982. Dynamics of biofilm processes. J. Water Pollut. Control Fed. 54: 1288–1301.

Tsai, Y.-L., and B.H. Olson. 1992. Rapid method for separation of bacterial DNA from humic substances in sediments for polymerase chain reaction. Appl. Environ. Microbiol. 58: 2292–2295.

Tsai, Y.-L., C.J. Palmer, and L.R. Sangermano. 1993. Detection of *Escherichia coli* in sewage and sludge by polymerase chain reaction. Appl. Environ. Microbiol. 59: 353–357.

Tsien, H.-C., G.A. Brusseau, R.S. Hanson, and L.P. Wackett. 1989. Biodegradation of trichloroethylene by *Methylococcus trichosporium* OB3b. Appl. Environ. Microbiol. 55: 3155–3161.

Tucker, E.S., V.W. Saeger, and O. Hicks. 1975. Activated sludge primary biodegradation of polychlorinated biphenyls. Bull. Environ. Contam. Toxicol. 14: 705–713.

Tuovinen, O.H., and J.C. Hsu. 1982. Aerobic and anaerobic microorganisms in tubercles of the Columbus, Ohio, water distribution system. Appl. Environ. Microbiol. 44: 761–764.

Turner, A.P.F., I. Karube, and G.S. Wilson, Eds. 1987. *Biosensors: Fundamentals and Applications*, Oxford University Press, Oxford, U.K.

Uchiyama H., T. Nakajima, O. Yagi, and T. Nakahara. 1992. Role of heterotrophic bacteria in complete mineralization of trichloroethylene by *Methylocystis* sp. strain M. Appl. Environ. Microbiol. 58: 3067–3071.

Uhlmann, D. 1979. *Hydrobiology: A Text for Engineers and Scientists*. Wiley, New York.

Unz, R.F., and S.R. Farrah. 1976. Observations on the formation of wastewater zoogleae. Water Res. 10: 665–671.

Unz, R.F., and T.M. Williams. 1988. Effect of controlled pH on the development of rosette-forming bacteria in axenic culture and bulking activated sludge. In: *Proceedings of the International Conference on Water and Wastewater Microbiology*, Newport Beach, CA, Feb. 8–11, 1988, Vol. 1.

Urbain, V., J.C. Block, and J. Maneim. 1993. Bioflocculation in activated sludge: An analytic approach. Water Res. 27: 829–838.

U.S. DHEW. 1967. Policy statement on the use of ultraviolet process for disinfection of water. U.S. Department of Health, Education and Walfare, Washington. D.C.

U.S. DHEW. 1969. *Manual of Septic Tank Practices*. Public Health Service, Pub. 528, Washington, D.C.

U.S. EPA. 1975. *Process Design Manual for Nitrogen Control*. Office of Technology Transfer, Washington, DC.

U.S EPA. 1977. *Wastewater Treatment Facilities for Sewered Small Communities*. EPA-625/1–77–009.

U.S. EPA. 1979. Criteria for classification of solid waste disposal facilities and practices: Final rules. Fed. Reg. 44(179): 53438.

U.S. EPA. 1980. *Onsite Wastewater Treatment and Disposal Systems*. EPA-625/1–80–012.

U.S. EPA. 1981. *Process Design Manual for Land Treatment of Municipal Wastewater*. EPA-625/1–81–013, U.S. EPA, Cincinnati, OH.

U.S. EPA. 1982. *Estimating Microorganisms Densities in Aerosols from Spray Irrigation of Wastewa-*

ter. EPA-600/9–82–003, Center for Environmental Research Information, Cincinnati, OH.

U.S. EPA. 1983. *Process Design Manual for Land Application of Municipal Sludge*. EPA-625/1–83–016.

U.S. EPA. 1984. *Environmental Regulations and Technology: Use and Disposal of Municipal Wastewater Sludge*. EPA-625/10–84–003, Spt. 1984.

U.S. EPA 1985. *Odor and Corrosion Control in Sanitary Systems and Treatment Plants*. EPA-625/1-85–018.

U.S. EPA. 1986. *Septic Systems and Groundwater Protection: An Executive Guide*. Office of Groundwater Protection, Washington, DC.

U.S. EPA. 1987a. *The Causes and Control of Activated Sludge Bulking and Foaming*. EPA-625/8–87/012, Cincinnati, OH.

U.S. EPA. 1987b. *Design Manual: Phosphorus Removal*. EPA-625/1–87/001, Cincinnati, OH.

U.S. EPA. 1988. *Methods for Aquatic Toxicity Identification Evaluation—Phase I Toxicity Characterization Procedures*. EPA-600/3–88/034, Duluth, MN.

U.S. EPA. 1989a. *Methods for Aquatic Toxicity Identification Evaluation—Phase II Toxicity Identification Procedures*. EPA-600/3–88/035, Duluth, MN.

U.S. EPA. 1989b. *Methods for Aquatic Toxicity Identification Evaluation—Phase III Toxicity Confirmation Procedures*. EPA-600/3–88/036, Duluth, MN.

U.S. EPA. 1989c. Algal (*Selenastrum capricornutum*) growth test, pp. 147–174, in: *Short-Term Methods for Estimating the Chronic Toxicity of Effluents and Receiving Waters to Freshwater Organisms*. EPA-600/4–89/001, Environmental Monitoring Systems Laboratory, Cincinnati, OH.

U.S. EPA. 1989d. National primary drinking water regulations; filtration and disinfection; turbidity; *Giardia lamblia*; viruses, *Legionella*, and heterotrophic bacteria. Fed. Reg. 54: 27486–27541.

U.S. EPA. 1989e. Drinking Water Health Effects Task Force. *Health Effects of Drinking Water Treatment Technologies*. Lewis Pubs., Chelsea, MI.

U.S. EPA. 1989f. *Assessment of Needed Publicly Owned Wastewater Treatment Facilities in the United States*. 1988 Needs Survey Report to Congress. EPA-430/09–89–001.

Valcke, D., and W. Verstraete. 1983. A practical method to estimate the acetoclastic methanogenic biomass in anaerobic sludges. J. Water Pollut. Control Fed. 55: 1191–1195.

Vallom, J.K., and A.J. McLoughlin. 1984. Lysis as a factor in sludge flocculation. Water Res. 18: 1523–1528.

Valo, R.J., M.M. Haggblom, and M.S. Salkinoja-Salonen. 1990. Bioremediation of chlorophenol containing simulated ground water by immobilized bacteria. Water Res. 24: 253–258.

Vance, D.B. 1991. Onsite bioremediation of oil and grease contaminated soil. Nat. Environ. J. 1: 26–30.

Vandenbergh, P.A., A.M. Wright, and A.K. Vidaver. 1985. Partial purification and characterization of a polysaccharide depolymerase associated with phage-infected *Erwinia amylovora*. Appl. Environ. Microbiol. 49: 994–996.

Vander, A.J., J.H. Sherman, and D.S. Luciano. 1985. *Human Physiology: The Mechanisms of Body Function*. McGraw-Hill, New York.

Vandevivere, P., and P. Baveye. 1992. Effect of bacterial extracellular polymers on the saturated hydraulic conductivity of sand columns. Appl. Environ. Microbiol. 58: 1690–1698.

Varma, M.M., W.A. Thomas, and C. Prasad. 1976. Resistance to inorganic salts and antibiotics among sewage-borne Enteriobacteriaceae and Achromobacteriaceae. J. Appl. Bacteriol. 41: 347–349.

Vaughn, J.M., E.F. Landry, T.J. Vicale, and W.F. Penello. 1979. Survey of human enteroviruses occurring in fresh and marine surface waters on Long Island. Appl. Environ. Microbiol. 38: 290–296.

Vaughn, J.M., E.F. Landry, C.A. Beckwith, and M.Z. Thomas. 1981. Virus removal during groundwater recharge: Effects of infiltration rate on adsorption of poliovirus to soil. Appl. Environ. Microbiol. 41: 139–143.

Vaughn, J.M., E.F. Landry, and M.Z. Thomas. 1983. Entrainment of viruses from septic tank leach fields through a shallow, sandy soil aquifer. Appl. Environ. Microbiol. 45: 1474–1480.

Vaughn, J.M., and J.F. Novotny. 1991. Virus inactivation by disinfectants, pp. 217–241, in: *Modeling the Environmental Fate of Microorganisms*, C.J. Hurst, Ed. American Society for Microbiology, Washington, DC.

Venkobachar, C., Z. Invegar, and A.V.S. Prabhakara Raj. 1977. Mechanism of disinfection: Effect of chlorine on cell membrane functions. Water. Res. 11: 727–729.

Venosa, A.D. 1986. Detection and significance of pathogens in sludge, pp. 1–14, in: *Control of Sludge Pathogens*, C.A. Sorber, Ed. Water Pollution Control Federation, Washington, DC.

Venosa, A.D., A.C. Petrasek, D. Brown, H.L. Sparks, and D.M. Allen. 1984. Disinfection of secondary effluent with ozone/UV. J. Water Pollut. Control Fed. 56: 137–142.

Verstraete, W., and M. Alexander. 1972. Heterotrophic nitrification by *Arthrobacter* sp. J. Bacteriol. 110: 955–961.

Verstraete, W., and E. van Vaerenbergh. 1986. Aerobic activated sludge, pp. 43–112, in: *Biotechnology*, H.J. Rehm and G. Reed, Eds.; Vol. 8, *Aerobic Degradations*, W. Schonborn, Vol. Ed. VCH, Germany.

Vesper, S.J., W. Davis-Hoover, and L.C. Murdoch. 1992. Oxygen sources for *in situ* subsurface bioremediation. Abstr. Annu. Meet. Am. Soc. Microbiol., New Orleans.

Villacorta-Martinez de Maturana, I., M.E. Ares-Mazas, D. Duran-Oreiro, and M.J. Lorenzo-Lorenzo. 1992. Efficacy of activated sludge in removing *Cryptosporidium parvum* from sewage. Appl. Environ. Microbiol. 58: 3514–3516.

Vissier, F.A., J.B. van Lier, A.J.L. Macario, and E. Conway de Macario. 1991. Diversity and population dynamics of bacteria in a granular consortium. Appl. Environ. 57:1728–1734.

Vogel, T.M., and P.L. McCarty. 1985. Biotransformation of tetrachloroethylene to trichloroethylene, dichloroethylene, vinyl chloride and carbon dioxide under methanogenic conditions. Appl. Environ. Microbiol. 49: 1080–1083.

Vogels, G.D., J.T. Keltjensand, and C. van der Drift. 1988. Biochemistry of methane production, pp. 707–770, in: *Biology of Anaerobic Microorganisms*, A.J.B. Zehnder, Ed. Wiley, New York.

Volskay, V.T., Jr., and C.P.L Grady, Jr. 1990. Respiration inhibition kinetics analysis. Water Res. 24: 863–874.

Volskay, V.T., Jr., and C.P.L. Grady, Jr. 1988. Toxicity of selected RCRA compounds to activated sludge microorganisms. J. Water Pollut. Control Fed. 60: 1850–1856.

Volskay, V.T., Jr., C.P.L Grady, Jr., and H.H. Tabak. 1990. Effect of selected RCRA compounds on activated sludge activity. J. Water Pollut. Control Fed. 62: 655–664.

Wachinski, A.M. 1988. Waste management in U.S. space program. J. Water Pollut. Control Fed. 60: 1790–1797.

Wada, S., H. Ichikawa, and K. Tatsumi. 1992. Removal of phenols with tyrosinase immobilized on magnetite. Water Sci. Technol. 26: 2057–2059.

Wadowsky, R.M., and R.B. Yee. 1985. Effect on non-Legionellaceae bacteria on the multiplication of *Legionella pneumophila* in potable water. Appl. Environ. Microbiol. 49: 1206–1210.

Walker, G.S., F.P. Lee, and E.M. Aieta. 1986. Chlorine dioxide for taste and odor control. J. Am. Water Works Assoc. 78: 84–93.

Walker, J.M., and G.B. Wilson. 1973. Composting sewage sludge: Why? Compost Sci. 14: 10–12.

Wallis, C., C.H. Stagg, and J.L. Melnick. 1974. The hazards of incorporating charcoal filters into domestic water systems. Water Res. 8: 111–113.

Wallis, P.M., and D.L. Lehmann, Eds. 1983. *Biological Health Risks of Sludge Disposal to Land in Cold Climates*. University of Calgary Press, Calgary, Canada.

Walsh, G.E., and R.L. Garnas. 1983. Determination of bioactivity of chemical fractions of liquid wastes using freshwater and saltwater algae and crustaceans. Environ. Sci. Technol. 17: 180–182.

Walter, M.V., and J.W. Vennes. 1985. Occurrence of multiple-antibiotic-resistant enteric bacteria in domestic sewage and oxidation lagoons. Appl. Environ. Microbiol. 50: 930–933.

Wang, Y.-T., H.D. Gabbard, and P.-C. Pai. 1991. Inhibition of acetate methanogenesis by phenols. J. Environ. Eng. 117: 487–496.

Wang, X., X. Yu, and R. Bartha. 1990. Effect of bioremediation on polycyclic aromatic hydrocarbon residues in soil. Environ. Sci. Technol. 24: 1086–1089.

Wanner, J., J. Chudoba, K. Kuckman, and L. Proske. 1987a. Control of activated sludge filamentous bulking. VII. Effect of anoxic conditions. Water Res. 21: 1447–1451.

Wanner, J., and P. Grau. 1989. Identification of filamentous microorganisms from activated sludge: A compromise between wishes and possibilities. Water Res. 23: 883–891.

Wanner, J., K. Kuckman, V. Ottova, and P. Grau. 1987b. Effect of anaerobic conditions on activated sludge filamentous bulking in laboratory systems. Water Res. 21: 1541–1546.

Wanner, O., and W. Gujer. 1984. Competition in biofilms. Paper presented at the 12th Conference of the International Association on Water Pollution Research, Amsterdam.

Ward, N.R., R.L. Wolfe, and B.H. Olson. 1984. Effect of pH, application technique and chlorine-to-nitrogen ratio on disinfectant activity of inorganic chloramines with pure culture bacteria. Appl. Environ. Microbiol. 48: 508–514.

Ward, R.L., and C.S. Ashley. 1977a. Identification of the viricidal agent in wastewater sludge. Appl. Environ. Microbiol. 33: 860–864.

Ward, R.L., and C.S. Ashley. 1977b. Discovery of an agent in wastewater sludge that reduces the heat required to inactivate reovirus. Appl. Environ. Microbiol. 34: 681–688.

Ward, R.L., C.S. Ashley, and R.H. Moseley. 1976. Heat inactivation of poliovirus in wastewater sludge. Appl. Environ. Microbiol. 32: 339–346.

Ward, R.L., D.R. Knowlton, J. Stober, W. Jakubowski, T. Mills, P. Graham, and D.E. Camann. 1989. Effect of wastewater spray irrigation on rotavirus infection rates in an exposed population. Water Res. 23: 1505–1509.

Ward, R.L., J.G. Yeager, and C.S. Ashley. 1981. Response of bacteria in wastewater sludge to moisture loss by evaporation and effect of moisture content on bacterial inactivation by ionizing radiation. Appl. Environ. Microbiol. 41: 1123–1127.

Washington, B., C. Lue-Hing, D.R. Zenz, K.C. Rao, and A.W. Kobayashi. 1983. Exertion of 5-day nitrogenous oxygen demand in nitrifying wastewater. J. Water Pollut. Control Fed. 55: 1196–1200.

Watanabe, I. 1973. Isolation of pentachlorophenol decomposing bacteria from soil. Soil Sci. Plant Nutr. 19: 109–116.

Water Pollution Control Federation. 1979. *Odor Control for Wastewater Facilities*. Manual of Practice No. 22. WPCF, Washington, DC.

Wattie, E., and C.T. Butterfield. 1944. Relative resistance of *Escherichia coli* and *Escherichia typhosa* to chlorine and chloramines. Public Health Rep. 59:1661–1665.

Wattie, E., and C.W. Chambers. 1943. Relative resistance of coliform organisms and certain enteric pathogens to excess-lime treatment. J. Am. Water Works Assoc. 35: 709–720.

Watkins, W.D., S.R. Rippey, C.R. Clavet, D.J. Kelley-Reitz, and W. Burkhardt III. 1988. Novel compound for identifying *E. coli*. Appl. Environ. Microbiol. 54: 1874–1875.

Way, J.S., K.L. Josephson, S.D. Pillai, M. Abbaszadegan, C.P. Gerba, and I.L. Pepper. 1993. Specific detection of *Salmonella* spp. by multiplex polymerase chain reaction. Appl. Environ. Microbiol. 59: 1473–1479.

Webb, C. 1987. Cell immobilization, pp. 347–376, in: *Environmental Biotechnology*, C.F. Forster and D.A.J. Wase, Eds. Ellis Horwood, Chichester, U.K.

Weber, W.J., M. Pirbazari, and G.L. Melson. 1978. Biological growth on activated carbon: An investigation by scanning electron microscopy. Environ. Sci. Technol. 12: 817–819.

Weddle, C.L., and D. Jenkins, 1971. The viability and activity of activated sludge. Water Res. 5: 621–640.

von Wedel, R.J., J.F. Mosquera, C.D. Goldsmith, G.R. Hater, A. Wong, T.A. Fox, W.T. Hunt, M.S. Paules, J.M. Quiros, and J.W. Wiegand. 1988. Bacterial biodegradation of petroleum hydrocarbons in groundwater: *In situ* augmented bioreclamation with enrichment isolates in California. In: *Proceedings of the International Conference on Water and Wastewater Microbiology*, Newport Beach, CA, Feb. 8–11, 1988, Vol. 2.

van der Wende, E., and W.G. Characklis. 1990. Biofilms in potable water distribution systems, pp. 249–268, in: *Drinking Water Microbiology*,. G.A. McFeters, Ed. Springer-Verlag, New York.

Weng, C.N., and A.H. Molof. 1974. Nitrification in the biological fixed film RBC. J. Water Pollut. Control Fed. 46: 1674–1685.

Wentzel, M.C., L.H. Lotter, R.E. Loewenthal, and G.V.R. Marais. 1986. Metabolic behaviour of *Acinetobacter* spp. in enhanced biological phosphorus removal: A biochemical model. Water S Afr 12: 209–224.

Werner, M., and R. Kayser. 1991. Denitrification with biogas as external carbon source. Water Sci. Technol. 23: 701–708.

Werner, P. 1985. Eine Methode zur Bestimmung der Veirkemungsneigung von Trinkwasser. Von Wasser 65: 257–262. Cited by Huck, 1990.

Westphal, P.A., and G.L. Christensen. 1983. Lime stabilization: effectiveness of two process modifications. J. Water Pollut. Control Fed. 55: 1381–1386.

Wetzler, T.F., J.R. Rea, G.J. Ma, and M. Glass. 1979. Non- association of *Yersinia* with traditional coliform indicators. Paper presented at the American Water Works Association, Denver.

Wheeler, D., J. Bartram, and B.J. Lloyd. 1988. The removal of viruses by filtration through sand, pp. 207–229, in: *Slow Sand Filtration: Recent Developments in Water Treatment Technology*, N.J.D. Graham, Ed. Ellis Horwood, Chichester, U.K.

Whitby, G.E., G. Palmateer, W.G. Cook, J. Maarschalkerweerd, D. Huber, and K. Flood. 1984. Ultraviolet disinfection of secondary effluent. J. Water Pollut. Control Fed. 56: 844–850

White, D.C., R.J. Bobbie, J.D. King, J. Nickels, and P. Amoe. 1979. Lipid analysis of sediments for microbial biomass and community structure, pp. 87–103, in: *Methodology for Biomass Determinations and Microbial Activities in Sediments*, C.D. Litchfield and P.L Seyfried, Eds. ASTM STP 673. American Society for Testing and Materials, Philadelphia.

White, J.M., D.P. Labeda, M.P. LeChevallier, J.R. Owens, D.D. Jones, and J.L. Gauthier. 1986. Novel actinomycete isolated from bulking industrial sludge. Appl. Environ. Microbiol. 52: 1324–1330.

Whitman, W.B., and R.S. Wolfe. 1980. Presence of nickel in factor F_{430} from *Methanobacterium bryantii*. Biochem. Biophys Res. Commun. 92: 1196–1201.

Whitmore, T.N., and S. Denny. 1992. The effect of disinfectants on a geosmin-producing strain of *Streptomyces griseus*. J. Appl. Bacteriol. 72: 160–165.

Wickramanayake, G.B., A.J. Rubin, and O.J. Sproul. 1985. Effect of ozone and storage temperature on *Giardia* cysts. J. Am. Water Works Assoc. 77: 74–77.

Widdel, F. 1988. Microbiology and ecology of sulfate- and sulfur-reducing bacteria, pp. 469–585, in: *Biology of Anaerobic Microorganisms*, A.J.B. Zehnder, Ed. Wiley, New York.

Wiedenmann, A., B. Fischer, U. Sraub, C.-H Wang, B. Flehmig, and D. Schoenen. 1993. Disinfection of hepatitis A virus and MS-2 coliphage in water by ultraviolet irradiation: Comparison of UV-susceptibility. Water Sci. Technol. 27: 335–338.

Wilcox, D.P., E. Chang, K.L. Dickson, and K.R. Johansson. 1983. Microbial growth associated with granular activated carbon in a pilot water treatment facility. Appl. Environ. Microbiol. 45: 406–416.

Willcomb, G.E. 1923. Twenty years of filtration practice at Albany. J. Am. Water Works Assoc. 10: 97–103.

Williams, C.M., J.C.H. Shih, and J.W. Spears. 1986. Effect of nickel on biological methane generation from a laboratory poultry waste digester. Biotechnol. Bioeng. 28: 1608–1610.

Williams, F.P., and E.W. Akin. 1986. Waterborne viral gastroenteritis. J. Am. Water Works Assoc. 78: 34–39.

Williams, R.B., and G.L. Culp. 1986. *Handbook of Public Water Systems*. Van Nostrand Reinhold Co., New York.

Williams, R.T., P.S. Ziegenfuss, G.B. Mohrman, and W.E. Sisk. 1989. Composting of explosives and propellant contaminated sediments, pp. 269–281, in: *Hazardous Waste Treatment: Biosystems for Pollution Control*. Air and Waste Management Association, Pittsburgh, PA.

Williams, T.M., and R.F. Unz. 1983. Environmental distribution of *Zoogloea* strains. Water Res. 17: 779–787.

Williams, T.M., and R.F. Unz, 1985a. Filamentous sulfur bacteria of activated sludge characterization of *Thiothrix*, *Beggiatoa*, and Eikelboom "type 021N" strains. Appl. Environ. Microbiol. 49: 887–898.

Williams, T.M., and R.F. Unz. 1985b. Isolation and characterization of filamentous bacteria present in bulking activated sludge. Appl Microbiol. Biotechnol. 22: 273–282.

Williams, T.M., and R.F. Unz. 1989. The nutrition of *Thiothrix*, Type 021N, *Beggiatoa* and *Leucothrix* strains. Water Res. 23: 15–22.

Williams, T.M., R.F. Unz, and J.T. Doman. 1987. Ultrastructure of *Thiothrix* spp. and "type 021N" bacteria. Appl. Environ. Microbiol. 53: 1560–1570.

Williamson, K.J., and D.G Johnson. 1981. A bacterial bioassay for assessment of wastewater toxicity. Water Res. 15:383–390.

Williamson, K,. and P.L. McCarty. 1976. A model for substrate utilization by bacterial films. J. Water Pollut. Control Fed. 48: 9–24.

Wilson, B., P. Roessler, M. Abbaszadegan, C.P. Gerba, and E. van Dellen. 1992. UV dose bioassay using coliphage MS-2. Abstr. Annu. Meet. Am. Soc. Microbiol., New Orleans.

Wilson, J.T., et al. 1986. In situ biorestoration as a groundwater remediation technique. Ground Water Monit. Rev. 6: 56–64.

Wilson, J.T., D.H. Kampbell, S.R. Hutchins, D.A. Kovacs, W. Korreck, R.H. Douglass, and D.J. Hendrix. 1989. Performance of two demonstration projects for the bioremediation of an aquifer contaminated by fuel spills from underground storage tanks. Paper presented at the Tenth Annual Meeting of the Society for Environmental Toxicology and Chemistry, Toronto, Canada, Oct. 28 to Nov. 2, 1989.

Witherell, L.E., R.W. Duncan, K.M. Stone, L.J. Stratton, L. Orciani, S. Kappel, and D.A. Jillson. 1988. Investigation of *Legionella pneumophila* in drinking water. J. Am. Water Works Assoc. 80: 87–93.

Wolfe, R.L. 1990. Ultraviolet disinfection of potable water. Environ. Sci. Technol. 24: 768–773.

Wolfe, R.L., and B.H. Olson. 1985. Inability of laboratory models to accurately predict field performance of disinfectants, pp. 555–573, in: *Water Chlorination: Environmental Impact and Health Effects*, Vol. 5, R.L. Jolley, R.J. Bull, W.P. Davis, S. Katz, M.H. Roberts, and V.A. Jacobs, Eds. Ann Arbor Science, Ann Arbor, MI.

Wolfe, R.L., N.R. Ward, and B.H. Olson. 1984. Inorganic chloramines as drinking water disinfectants: A review. J. Am. Water Works Assoc. 76: 74–88.

Wolinsky, E. 1979. Nontuberculous mycobacteria and associated diseases. Am. Rev. Resp. Dis. 119: 107–159.

von Wolzogen-Kuhr, C.A.H., and A.S. van der Vlugt. 1934. The graphitization of cast iron as an electrobiochemichal process in anaerobic soils. Water 18: 147–165.

Won, W.D., and H. Ross. 1973. Persistence of virus and bacteria in seawater. J. Environ. Eng. Div. Am. Soc. Civ. Eng. 99: 205–211.

Woods, S.L., J.F. Ferguson, and M.M. Benjamin. 1989. Characterization of chlorophenol and chloromethoxybenzene biodegradation during anaerobic treatment. Environ. Sci. Technol. 23: 62–68.

World Health Organization. 1979. *WHO International Reference Center for Community Water Supply, Annual Report*. Rijswijk, The Netherlands.

World Health Organization (WHO). 1989. *Guidelines for Use of Wastewater in Agriculture and Aquaculture*. WHO Technical Report Series 778. WHO, Geneva.

Wu, W., J. Hu, X. Gu, Y. Zhao, and H. Zhang. 1987. Cultivation of anaerobic granular sludge with aerobic activated sludge as seed. Water Res. 21: 789–799.

Xu, H., and B.J. Dutka. 1987. ATP-TOX system: A new rapid sensitive bacterial toxicity screening system based on the determination of ATP. Toxicity Assess. 2:149–166.

Yaguchi, J., K. Chigusa, and Y. Ohkubo. 1991. Isolation of microorganisms capable of lysing the filamentous bacterium, type 021N (studies on lytic enzyme against the filamentous bacterium, type 021N, screening studies). Water Sci. Technol. 23: 955–962.

Yahya, M.T., and C.P. Gerba. 1990. Inactivation of bacteriophage MS-2 and poliovirus by copper, silver and low levels of free chlorine (Abstract). International Symposium on Health-Related Water Microbiology, Tubingen, West Germany, April 1–6, 1990.

Yahya, M.T., L.K. Landeen, S.M. Kutz, and C.P. Gerba. 1989. Swimming pool disinfection: An evaluation of the efficacy of copper:silver ions. J. Environ. Health 51: 282–287.

Yahya, M.T, and W.A Yanko. 1992. Comparison of a long-term enteric virus monitoring data base with bacteriophage reduction in full scale reclamation plants. Paper presented at the International Water Pollution Research Conference International Symposium, Washington, DC, May 26–29, 1992.

Yang, J., and R.E. Speece. 1985. Effects of engineering controls on methane fermentation toxicity response. J. Water Pollut. Control Fed. 57: 1134–1141.

Yang, J., and R.E. Speece. 1986. The effects of chloroform toxicity on methane fermentation. Water Res. 20: 1273–1279.

Yates, M.V. 1985. Septic tank density and ground water contamination. Ground Water 23: 586–591.

Yates, M.V., and C.P. Gerba. 1984. Factors controlling the survival of virus in groundwater. Water Sci. Technol. 17: 681–687.

Yates, M.V., C.P. Gerba, and L.M. Kelley. 1985. Virus persistence in groundwater. Appl. Environ. Microbiol. 49: 778–781.

Yates, M.V., and Y. Ouyang. 1992. VIRTUS, a model of virus transport in unsaturated soils. Appl. Environ. Microbiol. 58: 1609–1616.

Yates, M.V., and S.R. Yates. 1987. A comparison of geostatistical methods for estimating virus inactivation rates in ground water. Water Res. 21: 1119–1125

Yates, M.V., and S.R. Yates. 1989. Septic tank setback distances: A way to minimize virus contamination of drinking water. Ground Water 27: 202–208.

Yates, M.V., S.R. Yates, A.W. Warrick, and C.P. Gerba. 1986. Predicting virus fate to determine septic tank setback distances using geostatistics. Appl. Environ. Microbiol. 49: 479–483.

Yaziz, M.I., and B.J. Lloyd. 1979. The removal of Salmonellas in conventional sewage treatment plants. J. Appl. Bacteriol. 46: 131–142.

Yeager, J.G., and R.T. O'Brien. 1983. Irradiation as a means to minimize public health risks from sludge-borne pathogens. J. Water Pollut. Control Fed. 55: 977–983.

Yee, R.B., and R.M. Wadowsky. 1982. Multiplication of *Legionella pneumophila* in unsterilized tap water. Appl. Environ. Microbiol. 43: 1130–1134.

Yoda, M., M. Kitagawa, and Y. Miyaji. 1987. Long-term competition between sulfate-reducing and methane-producing bacteria for acetate in anaerobic biofilm. Water Res. 21: 1547–1556.

Yoda, M., M. Kitagawa, and Y. Miyaji. 1989. Granular sludge formation in the anaerobic expanded micro-carrier bed process. Water Sci. Technol. 21: 109–120.

Yoshpe-Purer, Y., and S. Golderman. 1987. Occurrence of *Staphylococcus aureus* and *Pseudomonas aeruginosa* in Israeli coastal water. Appl. Environ. Microbiol. 53: 1138–1141.

You, I.S., and R. Bartha. 1982. Stimulation of 3,4-dichloroaniline mineralization by aniline. Appl. Environ. Microbiol. 44: 678–681.

Young, J.C. 1983. Comparison of three forms of 2-chloro-6-(trichloromethyl) pyridine as a nitrification inhibitor in BOD tests. J. Water Pollut. Control Fed. 55: 415–416.

Young, J.C., and P.L. McCarty. 1969. The anaerobic filter for waste treatment. J. Water Pollut. Control Fed. 41: R160–173.

Young, L.Y., and M.M. Häggblom. 1989. The anaerobic microbiology and biodegradation of aromatic compounds, pp. 3–17, in: *Biotechnology and Biodegradation*, D. Kamely, A. Chackrabarty, and G.S. Ommen, Eds. Gulf Publishing Co., Houston, TX.

Zajik, J.E. 1971. *Water Pollution: Disposal and Reuse*. Marcel Dekker, New York.

Zaske, S.K, W.S. Dockins, and G.A. McFeters. 1980. Cell envelope damage in *E. coli* caused by short-term stress in water. Appl. Environ. Microbiol. 41: 386–390.

Zehnder, A.J.B., Ed. 1988. *Biology of Anaerobic Microorganisms*. Wiley, New York.

Zeikus, J.G. 1980. Chemical and fuel production by anaerobic 'bacteria. Annu. Rev. Microbiol. 34: 423–464.

Zeyer, J., P. Eicher, J. Dolfing, and R.P. Schwarzenbach. 1989. Anaerobic degradation of aromatic hydrocarbons. in: *Biotechnology and Biodegradation*, D. Kamely, A. Chakrabarty, and G.S. Omenn, Eds. Gulf Publishing Co., Houston, TX.

Ziebell, W.A., D.H. Nero, J.F. Deininger, and E. McCoy. 1975. Use of bacteria in assessing waste treatment and soil disposal systems. in: *Home Sewage Disposal*. Proc. Nat. Home Sewage Disposal Symp., 9–10 Dec. 1974, Am. Soc. Agric. Eng., Pub. # 175, St. Joseph, Mich.

Ziegler, M., M. Lange, and W. Dott. 1990. Isolation and morphological and cytological characterization of filamentous bacteria from bulking sludge. Water Res. 24: 1437–1451.

Zierler, S., R.A. Danley, and L. Feingold. 1987. Type of disinfectant in drinking water and patterns of mortality in Massachussetts. Environ. Health Perspect. 68: 275–287.

Zimmerman, N.J., P.C. Reist, and A.G. Turner. 1987. Comparison of two biological aerosol sampling methods. Appl. Environ. Microbiol. 53: 99–104.

Zinder, S.H., S.C. Cardwell, T. Anguish, M. Yee, and M. Koch. 1984. Methanogenesis in a thermophilic (58°C) anaerobic digestor: *Methanothrix* sp. as an important aceticlastic methanogen. Appl. Environ. Microbiol. 47: 796–807.

Ziogou, K., P.W.W. Kirk, and J.N. Lester. 1989. Behaviour of phthalic acid esters during batch anaerobic digestion of sludge. Water Res. 23: 743–748.

Zoeteman, B.C.J., G.J. Piet, and L. Postma. 1980. Taste as an indicator for drinking water quality. J. Am. Water Works Assoc. 72: 537–540.

Zyman, J., and C.A. Sorber. 1988. Influence of simulated rainfall on the transport and survival of selected indicator organisms in sludge-amended soils. J. Water Pollut. Control Fed. 60: 2105–2110.

INDEX